FUNDAMENTALS OF
CHEMISTRY

SECOND EDITION

FUNDAMENTALS OF
CHEMISTRY

SECOND EDITION

David E. Goldberg
Brooklyn College

WCB
McGraw-Hill

Boston, Massachusetts   Burr Ridge, Illinois   Dubuque, Iowa
Madison, Wisconsin   New York, New York   San Francisco, California   St. Louis, Missouri

WCB/McGraw-Hill

*A Division of The **McGraw·Hill** Companies*

INTRODUCTION TO CHEMISTRY, SECOND EDITION

Recycled paper/acid free paper

 This book is printed on recycled, acid-free paper containing 10% postconsumer waste.

1 2 3 4 5 6 7 8 9 0 QPD/QPD 0 9 8 7

Library of Congress Catalog Number: 96-86011

ISBN 0-697-29150-2

Editorial director: Kevin T. Kane
Publisher: James M. Smith
Sponsoring editor: Kent A. Peterson
Developmental editor: Russell L. Lidberg
Marketing manager: Martin J. Lange
Project manager: Marilyn M. Sulzer
Production supervisor: Sandy Hahn
Designer: K. Wayne Harms
Cover design: © Comstock
Photo research coordinator: Lori Hancock
Art editor: Joyce Watters
Compositor: Graphic World, Inc.
Typeface: 10/12 Times Roman
Printer: Quebecor, Inc.

1 2 3 4 5 6 7 8 9 0 QPD/QPD 0 9 8 7

When ordering this title, use ISBN 0-07-115290-3

http://www.mhcollege.com

To my students,
who have taught me
a great deal

Brief Contents

CHAPTER 1   Basic Concepts   1

CHAPTER 2   Measurement   21

CHAPTER 3   Atoms and Atomic Masses   63

CHAPTER 4   Electronic Configuration of the Atom   78

CHAPTER 5   Chemical Bonding   100

CHAPTER 6   Nomenclature   128

CHAPTER 7   Formula Calculations   148

CHAPTER 8   Chemical Reactions   165

CHAPTER 9   Net Ionic Equations   193

CHAPTER 10   Stoichiometry   204

CHAPTER 11   Molarity   231

CHAPTER 12   Gases   251

CHAPTER 13   Atomic and Molecular Properties   282

CHAPTER 14   Solids and Liquids, Energies of Physical and Chemical Changes   303

CHAPTER 15   Solutions   328

CHAPTER 16   Oxidation Numbers   345

CHAPTER 17   Reaction Rates and Chemical Equilibrium   368

CHAPTER 18   Acid-Base Theory   387

CHAPTER 19   Organic Chemistry   406

CHAPTER 20   Nuclear Reactions   433

APPENDICES   457

Contents

Preface   xiii
To the Student   xxix

CHAPTER 1

**Basic
Concepts   1**

1.1   Classification of Matter   2
1.2   Properties   5
1.3   Matter and Energy   8
1.4   Chemical Symbols   9
1.5   The Periodic Table   11
1.6   Laws, Hypotheses, and Theories   15

   Summary   16
   Problems   17

CHAPTER 2

Measurement   21

2.1   Factor Label Method   22
2.2   The Metric System   27
   Length or Distance   32
   Mass   33
   Volume   33
2.3   Significant Digits   35
   *Significant Digits in Calculated
      Results   39*
   Rounding Off   41
2.4   Exponential Numbers   43
   *Changing the Form of Exponential
      Numbers   46*
   *Multiplication and Division
      of Exponential Numbers   46*
   *Addition and Subtraction
      of Exponential Numbers   49*
   *Raising an Exponential Number
      to a Power   50*
2.5   Density   51

2.6   Time, Temperature, and Energy   53
       Time   53
       Temperature and Energy   53
       Temperature Scales   54
       Energy   55

       Summary   56
       Problems   57

CHAPTER 3

**Atoms and
Atomic Masses 63**

3.1   Laws of Chemical Combination   64
3.2   Dalton's Atomic Theory   67
3.3   Subatomic Particles   68
3.4   Atomic Mass   70
3.5   Development of the Periodic Table   71

       Summary   73
       Problems   74

CHAPTER 4

**Electronic
Configuration
of the Atom   78**

4.1   Bohr Theory   79
4.2   Quantum Numbers   81
4.3   Relative Energies of Electrons   83
4.4   Shells, Subshells, and Orbitals   86
4.5   Shapes of Orbitals   89
4.6   Energy Level Diagrams   89
4.7   Periodic Variation of Electronic
       Configuration   92

       Summary   95
       Problems   96

CHAPTER 5

**Chemical
Bonding   100**

5.1   Chemical Formulas   101
5.2   Ionic Bonding   105
       *Detailed Electronic Configurations
         of Anions   108*
       *Detailed Electronic Configurations
         of Cations   108*
5.3   Formulas for Ionic Compounds   109
5.4   Electron Dot Diagrams   112
5.5   Covalent Bonding   113
       *Systematic Method for Drawing
         Electron Dot Diagrams   115*
       Polyatomic Ions   118
       Non-Octet Structures   121

       Summary   121
       Problems   123

CHAPTER 6

Nomenclature 128

6.1   Binary Nonmetal-Nonmetal
       Compounds   129
6.2   Naming Ionic Compounds   131
       Naming Cations   131
       Naming Anions   134
       *Naming and Writing Formulas
         for Ionic Compounds   135*
6.3   Naming Acids and Acid Salts   136
       Naming Acids   137
       Naming Acid Salts   140
6.4   Hydrates   140

       Summary   141
       Problems   142

Contents

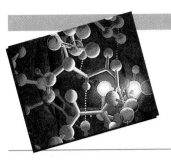

CHAPTER 7

Formula Calculations 148

7.1	Formula Masses	149
7.2	Percent Composition	150
7.3	The Mole	151
7.4	Empirical Formulas	155
7.5	Molecular Formulas	158

Summary 160
Problems 161

CHAPTER 9

Net Ionic Equations 193

9.1	Properties of Ionic Compounds in Aqueous Solution	194
9.2	Writing Net Ionic Equations	196

Summary 200
Problems 200

CHAPTER 8

Chemical Reactions 165

8.1	The Chemical Equation	166
8.2	Balancing Equations	167
8.3	Predicting the Products of Chemical Reactions	171
	Combination Reactions 171	
	Decomposition Reactions 173	
	Substitution Reactions 174	
	Double Substitution Reactions 176	
	Combustion Reactions 179	
8.4	Acids and Bases	180
	Properties of Acids and Bases 181	
	Acidic and Basic Anhydrides 184	
	Acid Salts 184	
	Carbonates and Acid Carbonates 185	

Summary 187
Problems 188

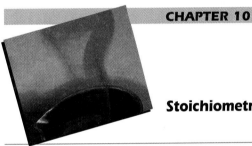

CHAPTER 10

Stoichiometry 204

10.1	Mole Calculations for Chemical Reactions	205
10.2	Mass Calculations for Chemical Reactions	207
10.3	Calculations Involving Other Quantities	211
10.4	Problems Involving Limiting Quantities	214
10.5	Theoretical Yield and Percent Yield	221
10.6	Calculations with Net Ionic Equations	223

Summary 224
Problems 225

CHAPTER 11

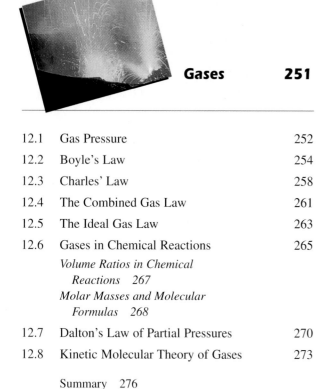

Molarity   231

11.1   Definition and Uses of Molarity   232
11.2   Molarities of Ions   238
11.3   Titration   242

Summary   246
Problems   247

CHAPTER 12

Gases   251

12.1   Gas Pressure   252
12.2   Boyle's Law   254
12.3   Charles' Law   258
12.4   The Combined Gas Law   261
12.5   The Ideal Gas Law   263
12.6   Gases in Chemical Reactions   265
       *Volume Ratios in Chemical
          Reactions   267*
       *Molar Masses and Molecular
          Formulas   268*
12.7   Dalton's Law of Partial Pressures   270
12.8   Kinetic Molecular Theory of Gases   273

Summary   276
Problems   278

CHAPTER 13

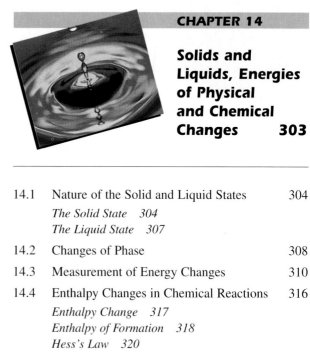

Atomic and Molecular Properties   282

13.1   Atomic and Ionic Sizes   283
13.2   Ionization Energy and Electron Affinity   286
13.3   Electronegativity and Bond Polarity   288
13.4   Molecular Shape   289
13.5   Polar and Nonpolar Molecules   292
13.6   Intermolecular Forces   295
       Dipolar Attractions   295
       van der Waals Forces   296
       Hydrogen Bonding   297

Summary   299
Problems   300

CHAPTER 14

Solids and Liquids, Energies of Physical and Chemical Changes   303

14.1   Nature of the Solid and Liquid States   304
       The Solid State   304
       The Liquid State   307
14.2   Changes of Phase   308
14.3   Measurement of Energy Changes   310
14.4   Enthalpy Changes in Chemical Reactions   316
       Enthalpy Change   317
       Enthalpy of Formation   318
       Hess's Law   320

Summary   322
Problems   323

Contents

CHAPTER 15

Solutions   328

15.1   The Solution Process   329

15.2   Saturated, Unsaturated,
        and Supersaturated Solutions   330

15.3   Molality   332

15.4   Mole Fraction   334

15.5   Colligative Properties   335

Vapor-Pressure Lowering   335
Freezing-Point Depression   336
Boiling-Point Elevation   338
Osmotic Pressure   339

Summary   340
Problems   342

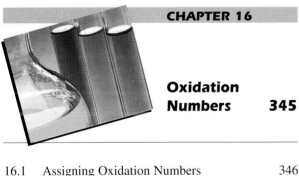

CHAPTER 16

**Oxidation
Numbers   345**

16.1   Assigning Oxidation Numbers   346

16.2   Using Oxidation Numbers in
        Naming Compounds   350

16.3   Periodic Variation of Oxidation Numbers   351

Predicting Oxidation Numbers   351
*Writing Formulas for Covalent
    Compounds   354*

16.4   Balancing Oxidation-Reduction Equations   354

16.5   Equivalents and Normality   359

Normality   360
Equivalent Mass   362

Summary   363
Problems   364

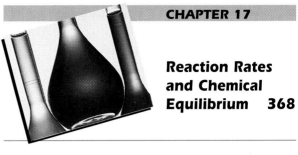

CHAPTER 17

**Reaction Rates
and Chemical
Equilibrium   368**

17.1   Rates of Reaction   369

17.2   The Condition of Equilibrium   370

17.3   LeChâtelier's Principle   372

17.4   Equilibrium Constants   375

*Finding Values of Equilibrium
    Constants   376*
*Calculations Using Equilibrium
    Constants   378*

Summary   381
Problems   382

CHAPTER 18

**Acid-Base
Theory   387**

18.1   The Brønsted Theory   388

18.2   Ionization Constants   392

18.3   Autoionization of Water   394

18.4   Buffer Solutions   398

Summary   402
Problems   403

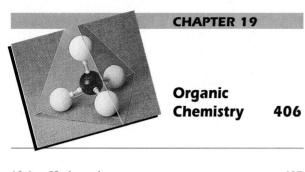

CHAPTER 19

Organic Chemistry 406

CHAPTER 20

Nuclear Reactions 433

19.1 Hydrocarbons   407
    The Alkanes   408
    The Alkenes   409
    The Alkynes   411
    The Aromatic Hydrocarbons   412

19.2 Isomerism   413

19.3 Some Other Classes of Organic Compounds   418
    Organic Halides   418
    Alcohols and Ethers   419
    Aldehydes and Ketones   421
    Organic Acids and Esters   422
    Amines and Amides   423

19.4 Polymers   424

19.5 Foods   426
    Fats   426
    Carbohydrates   427

    Summary   429
    Problems   430

20.1 Natural Radioactivity   434
    Radioactive Series   437
    Tracers   440

20.2 Half-Life   441

20.3 Nuclear Fission   448
    Chain Reactions   449
    Energetics of Nuclear Reactions   451

20.4 Nuclear Fusion   451

    Summary   453
    Problems   454

Appendix 1: Scientific Calculations   457

Appendix 2: Tables of Symbols, Abbreviations,
    and Prefixes and Suffixes   477

Appendix 3: Table of Basic
    Mathematical Equations   481

Appendix 4: Answers to Practice Problems   483

Appendix 5: Answers to Selected
    End-of-Chapter Problems   495

Glossary   540

Photo Credits   550

Index   552

Preface

Chemistry is a dynamic, rapidly changing field that comprises an extraordinarily interesting subject to study and a challenging one to teach. Today, as perhaps never before, the world is looking to the field of chemistry to provide answers to some of the most challenging problems that confront each of us as citizens of the global community. We live in a world in which we are confronted with products, problems, and changes that our parents and grandparents never dreamed of. And we are learning that what happens on the other side of the world can have a profound impact on us and our lives. Problems such as global climate change, loss of stratospheric ozone, and human-created ecological disasters such as the 1991 oil well fires in Kuwait have become subjects of daily discussion in newspapers and on television. Each discipline of science, but especially chemistry, will contribute to solving these problems, as well as others not yet encountered. In addition, the field of chemistry will continue to yield new processes and materials that will open up whole new vistas of opportunities and benefits.

As citizens of the global community, we need a solid foundation in scientific principles—including chemical principles—to help us better understand the world around us, as well as to contribute positively to that world. It is my hope that students using this book will develop a foundation of chemical principles with which they can begin to understand the processes that make up the world and underlie life and will use that foundation to succeed and prosper, not only in any subsequent chemistry courses but in life as well.

Audience and Philosophy

This book is intended primarily to serve beginning chemistry students, who typically have had no instruction, or limited instruction, in chemistry. These students often need help not only with mathematical manipulations, but also with reading and writing scientific material with precision. When the author was a child, he was told orally that Benedict Arnold was a traitor and at a different time that Lewis and Clark were traders. These two descriptive words sounded alike to me, but I knew from the context that Lewis and Clark were not held in the same disdain. If the two words had been presented together, there would have been no problem in understanding the difference. In this text, I have tried to put concepts and terms that might be confused by students together in the same problem so that the differences can be understood from the beginning.

The factor label method (dimensional analysis) is used help the student translate word problems into easily solved algebraic expressions. In many places, a problem is stated in parts to lead the student stepwise through a solution; later the same problem appears as a whole, worded as it might appear on an examination. There is a series of figures that build on preceding ones to reveal the fundamental unity of the concept of the mole. Many problems are worded so as to show the student that very different questions may sound similar and that the same question may be presented in very different words; these will encourage the student to try to understand concepts and not to memorize solutions. When different terms that look or sound alike are presented and explained together, the student can more easily learn both (see Problems 5.1, 5.2, 5.3, 5.5, 5.6, 5.14 and 5.16, for example).

Frequent use of analogies to daily life helps students understand that chemistry problems are not significantly different from everyday problems, even though they may seem more difficult because of their unfamiliarity. For example, calculations involving dozens of pairs of socks and moles of diatomic molecules may be carried out by the same methods (Problems 7.6 and 7.7). Oxidizing and reducing agents can be compared conceptually to dish towels and wet dishes (Examples 14.5 and 14.6). Specific heat calculations are like those involving room rates at a hotel (Example 16.13). The similarity of a catalyst to a marriage broker is presented in Problem 8.15.

Modern nomenclature is used throughout the text (the Stock system for inorganic compounds and IUPAC nomenclature for organic compounds), but common names for the simplest organic compounds are included and the older system for naming cations is mentioned. Classical group numbering is used in the periodic table since this numbering is an aid to learning many elementary concepts (the number of electrons in an atom's outermost shell, for example).

The second edition of this text has been changed to reflect requests by colleagues in several areas, but its major philosophy and teaching techniques have not been altered. The principal aim to teach the student as (s)he comes to the class has been maintained.

The second edition includes several major topics not covered in the first edition. Notable among these are Atomic and Ionic Sizes, Ionization Potential and Electron Affinity, Electronegativity and Bond Polarity, Molecular Shape, and Polar and Nonpolar Molecules, (all in Chapter 13) and Enthalpies of Chemical Reactions (Section 14.4). The added material gives instructors even more flexibility in the topics they choose to cover in their courses. (Practically no one covers the whole book.)

This edition has also been made more flexible by making optional Chapter 4, Detailed Electronic Configurations, and the parts of later chapters dependent on this chapter. The later material dependent on Chapter 4 has been highlighted with green background, so that instructors can easily tell their students which parts to omit if they so choose.

The number of end-of-chapter exercises has been increased about 50%, with selected problems highlighted in red that have solutions included in the text as Appendix 5.

Pedagogical Devices

The text includes a variety of pedagogical devices. These were chosen and designed to answer the question "If I were a student, what would help me organize and understand the material covered in this book?"

1. *Key Terms and Symbols.* Each chapter begins with a listing of the key terms and the symbols or abbreviations that are used in the chapter. Sectional references are provided. Students have the opportunity to use these lists to

review their understanding of the important terms and symbols before examinations.

2. *Chapter Outline.* There is also an outline at the beginning of each chapter. The outline allows students to tell at a glance how the chapter is organized and what major topics are included.

3. *Learning Objective.* At the start of each section, a learning objective is presented to alert students to the key concepts covered in the section. These objectives are another valuable study tool for students when they are reviewing chapter material for examinations.

4. *Boldfaced Key Terms.* Key terms appear in **boldface** when they are introduced within the text and are immediately defined in context. These terms are also listed, with sectional references, at the start of the chapter. All key terms are defined in the glossary.

5. *Items of Interest.* Throughout the text, boxes titled "Item of Interest" relate the subject matter to the real world.

6. *Marginal Comments.* Marginal comments are designed to alert the student to a key point, a helpful hint, or a safety caution.

7. *Tables.* Numerous strategically placed tables list and summarize important information, making it readily accessible for efficient study.

8. *Enrichment Boxes.* Throughout the text, boxes titled "Enrichment" highlight special topics that take the text material to a more extended level. Students will find them to be a lively and interesting feature as they investigate the processes of chemistry.

9. *Examples and Practice Problems.* The book has a wealth of Examples that show the student the step-by-step solution to the problem presented, which is directly related to the preceding textual information. The Practice Problems that follow most Examples give students the opportunity to solve a similar problem immediately. Solutions to Practice Problems are presented at the end of the book to provide immediate feedback.

10. *Photographs.* A wide array of visually appealing and informative photographs is used to help students understand chemical and physical phenomena and pique their interest.

11. *Illustrations.* Because a picture is worth a thousand words, each chapter is amply illustrated with accurate, colorful diagrams that clarify difficult concepts and enhance learning.

12. *Flow Diagrams.* To help students understand the steps in problem solving, flow diagrams have been included at key locations throughout the text. These diagrams allow students to visualize the process of solving a problem.

13. *Summary.* At the end of the chapter, there is a summary of the major concepts covered. Each section is reviewed in paragraph form. The summary, along with the chapter outline and section objectives, provides a complete overview of the chapter material.

14. *Items for Special Attention.* Appearing at the end of every chapter, this unique section highlights and emphasizes key concepts that often confuse students. This section anticipates students' questions and problem areas and helps them avoid many pitfalls.

15. *Self-Tutorial Problems.* This end-of-chapter section presents problems in simple form designed as teaching devices. Many are from everyday life, and they

emphasize the importance of identifying the information needed to answer questions, thus encouraging the advancement of students' analytical skills.

16. *Problems.* The problems in this end-of-chapter set are grouped under headings that match the chapter's section titles. This organization allows students to practice the problem-solving skills and methods associated with each important concept presented in the text.

17. *General Problems.* The final set of problems in each chapter is more difficult than the others and is not classified by topic. Many of these problems require knowledge of two or more concepts. Similar in scope to the type of questions students will be confronted with on tests, these problems provide students with an excellent means by which to judge their knowledge of the chapter's contents.

18. *Appendixes.* The book contains a complete set of appendixes, which include the solutions to the in-text Practice Problems (Appendix 4) and selected end-of-chapter problems (Appendix 5). A short review of scientific algebra and a unique presentation detailing the use of the electronic calculator (Appendix 1) will help students overcome any mathematical deficiencies. Lists of symbols, abbreviations, prefixes and suffixes, and mathematical equations (Appendixes 2 and 3) make the book more user-friendly.

19. *Glossary.* A complete glossary of all important terms is found at the end of the text.

Supplemental Materials

An extensive supplemental package has been designed to support this book. It includes the following elements:

1. *Instructor's Manual.* The instructor's manual contains the printed test item file, a list of transparencies, and suggestions on how to organize the course.

2. *Student Study Guide.* The student study guide offers students a variety of exercises, self-tests, and hints to promote their comprehension of the basics as well as the more difficult concepts.

3. *Transparencies.* A set of 50 color transparencies is available to help the instructor coordinate the lecture with key illustrations from the text.

4. *Customized Transparency Service.* If adopters are interested in acetates of text figures not included in the standard transparency set, those acetates will be custom-made upon request. Contact your local Wm. C. Brown Publishers sales representative for more information.

5. *TestPak.* This computerized classroom management system/service includes a database of test questions, reproducible student self-quizzes, and a grade-recording program. Disks are available for IBM and Apple computers, and no programming experience is required. If a computer is not available, instructors can choose questions from the Test Item File in the instructor's manual and phone or FAX Wm. C. Brown Publishers to request a printed exam, which will be returned within 48 hours.

6. *Laboratory Manual.* Written by Kathy Tyner of Southwestern College, *Lab Exercises in Preparatory Chemistry* features 62 class-tested experiments. The manual can be easily customized to suit instructors' individual needs. The instructor can delete experiments, add his or her own experiments, or change the arrangement to create a custom manual to fit specific class needs.

7. *Laboratory Resource Guide.* This helpful prep guide contains the hints that the author has learned over the years to ensure success in the laboratory.

8. *ChemSkill Builder, Personalized Problem-Sets for General Chemistry.* Developed by James D. Spain and Harold J. Peters of Electronic Homework Systems, Inc., *ChemSkill Builder* software challenges your students knowledge of introductory chemistry with an array of individualized problems. Organized to accompany any introductory chemistry text, this student-oriented software generates questions for students in a randomized fashion with a constant mix of variables. No two students will receive the same electronic homework problems-ensuring an accurate test of students' knowledge. This unique software program records grades on the quizzes. These grades can easily be transferred to an instructor's record keeping file.

9. *Videotapes.* Narrated by Ken Hughes of the University of Wisconsin—Oshkosh, the tapes provide six hours of laboratory demonstrations. Many of the demonstrations are of high-interest experiments, too expensive or too dangerous to be performed in the typical introductory laboratory. Contact your local Wm. C. Brown Publishers sales representative for more details.

10. *How to Study Science.* Written by Fred Drews of Suffolk County Community College, this excellent workbook offers students helpful suggestions for meeting the considerable challenges of a science course. It offers tips on how to take notes and how to get the most out of laboratories, as well as how to overcome science anxiety. The book's unique design helps to stir critical thinking skills, while facilitating careful note taking on the part of the student.

Acknowledgments

The preparation of a textbook is a family effort, and the quality of the final product is a reflection of the dedication of all the family members. First, I would like to thank my own family, without whose patience and support this project would not have been possible. Second, I would like to thank the scores of my fellow chemists who have taught me much in the past and continue to do so. Learning is a never-ending process, and I continue to learn from my colleagues and students. I would also like to thank the members of my extended family at Wm. C. Brown Publishers, without whom there would not have been a text: my Developmental Editors, Brittany Rossman, Russ Lidberg, and Bob Fenchel, and my Acquisitions Editor, Craig Marty. I gratefully acknowledge the invaluable help of the following dedicated reviewers, who provided expert suggestions and the needed encouragement to improve the text:

Caroline Ayers
East Carolina University

Stan Grenda
University of Nevada, Las Vegas

A. Kurtz Carpenter
Lower Columbia College

Susan A. Herking
Vincennes University–Jasper Center

Pamela Coffin
University of Michigan–Flint

Narayan S. Hosmane
Southern Methodist University

Donald Gauntlett
Lehigh County Community College

T. G. Jackson
University of South Alabama

To the Student

This book is designed to help you learn the fundamentals of chemistry. To be successful, you must master the concepts of chemistry and acquire the mathematical skills that are necessary to solve problems in this quantitative science. If your algebra is rusty, you should polish it up. Appendix 1 reviews the algebra used in basic chemistry and also shows how to avoid mistakes while solving chemistry problems with your scientific calculator. The factor label method is introduced in Chapter 2 to show you how to use units to help with problem solutions. You can help yourself by using the standard symbols and abbreviations for various quantities (such as m for mass, m for meter, mol for moles, and M for molarity). Always use the proper units with your numerical answers; it make a big difference whether your roommate's pet is 3 inches long or 3 feet long!

Many laws, generalizations, and rules are presented in the study of basic chemistry. Most students can master these. Successful students, however, not only know them, but also know *when to use each one.* Word problems are the biggest hurdle for most students who do have difficulty with chemistry. The best way to learn to do word problems is to practice intensively. Review the Examples and do the Practice Problems until you feel confident that you understand the concepts and techniques involved. (Do not try to memorize solutions; there are too many different ways to ask the same questions, and many similar-sounding questions are actually quite different.) Then, do as many of the end-of-chapter problems as you possibly can to see whether you have mastered the material.

You should not try to speed-read chemistry. Mere reading of a section will not generally allow full comprehension of the material. You must be able to solve the problems to be sure that you have really mastered the concepts. Many of the problems sound alike but are very different (for example, Problems 5.5, 7.6, 7.7, and 11.7), and many others sound different but are essentially the same (for example, 3.4, 5.13, 8.2, and 8.17). These will help you develop careful reading habits and prepare you for the questions asked on examinations.

Problems from everyday life that are analogous to scientific problems are included to help you understand certain points better. Other problems are first presented in parts to help you work through the solution and later appear as a single question, as is more likely to occur on examinations. Some of the problems are very easy; these are generally intended to emphasize an important point. After solving one of these problems, ask yourself why such a question was asked. Make sure you understand the point.

Make sure you understand the scientific meaning of each new term introduced. For example, the word *significant* as used in Chapter 2 means something entirely different from its meaning in everyday conversation; be sure you understand the difference. Key terms are **boldfaced** when they are first introduced in the text. A list of these terms is given at the beginning of each chapter. A complete glossary of all important terms is provided at the end of the book.

Other materials to aid your study include lists of standard symbols and abbreviations for variables, units, and subatomic particles, found in Appendix 2. A summary of the mathematical equations used in the book is presented in Appendix 3. The solutions to all Practice Problems and selected end-of-chapter problems are provided in Appendices 4 and 5, respectively. The selected end-of-chapter problem numbers are printed in red. A periodic table is printed inside the front cover of the book, and a table of the elements appears inside the back cover. Let these tools help you succeed!

1 Basic Concepts

- 1.1 Classification of Matter
- 1.2 Properties
- 1.3 Matter and Energy
- 1.4 Chemical Symbols
- 1.5 The Periodic Table
- 1.6 Laws, Hypotheses, and Theories

■ Key Terms *(Key terms are defined in the Glossary.)*

alkali metal (1.5)
alkaline earth metal (1.5)
analytical chemistry (intro)
biochemistry (intro)
chemical change (1.1)
chemical reaction (1.1)
chemistry (1.3)
coinage metal (1.5)
compound (1.1)
definite composition (1.1)
dissolve (1.2)
ductile (1.5)
element (1.1)
energy (1.3)
extensive property (1.2)
family (1.5)
formula (1.4)

group (1.5)
halogen (1.5)
heterogeneous mixture (1.1)
homogeneous mixture (1.1)
hypothesis (1.6)
inner transition element (1.5)
inorganic chemistry (intro)
intensive property (1.2)
law (1.6)
law of conservation of energy (1.3)
law of conservation of mass (1.6)
main group element (1.5)
malleable (1.2)
mass (1.3)
matter (1.3)
metal (1.5)

metalloid (1.5)
mixture (1.1)
noble gas (1.5)
nonmetal (1.5)
organic chemistry (intro)
period (1.5)
periodic table (1.5)
physical change (1.1)
physical chemistry (intro)
property (1.2)
quantitative property (1.2)
solution (1.1)
substance (1.1)
symbol (1.4)
theory (1.6)
transition element (1.5)

Chemistry is the study of matter and energy. Matter includes all the material things in the universe. In Section 1.1, you will learn to classify matter into various types—elements, compounds, and mixtures—based on composition. Properties—the characteristics by which types of matter may be identified—are discussed in Section 1.2.

Energy may be defined as the ability to do work. We often carry out chemical reactions for the sole purpose of changing energy from one form to another—for example, when we burn fuel in our homes or cars. The relationship between energy and matter, an important one for chemists, is explored in Section 1.3.

Symbols, introduced in Section 1.4, are used to represent the elements. The periodic table, introduced in Section 1.5, groups elements with similar properties. Section 1.6 presents scientific laws, hypotheses, and theories that generalize and explain natural phenomena.

For convenience, chemistry is often divided into the following five subdisciplines: organic chemistry, inorganic chemistry, analytical chemistry, physical chemistry, and biochemistry. **Organic chemistry** deals with most compounds of carbon. These compounds are introduced systematically in Chapter 19. **Inorganic chemistry** deals with all the elements and with compounds that are not defined as organic. **Analytical chemistry** involves finding which elements or compounds are present in a sample and/or how much of each is present. **Physical chemistry** deals with the properties—especially quantitative (measurable) properties—of substances. **Biochemistry** deals with the chemistry of living things.

These subdivisions of chemistry are somewhat arbitrary. A chemist specializing in any one of the first four subdivisions uses all of the first four and often biochemistry as well. A biochemist uses all five specializations. For example, the modern organic chemist often uses inorganic compounds to convert starting materials to desired products and then analyzes the products and measures their properties. In addition, many organic chemists now are investigating compounds of biological interest.

1.1   Classification of Matter

All the materials in the world are composed of a few more than a hundred elements. **Elements** are the simplest form of matter and cannot be broken down chemically into simpler, stable substances. They can be thought of as building blocks for everything in the universe. The same elements that make up the earth also make up the moon, as shown by actual analysis of moon samples. Moreover, indirect evidence obtained from analysis of light from stars shows that the rest of the universe is composed of the same elements.

Clearly the number of different combinations of elements must be huge to get all the varieties of materials in the universe. But elements can combine in only two fundamentally different ways: by physical changes to form mixtures or by chemical changes to produce compounds. **Chemical changes,** also called **chemical reactions,** change the composition (or structure) of a substance. **Physical changes** do not alter the composition. The breaking of glass into small pieces is an example of a physical change. The glass still has the same composition and the same properties as before, but its external form is changed. The burning of charcoal (mostly carbon) in air (or in pure oxygen) to get carbon dioxide, a colorless gas, is an example of a chemical reaction. Not only the form of the material but also its composition has changed. The gas has both carbon and oxygen in it, but the charcoal had no oxygen.

If a sample of matter cannot be broken down into simpler substances by ordinary chemical means, the sample is an element. An element has a definite set of properties. A **compound** is a chemical combination of elements that has its own set of properties and a **definite composition.** For example, pure water obtained from any natural source contains 88.8% oxygen and 11.2% hydrogen by mass. Compounds can be separated into their constituent elements only by chemical means.

■ EXAMPLE 1.1

The percentage of sodium in a 1.00-ounce sample of pure baking soda (sodium hydrogen carbonate) is 27.3%. (a) Is baking soda an element or a compound? (b) What is the percentage of sodium in a 23.1-ounce sample of the same substance?

Solution

(a) Baking soda is a compound; it contains more than one element.

(b) The larger sample is also 27.3% sodium because a given compound always contains the same percentage of each of its elements, no matter what its size. ■

Elements and compounds are the two types of **substances,** often referred to as pure substances.

Two or more substances—elements, compounds, or both—can combine physically to produce a mixture. A mixture can be separated into its components by physical means. **Mixtures** are physical combinations of substances that have properties related to those of their components but that do not have definite compositions. They can be either **heterogeneous mixtures** or **homogeneous mixtures.** In heterogeneous mixtures, two or more different types of matter can be seen with the naked eye or a good optical microscope. Homogeneous mixtures, also called **solutions,** look alike throughout, even with a microscope.

Both types of pure substances are homogeneous.

ITEM OF INTEREST

The difference between elements and compounds is illustrated in human nutrition. *Vitamins* are complex compounds of carbon, hydrogen, and several other elements. A vitamin owes its activity to the nature of the compound as a whole, and any slight change in it can destroy its nutritional value. About 20 elements are called *minerals*. They play a role in human nutrition. The minerals known to be essential for good health are calcium, phosphorus, potassium, sulfur, sodium, chlorine, magnesium, iron, manganese, copper, iodine, cobalt, fluorine, and zinc. Traces of silicon, boron, arsenic, strontium, aluminum, bromine, and nickel may also be required. These elements are eaten in the form of their compounds, but it does not matter much which compounds.

Heating a vitamin will destroy its potency by breaking the compound into other compounds. In contrast, heating a compound that contains one of the essential minerals might destroy the compound, but it will not change the mineral into another element. For example, calcium citrate can be changed into another calcium-containing compound, but the calcium is still present.

Figure 1.1 Classification of Matter

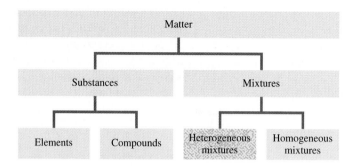

The word *homogenize* is related to the term *homogeneous,* but as used in everyday conversation, it does not mean exactly the same thing. For example, homogenized milk is not really homogeneous; you can see individual particles of cream under a microscope. Truly homogeneous liquids are transparent (though not always colorless). If you cannot recognize objects viewed through a thin layer of liquid, the liquid is not homogeneous.

Table 1.1 Classification of Matter

Pure substances
    Elements
    Compounds
Mixtures
    Heterogeneous mixtures
    Homogeneous mixtures
      (solutions)

The entire classification scheme for matter discussed in this section is outlined in Table 1.1 and Figure 1.1.

■ **EXAMPLE 1.2**

If you stir a teaspoon of salt into a glass of water and a teaspoon of mud into another glass of water, the salt will disappear (dissolve) into the water, but the mud will not (see Figure 1.2). Which mixture is a solution?

Solution

The salt forms a solution—a homogeneous mixture—with the water. The mud and water form a heterogeneous mixture. Particles of mud are easy to see in the

Figure 1.2 Salt Plus Water, and Mud Plus Water

(a) The salt dissolves in the water and is not distinguishable from the water; a solution is formed. (b) The mud does not dissolve in the water; a heterogeneous mixture is formed.

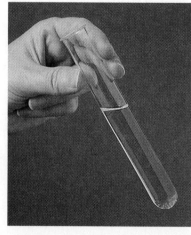 

(a) (b)

mud-water mixture, but seeing any salt particles in the salt-water solution is impossible, no matter how hard you look (even with a microscope).

Practice Problem 1.2   When solid iodine is added to ethyl alcohol, a colorless liquid, it forms a transparent liquid mixture with a deep color. Is the mixture homogeneous or heterogeneous? ■

■ EXAMPLE 1.3

Classify each of the following statements as true or false:

(a) Every substance is a compound.

(b) Every compound is a substance.

(c) Every compound contains two or more elements.

(d) Every mixture contains two or more compounds.

(e) Every mixture contains two or more substances.

(f) Every mixture contains two or more free elements.

(g) All substances are homogeneous.

(h) All mixtures are homogeneous.

(i) All heterogeneous samples are mixtures.

Solution

(a) False (Some are free elements.)

(b) True

(c) True

(d) False (They may contain free elements or compounds or both.)

(e) True

(f) False (They may contain compounds and only one or no free elements.)

(g) True

(h) False (Some are heterogeneous.)

(i) True

Practice Problem 1.3   Classify each of the following statements as true or false:

(a) Every substance contains two or more elements.

(b) Every mixture contains two or more elements.

(c) All mixtures are heterogeneous.

(d) All homogeneous samples are solutions. ■

1.2 Properties

OBJECTIVE

■ to use properties to help identify substances

Every substance has a definite set of properties. **Properties** are the characteristics by which we can identify something. For example, we know that pure water is a colorless, odorless, tasteless substance that is a liquid under the conditions usually found in an ordinary room. Water puts out fires, and it dissolves sugar and salt. Liquid water can be changed into a gas (called water vapor or steam) by heating it or into a solid (ice) by cooling it. Salt has a different set of properties than water does; sugar has yet another set.

Some properties of a sample of a substance depend on the quantity of the sample. These properties are called **extensive properties.** For example, the volume of a solid sample depends on how much substance is present. Other properties, such as color and taste, do not depend on how much substance is present. These properties are known as **intensive properties.** Intensive properties are much more useful for identifying substances.

■ EXAMPLE 1.4

(a) You have one sample that weighs a pound and a second sample that weighs an ounce. Can you tell which sample is aluminum and which is sand?

(b) You have one sample that is shiny and another sample that is a dull brown. Can you tell which sample is aluminum and which is sand?

Solution

(a) The weight of a sample is an extensive property that does not tell you anything about the material's identity.

(b) The intensive properties described allow you to tell which of the two samples is aluminum (the shiny one) and which is sand (the brown one).

Practice Problem 1.4   A chemistry professor drove 24 miles to work one day. Can you tell how fast the professor drove, or how long it took? ■

Some of the most important intensive properties that chemists use to identify substances are ones that they measure; they are called **quantitative properties.** Two such properties are the freezing point and the normal boiling point of a substance, which are the temperatures at which a liquid freezes to form a solid and boils to form a gas under normal atmospheric conditions. We will discuss quantitative properties in more detail in Chapter 2.

We can distinguish compounds from mixtures because of compounds' characteristic properties. Mixtures have properties like those of their constituents. The more of a given component present in a mixture, the more the mixture's properties will resemble those of that component. For example, the more sugar we put into a glass of water, the sweeter is the solution that will be produced.

An experiment will illustrate how properties are used to distinguish between a compound and a mixture. We place small samples of iron filings and powdered sulfur on separate watch glasses to investigate their properties (Figure 1.3a). We note that both are solids. We place the samples in separate test tubes and then hold a magnet beside the first tube (Figure 1.3b). We find that the iron is attracted to the magnet. When we hold the magnet next to the tube with the sulfur, nothing happens; the sulfur is not attracted by the magnet.

When we pour carbon disulfide, a colorless, flammable liquid, on the sulfur sample, the solid sulfur disappears, and the liquid turns yellow. The sulfur has **dissolved,** forming a solution with the carbon disulfide. When we pour carbon disulfide on the iron, nothing happens; the iron stays solid, and the liquid stays colorless. If we had large pieces of each element, we could pound them with a hammer and find that the sulfur is brittle and easily powdered but that the iron

Caution: Carbon disulfide is both explosive in air and poisonous.

Figure 1.3 *Iron, Sulfur, and a Mixture of the Two*

(a) Iron filings (black) and powdered sulfur (yellow).
(b) The iron is attracted by the magnet, but the sulfur is not.
(c) The iron filings in a mixture of iron and sulfur are still attracted by the magnet. Some of the powdered sulfur sticks to the iron filings, but the sulfur is not attracted by the magnet.

(a)

(b)

(c)

does not easily break into small pieces. Iron is **malleable**—that is, it can be pounded into various shapes. Table 1.2 lists the properties discussed so far of the two elements.

Next we pour some iron filings and some powdered sulfur into a large test tube and stir them together. The sample appears to be a dirty yellow, but if we look closely, we can see yellow specks and black specks. If we hold a magnet next to the test tube (Figure 1.3c), mostly black particles with some yellow particles clinging to them are attracted by the magnet. When we pour some carbon disulfide on the sample, the liquid turns yellow. We pour off that liquid and pour on more carbon disulfide until no yellow solid remains in the sample. When we evaporate the carbon disulfide in a fume hood, we get a yellow solid again. If we place a magnet next to the black material left in the large test tube, we find that it is attracted to the magnet. It seems that mixing the two samples of elements has not changed their properties. The sulfur is still yellow and still soluble in carbon

Table 1.2 *Some Properties of Iron, Sulfur, and an Iron-Sulfur Compound*

Iron	Sulfur	Iron-Sulfur Compound
Solid	Solid	Solid
Black	Yellow	Black
Malleable	Brittle	Brittle
Shiny	Dull	Dull
Magnetic	Not magnetic	Not magnetic
Insoluble in carbon disulfide	Soluble in carbon disulfide	Insoluble in carbon disulfide

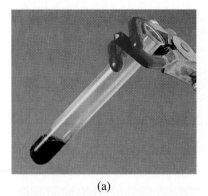

(a)

(b)

Figure 1.4 Reaction of Iron and Sulfur

(a) When a mixture of iron and sulfur is heated, the two elements react. Some sulfur is vaporized and then deposits on the test tube wall.
(b) The pulverized product of the reaction is not attracted by a magnet.

disulfide; the iron is still black and still magnetic. The two elements have retained their properties and their identities; they are still elements. The combination of the two is a mixture. A mixture does not have a definite composition, and it has properties related to the properties of its components.

Now we place two new, carefully measured samples of iron filings and powdered sulfur in another large test tube and heat the mixture strongly with a Bunsen burner. After a time, a red glow appears in the bottom of the tube and gradually spreads throughout the sample. Some sulfur escapes into the gas phase because of the heat and then deposits on the test tube wall (Figure 1.4a). A black solid results from the chemical reaction. When we remove the solid from the test tube (we may have to break the tube to get it out), we can pulverize the solid with a hammer—that is, it is brittle. If we try to dissolve the material in carbon disulfide, it does not dissolve. If we bring the magnet close to it, it is not attracted (Figure 1.4b). This material has its own set of properties: a dull black color, brittleness, insolubility in carbon disulfide, lack of attraction to a magnet (see Table 1.2). It is a compound—a chemical combination of iron and sulfur.

■ EXAMPLE 1.5

After 10.0 grams of a certain substance is heated in air until no further reaction takes place, 9.32 grams of a metal is left. After 10.0 grams of another substance is heated in air, 14.0 grams of white powder is left. Can you tell whether the reactants and the products are elements or compounds?

Solution

The first substance is a compound. When it is heated, it decomposes into a metallic material that is left behind and some gaseous product that escapes into the air. Since the metal has less mass than the original substance, it is simpler. The original substance is decomposable—it is not an element. The metal product might or might not be decomposable, so you cannot tell from the information given whether it is an element or a compound.

The second substance combined with something in the air; it gained mass. The powdery product is therefore a combination of substances and cannot be an element. You do not know if the original substance can be decomposed (it was not decomposed in this experiment), so you cannot tell if it is an element or a compound.

Practice Problem 1.5　A 7.00-gram sample of a certain shiny substance is heated in air. Afterward, 10.80 grams of a fluffy, white powder is present. Is the powder an element? ■

1.3　Matter and Energy

■ to distinguish between matter and energy, as well as among matter, mass, and weight

Matter is anything that has mass and occupies space. All the material things in the universe are composed of matter, including anything you can touch as well as the planets in the solar system and all the stars in the sky.

The **mass** of an object measures how much matter is in the object. Mass is directly proportional to weight at any given place in the universe. If you leave the surface of the earth, your mass remains the same, but your weight changes. An astronaut positioned between two celestial bodies such that their gravitational attractions pull equally in opposite directions is weightless, but the astronaut's mass remains the same as it was on earth. Since chemists ordinarily do their work on

ENRICHMENT

In 1905, Albert Einstein (1879–1955) published his theory that the mass of a sample of matter is increased as the energy of the sample is increased. For example, a baseball in motion has a very slightly greater mass than the same baseball at rest. The difference in mass is given by the famous equation

$$E = mc^2$$

In this equation, E is the energy of the object, m is the *mass difference,* and c^2 is a very large constant—the square of the velocity of light:

$$c^2 = (186\ 000\ \text{miles/second})^2$$
$$= 34\ 600\ 000\ 000\ \text{miles}^2/\text{second}^2$$

For macroscopic bodies such as a baseball, the increase in mass because of the added energy is so small that it is not measurable. It was not even discovered until the beginning of the twentieth century. At atomic and subatomic levels, however, the conversion of a small quantity of matter into energy is very important. It is the energy source of the sun and the stars, the atomic bomb, the hydrogen bomb, and nuclear power plants.

Table 1.3   Forms of Energy

Heat

Mechanical
  Kinetic (energy of motion)
  Potential (energy of position)

Electrical

Sound

Chemical

Nuclear

Electromagnetic (light)
  Visible light
  Ultraviolet
  Infrared
  X rays
  Solar*
  Radio waves
  Microwaves

*Solar energy is a combination of several forms of light.

the earth's surface and since mass and weight are directly proportional here, many chemists use the terms *mass* and *weight* interchangeably, but you should remember that they differ.

Energy is the capacity to do work. You cannot hold a sound or a beam of light in your hand; they are not forms of matter but forms of energy. The many forms of energy are outlined in Table 1.3. Energy cannot be created or destroyed, but it can be converted from one form to another. This statement is known as the **law of conservation of energy.**

■ EXAMPLE 1.6

What desired energy conversion is exhibited by (a) a neon sign in operation and (b) a car battery starting a car?

Solution

(a) Electrical energy is converted to light.

(b) The chemical energy stored in the battery is converted to electrical energy, which is then converted to mechanical energy by the starter.

Practice Problem 1.6   What desired energy conversion is exhibited by (a) an alternator in a car and (b) an elevator going up? ■

Chemistry is the study of the interaction of matter and energy and the changes that matter undergoes. (In nuclear reactions, tiny quantities of matter are actually converted to relatively large quantities of energy. See Chapter 20.)

1.4 Chemical Symbols

OBJECTIVE

■ to write symbols for the elements and names of elements from the symbols

Since the elements are the building blocks of all materials in the universe, we need an easy way to identify and refer to them. For this purpose, each chemical element is identified by an internationally used **symbol** consisting of one or two letters. (Certain recently created elements have three-letter symbols by international agreement, but we will not consider these elements at all in this course.) The first

Figure 1.5 *Elements Whose Names and Symbols Should Be Learned*

Legend:
- Most important elements in this course
- Other important elements in this course

1																	
H Hydrogen																	2 **He** Helium
3 **Li** Lithium	4 **Be** Beryllium											5 **B** Boron	6 **C** Carbon	7 **N** Nitrogen	8 **O** Oxygen	9 **F** Fluorine	10 **Ne** Neon
11 **Na** Sodium	12 **Mg** Magnesium											13 **Al** Aluminum	14 **Si** Silicon	15 **P** Phosphorus	16 **S** Sulfur	17 **Cl** Chlorine	18 **Ar** Argon
19 **K** Potassium	20 **Ca** Calcium	21 **Sc** Scandium	22 **Ti** Titanium	23 **V** Vanadium	24 **Cr** Chromium	25 **Mn** Manganese	26 **Fe** Iron	27 **Co** Cobalt	28 **Ni** Nickel	29 **Cu** Copper	30 **Zn** Zinc	31 **Ga** Gallium	32 **Ge** Germanium	33 **As** Arsenic	34 **Se** Selenium	35 **Br** Bromine	36 **Kr** Krypton
37 **Rb** Rubidium	38 **Sr** Strontium								46 **Pd** Palladium	47 **Ag** Silver	48 **Cd** Cadmium		50 **Sn** Tin	51 **Sb** Antimony	52 **Te** Tellurium	53 **I** Iodine	54 **Xe** Xenon
55 **Cs** Cesium	56 **Ba** Barium		74 **W** Tungsten						78 **Pt** Platinum	79 **Au** Gold	80 **Hg** Mercury		82 **Pb** Lead	83 **Bi** Bismuth			86 **Rn** Radon
87 **Fr** Francium	88 **Ra** Radium																

92 **U** Uranium

Figure 1.5 Elements Whose Names and Symbols Should Be Learned

The elements shown with a red background are most important in this course. Those with a blue background are also important.

Table 1.4 *Elements Whose Names and Symbols Begin with Different Letters*

Name	Symbol
Antimony	Sb
Gold	Au
Iron	Fe
Lead	Pb
Mercury	Hg
Potassium	K
Silver	Ag
Sodium	Na
Tin	Sn
Tungsten	W

letter of an element's symbol is always capitalized. If there is a second letter in the symbol, it is a lowercase (small) letter. The symbol is an abbreviation of the element's name, but some symbols represent names in languages other than English. The 10 elements whose symbols and names have different first letters are listed in Table 1.4. A list of the names and symbols of the first 103 elements, along with some other information, is presented in a table inside the back cover of this book. In that table, the elements are alphabetized according to their names, but duplicate entries appear under the initial letter of the symbols for the elements in Table 1.4.

The most important symbols for beginning students to learn are given in Figure 1.5. The names of the elements indicated and their symbols must be memorized. Don't bother to memorize the numbers shown in the boxes with the elements. The elements indicated by red shading should be learned first.

Chemists write symbols together in **formulas** to identify compounds. For example, the letters CO represent a compound of carbon and oxygen. Be careful to distinguish the formula CO from the symbol Co, which represents the element cobalt. The capitalization of letters is very important! Formulas are sometimes written with subscripts to tell the relative proportions of the elements present. For example, H_2O represents water, which has two atoms of hydrogen for every atom of oxygen present. More about formulas will be presented in Section 5.1.

1.5 The Periodic Table

OBJECTIVE

■ to begin to classify the elements in a systematic manner. To identify periodic groups, periods, and sections of the periodic table by name and/or number

In Section 1.2, you learned some of the properties of sulfur and iron. Do you have to learn the properties of all 100 or so elements individually, or are there some ways to ease that burden? For over 150 years, chemists have arranged the elements in groups with similar chemical characteristics, which makes it easier to learn their properties. This grouping of the elements has been refined to a high degree, and the modern **periodic table** is the result. A full periodic table is shown inside the front cover of this book. We will explore several uses for the periodic table in this section, as well as a number of terms used with it. This table will be used extensively throughout the rest of this course, and in subsequent chemistry courses.

All the elements in any horizontal row of the periodic table are said to be in the same **period.** There are seven periods, the first consisting of just 2 elements. The second and third periods contain 8 elements each, and the next two contain 18 elements each. The sixth period has 32 elements (including 14 inner transition elements numbered 57 through 71, located at the bottom of the table), and the last period is not complete. The periods are conventionally numbered with the Arabic numerals 1 through 7 (see Figure 1.6).

■ EXAMPLE 1.7

Which element begins the fourth period of the periodic table? Which element ends it? How many elements are in that period?

Solution

Potassium (K) begins the period, krypton (Kr) ends it, and there are 18 elements in the period.

Practice Problem 1.7   Which element begins the sixth period of the periodic table? Which element ends it? How many elements are in that period? ■

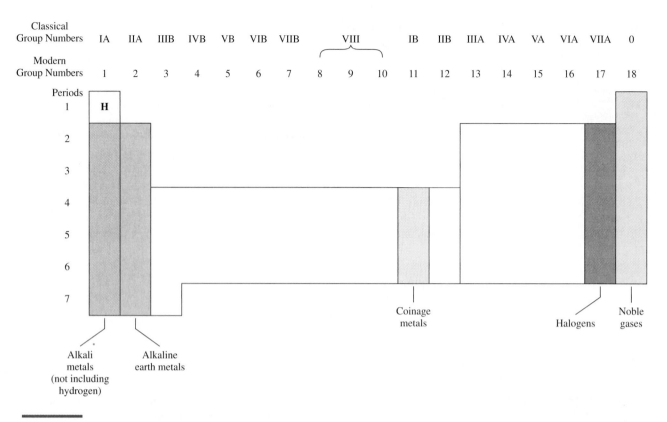

Figure 1.6 Groups and Periods

The elements in any vertical column in the periodic table are in the same **group,** or **family.** They have similar chemical properties, which change gradually from each one to the one below it; in some groups, the elements are very similar. The groups have two sets of group numbers (shown in Figure 1.6). The classical group numbers are Roman numerals followed by a letter A or B. These are more useful for beginning students in learning about atomic structure and bonding. The elements in two groups having the same number have some chemical similarities also, especially in the formulas of some of their compounds. For example, Na_2S and Cu_2S are formulas for sulfur compounds with elements in group IA and group IB, respectively. You can see that the formulas are similar. Another example is the pair of compounds Na_2SO_4 and Na_2CrO_4, in which elements from groups VIA and VIB, respectively, are present. The modern group numbers are given as Arabic numerals. The classical group numbers will be used throughout this book, with the modern group numbers sometimes added in parentheses afterward.

Five groups have family names (see Figure 1.6). The **alkali metals** include all the elements of group IA (1) except hydrogen. The **alkaline earth metals** are the elements of periodic group IIA (2), and the **coinage metals** are those of group IB (11). The **halogens** form group VIIA (17), and the **noble gases** constitute group 0 (18).

■ **EXAMPLE 1.8**

Is each of the following sets of elements in the same periodic group or in the same period? Which set has the more similar chemical properties?

(a) Be, Mg, Ca, Sr   (b) Si, P, S, Cl

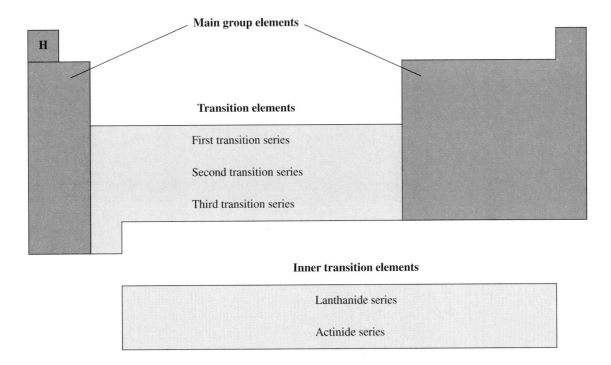

Figure 1.7 Main Group Elements, Transition Elements, and Inner Transition Elements

Solution

Set (a) is in the same group; set (b) is in the same period. Set (a) has the more similar chemical properties.

Practice Problem 1.8   Is calcium more likely to be similar to potassium or magnesium in its chemical properties? ■

Another major classification of the elements in terms of the periodic table is shown in Figure 1.7. Three areas are defined and named the **main group elements,** the **transition elements,** and the **inner transition elements.** The main group elements are the simplest to learn about, and they will be studied first. The transition elements include some of the most important elements in our everyday lives, such as iron and copper. The transition elements are often divided into three rows of elements, called the first transition series, the second transition series, and the third transition series. The two inner transition series fit into the periodic table in periods 6 and 7, right after lanthanum (La) and actinium (Ac), respectively. The inner transition elements include a few important elements, including uranium and plutonium. The first series of inner transition elements is called the lanthanide series, after lanthanum, the element that precedes them; the second series is called the actinide series, after actinium, the element that precedes them. These elements are conventionally placed below the others so as not to make the periodic table too wide. None of the actinide elements to the right of uranium has been found in nature; all of these elements were created by humans. All of the elements in the actinide series are radioactive.

■ EXAMPLE 1.9

How many transition elements are there (not counting the synthetic elements with three-letter symbols)?

Solution

There are 31.

Practice Problem 1.9   How many inner transition elements are there? ■

■ EXAMPLE 1.10

In what period are the lanthanide elements found?

Solution

The lanthanide elements, 58–71, follow element 57 and therefore are in period 6.

Practice Problem 1.10   In what period are the actinide elements found? ■

We can also divide the elements into metals and nonmetals because each of these classes has some distinctive properties common to all their members. For example, metals generally have a metallic luster (a glossy or shiny appearance) and are generally malleable (can be pounded into thin sheets) and **ductile** (can be drawn into a wire); nonmetals are generally brittle. Metals conduct electricity; most nonmetals do not.

In the periodic table, the **metals** are to the left of a stepped line starting to the left of boron (B) and continuing downward and to the right, ending to the left of astatine (At) (see Figure 1.8). Except for hydrogen, all the **nonmetals** are to the right of this line. The metallic elements greatly outnumber the nonmetallic elements. The properties of the elements vary gradually across the periodic table. Several of the elements near the stepped line have some properties of metals and some properties of nonmetals; they are sometimes called **metalloids.**

Figure 1.8 Metals and Nonmetals

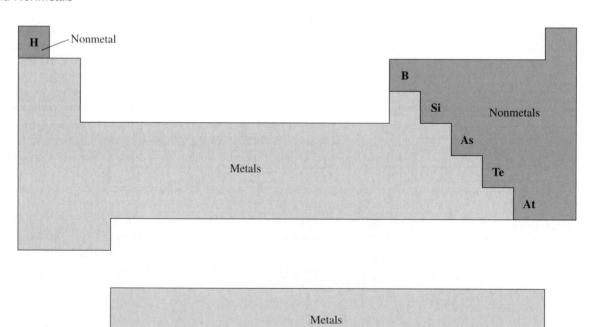

■ EXAMPLE 1.11

Which of the following elements are metals, and which are nonmetals?

(a) Aluminum   (b) Silver

(c) Nitrogen   (d) The carbon in a "lead" pencil

Solution

(a) Aluminum is a metal.   (b) Silver is a metal.

(c) Nitrogen is a nonmetal.   (d) Carbon is a nonmetal.

Practice Problem 1.11   Classify each of the following as metal or nonmetal:

(a) The carbon in a diamond   (b) Copper ■

Hydrogen is unique in its properties. It is placed on the side of the stepped line with the metals because it has some chemical properties similar to those of metals. In some periodic tables, it is also placed in another position, above the halogens, because of its nonmetallic chemical and physical properties. It actually does not fit comfortably in either position, since it is neither an alkali metal nor a halogen. To reflect its unique properties, it is placed in the center of still other periodic tables.

■ EXAMPLE 1.12

Use the periodic table to identify each of the following:

(a) The second element of the first transition series

(b) The third lanthanide

(c) The element of the fourth period that is also in group IIIB

(d) The fifth transition element   (e) The sixth inner transition metal

(f) The first element of group VIII   (g) The second halogen

(h) The last alkali metal   (i) The second coinage metal

Solution

(a) Ti   (b) Nd   (c) Sc   (d) Mn   (e) Eu   (f) Fe

(g) Cl   (h) Fr   (i) Ag

Practice Problem 1.12   Identify the first alkali metal. ■

1.6 Laws, Hypotheses, and Theories

OBJECTIVE

■ to distinguish among laws, hypotheses, and theories

So many facts are available to scientists as they do experiments and observe natural phenomena that the data must be classified so that they can be learned and understood. When a large group of scientific observations is generalized into a single statement, that statement is called a **law.** For example, when you drop a pencil, it falls downward. When you drop a book, it falls downward. These and millions of other such observations are grouped together and generalized as the law of gravity. A law is a general statement about observable facts.

After organizing observed data into a law, scientists try to explain the law. A statement that attempts to explain why a law is true is called a **hypothesis.** If the hypothesis becomes generally accepted, it becomes a **theory.** Einstein explained the law of gravity with his theory of relativity. Laws and theories are necessary because learning or remembering all the data that have been observed over the ages is impossible.

One of the most important laws in chemistry is the **law of conservation of mass.** This law states that, in any chemical reaction or physical change, the total mass present after the change is equal to the total mass present before the change. Chapter 3 will present John Dalton's explanation of this law, in which he proposed that the particles that make up matter can rearrange themselves in various ways but cannot be created or destroyed. That explanation is a theory; it explains the law. If the particles that make up the materials before and after the change are the same, the total mass must also be the same.

■ ■ ■ ■ ■ ■ ■ SUMMARY ■ ■ ■ ■ ■ ■ ■

Matter includes every material thing in the universe, and to be able to understand such a wide variety of items, we must classify matter. Matter is divided into pure substances and mixtures. Pure substances may be elements or compounds. Mixtures may be either heterogeneous or homogeneous. Elements are the fundamental building blocks of matter and cannot be broken down to simpler substances by chemical or physical means. Compounds are chemical combinations of elements; they have their own sets of properties and definite compositions. A physical combination of substances results in a mixture, whose components retain most of their properties. Mixtures do not have definite compositions. Homogeneous mixtures, called solutions, look alike throughout, but some parts of a heterogeneous mixture can be seen to be different from other parts. (Section 1.1)

Properties are the characteristics by which we can identify samples of matter. Intensive properties, such as color and brittleness, do not depend on the size of the sample, but extensive properties, such as volume, do. Intensive properties are more important in identifying substances. We can determine whether a combination of substances is a mixture or a compound by its properties. When we combine samples of matter, the result has more matter present than any of the original samples. When we break down a sample, each of the resulting products is composed of less matter than the starting sample. (Section 1.2)

Matter is anything that has mass and occupies space. Mass is a measure of the quantity of matter in a sample. The mass of an object does not change with its position in the universe. On the surface of the earth, mass is directly proportional to weight, and we determine the mass of an object by "weighing" it. Energy is the ability to do work and comes in many forms (Table 1.3). Energy cannot be created or destroyed, but it can be converted from one form to another. Chemistry is the study of the interaction of matter and energy and the changes that matter undergoes. (Section 1.3)

Each of the 100 or so elements important for this course has a chemical symbol consisting of one or two letters. The first letter (or the only one) is always written as a capital letter; the second, if present, is always written as a lowercase (small) letter. Associating the names of the most important elements (shown in Figure 1.5) with their symbols, and their symbols with their names, is a necessary skill. (Section 1.4)

The periodic table is a classification scheme for elements that is tremendously useful in learning the properties of the elements. It consists of seven periods and 16 classical groups, or families (18 in a more modern but less useful version). Several of the groups have names, which beginning students need to learn. The elements are separated into metals and nonmetals on the periodic table. They are also subdivided into main group elements, transition elements, and inner transition elements. (Section 1.5)

A statement that summarizes innumerable scientific facts and allows scientists to predict what will happen in a certain type of situation in the future is called a law. (For example, the law of gravity allows us to predict that if we drop something, it will fall downward. This law resulted from innumerable observations.) One of the most important chemistry is the law of conservation of mass, which states that mass cannot be created or destroyed in any chemical reaction or physical change. An explanation that is proposed to explain why a law works is called a hypothesis. If the explanation is accepted by the scientific community, it is known as a theory. (Section 1.6)

Items for Special Attention

■ Be sure to use the correct capitalization and abbreviations throughout your study of chemistry. Small differences can completely change the meaning of a term. For example, Co and CO are different substances.

■ The word *homogeneous* does not necessarily refer to a homogeneous mixture. Pure substances are also homogeneous.

■ The elements in a given group of the periodic table have similar properties. This fact can help you learn a great deal of chemistry with less effort than would otherwise be required.

■ Like some groups of the periodic table, some portions of periods have special names. For example, the first transition series (elements 21–30) is part of the fourth period.

Self-Tutorial Problems

1.1 Explain the difference between the results of hitting a piece of aluminum foil on a hard surface with a hammer and similarly hitting a pane of glass. Use the word *brittle* in your explanation.

1.2 Cm is the chemical symbol for curium, named after the famous Madame Curie. Why wasn't the symbol C, Cu, or Cr used instead?

1.3 Which classical periodic group number is used for each of the following families?

(a) Coinage metals

(b) Alkali metals

(c) Alkaline earth metals

(d) Noble gases

(e) Halogens

1.4 Write the symbols from the names for the following:

(a) The first 18 elements in Figure 1.5

(b) The second 18 elements in the figure

(c) The rest of the elements shown

1.5 Write the names from the symbols for the following:

(a) The first 18 elements in Figure 1.5

(b) The second 18 elements in the figure

(c) The rest of the elements shown

1.6 How many elements are present in each of the following?

(a) No and NO   (b) HF and Hf

(c) $PoCl_2$ and $POCl_3$ (d) Si and SI_2

1.7 Would it be considered unusual for an organic chemist to use an elemental metal to help prepare a new compound, even though the metal is considered inorganic?

1.8 Can you tell how far a person drove to work in the morning if you know that the person drove at 40 miles per hour? Is speed intensive or extensive?

1.9 All brands of pure aspirin are the same compound. If you need aspirin, how should you choose a brand to buy?

1.10 Are the nonmetals main group elements, transition elements, or inner transition elements?

1.11 Do elements in the same period or elements in the same group have similar chemical properties?

1.12 Which classical transition group has the most elements?

1.13 Which of the following are samples of matter, and which are samples of energy?

(a) A hot potato   (b) Cold turkey

(c) A beam of light   (d) The sound of a ringing bell

1.14 Does each main group have more or fewer elements than a typical transition group?

1.15 (a) In which group of the periodic table is Al?

(b) In which period is Al?

(c) What type of element is Al—a main group element, a transition element, or an inner transition element?

1.16 A chemist in which branch of chemistry deals with the chemistry of compounds of the transition metals with oxygen?

■ ■ ■ PROBLEMS ■ ■ ■

1.1 Classification of Matter

1.17 (a) When pure water is cooled below 32°F (0°C), it freezes (solidifies). When the solid is warmed above that temperature, it melts again. Its composition does not change during the entire process. Are these chemical or physical changes?

  (b) When gaseous ethene is treated with a tiny quantity of a certain other substance, it solidifies. It is difficult to cause the solid to reform a gas. Is the solidification a chemical or a physical change?

1.18 When nitrogen dioxide, a brown gas, is cooled, it turns into a colorless liquid. Is this a chemical or a physical change?

1.19 When some table salt is added to water, a solution is formed. State several ways in which you can tell that the combination is a solution rather than a new compound.

1.20 In the iron-sulfur experiment described in Section 1.2, heat was used to start a chemical reaction, which gave off more heat. Can you think of another example of a reaction that is started by heating and then gives off more heat?

1.21 Classify the following materials as homogeneous or heterogeneous:

  (a) White paint

  (b) Milk

  (c) Ammonia gas dissolved in water

  (d) A teaspoon of sugar in a glass of warm water after having been stirred thoroughly

  (e) A glass of pure water containing an ice cube (also pure water)

  (f) A cola drink with no bubbles visible

  (g) A cola drink with bubbles

1.22 Classify each of the following as a chemical change or a physical change:

  (a) Spreading salt on an icy sidewalk to melt the ice

  (b) Striking a match

  (c) Breaking a piece of metal by bending it back and forth

  (d) Baking a cake

  (e) Using a tea bag

  (f) Cooking a hamburger

  (g) Rubbing your hands together to get them warm

1.23 What kind of change—chemical or physical—accompanies each of the following?

  (a) The conversion of a compound to two elements

  (b) The conversion of two compounds into a solution

  (c) The separation of a mixture into its components

  (d) The conversion of two elements to a compound

1.24 When two separate samples of matter are mixed, a great deal of heat is generated. Is this more likely to be a chemical or a physical change?

1.2 Properties

1.25 Both nitrogen and hydrogen are odorless.

  (a) What is the odor of a mixture of the two gases?

  (b) Explain why ammonia, a compound of nitrogen and hydrogen, smells so strongly.

1.26 Classify each of the following as an element, a compound, or a mixture:

  (a) A homogeneous combination of iodine and alcohol (tincture of iodine), which retains a dark color and the liquid state of the alcohol

  (b) A homogeneous combination of sodium and iodine, which is a white solid

  (c) Solid iodine (a dark violet solid)

1.27 Classify each of the following as a compound or a mixture:

  (a) Salt water

  (b) Carbonated water

  (c) The liquid formed by a certain combination of oxygen and hydrogen gases

1.28 Bromine melts at −7.2°C; sodium melts at 97.8°C. A certain combination of the two melts at 747°C. Is the combination a mixture or a compound?

1.29 Classify each of the following as a compound or mixture. If it is impossible to tell, explain why.

  (a) A material containing only hydrogen and oxygen that is a gas under ordinary room conditions

  (b) A material containing 88.8% oxygen and 11.2% hydrogen

  (c) A material that is explosive and that contains 88.8% oxygen and 11.2% hydrogen

  (d) A solid combination of iron and oxygen, no part of which is attracted by a magnet

  (e) A material that consists of red particles and green particles

1.30 Which of the following properties are extensive and which are intensive?

  (a) Length   (b) Speed   (c) Color

  (d) Mass   (e) Volume   (f) Freezing point

  (g) Total cost   (h) Price per unit

1.31 Which types of matter in Table 1.1 are homogeneous?

1.32 Electricity is passed through 73.0 grams of a pure substance, and 32.1 grams of one material and 40.9 grams of another material are produced. Is the original substance an element or a compound?

1.33 List four or five properties you could use to distinguish between iron and aluminum.

1.34 List four or five characteristics that allow you to distinguish between water and gasoline.

1.35 A 7.0-ounce sample of a solid substance is heated under a stream of hydrogen gas, and 2.5 ounces of solid remains after the treatment. Further treatment with hydrogen causes no further change. Is the original substance an element or a compound?

1.36 If 2.00 ounces of gold costs $1200 and 10.00 ounces of gold costs $6000, is the price of gold intensive or extensive? Is the cost intensive or extensive?

1.37 A sample of a liquid is homogeneous. When it is cooled to 15°C, part of the liquid solidifies. The solid part is removed, and the liquid part is cooled further, but no other change takes place. Is the original liquid a compound or a solution?

1.3 Matter and Energy

1.38 Name a device commonly found on a car that changes:
 (a) Mechanical energy to electrical energy
 (b) Electrical energy to chemical energy
 (c) Chemical energy to electrical energy
 (d) Electrical energy to sound
 (e) Chemical energy to mechanical energy
 (f) Electrical energy to mechanical energy
 (g) Electrical energy to light

1.39 What two changes in energy accompany the use of a flashlight?

1.40 Name one common device, not on a car, that performs each of the following conversions:
 (a) Electricity to light
 (b) Electrical energy to sound
 (c) Chemical energy to heat
 (d) Chemical energy to mechanical energy

1.41 For a given quantity of energy, the electricity produced by a battery is much more expensive than that provided by the electric company. Why do we still use batteries?

1.42 Explain the advantages and disadvantages of house current versus batteries for use in a home smoke detector.

1.43 List as many kinds of energy as you can think of without consulting the text.

1.4 Chemical Symbols

1.44 Calculate the percentage of all elements whose names start with the letter A.

1.45 How many elements are present in each of the following?
 (a) H_2O (b) Ho   (c) Ni   (d) $COCl_2$

1.46 Beginning students often mix up the following elements. Give the name for each element.
 (a) Mg and Mn   (b) K and P   (c) Na and S
 (d) Cu and Co

1.47 Without consulting any tables, write the symbols for the following elements:
 (a) Magnesium and manganese
 (b) Boron, barium, and bismuth
 (c) Carbon, cadmium, and calcium
 (d) Cobalt and copper
 (e) Sodium and sulfur
 (f) Potassium and phosphorus

1.48 Without consulting any tables, write the names of the following elements:
 (a) Na and N   (b) F and Fe   (c) Ag and Au
 (d) Sn and Si   (e) K and P

1.49 Write the symbol for each of the following elements:
 (a) Tungsten   (b) Sodium   (c) Copper
 (d) Potassium   (e) Lead   (f) Iron
 (g) Mercury   (h) Silver   (i) Gold
 (j) Antimony

1.5 The Periodic Table

1.50 Which of the following neighbors of chlorine in the periodic table has properties most like those of chlorine—F, S, or Ar?

1.51 How many elements are in each of the first four periods of the periodic table?

1.52 Would you expect sodium or bismuth to act more like a typical metal?

1.53 Which two elements are most like sodium in chemical properties?

1.54 Which element of periodic group IA is *not* an alkali metal?

1.55 Using the table inside the back cover of the text, determine:
 (a) How many elements whose names start with the letter A are transition elements?
 (b) How many elements whose symbols start with the letter A are transition elements?

1.56 Using the table inside the back cover of the text, determine:
 (a) How many elements whose names start with the letter H are transition elements?
 (b) How many elements whose symbols start with the letter H are transition elements?

1.57 Which element is in group IV of the second transition series?

1.58 Which element with a two-letter symbol is in the fourth transition series?

1.59 State the group number and period number of each of the following elements:

(a) Co   (b) H   (c) Ca

1.60 Name the periodic group and state the group number of each of the following elements:

(a) Xe   (b) Na   (c) Ag

(d) Br   (e) Mg

1.61 (a) What two elements are in group IV of period 4?

(b) What element(s) is (are) in group III of period 3?

1.6 Laws, Hypotheses, and Theories

1.62 Would an accepted generalization that explains why active metals react with acids be referred to as a law, hypothesis, or theory?

1.63 Suppose that you are a consultant to the National Science Foundation, an agency of the U.S. Government. In a proposal for a $1 million grant, a claim is made that a method will be developed to make 20 grams of gold from 10 grams of gold and no other ingredients. Would you recommend that government money be spent on this proposed research? Explain your reasoning.

■ ■ ■ GENERAL PROBLEMS ■ ■ ■

1.64 (a) Predict the color of a solution of a light yellow substance dissolved in a light blue substance.

(b) Can you predict the color of a compound of a light yellow substance and a light blue substance?

(c) Explain your answers.

1.65 (a) Count the number of each of the following types of elements in Figure 1.5: main group elements, transition elements, inner transition elements. Now calculate the percentage of each type important enough for you to learn of all the elements in that type. For example, of all the main group elements, what percentage is important for you to learn (from Figure 1.5)? (In your calculations, omit the elements above 103.)

(b) Which type of element do you think will be most important in this course? Which will be second most important?

(c) Calculate the percentage of the elements in Figure 1.5 of all 103 elements considered in the course.

1.66 A chemist uses a compound of carbon, hydrogen, and oxygen to separate a metal from the rest of a sample to determine the metal's percentage in the sample. What branch of chemistry is the chemist practicing?

1.67 (a) A chemist in which branch of chemistry is most likely to determine the number of parts per million of an impurity in a city's drinking water?

(b) A chemist in which branch of chemistry determines the electrical conductivity of a metal already prepared by another chemist?

1.68 What kind of electrical device has the advantage of portability like a dry cell but better economy?

1.69 A nutritionist recommends more iron and less sodium in the diet of a patient with a blood problem. Does the nutritionist advocate eating iron metal, but not sodium metal? Explain.

1.70 Ratios are intensive. Explain why.

1.71 Explain the following statements sometimes made in everyday conversation:

(a) "Oil and water do not mix."

(b) "Gasoline and alcohol do not mix."

1.72 $NaClO_4$ is a formula for a certain chlorine compound. Which of the following formulas is most likely to be the formula for a manganese compound?

(a) Na_3MnO_4 (b) $NaMnO_4$ (c) $NaMnO_5$

2 Measurement

- 2.1 Factor Label Method
- 2.2 The Metric System
- 2.3 Significant Digits
- 2.4 Exponential Numbers
- 2.5 Density
- 2.6 Time, Temperature, and Energy

■ Key Terms (*Key terms are defined in the Glossary.*)

accuracy (2.3)
base (2.4)
calorie (2.6)
Celsius scale (2.6)
centigrade scale (2.6)
coefficient (2.4)
conversion factor (2.1)
cubic meter (2.2)
density (2.5)
exponent (2.4)
exponential notation (2.4)
exponential part (2.4)

factor label method (2.1)
Fahrenheit scale (2.6)
gram (2.2)
heat (2.6)
joule (2.6)
Kelvin scale (2.6)
length (2.2)
liter (2.2)
meter (2.2)
metric system (2.2)
percent (2.1)
precision (2.3)

rounding (2.3)
scientific notation (2.4)
second (2.6)
SI (2.2)
significant digit (2.3)
significant figure (2.3)
standard (2.2)
standard exponential form (2.4)
temperature (2.6)
unit (2.1)
volume (2.2)

■ Symbols

c (centi-) (2.2)
d (density) (2.5)
g (gram) (2.2)
k (kilo-) (2.2)

L (liter) (2.2)
m (mass) (2.2)
m (meter) (2.2)

m (milli-) (2.2)
s (second) (2.6)
V (volume) (2.2)

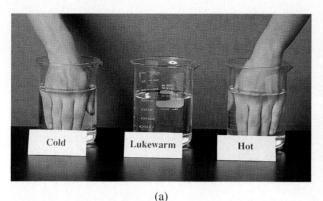 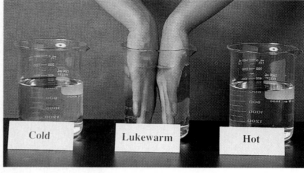

(a)   (b)

Figure 2.1 Hot
and Cold Experiment

When the hands are moved from
position (a) to position (b), the right
hand feels hot and the left hand feels
cold, even though they are both in
water of the same lukewarm
temperature. Try it.

Measurement is the heart of modern science, and even social scientists are becoming more quantitative. Measurements make identifications of substances more precise and allow more scientific generalities to be made. For example, even ancient peoples knew that when objects were dropped, they fell downward. The law of gravity was extended by measurements, allowing Sir Isaac Newton (1642–1727) to determine that the same laws of gravity that govern the fall of an object here on earth also govern the motions of the moon and the planets in the solar system.

A simple do-it-yourself project will convince you that measuring things quantitatively tells more than qualitative estimates, especially those made using the human senses: Fill one beaker with cold water, a second beaker with hot water, and a third beaker with a mixture of equal amounts of hot and cold water. Place one hand in the cold water and the other hand in the hot water at the same time (Figure 2.1). Leave them there for 2 minutes. Then place both hands in the mixed water. That water will feel hot to the hand originally in the cold water but cold to the hand originally in the hot water, even though both hands are now in the same water!

Several aspects of measurement will be considered in this chapter. First, Section 2.1 presents the factor label method, which makes calculations with measured quantities easier. This method will be used in the sections that follow and throughout the book. Section 2.2 introduces the metric system, a system of weights and measures designed to make calculations as easy as possible. In Section 2.3, we discuss the accuracy of measurements and how that accuracy should be reported, using the proper number of significant digits. Next, in Section 2.4, we consider how to calculate with extremely large and extremely small numbers, using exponential notation. The concept of density, considered in Section 2.5, not only is useful in itself, especially for identifying substances, but also allows us to apply the concepts presented in previous sections. Finally, Section 2.6 briefly discusses time, temperature, and energy.

2.1 Factor Label Method

■ to use the units of a
measurement to help to do
calculations involving that
measurement

Every measurement results in a number and a **unit.** Reporting the unit is just as important as reporting the number. For example, it makes quite a bit of difference whether your pet is 3 *inches* tall or 3 *feet* tall! The units are an integral part of any measurement, and from the outset, you must get used to stating the units for every measured quantity and for every quantity calculated from measured data. Always use full spellings or standard

Always use full spellings or
standard abbreviations for all
units.

abbreviations for all units. In many cases, you can use the units as a clue
to which operation—multiplication or division—to perform in calculations with
measured quantities.

The units of measurement can be treated as algebraic quantities in calculations. For example, we can calculate the total wages of a student aide who has
earned 7 dollars per hour for 15 hours of work, as follows:

$$\text{Total wages} = (\text{hours worked})(\text{hourly rate})$$

$$= 15 \text{ hours} \left(\frac{7 \text{ dollars}}{1 \text{ hour}} \right) = 105 \text{ dollars}$$

The unit *hours* in the time cancels the unit *hour* in the rate, leaving the unit *dollars* in the answer. Each unit is treated as a whole, no matter how many letters
it consists of. Moreover, for the units to cancel, it does not matter if the unit is
singular (such as hour) or plural (such as hours). If we did not know the equation to calculate the total wages, we could have put down the time with the unit
hours and multiplied by the rate of pay, which has the unit *hour* in its denominator. The units tell us that we must multiply!

The previous calculation is an example of the use of the **factor label
method,** in which a quantity is multiplied by a factor equal to 1. The units included in the factor are the labels. In the previous example, 7 dollars is equivalent to 1 hour, and the calculation changes the number of hours worked to
the equivalent number of dollars. To use the factor label method, you first put
down the given quantity, then multiply by a **conversion factor** (a rate or
ratio) that will change the units given to the units desired for the answer.
The factor may be a constant known to you or a value given to you in the
problem.

Small diagrams that show the initial units and the final units connected by
the conversion factor can be used to show how to change a quantity from one of
the units to the other. For example, for calculating the student aide's total wages,
we could use the following diagram:

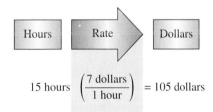

Diagrams like this will accompany many of the solutions to the in-text examples.
When you solve the practice problems and the problems at the end of the chapter, you may want to make your own diagrams.

■ EXAMPLE 2.1

Change 456 minutes to hours.

Solution

You put down the quantity given and then multiply it by a factor (which in
this case you know) that changes minutes to hours. The factor should have
the unit *minutes* in the denominator to cancel the *minutes* in the quantity

given. It should also have the unit *hour* in the numerator so that the answer is in hours:

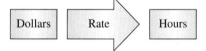

$$456 \text{ minutes} \left(\frac{1 \text{ hour}}{60 \text{ minutes}} \right) = 7.60 \text{ hours}$$

Any units in the denominator are divided into units in the numerator, just as any number in the denominator is. Any units and any numbers in the numerator are multiplied. (If a quantity, such as 456 minutes, is given with no denominator, the quantity is considered to be in a numerator.)

Practice Problem 2.1   Change 456 seconds to minutes. ■

A ratio or rate may be inverted (turned upside down) if the units that need to be canceled call for that.

■ EXAMPLE 2.2

Calculate the time required for a student aide to earn 350 dollars at 7 dollars per hour.

Solution

First, put down the quantity given; then multiply it by a factor involving the rate:

$$350 \text{ dollars} \left(\frac{1 \text{ hour}}{7 \text{ dollars}} \right) = 50.0 \text{ hours}$$

In this case, the inverse of the rate of pay (the factor used previously to calculate total wages) is employed. Rates or ratios can be used either right side up or upside down; getting the units to cancel properly will tell which form to use. Just be sure that the number in the rate (such as 7) stays with the proper unit (dollars).

Practice Problem 2.2   Calculate the time required to travel 7.00 miles at 35.0 miles per hour. ■

■ EXAMPLE 2.3

Explain why the factor label method works, using the conversion of $6.65 to cents as an example.

Solution

Consider this equality:

$$1 \text{ dollar} = 100 \text{ cents}$$

Dividing both sides of this equation by 1 dollar yields

$$1 = \frac{100 \text{ cents}}{1 \text{ dollar}}$$

Anything divided by itself is equal to 1, so the left side of this equation is 1. The right side of the equation is thus equal to 1 and therefore may be used to multiply any quantity to change its *form* without changing its *value*.

$$6.65 \text{ dollars} \left(\frac{100 \text{ cents}}{1 \text{ dollar}} \right) = 665 \text{ cents}$$

In any factor, the numerator is equal or equivalent to the denominator, so the value of the number multiplied is not changed, even though the units are. ■

A percentage can be used as a factor, since it represents the ratio of the number of one particular item to the total number of items in a group. **Percent** means the number *per hundred* total. Thus, 65% men in a class represents

$$\frac{65 \text{ men}}{100 \text{ people total}}$$

A ratio representing a percentage can also be inverted:

$$\frac{100 \text{ people total}}{65 \text{ men}}$$

■ EXAMPLE 2.4

Calculate the total number of people in a class containing 25% women if 17 women are in the class.

Solution

Put down the number of women (the quantity given) first. Then multiply by a ratio reflecting the percentage so that the unit given will be canceled and that desired will remain:

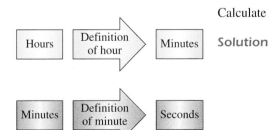

$$17 \text{ women} \left(\frac{100 \text{ people}}{25 \text{ women}} \right) = 68 \text{ people}$$

There are 68 people in the class.

Practice Problem 2.4   Calculate the number of men in a class of 400 people that has 53% men. ■

It may be necessary to use more than one factor to get a desired answer. The factors may be used in separate steps or may be combined in a single step.

■ EXAMPLE 2.5

Calculate the number of seconds in 3.55 hours.

Solution

$$3.55 \text{ hours} \left(\frac{60 \text{ minutes}}{1 \text{ hour}} \right) = 213 \text{ minutes}$$

$$213 \text{ minutes} \left(\frac{60 \text{ seconds}}{1 \text{ minute}} \right) = 12\,780 \text{ seconds}$$

Alternatively,

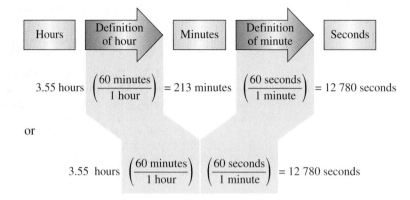

$$3.55 \text{ hours} \left(\frac{60 \text{ minutes}}{1 \text{ hour}} \right) = 213 \text{ minutes} \left(\frac{60 \text{ seconds}}{1 \text{ minute}} \right) = 12\,780 \text{ seconds}$$

or

$$3.55 \text{ hours} \left(\frac{60 \text{ minutes}}{1 \text{ hour}} \right) \left(\frac{60 \text{ seconds}}{1 \text{ minute}} \right) = 12\,780 \text{ seconds}$$

In this particular problem, it does not matter if you press the equal key on your calculator after entering the first 60. Similarly, you can write down the 213 minutes or not write it down, as you please, but the final answer is still the same.

Practice Problem 2.5   Calculate the number of seconds in exactly 3 weeks.   ■

A factor can be raised to a power if the units to be converted require that. Remember that when a ratio in parentheses is raised to a power, all the numbers and all the units within the parentheses must be raised to that power.

■ EXAMPLE 2.6

How many square feet are in 3.10 square yards?

Solution

$$3.10 \text{ yards}^2 \left(\frac{3 \text{ feet}}{1 \text{ yard}} \right)^2$$

$$= 3.10 \text{ yards}^2 \left(\frac{9 \text{ feet}^2}{1 \text{ yard}^2} \right) = 27.9 \text{ feet}^2$$

The second factor can be derived as follows:

$$3 \text{ feet} = 1 \text{ yard}$$
$$(3 \text{ feet})^2 = (1 \text{ yard})^2$$
$$3^2 \text{ feet}^2 = 1^2 \text{ yard}^2$$
$$9 \text{ feet}^2 = 1 \text{ yard}^2$$

Note that the number 3 is squared, the unit *feet* is squared, the number 1 is squared, and the unit *yard* is squared. There are 9 feet² in 1 yard² (see Figure 2.2).

Practice Problem 2.6   How many cubic feet of cement can be held in a cement mixer with a capacity of 3.10 cubic yards?   ■

A ratio may be changed to an equivalent ratio with different units by applying the factor label method.

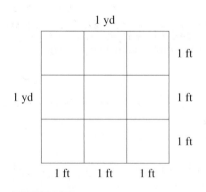

Figure 2.2 Number of Square Feet in a Square Yard

(not drawn to scale)

■ **EXAMPLE 2.7**

Change 30 miles per hour to feet per second.

Solution

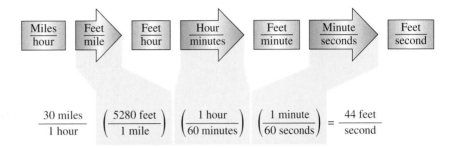

$$\frac{30\ miles}{1\ hour}\left(\frac{5280\ feet}{1\ mile}\right)\left(\frac{1\ hour}{60\ minutes}\right)\left(\frac{1\ minute}{60\ seconds}\right) = \frac{44\ feet}{second}$$

Practice Problem 2.7 Calculate the speed in miles per hour of a runner who runs the 100-yard dash in 9.40 seconds. ■

To summarize the steps of the factor label method:

1. Put down the *quantity* given (or, occasionally, a ratio to be converted).
2. Multiply the quantity by one or more factors—rates or ratios—which will change the units *given* to those *required* for the answer. The conversion factors may be given in the problem, or they may be constants of known value.

To use the factor label method effectively, you must know the units of all the quantities being dealt with.

Many more examples of the use of the factor label method will be presented in the sections that follow, where we will work problems involving quantities that are directly proportional to each other.

2.2 The Metric System

OBJECTIVE

■ to use the basic elements of the metric system—a system of units and prefixes designed to make scientific calculations as easy as possible

The **metric system** and its more modern counterpart **SI** (for S̲ystème Interna-tional d'Unités) are systems of units designed to make calculations as easy as possible. If you were to design a system of units yourself, you would probably make every word mean one and only one thing. You would make subdivisions and multiples of units be powers of 10 times a primary unit. You would make pre-fixes mean the same thing, no matter what unit they were attached to. You would make the abbreviations for the quantities and prefixes easy to remember. All these features have been built into the metric system.

Learning the following six words is essential to understanding the metric system:

1. meter   4. centi-
2. gram   5. milli-
3. liter   6. kilo-

(A few more words will be added as we progress.) Meter, gram, and liter are the units of length, mass, and volume, respectively, in the metric system. Just as the

Table 2.1   *Metric Prefixes**

Prefix	Abbreviation	Meaning	
Mega-	M	One million	1 000 000
Kilo-	**k**	**One thousand**	**1000**
Deci-	d	One-tenth	0.1
Centi-	**c**	**One-hundredth**	**0.01**
Milli-	**m**	**One-thousandth**	**0.001**
Micro-	μ	One-millionth	0.000 001
Nano-	n	One-billionth	0.000 000 001
Pico-	p	One-trillionth	0.000 000 000 001

*The most important prefixes are given in **boldface** type.

English system has subdivisions of its primary units (12 inches in a foot, for example), so does the metric system. But the metric system uses prefixes that mean the same thing no matter what primary unit they are used with. Centi-, milli-, and kilo- are prefixes that indicate certain multiples or divisions of any primary unit. Other prefixes are given in Table 2.1.

The **meter** is the primary unit of length in the metric system. Its abbreviation is m. The meter is defined in such a way that it can be duplicated precisely in any well-equipped laboratory in the world. Its original definition was one ten-millionth of the distance from the North Pole to the equator along an imaginary line running through Paris, France. Later, it was defined as the distance between two marks on a metal bar kept at the Bureau of Weights and Measures in Paris. (It now has an even more precise definition.) A meter is about 39.37 inches long, to give you an idea of its length.

The **gram** is the primary unit of mass in the metric system. The gram, abbreviated g, is such a small mass that the kilogram has been chosen as the legal **standard** of mass in the United States and as the worldwide standard in SI. Mass is measured by comparison with standard masses. The kilogram (kg) is a mass equivalent to about 2.2 pounds (Figure 2.3).

The **cubic meter** is the primary unit of volume in SI. A smaller unit, the **liter,** is the primary unit of volume in the metric system. The abbreviation for liter is L. You need to know both the cubic meter and the liter. Table 2.2 summarizes the primary metric units of distance, mass, and volume.

The prefix **centi-** means one-hundredth of any primary unit. For example, a centimeter is 0.01 meter, and a centigram is 0.01 gram.

Table 2.2   *Primary Metric Units*

	Unit	Symbol	Equivalencies
Distance	Meter	m	
Mass	Gram	g	
Volume	Liter	L	
Volume	Cubic meter	m^3	1000 L = 1 m^3
Volume	Cubic centimeter	cm^3	1000 cm^3 = 1 L

Figure 2.3 Measurement
of Mass

(a) The mass of a jelly bean is
comparable to a 16-g mass.
(b) This chicken has a mass of 2 kg.

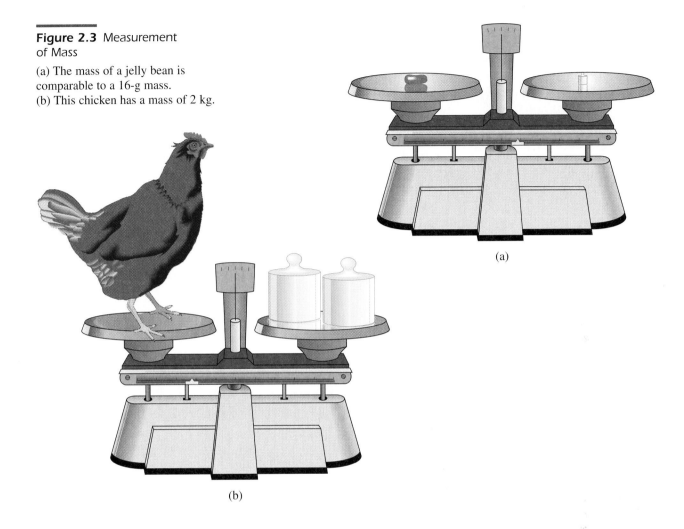

(a)

(b)

The prefix **milli-** means one-thousandth. No matter which primary unit it is
used with, it always means 0.001 times that unit. A millimeter is 0.001 meter, a
milliliter is 0.001 liter, and so on.

The prefix **kilo-** means 1000 times the primary unit, no matter which primary
unit it is used with. For example, a kilogram is 1000 grams, and a kilometer is
1000 meters.

The metric system is easier to use than the English system.

■ **EXAMPLE 2.8**

(a) How many meters are in 3.300 km?

(b) How many yards are in 3.300 miles?

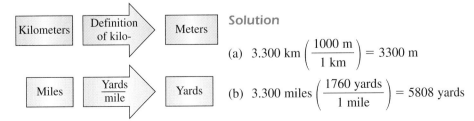

Solution

(a) $3.300 \text{ km} \left(\dfrac{1000 \text{ m}}{1 \text{ km}} \right) = 3300 \text{ m}$

(b) $3.300 \text{ miles} \left(\dfrac{1760 \text{ yards}}{1 \text{ mile}} \right) = 5808 \text{ yards}$

Figure 2.4 English and Metric Usage

The metric system problem, part (a), can be solved in your head—by moving the decimal point in 3.300 three places to the right. The English system conversion, part (b), requires that you remember the number of yards per mile (harder than the 1000 m/km metric conversion factor) and that you use pencil and paper or a calculator to do the arithmetic. The conversion factor 1000 is used for kilograms, kiloliters, kilowatts, and any other factor involving the prefix *kilo-*. The English conversion factor 1760 yards/mile is not used in any other conversion.

Practice Problem 2.8 (a) How many centimeters are in 29.5 m? (b) How many inches are in 29.5 feet? (c) For which of these two conversions do you need to use a calculator? ■

To convert a value expressed in a primary metric unit to its equivalent in a subunit, or vice versa, use a conversion factor with a 1 in front of the subunit and the equivalent value in front of the main unit. Note that you will have either the prefix abbreviation or its equivalent in front of the symbol for the primary unit:

Prefix	Equivalent value
Kilo	1000
Centi	0.01
Milli	0.001

For example, either of the following conversion factors is correct:

$$\left(\frac{1 \text{ centi meter}}{0.01 \text{ meter}} \right) \quad \left(\frac{0.01 \text{ meter}}{1 \text{ centi meter}} \right)$$

Thus, to convert 231 cm to meters:

$$231 \text{ cm} \left(\frac{0.01 \text{ m}}{1 \text{ cm}} \right) = 2.31 \text{ m}$$

To convert 4.73 m to centimeters:

Note that **1 centi-** (1 c) means **0.01**.

$$4.73 \text{ m} \left(\frac{1 \text{ cm}}{0.01 \text{ m}} \right) = 473 \text{ cm}$$

■ **EXAMPLE 2.9**

Convert 45.6 g to (a) kilograms and (b) milligrams.

Solution

(a) $45.6 \text{ g} \left(\dfrac{1 \text{ kg}}{1000 \text{ g}} \right) = 0.0456 \text{ kg}$

(b) $45.6 \text{ g} \left(\dfrac{1 \text{ mg}}{0.001 \text{ g}} \right) = 45\,600 \text{ mg}$

Practice Problem 2.9 Convert 3.59 mL to liters. ■

Some conversions between English and metric system units are presented in Table 2.3. Engineers must know how to do such conversions, since they still use some English system units. However, scientists rarely use English system units, and therefore, these conversions are less important for them. (A use of metric

Table 2.3 English-Metric Conversions

Length	Mass	Volume
1 meter = 39.37 inches	1 kilogram = 2.2045 pounds	1 liter = 1.059 quarts
2.540 centimeters = 1 inch	453.6 grams = 1 pound	29.57 milliliters = 1 fluid ounce
1.609 kilometers = 1 mile	28.35 grams = 1 ounce (avoirdupois)	
= 1760 yards		
= 5280 feet		

units that is becoming familiar to the general public is shown in Figure 2.4. Note that 80 km/h is about 50 miles/hour.)

When we add or subtract measured quantities, we treat the units just as we treat variables (such as x, y, and z) in algebraic manipulations (see Appendix 1). The units must be the same for the addition or subtraction of numbers that represent measurements. A sum or difference will have the same units as the quantities being added or subtracted.

> The units must be the same for the addition or subtraction of numbers that represent measurements.

■ **EXAMPLE 2.10**

Add (a)   2.00 m + 0.35 m   and   (b) 2.00 m + 35 cm.

Solution

(a)   The units are the same, so you simply add:

$$2.00 \text{ m} + 0.35 \text{ m} = 2.35 \text{ m}$$

(b)   You must either change 2.00 m to centimeters or 35 cm to meters, and then add:

$$35 \text{ cm}\left(\frac{0.01 \text{ m}}{1 \text{ cm}}\right) = 0.35 \text{ m}$$

$$2.00 \text{ m} + 0.35 \text{ m} = 2.35 \text{ m}$$

Practice Problem 2.10   Add the following algebraic quantities:

(a) $200x + 35x$ (b) $200x + 35.00y$, where $x = 0.01y$ ■

To multiply or divide measured quantities of the same type, such as two lengths, you may have to convert the units so that they are the same. For quantities of different types, the units cannot be the same.

■ **EXAMPLE 2.11**

Multiply (a) 2.50 cm × 3.00 cm   and   (b) 2.00 m × 25 cm.

Solution

(a)   The units are already the same, so you just multiply:

$$2.50 \text{ cm} \times 3.00 \text{ cm} = 7.50 \text{ cm}^2$$

Note that both the numbers and the units are multiplied.

(b) The 2.00 m could be changed to centimeters, or the 25 cm could be changed
to meters:

$$25 \text{ cm} \left(\frac{0.01 \text{ m}}{1 \text{ cm}} \right) = 0.25 \text{ m}$$

$$2.00 \text{ m} \times 0.25 \text{ m} = 0.50 \text{ m}^2$$

Practice Problem 2.11   Multiply the following algebraic quantities:

$$(2.50x)(3.00x).$$

How does the x in this problem resemble the unit in Example 2.11(a)? ■

■ **EXAMPLE 2.12**

The cost of a certain 15.00-inch gold chain is $260.00. What is the cost per inch?

Solution

$$\frac{260.00 \text{ dollars}}{15.00 \text{ inches}} = 17.33 \text{ dollars/inch}$$

Practice Problem 2.12   Divide the algebraic quantities: $(100x)/(20z^3)$. ■

Length or Distance

The primary unit of **length** in the metric system is the meter, which is approximately 10% longer than a yard. The same prefixes are used with the meter as with all other metric units.

■ **EXAMPLE 2.13**

Olympic divers use 3-m boards and 10-m boards. Calculate these heights in centimeters.

Solution

You can use the factor label method (Section 2.1) to do these metric calculations. A centimeter is 0.01 m, just as a cent is 0.01 dollar. There are 100 cm in 1 m, so in exactly 3 m, there are 300 cm:

$$3.00 \text{ m} \left(\frac{1 \text{ cm}}{0.01 \text{ m}} \right) = 300 \text{ cm}$$

In exactly 10 m, there are 1000 cm:

$$10 \text{ m} \left(\frac{1 \text{ cm}}{0.01 \text{ m}} \right) = 1000 \text{ cm}$$

Practice Problem 2.13   Calculate the heights of Olympic diving boards in millimeters. ■

■ **EXAMPLE 2.14**

How many kilometers are in 21.7 m?

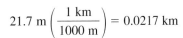

Solution

There are 1000 m in 1 km, by definition.

$$21.7 \text{ m} \left(\frac{1 \text{ km}}{1000 \text{ m}} \right) = 0.0217 \text{ km}$$

Practice Problem 2.14 How many millimeters are in 73.2 cm? ■

Mass

As stated earlier, the primary unit of mass in the metric system is the gram. Because the gram is so small, however, the standard mass in SI and the legal standard in the United States is the kilogram.

■ **EXAMPLE 2.15**

How many grams are in 1 kilogram (1 kg)?

Solution

The prefix *kilo-* means 1000 of whatever it is attached to. Therefore, 1 kg is 1000 g.

Practice Problem 2.15 How many grams are in 1 milligram (1 mg)? ■

■ **EXAMPLE 2.16**

Airlines sometimes have a 15-kg free-baggage allowance for overseas travelers. (a) How many grams does this allowance represent? (b) How many pounds?

Solution

(a)

$$15 \text{ kg} \left(\frac{1000 \text{ g}}{1 \text{ kg}} \right) = 15\ 000 \text{ g}$$

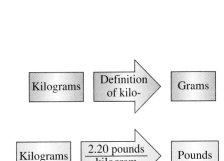

(b) From Table 2.3, you see that 1 kilogram = 2.20 pounds:

$$15 \text{ kg} \left(\frac{2.20 \text{ pounds}}{1 \text{ kg}} \right) = 33 \text{ pounds}$$

Practice Problem 2.16 Calculate the mass of a 172-pound person in kilograms. ■

Volume

Volume can be measured in two ways (Figure 2.5): (1) using the capacity of a certain container, and (2) using the space defined by a cube of length l on each side. The second method uses the cube of a length (and thus the unit for volume is the cube of a length unit). The volume of a rectangular solid is given by

$$\text{Volume} = \text{length} \times \text{width} \times \text{height}$$

$$V = l \times w \times h$$

A cube is a special case for which $l = w = h$, so the volume of a cube is $V = l^3$.

Figure 2.5 Two Methods to Determine Volume

(not drawn to scale)
Note that 10 cm × 10 cm × 10 cm = 1000 cm³.

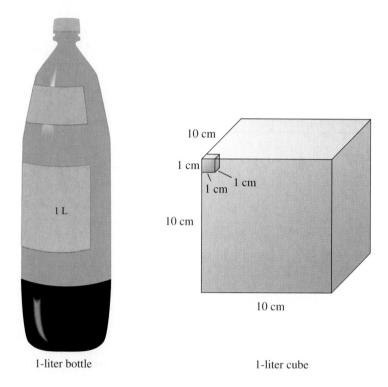

1-liter bottle   1-liter cube

The metric system unit of volume is the liter (L), originally defined as the volume occupied by a cube exactly 10 cm on each side (see Figure 2.5). In SI, the cubic meter is the standard. Since the cubic meter is a rather large volume (about half the capacity of a small cement truck), the liter is favored by chemists (see Figure 2.6).

$$1.00 \text{ L} = (10 \text{ cm})^3 = 1000 \text{ cm}^3 = 1000 \text{ mL}$$

Thus,

$$1.00 \text{ cm}^3 = 1.00 \text{ mL}$$

■ **EXAMPLE 2.17**

(a) How many 1-L cubes fit along the top front edge of the cubic meter pictured in Figure 2.6?

(b) How many fit on the front face?

(c) How many such vertical layers are in the entire cube?

(d) How many liters are in 1 m³?

Solution

(a) Ten 1-L cubes fit along the edge.

(b) One hundred 1-L cubes fit in the 10 rows on the front face.

(c) There are 10 layers from front to back.

(d) One thousand (10 × 10 × 10) 1-L cubes fit into 1 m³. Thus,

$$1 \text{ m}^3 = 1000 \text{ L} = 1 \text{ kL}$$

Practice Problem 2.17   How many cubic centimeters are in 1 L? ■

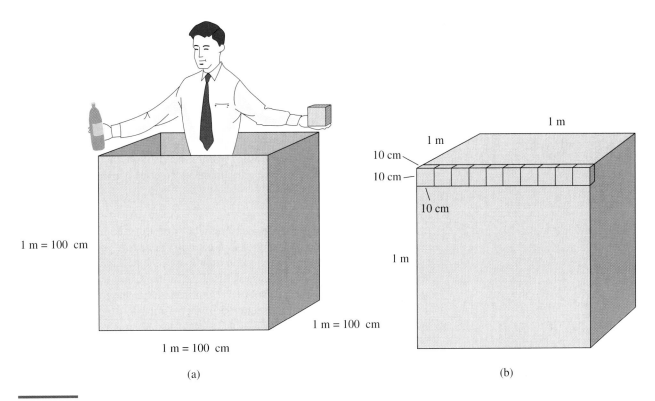

Figure 2.6 Cubic Meter and Liter

(a) The man in the cubic meter box is holding two objects, each of which is 1 L in volume. Note the difference in size between 1 L and 1 m³. (b) A cubic meter has edges that are 1 m long. Ten boxes with 10-cm edges fit along each such edge. Ten rows of those 10-cm boxes would fit in a layer covering the whole top surface of the cubic meter, and ten layers would fill the whole volume. There are 1000 liters in a cubic meter.

2.3 Significant Digits

OBJECTIVE

■ to use the correct number of digits so as to indicate the accuracy of a measurement or a calculated result

No matter how accurate your measuring tool, the accuracy of your measurements is limited. For example, an automobile odometer has divisions of 0.1 mile (or 0.1 km), and you can estimate to one-tenth of that smallest scale division, but you cannot measure 1 inch or even 1 foot with an odometer. Similarly, you cannot measure the thickness of a piece of paper with a ruler marked off in centimeters.

Scientific measurements are often repeated three or more times. The average value of the measurements is probably closer to the true value than any one of them. The **accuracy** is the closeness of the average of a set of measurements to the true value. The **precision** is the closeness of all of a set of measured values to one another. A set of measurements may be precise without being accurate or accurate without being precise, but the best measurements are both accurate and precise.

■ **EXAMPLE 2.18**

A set of measurements is done with a ruler that was made improperly. The ruler starts at 1 cm instead of at 0 cm (Figure 2.7). (Defective instruments do get

Figure 2.7 Measurement with a
Faulty Ruler

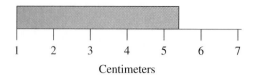

Centimeters

produced from time to time; you should check your instruments before using them
to make sure that they are not faulty.) Is it possible to measure a 4-cm bar with this
ruler and get close to the same reading every time? Is it possible to get true measurements with this ruler? Are the measurements precise? Are they accurate?

Solution

It is possible to get reproducible measurements with this ruler and, therefore, to
get good precision. However, you cannot get true readings if you start at the end
of the ruler because the ruler was made improperly; the accuracy is not good. (If
you calibrate the ruler—that is, make allowance for the faulty scale—you can get
accurate readings, too. You just subtract 1 cm from the readings to get the values
to report.) With a good ruler, scattered results—some too high and some too
low—are possible. The average of such a set of values might be near the true
value, and therefore be accurate, without being precise. Of course, readings that
are very close to one another and very close to the true value are best.

Practice Problem 2.18   A machine wraps sticks of butter with paper that is
printed with lines marking 1-ounce portions. If the machine wraps the butter one-
eighth of an inch away from the correct position (Figure 2.8), the first portion
might be too small and the last too large. If you buy 12 sticks of butter all
wrapped by the same machine in the same way: (a) Will you get the same mass
of butter at the first mark each time? (b) Will each portion be 1 ounce in mass?
(c) Is the wrapping more precise or more accurate? ■

The accuracy with which you can measure must be indicated when you report a measurement. When you use a measuring instrument, you should estimate
to one digit beyond the smallest scale division, if possible. For example, see Figure 2.9. If you measure the length of the bar with the top ruler, calibrated in centimeters, you can see that the bar is between 3 and 4 cm long and can estimate
that it extends 0.7 cm past 3 cm, for a reading of 3.7 cm. In contrast, if you use

■ ■ ■ ■ ■ ■

Two examples in everyday life of the use of measurements without due regard for the principles of accuracy and precision are first-down calls in football and congressional reapportionment based on census data. The football
is "spotted" by eye by an official from as far as 25 yards away from the
play—a very imperfect measurement. Then a much more exact measurement (to 0.001 yard, perhaps) is made with a metal chain to see if 10 000
yards or more has been gained. The census is taken every 10 years by asking people "Who lives here?" The answers in many cases are inaccurate because some people want to hide illegal immigration or building code viola-
tions. Yet, the results are used as though they were exact to divide the
congressional representation between states and within each state.

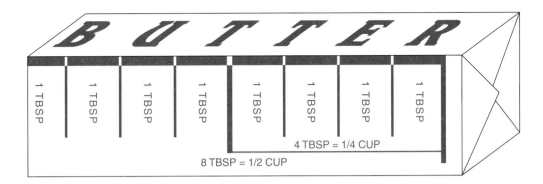

Figure 2.8 Precise but
Inaccurate Measurement

If the wrapping machine regularly
places the wrapper too close to
one end of the stick of butter, the
end piece might be the same weight
in each stick but still be far from
1 tablespoon.

the bottom ruler, calibrated in tenths of centimeters—that is, millimeters—you can see that the length of the bar is between 3.7 and 3.8 cm. You can estimate that it is 3.73 cm. The last digit you use to report this measurement tells anyone reading the result that you used a ruler with a millimeter scale.

Suppose that the bar extended exactly to the 4.4 line on the millimeter ruler. How would you report the result? You should report 4.40 cm. If you omit the zero, someone reading the result might think that you used a ruler calibrated only in centimeters. The third digit indicates that the result was obtained on a more accurate ruler, but just happened to be a value ending in zero.

■ **EXAMPLE 2.19**

About what fraction of measurements reported should be values ending in a zero?

Solution

About one time in ten the last digit of a reported measurement should be a zero. (There is an equal possibility of each digit, 0–9, being the last, so one-tenth of the time it should be a zero, and one-tenth of the time it should be a one, and one-tenth of the time a two, and so on.) ■

Scientists report the accuracy of their measurements every time they write one down. The number of digits they use always consists of the absolutely certain digits plus one estimated digit. Every digit that reflects the accuracy of the measurement is called a **significant digit,** or **significant figure.** Note that the word *significant* has a different meaning here than in everyday conversation, where it means "important."

Sometimes, zeros are used merely to indicate the magnitude of a number (how big or small the number is). If the purpose of a zero is *only* to establish the magnitude of the number, that zero is not significant. Determining which zeros are significant in a properly reported measurement is important.

Figure 2.9 Measurements
of Different Accuracies

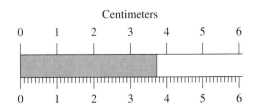

■ EXAMPLE 2.20

A measurement of a bar yields a length of 2.42 cm. How many meters is that? How many significant digits are in the number of meters?

Solution

$$2.42 \text{ cm} \left(\frac{0.01 \text{ m}}{1 \text{ cm}} \right) = 0.0242 \text{ m}$$

Since there are three significant digits in the number of centimeters, there are three significant digits in the number of meters. The calculation changing the value from centimeters to meters does not change the accuracy with which the measurement was made. (That 0.01 m equals 1 cm is a *definition* not a *measurment*.) The zeros in 0.0242 m are not significant; they merely show the magnitude of the number. (If they were not there, the value of the number would be different. They are important, but they are not significant; they do not tell anything about the accuracy of the measurement.)

Practice Problem 2.20   How many significant digits does the measurement 2.4 m have? If the measurement is changed to centimeters, how many significant digits will be in that value? ■

The following rules allow chemists to tell whether zeros in a number are significant or not:

1. Any zeros to the left of all nonzero digits (for example, in 0.003) are not significant.
2. Any zeros to the right of all nonzero digits and to the right of the decimal point (for example, in 4.00) are significant.
3. Any zeros between significant digits (for example, in 107) are significant.
4. Any zeros to the right of all nonzero digits in an integer (for example, in 500) are uncertain. If they only indicate the magnitude of the measurement, they are not significant. However, if they also show something about the accuracy of the measurement, they are significant. You cannot tell whether they are significant merely by looking at the number.

■ EXAMPLE 2.21

Underline the significant digits in each of the following measurements. If a digit is uncertain, place a question mark under it.

(a) 1.20 m   (b) 0.020 m   (c) 1.002 m   (d) 800 m

Solution

(a) 1.20 m
     The zero to the right of the two is significant (rule 2).
(b) 0.020 m
     The zeros to the left of the two are not significant (rule 1), but the one to the right is (rule 2).
(c) 1.002 m
     Zeros between significant digits are significant (rule 3).
(d) 800 m
        ??

The zeros to the right of all other digits in an integer are uncertain; they may reflect the accuracy or just the magnitude of the number. Without further information, it is impossible to tell (rule 4).

Practice Problem 2.21   Underline the significant digits in each of the following measurements. If a digit is uncertain, place a question mark under it.

(a) 20.0 cm   (b) 101.20 cm   (c) 3.002 cm   (d) 4000 cm ■

■ EXAMPLE 2.22

How many significant digits are in each of the following measurements?

(a) 1.2 m   (b) 1.20 m   (c) 1.200 m

Solution

(a) Two   (b) Three   (c) Four

The zeros in the numbers of parts (b) and (c) do not affect the magnitude; they must therefore show something about the accuracy of the measurements (rule 2). ■

■ EXAMPLE 2.23

Change each of the measurements given in meters in Example 2.22 to millimeters. Can you tell how many significant digits are in each result? How?

Solution

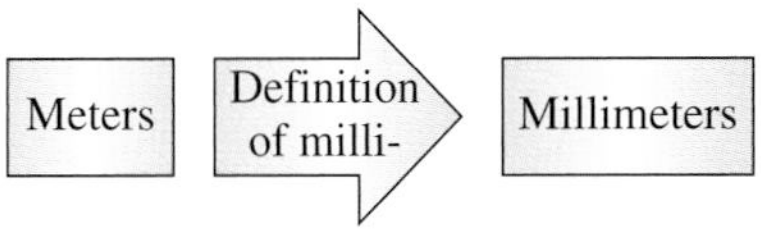

(a) $1.2 \text{ m} \left(\dfrac{1 \text{ mm}}{0.001 \text{ m}} \right) = 1200 \text{ mm}$

(b) $1.20 \text{ m} \left(\dfrac{1 \text{ mm}}{\phantom{0.001 \text{ m}}} \right) = 1200 \text{ mm}$

(c) $1.200 \text{ m} \left(\dfrac{1 \text{ mm}}{\phantom{0.001 \text{ m}}} \right) = 1200 \text{ mm}$

There are two significant digits in the result for part (a), three in the result for part (b), and four in the result for part (c). You know these numbers of significant digits only because you know the numbers of significant digits in the original numbers of meters given in Example 2.22. However, if you just look at the results, they all look the same! You cannot tell just by looking whether the zeros reflect the accuracy of the measurement or not.

Significant Digits in Calculated Results

In this electronic age, we have come to depend on our electronic calculators for solving arithmetic problems. However, electronic calculators usually do not give the correct number of significant digits. (A calculator may give the correct number of significant digits just by chance.) *You* are responsible for making sure that the number of significant digits in a calculated answer is correct.

Not only must your recorded data reflect the accuracy of the measurements, but any results calculated from the data must also reflect that accuracy. Certain rules govern how many significant digits are permitted in calculated results and how to get that many digits.

In the answers to addition and subtraction problems, the estimated digit that is farthest to the left is the last digit that can be retained. For example, let's add 3.2 cm and 8.007 cm:

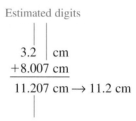

$$
\begin{array}{r}
3.2 \quad \text{cm} \\
+8.007 \text{ cm} \\
\hline
11.207 \text{ cm} \rightarrow 11.2 \text{ cm}
\end{array}
$$

The left-most estimated digit in the answer is the last digit that can be retained.

The digit 2 in 3.2 cm is an estimated digit; it has some uncertainty in it. Therefore, the digit 2 in the answer is also uncertain, and 0 and 7 are completely unknown. We cannot report the value 11.207 cm, or the reader will believe that 7 is the only uncertain digit. We must reduce the number of digits to leave 2 as the last digit.

Thus, for addition or subtraction, we retain digits in an answer only as far to the right as the left-most uncertain digit in any of the numbers being added or subtracted. Note that the *number* of significant digits does not matter for addition or subtraction; what matters is *where the last digits lie.* In the previous calculation, there are two significant digits in the first number and four in the second, but the answer has three.

■ **EXAMPLE 2.24**

Calculate the sum of 33.172 cm + 2.1 cm + 7.92 cm.

Solution

$$
\begin{array}{r}
33.172 \text{ cm} \\
2.1 \quad\ \text{cm} \\
7.92 \quad \text{cm} \\
\hline
43.192 \text{ cm} \rightarrow 43.2 \text{ cm}
\end{array}
$$

Practice Problem 2.24   Calculate the answer to the proper accuracy:

$$120.2 \text{ cm} - 34.567 \text{ cm} - 0.011 \text{ cm} \quad ■$$

For multiplication and division, the number of significant digits in the factor with the fewest significant digits limits the number of significant digits in the answer. For example, let's multiply 3.2 cm by 8.007 cm:

$3.2 \text{ cm} \times 8.007 \text{ cm} = 25.6224 \text{ cm}^2$ (Incorrect number of significant digits)

If we just leave the answer the way our electronic calculator gives it to us, anyone could assume that the measurement had been carried out with a precision of 1 part in 256 224, which is not true. We must reduce the number of significant digits in the answer to two because the factor with fewer significant digits has two. Thus, we change the answer to 26 cm². (See the discussion of rounding off in the next subsection.)

■ EXAMPLE 2.25

Do the following calculations, and report the answers to the correct number of significant digits:

(a) 2.00 cm × 3.051 cm   (b) 5.05 g/2.02 cm³

Solution

(a) The answer is 6.10 cm². You have to reduce the number of digits in 6.102 cm² to three significant digits, because the first factor has only three significant digits.

(b) The answer is 2.50 g/cm³. It must have three significant digits because both the dividend (5.05) and the divisor (2.02) have three significant digits. In this case, you need to *add a zero* to the answer given by your electronic calculator (2.5) to get the correct number of significant digits.

Practice Problem 2.25   Perform the following calculations and limit the answers to the correct number of significant digits:

(a) 1.6 cm × 2.00 cm × 3.051 cm

(b) 5.05 g/(2.02 cm × 3.0 cm × 3.12 cm) ■

Numbers that are definitions and not measurements, such as the number of centimeters in a meter (100) or the number of radii in the diameter of a circle (2), are exact numbers. They do not limit the number of significant digits in a calculated result.

■ EXAMPLE 2.26

The radius of a circle is 2.22 cm. Calculate the diameter of the circle to the correct number of significant digits.

Solution

$$d = 2r = 2(2.22 \text{ cm}) = 4.44 \text{ cm}$$

The *measurement* with the fewest significant digits is 2.22 cm. (It is the only measurement in the problem.)

Practice Problem 2.26   Calculate the number of centigrams in 5.4321 g. ■

Rounding Off

Reducing the number of digits to the number permitted involves a process called *rounding off,* often referred to simply as **rounding.** The process generally involves dropping one or more digits to the right of the decimal point and adjusting the last remaining digit if necessary. If the left-most digit to be dropped is greater than or equal to 5 with any nonzero digit anywhere to its right, we increase the last retained digit by 1 without regard to the sign of the number. If the left-most digit to be dropped is less than 5, we do not change the final digit we keep. If the digit(s) that is (are) dropped consist of a 5 alone or followed only by zeros, we raise the last digit kept to the next higher even digit if it is odd or leave it alone if it is even.

■ EXAMPLE 2.27

Round off each of the following numbers to an integer:

(a) 1.4   (b) 1.6   (c) 2.5   (d) 1.5   (e) 2.5001   (f) 1.5000

Solution

(a) 1; since 4 is less than 5, the integer 1 is not changed.

(b) 2; since 6 is greater than 5, the last digit retained is increased by 1 to 2.

(c) 2; the digit to be dropped is exactly 5, and the integer is left as 2 because it is already even.

(d) 2; the digit to be dropped is again exactly 5, and the integer 1 is rounded up to the next higher (even) integer, 2.

(e) 3; the digit 5 is followed by a 1 to its right, so the last digit retained is rounded up.

(f) 2; the digits to be dropped consist of a 5 followed by zeros, so 1 is raised to the next even integer, 2.

Practice Problem 2.27   Round off the following changes in mass to integers:

(a) −1.6 g   (b) −1.4 g   (c) −2.5 g   (d) −1.5 g ■

■ EXAMPLE 2.28

Round off each of the following lengths to retain one decimal place:

(a) 1.23 m   (b) 1.27 m   (c) 1.25 m

Solution

(a) 1.2 m   (b) 1.3 m   (c) 1.2 m

Where a rounded digit lies in relation to the decimal point is immaterial, as long as it is to the right of the decimal point.

Practice Problem 2.28   Round off each of the following lengths to retain two decimal places:

(a) 1.273 m   (b) 1.277 m   (c) 1.275 m ■

If the digits to be rounded are to the left of the decimal point rather than to the right, they are changed to (insignificant) zeros rather than being dropped. However, quantities greater than 10 are better expressed in scientific notation (Section 2.4).

■ EXAMPLE 2.29

Round off each of the following values to two significant digits:

(a) 124 cm   (b) 126 cm   (c) 125 cm

Solution

(a) 120 cm   (b) 130 cm   (c) 120 cm

In each case, the digit rounded is changed to an insignificant zero, not dropped.

Practice Problem 2.29 The state of Colorado is essentially a rectangle, measuring 623 km from east to west and 444 km from north to south. Calculate the area of Colorado to three significant digits. ■

2.4 Exponential Numbers

O B J E C T I V E

■ to use exponential notation to work with very large and very small numbers

Objects of scientific interest range from incredibly tiny to almost unimaginably large. The number of iron atoms that would fit side by side on a line 1 cm in length is about 80 million (Chapter 3). The number that could be packed into a volume of 1 cm^3 is 80 million cubed—about 500 thousand billion billion! Each iron atom is almost unimaginably small.

■ **EXAMPLE 2.30**

To get an idea of how large a number 1 billion is, calculate the number of years it would take to spend $1 billion if you spent $1000 per day. (Assume that there is no interest or other addition to the $1 billion.)

Solution

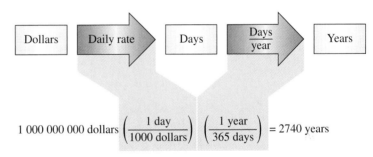

$$1\ 000\ 000\ 000\ \text{dollars} \left(\frac{1\ \text{day}}{1000\ \text{dollars}} \right) \left(\frac{1\ \text{year}}{365\ \text{days}} \right) = 2740\ \text{years}$$

It would take over 2700 years to spend a billion dollars by spending $1000 a day! Just think how large the number 100 billion or 1 billion billion is. Some numbers common in science are even larger than these.

Practice Problem 2.30 Calculate the amount of money that you would have to spend *per second* to use up $200 billion in 2500 years. ■

Scientists handle large and small numbers using **exponential notation**. A number written in this format has the following parts:

$$\underbrace{1.23}_{\text{Coefficient}} \times 10^4$$

Coefficient
Base Exponent
Exponential part

The **coefficient** is an ordinary number that may or may not include a decimal point. It is multiplied by an **exponential part,** consisting of a **base** and an

Table 2.4 Important Exponential Parts
and Their Meanings

Exponential Part	Value	Meaning
10^{-9}	0.000 000 001	One billionth
10^{-6}	0.000 001	One millionth
10^{-3}	0.001	One thousandth
10^{-2}	0.01	One hundredth
10^{-1}	0.1	One tenth
10^{0}	1	One
10^{1}	10	Ten
10^{2}	100	One hundred
10^{3}	1000	One thousand
10^{6}	1 000 000	One million
10^{9}	1 000 000 000	One billion

exponent. In scientific work, the base is usually 10, and the exponent is an integer (a whole number). The coefficient is multiplied by the base the number of times given by the exponent. That is, the number in the example is 1.23 multiplied four times by 10:

$$1.23 \times 10^4 = 1.23 \times 10 \times 10 \times 10 \times 10 = 12\ 300$$

Table 2.4 lists important exponential parts and their meanings.

▪ **EXAMPLE 2.31**

How would you write 1 billion in exponential notation?

Solution

$$1\ 000\ 000\ 000 = 1 \times 10 \times 10 \times 10 \times 10 \times 10 \times 10 \times 10 \times 10 \times 10$$
$$= 1 \times 10^9$$

Practice Problem 2.31   How would you write 20 billion in exponential notation? ▪

Scientists generally report numbers in exponential notation with coefficients that have one and only one integer digit, which is not zero. That is, the coefficient is a number that is greater than or equal to 1 and less than 10. Numbers in this format are said to be written in **standard exponential form,** or **scientific notation.** A scientific calculator gives exponential numbers in this form, unless "engineering format" is selected.

> In standard exponential notation, the coefficient is 1 or more but less than 10.

▪ **EXAMPLE 2.32**

Which of the following numbers is written in scientific notation?

(a) 20×10^7 (b) 0.88×10^3 (c) 9.876×10^1

Solution

(a) The number 20×10^7 is not in scientific notation because its coefficient is a two-digit integer.

(b) The number 0.88×10^3 is not in scientific notation because the integer digit of its coefficient is zero.

(c) The number 9.876×10^1 is in scientific notation because the coefficient has only one integer digit.

Practice Problem 2.32   Which of the following numbers is in scientific notation?

(a) 0.089×10^7 (b) 39×10^3 (c) 1.0000×10^{-3} ■

An advantage to scientific notation is that all digits of the coefficient of a number in scientific notation are significant. The exponent determines the magnitude of the number, so any zeros present in the coefficient must be significant.

■ EXAMPLE 2.33

How can you resolve the difficulty presented in Example 2.23?

Solution

One way to resolve this problem is to report the values in scientific notation, where all digits of the coefficient are significant: (a) 1.2×10^3 mm, (b) 1.20×10^3 mm, (c) 1.200×10^3 mm.

Practice Problem 2.33   Report the number of grams in each of the following quantities in such a way that the proper number of significant digits is obvious:

(a) 2.4 kg   (b) 2.40 kg   (c) 2.400 kg ■

To enter a number in exponential notation on an electronic calculator, you enter the coefficient, press the EXP or EE key, then enter the exponent. *Do not* press the multiplication key ⊠. See Appendix 1. If a number is given in exponential notation without a coefficient, a coefficient of 1 is assumed. Thus, 10^9 is 1 billion. (Some electronic calculators require a coefficient of 1 to be entered.)

■ EXAMPLE 2.34

How many times should you press the multiplication key to solve the following problem:

$$(1.0 \times 10^2)(2.0 \times 10^4)$$

Solution

Once, for the multiplication process. (You press the EXP or EE key twice.) ■

Changing the Form of Exponential Numbers

The *form* of an exponential number may be changed without changing its *value*. For example, 1.23×10^4 can be changed to another number times 10^3 or a different number times 10^2, and so on:

$$
\begin{aligned}
1.23 \times 10^4 &= 1.23 \times 10 \times 10 \times 10 \times 10 \\
&= 1.23 \times 10 \times (10 \times 10 \times 10) = 12.3 \times 10^3 \\
&= 1.23 \times 10 \times 10 \times (10 \times 10) = 123 \times 10^2 \\
&= 1.23 \times 10 \times 10 \times 10 \times (10) = 1230 \times 10^1 \\
&= 1.23 \times 10 \times 10 \times 10 \times 10 \ = 12\,300 \times 10^0 = 12\,300
\end{aligned}
$$

In the first conversion, we multiplied the coefficient by one of the tens and ended up with one fewer ten in the exponential portion of the number. The *values* of all these numbers are the same; only their *format* is different. You may need to change to different formats when you add or subtract exponential numbers, unless you use a scientific calculator.

We can *increase* either the coefficient or the exponential part of a number by any factor without changing the number's overall value if we *reduce* the other part by the same factor. A simple working rule allows changing the format of a number in exponential notation: Move the decimal point in the coefficient *to the right* n *places and reduce the exponent* n *units*, or move the decimal point in the coefficient *to the left* n *places and increase the exponent* n *units*.

> Move the decimal point in the coefficient to the right *n* places and reduce the exponent *n* units,
>
> or
>
> move the decimal point in the coefficient to the left *n* places and increase the exponent *n* units.

■ EXAMPLE 2.35

Change the format of each of the following numbers to scientific notation:

(a) 20.0×10^7 (b) 0.0490×10^3 (c) 303×10^1

Solution

(a) The decimal point must be moved one place to the *left*, so the exponent is *increased* by 1: 2.00×10^8.

(b) The decimal point has to be moved two places to the *right*, so the exponent is *reduced* by 2: 4.90×10^1.

(c) The coefficient is reduced by a factor of 100 (equal to 10^2), and the exponential part is increased by the same factor: 3.03×10^3.

Practice Problem 2.35   Change the format of each of the following numbers to scientific notation:

(a) 0.0202×10^7 (b) 400.0×10^3 (c) 0.670×10^1 ■

Multiplication and Division of Exponential Numbers

To multiply numbers in exponential format, we multiply the coefficients and the exponential parts separately. To multiply exponential parts, the exponents are *added*. For example, let's multiply 4.0×10^4 and 2.0×10^3:

$$(4.0 \times 10^4) \times (2.0 \times 10^3) = (4.0 \times 2.0) \times (10^4 \times 10^3)$$

$$= 8.0 \times 10^{4+3} = 8.0 \times 10^7$$

It's easy to see that if you multiply four tens by three tens, you get seven tens:

$$10^4 \times 10^3 = (10 \times 10 \times 10 \times 10) \times (10 \times 10 \times 10)$$
$$= 10 \times 10 \times 10 \times 10 \times 10 \times 10 \times 10 = 10^7$$

■ **EXAMPLE 2.36**

Multiply the following numbers, and express the answers in scientific notation:

(a) $(3.0 \times 10^6) \times (5.0 \times 10^3)$
(b) $(4.0 \times 10^2) \times (7.7 \times 10^4)$
(c) $(4.4 \times 10^7) \times (3.6 \times 10^1)$

Solution

(a) $(3.0 \times 10^6) \times (5.0 \times 10^3) = 15 \times 10^9 = 1.5 \times 10^{10}$
(b) $(4.0 \times 10^2) \times (7.7 \times 10^4) = 31 \times 10^6 = 3.1 \times 10^7$
(c) $(4.4 \times 10^7) \times (3.6 \times 10^1) = 16 \times 10^8 = 1.6 \times 10^9$

Practice Problem 2.36 Multiply the following numbers and express the answers in scientific notation:

(a) $(7.5 \times 10^9) \times (6.0 \times 10^5)$
(b) $(2.0 \times 10^3) \times (9.7 \times 10^3)$
(c) $(1.4 \times 10^6) \times (4.6 \times 10^7)$ ■

To divide exponential numbers, we divide the coefficients and the exponential parts separately. To divide exponential parts, we *subtract* the exponents.

■ **EXAMPLE 2.37**

Divide 6.0×10^5 by 2.0×10^2.

Solution

$$(6.0 \times 10^5)/(2.0 \times 10^2) = (6.0/2.0) \times 10^{5-2} = 3.0 \times 10^3$$

You can see that this procedure is correct:

$$\frac{6.0 \times 10^5}{2.0 \times 10^2} = \frac{6.0 \times 10 \times 10 \times 10 \times 10 \times 10}{2.0 \times 10 \times 10}$$
$$= 3.0 \times 10 \times 10 \times 10 = 3.0 \times 10^3$$

Practice Problem 2.37 Divide 2.0×10^5 by 4.0×10^2. ■

We can apply this procedure to calculate the quotient of two exponential numbers even when the denominator has a larger magnitude than the numerator. For example, let's divide 6.0×10^6 by 2.0×10^8. The rule for dividing exponential numbers gives the following result:

$$(6.0 \times 10^6)/(2.0 \times 10^8) = 3.0 \times 10^{6-8} = 3.0 \times 10^{-2}$$

What does the negative exponent mean? Writing out all the expressions allows us to see:

$$\frac{6.0 \times 10^6}{2.0 \times 10^8} = \frac{6.0 \times 10 \times 10 \times 10 \times 10 \times 10 \times 10}{2.0 \times 10 \times 10 \times 10 \times 10 \times 10 \times 10 \times 10 \times 10}$$

$$= 3.0 \times \left(\frac{1}{10 \times 10}\right) = 0.030 = 3.0 \times 10^{-2}$$

The negative exponent means to *divide* the coefficient by the base a certain number of times.

■ **EXAMPLE 2.38**

Calculate the quotient of $(6.0 \times 10^3)/(2.0 \times 10^3)$.

Solution

By the rule for division of exponential numbers:

$$\frac{6.0 \times 10^3}{2.0 \times 10^3} = 3.0 \times 10^0$$

By cancellation:

$$\frac{6.0 \times 10^3}{2.0 \times 10^3} = \frac{6.0 \times 10 \times 10 \times 10}{2.0 \times 10 \times 10 \times 10} = 3.0$$

Since $3.0 = 3.0 \times 10^0$, it is apparent that 10^0 is equal to 1.

Practice Problem 2.38   What is the decimal value (the value with no power of 10 shown) of 5.65×10^0? ■

■ **EXAMPLE 2.39**

Change the format of each of the following numbers to scientific notation:

(a) 10.0×10^{-6} (b) 0.050×10^{-3} (c) 303×10^{-1}

Solution

(a) The decimal point must be moved one place to the *left,* so the exponent is *increased* by 1: 1.00×10^{-5}.

(b) The decimal point has to be moved two places to the *right,* so the exponent is *reduced* by 2: 5.0×10^{-5}.

(c) The coefficient is reduced by a factor of 100 (equal to 10^2), and the exponential part is increased by the same factor: 3.03×10^1.

Practice Problem 2.39   Change the format of each of the following numbers to scientific notation:

(a) 0.0101×10^{-6} (b) 200.0×10^{-3} (c) 0.300×10^0 ■

The rules for multiplication and division need to be stated slightly differently to allow for negative exponents. To multiply exponential parts, add the exponents *algebraically.* To divide exponential parts, subtract the exponents *algebraically.* The word *algebraically* means "with due regard for the signs."

■ EXAMPLE 2.40

Divide 6.0×10^5 by 2.0×10^{-3}.

Solution

$$\frac{6.0 \times 10^5}{2.0 \times 10^{-3}} = 3.0 \times 10^{5-(-3)} = 3.0 \times 10^8$$

Instead of dividing a negative exponent by changing its sign and adding, you may transfer the *exponential part* of a number from numerator to denominator or from denominator to numerator if you simply change the sign of the exponent:

$$\frac{6.0 \times 10^5}{2.0 \times 10^{-3}} = \frac{6.0 \times 10^5 \times 10^{+3}}{2.0} = 3.0 \times 10^8$$

Practice Problem 2.40   Divide 2.0×10^{-5} by 8.0×10^{-3}. ■

Addition and Subtraction of Exponential Numbers

When we add or subtract numbers in exponential notation, *the exponents must be the same.* (This rule is related to the rule that requires numbers being added or subtracted to have their decimal points aligned.) The answer is then the sum or difference of the coefficients times the same exponential part as in each number being added or subtracted. (The calculator does this operation automatically, but you must know what is happening in order to report the proper number of significant digits [see Section 2.3].)

■ EXAMPLE 2.41

Add (a) $1.23 \times 10^4 + 4.56 \times 10^4$ and   (b) $1.23 \times 10^4 + 4.5 \times 10^3$.

Solution

(a) 1.23×10^4
     $+4.56 \times 10^4$ $\rightarrow$ Same exponent
     $\underline{5.79 \times 10^4}$
     Sum of coefficients

Since the exponents are the same, the coefficients are simply added. The answer has the same exponential part as each number added.

(b) 1.23×10^4
     $+0.45 \times 10^4$ Coefficient and exponent amended to allow addition
     $\overline{1.68 \times 10^4}$

Since the exponents are not the same, one of them must be changed to equal the other. Of course, you cannot change only the exponent because that would change the value of the number. You can move the decimal point one place to the left in 4.5×10^3 (making the coefficient smaller) and increase the exponent by one (making the exponential part larger). The value of the number is unchanged, but its format is now suitable for the addition you want to do.

Practice Problem 2.41 Subtract (a) $1.11 \times 10^4 - 6.66 \times 10^4$ and (b) $1.23 \times 10^4 - 9.25 \times 10^5$. ■

Raising an Exponential Number to a Power

To raise an exponential number to a power, we raise both the coefficient and the exponential part to the power. We raise an exponential part to a power by *multiplying* the exponent by the power.

■ **EXAMPLE 2.42**

Calculate the cube of 2.0×10^3 cm.

Solution

$$(2.0 \times 10^3 \text{ cm})^3 = (2.0)^3 \times (10^3)^3 \text{ cm}^3 = (2.0)^3 \times 10^{3 \times 3} \text{ cm}^3$$
$$= 8.0 \times 10^9 \text{ cm}^3$$

Note that the unit must be raised to the proper power. It is easily shown that this procedure is correct:

$$(2.0 \times 10^3 \text{ cm})^3 = (2.0 \times 10^3 \text{ cm})(2.0 \times 10^3 \text{ cm})(2.0 \times 10^3 \text{ cm})$$
$$= 8.0 \times 10^9 \text{ cm}^3$$

Practice Problem 2.42 Calculate the value of $(3.0 \times 10^{-3} \text{ cm})^2$. ■

Taking the square root of a number is equivalent to raising the number to the 1/2 power. In general, the *n*th root of a number is the number to the 1/*n* power.

■ **EXAMPLE 2.43**

Calculate the square root of 4.00×10^{-8}.

Solution

$$\sqrt{4.00 \times 10^{-8}} = 2.00 \times \sqrt{10^{-8}} = 2.00 \times (10^{-8})^{1/2} = 2.00 \times 10^{-4}$$

Practice Problem 2.43 Calculate the cube root of 8.00×10^9. ■

■ **EXAMPLE 2.44**

A *unit cell* is a small portion of a substance that, when repeated very many times, builds up an entire sample. A unit cell of a certain substance is cubic, with edge length 4.07×10^{-10} m. How many unit cells does it take to occupy 1.00 cm³?

Solution

The volume of the unit cell is

$$(4.07 \times 10^{-10} \text{ m})^3 = (4.07 \times 10^{-8} \text{ cm})^3 = 6.74 \times 10^{-23} \text{ cm}^3$$

The number of unit cells is therefore

$$1.00 \text{ cm}^3 \left(\frac{1 \text{ unit cell}}{6.74 \times 10^{-23} \text{ cm}^3} \right) = 1.48 \times 10^{22} \text{ unit cells} \quad ■$$

2.5 Density

Density is defined as *mass per unit volume:*

$$\text{Density} = \text{mass/volume}$$

In symbols,

$$d = m/V$$

The dimensions (combination of units) of density involve a mass unit divided by a volume unit, such as g/mL or g/cm³. Thus, to get the density of an object, we simply divide the mass of the object by its volume. Problems involving density usually involve finding one of the variables—*d*, *m*, or *V*—having been given the other two. Either the equation or the factor label method can be used to solve density problems. The equation is most often used when mass and volume are given. The factor label method is perhaps easier when density and one of the other two variables are given and the third variable is sought.

Densities of some common substances are given in Table 2.5. Note that scientists generally put the units at the heads of the columns when reporting data in a table. You should remember that the density of liquid water is about 1.00 g/mL = 1.00 g/cm³.

Table 2.5 Densities of Some Common Substances

Substance	Density (g/mL)
Aluminum	2.702
Copper	8.92
Gold	19.3
Iron	7.86
Lead	11.3
Magnesium	1.74
Mercury	13.6
Octane	0.7025
Platinum	21.45
Salt (NaCl)	2.165
Sugar (sucrose)	1.56
Water (at 4°C)	1.000

■ **EXAMPLE 2.45**

Calculate the density of the wood in a desk if its mass is 30.0 kg and its volume is 35.0 L.

Solution

$$d = m/V = (30.0 \text{ kg})/(35.0 \text{ L}) = 0.857 \text{ kg/L}$$

Practice Problem 2.45 Calculate the density of a rectangular metal bar that is 5.00 cm long, 2.00 cm wide, and 1.00 cm thick and has a mass of 113 g. ■

■ **EXAMPLE 2.46**

Calculate the mass of 245 mL of mercury (density = 13.6 g/mL).

Solution

$$245 \text{ mL} \left(\frac{13.6 \text{ g}}{1 \text{ mL}} \right) = 3330 \text{ g} = 3.33 \text{ kg}$$

Practice Problem 2.46   Calculate the volume of 668 g of mercury (density = 13.6 g/mL). ■

■ EXAMPLE 2.47

Calculate the volume in liters of 1012 g of mercury. (Hint: see Table 2.5.)

Solution

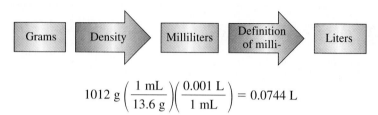

$$1012 \text{ g} \left(\frac{1 \text{ mL}}{13.6 \text{ g}} \right) \left(\frac{0.001 \text{ L}}{1 \text{ mL}} \right) = 0.0744 \text{ L}$$

Practice Problem 2.47   Calculate the mass of 1.000 L of aluminum. ■

Substances expand when heated, and the resulting change in volume causes some change in density. Within reasonable temperature ranges, the density of a substance is relatively constant. For example, water varies from 0.99979 g/mL at 0°C to 1.0000 g/mL at 4°C to 0.95838 g/mL at 100°C. We will usually ignore such slight differences, especially since we will work most often with densities measured to only three significant digits.

Density is an intensive property, useful in identifying substances. For example, gold can be distinguished from iron pyrite by their greatly differing densities—19.3 g/cm³ for gold and 5.0 g/cm³ for iron pyrite. Iron pyrite is known as "fool's gold" because of its striking visual resemblance to gold. Many prospectors in the western United States in gold-rush days were terribly disappointed when the test in the assay office showed that they had found iron pyrite rather than gold.

Relative densities determine whether an object will float in a given liquid in which it does not dissolve. An object will float if its density is less than the density of the liquid. For example, the density of liquid water is 1.00 g/mL and that of a particular kind of wood is 0.800 g/mL. The wood will float in water, since it has a lower density.

■ EXAMPLE 2.48

An 8.00 cm × 2.00 cm × 2.00 cm rectangular metal bar has a mass of 362 g. Will the bar float in water or in mercury (density = 13.6 g/mL)?

Solution

The volume of the bar is 32.0 cm³, and its density is therefore

$$d = (362 \text{ g})/(32.0 \text{ cm}^3) = 11.3 \text{ g/cm}^3$$

The bar will sink in water (density = 1.00 g/cm³), but it will float in mercury.

Practice Problem 2.48   A 9.00 cm × 4.00 cm × 2.00 cm rectangular solid has a mass of 566 g. Will the object float in water? ■

■ EXAMPLE 2.49

Identify the metal in the bar of Example 2.48.

Solution

The density is that of lead (Table 2.5).

Practice Problem 2.49   Identify the substance in Practice Problem 2.48.   ■

2.6 Time, Temperature, and Energy

OBJECTIVE

■ to use the units associated with time, temperature, and energy, and to distinguish between temperature and energy

Time

The units of time in the metric system are the units we use in everyday life. The primary SI unit is the **second** (s). Other time units are minute (min), hour (h), day (d), week (wk), and year (y).

■ **EXAMPLE 2.50**

How many seconds are in 1.00 h?

Solution

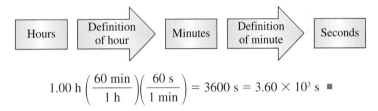

$$1.00 \text{ h} \left(\frac{60 \text{ min}}{1 \text{ h}} \right) \left(\frac{60 \text{ s}}{1 \text{ min}} \right) = 3600 \text{ s} = 3.60 \times 10^3 \text{ s} \quad ■$$

Temperature and Energy

Although they are related to each other, temperature and energy are not the same. The **temperature** of a body is a measure of the *intensity* of energy in the body. As energy is added to a body, in the form of **heat,** for example, the body generally gets warmer. However, adding the same amount of energy to two different-size samples of the same substance will cause the smaller sample to rise in temperature more than the larger sample (Figure 2.10).

Figure 2.10 Experiment Showing the Difference between Temperature and Energy

(a) First, 50 mL of water at 20°C is heated over a Bunsen burner for exactly 2 minutes. (b) Next, 150 mL of water at 20°C is heated for exactly 2 minutes over the same flame. The final temperature of the 50-mL sample will be higher than that of the 150-mL sample, even though approximately the same amount of heat will have been added to each sample.

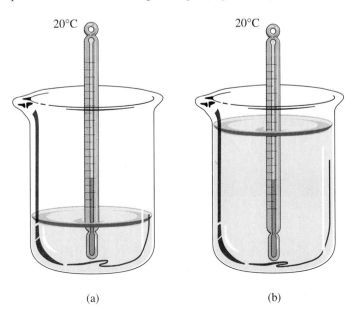

(a)   (b)

The temperatures of two bodies placed in contact with one another determine the direction of spontaneous heat flow between them. Heat will flow from the hotter body to the colder one (unless some energy is used to make the heat flow in the other direction).

■ **EXAMPLE 2.51**

If a hot piece of metal is placed in cold water, what happens to the temperatures of the metal and the water?

Solution

Heat flows from the hot metal to the cold water; the water gets warmer, and the metal gets colder.

Practice Problem 2.51   What happens to the temperature of hot coffee when you add some cold milk to it? ■

Temperature Scales

In the United States, it is necessary to know three different temperature scales. The scale used in everyday American life is the **Fahrenheit scale,** on which the temperature of freezing water is defined as 32°F, and the temperature of water boiling under normal conditions is defined as 212°F. Scientists do not use the Fahrenheit scale (although American engineers sometimes do). Instead, they use the **Celsius scale,** the metric scale for temperature, which is sometimes called the **centigrade scale.** On the Celsius scale, the temperature of freezing water is defined as 0°C, and the temperature of water boiling under normal conditions is defined as 100°C. The Celsius scale is used in most other countries of the world for everyday measurements as well as scientific ones. The **Kelvin scale** for measuring temperatures is important for work with gases (Chapter 12) and in other advanced work. On the Kelvin scale, the temperature of freezing water is 273 K, and the temperature of water boiling under normal conditions is 373 K. The degree sign (°) is not used with the Kelvin scale, and the units are called kelvins rather than degrees. The three scales are pictured in Figure 2.11.

To convert from degrees Fahrenheit (F) to degrees Celsius (C), or vice versa, we use the following equations:

$$C = \tfrac{5}{9}(F - 32) \qquad \text{and} \qquad F = \tfrac{9}{5}C + 32°$$

The 32° subtracted is a definition and can be expanded to 32.0°, 32.00°, etc. To convert from degrees Celsius to kelvins, we use the following equation:

$$K = C + 273$$

■ **EXAMPLE 2.52**

Convert 98.6°F (normal body temperature) to a Celsius temperature.

Solution

$$C = \tfrac{5}{9}(F - 32) = \tfrac{5}{9}(98.6 - 32.0) = 37.0°C \quad ■$$

Figure 2.11 Comparison of Temperature Scales

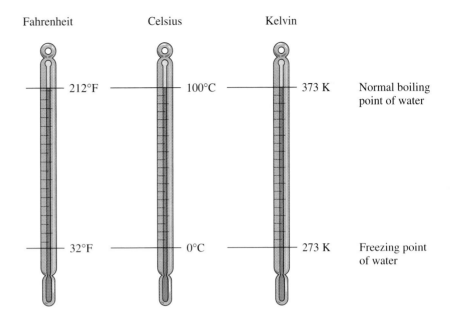

Fahrenheit   Celsius   Kelvin

212°F —————— 100°C —————— 373 K   Normal boiling
                                                   point of water

32°F —————— 0°C —————— 273 K   Freezing point
                                                   of water

Energy

Energy comes in many different forms, but each form can be measured in the same units. Several energy units are in common use; we shall use the metric system unit, which is the calorie (cal), and the SI unit, which is the joule (J). The **calorie** is the quantity of heat required to raise the temperature of 1 gram of water 1°C. (The kind of calorie counted by someone on a diet is actually a *kilocalorie*—enough energy to raise the temperature of a *kilogram* of water 1°C.) Electrical energy is the form of energy that is easiest to measure with great precision, and since it is usually measured in joules, the **joule** is the primary unit of energy in SI. Exactly 4.184 J is equal to 1.000 cal.

The kinetic energy of a body in motion is equal to

$$KE = \tfrac{1}{2}mv^2$$

where v is the velocity of the body and m is its mass. The units associated with this equation are

$$1 \text{ J} = 1 \text{ kg} \cdot \text{m}^2/\text{s}^2$$

■ EXAMPLE 2.53

How much kinetic energy does a 2.00-kg body have if it is moving at 3.00 m/s?

Solution

$$KE = \tfrac{1}{2}mv^2 = \tfrac{1}{2}(2.00 \text{ kg})(3.00 \text{ m/s})^2 = 9.00 \text{ kg} \cdot \text{m}^2/\text{s}^2 = 9.00 \text{ J} \quad ■$$

■ ■ ■ ■ ■ ■ ■ SUMMARY ■ ■ ■ ■ ■ ■ ■

Measurement is the key to quantitative physical science. The results of every measurement must include both a numeric value and a unit (or set of units). Be sure to use standard abbreviations for all units. The factor label method is used to convert a quantity from one set of units to another without changing its value. The original quantity is multiplied by a factor equal to 1. (The numerator and denominator of the factor are equal to each other in value but different in form.) To use the factor label method: (1) write down the quantity given, (2) multiply by a factor that will yield the desired units, (3) cancel the units, (4) multiply all numbers in the numerators and divide by the number(s) in the denominator(s). Sometimes, it is necessary to multiply by more than one factor. If you do so, you may solve for the intermediate answers, but you do not have to. (Section 2.1)

The metric system and its newer counterpart, SI, use subunits and multiples of units that are equal to powers of 10, and they also use the same prefixes to mean certain fractions or multiples, no matter what primary unit is being modified. The meter is the primary unit of length; the gram is the primary unit of mass; and the liter (the cubic meter in SI) is the primary unit of volume. The prefixes *centi-* (0.01), *milli-* (0.001), and *kilo-* (1000) are used with any of these or with any other metric unit. Conversions between English and metric units are necessary for American engineers and may be helpful to get you familiar with the relative sizes of the metric units. To add, subtract, multiply, or divide measured quantities of the same type, first be sure that they all have the same units. (Section 2.2)

The number of digits reported for a measurement or for the result of a calculation involving measurements is a means of showing how accurately the measurement(s) was (were) made. The last digit of reported measurements is usually an estimate based on tenths of the smallest scale division of the measuring instrument. Any digit from 1 through 9 in a properly reported result is significant. (The word *significant* as used here refers to the accuracy of the measurement and does not mean "important.") Zeros may or may not be significant; if they merely show the magnitude of the number, they are not significant.

When measurements are added or subtracted, the one that has significant digits least far to the right is the one that limits the number of significant digits in the answer. When measurements are multiplied or divided, the one with the fewest significant digits limits the number of significant digits in the answer. If too many digits are present in a calculated answer, the answer is rounded off: Some digits are dropped if they occur to the right of the decimal point or changed to nonsignificant zeros if they occur to the left of the decimal point. If too few digits are present, significant zeros are added. Note that, although a calculator gives the correct magnitude when used properly, you must understand the calculation processes to be able to determine the number of significant figures to report. (Section 2.3)

Exponential notation allows easy reporting of extremely large and extremely small numbers. A number in scientific notation consists of a coefficient times 10 to an integral power, where the coefficient is equal to or greater than 1 but less than 10. You must know how to convert numbers from exponential notation to ordinary decimal values, and vice versa, and also how to use exponential numbers in calculations. You should be able to use effectively an electronic calculator with exponential capability (see Appendix 1). All digits in the coefficients of numbers in scientific notation are significant. (Section 2.4)

Density, an intensive property, is defined as mass per unit volume. It can be calculated by dividing the mass of a sample by its volume. If a density is given, it may be used as a factor to solve for mass or volume. Samples of lower density float in liquids of higher density. (Section 2.5)

Time, temperature, and energy are familiar quantities used extensively in scientific work. Temperature is a measure of the intensity of energy in a sample. Scientists use two different temperature scales—Celsius and Kelvin. Their relationship to each other and to the more familiar Fahrenheit scale is shown in Figure 2.11. (Section 2.6)

Items for Special Attention

- A percentage can be used as a factor label.

- The metric units used to report volumes may involve liters or the cube of a length unit.

- Adding a positive value to a negative exponent either makes the exponent positive or makes the exponent less negative (gives the exponent a smaller magnitude).

- Do not key in 10 for the base of a number in scientific notation on your electronic calculator. The EXP or EE key means "× 10 to a power."

- Significant digits and decimal places are two different things. Do not confuse them. For example, 3.01 has three significant digits and two decimal places.

- The expressions that follow are simplified in the same way. The algebraic quantities x and y and units such as meters and seconds (m and s) or miles and hours are treated the same.

$$4y\left(\frac{50x}{y}\right) = 200x \qquad 4\text{ s}\left(\frac{50\text{ m}}{1\text{ s}}\right) = 200\text{ m}$$

$$4\text{ hours}\left(\frac{50\text{ miles}}{1\text{ hour}}\right) = 200\text{ miles}$$

- When measurements expressed in exponential notation are added or subtracted, *both the units and the exponents* must be the same. When such measurements are multiplied or divided, units and exponents can be different.

Self-Tutorial Problems

2.1 (a) Which is bigger, a cent or a dollar? Which would you need more of to buy a certain textbook?

    (b) Which is bigger, a centimeter or a meter? Which does it take more of to measure the length of a certain textbook?

2.2 Identify each of the following as a quantity or rate or ratio:

    (a) Minimum wage   (b) Amount of pay

    (c) Number of hours worked

2.3 Which of the following numbers have values less than zero? Which have magnitudes less than one? Which have values less than one?

    (a) 1.0×10^{-3} (b) -2.5×10^{-2}

    (c) -3.7×10^{7} (d) 4.2×10^{4}

2.4 Which is bigger—1 kg or 1 mg? Which would there be more of in a measurement of your own mass?

2.5 Which of the following are units of length?

    kg mg mL mm g m

2.6 Use your electronic calculator to find the value of this expression:

$$1 \times 10^{0} + 0$$

2.7 Use your electronic calculator to do the following calculation to check your calculator procedure:

$$\frac{6 \times 3}{2 \times 3}$$

2.8 Which of the following animals is most likely to have a mass of 1 kg?

    (a) Chicken   (b) Elephant

    (c) Fly   (d) Saint Bernard dog

2.9 What factor is used to convert a measurement in meters to millimeters?

2.10 (a) If you multiply a certain number by 10 and then divide the result by 10, what is the relationship of the final answer to the original number?

    (b) If you multiply the coefficient of an exponential number by 10 and divide the exponential part by 10, what is the effect?

2.11 What is the exponential equivalent of each of the following metric prefixes?

    (a) milli-   (b) kilo-   (c) centi-

2.12 (a) How can a company raise the price of coffee at the supermarket without charging more for each can?

    (b) How can you reduce the sweetness of a cup of coffee without taking out any sugar?

    (c) What happens to the average speed for a trip if the time spent traveling remains unchanged but the distance is decreased?

2.13 What is the difference between the masses 1 mg and 1 Mg?

2.14 How many square centimeters are in a square measuring 4.0 cm along each edge?

2.15 What is the difference in density, if any, among the following?

    6.7 grams per milliliter

    6.7 g/mL

    6.7 g in exactly 1 mL

    6.7 g in 1.0 mL

    6.7 g divided by 1.0 mL

2.16 Write each of the following numbers in exponential notation:

    (a) 3 thousand   (b) 1 million

    (c) 200 million   (d) 7 billion

■ ■ ■ PROBLEMS ■ ■ ■

2.1 Factor Label Method

2.17 Assume that donuts are $5.00 per dozen. Use the factor label method to answer each of the following:

(a) How much do 6.00 dozen cost?

(b) How many dozens can you buy with $12.50?

(c) How many donuts can you buy with $12.50?

2.18 Determine the cost of 1600 pencils if the price is $1.50 per dozen.

2.19 Calculate the number of seconds in 7.75 hours.

2.20 There is 60.0% oxygen by mass in a compound of sulfur and oxygen. Percent by mass is a ratio of the number of grams of a particular component to 100 grams of the total sample. How many grams of sulfur are in a 33.5-g sample of the compound?

2.21 Calculate the number of square yards of rug required to cover a living-room floor that is 12.0 ft wide and 15.0 ft long.

2.22 Calculate the pay earned by a student who worked 15 hours per week for 35 weeks at $6.50 per hour.

2.23 Calculate the pay received for 1.00 hour of work by a junior executive who works 40 hours per week and earns $52,000 per year for 50 weeks of work.

2.2 Metric System

2.24 Convert:

(a) 1.23 m to centimeters

(b) 1.24 m to millimeters

(c) 1.25 m to kilometers

2.25 Which of the following is the smallest container that could hold 1 m^3 of liquid?

(a) Drinking glass (b) Thimble

(c) Swimming pool (d) Soda bottle

2.26 (a) How many milligrams are in 3.50 g?

(b) How many centimeters are in 3.50 m?

(c) How many kilograms are in 3.50 g?

2.27 Which of the following are units of volume?

mg mL mm^3 g m^3 cm cm^2 kL

2.28 Which of the following is the most probable distance between a dormitory room and the chemistry lecture room?

(a) 1 mm (b) 1 cm (c) 1 m (d) 1 km

2.29 How many milliliters are in 4.75 L?

2.30 The author of this text is of average build. Fill in the metric units in the following description:

Height: 172 _____

Mass: 65 _____

Total volume of blood in his system: 4 _____

2.31 Calculate the number of liters in

(a) 0.0203 m^3 (b) 303 cm^3

(c) 403 mL (d) 503 mm^3

2.32 What is the volume of a rectangular solid that is 4.5 m long, 2.5 cm wide, and 0.000020 km thick?

2.33 How many liters are in 7.23 m^3?

2.34 How many cubic millimeters are in 9.95 mL?

2.3 Significant Digits

2.35 Round off the following measurements to three significant digits:

(a) 144.6 mL (b) 1446 mL

(c) Are the answers the same?

2.36 How many significant digits are present in each of the following measurements?

(a) 0.0107 kg (b) 1007 cm^3

(c) 10.00 mL (d) 17.12 m

2.37 Underline the significant digits in each of the following measurements. If a digit is uncertain, place a question mark below it.

(a) 1.00 mm (b) 0.010 cm

(c) 100 m (d) 10.0 km

2.38 Underline the significant digits in each of the following measurements. If a digit is uncertain, place a question mark below it.

(a) 100 cm (b) 702.0 cm

(c) 70.0 cm (d) 0.0700 m

(e) 700 mm

2.39 How many significant digits are present in each of the following measurements?

(a) 0.0109 kg (b) 100.7 cm^3

(c) 210.0 mL (d) 0.07000 m

2.40 Underline the significant digits in each of the following measurements. If a digit is uncertain, place a question mark under it.

(a) 7.65 cm (b) 0.00730 cm

(c) 7.20 cm (d) 7.07 cm

2.41 Report the length of the shaded bar, using each of the rulers shown:

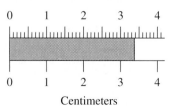

Centimeters

2.42 Add the following quantities, and report the answers to the proper number of significant digits:

(a) 6.17 g + 2.111 g

(b) 4.22 mL + 0.0112 mL

(c) 0.088 mL + 2.01 mL

(d) 1004 km + 21.02 km

2.4 Exponential Numbers

2.43 Convert each of the following numbers to standard exponential notation:

(a) 70.0 m   (b) 700.01 mg

(c) 830.1 g   (d) 0.00405 L

2.44 Convert each of the following values to ordinary (decimal) notation:

(a) 7.000×10^3 cm   (b) 7.00×10^3 cm

(c) 7.0×10^3 cm   (d) 7×10^3 cm

2.45 Convert each of the following numbers to centimeters, and express the answer in standard exponential notation:

(a) 60.2 mm   (b) 61.3 m

(c) 60.8 km   (d) 44.4×10^2 mm

2.46 Convert each of the following numbers to decimal format:

(a) 8.27×10^2 (b) 9.19×10^{-3}

(c) 2.00×10^1 (d) 3.00×10^0

2.47 Convert each of the following distances to meters, and express the results as ordinary numbers (not in exponential notation). Be sure to use the proper number of significant digits.

(a) 5.00×10^4 mm

(b) 5.00×10^4 cm

(c) 5.00×10^{-4} km

2.48 Write each of the following numbers in scientific notation:

(a) 2723

(b) 0.444

(c) 602 000 000 000 000 000 000 000

(d) 0.000 000 000 20

2.49 Express each of the following answers in scientific notation:

(a) $(3.0 \times 10^4)/(6.0 \times 10^4)$

(b) $(3.0 \times 10^2) + (3.3 \times 10^3)$

(c) $(4.0 \times 10^5) - (5.0 \times 10^4)$

(d) $(5.0 \times 10^2)(6.0 \times 10^3)$

(e) $(6.0 \times 10^{-3})(7.0 \times 10^{-2})$

2.50 Subtract:

$$(4.0 \times 10^{-3} \text{ cm}) - (-3.5 \times 10^{-2} \text{ cm})$$

2.51 The edge of a cube is 4.50×10^2 cm. What is the volume of the cube in cubic meters?

2.52 Express each of the following measurements as a decimal value:

(a) 3.00×10^{-3} L   (b) 3.0×10^{-3} L

(c) 3×10^{-3} L

2.53 Express each of the following measurements as a decimal value. State how many significant digits are in each result. Could you tell just from looking at the results without knowing the original values?

(a) 4.00×10^3 mL   (b) 4.0×10^3 mL

(c) 4×10^3 mL

2.54 Underline the significant digits in each of the following measurements. If a digit is uncertain, place a question mark below it.

(a) 1.0×10^2 cm   (b) 7.02×10^2 cm

(c) 7.0×10^2 cm   (d) 7.00×10^{-2} m

(e) 700 mm

2.55 Express each of the following volumes in milliliters:

(a) 5.00×10^{-3} L   (b) 5.0×10^{-3} L

(c) 5×10^{-3} L

2.56 Express each of the following measurements in liters:

(a) 6.00×10^3 mL   (b) 6.0×10^3 mL

(c) 6×10^3 mL

2.57 Add the following quantities, and report the answers to the correct number of significant digits:

(a) $2.71 \text{ kg} + (3.69 \times 10^2 \text{ g})$

(b) $0.0127 \text{ kg} + 21.7 \text{ g}$

(c) $0.019 \text{ kg} + 201.7 \text{ g}$

(d) $(2.00 \times 10^3 \text{ cm}) + (4.14 \times 10^2 \text{ cm})$

2.58 Multiply the following quantities, and report the answers to the correct number of significant digits:

(a) $1.79 \text{ cm} \times (2.61 \times 10^2 \text{ cm})$

(b) $0.9103 \text{ cm} \times 1.82 \text{ cm}$

(c) $0.1107 \text{ cm} \times 10.00 \text{ cm}$

(d) $(2.00 \times 10^3 \text{ cm}) \times (4.73 \times 10^2 \text{ cm})$

2.59 Divide the following quantities, and report the answers to the proper number of significant digits:

(a) $(6.62 \text{ g})/(9.170 \times 10^2 \text{ cm}^3)$

(b) $(0.00733 \text{ cm}^2)/(218 \text{ cm})$

(c) $(0.1033 \text{ cm}^3)/(400.0 \text{ cm})$

(d) $(2.72 \times 10^3 \text{ kg})/(4.31 \times 10^2 \text{ cm})^3$

2.60 Perform the following additions and explain the results:

(a) $(6.000 \times 10^{-3} \text{ mm}) + (6.0 \times 10^{-1} \text{ cm})$

(b) $(5.02 \times 10^2 \text{ g}) + 16.70 \text{ mg}$

2.61 In Problem 2.53, how can you report the proper number of significant digits in the results without using exponential notation?

2.62 In Problem 2.55, can you tell how many significant digits are in each measurement as it is given in exponential form in the problem? Can you tell in decimal form in the answer?

2.63 Add the following quantities, and report the answers to the proper number of significant digits:

(a) $(8.000 \times 10^{-3}\,\text{g}) + (8.0 \times 10^{-1}\,\text{g})$

(b) $(8.00 \times 10^{-1}\,\text{g}) + (2.72 \times 10^{-4}\,\text{g})$

2.5 Density

2.64 Using the data of Table 2.5, explain why aluminum is preferable to steel (mostly iron) for building airplanes.

2.65 Calculate the density of an object that has a volume of 4.75 mL and a mass of 32.8 g.

2.66 Using the densities in Table 2.5, identify the metal in a 24.70 cm³ solid of mass 529.8 g.

2.67 Calculate the number of milliliters of mercury (density = 13.6 g/mL) having a mass of 719 g.

2.68 Calculate the density of a rectangular solid of mass 1004 g and dimensions:

(a) 33.50 cm × 4.070 cm × 1.503 cm

(b) 0.3350 m × 4.070 cm × 15.03 mm

2.69 Explain why gasoline floats on water. Is water good for putting out gasoline fires?

2.70 Calculate the number of kilograms of mercury (density = 13.6 g/mL) occupying 709 mL.

2.71 (a) Calculate the volume of a rectangular box 21.7 cm by 3.70 cm by 1.700 cm.

(b) Calculate the number of kilograms of mercury (density = 13.6 g/mL) that can fit in that box.

2.72 Calculate the mass of mercury (density = 13.6 g/mL) that fills a rectangular box 21.7 cm by 3.70 cm by 1.700 cm.

2.73 Calculate the depth of water in centimeters in a cubic box (not full) with 40.0-cm edges if the mass of the contents is 30.0 kg.

2.74 Calculate the number of milliliters of lead (density = 11.3 g/mL) having a mass of 23.6 kg.

2.75 Calculate the mass of water that occupies 7.26 L.

2.76 What quantity is obtained in each of the following cases?

(a) Mass is divided by volume.

(b) Mass is divided by density.

(c) Density is multiplied by volume.

2.77 Express the density 6.25 kg/L in grams per cubic centimeter.

2.78 Convert the density 6.67×10^3 kg/m³ to grams per cubic centimeter.

2.79 Does gold float in mercury? (Hint: See Table 2.5 if necessary.)

2.80 An object has a density of 8.85 g/mL. Convert this density to kilograms per cubic meter.

2.81 Calculate the density in grams per milliliter of an object that has a volume of 5.33 cm³ and a mass of 27.9 g.

2.82 Calculate the volume in cubic centimeters of an object with a density of 5.33 g/cm³ and a mass of 292 g.

2.6 Time, Temperature, and Energy

2.83 Calculate the temperature in degrees Celsius of each of the following:

(a) 100°F (b) 20.0°F (c) 77°F (d) 98.6°F

(e) 0°F (f) 32.0°F (g) 212°F

2.84 Calculate the temperature in degrees Fahrenheit of each of the following:

(a) 100°C (b) 0°C (c) 33.0°C (d) 25°C

(e) 10°C (f) −18°C (g) −273°C

2.85 Calculate the temperature in kelvins of each of the following:

(a) 90.0°C (b) 33.0°C (c) −10°C

(d) 37°C (e) 125°C

2.86 Calculate the temperature in degrees Celsius of each of the following:

(a) 298 K (b) 373 K (c) 310 K (d) 0 K

2.87 Explain why *month* is not a useful unit of time for calculations.

2.88 What is the purpose of a refrigerator? How is this purpose accomplished?

■ ■ ■ GENERAL PROBLEMS ■ ■ ■

2.89 Perform the following calculations, and report the results to the proper number of significant digits:

(a) (3.55 cm)(3.107 cm)(2.8 cm)

(b) (3.61 kg)/[(4.44 × 10² cm)(33 cm)(2.04 cm)]

(c) (3.1440 g − 0.084 g)/(4.5132 cm³)

2.90 Using the densities in Table 2.5, identify the metal in a cube with 2.00-cm edges and a mass of 154 g.

2.91 Calculate the density of a cube with each edge 3.52 cm and mass 252 g.

2.92 Calculate the density of a cube with each edge 0.0352 m and mass 0.252 kg.

2.93 Compare the sizes, masses, and densities of the cubes in Problems 2.91 and 2.92.

2.94 Convert 5.78×10^3 kg/m³ to grams per cubic centimeter. Compare the answer to the result of Problem 2.93.

2.95 Under a certain set of conditions, the density of water is 1.00 g/mL and that of oxygen gas is 1.25 g/L. Which will float on the other?

2.96 Change 4.02×10^3 cm to:

(a) meters   (b) kilometers   (c) millimeters

2.97 A certain road map of Maine shows part of Canada. The distances in Canada are shown in kilometers, and the legend states: "To convert kilometers to miles, multiply by 0.62." What is actually being converted?

2.98 Calculate the length of each edge of a cube that has a volume of 8.00 cm³. Be sure to use the proper units.

2.99 (a) A bank usually charges 16.0% interest for a certain type of loan. If the bank advertises a special 10.0% discount for that type of loan, what is the actual rate?

(b) The percentage of a certain ore in the rock from a mine is 21%. The percentage of iron in the ore is 72%. What is the percentage of iron in the rock?

2.100 Calculate the volume of a (spherical) atom that has a radius of 1.50×10^{-10} m. ($V = \frac{4}{3}\pi r^3$; use $\pi = 3.1416$ if π is not on your calculator.)

2.101 Express each of the following lengths in millimeters (to the proper number of significant digits):

(a) 2×10^{-3} m   (b) 2.0×10^{-3} m

(c) 2.00×10^{-3} m   (d) 2.000×10^{-3} m

2.102 What is the volume of a cube whose edge measures:

(a) 7.32 cm   (b) 7.52×10^{-3} cm

2.103 Calculate the volume of a (spherical) atom of chlorine, which has a radius of 1.05×10^{-10} m. ($V = \frac{4}{3}\pi r^3$; use $\pi = 3.1416$ if π is not on your calculator.)

2.104 A vitamin pill maker produces pills with a mass of 1.00 g each. If each of the following indicates the mass of active ingredient, what percentage of the pill is active in each case?

(a) 2.00 mg   (b) 0.200 mg   (c) 200 mg

2.105 A certain brand of vitamin pill contains 200.0 μg of the vitamin; another brand has 0.2000 mg of the vitamin per pill. Which is the better buy, all other factors being equal?

2.106 A nurse who is directed to give a patient a pill that has 0.300 cg of active ingredients has no pills with centigrams as units. What pill labeled in milligrams should the nurse administer?

2.107 Calculate the number of liters in:

(a) 203 mL   (b) 0.0204 m³

(c) 2.05 cm³   (d) 2.06×10^4 mm³

2.108 Calculate the density in grams per cubic centimeter of an average chlorine atom, which has a radius of $1.05 \times$

10^{-10} m and a mass of 5.89×10^{-23} g. ($V = \frac{4}{3}\pi r^3$; use $\pi = 3.1416$ if π is not on your calculator.)

2.109 Calculate the temperature in Celsius of each of the following: (a) 33°F, (b) 33.0°F, and (c) 0°F. How many significant digits should be reported for each?

2.110 Calculate the length of each side of a cube that has a volume of 73.6 cm³. (Be sure to use the proper units and the proper number of significant digits.)

2.111 Calculate the energy of a particle that has no mass at rest but has a mass of 1.3×10^{-35} kg in motion.

$$E = mc^2 \text{ (from Chapter 1)}$$
$$1 \text{ J} = 1 \text{ kg} \cdot \text{m}^2/\text{s}^2$$
$$c = 3.00 \times 10^8 \text{ m/s}$$

2.112 (a) Draw a figure showing the addition of the lengths of two line segments, 10.0 cm and 0.10 cm.

(b) Can you do the same with 10.0 cm and 0.010 cm? Explain.

2.113 Calculate the mass of gold in a bracelet that contains 62.5% gold by mass and that has a volume of 15.40 mL and a density of 14.7 g/mL.

2.114 How many grains of sand, each with a volume of 1.0 mm³, could be held in a volume approximately equal to that of the earth? (The earth's radius is 6.4×10^3 km. For a sphere, $V = \frac{4}{3}\pi r^3$; use $\pi = 3.1416$ if π is not on your calculator.)

2.115 Explain why a football referee, after two successive defensive offside penalties on a first down, rules *without a measurement* that a new first down has been achieved, but does not do so after a first-down running play for no gain followed by two offside penalties.

2.116 Calculate the radius of a sphere of volume 10.00 cm³. ($V = \frac{4}{3}\pi r^3$; use $\pi = 3.1416$ if π is not on your calculator.)

2.117 Calculate the average density of the earth, assuming it to be spherical with radius 6400 km and mass 6.1×10^{24} kg. ($V = \frac{4}{3}\pi r^3$)

2.118 Calculate the value of "2.00 kseconds" in (a) minutes and (b) hours.

2.119 Using the densities in Table 2.5, identify the substance in a sphere with a 2.00-cm radius and a mass of 52.3 g. ($V = \frac{4}{3}\pi r^3$)

2.120 Calculate the length of each side of a cube that has a volume of 8.00×10^6 cm³. (Be sure to use the proper units and the proper number of significant digits.)

2.121 Calculate the mass of a sphere of radius 2.00 cm and density 0.382 g/cm³. ($V = \frac{4}{3}\pi r^3$; use $\pi = 3.1416$ if π is not on your calculator.)

2.122 Calculate the density of a sphere of radius 2.00 cm and mass 125.0 g. ($V = \frac{4}{3}\pi r^3$; use $\pi = 3.1416$ if π is not on your calculator.)

2.123 Calculate the area of a circle of radius 2.00 cm. ($A = \pi r^2$; use $\pi = 3.1416$ if π is not on your calculator.)

2.124 The density of a solution 25.0% by mass sodium chloride (table salt) in water is 1.19 g/mL. Calculate the mass of sodium chloride in 121 mL of the solution.

2.125 Calculate the density of a 25.0% by mass sodium chloride solution in water if 182 g of sodium chloride is used to make 612 mL of the solution.

2.126 Calculate the length of a rectangular solid if its density is 3.38 g/cm³, its mass is 278 g, its width is 4.73 cm, and its thickness is 2.13 cm.

2.127 Calculate the sum of 1.00 m and 0.100 mm. Attempt to draw a picture of this sum to the proper scale. Explain the effect in terms of significant digits.

2.128 Calculate the temperature in kelvins of each of the following:

(a) 90°F   (b) −40°F

(c) 122°F   (d) 32°F

2.129 Calculate the temperature in degrees Fahrenheit of each of the following:

(a) 300 K   (b) 0 K   (c) 373 K   (d) 233 K

2.130 Calculate the density in grams per milliliter of an object with a volume of 0.0400 L and a mass of 25.5 g.

2.131 Calculate the approximate height in meters of the man in the cubic meter box (Figure 2.6). Use a ratio of the height of the man to the height of the box in the figure, which is equal to the ratio of the heights in real life.

(a) Use a metric ruler to measure the figure.

(b) Use an inch ruler.

(c) State another reason why the metric system is easier to use than the English system of measurement.

2.132 Repeat Problem 2.131, but estimate the man's width at his waist instead of his height.

2.133 If a patient has a blood count of 500 white corpuscles per cubic millimeter, what is the number of white corpuscles per cubic centimeter?

3 Atoms and Atomic Masses

- 3.1 Laws of Chemical Combination
- 3.2 Dalton's Atomic Theory
- 3.3 Subatomic Particles
- 3.4 Atomic Mass
- 3.5 Development of the Periodic Table

■ Key Terms (Key terms are defined in the Glossary.)

atom (3.2)
atomic mass (3.4)
atomic mass scale (3.4)
atomic mass unit (3.4)
atomic number (3.3)
atomic weight (3.4)
atomic weight scale (3.4)
Dalton's atomic theory (3.2)

electron (3.3)
isotope (3.3)
law of constant composition (3.1)
law of definite proportions (3.1)
law of multiple proportions (3.1)
mass number (3.3)
molecule (3.2)
neutral (3.3)

neutron (3.3)
nucleus (3.3)
postulate (3.2)
proton (3.3)
relative scale (3.4)
subatomic particle (3.3)
weighted average (3.4)

■ Symbols/Abbreviation

A (mass number) (3.3)
n (number of neutrons) (3.3)

p (number of protons) (3.3)
Z (atomic number) (3.3)

amu (atomic mass unit) (3.4)

The theory of the atom has had a long history. The ancient Greeks postulated that matter exists in the form of atoms, but they did not base their theory on experiments, nor did they use it to develop additional ideas about atoms. In 1803, John Dalton proposed the first modern theory of the atom, which was based on the experimentally determined laws of conservation of mass, definite proportions, and multiple proportions. Dalton suggested for the first time that atoms of different elements are different from each other. His theory generated a great deal of research activity, which brought forth additional laws and knowledge about atoms, and he is recognized as the father of the atomic theory.

Section 3.1 takes up the experimental laws on which Dalton based his atomic theory, and Section 3.2 discusses the theory itself. You will learn about some modern extensions of the theory, including subatomic particles and isotopes, in Section 3.3. Atomic mass is presented in Section 3.4, and the development of the periodic table is traced in Section 3.5. A more sophisticated theory of the atom will be presented in Chapter 4.

3.1 Laws of Chemical Combination

Antoine Lavoisier (1743–1794), called the father of modern chemistry, discovered the law of conservation of mass by showing that during a chemical reaction, matter is neither gained nor lost. His quantitative work (work involving measurements) enabled him to conclude that the mass of products generated during a chemical reaction is the same as the mass of the reactants used up (Figure 3.1). This was not an easy conclusion, since "anyone could see that the ashes left after a large log burned did not weigh as much as the log itself." However, when the oxygen from the air (also a reactant in the burning of wood) and the carbon dioxide, water vapor, and other products formed (in addition to the ash) were considered, the total mass of the reactants and the total mass of the products were found to be equal. Lavoisier's work led other chemists to measure their reactants and products to confirm his conclusions and to see if they could make other quantitative observations.

The **law of definite proportions,** or the **law of constant composition,** emerged after careful work by many investigators. This law states that any given compound is composed of definite proportions by mass of its elements. This law was difficult to prove because many samples of compounds contain impurities of other compounds that have the same elements. For example, carbon monoxide (CO) and carbon dioxide (CO_2) are two different compounds, but each is com-

Figure 3.1 *Illustration of the Law of Conservation of Mass*

In an eighteenth-century experiment, phosphorus in air is ignited by sunlight focused with a magnifying glass. The phosphorus reacts with the oxygen present in the air to produce an oxide of phosphorus, which dissolves in the water. The mass of the system after the reaction is the same as it was before, but the volume of gas trapped in the bell jar has obviously been reduced.

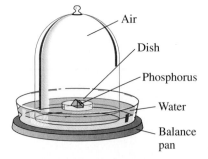

Before reaction: phosphorus in air trapped in bell jar

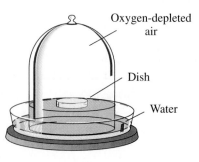

After reaction: an oxide of phosphorus dissolved in water

posed of only carbon and oxygen. The two compounds can form a homogeneous mixture in any proportions. Analysis of an impure sample of either gas could lead to a percentage of carbon anywhere between that in pure carbon monoxide and that in pure carbon dioxide. A pure sample of carbon monoxide or carbon dioxide, not an arbitrary mixture of the two, is necessary to measure a definite percent composition. Once chemists isolated and worked on pure compounds, it was apparent that the law of definite proportions was valid.

■ EXAMPLE 3.1

A 500-g sample of carbon dioxide is composed of 27.29% carbon and 72.71% oxygen by mass. What is the percent composition of a 10.5-g sample of carbon dioxide?

Solution

The 10.5-g sample is 27.29% carbon and 72.71% oxygen. *All* samples of carbon dioxide have the same percent composition, as required by the law of definite proportions.

Practice Problem 3.1   Carbon monoxide has a percent composition of 42.9% carbon and 57.1% oxygen. What possible percentages of carbon could be in a *mixture* of carbon monoxide and carbon dioxide? ■

■ EXAMPLE 3.2

Calculate the mass of carbon in a 5.000-g sample of carbon dioxide, using the percentages in Example 3.1.

Solution

$$5.000 \text{ g CO}_2\left(\frac{27.29 \text{ g C}}{100.0 \text{ g CO}_2}\right) = 1.364 \text{ g C}$$

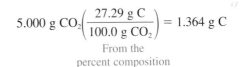

From the
percent composition

Practice Problem 3.2   Calculate the mass of carbon dioxide that contains 100 g of carbon. ■

The **law of multiple proportions** states that for two (or more) compounds composed of the same elements, for a given mass of one of the elements, the ratio of masses of any other element in the compounds is a small, whole-number ratio (Figure 3.2). For example, a compound of lead and oxygen contains 12.95 g of lead for each gram of oxygen present. Another compound of these elements contains 6.475 g of lead for each gram of oxygen present. For the fixed mass of oxygen (1 g in each case), the ratio of masses of lead is (12.95 g)/(6.475 g) = 2/1.

Sometimes, the ratio does not appear to be integral at first, but it can be converted to an integral ratio while keeping the value the same by multiplying both numerator and denominator by the same small integer. For example, a compound of manganese and oxygen contains 3.435 g of manganese per gram of oxygen, while a second compound contains 2.289 g of manganese per gram of oxygen. Per gram

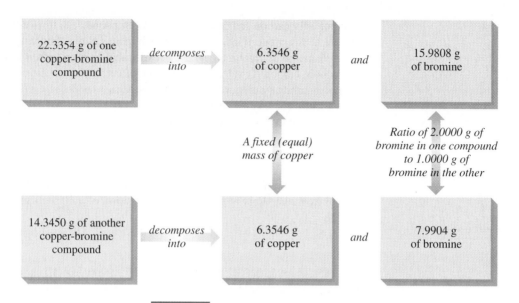

Figure 3.2 Example Illustrating the Law of Multiple Proportions

of oxygen, the ratio of masses of manganese is $(3.435 \text{ g})/(2.289 \text{ g}) = 1.500/1$. We can convert that ratio to an integral ratio by multiplying numerator and denominator by 2:

$$\frac{1.500 \times 2}{1 \times 2} = \frac{3}{2}$$

Converting ratios to integral ratios is discussed further in Appendix 1.

■ **EXAMPLE 3.3**

The percent compositions of carbon monoxide and carbon dioxide are as follows:

Carbon Monoxide	Carbon Dioxide
42.88% carbon	27.29% carbon
57.12% oxygen	72.71% oxygen

Show that these data follow the law of multiple proportions.

Solution

Per gram of carbon, the following mass of oxygen is present in each of the two compounds:

In Carbon Monoxide	In Carbon Dioxide
$\dfrac{57.12 \text{ g O}}{42.88 \text{ g C}} = \dfrac{1.332 \text{ g O}}{1 \text{ g C}}$	$\dfrac{72.71 \text{ g O}}{27.29 \text{ g C}} = \dfrac{2.664 \text{ g O}}{1 \text{ g C}}$

The ratio of grams of oxygen in carbon dioxide (per gram of carbon) to grams of oxygen in carbon monoxide (per gram of carbon) is

$$\frac{2.664 \text{ g O}}{1.332 \text{ g O}} = \frac{2.000}{1.000}$$

This ratio is, within limits of experimental error, equal to a small, whole-number ratio.

Figure 3.3 John Dalton

Note that it is *not* the ratio of mass of carbon to mass of oxygen that must be an integral ratio, according to the law of multiple proportions, but the ratio of the mass of oxygen in one compound to the mass of oxygen in the other compound (for the same mass of carbon in the two compounds).

Practice Problem 3.3   The percent compositions of two oxides of manganese are 77.45% Mn and 22.55% O for one oxide and 69.60% Mn and 30.40% O for the other. Show that these compounds obey the law of multiple proportions. ■

3.2   Dalton's Atomic Theory

O B J E C T I V E

■ to interpret the classical laws of chemical combination using Dalton's atomic theory

John Dalton (1766–1844) (see Figure 3.3) proposed the following **postulates** to explain the laws of chemical combination discussed in Section 3.1:

1. Matter is made up of very tiny, indivisible particles called **atoms.**

2. The atoms of each element all have the same mass, but the mass of the atoms of one element is different from the mass of the atoms of any other element.

3. Atoms combine to form **molecules.** When they do so, they combine in small, whole-number ratios.

4. Atoms of some pairs of elements can combine with each other in different ratios to form different compounds.

5. If atoms of two elements can combine to form more than one compound, the most stable compound has the atoms in a 1:1 ratio. (This postulate was quickly shown to be incorrect.)

The first three postulates have had to be amended, and the fifth was quickly abandoned altogether. But the postulates explained the laws of chemical combination known at the time, and they caused great activity among chemists, which led to more generalizations and further advances in chemistry.

The postulates of **Dalton's atomic theory** explained the laws of chemical combination very readily. The law of conservation of mass is explained as follows: Since atoms merely exchange "partners" during a chemical reaction and are not created or destroyed, their mass is also neither created nor destroyed. Thus, mass is conserved during a chemical reaction. The law of definite proportions is explained as follows: Since atoms react in definite integral ratios (postulate 3), and each kind of atom has a definite mass (postulate 2), the mass ratio of one element to the other(s) must also be definite. The law of multiple proportions is explained as follows: Since atoms combine in different ratios of small whole numbers (postulate 4), for a given number of atoms of one element, the number of atoms of the other element is in a small, whole-number ratio. A given number of atoms of the first element implies a given mass of that element, and a small, whole-number ratio for the atoms of the second element (each of the same mass) implies a small, whole-number ratio of masses of the second element (Figure 3.4). For example, consider water (H_2O) and hydrogen peroxide (H_2O_2), two compounds of hydrogen and oxygen. For a given number of hydrogen atoms (2), the numbers of oxygen atoms in the two compounds are 1 and 2. Stated another way, for a given mass of hydrogen (2.016 g), the ratio of masses of oxygen in the two compounds is 15.9994 g to 31.9988 g, a ratio of 1 to 2—a small, whole-number ratio.

We will discuss the ways in which the first three of Dalton's postulates have had to be amended after we learn more about the atom.

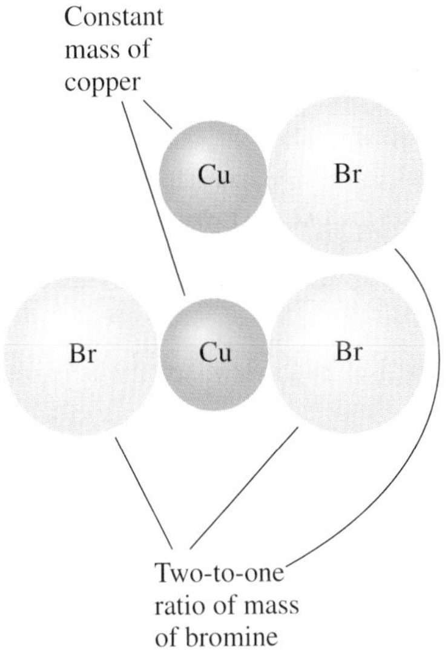

Constant mass of copper

Two-to-one ratio of mass of bromine

Figure 3.4 *Dalton's Explanation of the Law of Multiple Proportions*

Since the atoms of each element have a given mass, the fact that the atomic ratio is two atoms of bromine in one compound to one atom of bromine in the other (for one atom of copper in each) means that there is a 2:1 ratio of masses of bromine in the two compounds (for a given mass of copper).

3.3 Subatomic Particles

■ to use the properties of protons, electrons, and neutrons— subatomic particles— to determine atomic structure

The nucleus does not change during any ordinary chemical reaction.

The atomic number is equal to the number of protons in the atom.

The mass number is equal to the sum of the numbers of protons and neutrons in the atom.

The atom is composed of many **subatomic particles,** but only three types will be important in this course. **Protons** and **neutrons** exist in the atom's **nucleus,** and **electrons** exist outside the nucleus. The nucleus (plural, *nuclei*) is incredibly small, with a radius about one ten-thousandth of the radius of the atom itself. The nucleus does not change during any ordinary chemical reaction. The protons, neutrons, and electrons have the properties listed in Table 3.1. These properties are independent of the atom of which the subatomic particles are a part. Thus, the atom is the smallest unit that has the characteristic composition of an element, and in that sense, it is the smallest particle of an element.

Uncombined atoms are neither positive nor negative but electrically **neutral** and thus must have equal numbers of protons and electrons:

Number of protons = number of electrons   (For a neutral atom)

Since neutrons are neutral (see Table 3.1), the number of neutrons does not affect the charge on the atom. The number of protons in an atom determines the element's identity. All atoms having the same number of protons are atoms of the same element. The number of protons in an atom (p) is what differentiates each element from all others. It is called the **atomic number** (Z) of the element:

$$Z = p$$

The number of neutrons in the nuclei of atoms of the same element can differ. If two atoms have the same number of protons and different numbers of neutrons, they are atoms of the same element (they have the same atomic number). However, they have different masses because of the different numbers of neutrons. Such atoms are said to be **isotopes** of each other. Each isotope of an element is usually identified by its **mass number** (A), which is defined as the sum of the number of protons (p) and the number of neutrons (n) in the atom:

$$A = p + n = Z + n$$

Generally, the mass number for the isotopes rises as the atomic number rises, but the two are *not* directly proportional.

Symbols that represent the elements were introduced in Chapter 1. In addition, each of those symbols can be used to represent an atom of the element. Moreover, the symbols are used not only for the elements but also for their isotopes. An isotope is identified by the symbol of the element, with the mass num-

Table 3.1   Properties of Subatomic Particles

Particle	Charge (e)*	Mass (amu)†	Location in the Atom
Proton	1+	1.0073	In the nucleus
Neutron	0	1.0087	In the nucleus
Electron	1−	0.000549	Outside the nucleus

*The charges given are relative charges, based on the charge on the electron, e, as the fundamental unit of charge (1 e = 1.60 × 10⁻¹⁹ coulomb).
†The masses are given in atomic mass units (amu), described in Section 3.4.

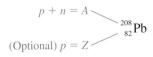

ber added as a superscript on the left side. For example, the isotope of lead with mass number of 208 is designated as ^{208}Pb. (Its name is lead-208.) Note that the number of neutrons is not given; the mass number is. The number of protons (the atomic number) may be shown as a subscript on the left, if desired, as in $^{208}_{82}$Pb. However, since the element's identity determines the atomic number, and vice versa, giving both the symbol and the atomic number is redundant—it identifies the element twice.

■ **EXAMPLE 3.4**

Two atoms have 17 protons each, but the first atom contains 18 neutrons and the second contains 20 neutrons. Show that their atomic numbers are the same but that their mass numbers differ.

Solution

The atomic numbers are the numbers of protons, in each case 17, so the atoms are both atoms of the same element—chlorine. (See the periodic table inside the front cover.) The mass number of the first atom is $17 + 18 = 35$, and the mass number of the second is $17 + 20 = 37$. Thus, the atoms have the same atomic number but different mass numbers. Their properties are essentially the same, since they are the same element, but their masses are somewhat different. They are isotopes of each other: ^{35}Cl and ^{37}Cl.

Practice Problem 3.4   Two atoms have mass number 127, but one has 74 neutrons and the other 75 neutrons. Are they isotopes of each other? ■

■ **EXAMPLE 3.5**

How many electrons are associated with each of the uncombined atoms in Example 3.4?

Solution

Each atom has 17 protons in its nucleus, so there is a 17+ charge on each nucleus. Since uncombined atoms are neutral, there must be 17 electrons, each with a 1− charge, to exactly balance the nuclear charge.

Practice Problem 3.5   How many electrons does each of the atoms in Practice Problem 3.4 have? ■

Dalton's first three postulates have had to be amended in light of information discovered after his work. The existence of subatomic particles means that atoms are not indivisible (postulate 1). Dalton thought that the mass differentiated the atoms of one element from those of another (postulate 2) because he believed that atoms were indivisible. However, atoms of *different* elements can have the *same mass number*. Atoms of each element have a distinctive atomic number—the number of protons in the nucleus—to distinguish them from atoms of other elements. In Chapter 5, postulate 3 will be shown to be only partially true. Only some combinations of atoms form molecules; other combinations form ionic compounds.

3.4 Atomic Mass

■ to calculate atomic masses— the relative masses of atoms— two ways: (1) from the ratios of masses of equal numbers of atoms, as was done histori- cally, and (2) from masses and abundances of the naturally occurring mixture of isotopes, the more modern method

Atoms are so tiny that, until recently, the masses of individual atoms could not be measured directly (Figure 3.5). However, since mass was so important in Dalton's theory, some measure of atomic masses was necessary. Therefore, a **relative scale**—the **atomic mass scale**—was set up. This scale is also called the **atomic weight scale.** On this scale, an average of the masses of all the atoms of the nat- urally occurring mixture of isotopes of a given element was measured relative to the mass of an atom of a *standard* isotope. The early pioneers of chemistry, try- ing to verify Dalton's atomic theory, could not measure the mass of individual atoms. The best they could do was to measure the masses of equal numbers of atoms of two (or more) elements at a time, to determine their relative masses. They established one element as a standard, gave it an arbitrary value of atomic mass, and used that value to establish the atomic mass scale. For example, since magnesium atoms have a mass about twice that of carbon atoms, the atomic mass of magnesium was set to a value about twice that of carbon.

The **atomic mass** of an element is the **weighted average** of the masses of the naturally occurring isotopes (not the mass numbers of the isotopes). The weighted average is obtained by multiplying the mass of each isotope by its percent abun- dance, summing all these products, and dividing the result by 100. Atomic masses are also called **atomic weights.** Modern practice defines the standard as the ^{12}C isotope, with a mass of *exactly* 12 **atomic mass units** (amu). The masses of atoms of all other elements are compared to the mass of ^{12}C. For example, a chlorine atom, on the average, has a mass about three times the mass of a ^{12}C atom, so the atomic mass of chlorine (35.453 amu) is about three times 12 amu. The atomic mass unit is tiny; it takes 6.02×10^{23} atomic mass units to make 1.00 gram.

Naturally occurring samples of an element have almost exactly the same mixture of isotopes, no matter what the source. For example, water from the rain forest of the Amazon, from an iceberg in the Arctic Ocean, or from the combus- tion of an oak tree in New York contains oxygen that is 99.759% ^{16}O, 0.037% ^{17}O, and 0.204% ^{18}O. Since the relative percentages of the isotopes in any naturally occurring element are remarkably constant, the average of the isotopic masses is also constant (to four, five, or even six significant digits). Thus, Dalton's postu- late of a constant mass for the atoms of an element explained the laws of chemi- cal combination because there is a constant *average* mass.

Figure 3.5 The Problem of Weighing Atoms

Try to weigh one grain of rice on a bathroom scale, and you will get an inkling of the difficulty of weighing atoms. The ratio of the mass of an atom to that of the smallest mass weighable on any balance is about 10^{-22} g/10^{-6} g, or 10^{-16}. This is much lower than the ratio of the mass of a grain of rice to that of a person, which is about 10^{-3} g/10^{5} g, or 10^{-8}.

■ EXAMPLE 3.6

Naturally occurring indium consists of 4.28% ^{113}In, which has a mass of 112.9043 amu, and 95.72% ^{115}In, which has a mass of 114.9041 amu. Calculate the atomic mass of indium.

Solution

The weighted average is given by the sum of the fraction of ^{113}In times its mass and the fraction of ^{115}In times its mass:

$$(0.0428)(112.9043 \text{ amu}) + (0.9572)(114.9041 \text{ amu}) = 114.8 \text{ amu}$$

Practice Problem 3.6   Naturally occurring antimony consists of 57.25% ^{121}Sb, with a mass of 120.9038 amu, and 42.75% ^{123}Sb, with a mass of 122.9041 amu. Calculate the atomic mass of antimony. ■

Figure 3.6 Dmitry Mendeleyev

How could those early chemists be sure that their samples of two elements had equal numbers of atoms? They made a compound of the elements in which the atomic ratio was 1:1. They did not need to know the exact number of atoms of each element, only that the atoms were present in a 1:1 ratio.

■ **EXAMPLE 3.7**

An equal number of men and women are at a party. The total weight of the men is 3000 pounds, and the total weight of the women is 1750 pounds. What is the ratio of the weight of the average man to that of the average woman?

Solution

The average man at the party weighs $(3000/x)$ pounds, and the average woman weighs $(1750/x)$ pounds, where x is the number of men or women. The ratio of the weights of the average man and the average woman is

$$\frac{3000/x}{1750/x} = 1.71$$

Even though you do not know the number of men or women, as long as you know the ratio of number of men to number of women (1:1) and the total weight ratio, you can calculate the ratio of the weights of the average man and the average woman.

Practice Problem 3.7   Three times as many men as women are at a party. The total weight of the men is 6300 pounds, and the total weight of the women is 1400 pounds. What is the ratio of the weight of the average man to that of the average woman at the party? ■

3.5 Development of the Periodic Table

OBJECTIVE

■ to repeat the thought processes of Mendeleyev and Meyer in the development of the periodic table

Many atomic masses were determined as a direct result of Dalton's postulates and the work that they stimulated, and several scientists attempted to relate the atomic masses of the elements to the elements' properties. This work culminated in the development of the periodic table by Dmitry Mendeleyev (1834–1907) (Figure 3.6) and independently by Lothar Meyer (1830–1895). Because Mendeleyev did more with his periodic table, he is often given sole credit for its development.

Mendeleyev put the known elements in order according to their atomic masses (atomic numbers had not yet been defined) and noticed that the properties of every seventh known element were similar. He arranged the elements in a table, with elements having similar properties in the same vertical group. At several points where an element did not seem to fit well in the position its atomic mass called for, he postulated that there was an undiscovered element for that position. For example, the next known element after zinc (Zn) by atomic mass was arsenic (As). However, since arsenic's properties were much more similar to those of phosphorus (P) than to those of aluminum (Al) or silicon (Si), Mendeleyev predicted that two elements that fit the positions under aluminum and silicon in his periodic table had not yet been discovered:

	B	C	N
	Al	Si	P
Zn	?	?	As
	In	Sn	Sb

He described their expected properties from those of the elements above and below them in his table. His predictions helped other chemists discover these elements, now known as gallium (Ga) and germanium (Ge).

Several other elements seemed out of order. For example, their atomic masses placed iodine (I) before tellurium (Te), but their chemical properties required the opposite order. Mendeleyev concluded that the atomic masses must have been determined incorrectly and put these two elements in positions reflecting their properties. We now know that the periodic properties of the elements are based on their atomic numbers, not their atomic masses, which explains Mendeleyev's difficulty with the placement of certain elements.

■ **EXAMPLE 3.8**

In the periodic table, locate two pairs of elements besides iodine and tellurium that are out of order, based on their atomic masses.

Solution

The elements argon and potassium and the elements cobalt and nickel are in reverse order with respect to their atomic masses.

Practice Problem 3.8   Are any elements in the periodic table out of order according to their atomic numbers? ■

An entire group of elements—the noble gases—was discovered after the periodic table was first formulated. These elements are colorless, odorless gases and almost totally inert. Their lack of combining capacity means that they are not found in any naturally occurring compound. If some compound had had a percentage of its mass unaccounted for, chemists would have known to look for the missing elements, but since the noble gases do not combine, there was no clue to their existence. (Helium was first discovered by its spectral lines in sunlight and was named for *helios,* the Greek word for sun; see Figure 3.7.)

That each element fits properly into place in a vertical column proves the fundamental correctness of arranging the elements according to their atomic numbers and chemical properties.

Atomic numbers and atomic masses are usually included in the boxes with the chemical symbols in the periodic table. The atomic number is the integer. (A mass number, which is also an integer, is given in parentheses for the most stable isotope of a few elements.)

The periodic table is a tremendous source of information for students who learn to use it well. In Chapter 4, you will learn to use the periodic table to predict

Figure 3.7 *Spectral Lines in Sunlight*

The dark lines on the visible spectrum of sunlight are due to energies of precise wavelengths of light being absorbed by atoms of elements in the outer layers of the sun. Bright lines of the same wavelengths are emitted when gaseous samples of those elements are excited electrically.

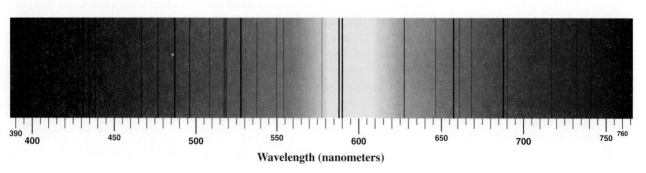

390   400   450   500   550   600   650   700   750 760

Wavelength (nanometers)

the electronic configuration of each of the elements, and in Chapter 5, you will use it to predict outermost electron shell occupancy. The table's numeric data is used in later chapters on formula calculations and stoichiometry, and its information on chemical trends is applied in the chapters on bonding and molecular structure.

■ ■ ■ ■ ■ ■ **SUMMARY** ■ ■ ■ ■ ■ ■

Lavoisier discovered the law of conservation of matter, which states that matter cannot be created or destroyed during chemical reactions. This generalization increased chemists' efforts to measure the masses of elements in compounds and resulted in two more laws. The law of definite proportions states that the percentage of each element in any sample of a pure compound is always the same. According to the law of multiple proportions, if the mass of one of the elements in two or more compounds of the same elements is held constant, the masses of each other element form a small, whole-number ratio. (Section 3.1)

Dalton suggested that the elements are composed of indivisible atoms and that the atoms of each element have a characteristic mass, different from the mass of any other element. He stated that the atoms combine to form molecules when the elements combine to form compounds. These postulates explained the laws of chemical combination known at that time, but most of them have been amended in light of later discoveries. However, the atom is still considered the fundamental particle of an element. (Section 3.2)

In the past 150 years, numerous experiments have shown that the atom is not indivisible but is composed of electrons plus a nucleus containing protons and neutrons. The nucleus does not change in any chemical reaction. The characteristics of the subatomic particles (Table 3.1) are important. The number of protons, called the atomic number, governs the number of electrons in the neutral atom. The sum of the numbers of protons and neutrons is called the mass number. All atoms of a given element have the same atomic number, which differs from the atomic numbers of other elements. Different atoms of the same element may have different numbers of neutrons and thus different mass numbers. Such atoms are called isotopes of each other. An isotope is identified by the symbol of the element, with the mass number as a superscript to the left. (Section 3.3)

The naturally occurring mixture of isotopes of any given element has about the same percentage of each isotope. Therefore, the average mass of all the atoms in any sample of the element is constant (to four or more significant digits). That weighted average is called the atomic mass (which is not the same as the mass number or the mass of an atom). Atomic masses are reported on a relative scale, with an atom of the ^{12}C isotope being defined as having a mass of exactly 12 amu. (Section 3.4)

When arranged in order of increasing atomic mass, the various elements, with few exceptions, have periodically recurring properties. Mendeleyev produced a periodic table based on this ordering. Later, it was learned that the atomic number is the basis for the chemical properties of an element, so the modern periodic table arranges the elements in order of increasing atomic number, with elements having similar properties arranged in vertical groups. The periodic table has many uses in the study of chemistry. (Section 3.5)

Items for Special Attention

■ Atomic mass and mass number are not the same. Atomic mass refers to the naturally occurring mixture of isotopes; mass number refers to an individual isotope. Atomic mass is an average and is never an exact integer; mass number is a sum (of the number of protons plus the number of neutrons) and is always an integer. Mass numbers are not given in the periodic table for most elements.

■ Atoms have masses between 1 and 250 atomic mass units, nowhere near as large as 1 gram. Be careful which units you use for the masses of individual atoms and which you use for the masses of weighable samples.

■ In the periodic table, the elements are arranged so that their atomic numbers are in increasing order and grouped vertically so that elements with similar chemical properties are in the same group (vertical column).

Self-Tutorial Problems

3.1 What is the difference between the symbol for an element and the symbol for an isotope of that element?

3.2 Which two types of subatomic particles must be present in equal numbers for an atom to be neutral?

3.3 The Brown family has triplet girls who weigh 93 pounds each and twin boys who weigh 75 pounds each.

(a) What is the total weight of the children?

(b) What is the average weight of the children?

(c) What is the average weight of the two children whose names begin with A—Ann and Andrew?

3.4 (a) What is the atomic number of nitrogen?

(b) How many protons are in a nitrogen atom?

(c) What is the number of positive charges on a nitrogen nucleus?

3.5 On what standard are all atomic masses based?

3.6 (a) What is the unit of atomic mass?

(b) What is the unit of electric charge used at the atomic level?

3.7 Of atomic number, atomic mass, and mass number, which two appear with most of the symbols for the elements in the periodic table?

3.8 Which of the following are synonyms?

mass of an atom

atomic weight

atomic number

mass number

atomic mass

3.9 What is the difference between the mass of an atom and the mass number of the atom?

3.10 In the periodic table (inside front cover), locate five elements for which mass numbers rather than atomic masses are given.

3.11 (a) Which element has atomic number 19?

(b) Which element has an atomic mass of 19.00 amu (to four significant figures)?

(c) Locate the elements of part (a) and part (b) on the periodic table.

3.12 At a racetrack, the winning horse paid "3 to 2." How much does a gambler win for each dollar bet on that horse? Is that ratio an integral ratio?

■ ■ ■ PROBLEMS ■ ■ ■

3.1 Laws of Chemical Combination

3.13 A 15.0-g sample of element A combines completely with a 10.3-g sample of element B. What is the total mass of the product?

3.14 A 150-g sample of an element combines completely with another element to make 233 g of a compound. What mass of the second element reacted?

3.15 A 100.0-g sample of a certain compound contains 79.89% carbon and 20.11% hydrogen.

(a) How much carbon is in a 5.000-g portion of this sample?

(b) How much carbon is in 5.000 g of a different sample of this compound?

3.16 A sample of a compound composed of only carbon and hydrogen contains 85.63% carbon. Show that this compound and the one in Problem 3.15 obey the law of multiple proportions.

3.17 A 43.5-g sample of a compound contains 39.35% sodium and 60.65% chlorine.

(a) Calculate the mass of sodium present.

(b) Calculate the mass of sodium present in a 100.0-g sample of the same compound.

3.18 A 5.555-g sample of calcium was burned, yielding 7.773 g of a calcium-oxygen compound. How many grams of oxygen was taken up in the reaction?

3.19 A 21.73-g sample of mercury(II) oxide was decomposed into mercury and oxygen, yielding 20.12 g of mercury.

(a) What mass of oxygen was obtained?

(b) What percentage of the compound was mercury?

3.20 If 6.667 mg of a compound containing only carbon and hydrogen is burned completely in oxygen and yields 19.55 mg of carbon dioxide and 12.01 mg of water, how much oxygen is used up?

3.21 When 4.41 mg of a compound containing only carbon and hydrogen was burned completely in 16.0 mg of oxygen, 13.2 mg of carbon dioxide and some water were formed. Calculate the mass of the water.

3.22 The ratio of masses of sulfur and oxygen in sulfur dioxide is 1.0 to 1.0. Is this fact a proof of the law of multiple proportions?

3.23 The ratio of the mass of carbon to the mass of oxygen in carbon monoxide is about 3:4. Does this fact confirm the law of multiple proportions?

3.24 Solve Example 3.3 (p. 66) again, this time using 1 g of oxygen in each compound. Is the law of multiple proportions still valid?

3.2 Dalton's Atomic Theory

3.25 According to Dalton's atomic theory, all atoms of the same element have the same mass. If an atom of carbon has a mass of 12 amu and an atom of oxygen has a mass of 16 amu:

(a) What is the mass ratio of one atom of oxygen to one atom of carbon?

(b) What is the total mass of 100 atoms of oxygen? What is the total mass of 100 atoms of carbon?

(c) What is the ratio of masses of 100 atoms of oxygen to 100 atoms of carbon?

(d) Choose an arbitrary, large number of atoms of oxygen. Then calculate the mass of that number and the mass of an equal number of carbon atoms. Calculate the ratio of the total masses.

(e) What can you conclude about the ratio of masses of equal numbers of oxygen and carbon atoms?

3.26 Would it make any difference in Problem 3.25 if average masses had been used? Explain.

3.27 What characteristic of an atom did Dalton think was the most important?

3.28 Restate Dalton's first three postulates in amended form, based on modern information.

3.29 What happens to a scientific hypothesis if experiments show it to be incorrect?

3.3 Subatomic Particles

3.30 Identify the only stable isotope that contains no neutrons.

3.31 Write the symbol for an isotope:

(a) For which the atomic number is 1 and the mass number is 1

(b) For which the atomic number is 1 and there is one neutron

(c) Containing one proton and two neutrons

(d) With a mass number of 3 and containing two neutrons

(e) With a mass number of 3 and containing one neutron

3.32 Deuterium (symbol: ^2D) is a special name for a certain isotope. It contains one proton and one neutron.

(a) Of what element is deuterium a part?

(b) Write the more familiar symbol for this isotope.

3.33 Isotopes of which element have:

(a) The smallest mass number

(b) The smallest atomic number

(c) The largest number of protons in the nuclei

3.34 Which isotope whose mass number is given in the periodic table on the inside front cover of the text has the:

(a) Smallest mass number

(b) Smallest number of neutrons

(c) Largest mass number

(d) Largest number of neutrons

3.35 Which two inner transition isotopes whose mass numbers are given in the periodic table on the inside front cover of the text have the greatest number of neutrons?

3.36 Complete the following table for neutral atoms of specific isotopes:

	Isotopic Symbol	Atomic Number	Mass Number	No. of Protons	No. of Neutrons	No. of Electrons
(a)	^{19}F	—	—	—	—	—
(b)		35	81	—	—	—
(c)	—	—	—	—	44	35
(d)	—	—	—	90	140	—
(e)	$^{56}_{26}$—	—	—	—	—	—
(f)	—	13	—	—	14	—
(g)	—	—	37	17	—	—

3.37 In the table in Problem 3.36, two pieces of quantitative information are given in each part.

(a) What two pieces of information are given in Problem 3.36(a)?

(b) Why would the atomic number and the number of protons not be sufficient for any part?

(c) Why would the number of protons and the number of electrons not be sufficient for any part?

3.38 Complete the following table for neutral atoms of specific isotopes:

	Isotopic Symbol	Atomic Number	Mass Number	No. of Protons	No. of Neutrons	No. of Electrons
(a)	^{34}S	—	—	—	—	—
(b)	—	52	127	—	—	—
(c)	—	—	102	45	—	—
(d)	—	—	—	—	104	71
(e)	—	83	—	—	126	—
(f)	—	—	133	—	—	55
(g)	—	6	—	—	8	—

3.4 Atomic Mass

3.39 ^{70}Ga does not occur naturally. Explain how gallium gets its atomic mass of 69.72 amu.

3.40 ^{108}Ag does not occur naturally. Explain how silver gets its atomic mass of 107.868 amu.

3.41 (a) What is the average of a 5.00-g mass and a 3.00-g mass?

(b) What is the weighted average of eight 5.00-g masses and two 3.00-g masses?

(c) What is the weighted average mass of three chlorine atoms with mass 35.0 amu each and one chlorine atom with mass 37.0 amu?

3.42 Calculate the atomic mass of oxygen if 99.759% of naturally occurring oxygen atoms have a mass of 15.9949 amu, 0.037% have a mass of 16.9991 amu, and 0.204% have a mass of 17.9992 amu.

3.43 Which of the following represent the mass of one atom (to three significant figures)?

(a) 4.00 g   (b) 16.0 amu   (c) 2.41×10^{24} amu

(d) 1.08×10^2 g

3.44 You can guess the mass number of the predominant isotope for many elements from the atomic mass of the element, but not in all cases. The mass numbers of the isotopes of selenium are 74, 76, 77, 78, 80, and 82. Explain why the atomic mass is so close to 79 amu.

3.45 (a) If 57.25% of the people in a Weight Watchers graduating class weigh 120.9 pounds each and the rest weigh 122.9 pounds each, what is the average mass of the class?

(b) If 57.25% of naturally occurring antimony atoms have an atomic mass of 120.9038 amu and the rest have an atomic mass of 122.9041 amu, what is the atomic mass of antimony?

3.46 Round off the atomic masses of the first 18 elements to two decimal places each.

3.47 After a calculation, a student reported the atomic mass of an element as 1.0×10^{-4} amu. The student later changed the value to $1.0 \times 10^{+4}$ amu. Which value, if either, is more probably correct?

3.5 Development of the Periodic Table

3.48 How important was it to the work of Mendeleyev that atomic mass and atomic number rise almost proportionally? Explain.

3.49 Could you use the average number of neutrons, instead of atomic number, to build a periodic table?

3.50 The following are the formulas for fluorides of certain second-period elements:

$$LiF \quad BeF_2 \quad BF_3 \quad CF_4 \quad NF_3 \quad OF_2$$

Predict the formula for a fluoride of each of the elements directly below these in the periodic table.

3.51 Why did Mendeleyev not use atomic numbers instead of atomic masses as the basis for his periodic table?

3.52 From the following properties of chlorine and iodine, predict the corresponding properties of bromine:

Chlorine	Iodine	Bromine
Gas under normal conditions	Solid under normal conditions	____
Light yellow	Deep violet	____
Reacts with metals	Reacts with metals	____
Reacts with oxygen	Reacts with oxygen	____
Does not conduct electricity	Does not conduct electricity	____

3.53 The following are the formulas for some oxides of third-period elements except argon:

$$Na_2O \quad MgO \quad Al_2O_3 \quad SiO_2 \quad P_2O_3 \quad SO_2 \quad Cl_2O$$

Predict the formula for an oxide of each of the elements directly below these in the periodic table.

■ ■ ■ GENERAL PROBLEMS ■ ■ ■

3.54 (a) Calculate the mass of oxygen in a 5.00-g sample of carbon dioxide, using the answer to Example 3.2.

(b) How much oxygen should be combined with the same mass of carbon as in Example 3.2 to form the compound carbon monoxide, assuming that there is twice the mass of oxygen per gram of carbon in carbon dioxide as there is in carbon monoxide?

(c) What is the percent composition of carbon monoxide?

3.55 ^{192}Ir does not occur naturally. Explain how iridium gets its atomic mass of 192.22 amu.

3.56 In a certain compound, 5.00 g of element A is combined with 8.20 g of element B. In another compound of A and B, 5.00 g of A can be combined with which of the following?

(a) 8.03 g of B (b) 16.4 g of B (c) 9.00 g of B
(d) 5.00 g of B (e) 10.0 g of B

3.57 Is there any possibility that the sulfur in the head of a match can combine with *all* the oxygen in the atmosphere of the earth to form a compound? Explain, using a law studied in this chapter.

3.58 A 15.0-g sample of element A reacts incompletely with a 5.00-g sample of element B. What is the total mass of the product and whatever parts of the reactants that did not react?

3.59 What factors limit the number of significant digits in the atomic mass of an element with two naturally occurring isotopes?

3.60 Three compounds of carbon, hydrogen, and sulfur have the following percent compositions. Show that these compounds obey the law of multiple proportions.

(a) 38.66% C, 9.734% H, 51.60% S

(b) 47.31% C, 10.59% H, 42.10% S

(c) 53.27% C, 11.18% H, 35.55% S

3.61 (a) Plot mass number versus atomic number for ^{1}H, ^{59}Ni, ^{118}Sn, ^{178}Hf, and ^{238}U.

(b) Are atomic number and mass number directly proportional?

(c) What can you say about the relationship of these two quantities?

3.62 At a racetrack, the winning horse paid $1.75 for each dollar bet. What odds were posted?

3.63 Explain why Mendeleyev could predict the existence of germanium but missed the entire group of noble gases.

3.64 Calculate the atomic mass of silver from the following data:

Isotope	Natural Abundance (%)	Relative Mass (amu)
^{107}Ag	51.82	106.90509
^{109}Ag	48.18	108.9047

3.65 The atomic mass of magnesium is 24.312 amu. Does any atom of any isotope of magnesium have a mass of 24.312 amu? Explain.

3.66 The atoms of element "Z" each have one-third the mass of a ^{12}C atom. Another element, "X," has atoms whose mass is four times the mass of "Z" atoms. A third element, "Q," has atoms with twice the mass of "X" atoms.

(a) Make a table of relative atomic masses based on ^{12}C as 12 amu.

(b) Identify the elements "Z," "X," and "Q."

3.67 All naturally occurring samples of potassium fluoride, when purified, contain the same percentage of potassium and the same percentage of fluorine. Naturally occurring fluorine consists of only one isotope. What do the constant percentages say about the three naturally occurring isotopes of potassium?

3.68 Calculate the atomic mass of selenium from the following data:

Isotope	Natural Abundance (%)	Relative Mass
^{74}Se	0.87	73.9205
^{76}Se	9.02	75.9192
^{77}Se	7.58	76.9199
^{78}Se	23.52	77.9173
^{80}Se	49.82	79.9165
^{82}Se	9.19	81.9167

3.69 If the radius of the nucleus is one ten-thousandth the radius of the atom as a whole, what is the ratio of their volumes? $(V = \frac{4}{3}\pi r^3)$

3.70 The formula for water is H_2O, signifying that there are two atoms of hydrogen for every atom of oxygen. If Dalton's fifth postulate had been true, what would the formula for water have been?

3.71 A typical atom has a radius of about 10^{-10} m. Estimate the radius of a typical nucleus.

3.72 Plot the mass number versus the atomic number for the last seven actinide elements. Can you see any relationship between these quantities for these elements that have very similar atomic numbers?

3.73 Naturally occurring bromine consists of 50.54% ^{79}Br, which has a mass of 78.9183 amu, and 49.46% ^{81}Br, which has a mass of 80.9163 amu. Calculate the atomic mass of bromine.

3.74 Sulfur dioxide (SO_2) has two atoms of oxygen per atom of sulfur, and sulfur trioxide (SO_3) has three atoms of oxygen per atom of sulfur. The mass ratio of sulfur to oxygen in SO_2 is 1.0:1.0. What is the mass ratio of sulfur to oxygen in SO_3?

3.75 Naturally occurring europium consists of 47.82% ^{151}Eu, which has a mass of 150.9196 amu, and 52.18% ^{153}Eu, which has a mass of 152.9209 amu. Calculate the atomic mass of europium.

3.76 Naturally occurring gallium consists of 60.4% ^{69}Ga, which has a mass of 68.9257 amu, and 39.6% ^{71}Ga, which has a mass of 70.9246 amu. Calculate the atomic mass of gallium.

4 Electronic Configuration of the Atom

- 4.1 Bohr Theory
- 4.2 Quantum Numbers
- 4.3 Relative Energies of Electrons
- 4.4 Shells, Subshells, and Orbitals
- 4.5 Shapes of Orbitals
- 4.6 Energy Level Diagrams
- 4.7 Periodic Variation of Electronic Configuration

Key Terms (Key terms are defined in the Glossary.)

angular momentum quantum number (4.2)
Bohr theory (4.1)
buildup principle (4.3)
degenerate (4.3)
discrete energy levels (4.1)
electronic configuration (4.4)
energy level diagram (4.6)
ground state (4.2)

Heisenberg uncertainty principle (4.5)
Hund's rule (4.6)
light absorption (4.1)
light emission (4.1)
lobe (4.5)
magnetic properties (4.6)
magnetic quantum number (4.2)
$n + \ell$ rule (4.3)
orbital (4.4)

orbital shape (4.5)
Pauli exclusion principle (4.2)
periodicity of electronic configuration (4.7)
principal quantum number (4.2)
quantum numbers (4.2)
shell (4.1)
spin quantum number (4.2)
subshell (4.4)

Symbols

d (subshell) (4.4)
f (subshell) (4.4)
ℓ (quantum number) (4.2)

m (quantum number) (4.2)
n (quantum number) (4.2)
p (subshell) (4.4)

s (quantum number) (4.2)
s (subshell) (4.4)

Figure 4.1 Niels Bohr

In Chapter 3, you learned that atoms owe their characteristics to their subatomic particles—protons, neutrons, and electrons. Electrons occur in shells outside the nucleus, and the electronic structure is responsible for all of the atom's chemical properties and many of its physical properties. The number of electrons in a neutral atom is equal to the number of protons in the nucleus. That simple description allows us to deduce much about atoms, especially concerning their interactions with one another (Chapter 5). However, a more detailed model of the atom allows even fuller explanations, including the reason for the differences between main group elements and elements of the transition and inner transition series.

Many details presented in this chapter are based on mathematics beyond the scope of this course, so some postulates must be accepted as "rules of the game." When the rules are followed, the explanations that result match the actual properties of the elements, which is assurance that the postulates are valid.

Section 4.1 describes how Niels Bohr deduced that electrons occur in shells having distinct energies. His theory was a milestone, but it does not explain the properties of atoms other than hydrogen. Section 4.2 introduces the quantum numbers, which provide a more satisfactory picture of electronic structure for atoms with more than one electron. The dependence of the energy of an electron on its quantum numbers is discussed in Section 4.3, and shells, subshells, and orbitals are covered in Section 4.4. The shapes of orbitals are described in Section 4.5, and diagrams depicting the energy levels of subshells are presented in Section 4.6. The electronic configuration of the atom is responsible for the chemical and physical properties of an element. The relationship between electronic configuration and position on the periodic table is developed in Section 4.7.

4.1 Bohr Theory

- to use the Bohr theory of energy levels in atoms to explain light emission and absorption by atoms

When gaseous atoms of a given element are heated, they emit light of specific energies. When gaseous atoms of that same element absorb light, they absorb those same energies. To explain these phenomena of **light emission** and **light absorption,** Niels Bohr (1885–1962) (Figure 4.1) postulated that the electrons in atoms are arranged in **shells,** each with a definite energy. The **Bohr theory** was the first to include the explanation that electrons in atoms have **discrete energy levels.**

When an atom absorbs energy, an electron is "promoted" to a higher energy level. Because each shell has a discrete energy level, the *difference* in energy between the shells is also definite. After an electron has been promoted to a higher energy level, it falls back to a lower energy level. When it falls back, light of energy equal to the difference in energy between the shells is emitted from the atom. In a different experiment, when light is absorbed by the atom, the electron is raised from one shell to another one. Since there is the same energy difference between the shells, the same energy of light is absorbed. An example of these effects is shown in Figure 4.2. Some of the possible electron transitions in a hydrogen atom are diagrammed in Figure 4.3.

■ **EXAMPLE 4.1**

The energy of the first shell of the hydrogen atom is -2.178×10^{-18} J, and that of the second shell is -5.445×10^{-19} J. (The negative value means that the electron in the atom has a lower energy than a free electron has.)

Hydrogen

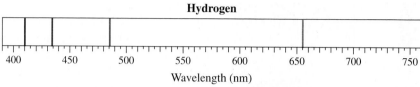

Wavelength (nm)

Helium

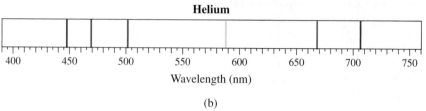

Wavelength (nm)

(a)

(b)

Figure 4.2 Emission of Light by Gaseous Atoms

(a) When neon atoms are excited by electrical energy, electron transitions between their subshells yield wavelengths of light that produce the familiar red color of a neon sign. Wavelengths corresponding to a large number of different transitions combine to yield that color. Other gases are used to produce other colors, although we refer to all such signs in everyday conversation as "neon" signs. (b) When hydrogen and helium, with fewer electrons than neon, are similarly excited, fewer transitions occur. These emission spectra are relatively simple. Hydrogen and helium are the major components in our sun.

(a)   What energy change takes place when an electron in a hydrogen atom moves from the first to the second shell?

(b)   What energy change takes place when an electron moves from the second to the first shell?

Solution

(a)   The difference in energy between the shells is 1.634×10^{-18} J. That much energy must be *absorbed* (for example, in the form of light, heat, or electricity) to get the electron promoted to the second shell. If light is absorbed, light of that particular energy and no other energy is involved.

(b)   The difference in energy between the two shells is still 1.634×10^{-18} J. In this case, the energy is *emitted* (given off) in the form of light. The energy of

Figure 4.3 Energy Levels and Some Possible Electron Transitions in the Hydrogen Atom

(not drawn to scale)

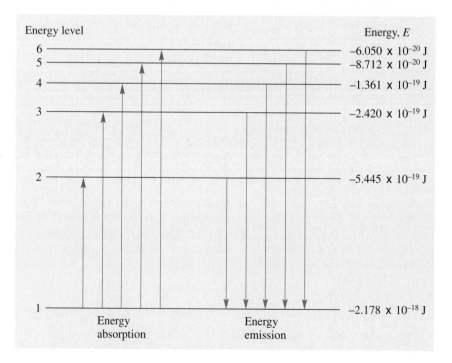

the light emitted in this case is equal to the energy of the light absorbed in part (a).

Practice Problem 4.1   The energy of the third shell of the hydrogen atom is -2.420×10^{-19} J. What energy change takes place when an electron moves from (a) the third to the second shell and (b) the third to the first shell? ■

Bohr postulated circular orbits for the electrons in an atom and developed a mathematical model to represent the energies of the orbits, as well as their distances from the atom's nucleus. His model worked very well for the hydrogen atom. It could be used to calculate the energy of the emitted and absorbed light, as well as the radius of the atom. However, the intensity of the various wavelengths of light involved was not explained well. Moreover, no other atom was explained well at all. While it has been replaced by a quantum mechanical model, Bohr's theory was a milestone because Bohr was the first to postulate energy levels in atoms.

4.2   Quantum Numbers

OBJECTIVE

■ to use quantum numbers to explain the electronic structures of the atoms in their most stable states

Each electron in an atom is associated with a set of four **quantum numbers.** The names of the quantum numbers, along with their symbols and permitted values, are given in Table 4.1.

The **principal quantum number** (n) can have any positive integral value, but the electrons in atoms in their most stable states have principal quantum numbers with values from 1 through 7. The most stable electronic state of an atom is called its **ground state.** The principal quantum number has the largest role in determining the energy of the electron, and it is also the main factor in determining how far the electron is, on average, from the nucleus. Thus, it is the most important quantum number.

For each value of n, the **angular momentum quantum number** (ℓ) for an electron can have integral values from zero to $(n - 1)$ but cannot be as large as n. The angular momentum quantum number also has a role in determining the energy of the electron, and it determines the shape of the volume of space that the electron can occupy (see Section 4.5).

Table 4.1   The Quantum Numbers

Name	Symbol*	Permitted Values	Examples
Principal quantum number	n	Any positive integer	1, 2, 3, . . .
Angular momentum quantum number	ℓ	Any integer from zero to $(n - 1)$	$0, \ldots,$ $(n - 1)$
Magnetic quantum number	m	Any integer from $-\ell$ to $+\ell$	$-\ell, \ldots, 0,$ $\ldots, +\ell$
Spin quantum number	s	$-\frac{1}{2}$ or $+\frac{1}{2}$	$-\frac{1}{2}, +\frac{1}{2}$

*The magnetic and spin quantum numbers are sometimes designated m_ℓ and m_s, respectively.

Figure 4.4 Baseball Ticket

No two tickets can have the same set of section designation, row number, seat number, and date.

■ **EXAMPLE 4.2**

What values for ℓ are permitted for an electron with (a) $n = 4$ and (b) $n = 1$?

Solution

(a) $\ell = 0, 1, 2,$ or 3, but no higher   (b) $\ell = 0$ (ℓ cannot be as high as n)

Note that the value of ℓ is not necessarily equal to $(n - 1)$ but can vary from zero up to $(n - 1)$.

Practice Problem 4.2   *How many* different values of ℓ are permitted for an electron with (a) $n = 4$ and (b) $n = 1$? ■

For each value of the angular momentum quantum number (ℓ), the **magnetic quantum number** (m) has values ranging from $-\ell$ through zero to $+\ell$ in integral steps. The value of m does not ordinarily affect the energy of an electron, but it does determine the orientation in space of the volume that can contain the electron (Section 4.5).

■ **EXAMPLE 4.3**

What values of m are permitted for an electron with (a) $\ell = 3$ and (b) $\ell = 0$?

Solution

(a) $m = -3, -2, -1, 0, 1, 2,$ or 3   (b) $m = 0$

Practice Problem 4.3   *How many* different values of m are permitted for an electron with (a) $\ell = 3$ and (b) $\ell = 0$? ■

■ **EXAMPLE 4.4**

Can you tell what values of m are permitted for an electron with $n = 2$?

Solution

The permitted values of m depend on the value of ℓ, not on the value of n. The most you can say in this case is that ℓ is limited to 0 or 1, so m must be 0 if $\ell = 0$, but may be $-1, 0,$ or $+1$ if $\ell = 1$.

Practice Problem 4.4   What is the lowest value of m permitted for any electron with $n = 4$? ■

The **spin quantum number** (s) may have a value of $-\frac{1}{2}$ or $+\frac{1}{2}$ only. The value of s does not depend on the value of any other quantum number. The spin value indicates that the electron is spinning on its axis in one direction or the opposite.

Another important limitation on the quantum numbers of electrons in atoms, in addition to those listed in Table 4.1, is the **Pauli exclusion principle.** This principle states that no two electrons in an atom can have the same set of four quantum numbers. This is like the business law that states that no two tickets to a baseball game can have the same set of date and section, row, and seat numbers (Figure 4.4). The row number may depend on the section number, and the seat number may depend on the row number, but the date does not depend on any of the other three.

Together with the $n + \ell$ rule, discussed in the next section, the Pauli exclusion principle determines the number of electrons in each of the shells in an atom.

4.3 Relative Energies of Electrons

The energies of the electrons in an atom are of paramount importance to the atom's properties. Electrons increase in energy as the sum $n + \ell$ increases. We call this the **$n + \ell$ rule.** Thus, we can make a list of sets of quantum numbers in order of their increasing energies by ordering the electrons according to increasing values of $n + \ell$. As a corollary, if two electrons have the same value of $n + \ell$, then the one with the lower n value is lower in energy. If the two n values are the same and the two ℓ values are the same, then the electrons are equal in energy. In an atom, electrons with the same energy are said to be **degenerate.**

■ **EXAMPLE 4.5**

Suppose that you must choose two integers with the lowest sum in the following ranges: For the first, you may choose any integer between 1 and 7. For the second, you may choose any nonnegative integer below the first. What integers must you select?

Solution

The lowest sum will come from the lowest possible first integer (1) and the lowest possible (and only possible) second integer (0). ■

Let's determine sets of four quantum numbers for the electrons of the ground states of the atoms of the first 10 elements. Hydrogen has only one electron. For that electron to be in its lowest energy state, it needs the lowest possible sum of n and ℓ, so we will choose the lowest value of n: $n = 1$. Then, referring to Table 4.1, we determine values for the other three quantum numbers:

With $n = 1$, the only permitted value of ℓ is 0.
With $\ell = 0$, the only permitted value of m is 0.
The value of s can be either $-\frac{1}{2}$ or $+\frac{1}{2}$.

The set of quantum numbers for hydrogen in its ground state can therefore be either of these:

$$
\begin{array}{ccc}
n = & 1 & \text{or} \quad n = \quad 1 \\
\ell = & 0 & \ell = \quad 0 \\
m = & 0 & m = \quad 0 \\
s = & -\frac{1}{2} & s = +\frac{1}{2}
\end{array}
$$

Since the n values and the ℓ values are the same in both of these sets of quantum numbers, these possible configurations represent the same energy. Thus, either set of quantum numbers could represent the electron of hydrogen.

A helium atom has two electrons, so we need two sets of quantum numbers. To represent the atom in its lowest energy state, we want each electron to have the lowest energy possible. If we let the first electron have the value of 1 for its principal quantum number, the set of quantum numbers for it will be the same as that given previously for the one electron of hydrogen. The other electron of helium can then have the other set of quantum numbers.

First Electron of Helium	Second Electron of Helium
$n = 1$	$n = 1$
$\ell = 0$	$\ell = 0$
$m = 0$	$m = 0$
$s = -\frac{1}{2}$	$s = +\frac{1}{2}$

Both of these electrons have the same energy, since they have the same n value and the same ℓ value. Either one could have been chosen as the "first" electron.

■ **EXAMPLE 4.6**

Could both electrons of helium have the value $s = -\frac{1}{2}$ with $n = 1$?

Solution

No. If n is 1, then ℓ and m must both have values of 0 (Table 4.1). If s were $-\frac{1}{2}$ for both electrons, they would have the same set of four quantum numbers, which is a violation of the Pauli principle.

Practice Problem 4.6   What is the maximum number of electrons an atom could have if the maximum value of n is 2 and the maximum value of ℓ is 0? ■

A lithium atom has three electrons. The first two of these can have the same sets of quantum numbers as the two electrons of helium. What should the set of quantum numbers for the third electron be? We cannot choose the lowest permitted value for n, which is 1, because ℓ and m would then both be 0. If we choose $-\frac{1}{2}$ as the value of s, the third electron would have a set of quantum numbers exactly the same as that of the first electron, and if we choose the value $s = +\frac{1}{2}$, the third electron would have the same set of quantum numbers as the second electron. Since neither of these situations is permitted by the Pauli principle, n cannot be 1 for the third electron. We must choose the next higher value, $n = 2$. With $n = 2$, the permitted values of ℓ are 0 and 1. Since $\ell = 0$ will give a lower value for the sum $n + \ell$, we choose that value for ℓ. With $\ell = 0$, m must be 0, and we can choose either $-\frac{1}{2}$ or $+\frac{1}{2}$ for s. The quantum numbers for the electrons of the lithium atom can thus be as follows:

First Electron of Lithium	Second Electron of Lithium	Third Electron of Lithium
$n = 1$	$n = 1$	$n = 2$
$\ell = 0$	$\ell = 0$	$\ell = 0$
$m = 0$	$m = 0$	$m = 0$
$s = -\frac{1}{2}$	$s = +\frac{1}{2}$	$s = -\frac{1}{2}$ (or $+\frac{1}{2}$)

With $s = -\frac{1}{2}$ for its third electron, the fourth electron of beryllium (Be) will have $n = 2$, $\ell = 0$, $m = 0$, and $s = +\frac{1}{2}$. For the fifth electron of boron (B), we cannot use the combination $n = 2$ and $\ell = 0$, because of the Pauli principle, so we use $n = 2$ and $\ell = 1$. There are three possible values for m with $\ell = 1$, and together

Table 4.2 Possible Sets of Quantum Numbers for the Ten Electrons of Neon

Quantum Number	First Electron	Second Electron	Third Electron	Fourth Electron	Fifth Electron	Sixth Electron	Seventh Electron	Eighth Electron	Ninth Electron	Tenth Electron
n	1	1	2	2	2	2	2	2	2	2
ℓ	0	0	0	0	1	1	1	1	1	1
m	0	0	0	0	-1	0	$+1$	-1	0	$+1$
s	$-\frac{1}{2}$	$+\frac{1}{2}$	$-\frac{1}{2}$	$+\frac{1}{2}$	$-\frac{1}{2}$	$-\frac{1}{2}$	$-\frac{1}{2}$	$+\frac{1}{2}$	$+\frac{1}{2}$	$+\frac{1}{2}$

with the two possible values for s, they yield six combinations of quantum numbers with $n = 2$ and $\ell = 1$.

The configurations of the first 10 electrons in a multi-electron atom are shown in Table 4.2. It must be emphasized that the value of m and the sign of the s value are arbitrary in some cases but not in others (see Problem 4.4 at the end of the chapter).

We can continue in this manner, building up the configuration of each element by adding a set of quantum numbers for one "last" electron to the configuration of the element before it. This process of adding one electron to the preceding element is called the **buildup principle.**

■ EXAMPLE 4.7

Write the sets of quantum numbers for the last eight electrons of argon, along with the sets given in Table 4.2 for the first 10 electrons.

Solution

The sets are shown in Table 4.3. Note that the combination $n = 3$, $\ell = 1$ has the same sum of n and ℓ as $n = 4$, $\ell = 0$. Since the sum is the same, the combination with the lower n value is used for the thirteenth through eighteenth electrons because it is lower in energy. ■

When we try to add the nineteenth electron to write the configuration for potassium (K), we encounter a new situation. The combination with the next lowest sum of n and ℓ is $n = 4$, $\ell = 0$. The combination $n = 3$, $\ell = 2$ is higher in energy. The nineteenth through twenty-first electrons can have the following sets of quantum numbers:

Table 4.3 Possible Sets of Quantum Numbers for the Last Eight Electrons of Argon

Quantum Number	Eleventh Electron	Twelfth Electron	Thirteenth Electron	Fourteenth Electron	Fifteenth Electron	Sixteenth Electron	Seventeenth Electron	Eighteenth Electron
n	3	3	3	3	3	3	3	3
ℓ	0	0	1	1	1	1	1	1
m	0	0	-1	0	$+1$	-1	0	$+1$
s	$-\frac{1}{2}$	$+\frac{1}{2}$	$-\frac{1}{2}$	$-\frac{1}{2}$	$-\frac{1}{2}$	$+\frac{1}{2}$	$+\frac{1}{2}$	$+\frac{1}{2}$

Quantum Number	Nineteenth Electron	Twentieth Electron	Twenty-first Electron
n	4	4	3
ℓ	0	0	2
m	0	0	-2
s	$-\frac{1}{2}$	$+\frac{1}{2}$	$-\frac{1}{2}$
$n + \ell$	4	4	5

The fact that electrons having quantum number values $n = 4$ and $\ell = 0$ are lower in energy than electrons with $n = 3$ and $\ell = 2$ is of extreme importance; it explains the existence and position on the periodic table of the transition metals. This point will be explained later.

■ EXAMPLE 4.8

Arrange the following electrons, identified only by their n and ℓ quantum numbers, in order of increasing energy from lowest to highest.

(a) $n = 4, \ell = 1$ (b) $n = 4, \ell = 0$

(c) $n = 3, \ell = 2$ (d) $n = 3, \ell = 1$

Solution

Calculate the sum $n + \ell$ for each electron:

(a) $n = 4, \ell = 1$, so $n + \ell = 5$ (b) $n = 4, \ell = 0$, so $n + \ell = 4$

(c) $n = 3, \ell = 2$, so $n + \ell = 5$ (d) $n = 3, \ell = 1$, so $n + \ell = 4$

The ones with lower values of $n + \ell$ are lower in energy, so electrons (d) and (b) are lower in energy than (c) and (a). The sum $n + \ell$ is the same for (d) and (b), so (d), the one with the lower n value, is lower in energy. The sum $n + \ell$ is the same for (c) and (a), so (c), the one with the lower n value, is lower in energy. Thus, the order of increasing energy is (d) < (b) < (c) < (a).

Practice Problem 4.8   Arrange the following electrons, identified only by their n and ℓ quantum numbers, in order of increasing energy from lowest to highest.

(a) $n = 4, \ell = 1$ (b) $n = 6, \ell = 0$ (c) $n = 5, \ell = 1$

(d) $n = 5, \ell = 2$ (e) $n = 5, \ell = 0$ ■

4.4   Shells, Subshells, and Orbitals

OBJECTIVE

■ to write electronic configurations in a shorter notation, using the concepts of shells, subshells, and orbitals

A shell is defined as a group of electrons in an atom all having the same principal quantum number. A **subshell** is defined as a group of electrons in an atom all having the same principal quantum number and also the same angular momentum quantum number. If two electrons in an atom have the same principal quantum number, the same angular momentum quantum number, and the same magnetic quantum number, the electrons are said to be in the same **orbital.**

■ EXAMPLE 4.9

Show that a maximum of two electrons can be in a given orbital.

Solution

By definition, the electrons in a given orbital have the same n value, the same ℓ value, and the same m value. According to the Pauli exclusion principle, they must therefore have different values for their spin quantum numbers (s). Since only two s values ($-\frac{1}{2}$ and $+\frac{1}{2}$) are permitted, the maximum number of electrons in a given orbital is two.

Practice Problem 4.9   What is the maximum number of electrons that will fit into the first subshell of the third shell of an atom? ■

Even though the m and s values do not affect the energy of the electron, it is still important to learn about them. The number of combinations of permitted values of these quantum numbers determines the maximum number of electrons in a given type of subshell. For example, in a subshell for which $\ell = 1$, m can have three different values (-1, 0, and $+1$), and s can have two different values ($-\frac{1}{2}$ and $+\frac{1}{2}$). The six different combinations of m and s allow a maximum of six electrons in any subshell for which $\ell = 1$.

■ EXAMPLE 4.10

How many electrons are permitted in a subshell for which (a) $\ell = 0$ and (b) $\ell = 2$?

Solution

(a) When $\ell = 0$, $m = 0$. Since m has only one permitted value and s has two, there are two different combinations of m and s for this subshell. Thus, it can be occupied by a maximum of two electrons.

(b) Five permitted m values (-2, -1, 0, $+1$, and $+2$) times two permitted s values ($-\frac{1}{2}$ and $+\frac{1}{2}$) makes 10 combinations. A maximum of 10 electrons can occupy this subshell. ■

Writing out each quantum number value for every electron in an atom is very time-consuming. A more efficient method is to group all the electrons in a given subshell. In this method, the following four lowercase letters represent the possible ℓ values:

Value of ℓ	Letter
0	s
1	p
2	d
3	f

Since only n and ℓ values affect the energies of electrons, the electrons with the same n value and the same ℓ value all have the same energy. In other words, all the electrons in a given subshell have the same energy. Each subshell is denoted by its principal quantum number and the letter designation for ℓ. For example, for neon, with atomic number 10, the sets of quantum numbers for the 10 electrons are listed in Table 4.2. We can group them as follows:

Value of n	Value of ℓ	Number of Electrons	Subshell Designation
1	0	2	1s
2	0	2	2s
2	1	6	2p

We write the **electronic configuration** by listing each subshell in order of increasing energy, with a superscript giving the number of electrons in that subshell. That is, the detailed electronic configuration for a neon atom is

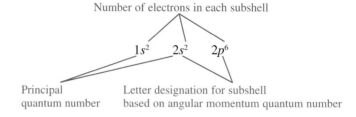

Number of electrons in each subshell

$$1s^2 \quad 2s^2 \quad 2p^6$$

Principal quantum number

Letter designation for subshell based on angular momentum quantum number

This configuration is read aloud as follows: "one ess two, two ess two, two pee six." (The superscripts are not exponents, so words such as *square* are not used.) The sum of the superscripts is the total number of electrons in the atom.

■ EXAMPLE 4.11

Using sets of quantum numbers from Table 4.2, write the detailed electronic configuration of oxygen.

Solution

Oxygen, with atomic number 8, has eight electrons. The first eight electrons shown in the table will fit into three subshells, as follows:

$$1s^2 \, 2s^2 \, 2p^4$$

Note that the $2p$ subshell can hold a maximum of six electrons, but in oxygen, only four electrons are left for that subshell.

Practice Problem 4.11   What element has the following electronic configuration?

$$1s^2 \, 2s^2 \, 2p^2 \quad ■$$

■ EXAMPLE 4.12

Write the detailed electronic structure of titanium (Ti). Comment on the relative energies of the "last" four electrons.

Solution

$$1s^2 \, 2s^2 \, 2p^6 \, 3s^2 \, 3p^6 \, 4s^2 \, 3d^2$$

Because the "last" two electrons are added to an inner shell (following the $n + \ell$ rule) instead of the outermost shell, the $3d$ subshell of a titanium atom must be higher in energy than the $4s$ subshell.

Practice Problem 4.12   What element has the following electronic configuration?

$$1s^2 \, 2s^2 \, 2p^6 \, 3s^2 \, 3p^6 \, 4s^2 \, 3d^8 \quad ■$$

4.5 Shapes of Orbitals

■ to draw the most common orbitals and to understand their spatial orientation and the uncertain nature of locating the electron in the atom

An orbital is an allowed energy state in an atom. Each orbital is designated by the three quantum numbers n, ℓ, and m. Since s is not specified, either value of s can be used, and a maximum of two electrons can occupy any given orbital in an atom.

Knowing exactly both the location and the momentum of an electron in an atom at the same time is impossible. This fact is known as the **Heisenberg uncertainty principle.** Therefore, scientists describe the *probable* locations of electrons. These locations describe the **orbital shapes,** which are important when the atom forms bonds with other atoms, because the orbital shapes are the basis of the geometry of the resulting molecule.

It is equally probable that s orbital electrons will be located in any direction about the nucleus. We say that an s orbital is spherically symmetrical. The $1s$ orbital is pictured in Figure 4.5(a). Since an electron with $\ell = 1$ has three possible m values, any p subshell has three orbitals. Each one lies along one of the coordinate axes—x, y, or z—as shown in Figure 4.5(b). Each p orbital consists of two three-dimensional **lobes** centered on one of the axes. An atom has five $3d$ orbitals, corresponding to the five possible m values (-2, -1, 0, $+1$, and $+2$) for a subshell with $\ell = 2$. Their orientations are shown in Figure 4.5(c).

4.6 Energy Level Diagrams

■ to represent pictorially the energies of subshells in atoms and of the electrons that occupy the subshells

Energy level diagrams are models for portraying electrons' occupancy of an atom's orbitals. They help chemists predict how many electrons are in each orbital of a subshell. Electron occupancy of the individual orbitals is important in determining an atom's magnetic properties. A line or a box or a circle is used to represent each orbital. An energy level diagram that could hold the electrons of any known atom is shown in Figure 4.6. The energy level diagram is like a graph in one dimension: The higher a subshell is placed, the higher is the energy of that subshell. The lines are spaced horizontally from left to right only to prevent crowding so that the diagram is easy to read.

The lowest line on the energy level diagram represents the orbital in the $1s$ subshell of the atom. Much higher in the diagram, indicating a much higher energy, lies the line for the $2s$ orbital. Somewhat higher than that are the three lines for the orbitals of the $2p$ subshell. The third shell lies at an even higher energy and consists of an s subshell, a p subshell, and a d subshell. Note that the $3d$ subshell lies at a slightly higher energy than the $4s$ subshell. The order of energy in the diagram is the same as that given by the $n + \ell$ rule.

We will often focus our attention on the portion of the energy level diagram containing the last electron added, in which we are most interested. The orbitals that lie above that portion are assumed to be empty, and any orbitals that lie below those pictured are completely filled when the atom is in its ground state.

We represent each electron with an arrow. Different electron spins (s value of $-\frac{1}{2}$ or $+\frac{1}{2}$) are indicated by arrows pointing downward or upward. Since each line represents one orbital, each line may hold a maximum of two arrows. If two

Figure 4.5 Shapes of Orbitals

(a) The 1*s* orbital.
(b) The three 2*p* orbitals.
(c) The five 3*d* orbitals.

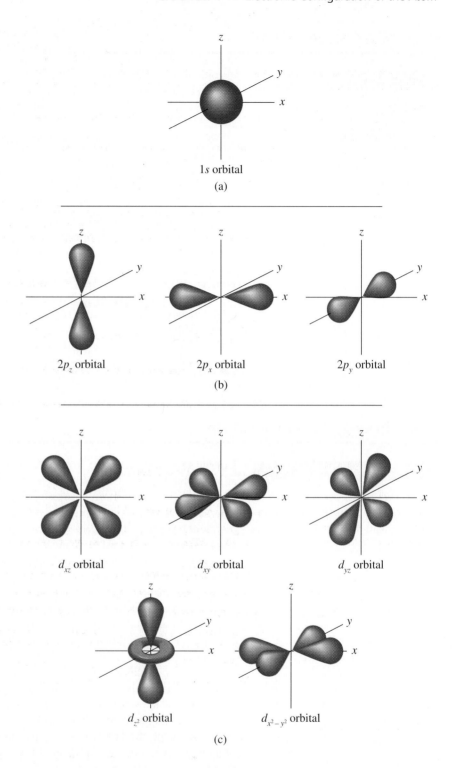

arrows are present, they must be pointing in opposite directions. The energy level diagram representing the neon atom is shown in Figure 4.7.

Hund's rule states that the electrons *within a given subshell* remain as *unpaired* as possible. Moreover, if there is more than one unpaired electron in a given subshell, they all must occupy different orbitals and have the same electron

Figure 4.6 Energy
Level Diagram

(not drawn to scale)

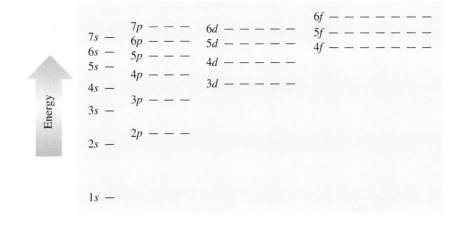

spin (all arrows representing unpaired electrons in a subshell point up or all point down). The energy level diagrams for the carbon, nitrogen, and oxygen atoms illustrate these rules:

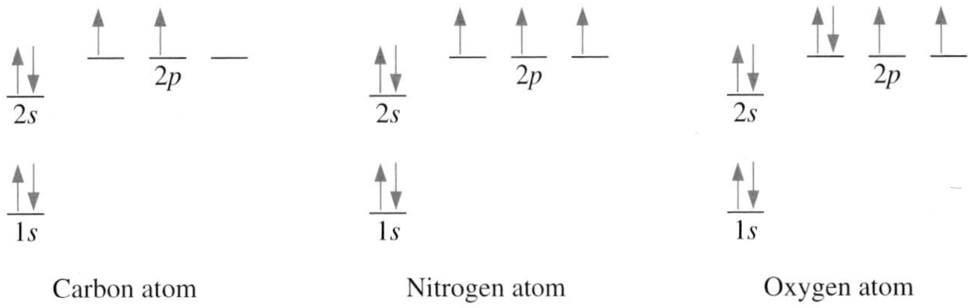

Carbon atom   Nitrogen atom   Oxygen atom

In the carbon atom, the lowest two subshells are filled; all electrons are paired in filled subshells. The $2p$ subshell has two electrons in the three orbitals, so each electron occupies a separate orbital. Moreover, both electrons have the same spin—both arrows point upward (alternatively, both could point downward). In the nitrogen atom, the $2p$ subshell is half filled. Each electron occupies a different orbital, and all arrows point in the same direction. In the oxygen atom, the $2p$ subshell is again partially filled. To get four electrons into the three orbitals requires the pairing of electrons in one orbital. In the other two, the electrons are unpaired and have the same spin; they are said to have **parallel spin**.

The **magnetic properties** of atoms allow us to tell if all the electrons in an atom are paired or, if not, how many electrons are unpaired. Atoms with all their electrons paired are repelled slightly from a magnetic field. If at least one electron per atom in a sample is unpaired, the sample tends to be drawn into a magnetic field. The greater the number of unpaired electrons, the greater the attraction into the magnetic field. (In elemental iron, cobalt, and nickel, the unpaired electrons in adjacent atoms reinforce one another, and a very much stronger attraction into a magnetic field results.)

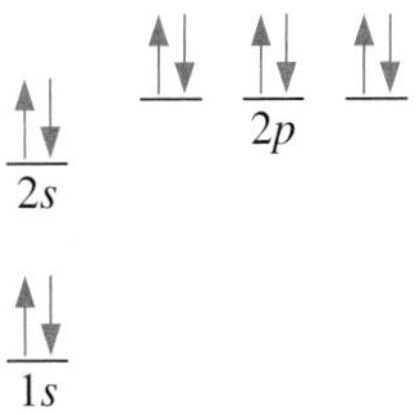

Figure 4.7 Electron Occupancy
of the Neon Atom

▪ **EXAMPLE 4.13**

How many unpaired electrons are in (a) a carbon atom and (b) a fluorine atom?

Solution

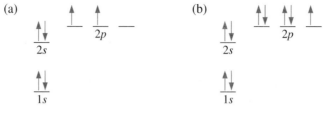

(a)   (b)

Carbon atom   Fluorine atom

The energy level diagrams show two unpaired electrons in a carbon atom and one in a fluorine atom.

Practice Problem 4.13   How many unpaired electrons are in (a) a neon atom and (b) a boron atom? ■

4.7   Periodic Variation of Electronic Configuration

There is a **periodicity of electronic configuration** of the elements. For example, if we examine the detailed electronic configurations of the alkali metals, we find that the outermost shell (specifically, the s subshell) of electrons contains only a single electron in each case. The alkaline earth metals have two outermost s electrons. The elements within each other group of the periodic table also have similarities in their outermost electronic configurations. We deduce that the outermost part of the electronic configuration is the main factor that determines the chemical properties of the elements, since the periodic table was constructed from data about the properties of the elements.

■ **EXAMPLE 4.14**

Write the electronic configurations of F, Cl, Br, and I. What feature makes them have similar chemical properties?

Solution

$$
\begin{array}{ll}
\text{F} & 1s^2\,2s^2\,2p^5 \\
\text{Cl} & 1s^2\,2s^2\,2p^6\,3s^2\,3p^5 \\
\text{Br} & 1s^2\,2s^2\,2p^6\,3s^2\,3p^6\,4s^2\,3d^{10}\,4p^5 \\
\text{I} & 1s^2\,2s^2\,2p^6\,3s^2\,3p^6\,4s^2\,3d^{10}\,4p^6\,5s^2\,4d^{10}\,5p^5
\end{array}
$$

The $ns^2\,np^5$ configuration of the outermost (nth) shell is common to all the halogens and different from the outermost configuration of the other elements. It is the cause of their similar chemical properties.

Practice Problem 4.14   Write the electronic configurations of the noble gases. State the feature that makes them have similar chemical properties. ■

The periodic table can be divided into blocks corresponding to the type of subshell occupied by the last electron added (Figure 4.8). The two groups at

Figure 4.8 Using the Periodic Table to Write Electronic Configurations

For the s and p blocks, the principal quantum number is equal to the period number. For the d block, the principal quantum number is equal to the period number minus 1. For the f block, the principal quantum number is equal to the period number minus 2.

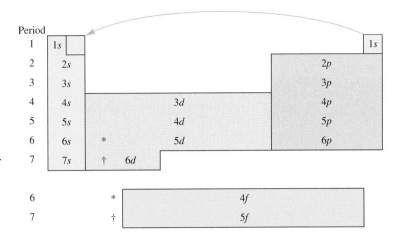

the left of the periodic table—the alkali metals and the alkaline earth metals—constitute the s block, since their last electrons occupy s subshells. Hydrogen and helium also are in this block, and we have to remember to shift helium to a place beside hydrogen for this purpose. The six periodic groups at the right of the table constitute the p block; their last electrons go into p subshells. The transition metals belong to the d block, and the f block consists of the inner transition metals.

Note the similarity between the number of elements in each period in a particular block and the maximum number of electrons permitted in the corresponding type subshell:

Type of Subshell or Block	Maximum Number of Electrons in Subshell	Number of Elements in Each Period in a Particular Block
s	2	2
p	6	6
d	10	10
f	14	14

> The electronic structure of atoms is the basis for the periodic behavior of the elements.

After each noble gas, a new shell of electrons is started, as is a new period of the periodic table. It turns out that electronic structure is the basis for the periodic behavior of the elements.

The three transition metal series arise because, for each of these elements, an electron has been added to the next-to-outermost shell. Addition of 10 electrons to the $3d$ subshell *after the completion of the 4s subshell* causes 10 elements to occur after calcium that have no precursors. The second and third transition series occur because the $4d$ and $5d$ subshells fill after the start of the fifth and sixth shells, respectively. The inner transition elements stem from the addition of electrons to f subshells two shells lower than the outermost shell of their atoms.

Since the periodic table reflects the electronic structures of the atoms, we can use it to deduce the configuration of any atom. We use the periodic table with its s, p, d, and f blocks, as shown in Figure 4.8. We imagine helium to be next to hydrogen in the $1s$ block. To determine the electronic configuration of an element, we start at hydrogen—element 1—and continue *in order of atomic numbers* until we get to the element in question. The subshells come from the blocks in the periodic table, and the numbers of electrons (the superscripts in the configuration) are the numbers of elements in the blocks. Thus, we determine the electronic configuration of aluminum as follows:

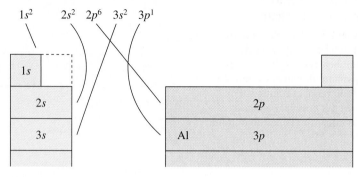

▪ **EXAMPLE 4.15**

Using the periodic table as an aid, write the detailed electronic configuration for each of the following elements:

(a) P   (b) Fe   (c) La

Solution

(a) $1s^2\ 2s^2\ 2p^6\ 3s^2\ 3p^3$

(b) $1s^2\ 2s^2\ 2p^6\ 3s^2\ 3p^6\ 4s^2\ 3d^6$

(c) $1s^2\ 2s^2\ 2p^6\ 3s^2\ 3p^6\ 4s^2\ 3d^{10}\ 4p^6\ 5s^2\ 4d^{10}\ 5p^6\ 6s^2\ 5d^1$

Practice Problem 4.15   Using the periodic table as an aid, write the detailed electronic configuration for each of the following elements:

(a) Sn   (b) Ga   (c) Se ▪

A more compact notation can sometimes be used to reduce the effort of writing long electronic configurations while retaining almost as much information. We are most interested in the outermost shell and the inner subshells having nearly the same energies. We can therefore write the detailed electronic configuration for just that shell and those subshells of the atom and use the symbol of the preceding noble gas in square brackets to represent all the other electrons. For example, the electronic configuration of Ba is denoted

$$\text{Ba}\qquad [\text{Xe}]\ 6s^2$$

▪ **EXAMPLE 4.16**

Write the electronic configuration for thallium (Tl) (element 81), using the shortened notation.

Solution

$$\text{Tl}\qquad [\text{Xe}]\ 6s^2\ 4f^{14}\ 5d^{10}\ 6p^1$$

The symbol for the preceding noble gas is written in square brackets. Then, starting at Cs, the alkali metal in the same period as thallium, you write $6s^2$ for Cs and Ba, $4f^{14}$ for the lanthanide elements, $5d^{10}$ for the elements from La to Hg, and $6p^1$ for Tl. You must be especially careful to follow the atomic numbers after elements 57 (La) and 89 (Ac), where the two inner transition series start. Note that the superscripts in the previous configuration plus the atomic number of Xe (54) add up to the atomic number of Tl (81).

Practice Problem 4.16   Write the electronic configuration for bismuth, using the shortened notation. ■

Using the periodic table as a mnemonic device has several advantages over relying on the $n + \ell$ rule and other rules: The periodic table is generally available for reference during examinations. The order of subshells is given "automatically." The maximum number of electrons in each subshell matches the number of elements in each block. To write a shortened notation for an element, you can start at the alkali metal in the same period.

■ ■ ■ ■ ■ ■ ■ **SUMMARY** ■ ■ ■ ■ ■ ■

The chemical properties of atoms depend on their electronic structures. The number of electrons in a neutral atom is equal to the number of protons in the nucleus—the atomic number of the element. Bohr first proposed the concept that electrons are arranged in discrete energy levels in the atom, which explained the emission of specific energies of light when gaseous atoms are heated. Although Bohr's theory could not explain many other details of the behavior of atoms, it was a milestone in relating electronic structure and properties of atoms. (Section 4.1)

The modern theory of electronic structure is based on the assignment of four quantum numbers to each electron in an atom. The principal quantum number, n, governs the energy of the electron and also its probable distance from the nucleus. The angular momentum quantum number, ℓ, also has an effect on the energy and determines the shape of the volume of space that the electron can occupy. The magnetic quantum number, m, determines the orientation in space of the volume occupied by the electron, and the spin quantum number, s, indicates the spin of the electron on its axis. The limits on the quantum numbers (Table 4.1) must be memorized. The Pauli exclusion principle states that no two electrons in an atom can have the same set of four quantum numbers. (Section 4.2)

The $n + \ell$ rule governs the order of increasing energy of the electrons in the atom. Subshells are filled with electrons in the order of increasing $n + \ell$, with due regard for the limitations on the quantum numbers and for the Pauli exclusion principle. In each case, the "last" electron can be added to the configuration of the element before, using a procedure known as the buildup principle. (Section 4.3)

Electrons in a given shell of an atom all have the same n value. Electrons in a given subshell of an atom

all have the same n value and the same ℓ value. Electrons in a given orbital of an atom all have the same n value, the same ℓ value, and the same m value. By convention, subshells are designated using lowercase letters that correspond to the various ℓ values, and the electronic configuration of an atom is written using superscripts for the numbers of electrons occupying the subshells (for example, $1s^2$ indicates that two electrons occupy the $1s$ subshell). (Section 4.4)

Electrons in the various orbitals occupy portions of space having specified shapes (Figure 4.5). (These are important when the shapes of molecules are considered in more advanced courses.) (Section 4.5)

Energy level diagrams portray electrons' occupation of the orbitals in an atom. Such diagrams are useful for understanding Hund's rule and atoms' magnetic properties. Hund's rule states that in partially filled subshells, the electrons occupy orbitals singly and have the same spins as far as possible. If all the electrons in a substance are paired (two electrons in each occupied orbital), the substance will be repelled slightly from a magnetic field. However, if at least one electron in each formula unit is unpaired, the substance will be drawn into a magnetic field. (Section 4.6)

The properties of the elements stem from their electronic configurations, and the properties place them in their locations in the periodic table. In each group, the elements have a characteristic outermost electronic configuration. The existence of the transition and inner transition elements stems from adding electrons to inner shells after outer shells have been started. Since the periodic table reflects the electronic structures of the atoms, it can be used as a mnemonic device when writing electronic configurations. The ability to write and understand such configurations is a very important skill. (Section 4.7)

Items for Special Attention

■ When electrons undergo transitions to higher shells, energy is *absorbed* by the atom; when electrons undergo transitions to lower shells, energy is *emitted* by the atom.

■ There is a difference between the questions "How many *m* values are possible?" and "What are the possible *m* values?" The number of *m* values is the number of orbitals in the subshell.

■ You can think of electrons in shells as being similar to small children on a ladder: They can never be between levels and are most stable at the lowest energy level possible.

■ The letter *s* is used in two ways—as the spin quantum number and as the letter designation for orbitals with $\ell = 0$.

■ The *p*, *d*, and *f* orbitals have more than one lobe each. Do not mistake each lobe for a separate orbital.

Self-Tutorial Problems

4.1 What values are possible for the principal quantum number *n* for electrons in the ground state of an atom of Lr, element 103?

4.2 (a) If you drop three marbles into an empty ice cream cone, how many will have the lowest position?

(b) If the cone is held steady, how many will have the lowest position possible under the circumstances?

(c) If you place three electrons in an atom in its ground state, will all three electrons have the same energy?

(d) Will all three have the lowest energy possible under the circumstances?

4.3 (a) What is the difference between an *s* subshell and an *s* orbital?

(b) What is the difference between a *p* subshell and a *p* orbital?

4.4 For the electrons of Table 4.2, is the sign of the *s* value arbitrary for (a) the first electron, (b) the second electron, (c) the third electron, (d) the tenth electron?

4.5 (a) Can two tickets to a concert have the same section, the same row, the same seat, and the same date?

(b) How many of these must be different to avoid seating problems?

(c) Can two electrons in the same atom have the same *n* value, the same ℓ value, the same *m* value, and the same *s* value?

(d) How many of these must be different to have a permissible situation?

4.6 For what element is the Bohr theory most useful?

4.7 Add the energies for the change of the electron in the hydrogen atom from the third shell to the second plus that from the second shell to the first (see Example 4.1 and Practice Problem 4.1 for data). Compare your answer to the energy for the change from the third shell to the first, and explain your result.

4.8 Explain why helium, with two outermost electrons, has the same inertness characteristic of neon and argon, each with eight outermost electrons.

■ ■ ■ PROBLEMS ■ ■ ■

4.1 Bohr Theory

4.9 Describe qualitatively the relationship between energy and the electron transitions occurring in the neon gas in a neon sign.

4.10 List the possible series of electron transitions for an electron descending from the fifth shell to the first in a hydrogen atom.

4.11 How many different energies of light would be emitted if many identical atoms underwent the changes described in Problem 4.10?

4.2 Quantum Numbers

4.12 What values of ℓ are permitted for an electron with $n = 5$?

4.13 What values of *s* are permitted for an electron with $n = 3$, $\ell = 2$, and $m = -1$?

4.14 What values of *m* are permitted for an electron with $\ell = 3$?

4.15 Make a chart showing all possible values of ℓ, *m*, and *s* for an electron with $n = 2$ and another with $n = 3$.

4.16 Which of the following sets of quantum numbers is (are) *not* permitted?

(a) $n = 2$ $\ell = 2$ $m = 0$ $s = -\frac{1}{2}$

(b) $n = 2$ $\ell = 1$ $m = +1$ $s = -1$

(c) $n = 3$ $\ell = -1$ $m = 0$ $s = -\frac{1}{2}$

(d) $n = 2$ $\ell = 1$ $m = +2$ $s = +\frac{1}{2}$

4.17 (a) How many 2*p* orbitals are present in any atom?

(b) What is the maximum number of electrons in the 2*p* subshell?

(c) How many electrons are present in the 2*p* subshell of a nitrogen atom?

(d) Explain why the subshell is not full in the nitrogen atom.

4.3 Relative Energies of Electrons

4.18 Compare the energies of the following electrons, identified by their quantum numbers only:

(a) $n = 2$, $\ell = 1$, $m = 0$, $s = -\frac{1}{2}$

(b) $n = 2$, $\ell = 1$, $m = 1$, $s = +\frac{1}{2}$

(c) $n = 2$, $\ell = 1$, $m = -1$, $s = -\frac{1}{2}$

4.19 What type (s, p, d, f) electron was the last electron added in the buildup process to

(a) An inner transition element

(b) A transition element

(c) A nonmetal

4.20 Arrange the following electrons, identified only by their n and ℓ quantum numbers, in order of increasing energy from lowest to highest.

(a) $n = 5$, $\ell = 1$ (b) $n = 5$, $\ell = 0$

(c) $n = 4$, $\ell = 2$ (d) $n = 4$, $\ell = 1$

4.21 Arrange the following electrons in order of increasing energy:

(a) $n = 3$, $\ell = 2$, $m = 0$, $s = -\frac{1}{2}$

(b) $n = 3$, $\ell = 2$, $m = +1$, $s = -\frac{1}{2}$

(c) $n = 3$, $\ell = 2$, $m = -1$, $s = +\frac{1}{2}$

(d) $n = 3$, $\ell = 2$, $m = 0$, $s = +\frac{1}{2}$

4.22 (a) What values of m are permitted for an electron with $\ell = 2$?

(b) How many different values of m are permitted for an electron with $\ell = 2$?

4.4 Shells, Subshells, and Orbitals

4.23 How many electrons are present in each of the following atoms? Assuming that each is a neutral atom, identify the element.

(a) $1s^2\, 2s^2\, 2p^4$

(b) $1s^2\, 2s^2\, 2p^6\, 3s^2\, 3p^6\, 4s^2\, 3d^8$

(c) $1s^2\, 2s^2\, 2p^6\, 3s^2\, 3p^6\, 4s^2\, 3d^{10}\, 4p^3$

4.24 What does the number of m values permitted for a given ℓ value have to do with the number of orbitals in a subshell?

4.25 What does the number of ℓ values permitted for a given n value have to do with the number of subshells in a shell?

4.26 What is the maximum number of electrons in a given atom that can have the following quantum numbers?

(a) $n = 5$, $\ell = 1$

(b) $n = 4$, $\ell = 0$

(c) $n = 2$, $\ell = 0$

(d) $n = 3$, $\ell = 1$, and $m = -1$

4.27 Explain why the helium atom is stable with only two electrons in its outermost shell, but beryllium is not.

4.28 Write detailed electronic configurations for Li, Na, K, and Rb, and deduce the outermost configuration for Cs and Fr.

4.29 (a) How many orbitals are in the $2p$ subshell?

(b) How many orbitals are in the $3p$ subshell?

(c) How many orbitals are in the $4p$ subshell?

(d) What is the maximum number of electrons permitted in a $3p$ subshell?

4.30 (a) What is the letter designation for $\ell = 2$?

(b) How many different m values are possible for an electron in a subshell for which $\ell = 2$?

(c) How many different orbitals are in an $\ell = 2$ subshell?

(d) What is the maximum number of electrons in an $\ell = 2$ subshell?

4.31 Write detailed electronic configurations for the following:

(a) Si   (b) Ca

(c) S   (d) Mg

(e) Al

4.32 How many electrons are permitted (a) in a p subshell and (b) in a p orbital?

4.33 Write detailed electronic configurations for F, Cl, Br, and I.

4.34 What principles or rules affect the energies of electrons in an atom?

4.35 (a) How many electrons are added to an atom in the buildup process before the start of the second shell? How many elements are in the periodic table before the start of the second period?

(b) How many electrons are added to an atom in the buildup process before the start of the third shell? How many elements are in the periodic table before the start of the third period?

(c) How many electrons are added to an atom in the buildup process before the start of the fourth shell? How many elements are in the periodic table before the start of the fourth period?

4.36 Write detailed electronic configurations for N, P, As, and Sb.

4.5 Shapes of Orbitals

4.37 (a) How many of the p orbitals pictured in Figure 4.5(b) are oriented along an axis? (b) How many of the d orbitals pictured in Figure 4.5(c) are oriented along axes?

4.38 According to Figure 4.5(c), which two $3d$ orbitals cannot have an electron in the xy plane?

4.39 According to Figure 4.5(b), which $2p$ orbital cannot have an electron in the xy plane?

4.40 How many d orbitals are pictured in Figure 4.5?

4.6 Energy Level Diagrams

4.41 How many unpaired electrons are present in the ground state of an atom if six electrons are present in each of the following subshells? There are no other unpaired electrons.

(a) $3p$ subshell

(b) $3d$ subshell

(c) $4f$ subshell

4.42 Draw an energy level diagram, and determine the number of unpaired electrons in an atom of each of the following:

(a) P　　(b) Ca　　(c) F　　(d) O

4.43 How many unpaired electrons are in an atom in the ground state, assuming that all other subshells are either completely full or empty, if its outermost p subshell contains (a) four electrons, (b) five electrons, (c) six electrons?

4.44 What is the maximum number of *unpaired* electrons in (a) a p subshell and (b) a d subshell?

4.45 Draw an energy level diagram for the iron atom.

4.46 How many unpaired electrons are in an atom in the ground state, assuming that all other subshells are either completely full or empty, if its outermost d subshell contains (a) four electrons, (b) five electrons, (c) six electrons?

4.47 Which of the following configurations represents the ground state of nitrogen?

(a)

$$\underset{2s}{\uparrow\downarrow} \qquad \underset{2p}{\uparrow\downarrow \quad \downarrow \quad \underline{}}$$

$$\underset{1s}{\uparrow\downarrow}$$

(b)

$$\underset{2s}{\uparrow\downarrow} \qquad \underset{2p}{\uparrow \quad \uparrow\downarrow \quad \underline{}}$$

$$\underset{1s}{\uparrow\downarrow}$$

(c)

$$\underset{2s}{\uparrow\downarrow} \qquad \underset{2p}{\uparrow \quad \uparrow \quad \uparrow}$$

$$\underset{1s}{\uparrow\downarrow}$$

(d)

$$\underset{2s}{\uparrow\downarrow} \qquad \underset{2p}{\uparrow \quad \downarrow \quad \uparrow}$$

$$\underset{1s}{\uparrow\downarrow}$$

(e)

$$\underset{2s}{\uparrow} \qquad \underset{2p}{\uparrow\downarrow \quad \uparrow \quad \uparrow}$$

$$\underset{1s}{\uparrow\downarrow}$$

(f)

$$\underset{2s}{\uparrow\downarrow} \qquad \underset{2p}{\downarrow \quad \downarrow \quad \uparrow}$$

$$\underset{1s}{\uparrow\downarrow}$$

4.7 Periodic Variation of Electronic Configuration

4.48 Use the periodic table to write the outer electronic configuration for each of the following elements:

(a) Rb　　(b) Ra　　(c) Sn

4.49 Locate in the periodic table (a) the element that has the first $3p$ electron and (b) the element that is the first to complete its $2p$ subshell.

■　■　■　■　**GENERAL PROBLEMS**　■　■　■　■

4.50 The energy of each of the first six shells of hydrogen is given in Figure 4.3. Calculate the energies emitted when the electrons in many hydrogen atoms descend from the fifth shell to the first. (Hint: See Problem 4.10.)

4.51 What is the maximum number of unpaired electrons in the ground state of an atom in which only the $1s$, $2s$, and $2p$ subshells have any electrons?

4.52 What is the maximum number of *unpaired* electrons in the $3d$ subshell of an atom?

4.53 (a) Convert the wavelength (λ) of the first line of the hydrogen spectrum, 410 nm, to meters.

(b) Calculate the energy of that transition, using the equation $E = hc/\lambda$, where $h = 6.61 \times 10^{-34}$ J $\cdot$ s and $c = 3.00 \times 10^8$ m/s.

4.54 The Bohr theory has been essentially replaced. Explain why any theory is ever rejected.

4.55 Which of the following configurations are *not* permitted for an atom in its ground state?

(a) $1s^2\, 2s^2$　　(b) $1s^1\, 2s^1$　　(c) $1s^2\, 2s^2\, 2p^2$

(d) $1s^6\, 2s^6\, 2p^6$

4.56 Figure 4.8 shows that the periodic table is based on the electronic structure of the atoms. Explain how Mendeleyev was able to create the periodic table without knowing about the electron at all.

4.57 Is the Bohr theory or the quantum mechanical theory better to describe the electronic arrangement of

(a) Helium　　(b) Nitrogen　　(c) Neon

4.58 What one color is produced by heating a mixture of hydrogen and helium gases in the proportions in which they are present in the sun?

4.59 Identify the element from each of the following *partial* configurations of neutral atoms:

(a) ... $4s^2\ 3d^5$ (b) ... $5s^2\ 4d^{10}\ 5p^2$

(c) ... $6p^3$ (d) ... $5s^1$

(e) ... $3d^{10}\ 4p^1$ (f) ... $6s^2\ 5d^1\ 4f^7$

4.60 Can you identify the following element from its inner electronic configuration? $1s^2\ 2s^2\ 2p^6$ ... Explain.

4.61 Does an electron gain or lose energy in each of the following transitions?

(a) From a $4f$ subshell to a $5s$ subshell

(b) From a $2s$ subshell to a $3p$ subshell

(c) From a $4s$ subshell to a $3d$ subshell

4.62 (a) Draw an energy level diagram for cobalt.

(b) Can you use this diagram for the electronic structure of carbon?

(c) Explain why one large energy level diagram is sufficient for all the elements.

4.63 What is wrong with each of the following ground state configurations?

(a) $1s^1\ 2s^1\ 2p^6$ (b) $1s^2\ 1p^6\ 2s^2\ 2p^5$

(c) $1s^2\ 2s^2\ 2p^4\ 3s^2$ (d) $1s^2\ 2p^6\ 3d^4$

(e) $1s^2\ 2s^2\ 2p^6\ 3s^2\ 3p^6\ 4s^2\ 3d^{11}$ (f) $[Xe]\ 6s^2\ 5d^{10}\ 4f^{16}$

(g) $1s^2\ 2s^2\ 2p^6\ 2d^2$ (h) $1s^2\ 2s^4\ 2p^6$

(i) $1s^2\ 2s^2\ 2p^6\ 3s^2\ 3p^6\ 4s^2\ 4d^{10}\ 4p^6$

4.64 In answering the question, "What is the maximum value for ℓ for any electron in the ground state of Lr, element 103?" several students gave the following answers and reasoning. Which one is correct?

(a) "The maximum $\ell = 6$ because the outermost shell has $n = 7$, and ℓ cannot be more than $n - 1$."

(b) "The maximum $\ell = 1$ because the outermost shell cannot have more than eight electrons, and $\ell = 1$ is the maximum ℓ for a filled octet."

(c) "The maximum $\ell = 3$ because the f subshell has an ℓ value of 3, and there is no subshell with a bigger ℓ value in Lr.

4.65 How many unpaired electrons are present in the ground state of an atom if there are five electrons in each of the following subshells? There are no other unpaired electrons.

(a) $3p$ subshell   (b) $3d$ subshell   (c) $4f$ subshell

4.66 Deduce the expected electronic configuration of (a) U, (b) Lu, and (c) Lr.

4.67 The orange line in the hydrogen spectrum (see Figure 4.2) is the change of the electron from the third shell to the second; the green line is the change from the fourth shell to the second; the two violet lines are the changes from the fifth and sixth shells to the second, respectively. Which color represents the most energy, and which represents the least? Is wavelength directly proportional to energy?

4.68 Which of the following sets of quantum numbers is (are) *not* permitted?

(a) $n = 2$ $\ell = 1$ $m = 1$ $s = -\frac{1}{2}$

(b) $n = 2$ $\ell = 2$ $m = 1$ $s = +\frac{1}{2}$

(c) $n = 3$ $\ell = 2$ $m = 0$ $s = -1$

(d) $n = 4$ $\ell = 3$ $m = 2$ $s = -\frac{1}{2}$

(e) $n = 4$ $\ell = 3$ $m = -3$ $s = +\frac{1}{2}$

4.69 Which metal in each of the following sets will be drawn into a magnetic field the most?

(a) Sc   Ti   V

(b) Ti   Mn   Zn

(c) Cu   Rh   Hg

(d) Mg   Mn   Cd

5 Chemical Bonding

- 5.1 Chemical Formulas
- 5.2 Ionic Bonding
- 5.3 Formulas for Ionic Compounds
- 5.4 Electron Dot Diagrams
- 5.5 Covalent Bonding

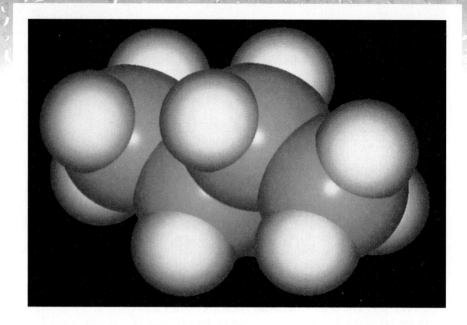

Key Terms (Key terms are defined in the Glossary.)

anhydrous (5.1)
anion (5.2)
anode (5.2)
binary compound (5.1)
cathode (5.2)
cation (5.2)
centered dot (5.1)
covalent bond (5.5)
diatomic molecule (5.1)
double bond (5.5)
duet (5.2)
electrode (5.2)
electron dot diagram (5.4)

electron sharing (5.5)
electronegativity (5.1)
elemental (5.1)
formula (5.1)
formula unit (5.1)
hydrate (5.1)
ion (5.2)
ionic bond (5.2)
lone pair (5.5)
macromolecule (5.5)
molecule (5.5)
monatomic ion (5.2)
noble gas configuration (5.2)

non-octet structure (5.5)
octet (5.2)
octet rule (5.2)
outermost shell (5.2)
polyatomic ion (5.5)
single bond (5.5)
sodium chloride structure (5.2)
structural formula (5.5)
subscript (5.1)
triple bond (5.5)
unshared pair (5.5)
valence electron (5.2)
valence shell (5.2)

Suffix

-ide (5.2)

The electronic structure of an uncombined atom, discussed in Chapter 4, determines the ability of that atom to combine with other atoms to produce molecules or ionic compounds. In this chapter, the fundamentals of chemical bonding are covered. To discuss compounds, chemical formulas are required. Moreover, when symbols for atoms are combined in a chemical formula, some type of bonding is implied. Therefore, chemical formulas are introduced first, in Section 5.1. (More information about chemical formulas will be presented in Chapter 7.) Ionic bonding, which occurs when electrons are transferred from one atom to another, is treated in Section 5.2. The number of electrons transferred from one atom to another, or the changes on the resulting ions, allow us to deduce the formulas for binary ionic compounds (Section 5.3). A convenient way to picture atoms with their outermost electrons—the electron dot diagram—is presented in Section 5.4. Atoms held together solely by covalent bonds form units called molecules (or much larger units called macromolecules). Covalent bonding, in which the sharing of electrons is the primary method of bonding, is introduced in Section 5.5, which also discusses combinations of ionic and covalent bonding.

> Ionic bonding involves transfer of electrons. Covalent bonding involves electron sharing. Both ionic and covalent bonding occur in many ternary compounds.

5.1 Chemical Formulas

OBJECTIVE

■ to interpret and write chemical formulas

Just as a symbol identifies an element, a **formula** is a combination of symbols that identifies a compound, an ion, or a molecule of an element. However, chemical formulas do much more. A formula also indicates the relative quantities of the elements contained in the compound or ion and implies some kind of chemical bonding between the atoms.

In the formula for a **binary compound** (one containing only two elements), the element that attracts electrons less is usually written first. The elements are assigned an **electronegativity** that reflects their affinity for electrons in chemical bonds. The elements that attract electrons most are said to have the highest electronegativities or to be the most electronegative. Fluorine, the most electronegative element, is assigned an electronegativity of 4.0, and the other elements have values relative to that of fluorine. The elements that attract electrons least are said to have the lowest electronegativities or to be the most electropositive.

Values for the electronegativities of the main group elements are presented in Figure 5.1. Except for those of the noble gases, the electronegativities of the elements increase toward the right and toward the top of the periodic table. Fluorine has the highest electronegativity of any element, and oxygen has the second highest value. The most electropositive element is francium (Fr). The metals are more electropositive than the nonmetals. Thus, the metal, if one is present, is written first in the formula of a binary compound. If no metal is present, the nonmetal closer to the metal portion of the periodic table is written first.

Formulas for binary compounds of hydrogen do not follow the rule just discussed. Hydrogen is written first in the formula if the compound is an acid (Chapter 8) and written later if the compound is not an acid. For example, HCl is hydrochloric acid, and NH_3 is ammonia. The position of the H in these formulas indicates that HCl is an acid and NH_3 is not.

Formulas are also used to identify molecules of free elements. Many free (uncombined) nonmetallic elements exist as molecules, such as H_2, N_2, O_2, F_2, Cl_2, Br_2, and I_2, as well as P_4 and S_8 (Figure 5.2). The formula S_8 indicates eight sulfur atoms bonded together. This formula does not represent a compound, since

Figure 5.1 Electronegativities of the Main Group Elements

H 2.1							He
Li 1.0	**Be** 1.5	**B** 2.0	**C** 2.5	**N** 3.0	**O** 3.5	**F** 4.0	**Ne**
Na 0.9	**Mg** 1.2	**Al** 1.5	**Si** 1.8	**P** 2.1	**S** 2.5	**Cl** 3.0	**Ar**
K 0.8	**Ca** 1.0	**Ga** 1.6	**Ge** 1.8	**As** 2.0	**Se** 2.4	**Br** 2.8	**Kr** 3.0
Rb 0.8	**Sr** 1.0	**In** 1.7	**Sn** 1.8	**Sb** 1.9	**Te** 2.1	**I** 2.5	**Xe** 2.6
Cs 0.7	**Ba** 0.9	**Tl** 1.8	**Pb** 1.9	**Bi** 1.9	**Po** 2.0	**At** 2.2	**Rn**
Fr 0.7	**Ra** 0.9						

Figure 5.2 Some Elements That Occur As Molecules

(a) Cl_2, chlorine.
(b) P_4, white phosphorus (the most stable form of elemental phosphorus).
(c) S_8, rhombic sulfur (the most stable form of elemental sulfur).

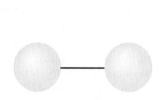

(a) Chlorine

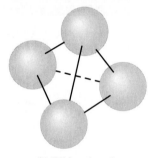

(b) White phosphorus

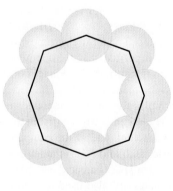

Top view

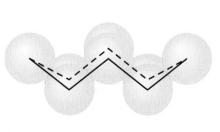

Side view

(c) Rhombic sulfur

Figure 5.3 The Seven Elements That Form Diatomic Molecules

Note that the shape formed by six of these elements in the periodic table is like a seven, starting at atomic number 7.

only one kind of atom is present. **Elemental** sulfur in its lowest-energy form occurs in such molecules.

Seven elements occur as **diatomic molecules** (molecules with two atoms) *when they are not combined with other elements.* Fortunately, these elements are easy to remember because, except for hydrogen, they form a shape like a seven in the periodic table, starting at the element with atomic number 7 (Figure 5.3).

The collection of atoms represented by a formula is called a **formula unit.** A chemical formula consists of symbols of element(s), often with **subscripts** that

> *Seven elements occur as diatomic molecules when they are not combined with other elements.*

ITEM OF INTEREST

Hydrogen molecules (H_2) are so much more stable than separated hydrogen atoms that the reaction of the atoms to form molecules produces a lot of heat:

$$2\,H \rightarrow H_2 + heat$$

Production of a given number of H_2 molecules from hydrogen atoms produces more heat than the production of the same number of CO_2 molecules from burning carbon (charcoal) in oxygen. Construction workers take advantage of the reaction of atomic hydrogen to weld steel pieces together in the absence of oxygen. That condition is desirable because oxygen might make the steel rust. Where do the hydrogen atoms come from? They are produced by electrical discharge in a welding torch, such as is diagrammed in Figure 5.4.

Figure 5.4 Welding Torch

Hydrogen gas is piped into a special tube in which many of the H_2 molecules are separated into atoms by an electrical discharge. The gas flow is adjusted to the rate at which most of the H atoms recombine into molecules just as they exit from the torch. The heat produced at that point is intense enough to weld pieces of steel together.

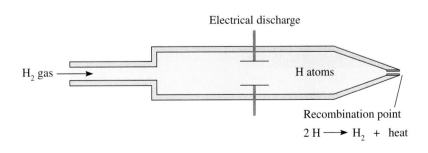

tell how many atoms of each element are present per formula unit. The subscript *follows* the symbol of the element it multiplies. Parentheses may be used in a formula to group bonded atoms together, and a subscript after the closing parenthesis tells how many of that group are present per formula unit. The following formulas convey the given information:

CO   One carbon atom and one oxygen atom are bonded in one formula unit.
CO_2 One carbon atom and two oxygen atoms are bonded in one formula unit.
$(NH_4)_2SO_4$ Two NH_4 groups, each containing one nitrogen atom and four hydrogen atoms, and one sulfur atom and four oxygen atoms are present in one formula unit. The atoms in the NH_4 groups are bonded in some way, as is the compound as a whole (Section 5.5).
I_2 Two atoms of iodine are bonded in one formula unit.
H_2O Two atoms of hydrogen and one atom of oxygen are bonded in one formula unit. This formula unit represents one molecule of water.

■ EXAMPLE 5.1

What is the difference between the formulas $(NH_4)_3PO_4$ and $N_3H_{12}PO_4$?

Solution

Although both formulas have the same ratios of atoms of each element to atoms of all the others and to formula units of the compound, the first formula states that the atoms of nitrogen and hydrogen are bonded together in some way in each of three NH_4 groups.

Practice Problem 5.1

(a) Write a formula that implies that a CH_3 group is connected to another CH_3 group by three CH_2 groups.

(b) Write a formula showing just the numbers of atoms of each element in this compound. ■

A few compounds have formulas that are written with a **centered dot,** as in $CuSO_4 \cdot 5H_2O$. In general, the dots connect the formulas of two or more compounds that could exist independently but are bonded in some way in a single compound. The coefficient (5) after the centered dot multiplies everything after it until the end of the formula. Thus, ten hydrogen atoms and nine oxygen atoms are in one formula unit of $CuSO_4 \cdot 5H_2O$. This particular formula could have been written $CuSO_4(H_2O)_5$. When water is one of the compounds in a formula with a centered dot, the combination is called a **hydrate.** The compound *without* the attached water is said to be **anhydrous.** Anhydrous $CuSO_4$ and its hydrate are pictured in Figure 5.5.

■ EXAMPLE 5.2

How many atoms of each element are present in one formula unit of each of the following compounds?

(a) $(NH_4)_3PO_4$ (b) $NaClO_2$ (c) $Al(ClO_4)_3$ (d) $BaCl_2 \cdot 2H_2O$

Solution

(a) 3 N, 12 H, 1 P, 4 O   (b) 1 Na, 1 Cl, 2 O

(c) 1 Al, 3 Cl, 12 O   (d) 1 Ba, 2 Cl, 4 H, 2 O

Figure 5.5 White, Anhydrous $CuSO_4$ and Its Blue Hydrate, $CuSO_4 \cdot 5H_2O$

Practice Problem 5.2   How many atoms of each element are present in one formula unit of each of the following compounds?

(a) $Co(NO_2)_3$ (b) $CoCO_3$ ■

In reading formulas aloud, the number is simply stated for any subscript following a symbol, as in H_2O: "H two O." To express parentheses followed by a subscript 2, the words "taken twice" are used; for a subscript 3, the words "taken three times" are used, and so on. The centered dot is read "dot." Here are some examples to illustrate these conventions:

$(NH_4)_2SO_4$	"N H four taken twice S O four"
$(NH_4)_3PO_4$	"N H four taken three times P O four"
$CH_3(CH_2)_4CH_3$	"C H three, C H two taken four times, C H three"
$CuSO_4 \cdot 5H_2O$	"C u S O four dot five H two O"

5.2 Ionic Bonding

OBJECTIVE

■ to write octet rule electronic structures for the formation of ionic compounds and to deduce the formulas of compounds of main group metals with nonmetals

Atoms tend to accept, donate, or share electrons to achieve the electronic structure of the nearest noble gas.

The electrons in atoms are arranged in groups having nearly the same energies. These energy levels are often referred to as shells. The first shell of any atom can hold a maximum of two electrons; the second shell can hold a maximum of eight electrons; and the other shells can hold a maximum of eight electrons when they are the outermost shell, but a greater number when they are not (Table 5.1). The **outermost shell** is the last shell that contains electrons.

The noble gases are composed of stable atoms; no reactions of the first three (He, Ne, Ar) have been discovered, and the others (Kr, Xe, Rn) are almost completely unreactive. The stability of the noble gases is due to the eight electrons in the outermost shell of each atom (two electrons in the case of helium). In fact, eight electrons in the outermost shell is a stable configuration for most atoms. Atoms other than those of the noble gases tend to form ionic and/or covalent bonds with other atoms to achieve this electronic configuration. The eight electrons in the outermost shell are called an **octet.** The tendency of atoms to be stable with eight electrons in the outermost shell is called the **octet rule.** In some compounds, one (or more) of the atoms does not obey the octet rule. Some exceptions to the octet rule will be mentioned in Section 5.5.

Since the maximum number of electrons in the first shell of an atom is two, helium is stable with two electrons in its only occupied shell. The other very light

Table 5.1   Maximum Electron Occupancy of Shells

Shell Number	Maximum Occupancy As the Outermost Shell	Maximum Occupancy As an Inner Shell
1	2	2
2	8	8
3	8	18
4	8	32
5	8	50*
6	8	72*
7	8*	98*

*More than the number of electrons available in any atom

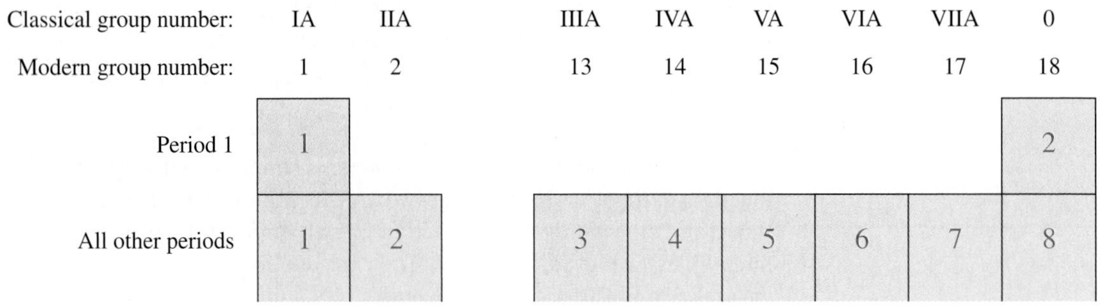

Classical group number:	IA	IIA		IIIA	IVA	VA	VIA	VIIA	0
Modern group number:	1	2		13	14	15	16	17	18

Period 1	1								2
All other periods	1	2		3	4	5	6	7	8

Figure 5.6 Numbers of Valence Electrons for Atoms of Main Group Elements

Na: $1s^2\ 2s^2\ 2p^6\ 3s^1$

Na$^+$: $1s^2\ 2s^2\ 2p^6\ 3s^0$

The number of electrons in the valence shell of an uncombined main group atom is equal to the classical periodic group number of the element.

Figure 5.7 Formula Units of Atoms, Molecules, and Ionic Compounds

The formula units Ne and Br$_2$ represent single atoms and diatomic molecules, respectively. Unlike the Br$_2$ molecule, in which one bromine atom is bonded to a specific other bromine atom, in the ionic compound NaCl, one Na$^+$ ion is bonded to six Cl$^-$ ions that are adjacent to it. Each of the Cl$^-$ ions is bonded to six Na$^+$ ions that are adjacent to it. (The fifth and sixth ions are in layers in front of and behind the layer shown here; see Figure 5.8.) The ratio of Na$^+$ ions to Cl$^-$ ions is therefore 1 : 1. Any pair of Na$^+$ and Cl$^-$ ions, such as those circled in red or the one circled in green, is a formula unit.

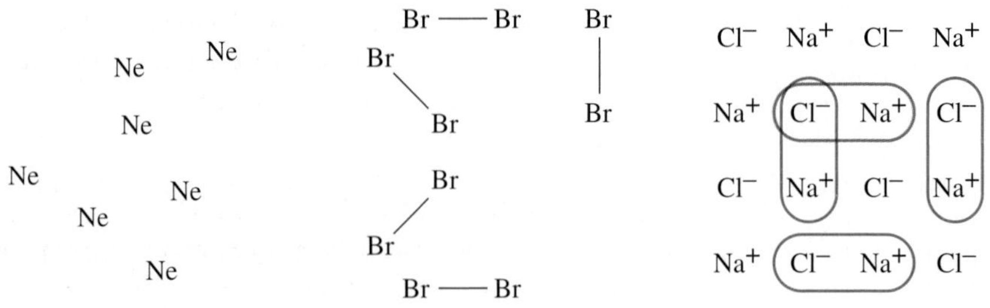

elements—hydrogen, lithium, and beryllium—tend to form stable states by achieving the two-electron configuration of helium. Having two electrons in the first shell, *when that is the only shell and therefore the outermost shell*, is a stable state, and the two electrons are sometimes called a **duet.** When there is only one shell, two electrons in that shell act like eight electrons in any other outermost shell. Therefore, an atom with two electrons in its outermost first shell is often said to obey the octet rule, although "duet rule" would be more precise.

The **valence shell** of electrons in an atom is the outermost shell of electrons of the *uncombined* atom. The electrons in that shell are called **valence electrons.** If all the electrons are removed from that shell, the next inner shell becomes the new outermost shell. For example, the sodium atom has two electrons in its first shell, eight electrons in its second shell (the maximum), and its last electron in its third shell. The valence shell is the third shell. If the one electron is removed from the third shell, the second shell becomes the outermost shell, containing eight electrons. The valence shell is still the (now empty) third shell. The number of electrons in the valence shell of an uncombined main group atom is equal to the classical periodic group number of the element (see Figure 5.6). The exceptions to this rule are that helium has two valence electrons and the other noble gases have eight valence electrons.

Metals react with nonmetals to form ionic compounds. Metals tend to transfer their valence electrons to nonmetals, and nonmetals tend to accept enough electrons from metals to achieve their octets. For example, a sodium atom has one electron in its valence shell, and a chlorine atom has seven electrons in its valence shell. When they react, the sodium atom transfers that one electron to the chlorine atom, forming two charged species called **ions.** Both ions have eight electrons in their outermost shells. (The sodium ion has eight electrons in its second shell, now its outermost shell.) *The electronic configurations of both ions are those of noble gas atoms* (the sodium ion has that of a neon atom, and the chlorine ion that of argon). The atoms have not been changed into noble gas atoms, however, because their nuclei have not changed.

The sodium ion has a single positive charge because the neutral atom has donated one negatively charged electron. The sodium ion is written as Na$^+$. The ion

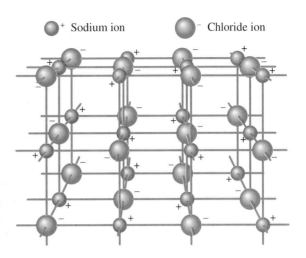

Figure 5.8 Sodium Chloride Structure

Each sodium ion is surrounded by six chloride ions, and each chloride ion is surrounded by six sodium ions. The ratio of sodium ions to chloride ions is 1:1, and the formula for sodium chloride is written NaCl. This shows only a small portion of the structure, which extends over thousands of ions or more in each direction.

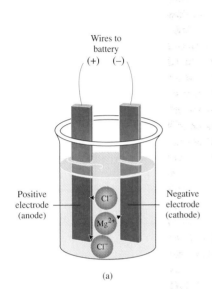

(a)

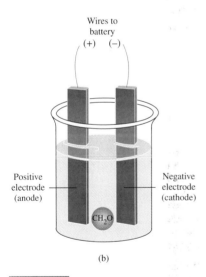

(b)

Figure 5.9 Conduction of Electricity by Ions

(a) When an ionic compound such as magnesium chloride, $MgCl_2$, is dissolved in water and a voltage is applied, the positive and negative ions carry the current. (b) When a compound such as CH_2O (formaldehyde), which is not ionic, is used, there are no charged particles to carry the current, and no flow of electricity is observed.

formed from the chlorine atom has a single negative charge, resulting from the gain of the electron by the neutral chlorine atom. This ion is written as Cl^-. Oppositely charged bodies attract each other, so Na^+ ions and Cl^- ions attract each other. In general, the transfer of electrons from one atom to another produces oppositely charged ions, which attract each other. The formula for the compound of sodium and chlorine is NaCl, which shows that one Na^+ ion is present for each Cl^- ion (Figure 5.7).

All ionic compounds have an overall net charge of zero because the electrons are transferred and do not disappear. The electrons that some atoms accept to form negative ions are donated by other atoms, which become positive ions. Positively charged ions are called **cations** (pronounced "cat′-ions"), and negatively charged ions are called **anions** (pronounced "an′-ions"). The sodium ion is positive; it is a cation. The ion produced by the chlorine atom accepting an extra electron is negative; it is an anion. If an anion is a **monatomic ion** (having only one atom), its name ends in *-ide,* so Cl^- is called the chloride ion.

Metallic and nonmetallic elements can react with each other to form compounds by transferring electrons from the metal atoms to the nonmetal atoms. The ions formed attract each other because of their opposite charges, and these attractions are called **ionic bonds.** However, in a solid ionic compound, a single pair of ions does not bond together; instead, an almost inconceivably huge number of both types of ions forms a lattice that extends in three dimensions. The three-dimensional nature of the **sodium chloride structure** (Figure 5.8) is typical of ionic solids.

The ionic nature of these compounds (the fact that charged particles are present) can be shown by experiments in which the ions are made to carry an electric current. Pure water does not conduct electricity well. However, if a compound that consists of ions is dissolved in water and the solution is placed between **electrodes** in an apparatus like that shown in Figure 5.9, the solution will conduct electricity when the electrodes are connected to the terminals of a battery. Each type of ion moves toward the electrode that has the *opposite* charge of that of the ion. That is, cations migrate to the negative electrode, called the **cathode,** and anions migrate to the positive electrode, called the **anode.** (The names *cation* and *anion* were derived from the words *cathode* and *anode.*) For electricity to be conducted, the ions must be free to move. In the solid state, an ionic compound will not conduct because the ions are trapped in the lattice. However, if the compound is heated until it melts or if it dissolves in water, the liquid compound or solution will conduct electricity because the ions are free to move.

Detailed Electronic Configurations of Anions (Optional)

The detailed electronic structures of monatomic ions may be deduced starting from the structures of the corresponding neutral atoms (presented in Chapter 4). Monatomic anions have simply added sufficient electrons to the outermost p subshell to complete that subshell. The $n + \ell$ rule can be used to deduce the structure of the ion as well as that of the neutral atom. For example, the electronic configuration of the oxide ion (the anion of oxygen) is deduced, starting with the configuration of oxygen:

$$\text{O atom:} \quad 1s^2\, 2s^2\, 2p^4$$

The oxide ion has a double negative charge, obtained by gaining two extra electrons. These electrons go into the $2p$ subshell:

$$\text{O}^{2-} \text{ ion:} \quad 1s^2\, 2s^2\, 2p^6$$

This electronic configuration is the same as that of neon. An ion with a configuration like that of a noble gas is said to have a **noble gas configuration.**

■ EXAMPLE 5.3

Write the electronic configuration for the bromide ion, Br^-.

Solution

$$\text{Br atom:} \quad 1s^2\, 2s^2\, 2p^6\, 3s^2\, 3p^6\, 4s^2\, 3d^{10}\, 4p^5$$

Adding an electron corresponding to the single negative charge yields the ion:

$$\text{Br}^- \text{ ion:} \quad 1s^2\, 2s^2\, 2p^6\, 3s^2\, 3p^6\, 4s^2\, 3d^{10}\, 4p^6$$

Practice Problem 5.3   Write the detailed electronic configuration of the nitride ion, N^{3-}. ■

Detailed Electronic Configurations of Cations (Optional)

To form monatomic cations, main group atoms lose electrons from the outermost subshell(s) of their valence shells. For example, the thallium atom loses its $6p^1$ electron to form Tl^+. The configuration is thus

$$\text{Tl}^+ \text{ ion:} \quad [\text{Xe}]\, 6s^2\, 5d^{10}\, 4f^{14}$$

■ EXAMPLE 5.4

Write the outermost electronic configuration for (a) the Pb^{2+} ion and (b) the Pb^{4+} ion.

Solution

$$\text{Pb atom:} \quad [\text{Xe}] \, 6s^2 \, 5d^{10} \, 4f^{14} \, 6p^2$$

(a) The outermost shell of the lead atom has $n = 6$, and the Pb^{2+} ion is formed by loss of the $6p^2$ electrons from that sixth shell:

$$Pb^{2+} \text{ ion:} \quad [\text{Xe}] \, 6s^2 \, 5d^{10} \, 4f^{14} \, 6p^0 \quad \text{or} \quad [\text{Xe}] \, 6s^2 \, 5d^{10} \, 4f^{14}$$

(b) The Pb^{4+} ion loses the $6p^2$ electrons plus two others—the $6s^2$ electrons. Those are the electrons in the sixth shell! Note that the $n + \ell$ rule does not apply to this case; the electrons in the outermost shell (that with the highest n value) are lost first.

$$Pb^{4+} \text{ ion:} \quad [\text{Xe}] \, 6s^0 \, 5d^{10} \, 4f^{14} \, 6p^0 \quad \text{or} \quad [\text{Xe}] \, 5d^{10} \, 4f^{14}$$

Practice Problem 5.4 Write the outermost electronic configuration for (a) Sn^{2+} and (b) Sn^{4+}. ■

The electronic configurations of transition metal ions, like those of main group ions, are determined by removal of the electrons from the shell of *highest n value first*. Next, electrons may be lost from the d subshell next to the valence shell. The capability of removing a variable number of electrons makes it possible for a transition metal to have ions of different charges.

■ EXAMPLE 5.5

Write detailed electronic configurations for Fe^{2+} and Fe^{3+}.

Solution

First, write the configuration for the iron atom:

$$\text{Fe atom:} \quad 1s^2 \, 2s^2 \, 2p^6 \, 3s^2 \, 3p^6 \, 4s^2 \, 3d^6$$

The outermost shell electrons of the iron atom are lost to form the $2+$ ion:

$$Fe^{2+} \text{ ion:} \quad 1s^2 \, 2s^2 \, 2p^6 \, 3s^2 \, 3p^6 \, 4s^0 \, 3d^6 \quad \text{or} \quad 1s^2 \, 2s^2 \, 2p^6 \, 3s^2 \, 3p^6 \, 3d^6$$

Then one $3d$ electron is lost to form the $3+$ ion:

$$Fe^{3+} \text{ ion:} \quad 1s^2 \, 2s^2 \, 2p^6 \, 3s^2 \, 3p^6 \, 4s^0 \, 3d^5 \quad \text{or} \quad 1s^2 \, 2s^2 \, 2p^6 \, 3s^2 \, 3p^6 \, 3d^5$$

Practice Problem 5.5 Write the electronic configurations of Cr^{2+} and Cr^{3+}. ■

5.3 Formulas for Ionic Compounds

OBJECTIVE

■ to learn how to deduce the formulas of compounds of main group metals with nonmetals and the formulas of any metal and nonmetal ion combination if we know the charges on the ions

Let's consider the ionic compound formed by the reaction of magnesium and chlorine. The magnesium atom has two electrons in its valence shell. (Magnesium is in periodic group IIA [2].) When the atom donates those two electrons to nonmetal atom(s), the positive ion formed has an octet like that of neon. However, the chlorine atom, with seven valence electrons, can accept only one additional electron. Therefore, it takes *two* chlorine atoms to accept the two electrons from *one* magnesium atom; so the formula for magnesium chloride is $MgCl_2$.

Mg²⁺ ion: $1s^2\,2s^2\,2p^6$

Cl⁻ ion: $1s^2\,2s^2\,2p^6\,3s^2\,3p^6$

Al atom: $1s^2\,2s^2\,2p^6\,3s^2\,3p^1$
Al³⁺ ion: $1s^2\,2s^2\,2p^6$

In compounds, the metals of periodic groups IA and IIA (1 and 2), as well as zinc, cadmium, aluminum, and silver, always form ions with positive charges equal to the element's classical periodic group number.

The charge on every monatomic anion (except H⁻) is equal to the classical group number of the element minus 8. The number of added electrons is the absolute value of that difference.

Na atom: $1s^2\,2s^2\,2p^6\,3s^1$
S atom: $1s^2\,2s^2\,2p^6\,3s^2\,3p^4$

Similarly, the formula of aluminum chloride is $AlCl_3$, since the aluminum atom has three valence electrons that it can donate to form the 3+ ion.

We can predict the charges on the ions of some elements but not others. In compounds, the metals of periodic groups IA and IIA (1 and 2), as well as zinc, cadmium, aluminum, and silver, always form ions with positive charges equal to the element's classical periodic group number. The charge on every monatomic anion (except H⁻) is equal to the classical group number of the element minus 8. The number of added electrons is the absolute value of that difference. (Not all nonmetals form monatomic anions, however.) Hydrogen can react with very active metals to form the hydride ion, H⁻, which has the two-electron configuration of helium. The maximum positive charge on a monatomic cation is 4+; the maximum negative charge on a monatomic anion is 3−. Charges on the most common monatomic ions are presented in Figure 5.10. In addition to the generalities just presented, note that all the elements of the first transition series except scandium form an ion with a 2+ charge, and most of them also form an ion having another charge. (The transition metals form ions having different charges by donating varying numbers—from 0 to 2—of their inner electrons to nonmetals.)

Since the overall charge on any ionic compound is zero, we can determine the formula of an ionic compound by balancing the charges on the cations and anions.

■ EXAMPLE 5.6

Determine the formula of the compound of sodium and sulfur.

Solution

Sodium, in group IA, has one valence electron, and sulfur, in group VIA, has six. Each sulfur atom needs two additional electrons to form S^{2-}, which has an octet in its valence shell, but each sodium atom can supply only one electron to form

Figure 5.10 Charges on Common Monatomic Ions

H 1+ 1−																	
Li 1+	Be 2+													N 3−	O 2−	F 1−	
Na 1+	Mg 2+											Al 3+		P 3−	S 2−	Cl 1−	
K 1+	Ca 2+	Sc 3+	Ti 2+ 3+	V 2+ 3+	Cr 2+ 3+	Mn 2+ 3+	Fe 2+ 3+	Co 2+ 3+	Ni 2+ 4+	Cu 1+ 2+	Zn 2+					Br 1−	
Rb 1+	Sr 2+								Pd 2+ 4+	Ag 1+	Cd 2+		Sn 2+ 4+			I 1−	
Cs 1+	Ba 2+								Pt 2+ 4+	Au 1+ 3+	Hg 2+ *		Pb 2+ 4+				
Fr 1+	Ra 2+																

*Mercury also forms a diatomic ion, Hg_2^{2+}.

Na⁺ ion: $1s^2 \, 2s^2 \, 2p^6$
S²⁻ ion: $1s^2 \, 2s^2 \, 2p^6 \, 3s^2 \, 3p^6$

Na⁺. Therefore, it takes two sodium atoms to supply the electrons for one sulfur atom, and the formula for sodium sulfide is Na_2S. Alternatively, we can say that it takes two monopositive sodium ions to balance the charge on one dinegative sulfide ion.

Practice Problem 5.6   Determine the formula of the compound of lithium and nitrogen. ■

■ EXAMPLE 5.7

Determine the formula of (a) the compound containing Co^{2+} and S^{2-} and (b) the compound containing Co^{3+} and O^{2-}.

Solution

The numbers of cations and anions in a compound depend only on their charges, and not on the identities of the ions.

(a) One Co^{2+} ion can balance the charge on one S^{2-} ion. Therefore, the ions bond in a 1:1 ratio to form CoS.

(b) It takes three O^{2-} ions to balance the charge on two Co^{3+} ions, so the formula is Co_2O_3.

Practice Problem 5.7   Determine the formula of (a) the compound containing Cu^{2+} and F^- and (b) the compound containing Co^{2+} and P^{3-}. ■

■ EXAMPLE 5.8

Write the formula for each type of ion in the following compounds:

(a) KBr   (b) MgO   (c) K₂S   (d) Al₂O₃

Solution

(a) K^+ Br^- (b) Mg^{2+} O^{2-} (c) K^+ S^{2-} (d) Al^{3+} O^{2-}

The charges on the cations are equal to the numbers of valence electrons originally in the atoms, and the charges on the anions are equal to 8 minus the number of valence electrons. The numbers of valence electrons for these elements are easily determined from their periodic group numbers. The two potassium ions in the compound of part (c) are not bonded to each other, since they are both positive and repel each other. They should not be written with a subscript (except in the formula for the compound, K_2S, in which they are bonded to the sulfide ion). If you want to show that two potassium ions are present, you must write $2\,K^+$.

Practice Problem 5.8   Write formulas for the ions in Ca_3P_2. ■

Since all compounds have overall charges of zero, we can deduce the charges on some metals' cations from the total charge on the anions bonded to them. For example, in $CuCl$ and $CuCl_2$, the charges on the copper ions are 1+ and 2+, respectively. The 1+ charge on the copper ion in $CuCl$ is required to balance the 1− charge on one Cl^- ion. The 2+ charge on the copper in $CuCl_2$ is required to balance the 1− charge on each of *two* Cl^- ions.

■ EXAMPLE 5.9

What is the charge on each cation in (a) CuO and (b) Cu_2O?

Solution

(a) The charge on the cation must be 2+, to balance the 2− charge on one O^{2-} ion.

(b) The charge on each cation must be 1+, since two cations are required to balance the charge on one O^{2-} ion.

Practice Problem 5.9　What is the charge on each cation in (a) AgCl and (b) $MnCl_2$? ■

5.4　Electron Dot Diagrams

<table>
<tr><td>

O B J E C T I V E

■ to write electron dot diagrams for keeping track of the valence electrons in compounds, especially those of main group elements

These four positions represent the four outermost *s* and *p* orbitals (Chapter 4).

Atomic orbitals hold a maximum of two electrons each.

</td><td>

The discussion in Section 5.2 showed that the valence electrons are very important in ionic bonding. Section 5.5 will show that they are also very important in covalent bonding. The **electron dot diagram,** which is a way to picture the transfer or sharing of valence electrons, aids in understanding both processes. Keep in mind, however, that electron dot diagrams are simplified representations of atoms and not true pictures.

In an electron dot diagram, the symbol of the element represents the nucleus of the atom plus its inner shells of electrons, and dots around the symbol stand for the valence electrons. The dots are placed to the left or right or above or below the symbol. In unbonded atoms, two dots, at most, are located in each position. For example, atoms of the elements sodium and sulfur may be represented as follows:

</td></tr>
</table>

$$\text{Na}\cdot \qquad :\overset{\cdot\cdot}{\underset{\cdot}{\text{S}}}:$$

■ EXAMPLE 5.10

Write four equivalent electron dot diagrams for the sodium atom.

Solution

$$\overset{\cdot}{\text{Na}}\cdot \qquad \overset{\cdot}{\text{Na}} \qquad \cdot\text{Na} \qquad \underset{\cdot}{\text{Na}}$$

Practice Problem 5.10

(a) Write four equivalent electron dot diagrams for the chlorine atom.

(b) Write six equivalent electron dot diagrams for the sulfur atom. ■

The reaction of sodium and sulfur to form Na_2S (see Example 5.6) can be visualized easily with electron dot diagrams:

$$
\begin{array}{cccc}
\text{Na}\cdot & & \text{Na}\cdot\searrow & \text{Na}^+ \\
& +\;\; :\overset{\cdot\cdot}{\underset{\cdot}{\text{S}}}: \;\; \rightarrow & :\overset{\cdot\cdot}{\underset{\cdot}{\text{S}}}: \;\; \rightarrow & + \left[:\overset{\cdot\cdot}{\underset{\cdot\cdot}{\text{S}}}:\right]^{2-} \\
\text{Na}\cdot & & \text{Na}\cdot\nearrow & \text{Na}^+
\end{array}
$$

Electrons from two sodium atoms are needed to allow one sulfur atom to attain its octet. We can write the reaction more simply as follows:

$$2\,Na\cdot\ +\ :\overset{\cdot\cdot}{\underset{\cdot\cdot}{S}}:\ \longrightarrow\ 2\,Na^+\ +\ \left[:\overset{\cdot\cdot}{\underset{\cdot\cdot}{S}}:\right]^{2-}$$

■ **EXAMPLE 5.11**

Use electron dot diagrams to picture the reaction of aluminum and sulfur.

Solution

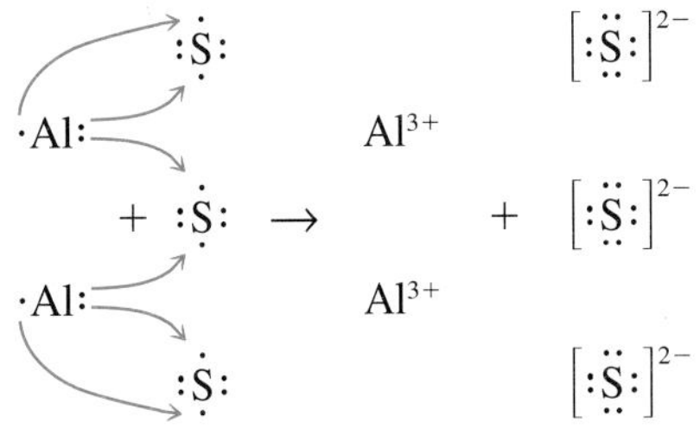

The six valence electrons of the two aluminum atoms are transferred to three sulfur atoms, yielding two Al^{3+} ions and three S^{2-} ions. The reaction can be written more simply as follows:

$$2\,\cdot Al:\ +\ 3\,:\overset{\cdot}{\underset{\cdot\cdot}{S}}:\ \longrightarrow\ 2\,Al^{3+}\ +\ 3\left[:\overset{\cdot\cdot}{\underset{\cdot\cdot}{S}}:\right]^{2-}$$

Practice Problem 5.11 Use electron dot diagrams to picture the reaction of calcium and phosphorus. ■

Electron dot diagrams are even more useful for understanding covalent bonding, as the next section shows.

5.5 Covalent Bonding

OBJECTIVE

■ to write electron dot diagrams for the compounds of two or more nonmetals bonded with shared electron pairs, and also for polyatomic ions

Metal atoms can donate electrons to nonmetal atoms, but nonmetal atoms cannot donate all their valence electrons to form octets and therefore do not form monatomic positive ions. (Single nonmetal atoms do not donate electrons at all, but some groups of nonmetal atoms can. This will be discussed later in this section.) Nonmetal atoms can accept electrons from metal atoms if such atoms are present; otherwise, they can attain an octet by **electron sharing. A covalent bond** consists of shared electrons. One pair of electrons shared between two atoms constitutes a *single* covalent bond, generally referred to as a **single bond.** An **unshared pair** of valence electrons is called a **lone pair.** Elements or compounds bonded only by covalent bonds form **molecules.**

Consider the hydrogen molecule, H_2. Each atom of hydrogen has one electron and would be more stable with two electrons (the helium configuration). There is no reason why one hydrogen atom would donate its electron and the other accept it. Instead, the two hydrogen atoms can *share* their electrons:

$$H\cdot\ +\ \cdot H \longrightarrow H{:}H$$

Electrons shared between atoms are counted toward the octets (actually duets) of *both* atoms. In the hydrogen molecule, each hydrogen atom has a total of two electrons in its first shell and, thus, a stable configuration.

▪ **EXAMPLE 5.12**

Draw an electron dot diagram for HCl. Label the single bond and the lone pairs.

Solution

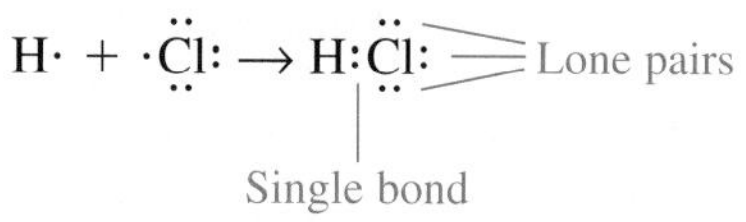

$$H\cdot \; + \; \cdot \overset{..}{\underset{..}{Cl}}\!: \; \rightarrow \; H\!:\!\overset{..}{\underset{..}{Cl}}\!: \;\; \text{Lone pairs}$$

Single bond

Practice Problem 5.12   Draw an electron dot diagram for F_2. Label the single bond and the lone pairs. ▪

In electron dot diagrams for atoms, the four areas around the symbol can hold a maximum of two electrons each. However, be aware that more than one pair of electrons can be placed between covalently bonded atoms.

Another representation of molecules is the **structural formula,** in which each electron pair being shared by two atoms is represented by a line or dash. Electrons not being shared may be shown as dots in such a representation. Structural formulas for H_2 and HCl are

$$H\!\!-\!\!H \qquad H\!\!-\!\!\overset{..}{\underset{..}{Cl}}\!:$$

Two atoms can share more than one pair of electrons to make an octet for each atom. Consider the nitrogen molecule:

$$:\!\overset{.}{N}\!\cdot \; + \; \cdot\overset{.}{N}\!: \; \rightarrow \; :\!N\!:\!:\!:\!N\!: \qquad \text{or} \qquad :\!N\!\!\equiv\!\!N\!:$$

In this case, three electron pairs are shared, and each nitrogen atom has an octet of electrons. There is one lone pair of electrons on each nitrogen atom. Three

ENRICHMENT

In addition to its stable elementary form—O_2—oxygen can also exist as O_3 molecules, a form called *ozone.* Ozone can be formed when an electrical discharge passes through oxygen gas, and it can also be formed in the upper atmosphere—the ozone layer—when high-energy rays from outer space bombard O_2 molecules. The ozone molecules in the atmosphere are important because they absorb harmful ultraviolet light from the sun. This prevents some of that light from reaching the earth's surface, where it could injure humans and other animals. The O_3 molecule is more reactive than O_2 and slowly decomposes spontaneously:

$$2\,O_3 \rightarrow 3\,O_2$$

Ozone is a powerful oxidizing agent. It is irritating and injurious in concentrations greater than two parts per million. It is used as a disinfectant and a bleach because of its oxidizing properties.

Some other free (uncombined) elements also occur in different forms. Such forms are called *allotropes.* Except for oxygen, the elements that form diatomic molecules when uncombined do not form allotropes, but many other nonmetals do. The allotropes of carbon—diamond and graphite—are perhaps best known to the general public.

pairs of electrons shared between the same two atoms constitute a **triple bond.** If two pairs of electrons are shared, a **double bond** results. Consider the carbon dioxide molecule:

$$:\ddot{O}: + \cdot\ddot{C}\cdot + :\ddot{O}: \rightarrow :\ddot{O}::C::\ddot{O}: \quad \text{or} \quad :\ddot{O}=C=O:$$

Double bond   Double bond

With a few simple rules, recognizing compounds that consist of molecules is fairly easy. All compounds that are gases or liquids at room temperature are molecular. (Solid compounds may be molecular.) Most compounds that do not have a metal atom or an ammonium ion (NH_4^+) in them are molecular. When not combined with other elements, most nonmetallic elements form molecules. (The noble gases have monatomic molecules; their atoms are uncombined.)

■ **EXAMPLE 5.13**

Which of the following formula units consist of uncombined atoms, which consist of molecules, and which consist of ions?

(a) Ne   (b) NaCl   (c) Cl_2 (d) NH_4Cl (e) CO_2

Solution

(a) Ne solution consists of uncombined atoms.

(b) NaCl consists of ions.

(c) Cl_2 consists of molecules.

(d) NH_4Cl is ionic, even though it contains no metal atoms.

(e) CO_2 consists of molecules.

Practice Problem 5.13   Which of the following formula units consist of uncombined atoms, which consist of molecules, and which consist of ions?

(a) ICl   (b) NaF   (c) Na (gaseous)   (d) FeF_2 ■

Some very large molecules, containing billions and billions of atoms, are called **macromolecules.** Diamond, graphite, and silica (sand) are examples (Figure 5.11). Formulas for macromolecules cannot state the number of atoms of each element in each molecule because there are too many. Therefore, these formulas give only the simplest ratio of atoms of one element to any others present. For example, the formula for both diamond and graphite is C and that for silica is SiO_2.

Systematic Method for Drawing Electron Dot Diagrams

Drawing electron dot diagrams for some compounds or ions can get complicated. If you cannot obtain a diagram in which each atom has an octet by using only single bonds, you may move unshared electrons in pairs to positions between atoms, forming double or triple bonds. However, for compounds or ions that obey the octet rule, you may wish to use a more systematic approach:

Step 1:   Determine the total number of valence electrons *available* from all the atoms in the formula unit.

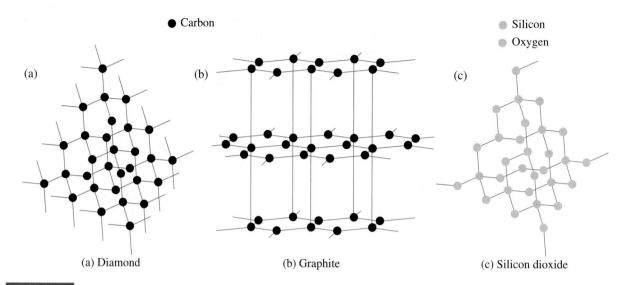

● Carbon

◐ Silicon
◐ Oxygen

(a)   (b)   (c)

(a) Diamond   (b) Graphite   (c) Silicon dioxide

Figure 5.11 Macromolecules of Diamond, Graphite, and Silica

All of the atoms pictured in each case, plus millions more, make up one giant molecule. (a) In diamond, carbon atoms are connected in a three-dimensional structure.
(b) In graphite, carbon atoms are connected in sheets or layers. The red lines do not represent bonds.
(c) Silica (silicon dioxide) has a structure somewhat like that of diamond, except that it contains silicon and oxygen atoms instead of carbon atoms, and an oxygen atom bridges each pair of silicon atoms.

Step 2: Determine the total number of electrons *required* to get eight electrons around each nonmetal atom except hydrogen and two electrons around each hydrogen atom. Note that hydrogen atoms need only two electrons in their outermost shells and that most main group metal ions need none.

Step 3: Subtract the number of electrons available from the number required to determine the number of shared electrons. (The shared electrons are counted for each atom; that is, they are counted twice to obtain the total number of electrons needed.)

Step 4: Distribute the shared pairs *between* adjacent atoms.

Step 5: Distribute the rest of the electrons to other positions.

■ EXAMPLE 5.14

Draw an electron dot diagram for $POCl_3$, in which the chlorine and oxygen atoms are all bonded to the phosphorus atom.

Solution

Atoms	Valence Electrons Available	Valence Electrons Required
P	$1 \times 5 = 5$	$1 \times 8 = 8$
O	$1 \times 6 = 6$	$1 \times 8 = 8$
3 Cl	$3 \times 7 = \underline{21}$	$3 \times 8 = \underline{24}$
Total	32	40

The number of electrons to be shared is $40 - 32 = 8$. Put one pair of electrons between each pair of adjacent atoms:

$$\overset{\text{O}}{\underset{\overset{\cdot\cdot}{\text{Cl}}}{\text{Cl}:\overset{\cdot\cdot}{\text{P}}:\text{Cl}}}$$

Adding the other available electrons (24) yields the complete electron dot diagram:

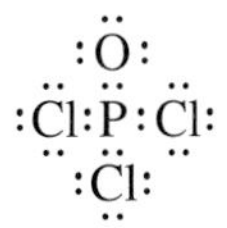

$$:\ddot{O}:$$
$$:\ddot{Cl}:\ddot{P}:\ddot{Cl}:$$
$$:\ddot{Cl}:$$

Be sure to check that all the atoms have the proper octets (duets for hydrogen atoms) and that all valence electrons and no other electrons are shown.

Practice Problem 5.14 Draw an electron dot diagram for H_2O_2, in which each oxygen atom is bonded to the other and also to one hydrogen atom. ■

■ **EXAMPLE 5.15**

Draw an electron dot diagram for formaldehyde (a biological preservative), CH_2O, in which the hydrogen and oxygen atoms are bonded to the carbon atom.

Solution

Atoms	Valence Electrons Available	Valence Electrons Required
2 H	$2 \times 1 = 2$	$2 \times 2 = 4$
C	$1 \times 4 = 4$	$1 \times 8 = 8$
O	$1 \times 6 = \underline{6}$	$1 \times 8 = \underline{8}$
Total	12	20

There are $20 - 12 = 8$ electrons shared. One double bond is needed for eight electrons to be shared between three pairs of atoms. The hydrogen atoms cannot be involved in a double bond because their maximum number of valence electrons is two. Therefore, the double bond must be between the carbon atom and the oxygen atom. The structure showing only the shared electrons is

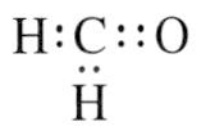

$$\text{H:C::O}$$
$$\ddot{H}$$

Then the unshared electrons are added:

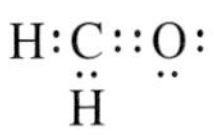

$$\text{H:C::}\ddot{O}:$$
$$\ddot{H}$$

There are eight electrons around the carbon atom and eight electrons around the oxygen atom, as well as two electrons around each hydrogen atom:

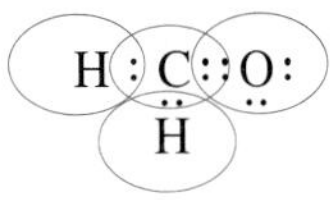

Practice Problem 5.15 Draw an electron dot diagram for $COCl_2$, in which the chlorine and oxygen atoms are bonded to the carbon atom. ■

The first problem in drawing an electron dot diagram for a complicated structure is to determine which atoms are bonded to which other atoms. Many common molecules and ions have one atom of one element and several atoms of another. The single atom of the one element is usually the central atom, with all the other atoms bonded to it.

■ **EXAMPLE 5.16**

Draw an electron dot diagram for (a) CH_4 and (b) NCl_3.

Solution

(a) The central atom is the carbon atom:

$$
\begin{array}{c}
H \\
H\ C\ H \\
H
\end{array}
$$

The number of shared electrons is $16 - 8 = 8$. The complete diagram is

$$
\begin{array}{c}
H \\
H:\overset{\cdot\cdot}{\underset{\cdot\cdot}{C}}:H \\
H
\end{array}
$$

Atoms	Valence Electrons Available	Valence Electrons Required
C	4	8
4 H	4	8
Total	8	16

(b) The central atom is the nitrogen atom:

$$
\begin{array}{c}
Cl \\
Cl\ N \\
Cl
\end{array}
$$

The number of shared electrons is $32 - 26 = 6$. The complete diagram is

$$
\begin{array}{c}
:\overset{\cdot\cdot}{\underset{}{Cl}}: \\
:\overset{\cdot\cdot}{\underset{\cdot\cdot}{Cl}}:\overset{}{\underset{\cdot\cdot}{N}}: \\
:\overset{}{\underset{\cdot\cdot}{Cl}}:
\end{array}
$$

Atoms	Valence Electrons Available	Valence Electrons Required
N	5	8
3 Cl	21	24
Total	26	32

Practice Problem 5.16   Draw an electron dot diagram for C_3H_8. ■

Polyatomic Ions

A great many compounds contain **polyatomic ions** ("many-atom" ions). There are many polyatomic anions but relatively few polyatomic cations. The most important polyatomic cation is the ammonium ion, NH_4^+ (compare with ammonia, NH_3). Some of the most important polyatomic anions are listed in Table 5.2.

The atoms *within* a polyatomic ion are bonded together with covalent bonds, but polyatomic ions as a whole are bonded to oppositely charged ions by the attraction of the opposite charges—by ionic bonding. For example, sodium chlorate, $NaClO_3$, contains sodium ions, Na^+, and chlorate ions, ClO_3^-. The Na^+ ions are attracted to the ClO_3^- ions by their opposite charges. The chlorine and oxygen atoms within each ClO_3^- ion are covalently bonded. The electron dot diagram for sodium chlorate is simply a combination of that for the sodium ion and that for the chlorate ion. The representation can be determined by the systematic process described previously. Note that the sodium ion is bonded ionically and that it shares *no* electrons with other atoms. For that reason, no electrons are allotted for its valence shell.

Table 5.2   Some Important Polyatomic Anions

Name	Formula
Hydroxide ion	OH^-
Cyanide ion	CN^-
Acetate ion	$C_2H_3O_2^-$
Chlorate ion	ClO_3^-
Bromate ion	BrO_3^-
Iodate ion	IO_3^-
Nitrate ion	NO_3^-
Sulfate ion	SO_4^{2-}
Carbonate ion	CO_3^{2-}
Phosphate ion	PO_4^{3-}

Atoms	Valence Electrons Available	Valence Electrons Required
Na	1	0
Cl	7	8
3 O	18	24
Total	26	32

The number of electrons to be shared is $32 - 26 = 6$. The structure, with only the shared electrons:

$$\text{Na}^+ \left[\text{O:Cl:O}\right]^-$$
$$\ddot{\text{O}}$$

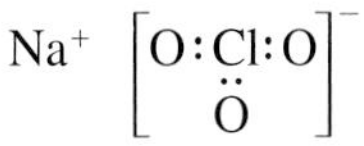

The rest of the electrons are added:

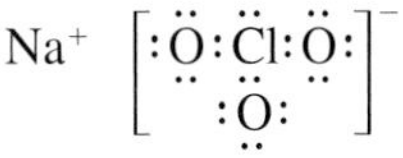

$$\text{Na}^+ \left[:\ddot{\text{O}}:\ddot{\text{Cl}}:\ddot{\text{O}}:\right]^-$$
$$:\ddot{\text{O}}:$$

Be sure to write the charge on each ion because the charge is an integral part of the formula. For example, there is a great difference between ClO_3 and ClO_3^-.

■ **EXAMPLE 5.17**

Draw an electron dot diagram for the chlorate ion, ClO_3^-.

Solution

The charge on the ion signifies the presence of an extra valence electron—from some other (unspecified) atom—which must be counted as available:

Atoms	Valence Electrons Available	Valence Electrons Required
Cl	7	8
3 O	18	24
Negative charge	1	
Total	26	32

The number of electrons to be shared is $32 - 26 = 6$. The structure, with only the shared electrons:

$$\left[\text{O:Cl:O}\right]^-$$
$$\ddot{\text{O}}$$

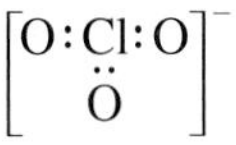

Adding the other electrons gives

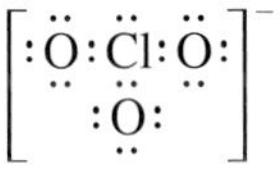

$$\left[:\ddot{\text{O}}:\ddot{\text{Cl}}:\ddot{\text{O}}:\right]^-$$
$$:\ddot{\text{O}}:$$

The structure of the chlorate ion is the same as when the ion was in sodium chlorate. In that case, the sodium atom donated its electron to the chlorate ion. In this case, you do not know where the extra electron came from, but it does not matter. The total number of valence electrons is still 26, and the number of electrons to be shared is still 6. The chlorate ion does not exist in isolation, even though we sometimes write it alone.

Practice Problem 5.17 Draw an electron dot diagram for the phosphate ion, PO_4^{3-}. ■

■ **EXAMPLE 5.18**

Draw an electron dot diagram for the CO_3^{2-} ion.

Solution

The number of shared electrons is $32 - 24 = 8$. Here is the structure with only the shared electrons shown:

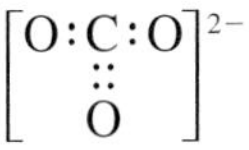

$$\left[\text{O:C:O}\right]^{2-}$$
$$::$$
$$\text{O}$$

Atoms	Valence Electrons Available	Valence Electrons Required
C	4	8
3 O	18	24
Negative charge	2	
Total	24	32

Here is the structure showing all the electrons:

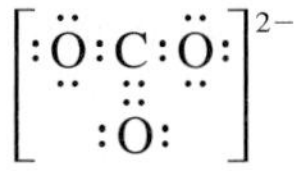

$$\left[\ddot{:}\ddot{O}:C:\ddot{O}\ddot{:}\right]^{2-}$$

The double bond was arbitrarily placed between the carbon atom and the oxygen atom below it. However, there is no difference between that oxygen and the one to the left of the carbon atom or the one to its right. You could draw the double bond between carbon and either of those atoms instead. All the following structures are equivalent:

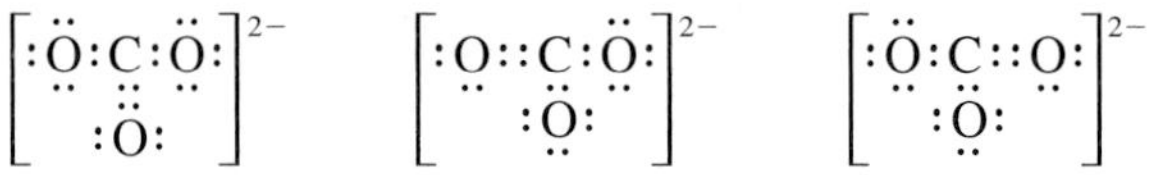

$$\left[\ddot{:}\ddot{O}::C:\ddot{O}\ddot{:}\right]^{2-}\qquad\left[\ddot{:}O::C:\ddot{O}\ddot{:}\right]^{2-}\qquad\left[\ddot{:}\ddot{O}:C::O\ddot{:}\right]^{2-}$$

Practice Problem 5.18 Draw an electron dot diagram for the NO_3^- ion. ■

Hydrogen atoms very often bond with oxygen atoms but seldom bond with oxygen atoms that are double-bonded to other atoms. Thus, if you have a choice between putting a hydrogen atom next to an oxygen atom connected by a single bond to another atom or putting it next to one connected by a double bond to another atom, choose the former.

■ **EXAMPLE 5.19**

Draw the electron dot diagram for HCO_3^-, in which the hydrogen atom is bonded to an oxygen atom.

Solution

Atoms	Valence Electrons Available	Valence Electrons Required
H	1	2
C	4	8
3 O	18	24
Negative charge	1	
Total	24	34

The number of shared electrons is $34 - 24 = 10$. The structure with the shared electrons is

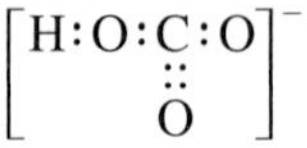

$$\left[H:O:C:O\atop\qquad\ddot{O}\right]^{-}$$

With all the electrons, it is

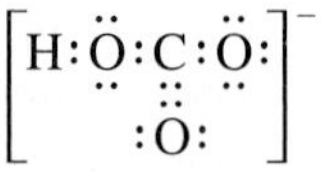

$$\left[H:\ddot{O}:C:\ddot{O}\ddot{:}\atop\qquad:\ddot{O}:\right]^{-}$$

The hydrogen atom can bond equally well to either of the oxygen atoms that are connected to the carbon atom by a single bond, but not to the double-bonded one.

Practice Problem 5.19 Draw an electron dot diagram for HNO_3. ■

With a little experience, you will recognize the familiar ions in formulas, which will allow you to deduce the formula of the other ion in the compound, even if it is unfamiliar to you.

■ **EXAMPLE 5.20**

Write formulas and electron dot diagrams for the ions in $NaHSeO_3$.

Solution

You recognize that sodium always exists in its compounds as the Na^+ ion. Thus, the other ion must contain all the other atoms and must have a single negative charge. The ions are Na^+ and $HSeO_3^-$. The electron dot diagrams are

$$Na^+ \quad \left[H\!:\!\overset{\cdot\cdot}{\underset{\cdot\cdot}{O}}\!:\!\overset{\cdot\cdot}{Se}\!:\!\overset{\cdot\cdot}{\underset{\cdot\cdot}{O}}\!: \atop :\overset{\cdot\cdot}{\underset{\cdot\cdot}{O}}: \right]^-$$

Practice Problem 5.20 Write formulas for the ions in $Na_2Cr_2O_7$. ■

Non-Octet Structures

Not all atoms in molecules or polyatomic ions obey the octet rule; those that do not are said to have **non-octet structures.** For example, boron, which is in the second period of the periodic table, is apt to have fewer than eight electrons in the valence shell of its atoms. Thus, the boron atom in BF_3 is represented as having only six electrons in its valence shell:

$$:\!\overset{\cdot\cdot}{\underset{\cdot\cdot}{F}}\!:\!B\!:\!\overset{\cdot\cdot}{\underset{\cdot\cdot}{F}}\!: \atop :\overset{\cdot\cdot}{\underset{\cdot\cdot}{F}}:$$

If the central element in a molecule or polyatomic ion is in the third period or higher and does not obey the octet rule, it is apt to expand its valence shell beyond eight electrons. The phosphorus atom in PF_5 has 10 electrons around it:

Such "expanded octets" are discussed in more advanced texts.

■ ■ ■ ■ ■ ■ ■ SUMMARY ■ ■ ■ ■ ■ ■ ■

Chemical formulas identify compounds, ions, or molecules. In formulas for binary compounds, the more electropositive element is written first. The formula implies that the atoms are held together by some kind(s) of chemical bond(s). When they are not combined with other elements, hydrogen, nitrogen, oxygen, fluorine, chlorine, bromine, and iodine exist as diatomic molecules (Figure 5.3). A formula unit represents the collection of atoms in the formula. Subscripts in a formula indicate the numbers of atoms of the elements in each formula unit. For example, the formula unit H_2O has two hydrogen atoms and one oxygen atom. Formula units of uncombined elements, such as Ne, are atoms. Formula units of covalently bonded atoms are called molecules. Formula units of ionic compounds do not have any special names. In formulas, atoms bonded in special groups may be enclosed in parentheses. A subscript following the closing parenthesis multiplies everything within the parentheses. For example, a formula unit of $Ba(NO_3)_2$ contains one barium atom, two nitrogen atoms, and six oxygen atoms. Formulas for hydrates have a centered dot preceding a number and the formula for water, such as $CuSO_4 \cdot 5H_2O$. The number multiplies everything following it to the end of the formula. (Section 5.1)

Atoms of main group elements tend to accept, donate, or share electrons to achieve the electronic structure of the nearest noble gas. Metal atoms tend to donate electrons and thereby become positive ions. When combining with metals, nonmetal atoms tend to accept electrons and become negative ions. The number of electrons donated or accepted by each atom depends to a great extent on the periodic group number; each atom tends to attain a noble gas configuration. The attraction of oppositely charged ions is called an ionic bond. Transition and inner transition metal atoms donate their valence electrons first but ordinarily do not achieve noble gas configurations. Most of them can also lose electrons from an inner shell and thus can form two cations with different charges. (Section 5.2)

Formulas for ionic compounds may be deduced from the charges on the ions, since all compounds have zero net charge. Given the constituent elements, you can predict the formula for binary compounds of most main group metals. You cannot do so for most transition metals because of their ability to form ions of different charges. (Given the specific ions, you can write a formula for any ionic compound.) Conversely, given the formula of an ionic compound, you can deduce the charges on its ions. Writing correct formulas for compounds and identifying the ions in compounds from their formulas are two absolutely essential skills. (Section 5.3)

Electron dot diagrams can be drawn for atoms, ions, and molecules, using a dot to represent each valence electron. These diagrams are most useful for main group elements. The diagrams help in visualizing simple reactions and structures of polyatomic ions and molecules. (Section 5.4)

Nonmetal atoms can share electrons with other nonmetal atoms, forming covalent bonds. For electron dot diagrams, the shared electrons are counted as being in the outermost shell of *each* of the bonded atoms. A single bond consists of one shared electron pair; a double bond consists of two shared electron pairs; a triple bond consists of three shared electron pairs. Macromolecules result from covalent bonding of millions of atoms or more into giant molecules.

Drawing electron dot diagrams for structures containing only atoms that obey the octet rule can be eased by subtracting the number of valence electrons *available* from the number *required* to get an octet (or duet) around each nonmetal atom. The difference is the number of electrons to be shared in the covalent bonds. For an ion, you must subtract one available electron for each positive charge on the ion or add one available electron for each negative charge. Main group metal ions generally require no outermost electrons, but each hydrogen atom requires two, and each other nonmetal atom requires eight. Some atoms do not follow the octet rule. (Section 5.5)

Items for Special Attention

■ Since formulas are used to represent unbonded atoms, covalently bonded molecules (Section 5.5), and ionically bonded compounds (Section 5.2), a formula unit can represent an atom, a molecule, or the simplest unit of an ionic compound (Figure 5.7). For example, Ne represents an uncombined atom; Cl_2 represents a molecule of an element; CO_2 represents a molecule of a compound; and NaCl represents one pair of ions in an ionic compound.

■ The seven elements that occur in the form of diatomic molecules (Figure 5.3) form such molecules *only when these elements are uncombined with other elements.* When combined in compounds, they may have one, two, three, four, or more atoms per formula unit, depending on the compound.

■ Learning and using a generalization is easier than memorizing individual facts. For example, if you learn that *in their compounds,* the metals of periodic groups IA and IIA (1 and 2) form ions with charges equal to their group numbers (1+ and 2+, respectively), you do not have to learn the charges on twelve separate metal ions.

■ All compounds, whether ionic or covalent, are electrically neutral. The total positive charge on the cations of an ionic compound must therefore be balanced by the total negative charge on the anions.

■ An ion is a charged species. A single ion is just *part* of a compound. The charges on ions are integral parts of the formulas of the ions. You should always include the charges when you write ions alone. For example, it makes quite a difference whether you are referring to SO_3 (sulfur trioxide) or SO_3^{2-} (sulfite ion). Writing the symbol or formula for a single ion does not imply that it can exist alone, but only that the ion of opposite charge is not of immediate importance. You may write the charges on both ions in an ionic compound while you are determining the compound's formula, but never write the charge on one ion without writing the charge on the other. To finish the formula, rewrite it without the charges. For example, write NaCl for Na^+Cl^-.

■ Except for H^-, monatomic anions have charges equal to their classical group number minus 8. (Not all nonmetals form monatomic ions.) Polyatomic anions containing oxygen and another element are not quite that easy to predict charges for, but generally, the charge is *odd* if the periodic group of the central element is odd and *even* if the periodic group of the central element is even.

■ The great majority of ionic compounds are composed of only one type of cation and one type of anion.

■ Electron dot diagrams are most useful for main group elements, and the systematic procedure for drawing electron dot diagrams works only for species in which all atoms obey the octet rule.

■ Nonmetal atoms accept electrons from metal atoms if the metal atoms are available, or else they share electrons; they never donate electrons to form monatomic cations. The largest magnitude of charge on monatomic cations is 4+, and on monatomic anions, it is 3−.

■ The terms *single bond, double bond,* and *triple bond* refer to covalent bonds only.

■ Polyatomic ions are held together by covalent bonds and are attracted to oppositely charged ions by ionic bonds.

■ After a little experience, you should recognize the monatomic and polyatomic ions introduced in this chapter. For example, every time the symbol for an alkali metal or an alkaline earth metal appears *in a compound,* it represents the ion with a charge equal to 1+ or 2+, respectively.

■ Both *chloride ion* and *chloride* can be used to refer to the Cl^- ion. To refer to Na^+, however, you must always include the word *ion* because *sodium* can refer to the element, the atom, or the ion.

Self-Tutorial Problems

5.1 What is the difference between $CoBr_2$ and $COBr_2$?

5.2 Which of the following have ionic bonds, and which have covalent bonds?

    (a) Cl_2 (b) $MgCl_2$ (c) SCl_2

5.3 (a) Distinguish between *diatomic* and *binary.*

    (b) Distinguish between *valence shell* and *outermost shell.*

5.4 How many electrons are "available" to draw the electron dot diagram of Cl^-? Where do they come from?

5.5 (a) What is the charge on an aluminum atom?

    (b) What is the charge on an aluminum ion?

    (c) What is the charge on an aluminum nucleus?

5.6 (a) What is the difference between group IA *metals* and group IA *elements?*

    (b) Which of the following statements is correct?

      Group IA metals form ions with a 1+ charge only.

      Group IA elements form ions with a 1+ charge only.

5.7 (a) Is the electron dot diagram of H^+ like that of any noble gas?

    (b) Is that of H^-?

5.8 (a) Which metals form ions of only one charge?

    (b) Which metals form ions of 1+ charge?

5.9 What is the charge on (a) the potassium ion, (b) the aluminum ion, (c) the bromide ion, and (d) the nitride ion?

5.10 In which classical periodic groups are the atoms' valence electrons equal in number to the group number?

5.11 (a) Write the formula of the compound of Ni^{2+} and Cl^-.

    (b) Identify the ions present in $NiBr_2$.

    (c) Write the formula of the compound of Zn^{2+} and O^{2-}.

    (d) Identify the ions in ZnS.

5.12 What is the charge on calcium in each of the following?

    (a) $CaCl_2$ (b) $CaCr_2O_7$ (c) Ca

    (d) CaO (e) $Ca_3(PO_4)_2$

5.13 Determine the formula of each of the following compounds:

    (a) The compound of sodium and bromine

    (b) The compound of Na^+ and Br^-

    (c) The product of the reaction of sodium and bromine

    (d) Sodium bromide

5.14 What is the difference between BrO_2 and BrO_2^-?

5.15 Identify the type of bonding in each of the following:

    (a) K_2S (b) SO_2 (c) S_4N_4

5.16 Write the formulas for ammonia and for the ammonium ion.

5.17 Draw an electron dot diagram for each of the following:

    (a) Mg (b) Mg^{2+} (c) S (d) S^{2-}

5.18 Draw electron dot diagrams for (a) NaH and (b) CaH_2.

5.19 What is a valid generalization about the charges on monatomic anions? What is a valid generalization about the charges on polyatomic anions containing oxygen and another element?

5.20 Draw an electron dot diagram for each of the following. Since these species all have the same number of electrons, explain why the diagrams are not all the same.

    (a) H^- (b) Li^+ (c) He

5.21 (a) How many valence electrons, if any, are in a magnesium ion?

    (b) How many electrons, if any, should a magnesium atom share in its compounds?

5.22 (a) Write the formula of the compound of Ag^+ and S^{2-}.

    (b) Identify the ions present in Ag_2O.

■ ■ ■ PROBLEMS ■ ■ ■

5.1 Chemical Formulas

5.23 What information (from Section 5.1) is conveyed by the formula $Al_2(SO_3)_3$?

5.24 What is implied about bonding in the mercury(I) ion, Hg_2^{2+}?

5.25 How many atoms of each element are present in one formula unit of each of the following?

(a) Na_3As

(b) $Al(ClO_4)_3$

(c) $(NH_4)_4P_2O_7$

5.26 How many atoms of each element are present in one formula unit of each of the following?

(a) $UO_2(ClO_3)_2$

(b) $Co(NO_3)_3$

(c) $Cr(C_2H_3O_2)_3$

(d) $KHCO_3 \cdot MgCO_3 \cdot 4H_2O$

5.2 Ionic Bonding

5.27 Complete the following table:

	Symbol	Atomic Number	No. of Protons	No. of Electrons	Net Charge
(a)	Na^+	____	____	____	____
(b)	____	16	____	18	____
(c)	____	____	7	10	____
(d)	____	20	____	____	2+
(e)	____	____	4	____	2+

5.28 Complete the following table:

	Symbol	Atomic Number	No. of Protons	No. of Electrons	Net Charge
(a)	O^{2-}	____	____	____	____
(b)	____	27	____	24	____
(c)	____	____	13	10	____
(d)	____	53	____	____	1−
(e)	____	____	25	____	2+

5.29 What difference, if any, is there between Na^+Cl^- and NaCl?

5.30 (a) Which metals form ions with 1+ charge? (b) Which metals form ions with 1+ charge only?

5.31 (optional) Write a detailed electronic configuration for each of the following ions:

(a) Br^- (b) N^{3-} (c) S^{2-}

5.32 (optional) Write a detailed electronic configuration for each of the following ions:

(a) Li^+ (b) Mg^{2+} (c) Al^{3+}

5.33 (optional) Write a detailed electronic configuration for each of the following ions:

(a) Co^{2+} (b) Cd^{2+} (c) Cr^{2+}

5.3 Formulas for Ionic Compounds

5.34 Write the formula for the compound formed between each of the following:

(a) Li and H   (b) Ca and H.

5.35 Iron forms ions of 2+ and 3+ charges. Write formulas for (a) two chlorides of iron and (b) two oxides of iron.

5.36 Write the formula for the compound formed between each of the following pairs of ions:

(a) S^{2-} and K^+

(b) Cu^+ and S^{2-}

(c) Cu^{2+} and S^{2-}

5.37 Write the formula for the ion formed by each of the following metals in all of its compounds:

(a) Potassium   (b) Cadmium   (c) Lithium

(d) Barium   (e) Aluminum

5.38 Complete the following table by writing the formula of the compound formed by the cation on the left and the anion at the top:

	I^-	S^{2-}	P^{3-}
Na^+	____	____	____
Ba^{2+}	____	____	____
Al^{3+}	____	____	____

5.39 Complete the following table by writing the formula of the compound formed by the cation on the left and the anion at the top:

	ClO_3^-	SO_4^{2-}	PO_4^{3-}
Ag^+	____	____	____
Co^{2+}	____	____	____
Cr^{3+}	____	____	____
NH_4^+	____	____	____

5.40 Identify the individual ions in each of the following compounds:

(a) KBr   (b) CaO   (c) Li_2S

(d) Mg_3N_2 (e) Na_3P (f) Li_3N

(g) CrF_3

5.41 Identify the individual ions in each of the following compounds:

(a) $CaCl_2$ (b) LiH   (c) $(NH_4)_2SO_4$

(d) $NaClO_2$ (e) $Ba(ClO)_2$ (f) Hg_2Cl_2

(g) KO_2

5.42 For each of the following compounds, identify the individual ions, and indicate how many of each are present per formula unit:

    (a) NaBr   (b) BaS   (c) $BaBr_2$ (d) Na_2S

5.43 How many valence electrons does an Sn^{2+} ion have?

5.44 What individual ions are present in (a) Cu_2O and (b) CuO?

5.45 Identify the anion and *both* cations in each of the following pairs of compounds:

    (a) $CrCl_2$ and $CrCl_3$

    (b) SnO and SnO_2

    (c) NiO and NiO_2

5.46 Complete the following table by writing the formula of the compound formed by the metal at the left and the nonmetal at the top:

	Chlorine	Sulfur	Phosphorus
Potassium	————	————	————
Magnesium	————	————	————
Aluminum	————	————	————

5.47 Write the formula for the compound formed by each of the following pairs of elements:

    (a) Na and P   (b) Mg and O   (c) Ca and Cl

    (d) Li and N   (e) Mg and N   (f) Zn and Br

    (g) Al and S   (h) Ca and P   (i) Al and I

5.48 Write the formula for the compound formed by each of the following pairs of elements:

    (a) Magnesium and selenium

    (b) Barium and sulfur

    (c) Lithium and bromine

    (d) Silver and oxygen

    (e) Potassium and fluorine

    (f) Aluminum and oxygen

5.4 Electron Dot Diagrams

5.49 Draw an electron dot diagram for each of the following ions:

    (a) P^{3-} (b) S^{2-} (c) Br^-

5.50 Draw an electron dot diagram for each of the following ions:

    (a) K^+ (b) Al^{3+} (c) Ca^{2+} (d) Pb^{2+}

5.51 Draw electron dot diagrams for atoms of the following elements and the ions they produce when they combine:

    (a) Li and N   (b) Ba and Br   (c) Al and S

5.52 Draw electron dot diagrams for atoms of the following elements and the ions they produce when they combine:

    (a) Li and H   (b) Be and H   (c) Al and H

5.5 Covalent Bonding

5.53 Which of the following involve ionic bonding only, which involve covalent bonding only, and which involve both?

    (a) $(NH_4)_2SO_4$ (b) $MgCl_2$ (c) $Ca(ClO_4)_2$

    (d) I_2 (e) PCl_3 (f) CF_4

5.54 What similarities and differences are there between a molecule and a polyatomic ion?

5.55 Draw an electron dot diagram for each of the following:

    (a) N_2 (b) PBr_3 (c) CCl_4

    (d) NH_3 (e) $COBr_2$ (carbon is the central atom)

    (f) SO_3

5.56 Draw an electron dot diagram for each of the following:

    (a) Cl_2 (b) SCl_2 (c) $CaCl_2$

5.57 Explain why a hydrogen atom cannot be bonded with a double bond or to two other atoms at the same time.

5.58 Draw a structural formula for each of the following:

    (a) PI_3 (b) HCN

    (c) SO_2Cl_2 (sulfur is the central atom)

5.59 Draw an electron dot diagram for each of the following compounds. Indicate any double or triple bonds.

    (a) C_3H_8 (b) C_3H_6 (c) C_3H_4

5.60 Draw an electron dot diagram for each of the following compounds. In the first four compounds, the hydrogen atom is bonded to an oxygen atom.

    (a) $HBrO_4$ (b) $HBrO_3$ (c) $HBrO_2$

    (d) HBrO   (e) HBr

5.61 For each of the following compounds, identify the individual ions, and indicate how many of each are present per formula unit:

    (a) NaCl   (b) MgO   (c) Na_2SO_3

    (d) $Sr_3(PO_4)_2$ (e) $(NH_4)_3PO_4$ (f) Mg_3N_2

    (g) CrF_3

5.62 Draw structural formulas for the compounds in Problem 5.60.

5.63 Draw structural formulas for the compounds in Problem 5.59.

5.64 Draw an electron dot diagram for HOCN, in which the atoms are bonded in that order.

5.65 Complete the following table by writing the formula of the compound formed by each cation on the left with each anion at the top:

	NO_3^-	SO_4^{2-}	CO_3^{2-}	PO_4^{3-}
K^+	———	———	———	———
Mg^{2+}	———	———	———	———
Fe^{2+}	———	———	———	———
Fe^{3+}	———	———	———	———

5.66 Complete the following table by writing the formula of the compound formed by each cation on the left with each anion at the top:

	NO_3^-	SO_4^{2-}	CO_3^{2-}	PO_4^{3-}
Ag^+	____	____	____	____
Ca^{2+}	____	____	____	____
Al^{3+}	____	____	____	____

5.67 Identify the cation and the anion in each of the following compounds:

(a) $Na_2Cr_2O_7$ (b) $La(OH)_3$ (c) Li_2SO_4

(d) $VOSO_4$ (e) $LiNO_3$ (f) Na_3PO_4

(g) $KHCO_3$ (h) NH_4BrO_3 (i) $(NH_4)_2SO_4$

5.68 Write formulas for the ions in each of the following compounds:

(a) $Ca(C_2H_3O_2)_2$ (b) $KMnO_4$ (c) $KSCN$

(d) $(NH_4)_2Cr_2O_7$ (e) Na_2CrO_4 (f) $BaSeO_4$

(g) NH_4CN (h) $Sr(OH)_2$

5.69 What familiar ion do you recognize in each of the following compounds? Also write the formula for the other ion present.

(a) $Ba(NO_2)_2$ (b) $(NH_4)_2SO_4$ (c) K_2SO_3

5.70 Draw an electron dot diagram for each of the following:

(a) Ozone, O_3

(b) Rhombic sulfur, S_8 (Hint: See Figure 5.2.)

5.71 Write the formula for the compound of each of the following pairs of ions:

(a) S^{2-} and Li^+

(b) PO_4^{3-} and NH_4^+

(c) CO_3^{2-} and Cr^{3+}

5.72 Identify the ions in each of the following:

(a) $Ce_3(PO_4)_4$

(b) $(VO)_3(PO_4)_2$

5.73 Write the formula of both the familiar ion and the unfamiliar ion in each of the following compounds:

(a) $BaMnO_4$ (b) $Ce(SO_4)_2$

(c) $K_2S_2O_3$ (d) $Ba(OCN)_2$

(e) $(VO_2)_2SO_4$ (f) $K_4P_2O_7$

(g) $UO_2(NO_3)_2$ (h) $(CH_3NH_3)_2SO_4$

5.74 Write the formulas of the ions present in each of the following compounds:

(a) BaO (b) $AlPO_4$ (c) $NaClO_4$

■ ■ ■ GENERAL PROBLEMS ■ ■ ■

5.75 Draw an electron dot diagram for each of the following:

(a) $NaOCN$ (b) CH_3NH_2 (c) CH_3CH_2OH

5.76 Draw an electron dot diagram for SCN^-, in which the carbon atom is the central atom.

5.77 (optional) State the octet rule in terms of detailed electronic configurations.

5.78 Try to draw an electron dot diagram for the "ammonium molecule," NH_4, which does not exist. What do you find? Draw an electron dot diagram for the ammonium ion, NH_4^+.

5.79 Draw electron dot diagrams for (a) $CoBr_2$ and (b) $COBr_2$.

5.80 Briefly define each of the following terms:

(a) Ion   (b) Cation   (c) Monatomic ion

(d) Ozone   (e) Noble gas configuration

(f) Triple bond   (g) Lone pair   (h) Octet

5.81 Draw a structural formula for each of the following:

(a) SO_3 (b) SO_3^{2-}

(c) Na_2SO_3 (d) H_2SO_3.

5.82 What is the charge on the only monatomic cation of bismuth?

5.83 What is the difference between SeO_3 and SeO_3^{2-}? Draw an electron dot diagram for each.

5.84 A certain ionic compound contains three oxygen atoms, one calcium atom, and one sulfur atom per formula unit. Identify the ions that make up the compound.

5.85 Complete the following table:

Symbol	Atomic Number	No. of Protons	No. of Electrons	Net Charge
K	____	____	____	____
____	16	____	____	2−
____	____	35	36	____
Se^{2-}	____	____	____	____
____	16	____	18	____
____	____	37	____	1+

5.86 Write the formula for the compound composed of each of the following pairs of ions:

(a) VO_2^+ and HPO_4^{2-}

(b) UO_2^{2+} and AsO_4^{3-}

(c) VO^{2+} and PO_4^{3-}

(d) NH_4^+ and ClO_4^-

5.87 (a) How many valence electrons, if any, are in an ammonium ion?

(b) How many electrons, if any, should that ion share with other ions in its compounds?

5.88 Draw an electron dot diagram for each of the following pairs of elements and their compounds:

(a) Lithium and nitrogen

(b) Barium and bromine

Contrast this problem with Problem 5.51(a and b).

5.89 Complete the following table:

Symbol	Atomic Number	No. of Protons	No. of Electrons	Net Charge
Mn	——	——	——	——
——	7	——	——	3−
——	——	12	10	——
Cr^{2+}	——	——	——	——
——	17	——	18	——
——	——	39	——	3+

5.90 Write formulas for the ions represented in each of the following:

(a) $(VO)_3(PO_4)_2$ (b) $(NH_4)_2HPO_4$ (c) $K_2Cr_2O_7$

5.91 Write formulas for the two new compounds formed if each of the following pairs of compounds traded anions:

(a) NaCl and $AgNO_3$

(b) $BaCl_2$ and $AgNO_3$

(c) $Pb(NO_3)_2$ and Na_2S

(d) $AlCl_3$ and $MgSO_4$

5.92 The formulas that follow represent compounds with ionic bonds only, with X representing one of the main group elements. In each case, state whether X is a metal or nonmetal, and determine to which main group X belongs.

(a) Na_2X (b) XCl_2 (c) XF_4 (d) X_2O_3

5.93 Which of the following have any ionic bonds?

CH_4 Na_2S $(NH_4)_2SO_4$ Pure HCl

5.94 Which one of the following is ionic? $POCl_3$ $PoCl_2$

5.95 In which of the following are there any covalent bonds?

$(NH_4)_2SO_4$ MgI_2 CH_4 Pure HCl

5.96 How many atoms of each element are in one formula unit of (a) $Co_2(CO)_9$ and (b) $(CH_3NH_3)_3PO_4$?

5.97 Write formulas for *both* kinds of ions in each of the following compounds:

(a) Li_3N (b) Na_2SO_4 (c) $KBrO_2$

5.98 List several ions that have two atoms of the same element covalently bonded together.

5.99 (*optional*) (a) Write the outer electronic configuration of thallium (Tl).

(b) On the basis of its configuration, explain why thallium forms both a 1+ ion and a 3+ ion.

5.100 Identify the cation in each of the following compounds:

(a) CuBr   (b) $CoSO_4$ (c) $CrPO_4$

5.101 Draw a structural formula for (a) C_2Cl_4 and (b) CCl_4.

5.102 Draw a structural formula for cyclohexane, C_6H_{12}, in which the six carbon atoms are bonded in a ring and each has two hydrogen atoms bonded to it.

5.103 Relatively speaking, how many atoms are covalently bonded in a diamond crystal?

5.104 (a) How many oxygen atoms are covalently bonded to each silicon atom in SiO_2 (Figure 5.11)?

(b) How many silicon atoms are bonded to each oxygen?

(Hint: Look at the top silicon atom and the oxygen atoms attached below it.)

5.105 Consult Figure 5.11 to determine how many carbon atoms are bonded to a given carbon atom in (a) diamond and (b) graphite.

(Hint: Look in the middle of each figure, not at the edges.)

6 Nomenclature

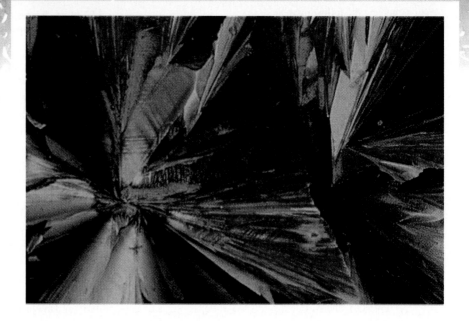

- 6.1 Binary Nonmetal-Nonmetal Compounds
- 6.2 Naming Ionic Compounds
- 6.3 Naming Acids and Acid Salts
- 6.4 Hydrates

■ Key Terms (Key terms are defined in the Glossary.)

acid (6.3)
acid salt (6.3)
ammonia (6.1)
ammonium ion (6.2)
base (6.3)

hydrogen (6.3)
ionizable hydrogen atom (6.3)
nomenclature (intro)
oxoacid (6.3)

oxoanion (6.2)
prefix (6.1)
salt (6.3)
Stock system (6.2)

■ Prefixes/Suffixes

-ate (6.2)
bi- (6.3)
deca- (6.1)
di- (6.1)
hepta- (6.1)
hexa- (6.1)
hydro- (6.3)

hypo- (6.2)
-ic (6.2)
-ic acid (6.3)
-ide (6.1)
-ite (6.2)
mono- (6.1)
nona- (6.1)

octa- (6.1)
-ous (6.2)
-ous acid (6.3)
penta- (6.1)
per- (6.2)
tetra- (6.1)
tri- (6.1)

So far in this book, we have used names for some simple chemical compounds, but we have not yet considered **nomenclature**—how to name compounds systematically. The great variety of compounds requires a systematic approach to naming them. Unfortunately, three or four different naming systems are used to name different types of compounds. Memorization of a few simple rules will allow naming of a great many compounds, but in addition to learning the rules, you must be sure to learn *when to use each one.* Learning generalities will help you to handle great quantities of information and to respond to specific questions.

This chapter covers the basic rules for naming many compounds and ions. Section 6.1 considers the naming of binary nonmetal-nonmetal compounds. The naming of ionic compounds is addressed in Section 6.2. First, the naming of cations and anions is discussed, leading into the naming of complete compounds. Section 6.3 covers the naming of acids and acid salts. Hydrates are considered briefly in Section 6.4. Tables and figures in the chapter summarize how to name compounds in a systematic way.

6.1   Binary Nonmetal-Nonmetal Compounds

OBJECTIVE

■ to name and write formulas for binary compounds of nonmetals

Figure 6.1 Household Ammonia

A solution of gaseous ammonia in water.

Except for compounds of hydrogen, the formulas for compounds of two nonmetals are written and named with the element farther to the left or lower in the periodic table given first. If one element is below and to the right of the other in the periodic table, the one to the left is given first, unless that element is oxygen or fluorine.

Binary compounds of hydrogen that are not acids are given special names. Two very important examples are water, H_2O, and **ammonia,** NH_3 (Figure 6.1). (Other much less important hydrogen-containing binary compounds are also known by common names. These include phosphine, PH_3, and arsine, AsH_3.) Hydrogen compounds that are acids in aqueous solution are named, and their formulas are written in special ways (see Section 6.3).

■ **EXAMPLE 6.1**

Which element is named first in a binary compound of each of the following pairs of elements?

(a) N and O   (b) P and N   (c) Se and I

(d) O and I   (e) F and Xe

Solution

The positions of the elements in the periodic table are used to determine the order of naming.

(a) Since nitrogen lies to the left of oxygen in the periodic table (in the same period), nitrogen is named first.

(b) Since phosphorus lies below nitrogen in the periodic table (in the same group), phosphorus is named first.

(c) Since selenium lies to the left of iodine in the periodic table, selenium is named first (despite being above iodine).

(d) Even though oxygen lies to the left of iodine in the periodic table, iodine is named first. Oxygen is an exception to the rule that a position toward the left is more important than a position lower in the table.

Table 6.1 Prefixes Used in Naming Binary Nonmetal-Nonmetal Compounds

Number of Atoms	Prefix
1	mono-
2	di-
3	tri-
4	tetra-
5	penta-
6	hexa-
7	hepta-
8	octa-
9	nona-
10	deca-

(e) Even though fluorine lies to the left of xenon in the periodic table, xenon is named first.

Practice Problem 6.1   Which element has its symbol written first in a binary compound of each of the following pairs of elements?

(a) F and I   (b) O and Xe   (c) F and O ■

Naming a binary compound of two nonmetals involves naming the first element and then using the root of the name of the second element with its ending changed to *-ide*. In addition, a **prefix** is used before the name of the *second* element to tell how many atoms of that element are present in each molecule of the compound. These prefixes are given in Table 6.1. When the name of the second element starts with a vowel and the prefix ends in *a* or *o*, that letter is usually dropped.

■ **EXAMPLE 6.2**

Write the name of each of the following compounds:

(a) BrF (b) CO (c) SO_2 (d) SO_3 (e) $SiCl_4$
(f) PF_5 (g) SCl_6 (h) XeO_2 (i) XeF_4 (j) BF_3

Solution

(a) Bromine monofluoride (Bromine is written first because it lies below fluorine in the periodic table. The ending of *fluorine* is changed to *-ide*. The prefix *mono-* is added to *fluoride* to show that only one fluorine atom is present in the molecule.)

(b) Carbon monoxide (Carbon is named first because it lies to the left of oxygen. The last *o* of the prefix *mono-* is dropped because the second element's name starts with *o*.)

(c) Sulfur dioxide   (d) Sulfur trioxide

(e) Silicon tetrachloride   (f) Phosphorus pentafluoride

(g) Sulfur hexachloride   (h) Xenon dioxide

(i) Xenon tetrafluoride   (j) Boron trifluoride

Practice Problem 6.2   Write the name of each of the following compounds:

(a) IF_5 (b) ICl_3 (c) SF_4 (d) PCl_5 (e) $AsBr_3$ (f) SeO_2 ■

For molecules that have more than one atom of the first element, such as P_4O_{10}, the prefixes from Table 6.1 can be used for both elements, except that the prefix *mono-* is not used for the first element of binary nonmetal-nonmetal compounds. The name for P_4O_{10} is tetraphosphorus decoxide. Classically, this compound was written P_2O_5 and named phosphorus pentoxide (without the prefix *di-*). Chemists know that two phosphorus atoms are needed for every five oxygen atoms, for reasons that will be presented in Chapter 16.

■ **EXAMPLE 6.3**

Name   (a) Cl_2O_3 (b) N_2O_4 (c) B_2O_3 (d) As_4O_{10}

Solution

(a) Dichlorine trioxide   (b) Dinitrogen tetroxide

(c) Diboron trioxide   (d) Tetraarsenic decoxide

Practice Problem 6.3   Write formulas for (a) tetrasulfur tetrafluoride and (b) dinitrogen trioxide. ■

6.2   Naming Ionic Compounds

OBJECTIVE

■ to name cations and anions and to name and write formulas for ionic compounds

In most cases, naming ionic compounds involves simply naming both ions. A huge majority of ionic compounds are made up of one type of cation plus one type of anion. (Alums are an exception.) Thus, to name most ionic compounds, we name the cation first and then the anion. The more difficult part of the process is learning to name cations and anions themselves.

The charges on the ions allow us to deduce the formula from the name of a compound, even though the numbers of each type of ion are not stated in the name. Writing formulas for ionic compounds requires deducing how many of each type of ion must be present to have a neutral compound.

Naming Cations

You learned in Chapter 5 that some metals always form monatomic ions having one given charge in all their compounds. In this book, we will call this type of ion the constant type. Other metals form monatomic ions with different charges (Figure 5.10). We will call this type the variable type. There are also some polyatomic cations, but only a few of these are important for this course. They are presented in Table 6.2. Thus, the first step in naming a cation is to decide which of these three types it is—polyatomic, constant type, or variable type. We name them in different ways.

Polyatomic cations are named as shown in Table 6.2 or a similar more extensive table from a more advanced text. The **ammonium ion** (NH_4^+) is very important, the hydronium ion (H_3O^+) is important in discussing acid-base equilibrium (Chapter 18), and the mercury(I) ion (Hg_2^{2+}) is fairly important.

Naming the constant type of cation involves naming the element and adding the word *ion*, unless a compound is being named. For example, Na^+ is the sodium ion, and Mg^{2+} is the magnesium ion; NaCl is sodium chloride. The alkali metals, the alkaline earth metals, zinc, cadmium, aluminum, and silver are the most important metals that form ions of the constant type. Each of these metals forms the same ion in any of its compounds, and the charge on the ion is equal to the classical periodic group number.

Naming ions of metals that form ions of more than one charge requires distinguishing between the possibilities. For example, iron forms Fe^{2+} and Fe^{3+} ions. We cannot call both of these "iron ion" because no one would know which of the two we meant. For *monatomic* cations of variable type, the *charge* in the form of a Roman numeral is attached to the element's name to indicate which ion we are talking about. For example, Fe^{2+} is called iron(II) ion and Fe^{3+} is called iron(III) ion. This system of nomenclature is called the **Stock system.**

Table 6.2   Names of Important Polyatomic Cations

Formula	Name
NH_4^+	Ammonium ion
H_3O^+	Hydronium ion
Hg_2^{2+}	Mercury(I) ion

CHAPTER 6 ■ Nomenclature

■ EXAMPLE 6.4

Name Cu^+ and Cu^{2+}.

Solution

Copper(I) ion and copper(II) ion, respectively.

■ EXAMPLE 6.5

How can you tell whether the copper ion in the compound CuCl has a 1+ or 2+ charge?

Solution

You know that the chloride ion has a 1− charge; the copper ion must have a 1+ charge to balance that charge and make the compound neutral. (See Section 5.3.)

Practice Problem 6.5 What is the charge on the anion in $(NH_4)_2MoO_4$? ■

■ EXAMPLE 6.6

Name (a) $CaCl_2$, (b) SCl_2, and (c) $CoCl_2$.

Solution

(a) Calcium chloride   (b) Sulfur dichloride

(c) Cobalt(II) chloride

Each of these compounds, whose formulas look so similar, is named in a different way. Calcium always forms a 2+ ion in its compounds, so the name *calcium ion* is sufficient for it. Sulfur forms a binary nonmetal-nonmetal compound with chlorine, so the second element is named with the prefix *di* (Section 6.1). Cobalt forms both Co^{2+} and Co^{3+} (Figure 5.10), so the name must have a Roman numeral to distinguish which of the two is present. Note that not only is it necessary to remember the rules for the different types of compounds but, just as important, *when to use each rule!*

> Not only is it necessary to remember the rules for the different types of compounds, but just as important, *when to use each rule!*

Practice Problem 6.6 Name (a) BCl_3, (b) $CrCl_3$, and (c) $AlCl_3$. ■

■ EXAMPLE 6.7

Name each of the following compounds:
(a) Cu_2S (b) CuS

Solution

(a) Copper(I) sulfide   (b) Copper(II) sulfide

In Cu_2S, the two copper ions are balanced by one sulfide ion with a 2− charge; the charge on each copper ion must be 1+. In CuS, only one copper ion is present to balance the 2− charge on the sulfide ion; the charge on the copper ion is 2+. Note that the Roman numerals in the names of monatomic cations

Table 6.3 Classical Names
of Some Common Cations

Periodic Group	Ion of Lower Charge	Ion of Higher Charge
VIB	Cr^{2+}, chromous	Cr^{3+}, chromic
VIIB	Mn^{2+}, manganous	Mn^{3+}, manganic
VIII	Fe^{2+}, ferrous	Fe^{3+}, ferric
VIII	Co^{2+}, cobaltous	Co^{3+}, cobaltic
VIII	Ni^{2+}, nickelous	Ni^{4+}, nickelic
IB	Cu^{+}, cuprous	Cu^{2+}, cupric
IB	Au^{+}, aurous	Au^{3+}, auric
IIB	Hg_2^{2+}, mercurous	Hg^{2+}, mercuric
IVA	Sn^{2+}, stannous	Sn^{4+}, stannic
IVA	Pb^{2+}, plumbous	Pb^{4+}, plumbic

denote the *charges on the ions.* The Arabic numerals appearing as subscripts in formulas denote the *number of atoms* of that element present per formula unit. Either of these numbers can be used to deduce the other, but they are not the same!

Practice Problem 6.7   Write the formula for (a) nickel(II) oxide and (b) nickel(IV) oxide. ■

An older nomenclature system (known as the classical system) uses suffixes to distinguish metal ions of the variable type. As Figure 5.10 shows, there are, at most, two possible monatomic cations for each metal listed. The ion with the higher charge is named with the ending changed to *-ic.* The ion of lower charge has its ending changed to *-ous.* For example, Ni^{2+} is called nickelous ion, and Ni^{4+} is called nickelic ion. For many elements, the Latin names are used instead of the English names. For example, Fe^{3+} is called ferric ion—from *ferrum,* the Latin for iron. Table 6.3 lists classical names for some important monatomic cations. This older system is more difficult to use in two ways: (1) you must remember the other possible charge on an ion in addition to the one given, and (2) you must remember a Latin name for many of the elements.

■ EXAMPLE 6.8

Name V^{3+} using the Stock system. Explain why use of the classical system would be harder.

Solution

The Stock system name—vanadium(III) ion—is easy. To use the classical system, you must know the answers to at least three questions: (1) What is the charge on the other monatomic ion of vanadium? (2) Is the Latin name for vanadium used in the classical system? (3) If the Latin name is used, what is that name? The Stock system was invented to make naming easier.

Practice Problem 6.8   Name Mn^{3+}. ■

Table 6.4　Names of Some Important Oxoanions*

Hypo ___ ite (Two Fewer Oxygen Atoms)	___ ite (One Fewer Oxygen Atom)	___ ate	Per ___ ate (One More Oxygen Atom)
ClO^-	ClO_2^-	ClO_3^-	ClO_4^-
BrO^-	BrO_2^-	BrO_3^-	BrO_4^-
IO^-	IO_2^-	IO_3^-	IO_4^-
PO_2^{3-}	PO_3^{3-}	PO_4^{3-}	
	NO_2^-	NO_3^-	
	SO_3^{2-}	SO_4^{2-}	
		CO_3^{2-}	

*The ions do not exist where there are spaces in the table.

Naming Anions

Just as for cations, there are three types of anions for naming purposes. Monatomic anions are easy to name. A second type, **oxoanions,** are anions that contain oxygen covalently bonded to another element. Table 6.4 presents some important oxoanions in a format designed to make their names easier to learn. Several other important anions, referred to as special anions in this book, are listed in Table 6.5.

All monatomic anions are named by changing the ending of the element's name to *-ide*. For example, Cl^-, H^-, and O^{2-} are called chloride ion, hydride ion, and oxide ion, respectively. (The names of a few special anions also end in *-ide;* among the most important are hydroxide and cyanide ions, listed in Table 6.5.) The charge on any monatomic anion is constant and, except for that on H^-, equal to the classical group number minus 8.

In many important anions, oxygen atoms are covalently bonded to a central atom. These ions have extra electrons from some source, which give them their negative charges. They are called oxoanions but were formerly known as oxyanions. For the seven most important oxoanions, the name is that of the central element with the ending changed to *-ate*. They are listed in the third column of Table 6.4. Once you learn the names and formulas of these ions, you can deduce the formulas of the corresponding ions with fewer or more oxygen atoms. Ions ending in *-ite* have one fewer oxygen atom than the corresponding *-ate* ions. In some cases, removal of two oxygen atoms from an ion ending in *-ate* results in an ion named with the prefix *hypo-* and the ending *-ite*. Addition of one oxygen atom to an ion with the ending *-ate* yields an ion named with the prefix *per-* and the ending *-ate*. Note in Table 6.4 that all the ions with a given central atom have the same charge. Note also that the charges are all odd for ions with a central element from an odd-numbered periodic group and all even for ions with a central element from an even-numbered periodic group.

Table 6.5　Names of Special Anions

Formula	Name
OH^-	Hydroxide
CN^-	Cyanide
O_2^{2-}	Peroxide
CrO_4^{2-}	Chromate
$Cr_2O_7^{2-}$	Dichromate
MnO_4^-	Permanganate
$C_2H_3O_2^-$	Acetate

■ **EXAMPLE 6.9**

Name Br^-, BrO_3^-, BrO_2^-, BrO^-, and BrO_4^-.

Solution

The names are bromide ion, bromate ion, bromite ion, hypobromite ion, and perbromate ion.

ITEM OF INTEREST

■ ■ ■ ■ ■ ■
The differences in the names of compounds can be life-and-death details. For example, physicians sometimes prescribe a barium sulfate slurry or barium sulfate enema for patients who are about to have stomach or intestinal X rays. The barium sulfate is opaque to X rays and outlines the stomach or colon clearly (Figure 6.2). However, barium ion is poisonous to humans. Barium sulfate is safe only because it is too insoluble to be harmful. However, if barium *sulfite* were given instead of barium *sulfate*, the compound would dissolve in the stomach or colon, and the patient might die. The one-letter difference in the name is critical.

Figure 6.2 Stomach X Ray
Barium sulfate, which is not soluble in water, is administered to humans to absorb X rays and outline organs. Barium salts that are soluble are poisonous.

Practice Problem 6.9 Write the formula for (a) iodite ion and (b) periodate ion. ■

■ EXAMPLE 6.10

What are the formulas for (a) sulfate ion and (b) bromate ion?

Solution

The formulas are (a) SO_4^{2-} and (b) BrO_3^{-}.

Practice Problem 6.10 Write the formula for each of the following ions:

(a) Nitrate ion   (b) Sulfite ion   (c) Phosphate ion   (d) Carbonate ion ■

Names for anions that contain oxygen but are not included in Table 6.4 may sometimes be determined because of a periodic relationship between their central element and that of an ion in that table. For example, MnO_4^{-} is analogous to ClO_4^{-} because both central elements are in periodic groups numbered VII. Its name is permanganate, which is analogous to perchlorate. Similarly, CrO_4^{2-} and SO_4^{2-} both have central atoms that are in periodic groups numbered VI. The name of CrO_4^{2-} is chromate, analogous to sulfate. (Not all such analogies are valid, however.)

■ EXAMPLE 6.11

Name AsO_4^{3-}.

Solution

Arsenic is just below phosphorus in the periodic table. You can guess that the AsO_4^{3-} ion is named analogously to the PO_4^{3-} ion. The name is arsenate ion.

Practice Problem 6.11 Name SiO_3^{2-}. ■

Naming and Writing Formulas for Ionic Compounds

Naming ionic compounds involves first naming the cation and then naming the anion. Therefore, the name of such a compound leads directly to its formula.

■ **EXAMPLE 6.12**

Name $AgClO_3$.

Solution

The cation is Ag^+, and the anion is ClO_3^-. The name of the compound is silver chlorate.

Practice Problem 6.12   Name $MgSO_4$. ■

■ **EXAMPLE 6.13**

Name (a) Cu_2S and (b) $(NH_4)_2SO_4$.

Solution

(a) Each cation is Cu^+; the anion is S^{2-}. It is important to recognize that each cation is a monatomic ion and that the two together do not make up a different ion. The compound is copper(I) sulfide. Note that the name of the compound does not explicitly mention that there are two copper(I) ions per sulfide ion.

(b) Each of the cations is NH_4^+, the ammonium ion (Table 6.2); the anion is the sulfate ion. The compound is ammonium sulfate.

Practice Problem 6.13   Name (a) $Cr(C_2H_3O_2)_3$ and (b) $Ca(ClO_3)_2$. ■

■ **EXAMPLE 6.14**

Write the formula for (a) lithium phosphate and (b) nickel(II) hypochlorite.

Solution

(a) The lithium ion is Li^+; the phosphate ion is PO_4^{3-}. The formula of the compound must balance positive charges and negative charges; it is Li_3PO_4.

(b) The nickel(II) ion is Ni^{2+}; the hypochlorite ion is ClO^-. The compound is $Ni(ClO)_2$. Parentheses are needed around the formula for the hypochlorite ion so that the subscript 2 indicates that two such ions are present. (If the parentheses were not written, the formula would appear to contain a chlorite ion, ClO_2^-.)

Practice Problem 6.14   Write the formula for (a) titanium(III) fluoride and (b) aluminum sulfate. ■

6.3 Naming Acids and Acid Salts

> **OBJECTIVE**
>
> ■ to name and write formulas for acids and acid salts

Acids are a special group of hydrogen-containing compounds whose properties will be covered more fully in Chapter 8. One of their most important properties is their reaction with **bases** to form **salts** (Section 8.4). Pure acids are covalent compounds, but they react to varying extents with water to form ions in solution. The hydrogen atoms that react with water to form ions are said to be **ionizable hydrogen atoms.** The formulas of acids have the ionizable hydrogen atoms written first. Thus, HCl is an acid with one ionizable hydrogen atom per molecule, and H_2SO_4 is an acid with two ionizable hydrogen atoms per molecule. However, CH_4 is not an acid at all, and NH_3 acts as a base in aqueous solution. In other words, the appearance of hydrogen first in a formula is not based on hydrogen's relative position in the periodic table, as is true for other elements, but only on whether the compound is an acid.

■ EXAMPLE 6.15

How many hydrogen atoms per molecule of acetic acid, $HC_2H_3O_2$, are ionizable?

Solution

One, represented by the first H, is ionizable. The other three hydrogen atoms of this compound are not ionizable, which is why they are written after the carbon atoms in the formula.

Practice Problem 6.15   How many hydrogen atoms per molecule of sebacic acid, $H_2C_{10}H_{16}O_4$, are ionizable? ■

Naming Acids

There are two types of acids for naming purposes: (1) binary compounds containing hydrogen and another nonmetal, and (2) **oxoacids** containing hydrogen, oxygen, and another nonmetal. The two types are named in different ways. The names of the oxoacids corresponding to the oxoanions in Table 6.4 can be remembered easily, since the ending -*ate* for an anion is simply changed to -*ic acid,* or the ending -*ite* for an anion is changed to -*ous acid.* (Sometimes, the stem is changed slightly.) If the anion has the prefix *hypo-* or *per-,* so does the acid. Binary hydrogen compounds that are acids include HF, HCl, HBr, HI, and H_2S. These compounds may be named by the usual rules for binary compounds; for example, HCl is hydrogen chloride. When these compounds are dissolved in water, they are usually named as acids. For acids related to monatomic anions, the -*ide* ending of the anion is changed to -*ic acid,* and the prefix *hydro-* is added. Thus, aqueous HCl is hydrochloric acid.

> Compounds named as acids do not include the word *hydrogen* in the name. The word *acid* implies the presence of hydrogen.

In summary, names of acids are related to the names of the corresponding anions as follows:

Name of Anion	*Name of Acid*
Per ____ ate	Per ____ ic acid
____ ate	____ ic acid
____ ite	____ ous acid
Hypo ____ ite	Hypo ____ ous acid
____ ide	Hydro ____ ic acid

■ EXAMPLE 6.16

Name the following acids:

(a) $HClO_3$ (b) H_3PO_3 (c) H_2SO_4 (d) HClO   (e) $HClO_4$ (f) HI

Solution

(a) Chloric acid. The ending -*ate* of the chlorate ion is changed to -*ic acid.*

(b) Phosphorous acid. The ending -*ite* of the phosphite ion is changed to -*ous acid.* In this case, the stem is also changed to *phosphor.*

(c) Sulfuric acid. The ending -*ate* is changed to -*ic acid,* and the stem is changed from *sulf* to *sulfur.*

(d) Hypochlorous acid. The ending *-ite* is changed to *-ous acid.* The prefix *hypo-* on the anion makes no difference; the prefix is included in the acid name.

(e) Perchloric acid. The ending *-ate* of perchlorate ion is changed to *-ic acid;* the prefix *per-* is not changed.

(f) Hydroiodic acid. The prefix *hydro-* distinguishes this binary acid from HIO_3.

Practice Problem 6.16   Name the following acids:

(a) $HBrO_3$ (b) HNO_2 (c) HNO_3 ■

Formulas for acids can be written by replacing every negative charge on the corresponding anion with one hydrogen atom. For example, SO_4^{2-} has two negative charges; therefore, sulfuric acid has two ionizable hydrogen atoms (and no charge): H_2SO_4.

■ **EXAMPLE 6.17**

Write the formula for each of the following acids:

(a) Phosphoric acid   (b) Hydrobromic acid   (c) Hypobromous acid

Solution

(a) H_3PO_4 (b) HBr (c) $HBrO$

Practice Problem 6.17   Write the formulas for (a) bromous acid and (b) bromic acid. ■

Figure 6.3 and Table 6.6 outline a systematic procedure for naming many compounds and ions. You may find these presentations very helpful, especially at first.

Table 6.6   Outline for Nomenclature

Is the compound (I) covalent or (II) ionic?

I. Covalent: Is the compound (A) an acid or (B) a binary compound of two nonmetals?
    A. Acid: Name the compound using a suffix and possibly a prefix related to the name of the analogous anion (II B). Add the word *acid.*
    B. Binary compound: Name the first element, and then name the second element with a prefix from Table 6.1 and with the ending changed to *-ide.*

II. Ionic: Name *both* (A) the cation and (B) the anion.
    A. Cation: Is the cation (1) polyatomic, (2) a metal forming ions with more than one charge, or (3) a metal with only one ion?
        1. Polyatomic: See Table 6.2.
        2. Variable: Use the name of the metal with a Roman numeral to indicate the charge.
        3. Constant: Use the name of the metal only.
    B. Anion: Is the anion (1) monatomic, (2) a tabulated oxoanion, or (3) something else?
        1. Monatomic: Change the ending of the element name to *-ide.*
        2. Oxoanion: See Table 6.4.
        3. Special: See Table 6.5.

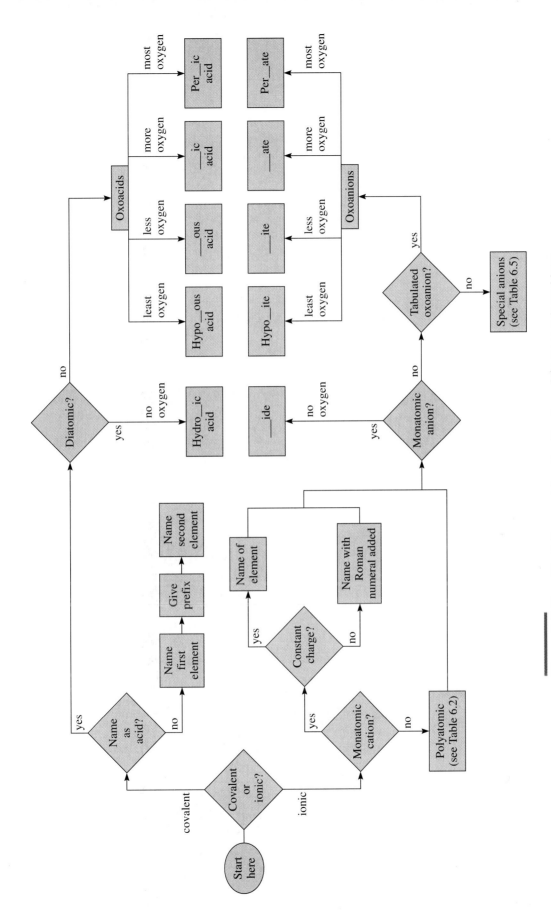

Figure 6.3 Flow Chart Summarizing the Naming of Compounds and Ions

Figure 6.4 Baking Soda

Ordinary baking soda is sodium hydrogen carbonate.

Acid salts, and their anions, have the word *hydrogen* in their names.

Naming Acid Salts

In Chapter 8, we will see that an acid with more than one ionizable hydrogen atom can react with bases in steps, with all but the last step yielding compounds called **acid salts.** Such salts consist of a cation, such as a sodium ion, plus an anion that has one or two hydrogen atoms still attached. Just as the hydrogen atoms are covalently bonded in the pure acid, the ones that are left in the acid salt are still covalently bonded. The anion is named with the word **hydrogen** followed by the name of the parent anion. For example, $NaHCO_3$ has a sodium cation, Na^+, and the hydrogen carbonate anion, HCO_3^-. The compound is sodium hydrogen carbonate. Acid salts of acids with three hydrogen atoms, such as phosphoric acid, require specification of how many hydrogen atoms are left. The prefixes *mono-* and *di-* are used for one and two hydrogen atoms, respectively. Thus, NaH_2PO_4 is sodium dihydrogen phosphate, and Na_2HPO_4 is sodium monohydrogen phosphate (or disodium hydrogen phosphate).

■ **EXAMPLE 6.18**

Name (a) NaHS and (b) HS^-.

Solution

(a) Sodium hydrogen sulfide   (b) Hydrogen sulfide ion

Including the word *ion* in the name in part (b) is important to distinguish this ion from hydrogen sulfide, H_2S.

Practice Problem 6.18 Write the formula for (a) sodium hydrogen sulfite and (b) magnesium hydrogen sulfite. ■

An older nomenclature system, still in use to some extent, uses the word *acid* to denote an acid salt. Also, the prefix *bi-* may be used for an acid salt of an acid with two ionizable hydrogen atoms. Thus, $NaHCO_3$ can be called sodium bicarbonate or sodium acid carbonate instead of sodium hydrogen carbonate (Figure 6.4).

6.4 Hydrates

OBJECTIVE

■ to name hydrates

Hydrates are stable crystalline compounds consisting of other compounds that are stable in their own right, with certain numbers of water molecules attached (Section 5.1). Naming and writing formulas for hydrates is easy. We simply name the compound first and then combine a Table 6.1 prefix that identifies the number of water molecules with the word *hydrate* to indicate the presence of the water molecules. For example, $CuSO_4 \cdot 5H_2O$ is called copper(II) sulfate pentahydrate. If we wish to emphasize that no water is attached, $CuSO_4$ may be called anhydrous copper(II) sulfate.

■ ■ ■ ■ ■ ■ ■ SUMMARY ■ ■ ■ ■ ■ ■ ■

Different systems are used for naming ionic compounds and binary covalent compounds, and acids are named still other ways. For binary nonmetal-nonmetal compounds (which are covalent), the left-most or lower element in the periodic table is named first, and then the other element is named. The ending of the second element is changed to -*ide,* and the number of atoms of that element in the molecule is indicated by a prefix (Table 6.1). If more than one atom of the first element is present per molecule, a prefix may be used for that element, too. (Section 6.1)

To name an ionic compound, name the cation first and then the anion. Monatomic cations of elements that form only one cation are named using just the name of the element. For monatomic cations of elements that can form more than one cation, the charge on the cation is indicated by a Roman numeral in parentheses added to the name of the element. Polyatomic cations have special names (Table 6.2).

The names of monatomic anions have the ending of the element's name changed to -*ide.* The charge on any monatomic anion (except H^-) is equal to the classical group number minus 8. The names of most familiar oxoanions end in -*ate* or -*ite,* depending on the relative number of oxygen atoms per ion. Ions with more oxygen atoms than those whose names end in -*ate* have the prefix *per-* added to the name; ions with fewer oxygen atoms than those whose names end in -*ite* have the prefix *hypo-* added to the name. Names

of other anions must be learned based on periodic table relationships or individually.

The charge on a cation is reported by a Roman numeral in parentheses in the *name* of the compound, and the number of ions is indicated by an Arabic numeral as a subscript in the *formula.* The charges allow you to deduce the numbers of ions, and vice versa, but the Roman numerals and the Arabic numerals do not represent the same quantities. (Section 6.2)

Acids can be recognized by the fact that the ionizable hydrogen atoms are written first in their formulas, but the word *hydrogen* does not appear in their names. The word *acid* implies the presence of the hydrogen. Oxoacids are named like the corresponding oxoanions, with the ending -*ate* changed to -*ic acid* or the ending -*ite* changed to -*ous acid.* Names of binary acids have the ending -*ide* of the corresponding anion changed to -*ic acid* and the prefix *hydro-* added. For example, Cl^- is chloride; HCl is hydrochloric acid.

Acid salts are named as ionic compounds, but the name of the anion has the word *hydrogen* in it (perhaps with a prefix) to indicate that at least one ionizable hydrogen atom is still left. For example, $NaHCO_3$ is sodium hydrogen carbonate. (Section 6.3)

Hydrates are named with a prefix from Table 6.1 before the word *hydrate,* to indicate the number of water molecules. For example, $CuSO_4 \cdot 5H_2O$ is named copper(II) sulfate pentahydrate. (Section 6.4)

Items for Special Attention

■ The following list summarizes the types of compounds and ions you have learned to name in this chapter:

Binary nonmetal-nonmetal compounds
Ionic compounds
    Cations
        Monatomic cations
            Variable charge
            Constant charge
        Polyatomic cations
    Anions
        Monatomic anions
        Oxoanions
        Special anions
Acids
Acid salts
Hydrates

■ The prefixes in Table 6.1 are used only for naming binary nonmetal-nonmetal compounds, acid salts, and hydrates.

■ It is critical to specify the charges in formulas for ions and to include the word *ion* if the name without that word means something else, such as sodium ion or hydrogen sulfide ion.

■ Roman numerals in names stand for charges, and subscripts in formulas represent numbers of atoms.

■ All monatomic anions have names ending in -*ide,* but not all anions with names ending in -*ide* are monatomic. Hydroxide ion, OH^-, and cyanide ion, CN^-, are important examples of diatomic ions with names ending in -*ide.*

■ Parentheses are used when two or more of a polyatomic ion are present in a given formula, as in $Ba(ClO)_2$. The parentheses indicate the number of atoms of each element present. In certain cases, the parentheses also distinguish between familiar ions, such as the ClO^- ions in $Ba(ClO)_2$ and the ClO_2^- ion in $NaClO_2$.

■ That hydrogen is present in an acid is stated in the name by the word *acid*, not by the word *hydrogen*. For example, HCl is hydrochloric *acid*. The word *hydrogen* is used in names of acid salts.

Self-Tutorial Problems

6.1 Name (a) CO and (b) Co.

6.2 Which metals form cations of the constant type? What are the charges on these cations?

6.3 What is the difference in the meanings of the prefixes *bi-* and *di-* (as used in this chapter)?

6.4 Classify each of the following as ionic or covalent, and name each:

(a) LiCl   (b) ICl   (c) PF_3 (d) AlF_3

6.5 What are the rules for remembering the charges on (a) monatomic anions and (b) oxoanions?

6.6 What is the difference between ClO_3 and ClO_3^-? Name each one.

6.7 Classify the metal in each of the following compounds as constant type or variable type, and then name each compound:

(a) $CaCl_2$ (b) $CuCl_2$ (c) AgCl

(d) $FeCl_2$

6.8 What is the difference between the two names for HCl—hydrogen chloride and hydrochloric acid?

6.9 Pure HCl is named as a binary nonmetal-nonmetal compound, whereas pure H_2SO_4 is named as an acid. Explain why H_2SO_4 is not named as HCl is named.

6.10 What can you tell from each of the following?

(a) The charges on the two ions making up a compound

(b) The fact that hydrogen is written first in a formula

(c) The fact that the name for a compound or ion ends in *-ate*

6.11 (a) What is the difference between hydrogen ion and hydride ion?

(b) Explain why H^+ is called the hydrogen ion rather than the hydrogen(I) ion, even though hydrogen can form two different ions.

6.12 Write formulas for (a) chloride ion, (b) chlorate ion, and (c) chlorite ion.

6.13 What are the differences in the following, as used in naming compounds?

(a) Hydro-   (b) Hydrogen   (c) Hypo-

6.14 Name (a) Na_3PO_4 and (b) H_3PO_4.

6.15 What is the difference between NH_3 and NH_4^+? Name each one.

6.16 What is the charge on each of the following?

(a) The hydrogen carbonate ion

(b) The dihydrogen phosphate ion

(c) The monohydrogen phosphate ion

6.17 Name each of the following acids:

(a) $HClO_4$ (b) $HClO_3$ (c) $HClO_2$

(d) HClO   (e) HCl

■ ■ ■ PROBLEMS ■ ■ ■

6.1 Binary Nonmetal-Nonmetal Compounds

6.18 Name each of the following compounds:

(a) IF_5 (b) H_2O (c) $AsCl_5$

(d) SO_3 (e) PBr_3

6.19 Write the formula for each of the following compounds:

(a) Sulfur dioxide

(b) Carbon tetrachloride

(c) Phosphorus pentachloride

(d) Arsenic trifluoride

(e) Ammonia

6.20 Name each of the following compounds:

(a) SiF_4 (b) SiO_2

(c) HCl (as the pure compound)

(d) H_2S (e) BrCl   (f) IF_3 (g) NBr_3

6.21 Name each of the following compounds:

(a) SF_2 (b) SF_4 (c) SF_6

6.22 Name each of the following:

(a) Cl_2O (b) Cl_2O_3 (c) Cl_2O_5 (d) Cl_2O_7

6.23 Write the formula for each of the following:

(a) Tetrasulfur tetranitride

(b) Diphosphorus pentoxide

(c) Tetraarsenic hexoxide

(d) Dichlorine trioxide

6.24 Write the formula for each of the following compounds:

(a) Carbon disulfide

(b) Sulfur difluoride

(c) Diphosphorus pentoxide

(d) Nitrogen trichloride

6.25 Name each of the following compounds:

(a) SeO_2 (b) HBr (c) P_2O_3

(d) BCl_3 (e) SO_3

6.2 Naming Ionic Compounds

6.26 Name each of the following cations:

(a) K^+ (b) Ba^{2+} (c) Cd^{2+}

6.27 Name each of the following anions:

(a) P^{3-} (b) O^{2-}

(c) N^{3-} (d) I^-

6.28 Name each of the following cations:

(a) Cr^{2+} (b) Cr^{3+}

(c) Al^{3+} (d) Ca^{2+}

6.29 Name each of the following anions:

(a) BrO_3^- (b) NO_3^- (c) CO_3^{2-}

6.30 Name each of the following anions:

(a) CrO_4^{2-} (b) $C_2H_3O_2^-$

(c) $Cr_2O_7^{2-}$ (d) MnO_4^-

(e) O_2^{2-} (f) CN^-

6.31 Explain why chemists often refer to Cl^- as "chloride" (without the word *ion*) but do not refer to Na^+ as "sodium" (without the word *ion*).

6.32 Name each of the following cations:

(a) Cu^{2+} (b) Ni^{2+} (c) Zn^{2+}

6.33 Write the formula for each of the following ions:

(a) Calcium ion (b) Manganese(II) ion

(c) Silver ion (d) Ammonium ion

(e) Mercury(I) ion

6.34 Write the formula for each of the following ions:

(a) Chlorate ion (b) Nitrite ion

(c) Phosphate ion (d) Chromate ion

(e) Cyanide ion (f) Hypobromite ion

6.35 Name each of the following compounds:

(a) K_2SO_4 (b) $Al(CN)_3$

(c) $(NH_4)_3PO_4$

6.36 Name each of the following compounds:

(a) CuS (b) CaS (c) $(NH_4)_2S$

6.37 Write the formula for each of the following compounds:

(a) Nickel(II) chlorate (b) Cobalt(III) hydroxide

(c) Magnesium sulfate (d) Copper(II) oxide

(e) Lithium cyanide (f) Ammonium carbonate

6.38 Write the formula for each of the following compounds:

(a) Copper(I) chloride (b) Silver acetate

(c) Barium carbonate (d) Nickel(II) nitrate

(e) Sodium peroxide

(f) Ammonium hydrogen carbonate

6.39 Complete the following table by writing the formula for each ionic compound whose cation is given on the left and whose anion is given at the top:

	Bromate	Sulfite	Phosphate	Acetate
Sodium	___	___	___	___
Chromium(II)	___	___	___	___
Iron(III)	___	___	___	___
Ammonium	___	___	___	___

6.40 Complete the following table by writing the formula for each ionic compound whose cation is given on the left and whose anion is given at the top:

	Fluoride	Oxide	Phosphate
Copper(II)	___	___	___
Cobalt(III)	___	___	___
Lead(IV)	___	___	___

6.41 Complete the following table by writing the formula for each ionic compound whose cation is given on the left and whose anion is given at the top:

	Hydroxide	Sulfide	Nitrate	Sulfate
Nickel(II)	___	___	___	___
Cerium(III)	___	___	___	___
Lead(II)	___	___	___	___
Sodium	___	___	___	___

6.42 Complete the following table by writing the formula for each ionic compound whose cation is given on the left and whose anion is given at the top:

	Hydrogen Sulfate	Sulfate	Acetate
Mercury(II)	___	___	___
Cobalt(II)	___	___	___
Iron(III)	___	___	___

6.43 An instructor tells the students in a class that Na^+ is the only stable ion of sodium and that Na^{2+} cannot be prepared in a solid. What name should the instructor use for Na^{2+}?

6.3 Naming Acids and Acid Salts

6.44 What is the difference between bromic acid and hydrobromic acid?

6.45 What is the difference between hypochlorous acid and hydrochloric acid?

6.46 Write the formula for each of the following acids:

(a) Hydrobromic acid   (b) Phosphoric acid

(c) Perchloric acid   (d) Sulfurous acid

6.47 What is the difference between the names *phosphorus* and *phosphorous?*

6.48 Name each of the following compounds as an acid and also as a binary compound of nonmetals:

(a) HCl   (b) H_2S (c) HBr

6.49 Name each of the following acids:

(a) H_2SO_3 (b) HClO   (c) H_3PO_4

6.50 What is the difference between phosphorous acid and hypophosphorous acid?

6.51 Classify each of the following as an acid, an acid salt, or a regular salt, and name each:

(a) $NaHSO_4$ (b) Na_2SO_4 (c) H_2SO_4

6.4 Hydrates

6.52 Name (a) $CdSO_4 \cdot 7H_2O$ and (b) $CaBr_2 \cdot 6H_2O$.

6.53 Write the formula for (a) calcium carbonate hexahydrate, (b) chromium(III) acetate monohydrate, and (c) cobalt(II) phosphate dihydrate.

■ ■ ■ GENERAL PROBLEMS ■ ■ ■

6.54 Which of the following compounds should be named using the prefixes of Table 6.1, which should be named with Roman numerals, and which should have neither?

(a) NaClO   (b) PCl_3 (c) CCl_4

(d) $BaCl_2$ (e) CdO   (f) $CoCO_3$

6.55 Name each of the following compounds:

(a) CF_4 (b) SiO_2

(c) HI (as the pure compound)   (d) H_2SO_3

(e) P_2O_5 (f) Cl_2O_5

(g) Na_2O_2 (h) $PbCr_2O_7$

6.56 Name each of the following compounds:

(a) SF_6 (b) SF_4 (c) SF_2

6.57 Name each of the following:

(a) SO_3 (b) SO_3^{2-} (c) Na_2SO_3

(d) H_2SO_3 (e) $CoSO_3$

6.58 Name (a) Cu_2O and (b) CuO.

6.59 Write the formula for each of the following compounds:

(a) Phosphorus trifluoride

(b) Titanium(IV) oxide

(c) Ammonium phosphate

(d) Hypochlorous acid

(e) Antimony(III) sulfide

6.60 Name (a) Na_2O_2 and (b) PbO_2. (Hint: Peroxide ion generally exists only in combination with metals in the form of their ion of highest charge.)

6.61 Write the formula for each of the following compounds:

(a) Iodine pentafluoride

(b) Titanium(III) sulfate

(c) Hypoiodous acid

(d) Chromium(III) sulfite

(e) Bismuth(III) chloride

6.62 Name (a) LiBr, (b) HBr (as an acid), and (c) NH_4Br.

6.63 Name each of the following compounds:

(a) Al_2O_3 (b) N_2O_3 (c) Co_2O_3

6.64 Name each of the following compounds:

(a) SF_2 (b) MgF_2 (c) CoF_2

6.65 Name each of the following ions:

(a) SO_4^{2-} (b) PO_4^{3-} (c) SeO_4^{2-} (d) AsO_4^{3-}

6.66 Name each of the following compounds:

(a) K_2O_2 (b) BaO   (c) BaO_2

6.67 Name the cation in each of the following compounds:

(a) K_2MoO_4 (b) $(NH_4)_2SeO_4$

(c) $CrPO_4$ (d) $CoHPO_4$

6.68 Name the anion in each of the following compounds:

(a) VO_2SO_4 (b) VO_2F (c) $Zn(H_2PO_4)_2$

6.69 Complete the following table by writing the formula and name of each compound formed from an anion at the top and a cation on the left:

	NO_3^-	CO_3^{2-}	PO_4^{3-}
NH_4^+	____	____	____
	____	____	____
Mn^{2+}	____	____	____
	____	____	____
Fe^{3+}	____	____	____
	____	____	____

6.70 Which transition metal ions have a charge of 1+?

6.71 Complete the following table by writing the formula and name of each compound formed from an anion at the top and a cation on the left:

	BrO_3^-	SO_3^{2-}	AsO_4^{3-}
Na^+	———	———	———
	———	———	———
Mg^{2+}	———	———	———
	———	———	———
Al^{3+}	———	———	———
	———	———	———

6.72 From Figure 6.3, give the route by which you would name (a) $Ca(ClO_3)_2$, (b) PCl_3, and (c) $(NH_4)_2Cr_2O_7$.

6.73 Give a more modern name for each of the following:

(a) Sodium bicarbonate   (b) Chromous chloride

(c) Ferrous nitrate   (d) Cobaltous sulfide

6.74 Write the formula for lithium bisulfide.

6.75 Write the formula for hydrosulfuric acid.

6.76 Name (a) $NaClO_3$ and (b) $Al(ClO_3)_3$.

6.77 Which of the following compounds have acid properties?

CH_4 $HC_2H_3O_2$ HCl $KHSO_3$ NH_3

6.78 Write formulas for hydrogen sulfide, hydrogen sulfide ion, and hydrosulfuric acid.

6.79 Which of the following pure compounds have covalent bonds only?

$KC_2H_3O_2$ CoF_2 CO   HCl   KCl

6.80 Write formulas for the following ions, as well as formulas and names for the corresponding acids and anions (with no hydrogen):

(a) Hydrogen sulfide ion

(b) Dihydrogen phosphate ion

(c) Hydrogen sulfite ion

(d) Hydrogen carbonate ion

6.81 Name each of the following compounds:

(a) CaS   (b) XeF_2 (c) P_2S_3 (d) CoF_3 (e) NH_3

6.82 Write formulas for (a) acetic acid, (b) ammonium selenate, and (c) lithium arsenate.

6.83 Name (a) NaH_2PO_4, (b) Na_2HPO_4, (c) H_3PO_4, and (d) Na_3PO_4.

6.84 Write formulas for (a) potassium monohydrogen phosphate and (b) potassium dihydrogen phosphate.

6.85 Name the following ions by the Stock system, using Table 6.3 if necessary:

(a) Nickelous ion   (b) Auric ion

(c) Mercuric ion   (d) Chromic ion

(e) Plumbous ion

6.86 Name and write formulas for the ions in (a) Na_2S and (b) Hg_2S.

6.87 (a) Azide ion has the formula N_3^-. Write the formula for the corresponding acid.

(b) What is the name of that acid? (Hint: Add the letter *o* to the stem.)

6.88 Name the following compounds. Identify the type of each, using the following symbols:

IV for compounds containing cations of variable type

IC for compounds containing cations of constant type

A for acids or acid salts

C for other binary covalent compounds

Formula	Name	Type
$BaCO_3$	———————	————
PCl_3	———————	————
H_2SO_4	———————	————
NH_4NO_3	———————	————
$MnSO_4$	———————	————
SF_4	———————	————
$Co_3(PO_4)_2$	———————	————
HNO_3	———————	————
CCl_4	———————	————
$FeCl_3$	———————	————
P_2S_3	———————	————
Cu_2S	———————	————
$Mg(OH)_2$	———————	————
BrF_3	———————	————
HCl	———————	————
BF_3	———————	————
AgBr	———————	————
H_3PO_4	———————	————
MnO_2	———————	————
CoF_3	———————	————
KOH	———————	————

6.89 Write formulas for the following compounds or elements. Identify the type of each, using the following symbols:

IV for compounds containing metal ions of variable type

IC for compounds containing metal ions of constant type

A for acids or acid salts

C for other binary covalent compounds

E for elements

Name	Formula	Type
Iron(III) oxide	—————	————
Nickel(II) sulfide	—————	————
Iodine trifluoride	—————	————
Lithium hydride	—————	————
Gold(I) oxide	—————	————
Calcium hydroxide	—————	————
Nitrous acid	—————	————
Ammonium sulfide	—————	————
Magnesium sulfide	—————	————
Phosphoric acid	—————	————
Hydrochloric acid	—————	————
Potassium sulfate	—————	————
Ammonium phosphate	—————	————

6.90 Write formulas for the following compounds or elements. Identify the type of each, using the following symbols:

IV for compounds containing metal ions of variable type

IC for compounds containing metal ions of constant type

A for acids or acid salts

C for other binary covalent compounds

E for elements

Name	Formula	Type
Potassium hydroxide		
Sodium sulfide		
Barium phosphate		
Nitric acid		
Potassium carbonate		
Cobalt(II) oxide		
Ammonium nitrate		
Chromium(II) phosphate		
Iron(II) sulfate		
Lead(IV) oxide		
Hydrobromic acid		
Sodium hydroxide		
Carbon disulfide		
Silver sulfide		
Aluminum sulfate		
Dinitrogen tetroxide		

6.91 Write formulas for the following compounds or elements. Identify the type of each, using the following symbols:

IV for compounds containing metal ions of variable type

IC for compounds containing metal ions of constant type

A for acids or acid salts

C for other binary covalent compounds

E for elements

Name	Formula	Type
Mercury(I) oxide		
Cadmium sulfide		
Boron trichloride		
Aluminum hydride		
Gold(I) sulfide		
Calcium phosphate		
Arsenic trichloride		
Magnesium nitride		
Zinc perchlorate		
Sodium acetate		
Copper(II) sulfide		
Nitrous acid		
Copper(I) sulfide		
Nitrogen		
Lead(II) chloride		
Barium hydroxide		
Calcium nitrate		
Nitrogen tribromide		
Potassium nitrate		
Hydrochloric acid		
Sodium phosphate		
Ammonium sulfate		

6.92 Name the following compounds. Identify the type of each, using the following symbols:

IV for compounds containing cations of variable type

IC for compounds containing cations of constant type

A for acids or acid salts

C for other binary covalent compounds

Formula	Name	Type
$KMnO_4$		
BaO_2		
$(NH_4)_2Cr_2O_7$		
$Ni(C_2H_3O_2)_2$		
$Mg(HCO_3)_2$		
SeO_3		
$FeAsO_4$		
P_4S_{10}		
IF_7		
$Co(H_2PO_4)_2$		
As_2O_5		
$HClO_3$		
NH_4ClO		
$HC_2H_3O_2$		

6.93 Name the following compounds:

Formula	Name
$Ca(OH)_2$	
HNO_2	
TiO	
$MgCrO_4$	
$Ni(NO_2)_2$	
$Na_2Cr_2O_7$	
$CrCl_2$	
NCl_3	
SF_6	
H_2SO_4	
$NaHSO_3$	
$BaCl_2$	
IF_7	
OF_2	
$LiCl$	
$Co(OH)_2$	
$AuCl_3$	
$(NH_4)_2SO_3$	
MnO	
$SnSO_4$	
$FeBr_2$	
NH_3	
MgF_2	
Li_3N	
$Ba(OH)_2$	
$FeCl_2$	
$(NH_4)_2S$	
XeF_2	
ICl	
CaH_2	
CI_4	
$AgCl$	
H_3PO_4	
$AgC_2H_3O_2$	

PbS   _____
$Zn(ClO_3)_2$ _____
CS_2 _____
$CoCO_3$ _____
$HClO_3$ _____
NaCN   _____

6.94 Write formulas for the following compounds or elements:

Name   *Formula*

Manganese(III) oxide   _____
Cobalt(II) sulfide   _____
Boron trifluoride   _____
Sodium hydride   _____
Gold(I) bromide   _____
Calcium hydride   _____
Carbon tetrachloride   _____
Lithium nitride   _____
Magnesium perchlorate   _____
Acetic acid   _____

Copper(II) sulfate   _____
Chlorous acid   _____
Copper(I) oxide   _____
Bromine   _____
Bromine trichloride   _____
Bromic acid   _____
Sodium hydrogen sulfate   _____
Calcium hydrogen sulfate   _____
Nitric acid   _____
Hydrochloric acid   _____
Lithium sulfite   _____
Ammonium dihydrogen
   phosphate   _____
Sodium hydroxide   _____
Calcium carbonate   _____
Nitrogen triiodide   _____
Potassium cyanide   _____
Chloric acid   _____
Sodium arsenate   _____
Ammonium selenate   _____

7 Formula Calculations

- 7.1 Formula Masses
- 7.2 Percent Composition
- 7.3 The Mole
- 7.4 Empirical Formulas
- 7.5 Molecular Formulas

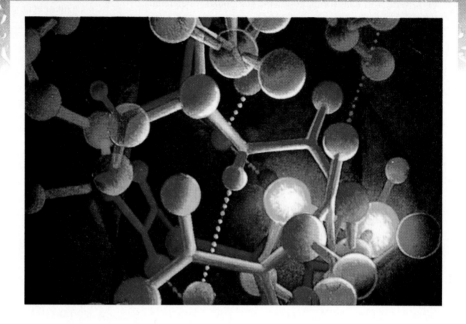

■ **Key Terms** *(Key terms are defined in the Glossary.)*

Avogadro's number (7.3)
empirical formula (7.4)
formula mass (7.1)
formula weight (7.1)

molar mass (7.3)
mole (7.3)
molecular formula (7.5)

molecular mass (7.1)
molecular weight (7.1)
percent composition (7.2)

■ **Abbreviations**

MM (molar mass) (7.3)

mol (mole) (7.3)

The meaning of a chemical formula was discussed in Chapter 5, and you learned how to interpret formulas in terms of the numbers of atoms of each element per formula unit. In this chapter, you will learn how to calculate the number of grams of each element in any given quantity of a compound from its formula and to do other calculations involving formulas. Formula masses are presented in Section 7.1, and percent composition is considered in Section 7.2. Section 7.3 discusses the mole—the basic chemical quantity of any substance. Moles can be used to count atoms, molecules, or ions and to calculate the mass of any known number of formula units of a substance. Section 7.4 shows how to use relative mass data to determine empirical formulas, and the method is extended to molecular formulas in Section 7.5.

7.1 Formula Masses

Since each symbol in a formula represents an atom, which has a given average atomic mass, the formula as a whole represents a collection of atoms with a given formula mass. The **formula mass,** or **formula weight,** is the sum of the atomic masses of all atoms of every element (not merely each type of atom) in a formula unit. Formula masses should be calculated to as many significant digits as are given in any data presented in a problem. If no data are given, at least three significant digits should be used. For example, we can calculate the formula mass for CO_2 as follows:

	Number of Atoms Per Formula Unit		Atomic Mass		
1 C	1	×	12.0 amu	=	12.0 amu
2 O	2	×	16.0 amu	=	32.0 amu
			Formula mass	=	44.0 amu

Since calculation of formula mass is essentially adding two or more numbers, the numbers to be added may be rounded to the same number of decimal places. Prior multiplication might affect the number of decimal places you retain in the atomic masses.

The three types of formula masses correspond to the three types of formula units: (1) atomic masses (also called atomic weights), (2) **molecular masses** (also called **molecular weights**), and (3) formula masses for ionic compounds (also called formula weights). The term *atomic mass* may be used whether an atom is combined or not, but it always refers to the mass of one atom of an element.

■ EXAMPLE 7.1

Why is it incorrect to refer to the molecular mass of NaF?

Solution

NaF is an ionic compound. It does not have molecules and thus does not have a molecular mass.

Practice Problem 7.1   Which of the following have molecular masses?

F_2 IF   LiF ■

■ EXAMPLE 7.2

Calculate the formula mass of (a) $(NH_4)_3PO_4$ and (b) $Cr(ClO_3)_2$.

Solution

(a) 3 N   3 × 14.01   =   42.03 amu
     12 H   12 × 1.008   =   12.10 amu
     1 P   1 × 30.97   =   30.97 amu
     4 O   4 × 16.00   =   64.00 amu
                  Formula mass = 149.10 amu

(b) 1 Cr   1 × 52.00   =   52.00 amu
     2 Cl   2 × 35.45   =   70.90 amu
     6 O   6 × 16.00   =   96.00 amu
                  Formula mass = 218.90 amu

Practice Problem 7.2   Calculate the formula mass of (a) $Ca(C_2H_3O_2)_2$ and (b) $Co(ClO_3)_3$. ■

7.2 Percent Composition

■ to calculate the percent
composition by mass from the
formula of a compound

If we know the total mass of each element in a formula unit and we also know the mass of the entire formula unit, we can calculate the **percent composition** of the compound. We simply divide the total mass of each element by the total mass of the formula unit and multiply each quotient by 100% to convert it to a percentage. All the percentages comprise the percent composition.

■ EXAMPLE 7.3

Calculate the percent composition of $Cr(ClO_3)_2$.

Solution

The chromium in one formula unit has a mass of 52.00 amu in a total mass of 218.90 amu, as calculated in Example 7.2(b). Chromium's percent by mass is 100% times the mass of a chromium atom divided by the formula mass:

$$\left(\frac{52.00 \text{ amu}}{218.90 \text{ amu}} \right) \times 100.00\% = 23.76\% \text{ Cr}$$

The percentages of chlorine and oxygen are calculated in the same way:

$$\text{Percentage of chlorine} = \left(\frac{70.90 \text{ amu}}{218.90 \text{ amu}} \right) \times 100.00\% = 32.39\% \text{ Cl}$$

$$\text{Percentage of oxygen} = \left(\frac{96.00 \text{ amu}}{218.90 \text{ amu}} \right) \times 100.00\% = 43.86\% \text{ O}$$

Notice that the *total* mass of chlorine or oxygen in the formula unit, not the atomic mass of chlorine or oxygen, is used in the calculation.

The sum of all the percentages of elements in any compound should be 100%:

$$23.76\% + 32.39\% + 43.86\% = 100.01\%$$

This is near enough to 100% to be correct. In general, the sum may not be exactly 100% because of prior rounding to the proper number of significant digits. If you get a total of 98% or 105%, you should look for an error. (Getting 100% does not guarantee that your percentages are correct, but getting a sum that is significantly different from 100% means that something is *incorrect.*)

Practice Problem 7.3   Calculate the percent composition of $(NH_4)_3PO_4$. ■

7.3   The Mole

The atomic mass unit (amu) is an extremely small unit, suitable for measuring masses of individual atoms and molecules. However, to measure masses on laboratory balances takes a huge number of atoms, molecules, or formula units. Chemists have to weigh a large collection of formula units, so that the total mass is measurable on a laboratory balance. (Try to weigh one grain of rice on a scale designed to weigh people, and you will get an inkling of the problem of measuring the mass of one atom or molecule. See Figure 3.5.)

The **mole** (abbreviated mol) is the standard chemical unit used to measure the quantity of a substance. A mole is defined as the number of ^{12}C atoms in *exactly* 12 g of ^{12}C. The mole is equal to 6.0225×10^{23} particles. This number is known as **Avogadro's number.** You should remember the value of this number to at least three significant digits.

Avogadro's number was set at 6.0225×10^{23} so that the atomic mass of each element and the number of grams per mole of that element have the same numeric value, although in different units. The atomic mass of ^{12}C is 12.00 amu, and that same number is the number of grams per mole of ^{12}C. The formula mass of any compound or element is also equal to its number of grams per mole. The formula mass of a substance in units of grams per mole is often called the **molar mass** of the substance. (Molar mass can be abbreviated MM.)

> The molar mass of any substance is equal to the number of grams per mole of that substance.

ENRICHMENT

It is also possible to think of a mole as the number of atomic mass units in 1 gram.

■ **EXAMPLE 7.4**

Use the mass of one ^{12}C atom (exactly 12 amu), the definition of a mole, and the value of Avogadro's number to calculate the number of atomic mass units per gram.

Solution

$$\underset{\text{Avogadro's number}}{\frac{6.0225 \times 10^{23}\ ^{12}C\ \text{atoms}}{1\ \text{mol}\ ^{12}C}} \left(\underset{\substack{\text{Mass of}\\ \text{1 mol}}}{\frac{1\ \text{mol}\ ^{12}C}{12.000\ \text{g}}} \right) \left(\underset{\substack{\text{Mass of}\\ \text{1 atom}}}{\frac{12.000\ \text{amu}}{1\ ^{12}C\ \text{atom}}} \right) = 6.0225 \times 10^{23}\ \text{amu/g}$$

The number of atomic mass units per gram is equal to Avogadro's number.

Practice Problem 7.4   Calculate the number of inches in 1 foot, knowing that a certain shoe box is 4.00 inches tall and that a stack of 1 dozen of these boxes is 4.00 feet tall. ■

■ EXAMPLE 7.5

Knowing that the molecular mass (formula mass) of H_2O is 18.0 amu, show that H_2O has a molar mass equal to 18.0 g/mol.

Solution

$$\frac{18.0 \text{ amu}}{1 \text{ molecule}}\left(\underbrace{\frac{6.02 \times 10^{23} \text{ molecules}}{1 \text{ mol}}}_{\text{Avogadro's number}}\right)\left(\underbrace{\frac{1.00 \text{ g}}{6.02 \times 10^{23} \text{ amu}}}_{\text{From Example 7.4}}\right) = \frac{18.0 \text{ g}}{1 \text{ mol}}$$

Practice Problem 7.5   The formula mass of $CaCO_3$ is 100.09 amu. What is its molar mass (in grams per mole)? ■

We can use Avogadro's *number* as a conversion factor to convert moles to *numbers* of formula units, and vice versa. We can use the molar *mass* to convert moles to *masses,* and vice versa. (See Figure 7.1.)

■ EXAMPLE 7.6

(a) Calculate the number of molecules in 3.00 mol of C_3H_8.

(b) Calculate the mass of 3.00 mol of C_3H_8.

Solution

(a) $3.00 \text{ mol C}_3\text{H}_8\left(\dfrac{6.02 \times 10^{23} \text{ molecules C}_3\text{H}_8}{1 \text{ mol C}_3\text{H}_8}\right) = 18.1 \times 10^{23} \text{ molecules C}_3\text{H}_8$

$= 1.81 \times 10^{24} \text{ molecules C}_3\text{H}_8$

(b) The molar mass of C_3H_8 is $3(12.0 \text{ g}) + 8(1.008 \text{ g}) = 44.1 \text{ g}$.

$$3.00 \text{ mol C}_3\text{H}_8\left(\frac{44.1 \text{ g C}_3\text{H}_8}{1 \text{ mol C}_3\text{H}_8}\right) = 132 \text{ g C}_3\text{H}_8$$

Practice Problem 7.6

(a) Calculate the number of moles of C_3H_6 in 17.0 g of C_3H_6.

(b) Calculate the number of moles of C_3H_6 in a sample containing 6.95×10^{24} C_3H_6 molecules. ■

■ EXAMPLE 7.7

Calculate the number of molecules of CH_4 in 34.56 g of CH_4.

Figure 7.1 Some Conversions Involving Moles

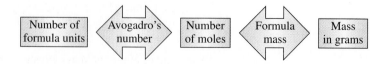

Solution

The molar mass of CH_4 is 12.01 g + 4(1.008 g) = 16.04 g. (The molar mass is calculated to four significant digits because the data in the problem are given to four significant digits.)

$$34.56 \text{ g CH}_4\left(\frac{1 \text{ mol CH}_4}{16.04 \text{ g CH}_4}\right) = 2.155 \text{ mol CH}_4$$

$$2.155 \text{ mol CH}_4\left(\frac{6.022 \times 10^{23} \text{ molecules}}{1 \text{ mol CH}_4}\right) = 1.298 \times 10^{24} \text{ molecules CH}_4$$

or

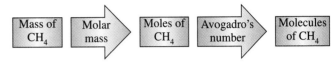

$$34.56 \text{ g CH}_4\left(\frac{1 \text{ mol CH}_4}{16.04 \text{ g}}\right)\left(\frac{6.022 \times 10^{23} \text{ molecules}}{1 \text{ mol CH}_4}\right)$$

$$= 1.298 \times 10^{24} \text{ molecules CH}_4$$

Practice Problem 7.7 Calculate the mass of 3.71×10^{22} molecules of CCl_4. ■

The chemical formula for a substance gives the ratio of atoms of each element in the substance to atoms of every other element in the substance. It also gives the ratio of dozens of atoms of each element in the substance to dozens of atoms of every other element in the substance. Moreover, it gives the ratio of *moles* of atoms of each element in the substance to *moles* of atoms of every other element in the substance. The mole ratio from the formula can be used as a factor to convert from moles of any element in the formula to moles of any other element or to moles of the formula unit as a whole. In Figure 7.2, these additional conversions have been added to those already presented in Figure 7.1.

Figure 7.2 More Conversions Involving Moles

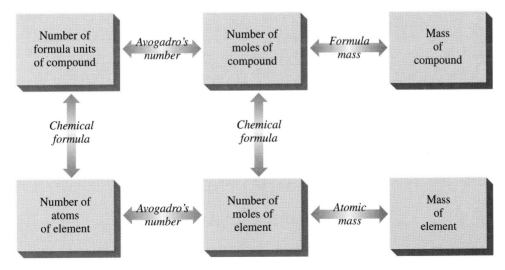

■ **EXAMPLE 7.8**

Calculate the number of moles of hydrogen atoms in 7.330 mol of H_2O.

Solution

$$7.330 \text{ mol } H_2O\left(\frac{2 \text{ mol } H}{1 \text{ mol } H_2O}\right) = 14.66 \text{ mol } H$$

Factor from
chemical formula

Practice Problem 7.8   Calculate the number of moles of $Mg(HCO_3)_2$ that contains 9.161×10^{21} oxygen atoms. ■

■ **EXAMPLE 7.9**

(a) Calculate the number of grams of fluorine in 10.05 g of CoF_2.

(b) Calculate the mass of CoF_2 that contains 10.05 g of fluorine.

Solution

(a) The molar mass (atomic mass in grams per mole) of fluorine is 19.00 g/mol;
     the molar mass of CoF_2 is 96.93 g/mol.

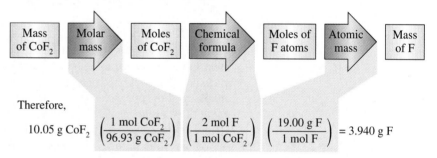

Therefore,

$$10.05 \text{ g } CoF_2 \left(\frac{1 \text{ mol } CoF_2}{96.93 \text{ g } CoF_2}\right)\left(\frac{2 \text{ mol } F}{1 \text{ mol } CoF_2}\right)\left(\frac{19.00 \text{ g } F}{1 \text{ mol } F}\right) = 3.940 \text{ g } F$$

(b)

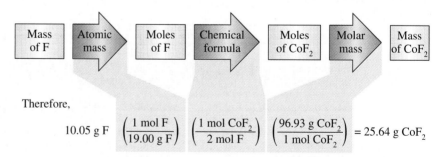

Therefore,

$$10.05 \text{ g } F \left(\frac{1 \text{ mol } F}{19.00 \text{ g } F}\right)\left(\frac{1 \text{ mol } CoF_2}{2 \text{ mol } F}\right)\left(\frac{96.93 \text{ g } CoF_2}{1 \text{ mol } CoF_2}\right) = 25.64 \text{ g } CoF_2$$

Practice Problem 7.9

(a) Calculate the mass of phosphorus in 50.0 g of $(NH_4)_3PO_4$, which is used in
     fertilizers.

(b) Calculate the mass of $(NH_4)_3PO_4$ that contains 50.0 g of phosphorus. ■

The percent composition of a compound can be calculated in terms of molar
masses instead of formula masses.

■ EXAMPLE 7.10

Calculate the percent composition of table sugar, $C_{12}H_{22}O_{11}$, using molar masses instead of formula masses.

Solution

The mass of a mole of $C_{12}H_{22}O_{11}$ is

$$12(12.01 \text{ g}) + 22(1.008 \text{ g}) + 11(16.00 \text{ g}) = 342.3 \text{ g}$$

The percentage of each element is given by

$$\left(\frac{12(12.01 \text{ g}) \text{ C}}{342.3 \text{ g } C_{12}H_{22}O_{11}} \right) \times 100.0\% = 42.10\% \text{ C}$$

$$\left(\frac{22(1.008 \text{ g}) \text{ H}}{342.3 \text{ g } C_{12}H_{22}O_{11}} \right) \times 100.0\% = 6.48\% \text{ H}$$

$$\left(\frac{11(16.00 \text{ g}) \text{ O}}{342.3 \text{ g } C_{12}H_{22}O_{11}} \right) \times 100.0\% = \underline{51.42\% \text{ O}}$$

$$\text{Total} = 100.00\%$$

Practice Problem 7.10 Calculate the percent composition of glucose, $C_6H_{12}O_6$. ■

7.4 Empirical Formulas

The **empirical formula** of a compound is the formula that gives the lowest whole-number ratio of atoms of all of the elements. For example, the empirical formula of the compound glucose, $C_6H_{12}O_6$, is CH_2O. The simplest ratio of carbon to hydrogen to oxygen atoms in glucose is 1 to 2 to 1. An empirical formula always has the smallest integral subscripts that give the correct ratio of atoms of the elements.

■ EXAMPLE 7.11

Write the empirical formulas for the compounds containing carbon and hydrogen in the following ratios:

(a) 2 mol carbon to 3 mol hydrogen

(b) 1.0 mol carbon to 1.5 mol hydrogen

(c) 1.902 mol carbon to 2.853 mol hydrogen

Solution

(a) The mole ratio is 2:3, so the empirical formula is C_2H_3.

(b) The mole ratio given is not integral, but you can multiply each value by 2 to get an integral ratio of 2:3. The empirical formula is again C_2H_3.

(c) This mole ratio is not integral, and this time it is more difficult to make it so. If you divide both values by the magnitude of the smaller one, you can get closer to an integral ratio:

$$\frac{1.902 \text{ mol C}}{1.902} = 1.000 \text{ mol C} \qquad \frac{2.853 \text{ mol H}}{1.902} = 1.500 \text{ mol H}$$

Now multiply by 2, as in part (b):

$$\frac{1.000 \text{ mol C}}{1.500 \text{ mol H}} = \frac{2 \text{ mol C}}{3 \text{ mol H}}$$

This ratio is also $2:3$, and again the empirical formula is C_2H_3.

Practice Problem 7.11   Write the empirical formula for a compound consisting of elements A and B for each ratio of A to B:

(a) $1:1$ (b) $1:1.5$ (c) $1:1.33$

(d) $1:1.25$ (e) $1:1.67$ (f) $1:1.75$ ■

We can find the empirical formula from percent composition data. The empirical formula represents a ratio; therefore, it does not depend on the size of the sample under consideration. Since the empirical formula reflects a *mole* ratio, and percent composition data are given in terms of *mass,* we have to convert the masses to moles. We then convert the mole ratio, which is unlikely to be an integral ratio, to the smallest possible whole-number ratio, from which we write the empirical formula.

The steps we take to obtain an empirical formula from percent composition data are as follows:

Step 1:   Change the percentages to numbers of grams (by assuming that 100.00 g of sample is present).

Step 2:   For each element, convert the number of grams to the number of moles.

Step 3:   Try to get an integral ratio by dividing *all* the numbers of moles by the magnitude of the smallest number of moles. This will make at least one number an integer.

Step 4:   If necessary, multiply *all* the numbers of moles by the same small integer to clear fractions. Round off the result to an integer only when the number of moles is within 1% of the integer. Always use at least three significant digits in empirical formula calculations; otherwise, rounding errors may produce an incorrect empirical formula.

■ **EXAMPLE 7.12**

Determine the empirical formula of a compound that has a percent composition of 43.66% O and 56.34% P.

Solution

Since the size of the sample does not matter in determining an empirical formula, you can assume a 100.00-g sample. That way, the percentages given are automatically equal numerically to the numbers of grams of the elements. For example:

$$100.00 \text{ g compound}\left(\frac{43.66 \text{ g O}}{\underset{\text{From the percentage}}{100.00 \text{ g compound}}}\right) = 43.66 \text{ g O}$$

You then convert the numbers of grams to moles:

$$43.66 \text{ g O}\left(\frac{1 \text{ mol O}}{16.00 \text{ g O}}\right) = 2.729 \text{ mol O}$$

$$56.34 \text{ g P}\left(\frac{1 \text{ mol P}}{30.97 \text{ g P}}\right) = 1.819 \text{ mol P}$$

You now have the mole ratio of phosphorus to oxygen, but it is not an integral ratio. The best way to try to get an integral ratio is to divide *each* of the numbers of moles by the magnitude of the lower number of moles:

$$\frac{2.729 \text{ mol O}}{1.819} = 1.500 \text{ mol O} \qquad \frac{1.819 \text{ mol P}}{1.819} = 1.000 \text{ mol P}$$

The numbers of moles are still not integers, but you can see that if you multiply *each* of them by 2, you will get an integral ratio:

$$2(1.500 \text{ mol O}) = 3.000 \text{ mol O} \qquad 2(1.000 \text{ mol P}) = 2.000 \text{ mol P}$$

The ratio is 3 mol O to 2 mol P, and the empirical formula is P_2O_3.

Note that you should *not* use the molecular mass of an oxygen *molecule* (or any other diatomic molecule) in empirical formula calculations because you are interested in the mole ratio involving oxygen *atoms*. Molecular oxygen, O_2, has nothing to do with empirical formula calculations.

Practice Problem 7.12 Determine the empirical formula of ethyl benzene, a compound containing 90.50% carbon and 9.50% hydrogen. ■

We can obtain an empirical formula from mass data instead of a percent composition.

■ EXAMPLE 7.13

Determine the empirical formula of a compound if a sample of the compound contains 20.47 g of iron and 7.821 g of oxygen.

Solution

Since the data are given in grams rather than percentages, you do not have to do the first step of changing to grams. You simply change the grams to moles:

$$20.47 \text{ g Fe}\left(\frac{1 \text{ mol Fe}}{55.85 \text{ g Fe}}\right) = 0.3665 \text{ mol Fe}$$

$$7.821 \text{ g O}\left(\frac{1 \text{ mol O}}{16.00 \text{ g O}}\right) = 0.4888 \text{ mol O}$$

Dividing by the smaller number of moles:

$$\frac{0.4888 \text{ mol O}}{0.3665} = 1.334 \text{ mol O} \qquad \frac{0.3665 \text{ mol Fe}}{0.3665} = 1.000 \text{ mol Fe}$$

Multiplying each number of moles by 3 yields 4 mol O and 3 mol Fe; the empirical formula is Fe_3O_4.

Practice Problem 7.13 Determine the empirical formula of a compound if a sample contains 35.88 g of sodium, 50.05 g of sulfur, and 37.46 g of oxygen. ■

A detective analyzes some white pills found in an aspirin bottle on an unidentified body. One pill is found to consist of 60.05% potassium, 18.44% carbon, and 21.51% nitrogen. A second pill contains 60.00% carbon, 4.48% hydrogen, and 35.53% oxygen and has a molar mass of 180 g/mol. Can these data help the detective deduce whether the death was accidental, murder, or suicide?

The percent composition shows that the first pill is KCN—potassium cyanide. The second pill could be aspirin—$C_9H_8O_4$. It looks like murder, since a cyanide pill is not likely to get into an aspirin bottle accidentally, and a person committing suicide is not likely to put a cyanide pill in an aspirin bottle. As in most human affairs, however, there is no certainty to this conclusion.

Most ionic compounds are identified by their empirical formulas, and such formulas are used for calculations involving these compounds. Such compounds as Na_2O_2, Hg_2Cl_2, $Na_2C_2O_4$, and $K_2S_2O_8$ are exceptions. For molecular substances, empirical formulas are used as a basis in determining molecular formulas, as described in the next section.

7.5 Molecular Formulas

OBJECTIVE

■ to determine the molecular formula from percent composition and molecular mass data or from the empirical formula and molecular mass data

The **molecular formula** gives not only the ratio of atoms of each element to atoms of every other element in a compound, but also the ratio of atoms of each element to molecules of the compound and the corresponding mole ratios. For example, C_2H_4 has a ratio of 2 moles of carbon atoms to 4 moles of hydrogen atoms, as well as a ratio of 2 moles of carbon atoms to 1 mole of C_2H_4 molecules. The molecular formula is always an integral multiple (1, 2, 3, . . .) of the empirical formula. Thus, the molecular formula gives all the information that the empirical formula gives plus the ratio of the number of moles of each element to the number of moles of the compound. Molecular formulas can be written only for compounds that exist in the form of molecules.

If we calculated the percent compositions of C_2H_2 and C_6H_6 (Figure 7.3), we would find that both have the same percentages of carbon and the same per-

Figure 7.3 Percent Compositions of Acetylene and Benzene

Consider a sample containing three molecules of acetylene, C_2H_2, and another sample containing one molecule of benzene, C_6H_6. Since both samples have the same number of carbon atoms (six) and both have the same number of hydrogen atoms (six), both obviously have the same percent composition. Since percent composition is an intensive property, the two compounds have the same percent composition, no matter how many molecules are present.

$$H—C\equiv C—H$$
$$H—C\equiv C—H$$
$$H—C\equiv C—H$$

Three molecules
of acetylene

One molecule
of benzene

centages of hydrogen (see Problem 7.42 at the end of the chapter). Both have the same empirical formula—CH. This result means that we cannot tell these two compounds apart from percent composition data alone. However, if we also have a molecular mass, we can use that information with the percent composition data to determine not only the empirical formula but also the molecular formula.

Determining the molecular formula of a compound involves first determining the empirical formula and then determining how many empirical formula units are in a molecule of the compound.

■ EXAMPLE 7.14

Determine the molecular formula of a hydrocarbon (a compound of carbon and hydrogen only) that contains 85.63% carbon and has a molar mass of 56.1 g/mol.

Solution

Since the total of the percentages must be 100.00%, the percentage of hydrogen in the compound must be 100.00% total − 85.63% C = 14.37% H.

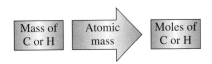

$$85.63 \text{ g C}\left(\frac{1 \text{ mol C}}{12.01 \text{ g C}}\right) = 7.130 \text{ mol C}$$

$$14.37 \text{ g H}\left(\frac{1 \text{ mol H}}{1.008 \text{ g H}}\right) = 14.26 \text{ mol H}$$

$$\frac{7.130 \text{ mol C}}{7.130} = 1.000 \text{ mol C} \qquad \frac{14.26 \text{ mol H}}{7.130} = 2.000 \text{ mol H}$$

The empirical formula is CH_2. Next, the molar mass (the number of grams per mole) is divided by the mass of a mole of empirical formula units to get the number of empirical formula units per molecule. The mass of 1.000 mol of CH_2 formula units is

$$12.01 \text{ g} + 2(1.008 \text{ g}) = 14.03 \text{ g}$$

Thus,

$$\frac{56.1 \text{ g/mol of molecules}}{14.03 \text{ g/mol of empirical formula units}} = \frac{4 \text{ mol of empirical formula units}}{1 \text{ mol of molecules}}$$

The empirical formula CH_2 is multiplied by 4 to get the molecular formula, C_4H_8.

Practice Problem 7.14   Determine the molecular formula of a hydrocarbon that contains 85.63% carbon and has a molar mass of 98.2 g/mol. ■

One important use of molecular formulas is to identify molecular compounds. If a chemist isolates a useful substance from a plant or animal source, the chemist wants to know the formula so that the compound can be made in the laboratory. Making a compound is often more convenient and more economical than obtaining it from its natural source. Certain vitamins and penicillin are examples.

■ ■ ■ ■ ■ ■ ■ SUMMARY ■ ■ ■ ■ ■ ■ ■

The formula mass (formula weight) of a substance is determined by adding the atomic masses (atomic weights) of each *atom* (not each element) in a formula unit. Molecular mass is one type of formula mass (for substances that form molecules) and is calculated in the same way as the formula mass for an ionic compound. For example, the formula mass of H_2O is 18.0 amu, the atomic mass of two hydrogen atoms plus that of one oxygen atom. Three or more significant digits should be used to report formula masses. (Section 7.1)

The percent composition is the percentage of each element in a compound. The percentage of an element in a compound is calculated by finding the ratio of the total mass of that element to the formula mass and multiplying by 100%. The percentages of all the elements in a compound should total 100% (within less than 1%). (Section 7.2)

The mole is defined as the number of ^{12}C atoms in exactly 12 g of ^{12}C, which is 6.02×10^{23}—Avogadro's number. The same number of moles of two (or more) different substances has the same number of formula units but *not* the same mass. The molar mass is the mass in grams of a mole of a substance. The number of grams per mole, or the molar mass, is a frequently used conversion factor, used for converting between grams and moles. (Section 7.3)

An empirical formula gives the lowest integral mole ratio of atoms of all the elements in a compound. An empirical formula is determined from a percent composition by changing the percentages to numbers of grams (by assuming a 100-g sample) and then dividing the number of grams of each element by its atomic mass in grams. The nonintegral mole ratio that results is converted to an integral mole ratio by dividing each of the numbers of moles by the smallest, and then, if necessary, multiplying every one of the quotients by the same small whole number. Never round off by more than 1% during this procedure. If data are given in grams rather than as a percent composition, simply omit the first step. (Section 7.4)

Molecular formulas give the same information as empirical formulas, as well as the ratio of the number of moles of each element to the number of moles of the compound. (Molecular formulas are used only for molecular substances, not ionic substances.) A molecular formula can be determined from the empirical formula of the compound and its formula mass: First, divide the formula mass by the mass in amu of one empirical formula unit, which will result in a small integer. Then, multiply each subscript of the empirical formula by that integer. (Section 7.5)

Items for Special Attention

■ A formula gives the mole ratio of one element to another. With a formula and atomic masses, you can calculate mass ratios—for example, percent by mass—rather easily.

■ To convert to or from moles, use Avogadro's *number* for *numbers* of formula units, the molar *mass* to convert to or from a *mass,* or the formula to convert from moles of atoms of an element to or from moles of atoms of another element or moles of entire formula units (see Figure 7.2). You must be sure to consider the units (for example, moles of atoms versus number of individual atoms versus mass of the atoms) as well as the species (for example, atoms of an element versus molecules of that element). These differences are apparent in the questions: "How many *atoms* of hydrogen are in 2.00 *mol* of hydrogen *molecules?*" and "What is the *mass* of the *atoms* of *hydrogen* in 1.00×10^{23} *molecules* of *water?*"

■ The most important skill learned in this chapter may be the ability to convert from grams to moles of a substance, and vice versa.

■ Do not use the term *molecule* or *molecular mass* when discussing ionic compounds, since they do not exist as molecules.

■ The unit usually used for the mass of a small number of atoms, molecules, or formula units is the amu, which is a very small fraction of a gram (about 10^{-24} g). The unit used for the mass of molar quantities of a substance is ordinarily the gram.

■ The percentages of the elements making up any compound must total 100%. If the percentages of all but one element are given, the percentage of that element can easily be calculated.

Self-Tutorial Problems

7.1 For each of the following, select the proper units from the following list: amu, grams, grams/mole.

(a) Formula mass   (b) Atomic mass

(c) Molar mass   (d) Mass

(e) Molecular mass

7.2 What small integer should you multiply each of the following ratios by to get a whole-number ratio? What ratio results in each case?

(a) $\dfrac{1.50}{1}$ (b) $\dfrac{1.33}{1}$ (c) $\dfrac{1.25}{1}$

(d) $\dfrac{1.67}{1}$ (e) $\dfrac{1.75}{1}$ (f) $\dfrac{2.25}{1}$

7.3 Which of the following are empirical formulas?

C_6H_{10} C_6H_{14} C_6H_{13} CH C_9H_{12}

7.4 A 100-g sample of a certain compound contains 87.0% carbon. What percentage of carbon is contained in a 14.0-g sample of the same compound?

7.5 What information do you need to determine each of the following?

(a) A formula mass   (b) An empirical formula

(c) A molecular formula

7.6 For this problem, assume that all the socks are identical.

(a) How many pairs of socks are in 3 dozen pairs of socks?

(b) How many socks are in 3 dozen socks?

(c) How many pairs of socks are in 3 dozen socks?

(d) How many socks are in 3 dozen pairs of socks?

(e) How many dozen pairs of socks are in 3 dozen socks?

(f) How many dozen socks are in 3 dozen pairs of socks?

7.7 (a) How many hydrogen molecules are in 3.00 mol of hydrogen molecules?

(b) How many hydrogen atoms are in 3.00 mol of hydrogen atoms?

(c) How many hydrogen molecules are in 3.00 mol of bonded hydrogen atoms?

(d) How many hydrogen atoms are in 3.00 mol of hydrogen molecules?

(e) How many moles of hydrogen molecules are in 3.00 mol of bonded hydrogen atoms?

(f) How many moles of hydrogen atoms are in 3.00 mol of hydrogen molecules?

7.8 Which of the following formulas identify ionic compounds but are not empirical formulas?

CaH_2 C_2H_4 Na_2SO_4 Na_2O_2 $Na_2S_2O_8$

7.9 What is the empirical formula of each of the following?

(a) C_3H_8 (b) $C_{10}H_8O_2$ (c) $C_{12}H_{22}O_{11}$ (d) $C_2H_4O_2$

7.10 (a) Which weighs more—a dozen grapes or a dozen watermelons? Which contains the greater number of fruits?

(b) Which weighs more—a mole of uranium atoms or a mole of lithium atoms? Which contains more atoms?

7.11 (a) Compare the mass of a dozen socks rolled into pairs with the mass of the same socks unrolled.

(b) Compare the mass of 1.00 mol of nitrogen atoms with that of the same atoms bonded into N_2 molecules.

(c) Compare the mass of 1.00 mol of nitrogen atoms with that of 1.00 mol of nitrogen molecules.

7.12 (a) What is the difference between the atomic mass of fluorine and the molecular mass of fluorine?

(b) Why is the phrase "molar mass of fluorine" ambiguous?

(c) To what does "molar mass of fluorine gas" refer?

7.13 State whether the percent composition, the empirical formula, or the molecular formula gives the information specified in each part.

(a) Ratio of masses of each element to mass of compound

(b) Ratio of moles of each element to each other element, and no more

(c) Ratio of moles of each element to moles of compound

7.14 What conversion factor is used to convert a number of moles of a substance to the number of grams of the substance?

7.15 What conversion factor(s) is (are) used to convert a number of formula units of a substance to a mass of that substance?

7.16 Which of the following substances have molecular masses? Which have molar masses?

AlF_3 NH_3 S_8 Ne

7.17 What can you determine from percent composition data?

7.18 What information is given in the formula $(NH_4)_2SO_3$?

7.19 (a) If a shirt box is 1.0 inch high, how many feet high is a stack of a dozen shirt boxes?

(b) If a shoe box is 4.0 inches high, how many feet high is a stack of a dozen shoe boxes?

(c) If a hat box is 8.0 inches high, how many feet high is a stack of a dozen hat boxes?

(d) Explain your results.

7.20 Calculate the formula mass of each of the following to two decimal places twice, first by rounding atomic masses and then by using the entire number of significant digits in the atomic masses and rounding the formula mass:

(a) C_4H_{10} (b) $NaCl$ (c) B_2H_6

■ ■ ■ PROBLEMS ■ ■ ■

7.1 Formula Masses

7.21 Calculate the formula mass of each of the following compounds to one decimal place:

    (a) Na_2SO_4 (b) $Ca(CNO)_2$

    (c) H_3PO_4 (d) $(NH_4)_2CO_3$

    (e) C_8H_{18} (f) LiH

7.22 Calculate the molar mass of each of the following compounds to three significant figures:

    (a) $(NH_4)_2SO_3$ (b) $Ba(NO_3)_2$ (c) $KMnO_4$

7.23 What is the smallest formula mass known for (a) any atom and (b) any molecule?

7.2 Percent Composition

7.24 Calculate the percent composition of nitroglycerin, $C_3H_5N_3O_9$.

7.25 Calculate the percent composition of vitamin D_2, $C_{28}H_{44}O$.

7.26 Calculate the percent composition of vitamin E, $C_{29}H_{50}O_2$.

7.27 Calculate the percent compositions of cyclobutadiene, C_4H_4, and ethylnaphthalein, $C_{12}H_{12}$. Compare the values, and explain the results.

7.28 Calculate the percent compositions of pentene, C_5H_{10}, and cycloheptane, C_7H_{14}. Compare the values, and explain the results.

7.29 Calculate the percent chlorine in DDT, $C_{14}H_9Cl_5$, an insecticide that has been discontinued because it does not biodegrade.

7.30 Calculate the percent composition of ethylene glycol, $C_2H_6O_2$, commonly used as a permanent antifreeze in cars.

7.31 Calculate the percent composition of aspirin, $C_9H_8O_4$.

7.32 Calculate the percent composition of vitamin A, $C_{20}H_{30}O$.

7.3 The Mole

7.33 How many moles of atoms of each element are present in 1.00 mol of each of the following compounds?

    (a) Na_2SO_4 (b) $CoCO_3$

    (c) Cu_2O (d) Hg_2Cl_2

7.34 How many moles of atoms of each element are present in 1.00 mol of each of the following compounds?

    (a) $Cr(ClO_4)_3$ (b) $(NH_4)_2SO_4$

    (c) BaS (d) $KHCO_3 \cdot MgCO_3 \cdot 4H_2O$

7.35 How many moles of atoms of each element are present in 1.00 mol of each of the following compounds?

    (a) $Al_2(Cr_2O_7)_3$ (b) $Ba(OH)_2$ (c) $(NH_4)_2S$

7.36 (a) Calculate the number of moles of C_2H_6 in 3.000 g of C_2H_6.

(b) Calculate the number of moles of carbon atoms in 3.000 g of C_2H_6.

(c) Calculate the number of individual carbon atoms in 3.000 g of C_2H_6.

7.37 Calculate the number of moles in 7.30 g of a covalent compound with a molar mass of 130.1 g/mol.

7.38 Calculate the number of moles of table sugar—sucrose, $C_{12}H_{22}O_{11}$—in 28.35 g (1.000 ounce).

7.39 Calculate the number of grams of formaldehyde, CH_2O, in 17.0 mol of formaldehyde.

7.40 Calculate the number of moles of methane (natural gas), CH_4, in 235 g of methane.

7.41 Calculate the mass of 2.50 mol of (a) unbonded chlorine atoms, (b) chlorine atoms bonded into Cl_2 molecules, and (c) Cl_2 molecules.

7.42 Calculate the percent compositions of acetylene, C_2H_2, and benzene, C_6H_6. Compare the values, and explain the results.

7.43 Calculate the percent compositions of pentene, C_5H_{10}, and cyclohexane, C_6H_{12}. Compare the values, and explain the results.

7.44 Calculate the number of carbon atoms in 3.000 mol of C_4H_{10}.

7.45 Calculate the number of hydrogen atoms in 22.20 mol of NH_3.

7.46 Calculate the number of moles of H_2SO_4 that contains 6.78×10^{22} oxygen atoms.

7.47 Calculate the number of moles of water that contains 7.77×10^{26} hydrogen atoms.

7.48 Calculate the number of grams of silver in 2.30 mol of $AgBr$.

7.49 Calculate the number of carbon atoms in 2.278 g of C_2H_6O.

7.50 Calculate the number of hydrogen atoms in 29.1 g of C_6H_6.

7.51 Calculate the mass of $Co(ClO_3)_2$ that contains 7.18×10^{20} chlorine atoms.

7.52 Calculate the mass of water that contains 1.27×10^{24} hydrogen atoms.

7.53 Calculate the number of grams of bromine in 7.29 mol of $AgBr$.

7.54 Calculate the number of molecules in 72.4 g of a compound with a molar mass of 90.5 g/mol.

7.55 Calculate the mass of the simple sugar glucose, $C_6H_{12}O_6$, containing 5.88×10^{24} molecules.

7.56 Calculate the number of grams of hydrogen in 22.7 mol of formaldehyde, CH_2O.

7.57 Calculate the mass of hydrogen in 7.00 g of CH_4, methane (natural gas).

7.58 Calculate the number of molecules of C_2H_4 containing 6.098×10^{24} C atoms.

7.59 Calculate the mass of Mg in 7.12×10^{22} formula units of Mg_3N_2.

7.60 Calculate the number of chlorine atoms in (a) 17.73 g of SCl_2 and (b) 17.73 g of $SOCl_2$.

7.61 Calculate the mass in grams of one ^{12}C atom.

7.4 Empirical Formulas

7.62 Decide whether or not each of the following is an empirical formula:

(a) Al_2Cl_6 (b) $NaClO_2$ (c) $MgCl_2O_4$

(d) Hg_2Cl_2 (e) S_8

7.63 Which of the following is an empirical formula?

$$C_4H_8 \qquad C_4H_9 \qquad C_4H_{10}$$

7.64 Find the empirical formula of glycine, the simplest amino acid, from its percent composition: 32.00% C, 6.71% H, 18.66% N, 42.63% O.

7.65 Determine the empirical formula of a compound whose percent composition is 40.04% S and 59.96% O. Use at least three significant digits in your calculations, and comment on why that is necessary.

7.66 TNT, trinitrotoluene, is composed of 37.01% carbon, 2.22% hydrogen, 18.50% nitrogen, and 42.26% oxygen. Determine its empirical formula.

7.67 Testosterone, a male hormone, is composed of 79.12% carbon, 9.79% hydrogen, and 11.10% oxygen. What is its empirical formula?

7.68 Determine an empirical formula from each of the following sets of percent composition data:

(a) 47.05% K, 14.45% C, 38.50% O

(b) 77.26% Hg, 9.25% C, 1.17% H, 12.32% O

(c) 43.64% P, 56.36% O

(d) 72.03% Mn, 27.97% O

(e) 87.73% C, 12.27% H

7.69 Polystyrene, a well-known plastic, is composed of 92.26% carbon and 7.74% hydrogen.

(a) How many grams of each element are in 100.0 g of polystyrene?

(b) How many moles of each element are in 100.0 g of polystyrene?

(c) What is the mole ratio in integers?

(d) What is the empirical formula?

7.70 Polystyrene, a well-known plastic, is composed of 92.26% carbon and 7.74% hydrogen. Determine its empirical formula.

7.71 Styrene, used in manufacturing a well-known plastic, is composed of 92.26% carbon and 7.74% hydrogen. Determine its empirical formula.

7.72 Determine an empirical formula from each of the following sets of percent composition data:

(a) 40.00% C, 6.71% H, 53.29% O

(b) 47.35% C, 10.60% H, 42.05% O

(c) 76.57% C, 6.43% H, 17.00% O

(d) 56.34% P, 43.66% O

(e) 36.85% N, 63.15% O

7.73 Determine an empirical formula from each of the following sets of percent composition data:

(a) 36.74% C, 5.14% H, 58.12% F

(b) 20.86% C, 2.19% H, 76.95% Cl

(c) 16.66% C, 1.40% H, 81.94% Cl

(d) 39.12% C, 8.76% H, 52.12% O

7.74 Calculate the empirical formula of each of the substances from the following analyses:

(a) 43.28 g C, 6.04 g H, 127.76 g Cl

(b) 73.67 g C, 4.94 g H, 93.22 g F

7.5 Molecular Formulas

7.75 Determine the molecular formula of a substance if its empirical formula is NO_2 and its molar mass is (a) 46.0 g/mol and (b) 92.0 g/mol.

7.76 Find the molecular formula of a substance if the empirical formula is CH_2 and its molar mass is (a) 56.0 g/mol, (b) 70.0 g/mol, (c) 98.0 g/mol, and (d) 294 g/mol.

7.77 Determine the molecular formula of glucose, a simple sugar, from its percent composition of 40.0% C, 6.67% H, 53.3% O, and its molar mass of 180 g/mol.

7.78 Calculate the molecular formula of each of the substances from the following analyses:

(a) 2.24 g H, 26.6 g C, 71.1 g O, MM = 90.0 g/mol

(b) 31.99 g C, 5.37 g H, 28.42 g O, MM = 74.1 g/mol

(c) 27.56 g C, 6.17 g H, 36.72 g O, MM = 92.0 g/mol

(d) 7.999 g C, 0.448 g H, 3.109 g N, 10.66 g O, MM = 200 g/mol

7.79 Find the molecular formula for toluene, which is composed of 91.25% carbon and 8.75% hydrogen and has a molar mass of 92.13 g/mol.

7.80 The most widely used antifreeze, ethylene glycol, is composed of 38.70% carbon, 9.74% hydrogen, and 51.56% oxygen. Its molar mass is 62.07 g/mol. Find its molecular formula.

7.81 Octane and heptane are two ingredients of gasoline. Octane has 84.12% carbon and 15.88% hydrogen, and heptane has 83.90% carbon and 16.10% hydrogen. Their molecular masses are 114 amu and 100 amu, respectively. What are their molecular formulas?

7.82 White phosphorus is one form of elemental phosphorus. Its molar mass is 123.9 g/mol. Calculate its molecular formula.

■ ■ ■ **GENERAL PROBLEMS** ■ ■ ■

7.83 Calculate the percent error in rounding off the atomic mass of each of the following elements to three significant digits:

    (a) Oxygen   (b) Carbon

    (c) Chlorine   (d) Hydrogen

7.84 (a) Vitamin B_{12} has one cobalt atom per formula unit. The compound is 4.348% Co. Calculate its molar mass.

    (b) Vitamin D_1 has two oxygen atoms per formula unit. The compound is 4.03% O. Calculate its molar mass.

7.85 Calculate the number of molecules of vitamin A, $C_{20}H_{30}O$, in 1000 mg of vitamin A.

7.86 Calculate the number of carbon atoms in 1.000 gallon of octane, C_8H_{18}, a major component of gasoline (1 gallon = 3.785 L; density = 0.7025 g/mL).

7.87 Calculate the number of carbon atoms in 75.0 g of a compound that contains 91.25% carbon and 8.75% hydrogen.

7.88 Calculate the number of molecules in 22.8 g of a compound that has a molar mass of 92.13 g/mol and contains 91.25% carbon and 8.75% hydrogen.

7.89 Calculate the number of hydrogen atoms in 9.92 g of a compound whose percent composition is 7.74% H and 92.26% C.

7.90 A sample is 40.0% NaCl by mass, and the rest is water. Calculate the number of molecules of water in 25.7 g of the sample.

7.91 A different method may be used to calculate the molecular formula from percent composition data plus a molar mass: First, calculate the mass of each element in 1.00 mol of compound. Next, calculate the number of moles of each of the elements in the mole of compound, and those results yield the molecular formula. Use this method to calculate the molecular formula of a hydrocarbon (a compound of carbon and hydrogen only) that contains 83.62% C and has a molar mass of 86.2 g/mol.

7.92 Calculate the molecular formula of a substance if its percent composition is 79.91% C and 20.09% H, and its molar mass is approximately 30 g/mol.

7.93 Calculate the percent composition of soluble saccharin, $C_7H_4NNaO_3S$.

7.94 Calculate the molecular formula of a substance if its percent composition is 85.63% C and 14.37% H, and its molar mass is approximately 41 g/mol.

7.95 A scientist isolates a pure substance from a newly discovered plant in the Amazon River basin. What data does the scientist need to start to determine whether the substance is a new compound and what its formula is?

7.96 How many moles of carbon atoms is present in the quantity of C_2H_6O that contains 23.10 g of hydrogen?

7.97 You will not be able to solve the following problem until you have studied Chapter 12: If 3.00 g of a gaseous compound composed of 80.0% carbon and 20.0% hydrogen occupies 4.00 L at 25°C and 1.00 atmosphere pressure, what is the molecular formula of the gas?

    (a) State what you *can* calculate.

    (b) What value can the rest of the data be used to calculate?

8 Chemical Reactions

- 8.1 The Chemical Equation
- 8.2 Balancing Equations
- 8.3 Predicting the Products of Chemical Reactions
- 8.4 Acids and Bases

■ Key Terms (Key terms are defined in the Glossary.)

acid (8.4)
acid carbonate (8.4)
acid salt (8.4)
acidic anhydride (8.4)
active (8.3)
anhydride (8.4)
Arrhenius theory (8.4)
balanced equation (8.1)
base (8.4)
basic anhydride (8.4)
carbonate (8.4)
catalyst (8.3)
coefficient (8.1)
combination reaction (8.3)

combustion reaction (8.3)
decomposition reaction (8.3)
displacement reaction (8.3)
double displacement reaction (8.3)
double substitution reaction (8.3)
equation (8.1)
indicator (8.4)
metathesis reaction (8.3)
molten (8.3)
neutral (8.4)
neutralization reaction (8.4)
precipitate (8.3)
product (8.1)

reactant (8.1)
reactive (8.3)
reagent (8.1)
salt (8.4)
solubility (8.3)
stability (8.3)
state (8.2)
strong acid (8.4)
strong base (8.4)
substitution reaction (8.3)
ternary compound (8.3)
weak acid (8.4)
weak base (8.4)

■ Symbols

(aq) (aqueous solution) (8.2)
(g) (gas) (8.2)

(l) (liquid) (8.2)
nr (no reaction) (8.3)

(s) (solid) (8.2)

In Chapter 7, you learned how to do numerical calculations for compounds, using their formulas as a basis. This chapter lays the foundation for doing similar calculations for chemical reactions, using the balanced equation as a basis. The chemical equation is introduced in Section 8.1, and methods for balancing equations are presented in Section 8.2. To write equations, you must often be able to predict the products of a reaction from a knowledge of the properties of the reactants. Section 8.3 shows how to classify chemical reactions into types to predict the products of thousands of reactions. An important type of reaction—the acid-base reaction—is discussed in Section 8.4.

8.1 The Chemical Equation

In a chemical reaction, the substances that react are called **reactants,** or sometimes, **reagents.** The substances that are produced are called **products.** During the reaction, some or all of the atoms of the reactants change their bonding. For example, two hydrogen molecules can react with an oxygen molecule to form two water molecules (Figure 8.1). In a less familiar example—the reaction of BaI_2 with AgF—the products are BaF_2 and AgI. The metal ions have merely traded nonmetal ions. All of the ions are still present after the reaction; they have just "changed partners." We can describe the reaction, pictured in Figure 8.2, in words:

Barium iodide plus silver fluoride yields barium fluoride plus silver iodide.

We use formulas to represent the substances involved in a reaction when we write a chemical equation. In an **equation,** the formulas for reactants are placed on the left side of the arrow and those for products are placed on the right side. Either substance may be written first on each side of the equation:

$$BaI_2 + AgF \rightarrow BaF_2 + AgI \qquad \textit{(Not balanced)}$$

Even better, we can write a **balanced equation,** which shows the relative numbers of atoms of each of the elements involved. The unbalanced equation just presented seems to indicate that an iodide ion has disappeared during the reaction and that a fluoride ion has appeared from nowhere. As written, that equation violates the law of conservation of mass. Thus, you must always write balanced equations for reactions. The word *equation* is related to the word *equal;* an equation must have equal numbers of atoms of each element on each side. Such an equation is said to be *balanced.*

Coefficients—numbers that are written before the formulas—tell the relative numbers of formula units of reactants and products involved in a reaction and bal-

Figure 8.1 Reaction of Hydrogen and Oxygen

The bonds in the diatomic molecules H_2 and O_2 are broken, and new bonds are formed between hydrogen and oxygen atoms.

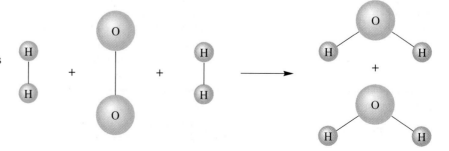

Figure 8.2 Reaction of Aqueous Barium Iodide with Aqueous Silver Fluoride

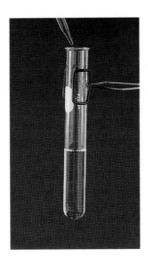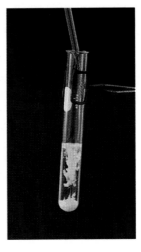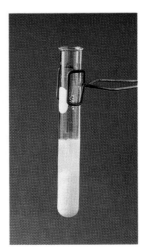

ance the number of atoms of each element involved. *The coefficient multiplies everything in the formula:*

$$BaI_2 + 2\,AgF \rightarrow BaF_2 + 2\,AgI \qquad \text{(Balanced)}$$

Coefficients

In a balanced chemical equation, the absence of a coefficient before a formula implies a coefficient of 1. The two formula units of AgF are composed of two Ag^+ ions and two F^- ions. The two F^- ions in one formula unit of BaF_2 come from the two formula units of AgF.

The balanced equation for the reaction of hydrogen and oxygen, illustrated in Figure 8.1, is

$$2\,H_2 + O_2 \rightarrow 2\,H_2O$$

> The coefficients in a balanced equation give the ratio of moles of each reactant and product to moles of any other reactant or product.

The coefficients in a balanced equation give the *ratio of moles* of each reactant and product to *moles* of any other reactant or product. They also give the ratio of formula units of each reactant and product to formula units of any other reactant or product. The balanced chemical equation is the cornerstone from which we can calculate how much of one substance reacts with or is produced by a certain quantity of another substance (to be covered in Chapter 10).

8.2 Balancing Equations

OBJECTIVE

■ to balance chemical equations—that is, to get the same number of atoms of each element on each side

The first major task of this chapter is to learn to balance equations for chemical reactions. Balancing simple equations will be covered in this chapter; equations for a more complex type of reaction will be considered in Chapter 16.

The first step in writing a complete and balanced equation for a chemical reaction is to write correct formulas for the reactants and products. To help you as you learn, you might write the equation in words and then write the formulas. Correct formulas cannot be changed to make an equation balance! Only after the correct formulas have been written can we go on to the next step. Then, we use coefficients to change the numbers of formula units to get the same number of atoms of each element on the two sides of the equation.

For example, the unbalanced equation for the reaction of carbon monoxide, CO, with oxygen, O_2, to give carbon dioxide, CO_2, is

$$CO + O_2 \rightarrow CO_2 \qquad \textit{(Not balanced)}$$

With one molecule of each substance, the numbers of oxygen atoms on the two sides of the equation are not equal, so the equation is not balanced. We can balance the equation by inserting proper coefficients in front of the formulas:

$$2\,CO + O_2 \rightarrow 2\,CO_2 \qquad \textit{(Balanced)}$$

The number 2 before the CO indicates that there are two carbon monoxide molecules, containing two carbon atoms and two oxygen atoms. There are two more oxygen atoms in the O_2 molecule. Since there are two carbon atoms and four oxygen atoms in the two CO_2 molecules, the equation is now balanced. You should always check an equation after you balance it to make sure that the numbers of atoms of each element on each side of the arrow are equal.

One of the problems encountered by students just learning to balance equations is that the absence of a coefficient in a balanced equation means a coefficient of 1, but the absence of a coefficient before the equation is fully balanced might mean that this substance has not yet been considered. To avoid any confusion, you can place a question mark before each formula when you start to balance an equation.

■ EXAMPLE 8.1

Balance the equation for the reaction of barium chloride and sodium carbonate to give barium carbonate and sodium chloride.

Solution

The first step is to write correct formulas for the reactants and products, as discussed in Chapter 5. Do not write incorrect formulas to make balancing easier; write correct formulas!

$$BaCl_2 + Na_2CO_3 \rightarrow BaCO_3 + NaCl \qquad \textit{(Not balanced)}$$

Insert question marks before each formula to avoid confusion later:

$$?\,BaCl_2 + ?\,Na_2CO_3 \rightarrow ?\,BaCO_3 + ?\,NaCl$$

Next, replace the question mark before one of the reactants or products with a 1; choosing the most complicated-looking formula is often the easiest approach.

$$?\,BaCl_2 + 1\,Na_2CO_3 \rightarrow ?\,BaCO_3 + ?\,NaCl$$

The one CO_3^{2-} in 1 Na_2CO_3 yields one CO_3^{2-} in $BaCO_3$; the two Na^+ ions in 1 Na_2CO_3 yield 2 NaCl:

$$?\,BaCl_2 + 1\,Na_2CO_3 \rightarrow 1\,BaCO_3 + 2\,NaCl$$

The $BaCl_2$ is balanced by considering the two Cl^- ions in 2 NaCl or the one Ba^{2+} ion in 1 $BaCO_3$.

$$1\,BaCl_2 + 1\,Na_2CO_3 \rightarrow 1\,BaCO_3 + 2\,NaCl$$

The equation is now balanced, but you need to delete any coefficients equal to 1:

$$BaCl_2 + Na_2CO_3 \rightarrow BaCO_3 + 2\,NaCl$$

Check: 1 Ba, 2 Cl, 2 Na, 1 C, 3 O on each side.

Practice Problem 8.1 Balance the equation for the reaction of barium hydroxide with hydrobromic acid to yield barium bromide and water. ■

If the initial placement of the coefficient 1 yields fractional coefficients in the equation, you can get integer values by simply multiplying every coefficient (including the coefficients equal to 1) by a small integer that will clear the fractions.

■ **EXAMPLE 8.2**

Balance the equation for the reaction of $FeCl_2$ and Cl_2, which produces $FeCl_3$.

Solution

$$1\,FeCl_2 + ?\,Cl_2 \rightarrow ?\,FeCl_3$$

Balancing iron:	$1\,FeCl_2 + ?\,Cl_2 \rightarrow 1\,FeCl_3$
Balancing chlorine:	$1\,FeCl_2 + \frac{1}{2}\,Cl_2 \rightarrow 1\,FeCl_3$
Clearing the fraction:	$2\,FeCl_2 + 1\,Cl_2 \rightarrow 2\,FeCl_3$
Final equation:	$2\,FeCl_2 + Cl_2 \rightarrow 2\,FeCl_3$

Check: 2 Fe, 6 Cl on each side.

Practice Problem 8.2 Balance an equation for the reaction of CH_2O with O_2 to yield CO and H_2O. ■

When any element appears in more than one substance on the same side of the equation, you should balance that element last.

■ **EXAMPLE 8.3**

Write a balanced equation for the following reaction:

$$NaBrO_3 + NaBr + HCl \rightarrow Br_2 + H_2O + NaCl$$

Solution

Place a 1 before $NaBrO_3$, the most complicated-looking formula, and a question mark before each of the other formulas:

$$1\,NaBrO_3 + ?\,NaBr + ?\,HCl \rightarrow ?\,Br_2 + ?\,H_2O + ?\,NaCl$$

Since sodium and bromine appear in two compounds on the left side of the equation, start working with oxygen:

$$1\,NaBrO_3 + ?\,NaBr + ?\,HCl \rightarrow ?\,Br_2 + 3\,H_2O + ?\,NaCl$$

Knowing how many hydrogen atoms are on the right side of the equation, you can balance hydrogen next:

$$1\,NaBrO_3 + ?\,NaBr + 6\,HCl \rightarrow ?\,Br_2 + 3\,H_2O + ?\,NaCl$$

Now balance chlorine:

$$1 \text{ NaBrO}_3 + ? \text{ NaBr} + 6 \text{ HCl} \rightarrow ? \text{ Br}_2 + 3 \text{ H}_2\text{O} + 6 \text{ NaCl}$$

Balance the sodium, being careful to note that there is already one Na^+ in the 1 NaBrO_3:

$$1 \text{ NaBrO}_3 + 5 \text{ NaBr} + 6 \text{ HCl} \rightarrow ? \text{ Br}_2 + 3 \text{ H}_2\text{O} + 6 \text{ NaCl}$$

Finally, balance the bromine:

$$1 \text{ NaBrO}_3 + 5 \text{ NaBr} + 6 \text{ HCl} \rightarrow 3 \text{ Br}_2 + 3 \text{ H}_2\text{O} + 6 \text{ NaCl}$$

The coefficient 1 may now be deleted:

$$\text{NaBrO}_3 + 5 \text{ NaBr} + 6 \text{ HCl} \rightarrow 3 \text{ Br}_2 + 3 \text{ H}_2\text{O} + 6 \text{ NaCl}$$

Check: 6 Na, 6 Br, 3 O, 6 H, 6 Cl on each side.

Practice Problem 8.3   Balance an equation for the reaction of CCl_4 with O_2 to yield COCl_2 and Cl_2. ■

■ **EXAMPLE 8.4**

Write a balanced equation for the reaction of sodium hydroxide with phosphoric acid to produce disodium monohydrogen phosphate and water.

Solution

First, write the correct formulas for the reactants and products, and place a question mark before each formula:

$$? \text{ NaOH} + ? \text{ H}_3\text{PO}_4 \rightarrow ? \text{ Na}_2\text{HPO}_4 + ? \text{ H}_2\text{O}$$

Replace the question mark before the most complicated-looking formula with a 1:

$$? \text{ NaOH} + ? \text{ H}_3\text{PO}_4 \rightarrow 1 \text{ Na}_2\text{HPO}_4 + ? \text{ H}_2\text{O}$$

Balance the sodium and phosphorus atoms first, since oxygen and hydrogen appear in two substances on each side of the equation:

$$2 \text{ NaOH} + 1 \text{ H}_3\text{PO}_4 \rightarrow 1 \text{ Na}_2\text{HPO}_4 + ? \text{ H}_2\text{O}$$

Balance the hydrogen atoms:

$$2 \text{ NaOH} + 1 \text{ H}_3\text{PO}_4 \rightarrow 1 \text{ Na}_2\text{HPO}_4 + 2 \text{ H}_2\text{O}$$

The oxygen atoms have been balanced in the process. Finally, eliminate the coefficients equal to 1:

$$2 \text{ NaOH} + \text{H}_3\text{PO}_4 \rightarrow \text{Na}_2\text{HPO}_4 + 2 \text{ H}_2\text{O}$$

Check: 2 Na, 6 O, 5 H, 1 P on each side.

Practice Problem 8.4   Balance an equation for the reaction of sodium sulfide with iron(II) chloride to yield iron(II) sulfide and sodium chloride. ■

Any polyatomic ion that maintains its composition through an entire reaction, such as the carbonate ion in Example 8.1, can be balanced as a group of atoms, instead of balancing the individual elements.

Information about the **state** of a reactant or product (whether it is present as a solid, liquid, gas, or solute) may be given in a chemical equation. The following abbreviations are used: solid (s), liquid (l), gas (g), and solute in aqueous solution (aq). Older practice was to use a downward-pointing arrow or underlining to show that a solid is formed from reactants in solution; an upward-pointing arrow or an overbar indicated formation of a gas. Thus, the reaction of silver nitrate with sodium chloride can be represented by any of the following equations:

$$AgNO_3(aq) + NaCl(aq) \rightarrow AgCl(s) + NaNO_3(aq)$$

$$AgNO_3 + NaCl \rightarrow AgCl\downarrow + NaNO_3$$

$$AgNO_3 + NaCl \rightarrow \underline{AgCl} + NaNO_3$$

The reaction of aqueous sodium carbonate with a solution of hydrochloric acid can be represented by any of the following equations:

$$Na_2CO_3(aq) + 2\ HCl(aq) \rightarrow 2\ NaCl(aq) + H_2O(l) + CO_2(g)$$

$$Na_2CO_3 + 2\ HCl \rightarrow 2\ NaCl + H_2O + CO_2\uparrow$$

$$Na_2CO_3 + 2\ HCl \rightarrow 2\ NaCl + H_2O + \overline{CO_2}$$

8.3 Predicting the Products of Chemical Reactions

- to predict the products of thousands of chemical reactions by categorizing reactions and following a few simple rules

Simple chemical reactions can be divided into the following classes:

1. Combination reactions
2. Decomposition reactions
3. Substitution (or displacement) reactions
4. Double substitution (or double displacement) reactions

In addition, there are the reactions of most elements and of many compounds with oxygen:

5. Combustion reactions

Another, more complicated, class of reaction will be presented in Chapter 16, and other complex reactions are covered in more advanced chemistry courses.

Combination Reactions

Combination reactions involve the reaction of two (or more) reactants to form one product. Perhaps the easiest combination reaction to recognize is one in which two free elements (at least one of which is a nonmetal) react with each other. The elements can do little except react with each other (or not react at all). For example, if we treat magnesium metal with chlorine gas, the elements can combine to form magnesium chloride:

$$Mg(s) + Cl_2(g) \rightarrow MgCl_2(s)$$

The formula for the product of a combination reaction must be written according to the rules presented in Chapter 5. *After* the product has been represented by the proper formula, the equation is balanced, as shown in Section 8.2.

■ EXAMPLE 8.5

Complete and balance an equation for the reaction of sodium metal and chlorine gas.

Solution

Step 1:   Identify the product—sodium chloride.

Step 2:   Write correct formulas for reactant(s) and product(s). Remember that chlorine is one of the seven elements that occur as diatomic molecules when uncombined with other elements (Figure 5.3):

$$Na(s) + Cl_2(g) \rightarrow NaCl(s) \qquad \textit{(Not balanced)}$$

Step 3:   Balance the equation:

$$2\,Na(s) + Cl_2(g) \rightarrow 2\,NaCl(s) \qquad \textit{(Balanced)}$$

Practice Problem 8.5   Complete and balance an equation for the reaction of magnesium metal with nitrogen gas. ■

When two nonmetallic elements combine, the product formed often depends on the relative quantities of the reactants present. For example, when carbon combines with oxygen, either of two possible compounds may be produced—carbon monoxide or carbon dioxide. When the supply of oxygen is limited, carbon monoxide is produced, but when excess oxygen is available, carbon dioxide results:

$$2\,C(s) + O_2(g,\ \text{limited quantity}) \rightarrow 2\,CO(g)$$

$$C(s) + O_2(g,\ \text{excess}) \rightarrow CO_2(g)$$

Of course, not every pair of elements will react with each other. For example, you know that the noble gases are quite stable in their elemental forms.

■ EXAMPLE 8.6

Predict what will happen when neon is treated with fluorine gas.

Solution

Nothing will happen; neon is too stable to react. In equation format,

$$Ne + F_2 \rightarrow nr \qquad \textit{(nr stands for "no reaction")}$$

Practice Problem 8.6   Predict the products of the reaction, if any, of calcium metal with potassium metal. ■

In another type of combination reaction, a compound may be able to combine with a particular free element to form another compound as the only product. This occurs most often when the free element is the same as one of the elements in the original compound. An example of such a combination reaction is

$$2\,FeCl_2(s) + Cl_2(g) \rightarrow 2\,FeCl_3(s)$$

ITEM OF INTEREST

What happens in the atmosphere to the carbon monoxide generated in automobile engines? The carbon monoxide gas reacts slowly with the oxygen gas in the air to produce carbon dioxide gas:

$$2\,CO(g) + O_2(g) \rightarrow 2\,CO_2(g)$$

Here, the element chlorine combines with a compound of iron and chlorine—iron(II) chloride—to form another compound of iron and chlorine—iron(III) chloride—in which the iron ion has a different positive charge.

In yet another type of combination reaction, two compounds containing the same element may be able to combine to form a single, more complex compound. The element the reactants have in common is very often oxygen:

$$CaO(s) + CO_2(g) \rightarrow CaCO_3(s)$$

$$MgO(s) + H_2O(l) \rightarrow Mg(OH)_2(s)$$

Decomposition Reactions

Decomposition reactions have the opposite effect from combination reactions. In a decomposition reaction, a single compound can decompose to two elements, to an element and a simpler compound, to two simpler compounds, or (rarely) to another combination of products. **Ternary compounds,** compounds containing three elements, do not decompose into three uncombined elements.

Input of energy in some form is often required to get a compound to decompose:

$$2\,H_2O(l) \xrightarrow[\text{Na}_2\text{SO}_4(aq)]{\text{Electricity}} 2\,H_2(g) + O_2(g) \quad \text{Two elements}$$

$$2\,H_2O_2(l) \xrightarrow{\text{Heat}} 2\,H_2O(l) + O_2(g) \quad \text{A simpler compound and an element}$$

$$CaCO_3(s) \xrightarrow{\text{Heat}} CaO(s) + CO_2(g) \quad \text{Two simpler compounds}$$

To get compounds to decompose using electricity, ions must be present, and the sample must be in some liquid form. (Electricity does not pass through solid ionic compounds, even though they are composed of positive and negative ions.) The ions in a liquid are free to move and thus conduct the current. The liquid can be a **molten** (melted) ionic substance or a solution of an ionic substance in water or another liquid (Figure 8.3). If a solution is used, the ionic compound must decompose more easily than the water or other liquid used to form the solution.

Note in the previous equations that the formulas of elemental hydrogen and oxygen are written as diatomic molecules—H_2 and O_2. Before equations are balanced, the formulas for all products must be written according to the rules given in Chapter 5.

Decomposition reactions are often used to prepare elements. Joseph Priestley (1733–1804), the discoverer of oxygen, used the decomposition of mercury(II) oxide, HgO, to prepare elemental oxygen (and free mercury):

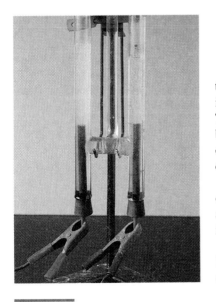

Figure 8.3 Electrolysis Reaction

$$2\,HgO(s) \xrightarrow{\text{Heat}} 2\,Hg(l) + O_2(g)$$

Table 8.1   Most Common Types of Decomposition Reactions

Reactant	Products	Example
Binary compound	→ Two elements	$2\,H_2O \xrightarrow{\text{Electricity}} 2\,H_2 + O_2$
Binary compound	→ Compound + element	$2\,H_2O_2 \rightarrow 2\,H_2O + O_2$
Ternary compound	→ Compound + element	$2\,NaNO_3 \xrightarrow{\text{Heat}} 2\,NaNO_2 + O_2$
Ternary compound	→ Two compounds	$CaCO_3 \xrightarrow{\text{Heat}} CaO + CO_2$

Figure 8.4 Decomposition of Potassium Chlorate

Students often decompose potassium chlorate to produce oxygen in the laboratory (Figure 8.4). This reaction is usually carried out by heating that compound in the presence of manganese(IV) oxide, MnO_2. The MnO_2 is a **catalyst**—a substance that changes the speed of a chemical reaction without undergoing a permanent change in its own composition. A catalyst is conventionally written above or below the reaction arrow:

$$2\,KClO_3(s) \xrightarrow[\text{Heat}]{MnO_2} 2\,KCl(s) + 3\,O_2(g)$$

Table 8.1 summarizes the most common types of decomposition reactions.

Substitution Reactions

The reaction of a free element with a compound of two (or more) other elements may result in the free element displacing one of the elements originally in the compound. A free metal can generally displace another metal in a compound; a free nonmetal can generally displace another nonmetal in a compound:

$$Zn(s) + CuCl_2(aq) \rightarrow ZnCl_2(aq) + Cu(s)$$
Zinc (metal) displaces copper (metal).

$$Cl_2(g) + CuBr_2(aq) \rightarrow CuCl_2(aq) + Br_2(aq)$$
Chlorine (nonmetal) displaces bromine (nonmetal).

In this class of reaction, called a **displacement reaction,** or **substitution reaction,** elements that are inherently more reactive can displace less reactive elements from their compounds, but the opposite process does not occur:

$$Cu(s) + ZnCl_2(aq) \rightarrow nr \quad \text{\textit{(nr stands for "no reaction")}}$$

Chemicals tend to react to go to a more stable, lower-energy state. When zinc reacts with copper(II) chloride, $CuCl_2$, the system goes to a lower-energy state. When an aqueous solution of zinc(II) chloride, $ZnCl_2$, is treated with copper metal, these chemicals are already in the lower-energy state, so they have no tendency to produce zinc metal and copper(II) chloride. We say that zinc is more **active,** or more **reactive,** than copper, which indicates that it has a greater tendency to leave its elemental state and form compounds. This is due to atoms of zinc having a greater tendency to lose electrons than those of copper do.

Figure 8.5 Reaction of Zinc
Metal with Hydrochloric Acid

Table 8.2 Relative Reactivities of Uncombined Elements

	Metals	**Nonmetals**	
Most active	Alkali metals and alkaline earth metals	F_2	*Most active*
		O_2	
	Al	Cl_2	
	Mn		
	Zn		
	Cr		
	Fe	Br_2	
	Sn		
	Pb		
	H*	I_2	*Less active*
	Cu		
	Ag		
Least active	Au		

*Hydrogen is included because it can be displaced from aqueous acids by reactive metals.

To predict which substitution reactions will occur, you need to know a little about the *relative* reactivities of some of the important metals and nonmetals. Some metals and a few nonmetals are listed in Table 8.2 in order of decreasing reactivity. Hydrogen is included in the list of metals because it can be displaced from aqueous acids by reactive metals (Figure 8.5) and can displace less active metals from their compounds:

$$Zn(s) + 2\ HCl(aq) \rightarrow H_2(g) + ZnCl_2(aq)$$

$$H_2(g) + CuCl_2(aq) \rightarrow Cu(s) + 2\ HCl(aq)$$

Very active metals can even displace hydrogen from water:

$$Ba(s) + 2\ H_2O(l) \rightarrow Ba(OH)_2(aq) + H_2(g)$$

■ **EXAMPLE 8.7**

Using data from Table 8.2, predict which of the following pairs of chemicals will react. If they will react, write a balanced equation for the reaction. If they will not react, write nr on the right-hand side of the arrow.

(a) Fe(s) and $Cu(NO_3)_2(aq)$ (b) Ag(s) and $Fe(NO_3)_2(aq)$

(c) HCl(aq) and Al(s)   (d) Au(s) and $AgNO_3(aq)$

(e) Cu(s) and $HClO_3(aq)$

Solution

(a) $Fe(s) + Cu(NO_3)_2(aq) \rightarrow Fe(NO_3)_2(aq) + Cu(s)$

(b) $Ag(s) + Fe(NO_3)_2(aq) \rightarrow$ nr   (Ag is less active than Fe.)

(c) $6\ HCl(aq) + 2\ Al(s) \rightarrow 3\ H_2(g) + 2\ AlCl_3(aq)$

(d) $Au(s) + AgNO_3(aq) \rightarrow$ nr   (Au is less active than Ag.)

(e) $Cu(s) + HClO_3(aq) \rightarrow$ nr   (Cu is less active than H.)

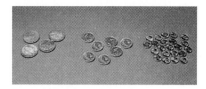

Figure 8.6 Gold, Silver, and Copper Coins

Silver tarnishes slowly in air to form silver sulfide, Ag_2S; it is more active than gold. Although copper is also used in coins, it is not as stable as the other coinage metals. Copper coins tarnish or corrode relatively quickly.

In a double substitution reaction, if the ions are not converted to covalent compounds, their charges do not change when they are converted from reactants to products.

Practice Problem 8.7   Predict whether each of the following pairs of substances will react. If they will react, write a balanced equation for the reaction. If they will not react, write nr on the right-hand side of the arrow.

(a) $Al(s)$ and $Pb(NO_3)_2(aq)$ (b) $HCl(aq)$ and $Ca(s)$ ■

Copper, silver, and gold—the coinage metals—have long been prized for their **stability,** or lack of reactivity (Figure 8.6). They can even occur uncombined in nature. Active metals do not occur naturally as free elements.

Double Substitution Reactions

The reaction of two compounds may yield two new compounds. Many reactions that occur in aqueous solution involve two ionic compounds trading anions. This class of reactions is called **double substitution reactions, double displacement reactions,** or **metathesis reactions.** As usual, the correct formulas must be written for the products before the equation is balanced. In a double substitution reaction, if the ions are not converted to covalent compounds, their charges do not change when they are converted from reactants to products.

■ **EXAMPLE 8.8**

Predict the products for reactions of the following pairs of reactants, and write balanced equations:

(a) Aqueous sodium chloride and aqueous silver nitrate

(b) Aqueous iron(II) chloride and aqueous silver nitrate

(c) Solid iron(III) hydroxide and aqueous hydrochloric acid

Solution

First, write the names of the products:

(a) Sodium chloride plus silver nitrate yields sodium nitrate plus silver chloride. The Na^+ and Ag^+ cations trade anions.

(b) Iron(II) chloride plus silver nitrate yields iron(II) nitrate plus silver chloride. Here, the cations also trade anions; however, the Fe^{2+} ion requires two singly charged anions to satisfy its dipositive charge. Note that the iron ion had a 2+ charge and still has that charge. Iron(III) nitrate is not expected as a product, because the Fe^{2+} ion does not change to an Fe^{3+} ion in this type of reaction.

(c) Iron(III) hydroxide plus hydrochloric acid yields iron(III) chloride plus water. The charge on the iron(III) ion remains the same throughout the reaction; the charge on the hydrogen ion changes, since the water formed is covalent, not ionic.

Next, write correct formulas for all reactants and products:

(a) $NaCl(aq) + AgNO_3(aq) \rightarrow NaNO_3(aq) + AgCl(s)$

(b) $FeCl_2(aq) + AgNO_3(aq) \rightarrow Fe(NO_3)_2(aq) + AgCl(s)$

(c) $Fe(OH)_3(s) + HCl(aq) \rightarrow FeCl_3(aq) + H_2O(l)$

Finally, balance the equations:

(a) $NaCl(aq) + AgNO_3(aq) \rightarrow NaNO_3(aq) + AgCl(s)$

(b) $FeCl_2(aq) + 2\,AgNO_3(aq) \rightarrow Fe(NO_3)_2(aq) + 2\,AgCl(s)$

(c) $Fe(OH)_3(s) + 3\,HCl(aq) \rightarrow FeCl_3(aq) + 3\,H_2O(l)$

Practice Problem 8.8   Complete and balance the equation for each of the following reactions:

(a) $Ba(OH)_2(aq) + HNO_3(aq) \rightarrow$ (b) $KCl(aq) + AgC_2H_3O_2(aq) \rightarrow$ ■

In aqueous solution, neither H_2CO_3 (carbonic acid) nor NH_4OH (ammonium hydroxide) is stable; they decompose to yield water and either CO_2 or NH_3. If either H_2CO_3 or NH_4OH appears to be an expected product of a double substitution reaction, CO_2 plus H_2O, or NH_3 plus H_2O, will be produced instead. Other unstable compounds are encountered much less frequently.

■ EXAMPLE 8.9

Predict the products of the following reactions:

(a) $Na_2CO_3(aq) + HCl(aq) \rightarrow$ (b) $NH_4Cl(aq) + NaOH(aq) \rightarrow$

Solution

(a) Ordinarily, you would predict that the ions will trade partners to yield NaCl and H_2CO_3. However, water and carbon dioxide are produced instead of H_2CO_3:

$$Na_2CO_3(aq) + 2\,HCl(aq) \rightarrow 2\,NaCl(aq) + H_2O(l) + CO_2(g)$$

(b) Ordinarily, you would predict that the ions will trade partners to yield NaCl and NH_4OH. However, instead of ammonium hydroxide, water and ammonia are produced:

$$NH_4Cl(aq) + NaOH(aq) \rightarrow NaCl(aq) + H_2O(l) + NH_3(aq)$$

Practice Problem 8.9   Complete and balance the following equations:

(a) $(NH_4)_2CO_3(aq) + HCl(aq) \rightarrow$ (b) $(NH_4)_2CO_3(aq) + NaOH(aq) \rightarrow$ ■

The driving force behind double substitution reactions is the formation of an insoluble ionic compound or a covalent compound (such as water or a gaseous compound) from ions in solution. A solid formed from ions in solution is called a **precipitate.** We can thus predict that a reaction will occur if soluble ionic compounds yield at least one insoluble ionic compound or one covalent compound. You learned in Chapter 5 that covalent compounds have no metallic element and no ammonium ion in them. In addition, you need to be familiar with the **solubilities** of some common ionic compounds in water. Some types of ionic compounds that are soluble or insoluble in water are listed in Table 8.3. A more comprehensive tabulation of solubilities is presented in Table 8.4.

Table 8.3 Water Solubility of Some Common Ionic Compounds

Soluble in Water	Insoluble in Water
All chlorates	$BaSO_4$
All nitrates	Most oxides
All acetates	Most sulfides
All compounds of alkali metals	Most phosphates
All compounds containing the ammonium ion	
All chlorides except those listed in the next column	$AgCl$, $PbCl_2$, Hg_2Cl_2, and $CuCl$

■ EXAMPLE 8.10

Complete and balance the equation for each of the following reactions:

(a) $BaCl_2(aq) + Na_2SO_4(aq) \rightarrow$ (b) $HClO_3(aq) + NaOH(aq) \rightarrow$

Table 8.4   *Tabulation of Solubilities*

	ClO_3^- NO_3^- $C_2H_3O_2^-$	Cl^- Br^- I^-	SO_4^{2-}	CO_3^{2-} SO_3^{2-} PO_4^{3-} CrO_4^{2-} BO_3^{3-}	S^{2-}	OH^-	O^{2-}
Pb^{2+}	s	ss-i	i	i	i	i	i
Na^+, K^+, NH_4^+	s	s	s	s	s	s	d
Hg_2^{2+}	s	i	i	i	i	i	i
Hg^{2+}	s	s-i†	s	i	i	i	i
Ag^+	s	i	ss	i	i	i	i
Mg^{2+}	s	s	s	s	s	i	i
Ca^{2+}	s	s	i	i	s	s	d
Ba^{2+}	s	s	i	i	s	s	d

Key: s = soluble (greater than about 1 gram solute/100 grams of water).
    ss = slightly soluble (approximately 0.1–1 gram solute/100 grams of water).
     i = insoluble (less than about 0.1 gram solute/100 grams of water).
     d = decomposes in water.
     †$HgCl_2$ is soluble, $HgBr_2$ is less soluble, and HgI_2 is insoluble.

Solution

(a) You see from Table 8.3 that $BaSO_4$ is insoluble in water. Thus, the following reaction will occur:

$$BaCl_2(aq) + Na_2SO_4(aq) \rightarrow BaSO_4(s) + 2\ NaCl(aq)$$

Barium sulfate precipitates.

ITEM OF INTEREST

An industrial process called the *Solvay process* uses the following set of reactions to produce Na_2CO_3 (known as washing soda). The reactants are inexpensive, and Na_2CO_3 is a very important industrial compound used in the manufacture of soap, glass, paper, detergents, and other chemicals.

$$CaCO_3(s) \xrightarrow{\text{Heat}} CaO(s) + CO_2(g)$$

$$2\ CO_2(g) + 2\ H_2O(l) + 2\ NH_3(aq) \rightarrow 2\ NH_4HCO_3(aq)$$

$$2\ NaCl(aq) + 2\ NH_4HCO_3(aq) \rightarrow 2\ NaHCO_3(s) + 2\ NH_4Cl(aq)$$
Very concentrated solutions

$$2\ NaHCO_3(s) \xrightarrow{\text{Heat}} Na_2CO_3(s) + CO_2(g) + H_2O(l)$$

$$CaO(s) + H_2O(l) \rightarrow Ca(OH)_2(aq)$$

$$2\ NH_4Cl(aq) + Ca(OH)_2(aq) \rightarrow 2\ NH_3(aq) + 2\ H_2O(l) + CaCl_2(aq)$$

If we add all the reactants and all the products in these equations and then delete the compounds that appear on both sides, we get the following overall equation:

$$CaCO_3 + 2\ NaCl \rightarrow Na_2CO_3 + CaCl_2$$

(b) In this reaction, a covalent compound, water, is formed from ions in solution:

$$HClO_3(aq) + NaOH(aq) \rightarrow NaClO_3(aq) + H_2O(l)$$

Practice Problem 8.10   Complete and balance the equation for each of the following reactions:

(a) $Li_2O(s) + H_2SO_4(aq) \rightarrow$ (b) $LiHCO_3(aq) + HCl(aq) \rightarrow$ ■

■ EXAMPLE 8.11

Can industrial chemists simply combine $NaCl$ and $CaCO_3$ to get Na_2CO_3 and $CaCl_2$?

Solution

The proposed reactants are more stable than the desired products because $CaCO_3$ is insoluble in water, and Na_2CO_3 and $CaCl_2$ are both soluble. Thus, this reaction is not feasible. ■

■ EXAMPLE 8.12

What type of reaction is each of the steps of the Solvay process?

Solution

In the order in which they are shown, the six reactions are classed as (1) decomposition, (2) combination, (3) double substitution, (4) decomposition, (5) combination, and (6) double substitution followed by decomposition. ■

Combustion Reactions

Reactions of oxygen gas with all types of substances are called **combustion reactions.** Combustion reactions of some elements can also be classified as combination reactions; the *type* of reaction is not as important as the *products*. For example, we can refer to the following reactions as combination reactions or combustion reactions:

$$S(s) + O_2(g) \rightarrow SO_2(g)$$

$$C(s) + O_2(g) \rightarrow CO_2(g)$$

The combustion reactions of hydrocarbons—compounds composed of carbon and hydrogen only—are especially important as sources of useful energy. We burn methane, CH_4, called natural gas, in our homes to provide heat, and we combust octane, C_8H_{18}, in our cars to provide mechanical energy:

$$CH_4(g) + 2\ O_2(g) \rightarrow CO_2(g) + 2\ H_2O(g)$$

$$2\ C_8H_{18}(l) + 17\ O_2(g) \rightarrow 16\ CO(g) + 18\ H_2O(g)$$

In such reactions, either carbon monoxide or carbon dioxide may be produced, in addition to water. If sufficient oxygen is present, carbon dioxide is produced. If the supply of oxygen is limited, carbon monoxide is the product. (With very limited oxygen, soot—a form of carbon—and water are produced.) *In any case, water is a product.*

■ EXAMPLE 8.13

Which of the following reactions was carried out in a limited supply of oxygen?

$$CH_4(g) + 2\ O_2(g) \rightarrow CO_2(g) + 2\ H_2O(g)$$

$$2\ C_8H_{18}(l) + 17\ O_2(g) \rightarrow 16\ CO(g) + 18\ H_2O(g)$$

Solution

The reaction of C_8H_{18} produced CO, so that reaction was run in limited oxygen. If the combustion reaction of C_8H_{18} is carried out in excess oxygen, the equation is written as follows:

$$2\ C_8H_{18}(l) + 25\ O_2(g) \rightarrow 16\ CO_2(g) + 18\ H_2O(g)$$

Practice Problem 8.13   What will be the effect if between 17 and 25 mol of oxygen is available for the combustion of 2 mol of octane? ■

■ EXAMPLE 8.14

Write a balanced equation for the reaction of methane in a limited oxygen supply.

Solution

$$2\ CH_4(g) + 3\ O_2(g) \rightarrow 2\ CO(g) + 4\ H_2O(g) \quad ■$$

The combustion reactions of compounds containing carbon, hydrogen, and oxygen (which include the carbohydrates we use for food) also produce either carbon monoxide or carbon dioxide, depending on the relative quantity of oxygen available.

■ EXAMPLE 8.15

Write a complete and balanced equation for the reaction of sucrose (table sugar), $C_{12}H_{22}O_{11}$, with excess oxygen.

Solution

$$C_{12}H_{22}O_{11}(s) + 12\ O_2(g) \rightarrow 12\ CO_2(g) + 11\ H_2O(g)$$

Practice Problem 8.15   Write a complete and balanced equation for the combustion reaction of sucrose, $C_{12}H_{22}O_{11}$, in a limited supply of oxygen. ■

8.4 Acids and Bases

OBJECTIVE

■ to predict the products of the reactions of acids with bases, and to use a specialized nomenclature for acid-base reactions

Most of the reactions of acids and bases fit into the substitution or double substitution classes discussed in Section 8.3. However, the reactions of acids and bases are so important that we need to take a closer look at them.

According to the most fundamental theory concerning acids and bases—the **Arrhenius theory**—an **acid** is a compound that furnishes hydrogen ions, H^+, to an aqueous solution, and a **base** is a compound that furnishes hydroxide ions, OH^-, to an aqueous solution. The hydrogen ion does not exist alone, as H^+, but

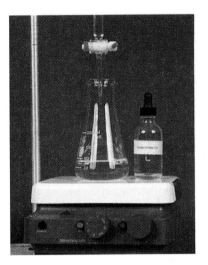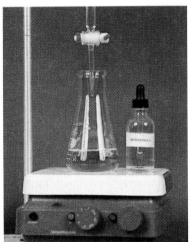

Figure 8.7 Neutralization Reaction

is stable in aqueous solution in the form H_3O^+, which is frequently represented as $H^+(aq)$.

In beginning courses, formulas for acids (and no other compounds except water) are written with the ionizable hydrogen atoms first, as in HCl.

$$HCl(g) \xrightarrow{H_2O} H^+(aq) + Cl^-(aq)$$

Methane, CH_4, ammonia, NH_3, and glucose, $C_6H_{12}O_6$, are examples of compounds that are not acids, since they do not provide hydrogen ions to aqueous solutions. Their hydrogen atoms are therefore not written first in their formulas. For certain acids, such as acetic acid, $HC_2H_3O_2$, only the hydrogen atoms(s) written first is (are) capable of being ionized; the other hydrogen atoms do not yield H^+ in solution.

Properties of Acids and Bases

Caution: Do not taste chemicals unless specifically directed to do so.

Acids in general have a sour taste, turn indicators certain colors, and react with bases to form salts. For example, the sour taste of lemon is the taste of citric acid, and the sour taste of vinegar is due mainly to acetic acid, its principal acid component.

Bases feel slippery, turn indicators certain colors that differ from those acids produce, and react with acids to form salts. You may experience the slipperiness of a base by putting your fingertips in some dilute ammonia-water.

Caution: Never put your skin into contact with concentrated solutions of strong bases, such as liquid Drano, because they are capable of dissolving the fat in your skin.

The most important reactions of acids and bases are their reactions with each other to form salts and water:

$$HNO_3(aq) + NaOH(aq) \rightarrow NaNO_3(aq) + H_2O(l)$$

An acid　　　A base　　　A salt　　　Water

A **salt** is any compound of a cation other than H^+ with an anion other than OH^- or O^{2-}. (The word *salt* in everyday conversation means sodium chloride, which is a salt under this definition.) Such reactions, actually specific examples of double substitution reactions, are called **neutralization reactions** (see Figure 8.7) because they produce products that are more neutral than acids or bases. **Neutral** means "neither acidic nor basic."

Many individual compounds that are acids have additional properties that make them dangerous. These dangerous compounds give acids a bad name to the general public. For example, LSD (lysergic acid diethylamide) is a mind-affecting hallucinogenic agent, but this property is in addition to any acid properties of the compound. Concentrated sulfuric acid, used in auto batteries, is a powerful oxidizing agent and dehydrating agent. A lump of sugar placed into concentrated sulfuric acid has the elements of water pulled from its molecules, leaving carbon (Figure 8.8):

$$C_{12}H_{22}O_{11}(s) \xrightarrow{\text{Concentrated } H_2SO_4} 12\ C(s) + 11\ H_2O(\text{in } H_2SO_4 \text{ solution}) + \text{heat}$$

Nitric acid, especially when concentrated but even in dilute solution, is another powerful oxidizing agent. In contrast, boric acid is such a weak acid that it is sometimes used in solution to bathe infected eyes.

An **indicator** is a dye that is intensely colored in acidic solution or basic solution or both. Its color is different, depending on the acidity or basicity of the solution, so an indicator can be used to tell when an acid or base has been neutralized in a reaction. Indicators must be intensely colored so that the color can be seen when the indicator is present in low concentration (a small quantity in a relatively large quantity of solution) (Figure 8.9). For example, the indicator called litmus is blue in basic solution and red in acidic solution. If you place a drop of a solution on a piece of paper treated with litmus and the paper turns red, the solution is acidic.

■ **EXAMPLE 8.16**

A common indicator, phenolphthalein, is colorless in acidic solution and red in basic solution. Describe the color changes when two drops of phenolphthalein solution are added to 100 mL of a colorless acidic solution and then the acid is gradually neutralized by adding drops of base.

Solution

The acidic solution is initially colorless and remains so when the indicator is added. As most of the base is added, no permanent color change takes place. When one drop or less of *excess* base is added, the solution will change to pink. The indicator, usually red in a basic solution, is present in very low concentration, and its red color looks pink in such a relatively large volume of solution.

Practice Problem 8.16   What happens to the color of the solution in Example 8.16 if acid is added after the last drop of base has been added? ■

Figure 8.8 Dehydration of Sugar by Sulfuric Acid

Figure 8.9 Action
of an Indicator

All hydrogen-containing acids are covalent compounds when they are not in solution; they ionize when they react with water:

$$HX(l \text{ or } g) + H_2O(l) \longrightarrow H^+(aq) + X^-(aq)$$

Those that react nearly 100% to form ions are called **strong acids.** Those that react only to a limited extent are called **weak acids.** The common strong acids are HCl, HBr, HI, $HClO_3$, $HClO_4$, HNO_3, and H_2SO_4. Practically all other acids are weak.

Bases provide hydroxide ions to aqueous solution. Soluble metal hydroxides, including those of the alkali metals and barium, are examples. The soluble metal hydroxides are ionic even when they are pure solids; *they remain ionic in water.* When they are dissolved in water, the hydroxide ions are totally separated from the metal ions. A soluble metal hydroxide is a **strong base.** A **weak base** is not 100% ionized. Ammonia, the most common weak base, reacts with water to a small extent to provide hydroxide ions:

$$NH_3(aq) + H_2O(l) \longrightarrow NH_4^+(aq) + OH^-(aq) \qquad \textit{(Usually from 0.1\% to 2\%)}$$

For example, if 1 mol of NH_3 is dissolved in a liter of water, only 0.004 mol of NH_4^+ and 0.004 mol of OH^- will be present. Almost all (0.996 mol) of the NH_3 remains in its molecular form.

Weak acids and weak bases react with water to a small extent but react with strong bases or acids essentially completely:

$$NaOH(aq) + HC_2H_3O_2(aq) \longrightarrow NaC_2H_3O_2(aq) + H_2O(l)$$

$$HCl(aq) + NH_3(aq) \longrightarrow NH_4Cl(aq)$$

A strong acid and a strong base react with each other completely to form a salt and water:

$$HCl(aq) + NaOH(aq) \longrightarrow NaCl(aq) + H_2O(l)$$

Acids usually react with insoluble bases to produce salts and water:

$$Ca(OH)_2(s) + 2\, HCl(aq) \longrightarrow CaCl_2(aq) + 2\, H_2O(l)$$

The driving force for double substitution reactions is formation of insoluble ionic compounds or covalent compounds from ions in solution. However, if an equation has an insoluble compound on one side and a covalent compound on the other side, which way does the reaction go? In many cases like this, the formation of covalent compounds is more important than the formation of insoluble ionic compounds, as shown by the reaction of $Ca(OH)_2$ with HCl.

The formation of a weak acid from ionic compounds is another example of the formation of a covalent compound. For example, in aqueous solution, the following reaction will occur:

$$NaC_2H_3O_2(aq) + HCl(aq) \longrightarrow \underset{\text{Weak acid}}{HC_2H_3O_2(aq)} + NaCl(aq)$$

Acids can react with metals more active than hydrogen (see Table 8.2) to produce a salt and hydrogen gas:

$$Zn(s) + 2\, HCl(aq) \longrightarrow ZnCl_2(aq) + H_2(g)$$

Extremely active metals, such as the alkali and alkaline earth metals, can even react with water to produce hydrogen gas plus the corresponding metal hydroxide. For example:

$$2\, Na(s) + 2\, H_2O(l) \longrightarrow 2\, NaOH(aq) + H_2(g)$$

Acidic and Basic Anhydrides

Most metal oxides in which the metal ion has a 1+ or 2+ charge are **basic anhydrides,** and most nonmetal oxides are **acidic anhydrides.** In general, an **anhydride** is any compound that can result from loss of water from another compound. If water is added to an acidic anhydride, the anhydride becomes an acid. For example, sulfur dioxide plus water yields sulfurous acid:

$$SO_2(g) + H_2O(l) \rightarrow H_2SO_3(aq)$$

If water is added to a basic anhydride, the anhydride becomes a base. For example, lithium oxide plus water yields lithium hydroxide:

$$Li_2O(s) + H_2O(l) \rightarrow 2\ LiOH(aq)$$

The first of the previous reactions is responsible for a good portion of the acid rain problem troubling the industrialized world. Sulfur, present in small quantities as an impurity in coal and oil, is converted to sulfur dioxide when the coal or oil is burned; then the sulfur dioxide reacts with the moisture in the air to produce sulfurous acid. Sulfurous acid can react with the oxygen in air to produce sulfuric acid. These acids are washed from the air by rain (or snow), and the solution can cause some corrosion of concrete and metal in buildings. Acids in the air and in the rain or snow also injure trees and other plants, as well as animals, including humans. In high concentrations, acids can make breathing difficult, especially for people who are already in poor health.

Acidic anhydrides can react directly with bases, and basic anhydrides can react directly with acids. The same salt is produced as would be produced by the acid and base:

$$CO_2(g) + 2\ NaOH(aq) \rightarrow Na_2CO_3(aq) + H_2O(l)$$

$$MgO(s) + 2\ HCl(aq) \rightarrow MgCl_2(aq) + H_2O(l)$$

An acidic anhydride and a basic anhydride can even react with each other in a combination reaction:

$$CO_2(g) + CaO(s) \rightarrow CaCO_3(s)$$

A few nonmetal oxides, including CO and N_2O, are not acidic anhydrides; they do not react with water to form acids or with bases to form salts:

$$CO(g) + H_2O(l) \xrightarrow[\text{Room temperature}]{} nr$$

$$CO(g) + NaOH(aq) \rightarrow nr$$

Acid Salts

Acids containing more than one ionizable hydrogen atom, such as H_2SO_4 and H_3PO_4, can be *partially* neutralized if less base is used than is needed for complete neutralization. The salt formed contains ionizable hydrogen atoms and therefore is still capable of reacting with bases:

$H_3PO_4(aq) + NaOH(aq) \rightarrow NaH_2PO_4(aq) + H_2O(l)$ *(Partial neutralization)*

$H_3PO_4(aq) + 2\ NaOH(aq) \rightarrow Na_2HPO_4(aq) + 2\ H_2O(l)$ *(Partial neutralization)*

$H_3PO_4(aq) + 3\ NaOH(aq) \rightarrow Na_3PO_4(aq) + 3\ H_2O(l)$ *(Complete neutralization)*

A substance produced by a partial neutralization, such as NaH_2PO_4 or Na_2HPO_4, is partially a salt and partially an acid. As the product of an acid and a base, it is a salt. However, it is capable of neutralizing more base, so it can also act as an acid:

$$NaH_2PO_4(aq) + 2\ NaOH(aq) \rightarrow Na_3PO_4(aq) + 2\ H_2O(l)$$

Such a substance is called an **acid salt.** Its name includes the word *hydrogen* to denote the fact that one or more ionizable hydrogen atoms remain. The prefix *mono-* or *di-* may be used when it is necessary to indicate how many hydrogen atoms are present:

NaH_2PO_4 Sodium dihydrogen phosphate

Na_2HPO_4 Sodium monohydrogen phosphate (or disodium hydrogen phosphate)

$NaHCO_3$ Sodium hydrogen carbonate

In an older nomenclature system, the word *acid* was used to denote an acid salt. The prefix *bi-* was used for a half-neutralized acid that originally contained two ionizable hydrogen atoms. Thus, sodium bicarbonate and sodium acid carbonate are other names that have been used for $NaHCO_3$.

In the anion of an acid salt, the number of hydrogen atoms plus the magnitude of the charge on the ion equals the magnitude of the charge on the oxoanion and also equals the number of hydrogen atoms in the acid:

PO_4^{3-} The charge on the phosphate ion is $3-$.

HPO_4^{2-} 1 hydrogen atom plus 2 negative charges equals 3.

$H_2PO_4^{-}$ 2 hydrogen atoms plus 1 negative charge equals 3.

H_3PO_4 3 hydrogen atoms plus 0 negative charges equals 3.

■ EXAMPLE 8.17

What is the charge on the hydrogen sulfate ion?

Solution

The charge is -1 in HSO_4^{-}.

Practice Problem 8.17   What is the charge on the hydrogen sulfide ion?   ■

Carbonates and Acid Carbonates

Carbonates are compounds containing the carbonate ion. **Acid carbonates** are compounds containing the hydrogen carbonate ion. Just as acid-base reactions are an important type of double substitution reaction, the reactions of carbonates and acid carbonates with acids are an important subtype of acid-base reaction.

Carbonates undergo double substitution reactions with acids to form carbon dioxide and water or acid carbonates, depending on the relative quantity of acid added:

$$Na_2CO_3(aq) + 2\ HCl(aq) \rightarrow 2\ NaCl(aq) + CO_2(g) + H_2O(l)$$

$$Na_2CO_3(aq) + HCl(aq) \rightarrow NaCl(aq) + NaHCO_3(aq)$$

The acid either totally or partially neutralizes the carbonate.

Carbon dioxide, an acidic anhydride, can react with a base to form a carbonate or an acid carbonate:

$$CO_2(g) + 2\,NaOH(aq) \rightarrow Na_2CO_3(aq) + H_2O(l)$$

$$CO_2(g) + NaOH(aq) \rightarrow NaHCO_3(aq)$$

The base either totally or partially neutralizes the carbon dioxide.

Carbon dioxide present in relatively high concentration in water can dissolve insoluble carbonates to yield soluble acid carbonates:

$$CO_2(g) + H_2O(l) + CaCO_3(s) \rightarrow Ca(HCO_3)_2(aq)$$

The reaction of limestone ($CaCO_3$) with water containing carbon dioxide in relatively high concentration can form natural caves, such as Luray Caverns in Virginia (Figure 8.10a). If the carbon dioxide concentration is lowered, the reverse reaction can occur:

$$Ca(HCO_3)_2(aq) \rightarrow CO_2(g) + H_2O(l) + CaCO_3(s)$$

Thus, water dripping from the ceiling of a cavern can deposit $CaCO_3$, a tiny particle at a time, and over long periods can form stalactites and stalagmites (Figure 8.10b).

(a)   (b)

Figure 8.10 Formation of Caves

(a) Huge underground chambers, such as the Luray Caverns in the Blue Ridge Mountains of Virginia, are formed over eons of time by the reaction of carbon dioxide dissolved in water with solid limestone, $CaCO_3$. (b) Water containing calcium hydrogen carbonate, $Ca(HCO_3)_2$, dripped from the ceiling of a limestone cavern and deposited solid calcium carbonate, $CaCO_3$, when the concentration of carbon dioxide was low. A droplet hanging from the ceiling formed a tiny portion of a stalactite; a droplet that hit the floor formed a tiny portion of a stalagmite.

Figure 8.11 Acid-Base
Reactions Involving Carbonates
and Acid Carbonates

Heating an acid carbonate, such
as sodium hydrogen carbonate,
produces the corresponding
carbonate plus carbon dioxide
and water. Mixing a carbonate,
such as sodium carbonate, with
carbon dioxide and water produces
the corresponding acid carbonate,
such as sodium hydrogen carbonate.

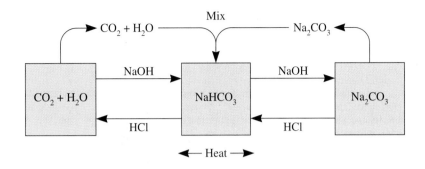

Acid carbonates undergo double substitution reactions with either acids or
bases, neutralizing either:

$$NaHCO_3(aq) + HCl(aq) \rightarrow NaCl(aq) + H_2O(l) + CO_2(g)$$

$$NaHCO_3(aq) + NaOH(aq) \rightarrow Na_2CO_3(aq) + H_2O(l)$$

These types of reactions are summarized in Figure 8.11.

■ ■ ■ ■ ■ ■ ■ SUMMARY ■ ■ ■ ■ ■ ■ ■

The balanced equation represents a chemical reaction. It not only identifies the reactants and the products, but also gives quantitative information on the ratios of all substances involved in the reaction. (Section 8.1)

To balance an equation—that is, to make the numbers of atoms of each of the elements the same on both sides of the equation—you place coefficients in front of each formula in the equation. The state of each substance may be indicated as gas (g), liquid (l), solid (s), or solute in aqueous solution (aq). (Section 8.2)

With a little experience, you can predict the products of simple reactions from the nature of the reactants. In writing formulas for the products, always use the rules given in Chapter 5; do not write incorrect formulas to make balancing an equation easier. Simple reactions can be divided into five types: combination reactions, decomposition reactions, substitution reactions, double substitution reactions, and combustion reactions. Identifying the type of reaction can help greatly in deducing the product(s). If two free elements are given, they can either combine or do nothing; they cannot be broken down into simpler substances. If only one compound is given, it probably will decompose, especially if energy is provided.

An element and a compound can react to give a new compound and another free element. Two ionic compounds can swap ions to produce two new compounds. The relative reactivity of the elements (Table 8.2) determines whether a substitution reaction can occur. Solubility in water (Table 8.3) often determines whether a double substitution reaction can occur. Reaction with oxygen is combustion; carbon-containing compounds react with limited oxygen to produce carbon monoxide or react with excess oxygen to give carbon dioxide. (Section 8.3)

Acids and bases react according to the rules in Section 8.3, but their reactions are so common that further details need to be learned. The double substitution reaction of an acid with a base is called a neutralization reaction. The products are water and a salt. An indicator is often used to signal that neutralization has been completed. Strong acids react with water completely to form ions, and weak acids react with water only slightly, but both kinds of acids react with bases to form salts. Substances that react with water to form acids or bases are called anhydrides. Acids containing more than one ionizable hydrogen atom can be partially neutralized, forming acid salts. Carbonates and acid carbonates react in some ways like bases. (Section 8.4)

Items for Special Attention

■ *Never* treat spilled acid or base with strong base or strong acid. Excess of the reagent might do more harm than the original acid or base, and the heat of the neutralization reaction might also cause problems. Instead, flood with water, and later treat with sodium hydrogen carbonate, $NaHCO_3$, which is almost neutral and produces safe reaction products. Called baking soda in everyday life, sodium hydrogen carbonate is as effective at home as it is in the laboratory.

■ You can often apply a generality to answer a specific question. For example, you can tell that Na_2MoO_4 is soluble in water even if you have never seen this formula before. According to Table 8.3, all alkali metal compounds are soluble, and this compound is an alkali metal salt.

Self-Tutorial Problems

8.1 Assign the following types to one of the five classes of reactions presented in Section 8.3:

Reactants	*Products*
(a) 2 elements	1 compound
(b) 1 compound	1 element + 1 compound
(c) 1 element + 1 compound	1 element + 1 compound
(d) 1 compound	2 elements
(e) 2 compounds	2 different compounds
(f) 1 element + 1 compound	1 compound
(g) 1 compound + O_2	2 or more compounds

8.2 What is the difference, if any, among (a) the reaction of barium with fluorine, (b) the combination of barium and fluorine, and (c) the formation of barium fluoride from its elements?

8.3 Which table in this chapter should be used when working with substitution reactions, and which one with double substitution reactions?

8.4 In a certain double substitution reaction, $FeCl_3$ is a reactant. Is $Fe(NO_3)_3$ or $Fe(NO_3)_2$ more likely to be a product?

8.5 What type of reaction is the following? What are the products?

$$C_4H_{10}(g) + O_2(g, \text{ excess}) \rightarrow$$

8.6 Classify each of the following as an acidic anhydride or a basic anhydride:

$$N_2O_3 \qquad CaO \qquad SO_2 \qquad Cl_2O_7$$

8.7 Are oxides of active metals or oxides of inactive metals more likely to decompose into their two elements when heated?

8.8 Which of the following compounds are acids?

$$PH_3 \qquad H_2Se \qquad NH_3 \qquad C_2H_6 \qquad H_2O \qquad HC_3H_5O_2$$

8.9 What type of substance can act as an acid but does not have hydrogen written first in its formula?

8.10 Explain how to recognize that O_2 and NiO will not react with each other in a substitution reaction.

8.11 Which, if any, of the common acids exist completely in the form of ions (a) in aqueous solution and (b) as a pure compound?

8.12 Write a balanced chemical equation for each of the following reactions:

(a) $SO_2(g) + Cl_2(g) \rightarrow SO_2Cl_2(l)$

(b) $SO_2(g) + PCl_5(s) \rightarrow SOCl_2(l) + POCl_3(l)$

8.13 Consider the reaction of aqueous bromine with aqueous potassium iodide.

(a) Identify the reaction type.

(b) Write correct formulas for all reactants and products.

(c) Write a balanced equation.

8.14 Give two reasons why the following reaction produces products:

$$Ba(OH)_2(aq) + H_2SO_4(aq) \rightarrow BaSO_4(s) + 2\,H_2O(l)$$

8.15 Explain how a catalyst resembles a marriage broker.

8.16 Rewrite the following equations with integral coefficients:

(a) $CrF_2(s) + \frac{1}{2} F_2(g) \rightarrow CrF_3(s)$

(b) $NH_3(g) + \frac{5}{4} O_2(g) \rightarrow NO(g) + \frac{3}{2} H_2O(g)$

(c) $CuCl(s) + \frac{1}{2} Cl_2(g) \rightarrow CuCl_2(s)$

(d) $\frac{2}{3} H_3PO_4(aq) + Ba(OH)_2(aq) \rightarrow$
$\frac{1}{3} Ba_3(PO_4)_2(s) + 2\,H_2O(l)$

8.17 What products are expected in each of the following cases?

(a) $KClO_3$ is heated in the presence of MnO_2 as a catalyst.

(b) $KClO_3$ is heated in the presence of MnO_2.

(c) $KClO_3$ and MnO_2 are heated together.

(d) $KClO_3$ is heated.

■ ■ ■ **PROBLEMS** ■ ■ ■

8.1 The Chemical Equation

8.18 (a) If two molecules of H_2O react with potassium metal according to the following equation, how many molecules of H_2 will be produced?

$$2 K(s) + 2 H_2O(l) \rightarrow 2 KOH(aq) + H_2(g)$$

(b) If 2 mol of H_2O reacts with potassium metal according to the equation, how many moles of H_2 will be produced?

8.19 (a) If one molecule of P_4 reacts with chlorine gas according to the following equation, how many molecules of PCl_5 will be produced?

$$P_4(s) + 10 Cl_2(s) \rightarrow 4 PCl_5(l)$$

(b) If 1 mol of P_4 reacts with chlorine gas according to the equation, how many moles of PCl_5 will be produced?

8.20 List the number of atoms of each element in the given number of formula units:

(a) 4 $NaClO_2$ (b) 8 NH_3 (c) 2 $CoCO_3$

(d) 3 $Ba(NO_2)_2 \cdot H_2O$ (e) 6 $(NH_4)H_2PO_3$

8.2 Balancing Equations

8.21 Write a balanced equation for the reaction at high temperature of oxygen gas and nitrogen gas to form gaseous N_2O.

8.22 Balance the equation for each of the following reactions:

(a) $C_3H_7OH(l) + O_2(g) \rightarrow CO_2(g) + H_2O(g)$

(b) $C_3H_8(g) + O_2(g) \rightarrow CO_2(g) + H_2O(g)$

(c) $BiCl_3(aq) + H_2O(l) \rightarrow BiOCl(s) + HCl(aq)$

(d) $Sb_2S_3(s) + O_2(g) \rightarrow Sb_2O_3(s) + SO_2(g)$

(e) $CH_2O(l) + O_2(g) \rightarrow CO(g) + H_2O(g)$

(f) $CO_2(g) + H_2(g) \rightarrow CO(g) + H_2O(g)$

(g) $ZnS(s) + O_2(g) \rightarrow ZnO(s) + SO_2(g)$

(h) $Cu_2S(s) + O_2(g) \rightarrow Cu(s) + SO_2(g)$

(i) $H_2O(l) + PCl_5(s) \rightarrow HCl(aq) + H_3PO_4(aq)$

(j) $O_2(g) + FeO(s) \rightarrow Fe_3O_4(s)$

(k) $Na_2SO_3(aq) + S(s) \rightarrow Na_2S_2O_3(aq)$

(l) $AlCl_3(aq) + NaOH(aq) \rightarrow$
$$NaAl(OH)_4(aq) + NaCl(aq)$$

(m) $CuSO_4 \cdot 5H_2O(aq) + NH_3(aq) \rightarrow$
$$CuSO_4 \cdot 4NH_3(aq) + H_2O(l)$$

8.23 Balance the equation for each of the following reactions:

(a) $B_2H_6(g) + O_2(g) \rightarrow B_2O_3(s) + H_2O(l)$

(b) $Zn(s) + NaOH(aq) + H_2O(l) \rightarrow$
$$Na_2Zn(OH)_4(aq) + H_2(g)$$

(c) $MnO_2(s) + H_2C_2O_4(aq) \rightarrow$
$$CO_2(g) + MnO(s) + H_2O(l)$$

(d) $NaAl(OH)_4(aq) + HCl(aq) \rightarrow$
$$AlCl_3(aq) + H_2O(l) + NaCl(aq)$$

(e) $Mn_3O_4(s) + O_2(g) \rightarrow Mn_2O_3(s)$

(f) $H_3AsO_4(aq) + NaOH(aq) \rightarrow Na_2HAsO_4(aq) + H_2O$

8.24 Write a balanced equation for the reaction of aqueous copper(II) chloride with aqueous potassium iodide to produce solid copper(I) iodide plus aqueous iodine plus aqueous potassium chloride.

8.25 Write a balanced chemical equation for each of the following reactions:

(a) Solid sulfur plus fluorine gas yields liquid sulfur hexafluoride.

(b) Aqueous calcium hydrogen carbonate plus hydrochloric acid yields calcium chloride plus carbon dioxide plus water.

(c) Aqueous barium hydroxide plus phosphoric acid yields barium phosphate plus water.

(d) Aqueous ammonium sulfate plus barium chloride yields barium sulfate plus ammonium chloride.

8.26 Balance the equation for each of the following reactions:

(a) $Li(s) + O_2(g) \rightarrow Li_2O(s)$
      Oxide

(b) $Na(s) + O_2(g) \rightarrow Na_2O_2(s)$
      Peroxide

(c) $K(s) + O_2(g) \rightarrow KO_2(s)$
      Superoxide

8.27 Balance the equation for each of the following reactions:

(a) $P(s) + O_2(g) \rightarrow P_2O_5(s)$

(b) $S(s) + O_2(g) \rightarrow SO_2(g)$

(c) $N_2(g) + O_2(g) \xrightarrow{\text{Lightning}} NO(g)$

(d) $SO_2(g) + O_2(g) \xrightarrow{\text{Catalyst}} SO_3(g)$

(e) $Mg(s) + O_2(g) \xrightarrow{\text{Heat}} MgO(s)$

(f) $Li(s) + Cl_2(g) \rightarrow LiCl(s)$

8.28 Write a balanced chemical equation for each of the following reactions:

(a) Lithium metal when heated with nitrogen gas reacts to produce solid lithium nitride.

(b) Water reacts with lithium metal to produce aqueous lithium hydroxide and hydrogen gas. *(Caution: This reaction is potentially explosive.)*

(c) Aqueous sodium hydroxide reacts with gaseous carbon dioxide to produce aqueous sodium carbonate and water.

(d) Solid magnesium sulfite decomposes on heating to produce solid magnesium oxide and sulfur dioxide gas.

(e) Ethane gas (C_2H_6) burns in excess oxygen to produce carbon dioxide and water.

8.3 Predicting the Products of Chemical Reactions

8.29 Write two balanced equations for the possible reactions of H_2SO_4 with NaOH.

8.30 Table 8.3 states that most sulfides are insoluble in water. Which sulfides are soluble?

8.31 Complete and balance a chemical equation for each of the following reactions:

(a) $C_4H_{10}(g) + O_2(g, \text{limited supply}) \rightarrow$

(b) $C_4H_{10}(g) + O_2(g, \text{excess}) \rightarrow$

(c) $C_8H_{18}(l) + O_2(g, \text{limited supply}) \rightarrow$

(d) $C_8H_{18}(l) + O_2(g, \text{excess}) \rightarrow$

8.32 Complete and balance each of the following equations:

(a) $SO_3(g) + CaO(s) \rightarrow$

(b) $SO_2(g) + NaOH(aq) \rightarrow$

(c) $SO_2(g) + CaO(s) \rightarrow$

8.33 Write two balanced equations for the possible reactions of benzene, C_6H_6, with oxygen.

8.34 Complete and balance each of the following equations:

(a) $C_5H_{12}O_2(l) + O_2(g, \text{excess}) \rightarrow$

(b) $C_6H_{12}O_2(l) + O_2(g, \text{limited}) \rightarrow$

8.35 Complete and balance an equation for each of the following chemical reactions:

(a) Sodium chloride plus silver nitrate

(b) Hydrochloric acid plus barium hydroxide

(c) Methane (CH_4) plus excess oxygen

(d) Production of aluminum chloride from its elements

(e) Production of aluminum oxide from its elements

8.36 Write a balanced equation for the reaction of chlorine with (a) an alkali metal and (b) an alkaline earth metal.

8.37 Write a balanced equation for the reaction of (a) iron with HCl(aq) to form an iron(II) compound and (b) iron with chlorine to form an iron(III) compound.

8.38 Consider the following pair of reactants:

$$Cr + CrCl_3 \rightarrow$$

(a) Adding chromium metal to the compound is equivalent to doing what with the chlorine?

(b) What other compound of chromium and chlorine exists?

(c) Complete and balance the preceding equation.

(d) Write the symbol for chromium surrounded by the symbols for three chlorine atoms, and write a second such set to the right of the first set. Add another chromium atom between two of the chlorine atoms, and encircle three sets of atoms to make the compound in part (b).

8.39 In which, if any, of the following systems is a reaction expected?

(a) $ZnCl_2(aq) + Cl_2(g)$

(b) $Ne(g) + O_2(g)$

(c) $MgCl_2(aq) + Al(s)$

(d) $Cu(s) + HCl(aq)$

8.40 In which of the following systems is a reaction expected? Write an equation for any reaction that occurs.

(a) $CrCl_2 + Cl_2 \rightarrow$

(b) $CrCl_3 + Cl_2 \rightarrow$

8.41 Complete and balance each of the following equations:

(a) $C_5H_{12}O(l) + O_2(g, \text{limited}) \rightarrow$

(b) $C_6H_{12}O(l) + O_2(g, \text{excess}) \rightarrow$

8.42 Complete and balance each of the following equations:

(a) $Li(s) + H_2O(l) \rightarrow$

(b) $F_2(g) + H_2O(l) \rightarrow$

8.43 Complete and balance each of the following equations:

(a) $HCl(aq) \xrightarrow{\text{Electricity}}$

(b) $H_2O(l) \xrightarrow[\text{Na}_2\text{SO}_4]{\text{Electricity}}$

(c) $NaCl(l) \xrightarrow{\text{Electricity}}$

8.44 Complete and balance each of the following equations:

(a) $BaCl_2(aq) + (NH_4)_2SO_4(aq) \rightarrow$

(b) $Ba(ClO_3)_2(aq) + MgSO_4(aq) \rightarrow$

(c) $NH_3(g) + HCl(aq) \rightarrow$

(d) $Ba(s) + FeCl_3(s) \xrightarrow{\text{Heat}}$

(e) $Cl_2(g) + AlI_3(aq) \rightarrow$

(f) $Ba(C_2H_3O_2)_2(aq) + Na_2CO_3(aq) \rightarrow$

8.45 Which type of reaction involving ionic compounds is most likely to occur without any change in the charges on the ions?

8.46 Complete and balance each of the following equations:

(a) $FeCl_2(s) + Cl_2(g) \rightarrow$

(b) $PCl_5(l) + H_2O(l) \rightarrow H_3PO_4(aq) +$

(c) $PCl_3(l) + Cl_2(g) \rightarrow$

(d) $Mg(s) + N_2(g) \rightarrow$

8.47 Complete and balance each of the following equations:

(a) $Fe_2(SO_4)_3(aq) + BaCl_2(aq) \rightarrow$

(b) $FeSO_4(aq) + BaCl_2(aq) \rightarrow$

8.48 If a compound decomposes without any external energy being added in some form, do you expect the compound to be very long-lasting? Explain.

8.49 The following reaction occurs in aqueous solution. What conclusions can you reach about the barium carbonate?

$$BaCl_2 + Na_2CO_3 \rightarrow BaCO_3 + 2\,NaCl$$

8.4 Acids and Bases

8.50 Complete and balance an equation for each of the following reactions. If no reaction occurs, write nr.

(a) $Zn(s) + HClO_4(aq) \rightarrow$

(b) $Cu(s) + HCl(aq) \rightarrow$

(c) $Al(s) + HI(aq) \rightarrow$

8.51 Complete and balance each of the following equations:

(a) $NH_4Cl(aq) + NaOH(aq) \rightarrow$

(b) $NaHCO_3(aq) + NaOH(aq) \rightarrow$

8.52 Complete and balance each of the following equations:

(a) $Na_2CO_3(aq) + HCl(aq, limited) \rightarrow$

(b) $Na_2CO_3(aq) + HCl(aq, excess) \rightarrow$

8.53 Solid $MgCO_3$ "dissolves" in excess $HCl(aq)$. Write an equation for the reaction. Describe what you would expect to see during this reaction.

8.54 Write an equation for the reaction of carbon dioxide and water with magnesium carbonate to produce a soluble product.

8.55 Complete and balance each of the following equations:

(a) $H_3PO_4(aq) + NaOH(aq, excess) \rightarrow$

(b) $H_2SO_4(aq, excess) + CaCO_3(s) \rightarrow$

(c) $HCl(aq) + NaHCO_3(aq) \rightarrow$

8.56 Complete and balance each of the following equations:

(a) $HClO_4(aq) + CaO(s) \rightarrow$

(b) $N_2O_3(g) + H_2O(l) \rightarrow$

(c) $SO_3(g) + NaOH(aq) \rightarrow$

8.57 Write balanced equations for two possible reactions of oxalic acid ($H_2C_2O_4$) with potassium hydroxide (limited and excess).

8.58 Complete and balance each of the following equations, assuming that an excess of the second reactant is present. Comment on why each reaction proceeds.

(a) $NaH_2BO_3(aq) + HCl(aq) \rightarrow$

(b) $BaCO_3(s) + HBr(aq) \rightarrow$

(c) $Mg(C_2H_3O_2)_2(aq) + HCl(aq) \rightarrow$

(d) $MgO(s) + HClO_4(aq) \rightarrow$

(e) $Mg(OH)_2(s) + HClO_4(aq) \rightarrow$

(f) $Ba(OH)_2(aq) + HNO_3(aq) \rightarrow$

■ ■ ■ GENERAL PROBLEMS ■ ■ ■

8.59 Is each of the following equations balanced? Is each correct?

(a) $AgCl(s) + KNO_3(aq) \rightarrow KCl(aq) + AgNO_3(aq)$

(b) $FeCl_2(aq) + Cu(s) \rightarrow CuCl_2(aq) + Fe(s)$

(c) $NaCl(aq) + KBr(aq) \rightarrow KCl(aq) + NaBr(aq)$

8.60 Balance the following equation:

$KI(aq) + Fe(NO_3)_3(aq) \rightarrow FeI_2(aq) + KNO_3(aq) + I_2(aq)$

8.61 Explain why the Solvay process is used instead of the following reaction:

$$CaCO_3 + 2\,NaCl \rightarrow CaCl_2 + Na_2CO_3$$

8.62 Complete and balance each of the following equations:

(a) $CrCl_2(aq) + AgNO_3(aq) \rightarrow$

(b) $CrCl_3(aq) + AgNO_3(aq) \rightarrow$

8.63 Complete and balance each of the following equations:

(a) $C_3H_6O(l) + O_2(g, excess) \rightarrow$

(b) $C_2H_2(g) + O_2(g, excess) \rightarrow$

(c) $C_{12}H_{26}(l) + O_2(g, excess) \rightarrow$
Kerosene

(d) $C_{12}H_{26}(l) + O_2(g, limited) \rightarrow$
Kerosene

8.64 How can you distinguish a combustion reaction from a displacement reaction, since each may involve an element and a compound?

8.65 Can a substitution reaction occur between an element and a compound of that same element?

8.66 Can a double substitution reaction occur between two compounds containing one ion in common?

8.67 Do the classes of reactions described in Section 8.3 include all possible types of chemical reactions?

8.68 Addition of aqueous ammonia to a solution of $Ca(HCO_3)_2(aq)$ causes a white solid to form. What is the formula of the solid?

8.69 What products are expected from the reaction of ammonium chloride and barium hydroxide? Write an equation for the reaction.

8.70 (a) Which class of reaction requires only one reactant?

(b) Does the addition of a catalyst change the answer to part (a)?

(c) How can you recognize a substance as a catalyst?

8.71 Give one example of each type of reaction in Problem 8.1.

8.72 Complete and balance each of the following equations:

(a) $FeBr_3(s) + Cl_2(g) \rightarrow$ (b) $FeCl_2(s) + Cl_2(g) \rightarrow$

(c) $FeBr_2(s) + Cl_2(g, excess) \rightarrow$

8.73 What is unusual about the following decomposition reactions?

(a) $NH_4HCO_3(s) \xrightarrow{\text{Heat}}$ (b) $Ca(HCO_3)_2(s) \xrightarrow{\text{Heat}}$

8.74 Explain the difference among the following three questions:

What is the product of the electrolysis of water containing dilute NaCl to carry the current?

What is the product of the electrolysis of water containing dilute NaCl?

What is the product of the electrolysis of dilute aqueous NaCl?

8.75 Consider the following pairs of reactants. For each, determine the possible reaction type, and write correct formulas for the products that could be produced. If the reaction can proceed, write a balanced equation.

(a) $CO(g)$ and $O_2(g)$

(b) $HCl(aq)$ and $Zn(s)$

(c) $NaNO_3(aq)$ and $AgCl(s)$

8.76 Inexpensive metal forks corrode rapidly if used in a delicatessen to remove pickles from the juice in which they are shipped. Explain the probable cause.

8.77 Assuming that water containing $Ca(HCO_3)_2$ deposits 1 mg of $CaCO_3$ per minute on the ceiling of a limestone cavern, how long will it take to produce a stalactite with a mass of 100 metric tons (1 metric ton = 1×10^6 g)?

8.78 Complete and balance each of the following equations:

(a) $HBr(aq) + NaHCO_3(aq) \rightarrow$

(b) $NaOH(aq) + NaHSO_3(aq) \rightarrow$

(c) $HC_2H_3O_2(aq) + Ba(OH)_2(aq) \rightarrow$

(d) $NaC_2H_3O_2(aq) + HCl(aq) \rightarrow$

9 Net Ionic Equations

- 9.1 Properties of Ionic Compounds in Aqueous Solution

- 9.2 Writing Net Ionic Equations

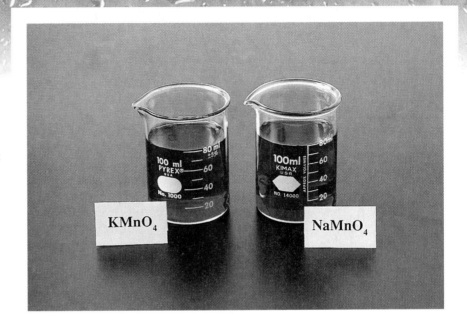

- Key Terms *(Key terms are defined in the Glossary.)*

ionic equation (9.2)
net ionic equation (9.2)

spectator ion (9.2)
total equation (9.2)

Section 9.1 describes the properties of ionic compounds in aqueous solution. Section 9.2 then explains how to write net ionic equations for many reactions in aqueous solution. These equations show the actual reactions that occur; ions that do not change at all during the reaction are not included. Each net ionic equation can summarize many equations involving complete compounds.

9.1 Properties of Ionic Compounds in Aqueous Solution

The properties of ionic compounds in solution are actually the properties of the individual ions themselves (Figure 9.1). For example, an aqueous solution of sodium chloride consists essentially of chloride ions and sodium ions in water. A similar solution of potassium chloride consists of chloride ions and potassium ions in water. If either solution is treated with a solution containing silver ions, the chloride ions will form silver chloride, which is insoluble. The chloride ions act independently of the cation that is also present, regardless of whether it is sodium ion or potassium ion. Since the properties of the compound are the properties of the component ions, you need to learn to write equations for only the ions that react, omitting the ions that remain unchanged throughout the reaction (Section 9.2).

Salts, strong acids, and strong bases (Table 9.1) all provide ions in solution, but the process by which these types of compounds form ions in solution differs. Pure strong acids are covalent compounds, and they undergo a chemical reaction with water to form ions in solution. This process, called *ionization,* will be discussed in more detail in Chapter 18. Salts and strong bases are ionic even when they are pure, and their interaction with water is more a physical process than a chemical reaction. The solution process for them is called *dissociation* because the ions dissociate from each other; that is, they get out of each other's sphere of influence and are able to move relatively independently of ions of the opposite charge.

An ionic compound *in aqueous solution* may be represented as separate ions, since the ions of each type are free to move about independently of the ions of the

(a)

(b)

(c)

Figure 9.1 Properties of Ions

(a) The purple color of these two solutions is due to the permanganate ion. (b) Solutions of potassium and sodium ions with different anions than the permanganate ion show that these cations are colorless. (c) The blue color of these solutions is characteristic of the copper(II) ion; the nitrate ion and the sulfate ion are colorless, as shown in part (b).

Table 9.1 Strong Acids and Bases

Strong Acids	Strong Bases
HCl	All soluble metal
HBr	hydroxides,
HI	such as NaOH,
$HClO_3$	KOH, and
$HClO_4$	$Ba(OH)_2$*
HNO_3	
H_2SO_4	

*Note that $Ba(OH)_2$ has limited solubility.

other type. However, an ionic solid that is not dissolved in water is not written as separate ions; the oppositely charged ions in the solid lattice of an ionic compound are not independent of each other (Figure 9.2).

Thus, compounds must be both soluble and ionic to be written in the form of their separate ions. A listing of water-soluble compounds was given in Table 8.3. In addition to the compounds listed there, all strong acids are water soluble. In summary, compounds that dissociate or ionize in aqueous solution include the following:

1. Strong acids (HCl, $HClO_3$, $HClO_4$, HBr, HI, HNO_3, H_2SO_4)

2. All soluble salts (compounds containing metal or ammonium ions)

3. All soluble metal hydroxides

All other compounds (for example, gases, other covalent compounds, and all solids) either contain no ions or have ions that are affected by the presence of the other ions. Such compounds are written using their regular formulas.

■ **EXAMPLE 9.1**

Write each of the following compounds to represent best how it acts in the presence of water:

(a) $NaClO_3(aq)$ (b) $Al(ClO_3)_3(aq)$ (c) $H_2O(l)$ (d) $CO_2(g)$
(e) $AgCl(s)$

Solution

(a) $Na^+(aq) + ClO_3^-(aq)$ (b) $Al^{3+}(aq) + 3 ClO_3^-(aq)$
(c) $H_2O(l)$ (d) $CO_2(g)$
(e) $AgCl(s)$ ■

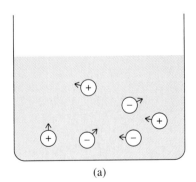
(a)

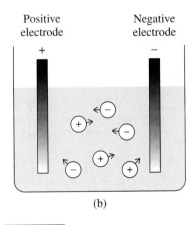
(b)

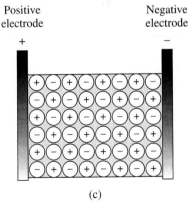
(c)

Figure 9.2 Mobility of Ions

(a) Ions in dilute solutions are free to move independently of other ions. In the absence of electrodes, they move in random directions. (b) Under the influence of the charges on electrodes, the ions move toward the electrode of opposite charge. (c) In contrast, even if charged electrodes are present, ions in solids cannot move because of the surrounding ions of opposite charge.

9.2 Writing Net Ionic Equations

■ to write net ionic equations for reactions in aqueous solution and to interpret such equations

When sodium chloride solution is added to silver nitrate solution, a precipitate of silver chloride is produced, and the solution contains sodium nitrate (Figure 9.3):

$$AgNO_3(aq) + NaCl(aq) \rightarrow AgCl(s) + NaNO_3(aq)$$

This type of equation can be called the **total equation.** (A total equation is sometimes referred to as a *molecular equation* because the compounds in it are written "as if they were molecules.")

Even more informative than a total equation is an **ionic equation.** An ionic compound *in aqueous solution* may be represented as separate ions, but an ionic solid that is not dissolved in water is written as a complete compound. We can write an ionic equation for the reaction of sodium chloride with silver nitrate in aqueous solution (Figure 9.3) as follows:

$$Ag^+(aq) + NO_3^-(aq) + Na^+(aq) + Cl^-(aq) \rightarrow$$
$$AgCl(s) + Na^+(aq) + NO_3^-(aq)$$

Since the Na^+ and NO_3^- ions appear on both sides of this equation (are unchanged by the reaction), they are called **spectator ions.** They may be eliminated from the equation:

$$Ag^+(aq) + Cl^-(aq) \rightarrow AgCl(s)$$

This equation is an example of a **net ionic equation.** All the spectator ions are omitted from a net ionic equation.

Thus, to write net ionic equations, we first write compounds that are both soluble and ionic in the form of their separate ions. All other compounds (for example, gases, other covalent compounds, and all solids) are written as complete compounds in an ionic equation. We then eliminate the ions that appear unchanged on both sides of the equation to obtain the net ionic equation.

Figure 9.3 Reaction of Silver Nitrate and Sodium Chloride

■ EXAMPLE 9.2

Write a net ionic equation for the reaction of potassium chloride and silver nitrate.

Solution

The total equation for the reaction is

$$AgNO_3(aq) + KCl(aq) \rightarrow AgCl(s) + KNO_3(aq)$$

The ionic equation is

$$Ag^+(aq) + NO_3^-(aq) + K^+(aq) + Cl^-(aq) \rightarrow AgCl(s) + K^+(aq) + NO_3^-(aq)$$

Eliminating the spectator ions produces the net ionic equation:

$$Ag^+(aq) + Cl^-(aq) \rightarrow AgCl(s)$$

This is the same as the net ionic equation given previously for the reaction of sodium chloride with silver nitrate because essentially the same reaction has taken place. Whether it was the sodium ions or the potassium ions that *did not react* is not important to us. In general, we can say that soluble ionic chlorides react with soluble silver salts to produce silver chloride. This statement does not mention the other ions present in the reactant solutions and may be represented by the net ionic equation, which similarly does not mention any other ions that may be present.

Practice Problem 9.2 Write a net ionic equation for the reaction of $AgC_2H_3O_2$ with $CuCl_2$. ■

Be careful not to misinterpret the name *net ionic equation*. It is not necessarily true that all the substances appearing in such an equation are ionic. Covalent compounds often occur in net ionic equations. Also, just because the formula for a complete compound is written in such an equation does not mean that the compound is not ionic; it might simply be insoluble.

An important type of reaction is the reaction of a strong acid with a strong base (Table 9.1) to produce a salt and water. In solution, strong acids and bases exist completely in the form of their anions and cations. All salts, the products of reactions of acids with bases, may also be regarded as completely ionic (but not all are water soluble). We may therefore write net ionic equations for this type of reaction. The reaction of hydrochloric acid and sodium hydroxide is typical:

$$HCl(aq) + NaOH(aq) \rightarrow NaCl(aq) + H_2O(l)$$

In solution, both of the reactants and the sodium chloride are ionic, but the water is covalent, of course. The ionic equation for the reaction is

$$H^+(aq) + Cl^-(aq) + Na^+(aq) + OH^-(aq) \rightarrow Na^+(aq) + Cl^-(aq) + H_2O(l)$$

Eliminating the spectator ions from both sides of this equation yields the net ionic equation:

$$H^+(aq) + OH^-(aq) \rightarrow H_2O(l)$$

Note that water, which is molecular, not ionic, is included in this net ionic equation.

■ EXAMPLE 9.3

Write a net ionic equation for the reaction of aqueous $Ba(OH)_2$ with aqueous HCl.

Solution

The total equation is

$$Ba(OH)_2(aq) + 2\ HCl(aq) \rightarrow BaCl_2(aq) + 2\ H_2O(l)$$

The ionic equation is

$$Ba^{2+}(aq) + 2\ OH^-(aq) + 2\ H^+(aq) + 2\ Cl^-(aq) \rightarrow$$
$$Ba^{2+}(aq) + 2\ Cl^-(aq) + 2\ H_2O(l)$$

Be careful with the coefficients of the ions in the ionic equation.

Eliminating the spectator ions yields

$$2\ OH^-(aq) + 2\ H^+(aq) \rightarrow 2\ H_2O(l)$$

This equation can be simplified by dividing each coefficient by 2:

$$OH^-(aq) + H^+(aq) \rightarrow H_2O(l)$$

Again, this is the same net ionic equation as that for the reaction of NaOH with HCl. In fact, the reaction of *any* aqueous strong acid with *any* aqueous strong base yields this net ionic equation (unless some precipitation occurs).

Figure 9.4 Silver Ion
Test Solution

A solution that can be used to test
whether another solution contains
silver ions can contain any soluble
ionic chloride. A bottle containing
such a solution may be labeled "Cl⁻
Solution" or "Ag⁺ Test Reagent."
That this solution also contains Na^+
ions or K^+ ions does not matter,
since these cations would not react
with the ions in a solution containing
silver ions.

Practice Problem 9.3   Write a net ionic equation for the reaction of KOH(aq)
with aqueous solutions of each of the following:

(a) HNO_3 (b) HBr   (c) $HClO_3$ (d) HI ■

Net ionic equations can also be written for reactions in which gases are produced. For example, sodium hydrogen carbonate reacts with nitric acid to produce sodium nitrate, carbon dioxide, and water:

$$NaHCO_3(aq) + HNO_3(aq) \rightarrow NaNO_3(aq) + CO_2(g) + H_2O(l)$$

The net ionic equation is

$$HCO_3^-(aq) + H^+(aq) \rightarrow CO_2(g) + H_2O(l)$$

What does a net ionic equation actually tell us? As an example, the net ionic
equation of Example 9.3 indicates that any strong acid in water reacts with any
soluble strong hydroxide to yield water as a product. The ions that do not react
are not of immediate concern. However, no aqueous solution contains only H^+
ions or only OH^- ions. The net ionic equation does not state that these ions occur
without ions of the opposite charge, only that the identities of the oppositely
charged ions are not important because they do not react (Figure 9.4).

Weak acids—any acids not listed in Table 9.1—are essentially covalent in
solution and should be written as molecular species.

■ **EXAMPLE 9.4**

Write a net ionic equation for the reaction of acetic acid, $HC_2H_3O_2$, with sodium
hydroxide, NaOH.

Solution

The total equation for the reaction is

$$HC_2H_3O_2(aq) + NaOH(aq) \rightarrow NaC_2H_3O_2(aq) + H_2O(l)$$

To write the ionic equation, you must remember that acetic acid is not a strong
acid; that is, it does not ionize completely in water. It is written as a covalent
compound. The ionic equation is

$$HC_2H_3O_2(aq) + Na^+(aq) + OH^-(aq) \rightarrow Na^+(aq) + C_2H_3O_2^-(aq) + H_2O(l)$$

The net ionic equation is written without the sodium ions:

$$HC_2H_3O_2(aq) + OH^-(aq) \rightarrow C_2H_3O_2^-(aq) + H_2O(l)$$

Note that this net ionic equation is *not* the same as the one for the reaction of
a strong acid and a strong base!

Practice Problem 9.4   Write a net ionic equation for the reaction of aqueous
NH_3 with aqueous HNO_3. ■

For a net ionic equation to be balanced, both the numbers of each type of
atom and the net charge must be the same on the two sides of the equation. For
example, we know that copper is more active than silver is (Table 8.2) and will
replace silver from its compounds. We start to write the net ionic equation for the
reaction in solution as follows:

$$Cu(s) + Ag^+(aq) \rightarrow Cu^{2+}(aq) + Ag(s) \qquad \text{(Not balanced)}$$

This net ionic equation is balanced only with regard to the numbers of copper and silver atoms. Since the charge is not balanced, however, the equation is not balanced. We can balance it by doubling the charge on the left side (with a 2 before the Ag^+) and keeping the number of silver atoms balanced (with a 2 before the Ag):

$$Cu(s) + 2\,Ag^+(aq) \rightarrow Cu^{2+}(aq) + 2\,Ag(s) \quad \text{(Balanced)}$$

■ EXAMPLE 9.5

Write a total equation corresponding to the net ionic equation for the reaction of silver ions and copper, using nitrate ions as the spectator ions. Explain why the charges in the net ionic equation have to be balanced.

Solution

$$Cu(s) + 2\,AgNO_3(aq) \rightarrow Cu(NO_3)_2(aq) + 2\,Ag(s)$$

Each positive charge in the net ionic equation represents one nitrate ion in this total equation. The charges must be balanced in the net ionic equation because the nitrate ions must be balanced in the total equation.

Practice Problem 9.5 Write a balanced total equation that is represented by the following net ionic equation:

$$Br_2(aq) + 2\,I^-(aq) \rightarrow I_2(aq) + 2\,Br^-(aq) \quad ■$$

■ EXAMPLE 9.6

Write a net ionic equation for the reaction of iron metal with aqueous iron(III) chloride to produce aqueous iron(II) chloride.

Solution

The total equation is

$$Fe(s) + 2\,FeCl_3(aq) \rightarrow 3\,FeCl_2(aq)$$

The ionic equation is

$$Fe(s) + 2\,Fe^{3+}(aq) + 6\,Cl^-(aq) \rightarrow 3\,Fe^{2+}(aq) + 6\,Cl^-(aq)$$

Eliminating the chloride ions from each side yields the net ionic equation:

$$Fe(s) + 2\,Fe^{3+}(aq) \rightarrow 3\,Fe^{2+}(aq)$$

Note that the cations are *not* eliminated because they are not the same on each side. There has been a change from iron(III) to iron(II). The uncharged metal atom has also changed and cannot be eliminated as a spectator ion. The net ionic equation indicates that iron metal will react with any soluble iron(III) compound to produce the corresponding soluble iron(II) compound.

> Be sure the total equation is balanced before attempting to write the ionic equation.

Practice Problem 9.6 Write a net ionic equation for the following reaction:

$$Cu(s) + CuBr_2(aq) \rightarrow 2\,CuBr(s) \quad ■$$

Net ionic equations are used extensively in chemistry. For example, equilibrium expressions for acid-base reactions, as well as for the ionization of water itself, are conventionally written in the form of net ionic equations. Many complex oxidation-reduction equations are balanced using net ionic equations. These topics are introduced in Chapters 16 through 18.

■ ■ ■ ■ ■ ■ ■ SUMMARY ■ ■ ■ ■ ■ ■ ■

In aqueous solutions of ionic compounds, the ions act independently of each other. Soluble ionic compounds are written as their separate ions. You must be familiar with the solubility rules presented in Chapter 8 and recognize that the following types of compounds are ionic: strong acids in solution, soluble metallic hydroxides, and salts. (Salts, which can be formed as the products of reactions of acids with bases, include all ionic compounds except strong acids and bases and metallic oxides and hydroxides.) Compounds must be *both ionic and soluble* to be written in the form of their separate ions. (Section 9.1)

A net ionic equation describes the actual reaction between ions of compounds in aqueous solution. Ions that do not change at all during the reaction are omitted from the equation; these ions are called spectator ions. One net ionic equation may describe the reactions of many compounds. For example, the net ionic equation

$$Ag^+(aq) + Cl^-(aq) \rightarrow AgCl(s)$$

summarizes all the reactions described by the statement: "Any soluble silver salt reacts with any soluble ionic chloride to produce (the insoluble) silver chloride." The equation also gives the mole ratios, which the statement does not.

Net ionic equations are balanced only if the numbers of atoms of each element and the net charge on each side of the equation are both balanced. (Section 9.2)

Items for Special Attention

■ Strong acids react completely with water to form ions in solution. Metal hydroxides and salts are ionic in the solid state, as well as in solution; however, in the solid state, such compounds are written as compounds because the ions are not independent of each other.

■ Most ionic compounds are composed of only one type of positive ion and one type of negative ion. (Of course, more than one of each type of ion may be present in each formula unit.)

■ You may be feeling confused about what should be included in net ionic equations. Easier to remember is what should be left out: *Only ions in solution that remain unchanged in solution* should be left out to produce net ionic equations; all other species must be included. Thus, insoluble compounds (ionic or not), covalent compounds, elements, and ions that change in any way between reactants and products are all included. Remembering what to omit—the spectator ions—is much easier!

Self-Tutorial Problems

9.1 Write formulas for the ions that constitute each of the following compounds:

(a) $FeCl_2$ (b) NH_4NO_3 (c) $NaClO_4$
(d) $Co(ClO)_2$ (e) $CuCl_2$ (f) $AlCl_3$

9.2 Write each of the following species (in aqueous solution, if soluble) as it should appear in an ionic equation:

(a) $HClO_3$ (b) $CaCl_2$
(c) HCl (d) $BaSO_4$
(e) $NaOH$ (f) $KClO_3$
(g) Zn (h) $(NH_4)_2SO_4$
(i) $Ni(ClO_3)_2$ (j) SO_2
(k) NH_3 (l) KOH
(m) $Al_2(SO_4)_3(aq)$ (n) $Mg(OH)_2(s)$
(o) H_2O (p) $CH_3OH(aq)$ (methyl alcohol)
(q) H_3PO_4 (r) $AgCl$
(s) $PbCl_2$ (t) $Zn(C_2H_3O_2)_2$
(u) $AgClO_3$ (v) NH_4I
(w) $K_2Cr_2O_7$ (x) $CH_2O(aq)$
(y) $KMnO_4$ (z) $HClO_4$

9.3 Write a net ionic equation for each of the following reactions:

(a) $Ba(OH)_2(aq) + 2\,HCl(aq) \rightarrow BaCl_2(aq) + 2\,H_2O(l)$
(b) $Ba(OH)_2(s) + 2\,HCl(aq) \rightarrow BaCl_2(aq) + 2\,H_2O(l)$

9.4 Write a net ionic equation for the reaction of (a) Ca with HCl, (b) Ca with any strong acid, and (c) Ca with a strong acid.

9.5 What is the difference in the nature of the bonding of the chlorine in the following species?

$$Cl_2 \qquad SrCl_2 \qquad SCl_2$$

9.6 For each of the following compounds, determine whether it is soluble in water, whether it is ionic in the pure state or in solution, and whether it should be written as a compound or as separate ions in an ionic equation. Then write the compound as it should be written in an ionic equation.

(a) H_2O (b) LiCl   (c) $BaSO_4$

(d) $HC_2H_3O_2$ (e) $HClO_3$ (f) CH_3OH
                                      (methyl alcohol)

9.7 Which, if any, of the common acids exist as ions (a) in the pure state and (b) in aqueous solution?

9.8 A bottle labeled "Sr^{2+} Test Reagent" in a chemistry lab is used to test the properties of the strontium ion. What does the bottle contain?

■ ■ ■ PROBLEMS ■ ■ ■

9.1 Properties of Ionic Compounds in Aqueous Solution

9.9 Assuming that each of the following acids is in aqueous solution, write its formula to best represent it:

(a) HCl   (b) HClO   (c) $HClO_2$

(d) $HClO_3$ (e) $HClO_4$

9.10 Assuming that each of the following compounds is in aqueous solution, write its formula to best represent it:

(a) NaCl   (b) NaClO   (c) $NaClO_2$

(d) $NaClO_3$ (e) $NaClO_4$

9.11 Write each of the following compounds to best represent it in the presence of water:

(a) CO_2 (b) $BaSO_4$ (c) $Cu(ClO_3)_2$

(d) $AlCl_3$ (e) $Ba(ClO_3)_2$ (f) SF_2

9.12 Assuming that each of the following acids is in aqueous solution, write its formula to best represent it:

(a) HNO_3 (b) $HC_2H_3O_2$

(c) HBr   (d) H_3PO_4

9.13 Assuming that each of the following compounds is in aqueous solution, write its formula to best represent it:

(a) $(NH_4)_2SO_4$ (b) $Mg(HCO_3)_2$ (c) NH_3

(d) $CaCl_2$ (e) $Fe(NO_3)_3$

9.14 Write each of the following compounds to best represent it in the presence of water:

(a) AgCl   (b) Li_2SO_4 (c) $Hg_2(NO_3)_2$

(d) CCl_4 (e) NH_4ClO_3

9.15 (a) Would NaF or HF be better for making a solution containing fluoride ion?

(b) Would NaCl or HCl be better for making a solution containing chloride ion?

(c) Would NH_4Cl or NH_3 be better for making a solution containing ammonium ion?

9.2 Writing Net Ionic Equations

9.16 Balance each of the following net ionic equations:

(a) $Cu_2O(s) + H^+(aq) \rightarrow Cu(s) + Cu^{2+}(aq) + H_2O(l)$

(b) $Co^{3+}(aq) + Co(s) \rightarrow Co^{2+}(aq)$

(c) $Ag^+(aq) + Cd(s) \rightarrow Cd^{2+}(aq) + Ag(s)$

(d) $I^-(aq) + Ce^{4+}(aq) \rightarrow Ce^{3+}(aq) + I_2(aq)$

9.17 Balance each of the following net ionic equations:

(a) $Zn(s) + Fe^{3+}(aq) \rightarrow Fe^{2+}(aq) + Zn^{2+}(aq)$

(b) $NH_3(aq) + Ag^+(aq) \rightarrow Ag(NH_3)_2^+(aq)$

(c) $Ce^{4+}(aq) + Br^-(aq) \rightarrow Br_2(aq) + Ce^{3+}(aq)$

(d) $PbO_2(s) + SO_4^{2-}(aq) + H^+(aq) + Br^-(aq) \rightarrow$
$$Br_2(aq) + PbSO_4(s) + H_2O(l)$$

9.18 Write a net ionic equation for each of the following reactions:

(a) $KHCO_3(aq) + KOH(aq) \rightarrow K_2CO_3(aq) + H_2O(l)$

(b) $KHCO_3(aq) + HNO_3(aq) \rightarrow$
$$KNO_3(aq) + H_2O(l) + CO_2(g)$$

(c) $KI(aq) + AgNO_3(aq) \rightarrow KNO_3(aq) + AgI(s)$

(d) $Ba(OH)_2(aq) + 2\ HClO_3(aq) \rightarrow$
$$Ba(ClO_3)_2(aq) + 2\ H_2O(l)$$

(e) $BaCl_2(aq) + Na_2SO_4(aq) \rightarrow 2\ NaCl(aq) + BaSO_4(s)$

(f) $2\ HNO_3(aq) + Ba(OH)_2(s) \rightarrow$
$$Ba(NO_3)_2(aq) + 2\ H_2O(l)$$

9.19 Write a net ionic equation for the reaction of an insoluble metal oxide, represented as M_2O, with a strong acid.

9.20 Write six total equations that correspond to the following net ionic equation and have an alkali metal ion and nitrate ion as spectator ions:

$$Pb^{2+}(aq) + 2\ Cl^-(aq) \rightarrow PbCl_2(s)$$

9.21 Write a net ionic equation for the reaction of aqueous barium nitrate with aqueous sodium carbonate to yield solid barium carbonate and aqueous sodium nitrate.

9.22 Write a net ionic equation for the reaction of silver nitrate with each of the following in aqueous solution:

(a) Iron(II) chloride

(b) Iron(III) chloride

(c) Zinc chloride

(d) Copper(II) chloride

(e) Nickel(II) chloride

(f) Cobalt(II) chloride

(g) Hydrochloric acid

9.23 (a) Write six total equations that correspond to the following net ionic equation and have an alkali metal ion and nitrate ion as spectator ions:

$$Ag^+(aq) + I^-(aq) \rightarrow AgI(s)$$

(b) If any of three anions were used as a spectator ion, how many total equations could be written?

9.24 Write nine total equations that are represented by the following net ionic equation and have spectator ions chosen from Na^+, NH_4^+, Fe^{2+}, ClO_3^-, NO_3^-, and $C_2H_3O_2^-$:

$$Ba^{2+}(aq) + SO_4^{2-}(aq) \rightarrow BaSO_4(s)$$

9.25 Write a net ionic equation for the reaction of

(a) a strong acid with a strong base

(b) a strong acid with ammonia (a weak base), and

(c) a weak acid, represented as HA, with a strong base.

9.26 Write a net ionic equation for the reaction of aqueous H_2SO_4 with aqueous $Ba(OH)_2$.

9.27 Write a net ionic equation for the reaction of aqueous silver chlorate with aqueous sodium carbonate to yield solid silver carbonate and aqueous sodium chlorate.

9.28 Write a net ionic equation for the reaction of sodium sulfide with each of the following in aqueous solution. An insoluble sulfide is formed in each case.

(a) Iron(II) chloride　　(b) Manganese(II) chloride

(c) Zinc chloride　　(d) Copper(II) chloride

(e) Nickel(II) chloride　　(f) Cobalt(II) chloride

9.29 Write a net ionic equation for the reaction of aqueous barium chlorate with aqueous sodium carbonate to yield solid barium carbonate and aqueous sodium chlorate.

9.30 Write a net ionic equation for each of the following reactions:

(a) $HClO_3(aq) + NaOH(aq) \rightarrow NaClO_3(aq) + H_2O(l)$

(b) $HClO_2(aq) + NaOH(aq) \rightarrow NaClO_2(aq) + H_2O(l)$

9.31 Barium hydroxide, $Ba(OH)_2$, has limited water solubility. If a small quantity of barium hydroxide is added to a given volume of water, it might dissolve. If a large quantity of barium hydroxide is added to the same volume of water, most of it will not dissolve. Thus, barium hydroxide might appear in an equation as either solid or aqueous. Write a net ionic equation for the reaction of barium hydroxide with HCl for each of these cases.

9.32 Balance the following equations. Then write a net ionic equation for each.

(a) $Zn(s) + HCl(aq) \rightarrow H_2(g) + ZnCl_2(aq)$

(b) $Zn(s) + H_2SO_4(aq) \rightarrow H_2(g) + ZnSO_4(aq)$

(c) $Zn(s) + HClO_3(aq) \rightarrow H_2(g) + Zn(ClO_3)_2(aq)$

9.33 Balance the following net ionic equations:

(a) $Ce^{4+}(aq) + Zn(s) \rightarrow Zn^{2+}(aq) + Ce^{3+}(aq)$

(b) $I^-(aq) + Cu^{2+}(aq) \rightarrow CuI(s) + I_2(aq)$

(c) $Al(s) + Sn^{2+}(aq) \rightarrow Sn(s) + Al^{3+}(aq)$

■ ■ ■ GENERAL PROBLEMS ■ ■ ■

9.34 Write a net ionic equation for the reaction of (a) a soluble carbonate with excess strong acid, (b) a soluble acid carbonate with a strong acid, and (c) a soluble acid carbonate with a strong base.

9.35 Explain why the reaction of HCl with NaOH and the reaction of $HClO_3$ with KOH yield the same quantity of heat per mole of water produced.

9.36 When $CH_3Cl(l)$ is treated with $AgNO_3(aq)$, no reaction occurs. Explain why.

9.37 Complete and balance the following net ionic equations:

(a) $H^+(aq) + OH^-(aq) \rightarrow$

(b) $H^+(aq) + CO_3^{2-}(aq) \rightarrow$

(c) $H^+(aq) + HCO_3^-(aq) \rightarrow$

(d) $Ba^{2+}(aq) + SO_4^{2-}(aq) \rightarrow$

(e) $Ag^+(aq) + Br^-(aq) \rightarrow$

9.38 Write five total equations that are represented by the following net ionic equation:

$$2\,H^+(aq) + CO_3^{2-}(aq) \rightarrow H_2O(l) + CO_2(g)$$

9.39 Write a complete equation and a balanced net ionic equation for each of the following reactions (there is a reaction in each case):

(a) $NH_4Cl(aq) + KOH(aq) \rightarrow$

(b) $NaC_2H_3O_2(aq) + HClO_4(aq) \rightarrow$

(c) $NaH_2PO_4(aq) + HNO_3(aq) \rightarrow$

(d) $Na_2CO_3(aq) + HClO_3(aq, excess) \rightarrow$

9.40 Write a net ionic equation for the reaction of sodium metal with water.

9.41 Write a complete equation and a balanced net ionic equation for each of the following reactions (there is a reaction in each case):

(a) $Zn(ClO_3)_2(aq) + (NH_4)_2S(aq) \rightarrow$

(b) $Hg(C_2H_3O_2)_2(aq) + Na_3PO_4(aq) \rightarrow$

(c) $CaCl_2(aq) + Na_2CO_3(aq) \rightarrow$

(d) $Ba(NO_3)_2(aq) + Na_3PO_4(aq) \rightarrow$

(e) $Hg_2(NO_3)_2(aq) + KCl(aq) \rightarrow$

(f) $CuCl_2(aq) + Na_2S(aq) \rightarrow$

9.42 Write a complete equation and a balanced net ionic equation for each of the following reactions (there is a reaction in each case):

(a) $BaCO_3(s) + H_2SO_4(aq) \rightarrow$

(b) $BaCO_3(s) + HCl(aq) \rightarrow$

(c) $MgCO_3(s) + CO_2(g) + H_2O(l) \rightarrow$

(d) $Mg(HCO_3)_2(aq) + HCl(aq) \rightarrow$

9.43 Write a complete equation and a balanced net ionic equation for each of the following reactions (there is a reaction in each case):

(a) $NH_4Cl(aq) + KOH(aq) \rightarrow$

(b) $NaC_2H_3O_2(aq) + HClO_4(aq) \rightarrow$

(c) $NaH_2PO_4(aq) + HNO_3(aq) \rightarrow$

10 Stoichiometry

- 10.1 Mole Calculations for Chemical Reactions
- 10.2 Mass Calculations for Chemical Reactions
- 10.3 Calculations Involving Other Quantities
- 10.4 Problems Involving Limiting Quantities
- 10.5 Theoretical Yield and Percent Yield
- 10.6 Calculations with Net Ionic Equations

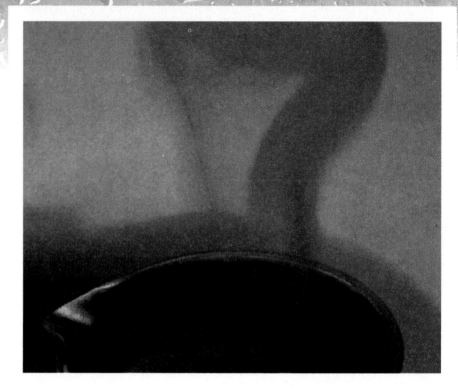

Key Terms *(Key terms are defined in the Glossary.)*

completion (10.4)
excess (10.4)
limiting quantity (10.4)

percent yield (10.5)
reacting ratio (10.1)

stoichiometry (10.1)
theoretical yield (10.5)

In Chapter 7, you learned to interpret chemical formulas in terms of the quantities of the elements involved. In Chapters 8 and 9, you learned to complete and balance chemical equations. This chapter shows how to interpret the quantities of substances involved in a chemical reaction, using the balanced chemical equation almost as the formula was used in Chapter 7.

Just as compounds have definite ratios of elements, chemical reactions have definite ratios of reactants and products. Those ratios are used in Section 10.1 to calculate the number of moles of other substances in a reaction from the number of moles of one of the substances. Section 10.2 combines information from Section 10.1, Chapter 7, and elsewhere to explain how to calculate the mass of any substance involved in a reaction from the mass of another. Section 10.3 demonstrates how to work with units other than moles or grams when finding quantities of reactants or products. Section 10.4 shows how to calculate the quantities of reactants and products involved in a reaction even if the quantities of reactants are not present in the mole ratio of the balanced equation. Section 10.5 covers the calculation of the percentage yield of a product from the actual yield and the theoretical yield, based on the amount(s) of reactant(s). Section 10.6 explains which types of calculations can and cannot be done for net ionic equations.

10.1 Mole Calculations for Chemical Reactions

Stoichiometry involves the calculation of quantities of any substances involved in a chemical reaction from the quantities of the other substances. The balanced equation gives the ratios of formula units of all the reactants and products of a chemical reaction. It also gives the corresponding ratios of moles of reactants and products. These relationships are shown in Figure 10.1. For example, the reaction of sodium metal with chlorine gas is governed by the equation

$$2\ Na(s) + Cl_2(g) \rightarrow 2\ NaCl(s)$$

This equation may be interpreted, as shown in Chapter 8, in two ways:

1. Two atoms of sodium react with one molecule of Cl_2 to produce two formula units of NaCl.

Figure 10.1 Mole Conversions for Stoichiometry Problems

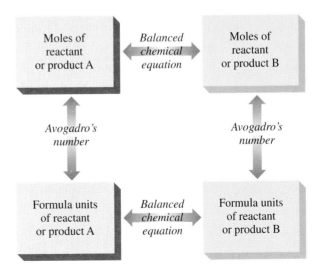

2. Two moles of sodium react with one mole of Cl_2 to produce two moles of NaCl.

Of course, the chemist is not required to place exactly 2 mol of Na and 1 mol of Cl_2 in a reaction flask. The equation gives the **reacting ratio.** Ratios of coefficients from balanced chemical equations may be used as conversion factors for solving problems.

▪ EXAMPLE 10.1

Write all the possible factors from the coefficients in the following balanced equation:

$$Cl_2(g) + 2\ Na(s) \rightarrow 2\ NaCl(s)$$

Solution

$$\frac{1\ mol\ Cl_2}{2\ mol\ Na} \quad \frac{1\ mol\ Cl_2}{2\ mol\ NaCl} \quad \frac{2\ mol\ Na}{2\ mol\ NaCl} \quad \frac{2\ mol\ Na}{1\ mol\ Cl_2} \quad \frac{2\ mol\ NaCl}{1\ mol\ Cl_2} \quad \frac{2\ mol\ NaCl}{2\ mol\ Na}$$

Practice Problem 10.1   Write all the possible factors from the coefficients in the following balanced equation:

$$3\ HCl(aq) + La(OH)_3(s) \rightarrow LaCl_3(aq) + 3\ H_2O(l)\ \ \blacksquare$$

The ratio that should be used in a particular problem will have the substance for which the number of moles is given in the denominator and the substance for which the number of moles is desired in the numerator.

▪ EXAMPLE 10.2

Calculate the number of moles of sodium atoms that will react with 2.27 mol of sulfur atoms to form sodium sulfide. The balanced equation is

$$2\ Na(s) + S(s) \rightarrow Na_2S(s)$$

Solution

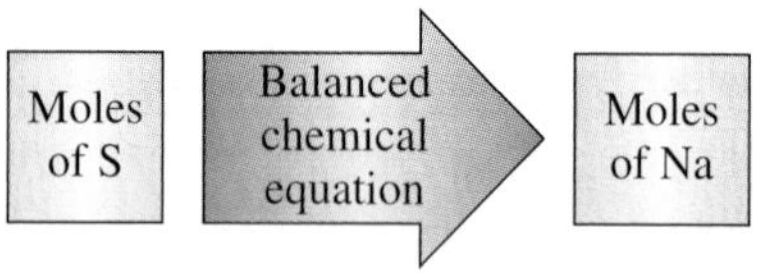

$$2.27\ mol\ S\left(\frac{2\ mol\ Na}{1\ mol\ S}\right) = 4.54\ mol\ Na$$

| Moles of S | Balanced chemical equation | Moles of Na |

Practice Problem 10.2   Calculate the number of moles of Na_2S that will be produced by the reaction in Example 10.2.   ▪

Essentially, most problems involving mole calculations are as simple as the one in Example 10.2. A problem may seem more difficult if you are required to write and balance an equation, but you learned how to do that in Chapter 8.

▪ EXAMPLE 10.3

Calculate the number of moles of hydrogen gas that can be produced by reaction of 7.18 mol of hydrochloric acid, HCl, with aluminum metal.

Solution

The first step, as in most stoichiometry problems, is to write a balanced equation for the reaction:

$$6 \ HCl(aq) + 2 \ Al(s) \longrightarrow 3 \ H_2(g) + 2 \ AlCl_3(aq)$$

Now the stoichiometry problem can be solved, as in Example 10.2:

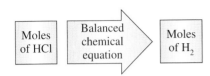

$$7.18 \ mol \ HCl\left(\frac{3 \ mol \ H_2}{6 \ mol \ HCl}\right) = 3.59 \ mol \ H_2$$

Practice Problem 10.3   Calculate the number of moles of aqueous H_3PO_4 that must react completely with aqueous NaOH to produce 2.74 mol of Na_3PO_4. ■

Remember that the quantities involved in mole calculations are the quantities that *react*, not necessarily the quantities that are *present*.

■ EXAMPLE 10.4

A sample of 0.2500 mol of solid $KClO_3$ is heated gently for a period of time, and 0.0678 mol of the compound decomposes. Calculate the number of moles of oxygen gas produced.

Solution

The equation (Section 8.3) is

$$2 \ KClO_3(s) \xrightarrow{\text{Heat}} 3 \ O_2(g) + 2 \ KCl(s)$$

In this experiment, even though 0.2500 mol of potassium chlorate is present, only 0.0678 mol reacts. The number of moles of oxygen gas produced depends on the number of moles of potassium chlorate that reacts:

$$0.0678 \ mol \ KClO_3\left(\frac{3 \ mol \ O_2}{2 \ mol \ KClO_3}\right) = 0.102 \ mol \ O_2$$

Practice Problem 10.4   In a certain reaction, 0.300 mol of H_2 gas reacts partially with N_2 gas to yield gaseous NH_3. If 0.096 mol of H_2 remains after the reaction is stopped, how many moles of NH_3 is produced? ■

10.2 Mas Calculations for Chemical Reactions

OBJECTIVE

■ to determine masses of substances involved in chemical reactions

In Section 10.1, you learned to calculate the number of moles of any substances involved in a chemical reaction from the number of moles of any other substance. You can solve problems that include mass calculations by simply changing the masses to moles or the moles to masses, as discussed in Chapter 7. In Figure 10.2, these conversions have been added to those shown in Figure 10.1.

You may need to review Chapter 6 to solve stoichiometry problems that give the names of compounds, rather than their formulas.

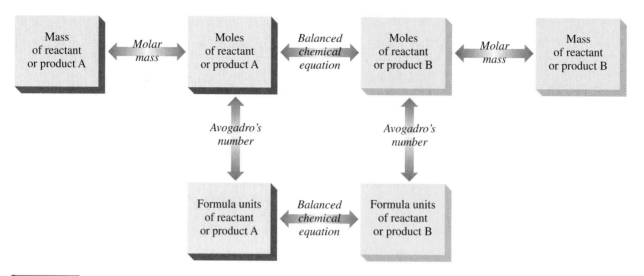

Figure 10.2 Mass
and Mole Conversions
for Stoichiometry Problems

■ **EXAMPLE 10.5**

Calculate the mass of sodium metal that will react with 241 g of chlorine gas to
form solid sodium chloride.

Solution

First, write the balanced chemical equation:

$$2 \text{ Na(s)} + \text{Cl}_2\text{(g)} \rightarrow 2 \text{ NaCl(s)}$$

Since the equation gives the *mole ratios,* you need to convert the mass of chlorine to the number of moles of chlorine, and then you can proceed as in Section 10.1:

$$241 \text{ g Cl}_2 \left(\frac{1 \text{ mol Cl}_2}{70.9 \text{ g Cl}_2} \right) = 3.40 \text{ mol Cl}_2$$

$$3.40 \text{ mol Cl}_2 \left(\frac{2 \text{ mol Na}}{1 \text{ mol Cl}_2} \right) = 6.80 \text{ mol Na}$$

Finally, convert the number of moles of sodium to the mass of sodium:

$$6.80 \text{ mol Na} \left(\frac{23.0 \text{ g Na}}{1 \text{ mol Na}} \right) = 156 \text{ g Na}$$

As usual, you could combine all these steps into a single calculation:

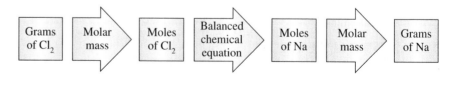

$$241 \text{ g Cl}_2 \left(\frac{1 \text{ mol Cl}_2}{70.9 \text{ g Cl}_2} \right) \left(\frac{2 \text{ mol Na}}{1 \text{ mol Cl}_2} \right) \left(\frac{23.0 \text{ g Na}}{1 \text{ mol Na}} \right) = 156 \text{ g Na}$$

Practice Problem 10.5   Calculate the mass of sodium chloride that will be
produced by the reaction in Example 10.5. ■

■ EXAMPLE 10.6

Electrolysis of concentrated aqueous sodium chloride solution (called brine) yields aqueous sodium hydroxide, hydrogen gas, and chlorine gas—three important industrial chemicals. Calculate the mass of chlorine that can be produced by electrolysis of 227 g of sodium chloride in concentrated aqueous solution:

$$2\ NaCl(aq) + 2\ H_2O(l) \xrightarrow{\text{Electricity}} 2\ NaOH(aq) + Cl_2(g) + H_2(g)$$

Solution

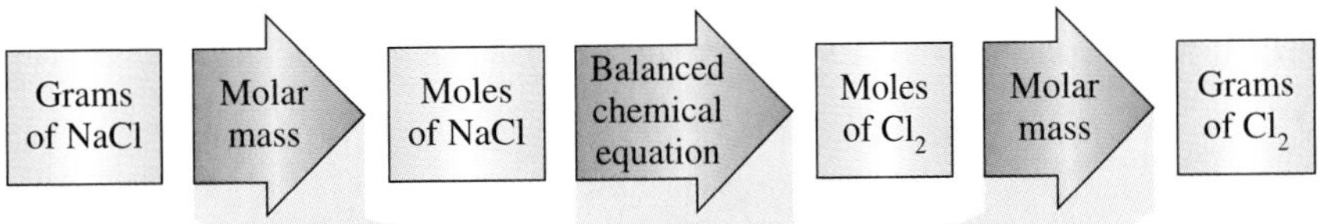

| Grams of NaCl | Molar mass | Moles of NaCl | Balanced chemical equation | Moles of Cl_2 | Molar mass | Grams of Cl_2 |

$$227\text{ g NaCl} \left(\frac{1\text{ mol NaCl}}{58.5\text{ g NaCl}}\right) \left(\frac{1\text{ mol }Cl_2}{2\text{ mol NaCl}}\right) \left(\frac{70.9\text{ g }Cl_2}{1\text{ mol }Cl_2}\right) = 138\text{ g }Cl_2$$

Practice Problem 10.6   The industrial process for the production of sodium metal and chlorine gas involves electrolysis of molten (melted) sodium chloride (in the absence of water). Calculate the mass of sodium that can be prepared by electrolysis of 227 g of sodium chloride. The balanced equation is

$$2\ NaCl(l) \xrightarrow{\text{Electricity}} 2\ Na(l) + Cl_2(g)\ ■$$

■ EXAMPLE 10.7

Sulfuric acid, H_2SO_4, is the chemical produced in the greatest tonnage worldwide. Calculate the number of metric tons of SO_2 gas required to prepare 10.0 metric tons of liquid H_2SO_4 (1 metric ton = 1×10^6 g). The balanced equation for the overall reaction, which is actually carried out in steps, is

$$2\ SO_2(g) + O_2(g) + 2\ H_2O(l) \rightarrow 2\ H_2SO_4(l)$$

Solution

The balanced equation can be used to calculate the quantity of a reactant from the quantity of any product, as well as vice versa. The same type of calculation is performed:

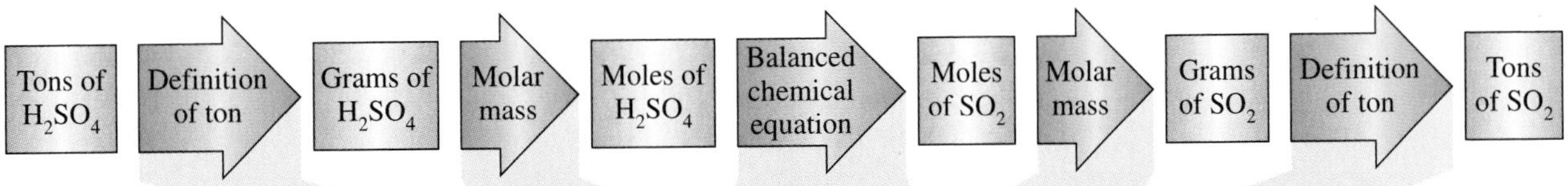

| Tons of H_2SO_4 | Definition of ton | Grams of H_2SO_4 | Molar mass | Moles of H_2SO_4 | Balanced chemical equation | Moles of SO_2 | Molar mass | Grams of SO_2 | Definition of ton | Tons of SO_2 |

$$10.0\text{ tons }H_2SO_4 \left(\frac{1 \times 10^6\text{ g}}{1\text{ ton}}\right) \left(\frac{1\text{ mol }H_2SO_4}{98.1\text{ g }H_2SO_4}\right) \left(\frac{2\text{ mol }SO_2}{2\text{ mol }H_2SO_4}\right) \left(\frac{64.1\text{ g }SO_2}{1\text{ mol }SO_2}\right) \left(\frac{1\text{ ton}}{1 \times 10^6\text{ g}}\right) = 6.53\text{ tons }SO_2$$

Note that the factors for the number of grams per metric ton cancel out.

Practice Problem 10.7   Calculate the mass of chlorine gas that must be treated with sodium metal to prepare 10.9 g of solid sodium chloride. ■

■ EXAMPLE 10.8

The industrial processing of copper(I) sulfide to produce copper metal involves roasting (heating) the solid ore in the presence of oxygen gas to produce the metal and sulfur dioxide gas. (The sulfur dioxide is used to make sulfuric acid.) Calculate the mass of copper produced by roasting 2.75 metric tons (1 metric ton = 1×10^6 g) of copper(I) sulfide.

Solution

$$Cu_2S(s) + O_2(g) \xrightarrow{\text{Heat}} 2\ Cu(s) + SO_2(g)$$

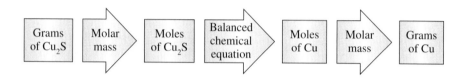

$$2.75 \times 10^6 \text{ g Cu}_2\text{S}\left(\frac{1 \text{ mol Cu}_2\text{S}}{159 \text{ g Cu}_2\text{S}}\right)\left(\frac{2 \text{ mol Cu}}{1 \text{ mol Cu}_2\text{S}}\right)\left(\frac{63.5 \text{ g Cu}}{1 \text{ mol Cu}}\right)$$
$$= 2.20 \times 10^6 \text{ g Cu} = 2.20 \text{ metric tons Cu}$$

Practice Problem 10.8   Copper(II) sulfide can also be roasted in the same way as copper(I) sulfide. Calculate the mass of copper produced by roasting 2.75 metric tons of copper(II) sulfide. ■

■ EXAMPLE 10.9

Excess hydrochloric acid was added to an aqueous solution of sodium carbonate, and the resulting solution was evaporated to dryness, which produced 2.12 g of solid product. Calculate the mass of sodium carbonate in the original solution.

Solution

The equation is

$$Na_2CO_3(aq) + 2\ HCl(aq) \rightarrow 2\ NaCl(aq) + CO_2(g) + H_2O(l)$$

After evaporation to dryness, the only solid remaining is sodium chloride. The carbon dioxide bubbled off during the reaction. The water produced by the reaction and the excess hydrochloric acid were evaporated, along with the water present to make the aqueous solution. Thus, the 2.12 g of solid is sodium chloride.

$$2.12 \text{ g NaCl}\left(\frac{1 \text{ mol NaCl}}{58.5 \text{ g NaCl}}\right)\left(\frac{1 \text{ mol Na}_2\text{CO}_3}{2 \text{ mol NaCl}}\right)\left(\frac{106 \text{ g Na}_2\text{CO}_3}{1 \text{ mol Na}_2\text{CO}_3}\right)$$
$$= 1.92 \text{ g Na}_2\text{CO}_3 \quad ■$$

10.3 Calculations Involving Other Quantities

Not only masses but quantities of substances in any units can be used for stoichiometry purposes. The quantities given must be changed to moles. Just as a mass is a measure of the number of moles of a reactant or product, the number of individual atoms, ions, or molecules involved in a chemical reaction may be converted to moles of reactant or product and used to solve problems. The number of moles of individual atoms or ions of a given element within a compound may also be used to determine the number of moles of reactant or product. The density of a substance may be used to determine the mass of a given volume of it and thus may be used to determine the number of moles present. Some of these additional relationships are illustrated in Figure 10.3.

■ **EXAMPLE 10.10**

Calculate the number of moles of solid mercury(II) oxide that can be produced by the reaction of oxygen gas with 15.0 mL of liquid mercury (density = 13.6 g/mL).

Solution

$$2 \, Hg(l) + O_2(g) \rightarrow 2 \, HgO(s)$$

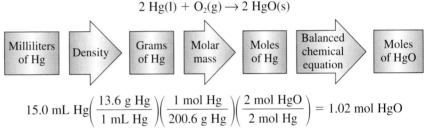

$$15.0 \text{ mL Hg}\left(\frac{13.6 \text{ g Hg}}{1 \text{ mL Hg}}\right)\left(\frac{1 \text{ mol Hg}}{200.6 \text{ g Hg}}\right)\left(\frac{2 \text{ mol HgO}}{2 \text{ mol Hg}}\right) = 1.02 \text{ mol HgO}$$

Figure 10.3 Mass, Mole, and Other Conversions

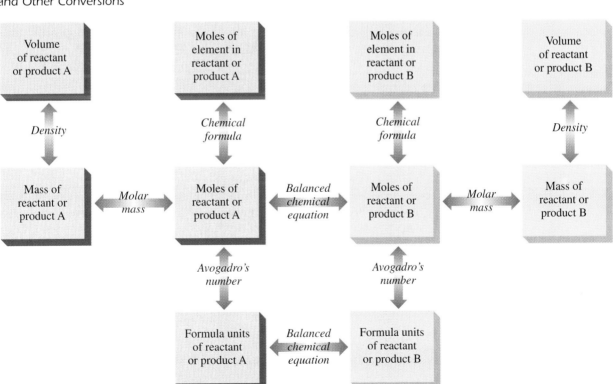

Practice Problem 10.10 Calculate the volume of liquid water (density = 1.00 g/mL) produced by burning 173 g of methane gas, CH_4, and condensing the gaseous water produced. ■

■ **EXAMPLE 10.11**

(a) A child takes 105 pennies from her piggy bank to buy jelly beans, which cost $3.50 per pound. If there are 80 jelly beans per pound, how many jelly beans can she buy?

(b) How many O_2 molecules does it take to produce 105 mmol of H_2O by a combination reaction with sufficient H_2 gas?

Solution

(a) Use the factor label method to obtain the answer from the quantities given:

$$105 \text{ cents}\left(\frac{1 \text{ dollar}}{100 \text{ cents}}\right)\left(\frac{1.00 \text{ pound}}{3.50 \text{ dollars}}\right)\left(\frac{80 \text{ jelly beans}}{1 \text{ pound}}\right) = 24 \text{ jelly beans}$$

(b) The balanced equation is

$$2 \text{ H}_2(g) + O_2(g) \rightarrow 2 \text{ H}_2O(l)$$

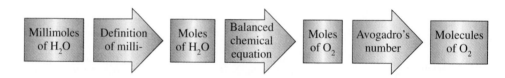

$$105 \text{ mmol H}_2O\left(\frac{1 \text{ mol H}_2O}{1000 \text{ mmol H}_2O}\right)\left(\frac{1 \text{ mol O}_2}{2 \text{ mol H}_2O}\right)\left(\frac{6.02 \times 10^{23} \text{ molecules O}_2}{1 \text{ mol O}_2}\right)$$
$$= 3.16 \times 10^{22} \text{ molecules O}_2$$

The main difference between parts (a) and (b) is that the values of more factors have to be stated in the problem in part (a) than in part (b).

Practice Problem 10.11 How many H_2 molecules does it take to combine with N_2 to produce 0.170 mol of NH_3? ■

■ **EXAMPLE 10.12**

Calculate the number of molecules of CO_2 that can be produced by complete combustion of 61.7 g of gaseous C_3H_6.

Solution

The balanced equation for the reaction is

$$2 \text{ C}_3H_6(g) + 9 \text{ O}_2(g) \rightarrow 6 \text{ CO}_2(g) + 6 \text{ H}_2O(g)$$

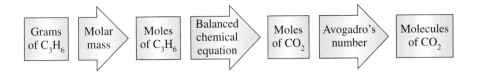

$$61.7 \text{ g } C_3H_6 \left(\frac{1 \text{ mol } C_3H_6}{42.1 \text{ g } C_3H_6} \right) \left(\frac{6 \text{ mol } CO_2}{2 \text{ mol } C_3H_6} \right) \left(\frac{6.02 \times 10^{23} \text{ molecules } CO_2}{1 \text{ mol } CO_2} \right)$$
$$= 2.65 \times 10^{24} \text{ molecules } CO_2$$

Practice Problem 10.12 Calculate the mass of H_2O produced by the complete combustion of 7.18×10^{23} molecules of C_3H_6. ■

■ **EXAMPLE 10.13**

Calculate the mass of H_2O that can be prepared by the reaction of 5.00×10^{23} O_2 molecules with hydrogen gas.

Solution

$$2 \, H_2(g) + O_2(g) \rightarrow 2 \, H_2O(l)$$

Since the equation gives the mole ratio, the number of molecules is changed to moles, which can then be converted to grams (mass) of product:

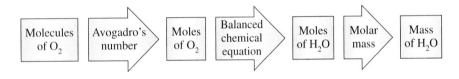

$$5.00 \times 10^{23} \, O_2 \text{ molecules} \left(\frac{1 \text{ mol } O_2}{6.02 \times 10^{23} \, O_2 \text{ molecules}} \right) \left(\frac{2 \text{ mol } H_2O}{1 \text{ mol } O_2} \right) \left(\frac{18.0 \text{ g } H_2O}{1 \text{ mol } H_2O} \right)$$
$$= 29.9 \text{ g } H_2O$$

Practice Problem 10.13 Calculate the number of individual atoms of sodium metal that can react when heated with solid zinc chloride to form solid sodium chloride and 121 g of zinc metal. ■

The number of moles of an element in a mole of compound may also be used to calculate the number of moles of the compound involved in a reaction. The ratio of the number of moles of an element within a compound to the number of moles of the compound is determined by the compound's chemical formula. Thus, the subscripts of the formula may be used to form a conversion factor.

■ **EXAMPLE 10.14**

The quantities of nitrogen, phosphorus, and potassium in a fertilizer are critical to the fertilizer's function in helping crops grow. Calculate the number of moles of nitrogen atoms in the ammonium phosphate, $(NH_4)_3PO_4$, produced by the reaction of excess aqueous ammonia with 22.2 mol of phosphoric acid.

Solution

The reaction is

$$3\ NH_3(aq) + H_3PO_4(aq) \rightarrow (NH_4)_3PO_4(aq)$$

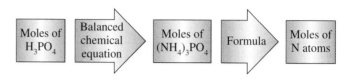

$$22.2\ \text{mol}\ H_3PO_4\left(\frac{1\ \text{mol}\ (NH_4)_3PO_4}{1\ \text{mol}\ H_3PO_4}\right)\left(\frac{3\ \text{mol}\ N}{1\ \text{mol}\ (NH_4)_3PO_4}\right) = 66.6\ \text{mol}\ N$$

From the
chemical formula

Practice Problem 10.14   Calculate the mass of nitrogen in the ammonium phosphate, $(NH_4)_3PO_4$, prepared by treating 3.00 kg of phosphoric acid with excess aqueous ammonia. ■

10.4 Problems Involving Limiting Quantities

In problems in the preceding sections, a quantity of one reactant was given, and it was assumed that enough of any other reactants was present. In Example 10.3, for instance, 7.18 mol of hydrochloric acid reacted with aluminum. If no aluminum is present, however, no reaction is possible—no matter how many moles of hydrochloric acid there are. In Example 10.3, we assumed that sufficient aluminum was present because nothing was stated about the quantity of aluminum used. In contrast, problems involving limiting quantities have the quantities of at least two reactants given. The reactant used up first limits the quantities of the products and is referred to as being present in **limiting quantity.** Any other reactant may be present in an amount that represents a greater number of moles than is required for the reaction and is said to be present in **excess.** The reaction is said to have gone to **completion** when the limiting quantity has been used up. An example from everyday life illustrates this principle.

■ **EXAMPLE 10.15**

(a) If cashew nuts cost $8.00 per pound, how many pounds of these nuts can be purchased with $256.00?

(b) How many pounds of cashew nuts can be purchased with $256.00 if the store has 28.0 pounds in stock?

Solution

(a) Assuming that the store has sufficient nuts,

$$256.00\ \text{dollars}\left(\frac{1\ \text{pound}}{8.00\ \text{dollars}}\right) = 32.0\ \text{pounds}$$

(b) Even though the amount of money is sufficient to purchase 32.0 pounds of nuts, the store does not have that much. The maximum quantity that can be purchased is the 28.0 pounds on hand. The nuts are said to be available in limited quantity. No matter how much money you have, you cannot buy more nuts than the store has.

Practice Problem 10.15 How many pounds of cashew nuts, at $8.00 per pound, can be purchased with $100.00 if the store has 25.0 pounds of the nuts? ■

■ EXAMPLE 10.16

(a) Calculate the quantity of zinc metal required to react with 4.16 mol of aqueous hydrochloric acid.

(b) Calculate the quantity of zinc that will react with 4.16 mol of HCl if 2.50 mol of Zn is present.

(c) Calculate the quantity of zinc that will react with 4.16 mol of HCl if 1.00 mol of Zn is present.

Solution

(a) The quantity of zinc is determined in the same way as the quantity of hydrogen was in Example 10.3:

$$Zn(s) + 2\ HCl(aq) \longrightarrow ZnCl_2(aq) + H_2(g)$$

$$4.16\ \text{mol HCl} \left(\frac{1\ \text{mol Zn}}{2\ \text{mol HCl}} \right) = 2.08\ \text{mol Zn}$$

(b) In part (a), you determined that 2.08 mol of Zn will react with 4.16 mol of HCl. Since more zinc is present in this case, 2.08 mol will be used up, and the remaining 0.42 mol will not react. Zinc is *in excess.*

(c) In part (a), you showed that 2.08 mol of Zn is required to react with 4.16 mol of HCl, but in this case, not that much zinc is present. Zinc is *in limiting quantity,* and the entire 1.00 mol of Zn will react with 2.00 mol of HCl:

$$1.00\ \text{mol Zn} \left(\frac{2\ \text{mol HCl}}{1\ \text{mol Zn}} \right) = 2.00\ \text{mol HCl}$$

The hydrochloric acid is in excess. The number of moles of HCl that will be left unreacted is the difference:

$$
\begin{array}{r}
4.16\ \text{mol HCl present} \\
-2.00\ \text{mol HCl reacts} \\
\hline
2.16\ \text{mol HCl unreacted}
\end{array}
$$

Practice Problem 10.16 Calculate the number of moles of hydrogen gas that will be produced in each part of Example 10.16. ■

The first task in doing a problem involving a limiting quantity is to recognize that it is such a problem. Fortunately, that is fairly easy: The quantities of two different reactants will be given. Then, follow these steps:

Step 1: Determine how many moles of each reactant is present, if this information is not already given.

Step 2: If the number of moles of one reactant that will react with all of the other reactants has been stated or calculated, then compare the numbers of moles present and required. In general, however, it is easier to determine the ratio of the numbers of moles of reactants present and the same ratio of the numbers of moles of the reactants required, which is given in the balanced equation. If the ratio of reactant A to reactant B *present* is greater than the ratio of reactant A to reactant B *in the balanced equation,* then reactant A is in excess. If the ratio present is lower than the ratio required, then reactant B is in excess.

You may find it helpful to put the reactant that is present in the greater number of moles in the numerator of the ratio present. Note, however, that the same reactant must be in the numerator in both the ratio present and the ratio required.

> The limiting quantity is used to determine the quantity of product(s) produced and also the quantity that will actually react of the reactant(s) present in excess.

For example, for Example 10.15(b), we could have compared the following ratios:

Present		*Required*
$\dfrac{256.00 \text{ dollars}}{28.0 \text{ pounds}} = \dfrac{9.14 \text{ dollars}}{1 \text{ pound}}$		$\dfrac{8.00 \text{ dollars}}{1 \text{ pound}}$

Note that *1 pound* is in the denominator of both ratios. Since the ratio *present* is larger, the quantity in the numerator (dollars) is in excess, and the number of pounds of nuts is the limiting quantity. The limiting quantity is used to determine the quantity of product(s) produced and also the quantity that will actually react of the reactant(s) present in excess.

> Caution: Use the limiting quantity present, not the calculated quantity in the ratio.

Caution: Use the limiting quantity present, not the calculated quantity in the ratio. In this example, use 28.0 pounds, not 1 pound.

■ EXAMPLE 10.17

(a) If 2.90 mol of chlorine gas is treated with 4.90 mol of sodium metal to produce solid sodium chloride, which reactant is in excess?

(b) How many moles of sodium chloride can be produced?

Solution

(a) The balanced chemical equation is

$$2 \text{ Na(s)} + \text{Cl}_2(\text{g}) \rightarrow 2 \text{ NaCl(s)}$$

The mole ratios are as follows:

Present		*Required*
$\dfrac{4.90 \text{ mol Na}}{2.90 \text{ mol Cl}_2} = \dfrac{1.69 \text{ mol Na}}{1 \text{ mol Cl}_2}$		$\dfrac{2 \text{ mol Na}}{1 \text{ mol Cl}_2}$

> Hint: On examinations, always state explicitly what is in limiting quantity.

Note that *1 mol* of the same substance (Cl_2) is in the denominator of *both* ratios. Since the ratio of moles of sodium present to moles of chlorine present (1.69) is lower than the required ratio from the balanced chemical equation (2), the chlorine is in excess, and the sodium is in limiting quantity.

(b) The number of moles of sodium present (the limiting quantity) is used to calculate the number of moles of NaCl produced:

$$4.90 \text{ mol Na}\left(\frac{2 \text{ mol NaCl}}{2 \text{ mol Na}}\right) = 4.90 \text{ mol NaCl}$$

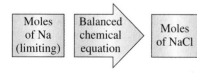

You can tabulate these quantities, all in moles, as follows:

	2 Na(s) +	Cl$_2$(g) →	2 NaCl(s)
Present initially	4.90	2.90	0.00
Change due to reaction	−4.90	−2.45	+4.90
Present finally	0.00	0.45	4.90

In such a tabulation, the numbers of moles in the row showing the changes due to the reaction will always be in the same ratio as the coefficients in the balanced equation.

Practice Problem 10.17

(a) If 12.9 mol of sodium metal is treated with 7.80 mol of solid sulfur to produce solid sodium sulfide, which reactant is in excess?

(b) How many moles of sodium sulfide can be produced? ■

■ EXAMPLE 10.18

Calculate the number of moles of $BaCl_2$(aq) that will be produced by the reaction of 2.50 mol of $Ba(OH)_2$(s) and 5.50 mol of HCl(aq).

Solution

You recognize this as a problem involving a limiting quantity, since the quantities of *two* reactants are given. You first write a balanced equation for the reaction:

$$2\ HCl(aq) + Ba(OH)_2(s) \rightarrow BaCl_2(aq) + 2\ H_2O(l)$$

Next, calculate the ratio of HCl to $Ba(OH)_2$ present and also the ratio required by the balanced equation:

Present   *Required*

$$\frac{5.50\ \text{mol HCl}}{2.50\ \text{mol Ba(OH)}_2} = \frac{2.20\ \text{mol HCl}}{1\ \text{mol Ba(OH)}_2} \qquad \frac{2\ \text{mol HCl}}{1\ \text{mol Ba(OH)}_2}$$

Since the ratio present is higher than that required, the HCl (which is in the numerator) is present in excess, and the $Ba(OH)_2$ is in limiting quantity. Since the quantity of $Ba(OH)_2$ is limiting, you use 2.50 mol of $Ba(OH)_2$ to calculate the number of moles of $BaCl_2$ that will be produced:

$$2.50\ \text{mol Ba(OH)}_2 \left(\frac{1\ \text{mol BaCl}_2}{1\ \text{mol Ba(OH)}_2} \right) = 2.50\ \text{mol BaCl}_2$$

You can tabulate these quantities, all in moles, as follows:

	2 HCl(aq) +	Ba(OH)$_2$(s) →	BaCl$_2$(aq) +	2 H$_2$O(l)
Present initially	5.50	2.50	0.00	
Change due to reaction	−5.00	−2.50	+2.50	
Present finally	0.50	0.00	2.50	

Practice Problem 10.18   Calculate the number of moles of $BaCl_2$(aq) that will be produced by the reaction of 2.50 mol of $Ba(OH)_2$(s) and 4.50 mol of HCl(aq). ■

If the number of moles present for each reactant is exactly the number required, then *both* reactants are in limiting quantity. *Either* quantity of reactant may be used to calculate the quantity of product.

■ EXAMPLE 10.19

Calculate the number of moles of $BaCl_2$ that will be produced by the reaction of 2.50 mol of $Ba(OH)_2$ and 5.00 mol of HCl.

Solution

Present		*Required*
$\dfrac{5.00 \text{ mol HCl}}{2.50 \text{ mol Ba(OH)}_2} = \dfrac{2.00 \text{ mol HCl}}{1 \text{ mol Ba(OH)}_2}$		$\dfrac{2 \text{ mol HCl}}{1 \text{ mol Ba(OH)}_2}$

Since the ratios are equal, neither reactant is in excess. Either quantity of reactant can be used to calculate the quantity of product. That is, the number of moles of $BaCl_2$ may be calculated by *either* of the following:

$$2.50 \text{ mol Ba(OH)}_2 \left(\frac{1 \text{ mol BaCl}_2}{1 \text{ mol Ba(OH)}_2} \right) = 2.50 \text{ mol BaCl}_2$$

or

$$5.00 \text{ mol HCl} \left(\frac{1 \text{ mol BaCl}_2}{2 \text{ mol HCl}} \right) = 2.50 \text{ mol BaCl}_2$$

You can see that these calculations yield the same answer. *Do not add these numbers of moles together!*

Tabulation, all in moles, shows the same result:

	2 HCl(aq) +	Ba(OH)$_2$(s) →	BaCl$_2$(aq) + 2 H$_2$O
Present initially	5.00	2.50	0.00
Change due to reaction	−5.00	−2.50	+2.50
Present finally	0.00	0.00	2.50

Practice Problem 10.19   (a) Calculate the number of moles of PbI_2 that can be produced by treating 2.00 mol of $Pb(NO_3)_2$ with 4.00 mol of NaI. (b) Calculate the number of moles of PbI_2 that can be produced by treating 4.00 mol of $Pb(NO_3)_2$ with 2.00 mol of NaI. ■

The number of moles of excess reactant can be calculated by subtracting the number of moles that react from the number of moles initially present.

■ EXAMPLE 10.20

How many moles of excess reactant are present after 3.10 mol of aluminum metal reacts with 4.50 mol of dilute aqueous sulfuric acid?

Solution

$$2 \text{ Al(s)} + 3 \text{ H}_2\text{SO}_4\text{(aq)} \rightarrow 3 \text{ H}_2\text{(g)} + \text{Al}_2(\text{SO}_4)_3\text{(aq)}$$

Present		*Required*	
$\dfrac{4.50 \text{ mol H}_2\text{SO}_4}{3.10 \text{ mol Al}} = \dfrac{1.45 \text{ mol H}_2\text{SO}_4}{1 \text{ mol Al}}$		$\dfrac{3 \text{ mol H}_2\text{SO}_4}{2 \text{ mol Al}} = \dfrac{1.5 \text{ mol H}_2\text{SO}_4}{1 \text{ mol Al}}$	

Since the ratio present is smaller, H_2SO_4 is in limiting quantity.

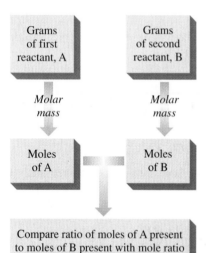
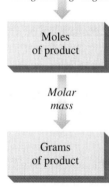

Figure 10.4 Procedure for Solving Problems Involving Limiting Quantities When Masses Are Given

$$4.50 \text{ mol } H_2SO_4 \left(\frac{2 \text{ mol Al}}{3 \text{ mol } H_2SO_4} \right) = 3.00 \text{ mol Al required}$$

3.10 mol Al present − 3.00 mol Al required = 0.10 mol Al excess

Again, you can tabulate these quantities, all in moles, as follows:

	2 Al(s) +	3 H₂SO₄(aq) →	3 H₂(g) +	Al₂(SO₄)₃(aq)
Present initially	3.10	4.50	0.00	0.00
Change due to reaction	−3.00	−4.50	+4.50	+1.50
Present finally	0.10	0.00	4.50	1.50

Practice Problem 10.20   How many moles of excess reactant are present after 2.50 mol of zinc metal reacts with 2.25 mol of aqueous phosphoric acid? ■

Problems involving limiting quantities may be stated in terms of masses, rather than moles, and a mass of product might be required. To solve, convert the masses of reactants to moles, perform the steps given earlier in this section, and convert the final number of moles of product to a mass, if required. Figure 10.4 summarizes the conversions and procedure.

■ **EXAMPLE 10.21**

What mass of NaCl will be formed by addition of 35.0 g of Na_2CO_3 in aqueous solution to an aqueous solution containing 25.5 g of HCl?

Solution

The reactants undergo the following reaction:

$$Na_2CO_3(aq) + 2 \text{ HCl}(aq) \rightarrow 2 \text{ NaCl}(aq) + CO_2(g) + H_2O(l)$$

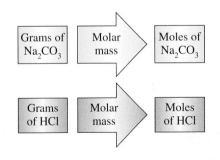

Change the given masses to moles:

$$35.0 \text{ g } Na_2CO_3 \left(\frac{1 \text{ mol } Na_2CO_3}{106 \text{ g } Na_2CO_3} \right) = 0.330 \text{ mol } Na_2CO_3 \text{ present}$$

$$25.5 \text{ g HCl} \left(\frac{1 \text{ mol HCl}}{36.5 \text{ g HCl}} \right) = 0.699 \text{ mol HCl present}$$

Compare the ratios:

Present		*Required*
$\dfrac{0.699 \text{ mol HCl}}{0.330 \text{ mol } Na_2CO_3} =$	$\dfrac{2.12 \text{ mol HCl}}{1 \text{ mol } Na_2CO_3}$	$\dfrac{2 \text{ mol HCl}}{1 \text{ mol } Na_2CO_3}$

Since the ratio present is greater than that required, the Na_2CO_3 (in the denominator) is in limiting quantity. You should base the calculation of the number of moles of NaCl formed on the number of moles of Na_2CO_3 present, and then convert to grams, since a mass is required:

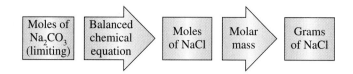

$$0.330 \text{ mol Na}_2\text{CO}_3\left(\frac{2 \text{ mol NaCl}}{1 \text{ mol Na}_2\text{CO}_3}\right)\left(\frac{58.5 \text{ g NaCl}}{1 \text{ mol NaCl}}\right) = 38.6 \text{ g NaCl}$$

Tabulation, again in *moles,* shows the same result:

	$\text{Na}_2\text{CO}_3(aq)$	+ 2 HCl(aq) →	2 NaCl(aq)	+ $\text{CO}_2(g)$	+ $\text{H}_2\text{O}(l)$
Present initially	0.330	0.699	0.000	0.000	
Change due to reaction	−0.330	−0.660	+0.660	+0.330	
Present finally	0.000	0.039	0.660	0.330	

$$0.660 \text{ mol NaCl}\left(\frac{58.5 \text{ g NaCl}}{1 \text{ mol NaCl}}\right) = 38.6 \text{ g NaCl}$$

Practice Problem 10.21   What mass of NaCl will be formed by addition of 26.0 g of NaHCO_3 to 15.3 g of HCl, all in aqueous solution? ■

■ **EXAMPLE 10.22**

(a) What mass of sodium perchlorate will result from the reaction of 25.0 g of aqueous sodium carbonate with 50.0 g of aqueous perchloric acid?

(b) What mass of excess reagent will remain unreacted?

Solution

(a) Write the balanced equation, and convert grams (given) to moles:

$$2 \text{ HClO}_4(aq) + \text{Na}_2\text{CO}_3(aq) \rightarrow 2 \text{ NaClO}_4(aq) + \text{CO}_2(g) + \text{H}_2\text{O}(l)$$

$$25.0 \text{ g Na}_2\text{CO}_3\left(\frac{1 \text{ mol Na}_2\text{CO}_3}{106 \text{ g Na}_2\text{CO}_3}\right) = 0.236 \text{ mol Na}_2\text{CO}_3 \text{ present}$$

$$50.0 \text{ g HClO}_4\left(\frac{1 \text{ mol HClO}_4}{100.5 \text{ g HClO}_4}\right) = 0.498 \text{ mol HClO}_4 \text{ present}$$

Compare ratios:

Present		*Required*
$\dfrac{0.498 \text{ mol HClO}_4}{0.236 \text{ mol Na}_2\text{CO}_3} = \dfrac{2.11 \text{ mol HClO}_4}{1 \text{ mol Na}_2\text{CO}_3}$		$\dfrac{2 \text{ mol HClO}_4}{1 \text{ mol Na}_2\text{CO}_3}$

Since the ratio present is greater, Na_2CO_3 (in the denominator) is in limiting quantity. Use the limiting quantity to find the mass of product:

$$0.236 \text{ mol Na}_2\text{CO}_3\left(\frac{2 \text{ mol NaClO}_4}{1 \text{ mol Na}_2\text{CO}_3}\right)\left(\frac{122.5 \text{ g NaClO}_4}{1 \text{ mol NaClO}_4}\right) = 57.8 \text{ g NaClO}_4$$

(b) The mass of $HClO_4$ that reacts is given by

$$0.236 \text{ mol } Na_2CO_3 \left(\frac{2 \text{ mol } HClO_4}{1 \text{ mol } Na_2CO_3} \right) \left(\frac{100.5 \text{ g } HClO_4}{1 \text{ mol } HClO_4} \right) = 47.4 \text{ g } HClO_4$$

The mass remaining unreacted is

$$50.0 \text{ g} - 47.4 = 2.6 \text{ g excess}$$

The tabulation still must be done in moles:

	$2 \text{ HClO}_4(aq)$	$+ \text{ Na}_2\text{CO}_3(aq) \rightarrow$	$2 \text{ NaClO}_4(aq)$	$+ \text{ CO}_2(g)$	$+ \text{ H}_2\text{O}(l)$
Present initially	0.498	0.236	0.000	0.000	
Change due to reaction	−0.472	−0.236	+0.472	+0.236	
Present finally	0.026	0.000	0.472	0.236	

(a) Converting to mass:

$$0.472 \text{ mol } NaClO_4 \left(\frac{122.5 \text{ g } NaClO_4}{1 \text{ mol } NaClO_4} \right) = 57.8 \text{ g } NaClO_4$$

(b) Converting to mass:

$$0.026 \text{ mol } HClO_4 \left(\frac{100.5 \text{ g } HClO_4}{1 \text{ mol } HClO_4} \right) = 2.6 \text{ g } HClO_4$$

Practice Problem 10.22   What mass of $K_2SO_4(aq)$ will result from the reaction of 42.7 g of $KOH(aq)$ and 21.1 g of $H_2SO_4(aq)$? ■

10.5 Theoretical Yield and Percent Yield

OBJECTIVE

■ to express the quantity of product obtained from a reaction as a percentage of what the reaction is theoretically capable of producing

When a quantity of product is calculated from a quantity or quantities of reactants, as was done in the earlier sections of this chapter, that quantity of product is called the **theoretical yield.** When a reaction is run, however, less product than the calculated amount is often obtained: Some of the product may stay in the solution in which the reaction was run; some side reaction may use up some of the reactants; or the reaction may be stopped before it is completed. No matter why, the fact is that many reactions produce less product than the calculated quantity; that is, the actual yield is less than the theoretical yield. No reaction can produce more than the theoretical yield. The **percent yield** is defined as 100% times the ratio of the actual yield to the theoretical yield:

$$\text{Percent yield} = \left(\frac{\text{Actual yield}}{\text{Theoretical yield}} \right) \times 100\%$$

■ EXAMPLE 10.23

Calculate the percent yield of a reaction if calculations indicated that 4.89 g of product could be obtained, but only 4.56 g of product was actually obtained.

Solution

$$\text{Percent yield} = \left(\frac{\text{Actual yield}}{\text{Theoretical yield}} \right) \times 100\% = \left(\frac{4.56 \text{ g}}{4.89 \text{ g}} \right) \times 100\% = 93.3\%$$

■ **EXAMPLE 10.24**

Calculate the percent yield if 5.18 g of solid PCl_5 is obtained in a certain experiment in which 3.45 g of liquid PCl_3 is treated with excess gaseous Cl_2.

Solution

The theoretical yield in grams is calculated as discussed in Section 10.2:

$$PCl_3(l) + Cl_2(g) \rightarrow PCl_5(s)$$

$$3.45 \text{ g } PCl_3\left(\frac{1 \text{ mol } PCl_3}{138 \text{ g } PCl_3}\right)\left(\frac{1 \text{ mol } PCl_5}{1 \text{ mol } PCl_3}\right)\left(\frac{208 \text{ g } PCl_5}{1 \text{ mol } PCl_5}\right) = 5.20 \text{ g } PCl_5$$

The percent yield is

$$\left(\frac{\text{Actual yield}}{\text{Theoretical yield}}\right) \times 100\% = \left(\frac{5.18 \text{ g}}{5.20 \text{ g}}\right) \times 100\% = 99.6\%$$

Practice Problem 10.24   Calculate the percent yield if 5.18 g of PCl_5 is obtained from treatment of 3.45 g of PCl_3 with 1.80 g of gaseous Cl_2. ■

■ **EXAMPLE 10.25**

Ozone, O_3, is a gas produced when O_2 molecules are subjected to electrical discharge or the action of cosmic rays in the upper atmosphere.

(a) Calculate the mass of ozone that could theoretically be produced by conversion of 15.0 g of O_2.

(b) If 1.55 g of O_3 is actually produced, what is the percent yield?

Solution

(a) $$3 \text{ } O_2(g) \rightarrow 2 \text{ } O_3(g)$$

$$15.0 \text{ g } O_2\left(\frac{1 \text{ mol } O_2}{32.0 \text{ g } O_2}\right)\left(\frac{2 \text{ mol } O_3}{3 \text{ mol } O_2}\right)\left(\frac{48.0 \text{ g } O_3}{1 \text{ mol } O_3}\right) = 15.0 \text{ g } O_3$$

This part of the problem could have been solved by simply applying the law of conservation of mass.

(b) The percent yield is

$$\left(\frac{1.55 \text{ g}}{15.0 \text{ g}}\right) \times 100\% = 10.3\% \text{ yield} ■$$

10.6 Calculations with Net Ionic Equations

■ to calculate the numbers of moles involved in net ionic equations, to calculate the masses of complete compounds involved in such equations, and to recognize the limitations of net ionic equations in calculating masses of individual ions

Net ionic equations (Chapter 9), like all other balanced chemical equations, give the mole ratios of reactants and products. Therefore, any calculations that require mole ratios may be done with net ionic equations as well as with total equations. However, a net ionic equation does not yield mass data directly because part of each soluble ionic compound is not given. For example, we can tell how many moles of silver ion are required to produce a certain number of moles of a product, but we cannot weigh out just the silver ions. The compound must contain some anions, too. The net ionic equation indicates that we are not interested in the anions because they do not react. However, the anions have some mass. We cannot tell how much of the mass of the compound is composed of silver ions and how much is composed of anions if we do not specify which anions are present. Thus, net ionic equations are often not directly useful for mass computations.

■ **EXAMPLE 10.26**

How many moles of silver ions and how many grams of silver nitrate are required to produce 212 g of silver chloride?

Solution

$$Ag^+(aq) + Cl^-(aq) \longrightarrow AgCl(s)$$

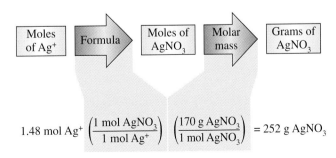

$$212 \text{ g AgCl} \left(\frac{1 \text{ mol AgCl}}{143.3 \text{ g AgCl}} \right) \left(\frac{1 \text{ mol Ag}^+}{1 \text{ mol AgCl}} \right) = 1.48 \text{ mol Ag}^+$$

You cannot weigh out 1.48 mol of Ag^+ if you do not know what anion is included in the compound. Even though the anion does not react, it still has some mass. Knowing that the silver ions are part of silver nitrate, you can calculate the mass of that compound that must be weighed out to provide the silver ions to react:

$$1.48 \text{ mol Ag}^+ \left(\frac{1 \text{ mol AgNO}_3}{1 \text{ mol Ag}^+} \right) \left(\frac{170 \text{ g AgNO}_3}{1 \text{ mol AgNO}_3} \right) = 252 \text{ g AgNO}_3$$

Practice Problem 10.26 How many moles of silver ions and how many grams of silver chlorate would be required to produce 212 g of silver chloride? ■

The small number of key terms for this chapter suggests that this chapter does not introduce many new concepts. However, this chapter may seem difficult because it draws extensively on background material from earlier chapters. Concepts presented in this chapter are extremely important because they are applied in later chapters on gas laws and equilibrium, among others.

■ ■ ■ ■ ■ ■ ■ SUMMARY ■ ■ ■ ■ ■ ■

The ratios of the numbers of moles of reactants and products involved in any chemical reaction are given by the coefficients in the balanced equation for the reaction. Each ratio of moles may be used as a factor to convert the number of moles of one reactant or product to the number of moles of any other. (Section 10.1)

If the quantity of any reactant or product is given in terms of mass instead of in moles, the mass must be changed to moles before calculating the number of moles of another substance in the reaction. If the mass of a reactant or product is required to answer a problem, its number of moles must be converted to a mass. (Conversions between mass and moles are presented in Chapter 7.) (Section 10.2) If some other measure of the quantity of a substance is given or required (for example, the number of molecules of a substance), an appropriate conversion factor is needed to convert to or from moles. (Section 10.3)

For problems in which the quantities of two (or more) reactants are given, you must determine if one of the reactants is present in a quantity less than, equal to, or greater than that required to react with *all* the other reactants. The ratio of the numbers of moles present of the two substances is compared to the ratio of the numbers of moles required. The ratio of numbers of moles required is equal to the ratio of coefficients from the balanced chemical equation. If the ratio present is greater than the corresponding ratio required, the reactant in the numerator is in excess, and the other is in limiting quantity. If the ratio present is less than the corresponding ratio required, the reactant in the numerator is in limiting quantity. The limiting quantity is used to calculate the quantities of the other substances that will react or be produced. If both ratios are equal, the reactants are present in precisely the right ratio, and either quantity given can be used in calculations. If masses are given, rather than moles, they are converted to moles before proceeding. (Section 10.4)

The theoretical yield is the quantity of product calculated from the quantity of reactant used (or the limiting quantity if more than one quantity is given). In some reactions, not all of the calculated product can be collected. The percent yield is the ratio of the actual yield to the theoretical yield, converted to a percentage:

$$\text{Percent yield} = \frac{\text{Actual yield}}{\text{Theoretical yield}} \times 100\%$$

(Section 10.5)

Net ionic equations can be used to calculate mole ratios but often cannot be used directly with masses. Although spectator ions do not react, they do have mass, and the mass of the compound cannot be determined if the other ions are not specified. (Section 10.6)

Items for Special Attention

■ The balanced chemical equation gives the *mole ratios* of reactants and products *involved in the reaction,* not the mass ratios and not the numbers of moles present.

■ Use the formulas of the substances involved, as well as the units, when applying the factor label method to solve stoichiometry problems. For example, write g NaCl or mol HCl, rather than just g or mol.

■ The substance present in limiting quantity is often present in a greater number of moles than the substance in excess (but always less than would be required to react with all of the substance in excess).

■ Get the limiting quantity from the *ratio of moles present:* from the *numerator* if the ratio present is less than the ratio required or from the *denominator* if the ratio present is greater than the ratio required.

Self-Tutorial Problems

10.1 Write all the possible conversion factors using the coefficients from the following equation:

$$PCl_5(s) + 4 H_2O(l) \rightarrow 5 HCl(aq) + H_3PO_4(aq)$$

10.2 Consider the following reaction:

$$Ca(s) + 2 HCl(aq) \rightarrow H_2(g) + CaCl_2(aq)$$

Since 1 Ca atom reacts with 2 formula units of HCl, how many Ca atoms will react with each of the following?

(a) 4 formula units of HCl

(b) 12 formula units of HCl

(c) 200 formula units of HCl

(d) 2 dozen formula units of HCl

(e) 2 mol of HCl

10.3 For the reaction

$$H_3PO_4(aq) + 3 KOH(aq) \rightarrow K_3PO_4(aq) + 3 H_2O(l)$$

one student placed 1.00 mol of KOH(aq) in a flask, a second student placed 2.00 mol of KOH(aq) in a flask, and a third student placed 4.00 mol of KOH(aq) in a flask. Which student(s) could carry out the reaction specified in the equation?

10.4 Consider the following balanced equation:

$$5 FeCl_2(aq) + KMnO_4(aq) + 8 HCl(aq) \rightarrow$$
$$5 FeCl_3(aq) + MnCl_2(aq) + 4 H_2O(l) + KCl(aq)$$

What is the ratio of moles of potassium permanganate to moles of iron(II) chloride?

10.5 (a) How many (two-slice) sandwiches can you make with 50 slices of bread?

(b) How many sandwiches can you make with 50 slices of bread and 30 hamburger patties?

(c) How many sandwiches can you make with 50 slices of bread and 20 hamburger patties?

(d) How can you recognize when a problem involves a limiting quantity?

10.6 Consider the following reaction:

$$Ba(OH)_2(aq) + 2 HCl(aq) \rightarrow BaCl_2(aq) + 2 H_2O(l)$$

(a) How many moles of $BaCl_2$ can you make with 50 mol of HCl?

(b) How many moles of $BaCl_2$ can you make with 50 mol of HCl and 30 mol of $Ba(OH)_2$?

(c) How many moles of $BaCl_2$ can you make with 50 mol of HCl and 20 mol of $Ba(OH)_2$?

10.7 Calculate the number of moles of each of the following that is necessary to produce 2.50 mol of AgCl by reaction with $AgNO_3$:

(a) KCl

(b) NH_4Cl

(c) CsCl

10.8 List what you would need to do to solve the following problem: Calculate the mass of solid produced by treatment of 14.5 g of aqueous sodium carbonate with 10.0 g of hydrochloric acid, followed by evaporation of the volatile components.

10.9 From each of the following pairs of ratios, determine whether reactant A or reactant B is in limiting quantity:

	Present	Required
(a)	$\dfrac{6.5 \text{ mol A}}{2.9 \text{ mol B}}$	$\dfrac{2 \text{ mol A}}{1 \text{ mol B}}$
(b)	$\dfrac{3.5 \text{ mol A}}{1.9 \text{ mol B}}$	$\dfrac{2 \text{ mol A}}{1 \text{ mol B}}$
(c)	$\dfrac{2.4 \text{ mol A}}{1.9 \text{ mol B}}$	$\dfrac{3 \text{ mol A}}{2 \text{ mol B}}$
(d)	$\dfrac{2.1 \text{ mol A}}{1.9 \text{ mol B}}$	$\dfrac{1 \text{ mol A}}{1 \text{ mol B}}$
(e)	$\dfrac{4.9 \text{ mol A}}{3.9 \text{ mol B}}$	$\dfrac{4 \text{ mol A}}{3 \text{ mol B}}$

10.10 How many moles of NaCl will be produced by the following reaction in each case?

$$2 Na(s) + Cl_2(g) \rightarrow 2 NaCl(s)$$

(a) 1 mol Na and 0 mol Cl_2

(b) 2 mol Na and 1 mol Cl_2

(c) 3 mol Na and 1 mol Cl_2

10.11 From each of the following pairs of ratios, determine whether reactant A or reactant B is in limiting quantity:

	Present	Required
(a)	$\dfrac{3.5 \text{ mol A}}{1.9 \text{ mol B}}$	$\dfrac{3 \text{ mol A}}{2 \text{ mol B}}$
(b)	$\dfrac{4.5 \text{ mol A}}{6.0 \text{ mol B}}$	$\dfrac{2 \text{ mol A}}{3 \text{ mol B}}$
(c)	$\dfrac{3.5 \text{ mol A}}{6.5 \text{ mol B}}$	$\dfrac{1 \text{ mol A}}{2 \text{ mol B}}$
(d)	$\dfrac{2.5 \text{ mol A}}{1.9 \text{ mol B}}$	$\dfrac{4 \text{ mol A}}{3 \text{ mol B}}$
(e)	$\dfrac{4.5 \text{ mol A}}{6.0 \text{ mol B}}$	$\dfrac{2 \text{ mol A}}{5 \text{ mol B}}$

10.12 Consider the following reaction:

$$La(OH)_3(s) + 3 HCl(aq) \rightarrow LaCl_3(aq) + 3 H_2O(l)$$

(a) How many moles of $LaCl_3$ can you make with 30 mol of HCl?

(b) How many moles of $LaCl_3$ can you make with 30 mol of HCl and 15 mol of $La(OH)_3$?

(c) How many moles of $LaCl_3$ can you make with 30 mol of HCl and 5 mol of $La(OH)_3$?

■ ■ ■ PROBLEMS ■ ■ ■

10.1 Mole Calculations for Chemical Reactions

10.13 Calculate the number of grams of each of the following that is necessary to produce 2.50 mol of AgCl by reaction with $AgNO_3$. (Compare with Problem 10.7.)

(a) KCl (b) NH_4Cl (c) CsCl

10.14 Which of the following samples of metal can produce the most hydrogen by reaction with HCl?

0.150 mol Ca 0.150 mol Al 0.150 mol Zn

10.15 Calculate the number of moles of H_3PO_4 that will react with 5.55 mol of $Ba(OH)_2$ to form solid $Ba_3(PO_4)_2$.

10.16 Calculate the number of moles of H_3PO_4 that will react with 5.55 mol of NaOH in aqueous solution to form Na_2HPO_4.

10.17 Calculate the number of moles of H_3PO_4 that will react with 5.55 mol of $Ba(OH)_2$ to form solid $BaHPO_4$.

10.18 Calculate the number of moles of H_3PO_4 that will react with 5.55 mol of $Ba(OH)_2$ to form $Ba(H_2PO_4)_2$.

10.19 How many moles of oxygen gas are required for the complete combustion of 2.96 mol of butene, C_4H_8?

10.20 How many moles of oxygen gas are required for the complete combustion of 14.0 mol of butane, C_4H_{10}?

10.21 How many moles of oxygen gas are required for the combustion of 2.00 mol of butane, C_4H_{10}, to yield CO and water?

10.22 How many moles of solid $BaCl_2$ are also produced when 9.12 mol of H_2O is formed from the reaction of $BaCO_3$ and HCl?

10.23 A 40.5-mmol sample of $KClO_3$ was partially decomposed by heating, and 40.5 mmol of O_2 was produced.

(a) Write the balanced equation for the reaction.

(b) Which of the numbers of millimoles given in the problem is governed by the balanced equation?

(c) Calculate the percentage of $KClO_3$ that decomposed.

10.24 How many moles of C_8H_{18} can be produced by the reaction of 2.15 mol of H_2 and sufficient C_8H_{14}? The balanced equation is

$$C_8H_{14}(l) + 2 H_2(g) \rightarrow C_8H_{18}(l)$$

10.25 How many moles of $BaCl_2$ are also produced when 12.50 mol of H_2O is formed from the reaction of $Ba(OH)_2$ and HCl?

10.26 (a) Calculate the number of millimoles of MnO_4^- that reacts with 10.00 mmol of Fe^{2+} according to the following equation:

$$5 Fe^{2+}(aq) + MnO_4^-(aq) + 8 H^+(aq) \rightarrow$$
$$5 Fe^{3+}(aq) + Mn^{2+}(aq) + 4 H_2O(l)$$

(b) How many millimoles of water are produced?

10.2 Mass Calculations for Chemical Reactions

10.27 Silver can be prepared from aqueous silver nitrate by reaction with copper metal. Copper(II) nitrate is the other product.

(a) Write a balanced equation for the reaction.

(b) Calculate the number of moles of copper in 9.97 g of copper.

(c) Calculate the number of moles of silver that can be produced from that number of moles of copper.

(d) Calculate the mass of that number of moles of silver.

(e) Combine the calculations for parts (b)–(d) into a factor label solution to determine how many grams of silver can be produced using 9.97 g of copper.

10.28 (a) Octane, C_8H_{18}, burns in excess oxygen to produce carbon dioxide and water. Write a balanced equation for the reaction.

(b) Calculate the number of moles of carbon dioxide in 983 g of carbon dioxide.

(c) Calculate the number of moles of octane required to produce that number of moles of carbon dioxide.

(d) Calculate the mass of octane in that number of moles.

(e) Combine the calculations for parts (b)–(d) into one factor label solution.

10.29 (a) Calculate the number of moles of H_2 that can be produced by the reaction of aqueous $HClO_3$ with 25.0 g of metallic Ba.

(b) Repeat the calculation of part (a) using 25.0 g of metallic Mg.

(c) Explain why the numbers of moles of H_2 produced in parts (a) and (b) differ greatly, even though the same number of grams of metal is used in each case.

10.30 (a) Calculate the number of moles of H_2 that can be produced by the reaction of aqueous HCl with 25.0 g of metallic Al.

(b) Compare the value with the result of Problem 10.29(b), and explain the difference.

10.31 Excess $AgNO_3(aq)$ was added to a sample of $FeCl_3(aq)$, and 17.2 g of AgCl(s) was produced. What mass of $FeCl_3$ was present initially?

10.32 Calculate the mass of product produced by each of the following combinations:

(a) 20.0 g of sodium with excess sulfur

(b) 30.0 g of magnesium with excess fluorine

(c) 40.0 g of aluminum with excess oxygen

10.33 Aluminum is produced commercially by high-temperature electrolysis of aluminum oxide ore dissolved in a nonaqueous melt. The electrodes are carbon. The reaction may be represented as follows:

$$Al_2O_3(\text{solution}) + 3\ C(s) \xrightarrow{\text{Electricity}} 2\ Al(l) + 3\ CO(g)$$

Calculate the mass of Al_2O_3 used to produce $1.50 \times 10^6\ g$ of aluminum by this process.

10.34 (a) Octane, C_8H_{18}, burns in limited oxygen supply to produce carbon monoxide and water. Write a balanced equation for the reaction.

(b) Calculate the number of moles of carbon monoxide in 427 g of carbon monoxide.

(c) Calculate the number of moles of octane required to produce that number of moles of carbon monoxide.

(d) Calculate the mass of octane in that number of moles.

(e) Combine the calculations for parts (b)–(d) into one factor label solution.

10.35 (a) Calculate the mass of NaCl that can be prepared with 165 g of Cl_2 and sufficient Na.

(b) Calculate the mass of NaCl that can be prepared with 165 g of HCl and sufficient NaOH.

(c) Which of parts (a) and (b) can be solved without any of the calculations presented in this chapter? Explain why the other part cannot be solved in the same manner.

10.36 The compound $(NH_4)_3PO_4$ is used as a fertilizer. Calculate the mass of $(NH_4)_3PO_4$ that can be produced by the reaction of $5.95 \times 10^5\ g$ of H_3PO_4 with sufficient NH_3:

$$3\ NH_3(aq) + H_3PO_4(aq) \rightarrow (NH_4)_3PO_4(aq)$$

10.37 Consider the following reaction:

$$2\ NaHCO_3(s) \xrightarrow{\text{Heat}} Na_2CO_3(s) + CO_2(g) + H_2O(g)$$

(a) What mass of Na_2CO_3 can be prepared by heating 25.0 g of $NaHCO_3$ until no further reaction takes place?

(b) What mass of solid product(s) can be prepared by heating 25.0 g of $NaHCO_3$ until no further reaction takes place?

(c) What is the difference, if any, in parts (a) and (b)?

10.38 Silver bromide can be "dissolved" by the action of aqueous $Na_2S_2O_3$ (called "hypo") after any silver that has been "activated" by exposure to light is reduced to metallic silver by another agent. This process is the basis for the development of black-and-white film. Calculate the mass of hypo necessary to dissolve 21.9 g of AgBr. The equation for the dissolving process is

$$AgBr(s) + 2\ Na_2S_2O_3(aq) \rightarrow$$
$$Na_3Ag(S_2O_3)_2(aq) + NaBr(aq)$$

10.39 In the softening of temporary hard water—water containing calcium hydrogen carbonate—the acid salt is converted to calcium carbonate, carbon dioxide, and water by heating. The process is described by the following equation:

$$Ca(HCO_3)_2(aq) \xrightarrow{\text{Heat}} CaCO_3(s) + CO_2(g) + H_2O(l)$$

Calculate the mass of $CaCO_3$ that can be produced from 729 g of $Ca(HCO_3)_2$.

10.40 Sulfur dioxide in the atmosphere contributes to acid rain. One method of controlling sulfur dioxide emission is to absorb the sulfur dioxide into a solution of a base. Calculate the mass of SO_2 that can be absorbed by 33.0 g of $Ba(OH)_2$ to make $BaSO_4$.

10.41 How many moles of KOH are required to completely neutralize 42.7 g of $H_2C_2O_4$?

10.42 Calculate the mass of NH_4NO_3 that can be produced by the reaction of 225 g of Zn according to the following balanced equation:

$$10\ HNO_3(aq) + 4\ Zn(s) \rightarrow$$
$$NH_4NO_3(aq) + 4\ Zn(NO_3)_2(aq) + 3\ H_2O(l)$$

10.43 Calculate the mass of silver metal that can be produced by the action of 424 g of zinc metal on excess aqueous silver nitrate.

10.44 Calculate the mass of NO_2 that can be produced by the reaction of 35.9 g of HNO_3 according to the following balanced equation:

$$4\ HNO_3(aq) + Cu(s) \rightarrow$$
$$2\ NO_2(g) + Cu(NO_3)_2(aq) + 2\ H_2O(l)$$

10.45 Calculate the mass of gaseous SO_2 that will be produced along with 2.50 kg of copper from the roasting of copper(II) sulfide. (Hint: See Practice Problem 10.8 and Example 10.8, if necessary.)

$$CuS(s) + O_2(g) \xrightarrow{\text{Heat}} Cu(s) + SO_2(g)$$

10.46 Powdered aluminum metal can be used to reduce iron(II) oxide to molten iron, usable for spot welding. Calculate the mass of iron that can be produced by the reaction of 468 g of aluminum.

$$2\ Al(s) + 3\ FeO(s) \xrightarrow{\text{Heat}} 3\ Fe(l) + Al_2O_3(s)$$

10.47 Excess $AgNO_3(aq)$ was added to a sample of $NiCl_2(aq)$, and 13.7 g of AgCl(s) was produced. What mass of $NiCl_2$ was present initially?

10.48 How many grams of C_8H_{18} can be produced by the reaction of 6.93 g of H_2 and sufficient C_8H_{14}? The balanced equation is

$$C_8H_{14}(l) + 2\ H_2(g) \rightarrow C_8H_{18}(l)$$

10.49 What mass of MgO can be "dissolved" by 121 g of HNO_3?

$$MgO(s) + 2\ HNO_3(aq) \rightarrow Mg(NO_3)_2(aq) + H_2O(l)$$

10.50 The compound HBF_4 is a useful reagent. Calculate the mass of HF required to make 255 g of HBF_4. It is prepared by the following reaction:

$$H_3BO_3(aq) + 4\ HF(aq) \rightarrow HBF_4(aq) + 3\ H_2O(l)$$

10.51 The discharge of a lead storage cell in an automobile battery can be represented by the following equation:

$$Pb(s) + PbO_2(s) + 2\ H_2SO_4(aq) \rightarrow$$
$$2\ PbSO_4(s) + 2\ H_2O(l)$$

Calculate the mass of sulfuric acid consumed when 672 g of lead is changed to lead(II) sulfate.

10.52 Calculate the mass of Cl_2 required to convert 72.4 g of $CrCl_2$ to $CrCl_3$.

10.53 What mass of tin can be prepared by the reaction of 27.6 g of iron and excess aqueous tin(II) nitrate? Iron(II) nitrate is the other product.

10.54 Calculate the mass of NH_3 that can be prepared by heating 2.50 mol of solid $(NH_4)_2CO_3$. The balanced equation is

$$(NH_4)_2CO_3(s) \xrightarrow{\text{Heat}} 2\ NH_3(g) + CO_2(g) + H_2O(g)$$

10.55 What mass of P_4O_{10} can be prepared by combustion of 6.173 g of P_4?

10.56 What mass of P_4O_6 can be prepared by combustion of 61.61 g of P_4?

10.57 Calculate the number of moles of oxygen gas required to convert 17.5 g of SO_2 to SO_3.

10.58 What mass of silver can be prepared by the reaction of 77.9 g of zinc and excess aqueous silver nitrate?

10.59 Iron ore is reduced to iron with coke (impure carbon). Calculate the mass of Fe_2O_3 that can be reduced to iron with 1055 g of carbon. The reaction may be represented as follows:

$$Fe_2O_3(s) + 3\ C(s) \xrightarrow{\text{Heat}} 2\ Fe(l) + 3\ CO(g)$$

10.60 Calculate the number of moles of NH_3 that can be prepared by heating 22.7 g of solid $(NH_4)_2SO_3$. The balanced equation is

$$(NH_4)_2SO_3(s) \xrightarrow{\text{Heat}} 2\ NH_3(g) + SO_2(g) + H_2O(g)$$

10.61 What mass of KOH is required to completely neutralize 29.7 g of H_2SO_4?

10.62 What mass of $CaCO_3$ can be "dissolved" by 17.6 g of HNO_3?

$$CaCO_3(s) + 2\ HNO_3(aq) \rightarrow$$
$$Ca(NO_3)_2(aq) + H_2O(l) + CO_2(g)$$

10.3 Calculations Involving Other Quantities

10.63 (a) How many moles of chlorine atoms do 7.22×10^{24} chlorine atoms represent?

(b) How many moles of PCl_5 contain that number of chlorine atoms?

(c) How many moles of $POCl_3$ can be prepared by treatment of that much PCl_5 with O_2?

$$2\ PCl_5(l) + O_2(g) \rightarrow 2\ POCl_3(g) + 2\ Cl_2(g)$$

(d) What mass of $POCl_3$ is that?

10.64 Calculate the number of chlorine atoms in the chromium(III) chloride prepared by treating chromium(II) chloride with 2.68 g of chlorine gas.

10.65 Calculate the number of moles of hydrogen atoms in 227 mL of water (density = 0.997 g/mL).

10.66 Calculate the number of moles of chlorine atoms in 73.2 mL of CCl_4 (density = 1.59 g/mL).

10.67 The hydrogen used for about 90% of the industrial synthesis of ammonia comes from the following reaction at high temperature:

$$CH_4(g) + H_2O(g) \xrightarrow{\text{Ni}} CO(g) + 3\ H_2(g)$$

Calculate the number of molecules of CH_4 required to produce 2.50 metric tons (2.50×10^6 g) of H_2.

10.68 A solid combustible material can sometimes be changed into a more useful fuel if it is converted to a gas before burning. The following reaction, known as the water gas reaction, is used to provide gaseous fuels under certain circumstances:

$$H_2O(g) + C(s) \xrightarrow{1200°C} CO(g) + H_2(g)$$

Calculate the number of moles of water required to convert 2.50×10^{28} carbon atoms to carbon monoxide and hydrogen.

10.69 Calculate the number of oxygen atoms in the sulfur trioxide prepared by treating sulfur dioxide with 72.9 g of oxygen gas.

10.70 How many nitrogen atoms are contained in the ammonium nitrate prepared by treatment of 75.0 g of aqueous ammonia with excess nitric acid?

$$NH_3(aq) + HNO_3(aq) \rightarrow NH_4NO_3(aq)$$

10.71 What mass of $COCl_2$ can be prepared by heating oxygen gas with the quantity of CCl_4 that contains 6.66×10^{23} Cl atoms? The balanced equation for the reaction is

$$2\ CCl_4(l) + O_2(g) \rightarrow 2\ COCl_2(g) + 2\ Cl_2(g)$$

10.72 What is the density of CCl_4 if 1.50×10^{24} molecules occupy a volume of 242 mL?

10.73 What mass of nitrogen is contained in the ammonium chlorate prepared by treatment of 25.0 g of aqueous ammonia with excess chloric acid?

$$NH_3(aq) + HClO_3(aq) \rightarrow NH_4ClO_3(aq)$$

10.74 What mass of carbon disulfide can be prepared by heating carbon with the quantity of sulfur that contains 5.00×10^{22} S atoms?

10.4 Problems Involving Limiting Quantities

10.75 The director of a summer baseball camp has 10 home plates and 27 bases. One home plate and 3 bases are needed for each baseball field.

(a) How many baseball fields can the director equip?

(b) How many extra pieces of equipment will there be?

10.76 Consider the following equation:

$$C_3H_4(g) + 2\ H_2(g) \xrightarrow{\text{Catalyst}} C_3H_8(g)$$

(a) Calculate the mass of C_3H_8 that can be prepared from 273.8 g of C_3H_4 and 40.18 g of H_2.

(b) Explain why this problem cannot be solved by applying the law of conservation of mass.

10.77 Calculate the mass of solid ZnO produced when 723 g of ZnS is treated with 212 g of O_2. SO_2 is the other product.

10.78 Calculate the number of moles of H_2O that will be produced by the reaction of 6.22×10^{23} molecules of H_2SO_4 and 179 g of $Ba(OH)_2$.

10.79 Calculate the number of moles of unreacted starting material that will be present when 37.17 mol of aqueous HNO_3 is treated with 1201 g of solid CaO.

10.80 Calculate the mass of unreacted starting material when 75.0 g of PCl_3 is treated with 22.8 g of H_2O.

$$PCl_3(l) + 3\ H_2O(l) \rightarrow H_3PO_3(l) + 3\ HCl(g)$$

10.81 Calculate the number of moles of unreacted starting material that will be present when 115 g of aqueous HNO_3 is treated with 52.0 g of solid MgO.

10.82 Calculate the mass of unreacted starting material when 25.0 g of SCl_4 is treated with 5.00 g of H_2O.

$$SCl_4(l) + 2\ H_2O(l) \rightarrow SO_2(l) + 4\ HCl(g)$$

10.83 Calculate the mass of solid Ag_2S produced when 2.50 g of $(NH_4)_2S$ is treated with 12.3 g of $AgNO_3$.

10.84 Calculate the number of moles of H_2O produced by the reaction of 9.53 g of $HClO_3$ and 2.00×10^{23} formula units of NaOH.

10.85 How many molecules of NO can be produced by the reaction of 3.50 mol of NH_3 with 4.75 mol of O_2, according to the following balanced equation?

$$4\ NH_3(g) + 5\ O_2(g) \rightarrow 4\ NO(g) + 6\ H_2O(g)$$

10.86 (a) Calculate the number of moles of H_2O produced when 0.197 mol of $Ba(OH)_2$ is treated with 2.27×10^{23} molecules of HCl.

(b) Calculate the number of moles of excess reactant.

10.87 Calculate the mass of zinc bromide produced when 72.7 g of zinc is treated with 95.1 g of hydrobromic acid.

10.88 (a) Calculate the number of moles of $AlCl_3$ that will react with 1.25 mol of $AgNO_3$.

(b) Calculate the number of moles of $AlCl_3$ that will react if 1.11 mol of $AlCl_3$ is treated with 1.25 mol of $AgNO_3$.

(c) Calculate the number of moles of $AlCl_3$ that will react if 0.411 mol of $AlCl_3$ is treated with 1.25 mol of $AgNO_3$.

10.89 Calculate the number of moles of AgCl that is produced in each part of Problem 10.88.

10.5 Theoretical Yield and Percent Yield

10.90 Calculate (a) the theoretical yield and (b) the percent yield, if 2.46 mol of liquid SO_2Cl_2 is obtained from the reaction of 2.48 mol of gaseous SO_2 and excess gaseous Cl_2.

10.91 Calculate the percent yield for an experiment in which 5.99 g of $SOCl_2$ was obtained by treatment of 5.89 g of SO_2 with excess PCl_5:

$$SO_2(g) + PCl_5(l) \rightarrow SOCl_2(l) + POCl_3(l)$$

10.92 Calculate (a) the theoretical yield and (b) the percent yield, if 121.2 g of liquid SO_2Cl_2 is obtained from the reaction of 60.2 g of gaseous SO_2 and excess gaseous Cl_2.

10.93 Calculate the percent yield for an experiment in which 5.50 g of $SOCl_2$ was obtained by treatment of 5.80 g of SO_2 with 40.1 g of PCl_5:

$$SO_2(g) + PCl_5(l) \rightarrow SOCl_2(l) + POCl_3(l)$$

10.6 Calculations with Net Ionic Equations

10.94 Calculate the number of moles of CO_2 that will be produced by the reaction of 0.575 mol of HCO_3^- with excess H^+.

10.95 How many moles of carbonate ion can be converted to carbon dioxide and water with 1.23 mol H^+?

10.96 Calculate the number of moles of solid Ag_2S that can be produced by the reaction of 0.508 mol of Ag^+ with excess S^{2-}.

10.97 Calculate the number of grams of CO_2 that will be produced by the reaction of 0.500 mol of CO_3^{2-} with 1.07 mol of H^+.

10.98 (a) Calculate the number of moles of Na^+ present in 0.654 mol of NaOH.

(b) How many moles of NaCl will be produced by the reaction of that quantity of NaOH with 0.407 mol of HCl?

(c) How many moles of NaOH will be present after the reaction?

(d) How many moles of Na^+ will be present in the final solution?

(e) Does your answer to part (d) confirm that Na^+ is a spectator ion?

10.99 Calculate the number of grams of $PbBr_2$ that can be produced by the reaction of 0.506 mol of Pb^{2+} with excess Br^-.

10.100 How many moles of sulfite ion can be converted to sulfur dioxide and water with 0.952 mol H^+?

10.101 Calculate the number of moles of SO_2 that will be produced by the reaction of 1.27 mol of HSO_3^- with excess H^+.

10.102 Calculate the number of grams of $BaSO_4$ that can be produced by the reaction of 1.74 mol of SO_4^{2-} with 1.99 mol of Ba^{2+}.

■ ■ ■ GENERAL PROBLEMS ■ ■ ■

10.103 Rewrite Problem 10.28 as a single-step problem.

10.104 In each case, calculate the mass of the product that is not water:

(a) 50.0 g of aqueous lithium hydroxide is treated with sufficient sulfuric acid so that the acid is completely neutralized

(b) 50.0 g of acetic acid is treated with excess aqueous potassium hydroxide

(c) 50.0 g of solid magnesium oxide is treated with excess nitric acid

(d) 50.0 g of dinitrogen pentoxide gas is treated with excess aqueous lithium hydroxide

10.105 List the steps necessary to do each of the following stoichiometry problems:

(a) Calculate the number of moles of aluminum metal required to prepare 6.99 mol of solid Al_2S_3.

(b) Calculate the number of grams of $Mg(OH)_2(s)$ required to prepare 6.99 mol of $Mg(NO_3)_2(aq)$ by reaction with $HNO_3(aq)$.

(c) Calculate the number of grams of $CO_2(g)$ that can be prepared from 16.2 g of $CO(g)$ and 5.67 g of $O_2(g)$.

10.106 What mass of nitrogen is contained in the ammonium nitrate that can be prepared by treatment of 25.0 g of aqueous ammonia with 40.0 g of nitric acid?

10.107 Calculate the number of moles of each substance (except water) in solution after 1.50 mol of $NaOH(aq)$ is added to 3.00 mol of $HC_2H_3O_2(aq)$.

10.108 Heating of solid sodium hydrogen carbonate is one step in the industrial process for production of washing soda—sodium carbonate. Carbon dioxide and water are also produced. Calculate the mass of solid produced when a 275-g sample of sodium hydrogen carbonate is heated.

10.109 What mass of barium sulfate can be produced by treatment in aqueous solution of 71.2 g of lithium sulfate with 244 g of barium nitrate?

10.110 What mass of barium carbonate can be produced by treatment in aqueous solution of 12.5 g of lithium carbonate with 27.0 g of barium nitrate?

10.111 (a) After 2.38 g of solid $KClO_3$ is heated for a brief time, 1.01 g of KCl has been produced. What mass of O_2 has been produced?

(b) Did all the $KClO_3$ decompose?

10.112 A 125-mmol sample of solid $KClO_3$ was partially decomposed by heating, and 124 mmol of O_2 was produced. Calculate the percentage of $KClO_3$ that decomposed.

10.113 Calculate the number of moles of each substance in solution after 3.00 mol of $NaC_2H_3O_2(aq)$ is added to 5.00 mol of $HCl(aq)$. The balanced equation is

$$NaC_2H_3O_2(aq) + HCl(aq) \rightarrow NaCl(aq) + HC_2H_3O_2(aq)$$

10.114 After 0.0500 mol of solid $KClO_3$ has been heated for a period of time, 0.0220 mol remains. How much KCl has been produced? How much O_2?

10.115 (a) A 5.00-g sample of a mixture of $KClO_3(s)$ and $MnO_2(s)$ was heated for a brief period of time, after which 4.82 g of solid remained. Write the balanced equation for the reaction, indicating which of the substances is the catalyst and including the states of all reactants and products.

(b) How many grams of oxygen were produced?

(c) How many grams of $KClO_3$ decomposed?

(d) How many grams of KCl were produced?

(e) What is the minimum number of grams of $KClO_3$ that was present originally?

10.116 A 5.00-g sample of a mixture of $KClO_3(s)$ and $MnO_2(s)$ was heated for a brief period of time, after which 4.82 g of solid remained. How many grams of KCl are present in the final mixture?

10.117 Sketch a diagram like Figure 10.3 that shows the steps necessary to solve Problem 10.70. What parts of your diagram are not found in Figure 10.3?

10.118 Calculate the answer to Problem 10.8.

10.119 (a) Sketch a diagram like Figure 10.2 that shows the conversion of moles of an element in a reactant to moles of an element in a product.

(b) Make the diagram of part (a) show the specific conversions needed to find the number of moles of NH_3 that can be made from a sample of NH_4Cl containing 5.92 mol of H atoms by reaction with NaOH.

10.120 For a store's going-out-of business sale, a set consisting of a card table and 4 chairs is advertised at $109.00. The store has 47 tables and 177 chairs. An outlet manager arrives with $3000. What is the maximum number of *sets* the outlet manager can buy?

10.121 Determine the number of moles of $MnCl_2$ that can be prepared by the reaction of 0.104 mol of $KMnO_4$, 0.502 mol of $FeCl_2$, and 1.82 mol of HCl, according to the following balanced equation:

$$KMnO_4(aq) + 5\ FeCl_2(aq) + 8\ HCl(aq) \rightarrow$$
$$MnCl_2(aq) + KCl(aq) + 5\ FeCl_3(aq) + 4\ H_2O(l)$$

11 Molarity

- 11.1 Definition and Uses of Molarity
- 11.2 Molarities of Ions
- 11.3 Titration

- Key Terms *(Key terms are defined in the Glossary.)*

buret (11.3)
concentration (11.1)
end point (11.3)
Erlenmeyer flask (11.3)

molar (11.1)
molarity (11.1)
pipet (11.3)

solute (11.1)
solvent (11.1)
titration (11.3)

- Symbols

M (molar) (11.1)
M (molarity) (11.1)

Solutions were introduced in Chapter 1, and quantities of substances in moles were presented in Section 7.3. We can easily measure the quantity of a solute in solution by measuring the volume of a solution if we can determine the concentration of the solute. In this chapter, we will limit our discussions to liquid solutions, mostly aqueous solutions.

Molarity, the most common measure of concentration used by chemists, is introduced in Section 11.1 and used to solve problems involving numbers of moles and volumes. The concentrations of individual ions in aqueous solutions of ionic substances are discussed in Section 11.2. The technique of titration, used to determine the concentrations of solutions experimentally, is presented in Section 11.3.

11.1 Definition and Uses of Molarity

OBJECTIVE

■ to define molarity—the most basic chemical measure of concentration—and to use molarity to solve problems

A **solute** is the substance dissolved in a **solvent,** which is the substance doing the dissolving. Most often, the solvent is the component present in greatest quantity, and the solutes are present in lower quantities. However, when water is a component, it is often regarded as the solvent, even if more of another component is present. For example, in a salt-water solution, the salt is the solute, and the water is the solvent.

Everyone is familiar with the concept of concentration. **Concentration** is quantity of solute in a given quantity of solvent or solution. For example, if you usually drink coffee with 2 teaspoons of sugar per cup, how much sugar would you use in half a cup of coffee to get the usual sweetness? The sweetness depends on the concentration—the amount of sugar *per given volume of solution.* You would use 1 teaspoon of sugar (half of your normal amount) in half a cup of coffee (half of your normal amount).

■ EXAMPLE 11.1

If you absentmindedly stir 3 teaspoons of sugar into a cup of coffee, which makes the coffee sweeter than you like, how can you make it less sweet?

Solution

Taking some sugar out of the solution is difficult, once the sugar has dissolved. The easy way to make the drink less sweet is to add more liquid. Increasing the volume makes less sugar *per unit volume,* so the coffee tastes less sweet.

■ EXAMPLE 11.2

A cup full of water contains two lumps of sugar; a second cup half full of water contains one lump of sugar.

(a) Which cup, if either, contains more sugar?

(b) Which cup, if either, contains the sweeter-tasting solution?

(c) What is the difference between *quantity* and *concentration?*

Solution

(a) The first cup contains more sugar—two lumps is more than one.

(b) The contents of both cups taste equally sweet—both solutions have the same concentration.

(c) Concentration is quantity divided by volume. ■

Molarity is defined as the *number of moles of solute per liter of solution:*

$$\text{Molarity} = \frac{\text{number of moles of solute}}{\text{number of liters of solution}}$$

This definition is often shortened to "moles per liter," but condensing it in this way does not change the fact that it is really the number of moles *of solute* per liter *of solution*. The unit of molarity is **molar,** symbolized M. An italic capital M is used as a symbol for molarity; note that a nonitalic capital M means molar.

■ EXAMPLE 11.3

Calculate the molarity of a solution containing 6.00 mol of CH_3OH in enough water to make 2.00 L of solution.

Solution

$$\text{Molarity} = \frac{6.00 \text{ mol}}{2.00 \text{ L}} = \frac{3.00 \text{ mol}}{1 \text{ L}} = 3.00 \text{ M}$$

Three different ways of stating this concentration are as follows:

The solution is 3.00 molar (3.00 M) in CH_3OH.
The CH_3OH is 3.00 M.
The molarity is 3.00 M.

Practice Problem 11.3 A solution is prepared using 3.90 L of water and 2.50 mol of a certain solute. The total volume is 4.00 L, and the number of moles of water is 217 mol. What is the molarity of the solute? ■

If quantities of solute and solution are given in units other than moles and liters, respectively, they can be changed to moles and liters to calculate the molarity.

■ EXAMPLE 11.4

Calculate the molarity of 750 mL of solution containing 70.0 g of CH_3OH.

Solution

Since molarity is defined in terms of moles of solute and liters of solution, the given quantities are converted to moles and liters, respectively:

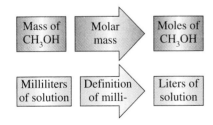

$$70.0 \text{ g } CH_3OH\left(\frac{1 \text{ mol } CH_3OH}{32.0 \text{ g } CH_3OH}\right) = 2.19 \text{ mol } CH_3OH$$

$$750 \text{ mL}\left(\frac{1 \text{ L}}{1000 \text{ mL}}\right) = 0.750 \text{ L}$$

$$M = \frac{2.19 \text{ mol } CH_3OH}{0.750 \text{ L}} = 2.92 \text{ M } CH_3OH$$

Practice Problem 11.4 Calculate the molarity of 2.17 L of solution containing 15.2 g of KBr. ■

■ EXAMPLE 11.5

Calculate the molarity of 20.0 mL of solution containing 3.75 mmol of solute.

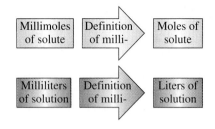

Solution

$$\frac{3.75 \ \cancel{mmol}\left(\dfrac{1 \ mol}{1000 \ \cancel{mmol}}\right)}{20.0 \ \cancel{mL}\left(\dfrac{1 \ L}{1000 \ \cancel{mL}}\right)} = \frac{3.75 \ mol}{20.0 \ L} = 0.188 \ M$$

Note that the number of millimoles per milliliter is equal to the number of moles per liter, and thus millimoles per milliliter is another way to define molarity. This equivalent definition is important, since quantities used in chemical laboratories are often measured in millimoles and milliliters.

Practice Problem 11.5   Calculate the molarity of a solution containing 58.5 mg of NaCl in 1.50 mL of solution. ▪

Since molarity is a ratio, like speed and density, it can be used as a conversion factor. Wherever it appears, the symbol M can be replaced by the ratio moles per liter (mol/L) or millimoles per milliliter (mmol/mL). For example, a concentration of 2.34 M can be used as any of the following factors:

$$\frac{2.34 \ mol}{1 \ L} \qquad \frac{1 \ L}{2.34 \ mol} \qquad \frac{2.34 \ mmol}{1 \ mL} \qquad \frac{1 \ mL}{2.34 \ mmol}$$

As another example, we can calculate the number of moles of solute present in 4.00 L of 2.00 M solution:

$$4.00 \ L\left(\frac{2.00 \ mol}{1 \ L}\right) = 8.00 \ mol$$

In this solution, the total quantity of solute is easy to visualize, as shown in Figure 11.1.

▪ **EXAMPLE 11.6**

Calculate the number of liters necessary to contain 1.75 mol of 1.25 M solute.

Solution

$$1.75 \ mol\left(\frac{1 \ L}{1.25 \ mol}\right) = 1.40 \ L$$

Practice Problem 11.6   Calculate the number of millimoles of $NaClO_3$ in 25.00 mL of 3.000 M solution. ▪

We can calculate the concentration of a solution that has been prepared by diluting a more concentrated solution by using the basic definition of molarity. For example, what concentration results when 2.00 L of 3.00 M NaCl is diluted with water to make 4.00 L of solution? When the solution is diluted with water, the *concentration* of NaCl is reduced, but the *quantity* of NaCl remains the same. We first calculate the original number of moles of NaCl and use that to calculate the final concentration:

$$2.00 \ L\left(\frac{3.00 \ mol}{1 \ L}\right) = 6.00 \ mol$$

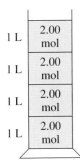

Figure 11.1 Moles and Concentration of Solute in a Solution

It is easy to see that 4.00 L of 2.00 M solution contains 8.00 mol of

Since only water—and no more solute—is added, there is 6.00 mol of solute in the final solution. Thus, the concentration of the final solution is

$$M = \frac{6.00 \text{ mol}}{4.00 \text{ L}} = 1.50 \text{ M}$$

It is easy to understand that when the volume is doubled, the concentration is halved, from 3.00 M to 1.50 M.

■ EXAMPLE 11.7

Calculate the final concentration of NaCl after 1.50 L of 0.200 M NaCl is diluted (a) to 3.50 L with water and (b) with 3.50 L of water.

Solution

The fraction of a mole of NaCl in the original solution is

$$1.50 \text{ L}\left(\frac{0.200 \text{ mol}}{1 \text{ L}}\right) = 0.300 \text{ mol}$$

The number of moles of NaCl is not changed by the addition of water.

(a) In this case, the final volume is 3.50 L, so the final concentration is

$$M = \frac{0.300 \text{ mol}}{3.50 \text{ L}} = 0.0857 \text{ M}$$

(b) In this case, the final volume is almost exactly 1.50 L + 3.50 L = 5.00 L, so the final concentration is

$$M = \frac{0.300 \text{ mol}}{5.00 \text{ L}} = 0.0600 \text{ M}$$

■ EXAMPLE 11.8

Calculate the final concentration after 1.50 L of 2.75 M NaCl is added to 2.25 L of 1.90 M NaCl and the resulting solution is diluted to 5.00 L.

Solution

Solute is contained in each solution, but not in the water used to dilute to 5.00 L. The number of moles of NaCl in the final solution is the sum of the numbers of moles in the two initial solutions:

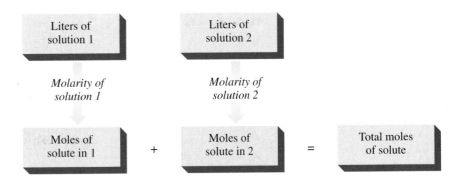

$$1.50 \text{ L}\left(\frac{2.75 \text{ mol}}{1 \text{ L}}\right) = 4.12 \text{ mol} \qquad 2.25 \text{ L}\left(\frac{1.90 \text{ mol}}{1 \text{ L}}\right) = 4.28 \text{ mol}$$

The total number of moles of NaCl is 4.12 mol + 4.28 mol = 8.40 mol. The final volume is 5.00 L, and the final concentration is

$$\frac{8.40 \text{ mol}}{5.00 \text{ L}} = 1.68 \text{ M}$$

Practice Problem 11.8 Calculate the final concentration after 150 mL of 2.72 M sugar solution is combined with 175 mL of 1.78 M sugar solution and the resulting solution is diluted to 600 mL. ■

Molarities and volumes may be used to calculate the numbers of moles involved in chemical reactions (Chapter 10). The conversions used are shown in Figure 11.2, where they have been added to those of Figure 10.2.

■ **EXAMPLE 11.9**

What volume of 1.75 M HCl is required to react with 44.1 mL of 1.57 M NaOH?

Solution

The reaction is

$$\text{NaOH(aq)} + \text{HCl(aq)} \longrightarrow \text{NaCl(aq)} + \text{H}_2\text{O(l)}$$

Molarity can be expressed as millimoles per milliliter.

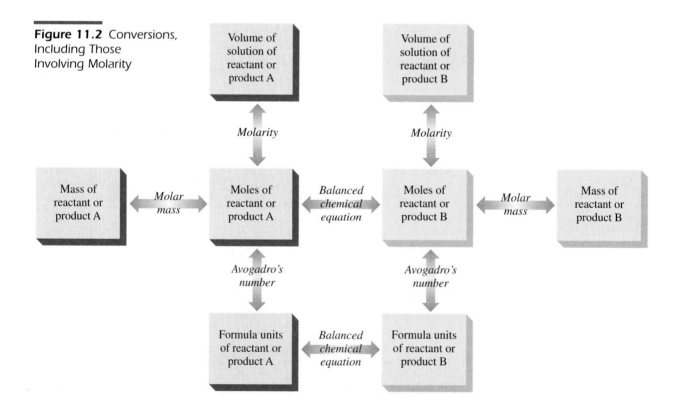

Figure 11.2 Conversions, Including Those Involving Molarity

The number of millimoles of base is

$$44.1 \text{ mL} \left(\frac{1.57 \text{ mmol NaOH}}{1 \text{ mL NaOH}} \right) = 69.2 \text{ mmol NaOH}$$

That number of millimoles of base is equal to the number of millimoles of acid, according to the balanced equation given previously. The volume of acid required is therefore

$$69.2 \text{ mmol HCl} \left(\frac{1 \text{ mL HCl}}{1.75 \text{ mmol HCl}} \right) = 39.5 \text{ mL HCl} \quad ■$$

■ **EXAMPLE 11.10**

What volume of 2.710 M HNO_3 is required to react with 31.70 mL of 0.2112 M $Ba(OH)_2$?

Solution

The equation is

$$Ba(OH)_2(aq) + 2 \text{ } HNO_3(aq) \rightarrow Ba(NO_3)_2(aq) + 2 \text{ } H_2O(l)$$

$$31.70 \text{ mL Ba(OH)}_2 \left(\frac{0.2112 \text{ mmol Ba(OH)}_2}{1 \text{ mL}} \right) = 6.695 \text{ mmol Ba(OH)}_2 \text{ present}$$

$$6.695 \text{ mmol Ba(OH)}_2 \left(\frac{2 \text{ mmol HNO}_3}{1 \text{ mmol Ba(OH)}_2} \right) = 13.39 \text{ mmol HNO}_3 \text{ required}$$

The volume of 2.710 M HNO_3 required is therefore

$$13.39 \text{ mmol HNO}_3 \left(\frac{1 \text{ mL HNO}_3}{2.710 \text{ mmol HNO}_3} \right) = 4.941 \text{ mL HNO}_3$$

Practice Problem 11.10   What volume of 0.2710 M $Ba(OH)_2$ is required to react with 31.70 mL of 0.2112 M HNO_3? ■

■ **EXAMPLE 11.11**

Calculate the number of moles of NaCl produced by reaction of 3.00 L of 1.75 M HCl and 4.00 L of 1.23 M NaOH.

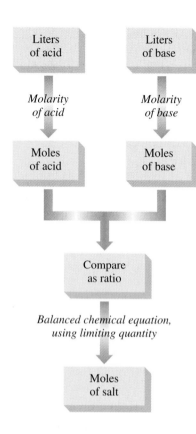

Solution

Since the quantities of two reactants are given, this problem involves a limiting quantity.

$$3.00 \text{ L HCl}\left(\frac{1.75 \text{ mol HCl}}{1 \text{ L HCl}}\right) = 5.25 \text{ mol HCl}$$

$$4.00 \text{ L NaOH}\left(\frac{1.23 \text{ mol NaOH}}{1 \text{ L NaOH}}\right) = 4.92 \text{ mol NaOH}$$

$$NaOH(aq) + HCl(aq) \rightarrow NaCl(aq) + H_2O(l)$$

Since the reactants react in a 1 : 1 mole ratio, HCl is in excess, NaOH is in limiting quantity, and 4.92 mol of NaCl will be produced.

Practice Problem 11.11   Calculate the number of grams of NaCl that can be produced by the reaction of 22.0 g of HCl and 553 mL of 3.00 M NaOH. ■

The maximum molarity possible in an aqueous solution of an ionic compound is about 40 M (with $LiClO_3$ as the solute). Pure water itself contains 55.6 mol/L. Sometimes, knowing the maximum molarity is important for determining whether your calculated answer is reasonable or for other purposes.

■ **EXAMPLE 11.12**

Two ionic solutes, called A and B, are dissolved in the same solvent. The ratio of the molarity of A to the molarity of B is 10^4. What can you tell from this fact?

Solution

Chemists customarily represent the concentration of a substance by enclosing its formula or symbol in square brackets. For example, [A] represents the molarity of A. Then

$$\frac{[A]}{[B]} = 1 \times 10^4$$

The concentration of B in its solution must be very low, since the concentration of A cannot be much greater than 10^1, at most.

Practice Problem 11.12   The ratio of molarities of A to B is 10^{-4}. What can you tell from this fact? ■

11.2 Molarities of Ions

As shown in Section 11.1, the molarities of ionic compounds can be calculated just as the molarities of covalent compounds are. The molarity of an ionic compound is the number of moles of the *compound* per liter of solution. However, as discussed in Chapter 9, it is often useful to describe ionic compounds in solution as the separate ions. The molarity of any ion is simply the number of moles of the *ion* per liter of solution.

■ EXAMPLE 11.13

(a) Calculate the molarity of $CaCl_2$ if 2.10 mol of that compound is dissolved in enough water to make 2.00 L of solution.

(b) Calculate the concentration of each type of ion in that solution.

Solution

(a)

$$\frac{2.10 \text{ mol } CaCl_2}{2.00 \text{ L}} = 1.05 \text{ M } CaCl_2$$

(b) The $CaCl_2$ is composed of Ca^{2+} ions and Cl^- ions. The formula indicates that there are twice as many Cl^- ions as Ca^{2+} ions. Thus,

$$2.10 \text{ mol } CaCl_2 \left(\frac{1 \text{ mol } Ca^{2+}}{1 \text{ mol } CaCl_2} \right) = 2.10 \text{ mol } Ca^{2+}$$

$$2.10 \text{ mol } CaCl_2 \left(\frac{2 \text{ mol } Cl^-}{1 \text{ mol } CaCl_2} \right) = 4.20 \text{ mol } Cl^-$$

The concentrations of ions are

$$\frac{2.10 \text{ mol } Ca^{2+}}{2.00 \text{ L}} = 1.05 \text{ M } Ca^{2+} \qquad \frac{4.20 \text{ mol } Cl^-}{2.00 \text{ L}} = 2.10 \text{ M } Cl^-$$

Practice Problem 11.13

(a) Calculate the molarity of $Al(NO_3)_3$ if 1.12 mol of that compound is dissolved in enough water to make 3.00 L of solution.

(b) Calculate the concentration of each ion in that solution. ■

When two different compounds containing an ion in common are placed in the same solution, if no reaction occurs, the numbers of moles of the common ion are added.

■ EXAMPLE 11.14

Calculate the concentration of each ion in a solution made by adding 2.00 L of 3.23 M NaCl to 1.50 L of 0.975 M $CaCl_2$ and diluting to 4.00 L.

Solution

When these two solutions are combined, there is no chemical reaction, but the final chloride ion concentration will have to include the chloride ions provided by both salts. The cation concentrations are not added, since the cations are different. The number of moles of each ion is calculated as follows:

$$2.00 \text{ L NaCl} \left(\frac{3.23 \text{ mol NaCl}}{1 \text{ L NaCl}} \right) = 6.46 \text{ mol NaCl}$$

$$1.50 \text{ L } CaCl_2 \left(\frac{0.975 \text{ mol } CaCl_2}{1 \text{ L } CaCl_2} \right) = 1.46 \text{ mol } CaCl_2$$

In 6.46 mol of NaCl, there are

$$6.46 \text{ mol } Cl^- \text{ and } 6.46 \text{ mol } Na^+$$

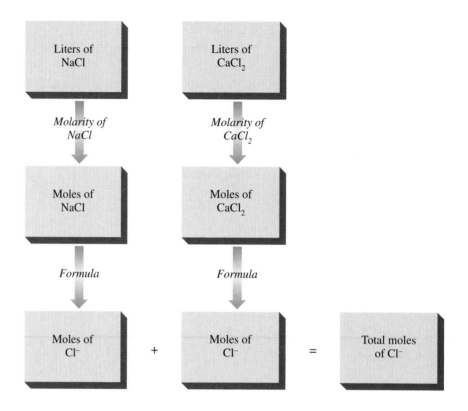

In 1.46 mol of $CaCl_2$, there are

$$2.92 \text{ mol Cl}^- \qquad \qquad \text{and } 1.46 \text{ mol Ca}^{2+}$$

The total numbers of moles are

$$9.38 \text{ mol Cl}^-, \quad 6.46 \text{ mol Na}^+, \text{ and } 1.46 \text{ mol Ca}^{2+}$$

The final concentrations are

$$\frac{6.46 \text{ mol Na}^+}{4.00 \text{ L}} = 1.62 \text{ M Na}^+ \qquad \frac{9.38 \text{ mol Cl}^-}{4.00 \text{ L}} = 2.34 \text{ M Cl}^-$$

$$\frac{1.46 \text{ mol Ca}^{2+}}{4.00 \text{ L}} = 0.365 \text{ M Ca}^{2+}$$

Practice Problem 11.14   Calculate the concentration of each ion in a solution made by adding 2.00 L of 1.72 M $(NH_4)_3PO_4$ to 2.50 L of 0.985 M NH_4Cl and diluting to 5.00 L. ▪

We can calculate the concentrations of ions in solution after a chemical reaction takes place. In doing so, we often use net ionic equations (Chapter 9).

▪ **EXAMPLE 11.15**

Calculate the concentration of each ion in a solution made by adding 2.00 L of 1.22 M HCl to 2.50 L of 0.285 M $Ba(OH)_2$ and diluting to 5.00 L.

Solution

When these two solutions are combined, there is a chemical reaction. The net ionic equation for the reaction is

$$H^+(aq) + OH^-(aq) \rightarrow H_2O(l)$$

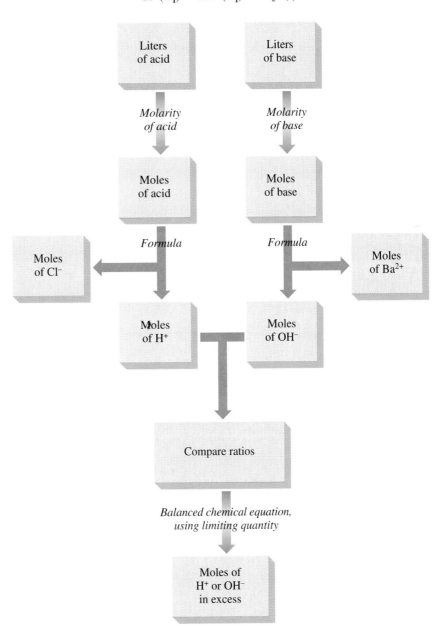

The numbers of moles of H^+ and OH^- in the two initial solutions are 2.44 mol H^+ and 1.42 mol OH^- (determined by calculations similar to those in Example 11.14). Since the OH^- ion is in limiting quantity, the 1.42 mol of OH^- will react with 1.42 mol of H^+, leaving 1.02 mol of H^+ in solution. The number of moles of water is of no interest; water is the solvent. The volume of water produced in the reaction is also of no interest. First, the volume of water produced is not significantly different from the sum of the volumes of H^+ and OH^- used up. Second, the final solution is diluted with water to 5.00 L. The Cl^- and Ba^{2+} ions are spectator ions; their numbers of moles do not change when the initial solutions are combined. Thus, there are 0.712 mol of Ba^{2+} and 2.44 mol of Cl^- in the final solution, along with the excess 1.02 mol of H^+. The concentrations are

$$\frac{1.02 \text{ mol } H^+}{5.00 \text{ L}} = 0.204 \text{ M } H^+ \qquad \frac{2.44 \text{ mol } Cl^-}{5.00 \text{ L}} = 0.488 \text{ M } Cl^-$$

$$\frac{0.712 \text{ mol } Ba^{2+}}{5.00 \text{ L}} = 0.142 \text{ M } Ba^{2+}$$

Practice Problem 11.15   Calculate the concentration of each ion in a solution made by adding 2.00 L of 3.23 M HCl to 2.50 L of 0.975 M NaOH and diluting to 5.00 L. ▪

11.3 Titration

OBJECTIVE

▪ to determine the concentration or the number of moles present of a substance by an experimental technique called titration

Titration is a technique to determine either the concentration of a solution of unknown molarity or the number of moles of a substance in a given sample. A chemical reaction is used for this purpose, and the reaction must be fast, be complete, and have a determinable end point. The reactions of strong acids and bases generally meet these criteria, and acid-base titrations are among the most important examples of this technique.

The purpose of a typical titration is the determination of the concentration of a solution of HCl, using a solution of NaOH of known concentration. Assume that 1 L of each solution is available (Figure 11.3a). Exactly 25.00 mL of the HCl solution is transferred with a **pipet** (Figure 11.3b) to a clean **Erlenmeyer flask** (Figure 11.3c), and two drops of an indicator are added to the solution. (Indicators were introduced in Chapter 8.) The indicator will show when the reaction is complete by changing color; that is, it will indicate when the **end point** has been reached. The NaOH solution is placed in a **buret** (Figure 11.3d). The tip of the buret is filled by allowing a small portion of the solution to run out the bottom, and then the level of the solution in the buret is read and recorded. Next, the NaOH solution is added from the buret to the Erlenmeyer flask—rapidly at first, then slowly as the reaction nears completion. When the last drop or half drop of NaOH solution completes the reaction, addition is stopped, and the final buret level is read and recorded. At this point, the number of moles of HCl originally in the solution and the number of moles of NaOH added to that solution are equal, just as they are in the balanced equation:

$$HCl(aq) + NaOH(aq) \rightarrow NaCl(aq) + H_2O(l)$$

The indicator allows us to tell when the reaction is complete because it has one color in the HCl solution and another in the NaOH solution (or it may be colorless in one solution and have a color in the other). As we add the NaOH solution from the buret, the color characteristic of that solution appears at the point at which the drop enters the HCl solution. As we swirl the flask rapidly to ensure mixing, that color disappears. As the reaction proceeds further toward completion, it takes longer for the color due to the added base to disappear, even though we are adding smaller portions of NaOH solution. When we add the last drop of

(a)   (b)   (c)   (d)

Figure 11.3 Apparatus for Performing an Acid-Base Titration (not shown to scale)

(a) Stock solutions of hydrochloric acid and sodium hydroxide. (b) A pipet is a piece of glassware shaped like a straw with an enlarged middle section. There is a mark about halfway up the top section. When the pipet is filled to the mark and then allowed to drain, the liquid delivered from the pipet will have the volume marked on the side. Common volumes are 5.00 mL, 10.00 mL, and 25.00 mL. The enlarged midsection allows more solution to be held, and the small top and bottom sections allow more-exact volumes to be delivered. (c) An Erlenmeyer flask is shaped so that the contents do not spill out when the flask is swirled. (d) A buret is a uniform-bore tube with volume calibrations marked on its side. Readings are taken to the bottom of the meniscus, the curved top surface of the liquid in the buret. The volume delivered from a buret is equal to the difference between the initial and final readings. The portion below the stopcock must be filled with liquid both before and after the titration to get an accurate volume.

NaOH solution, the associated color remains for at least 1 minute. We know that the end point of the titration has been reached.

We can now calculate the concentration of the HCl solution. The volume of NaOH solution used is the difference between the initial and final buret readings. The concentration of NaOH is known, and thus the number of millimoles (or moles) of NaOH can be calculated. From the way a titration is run and the fact that the mole ratio of HCl to NaOH in the balanced equation is 1:1, the number of millimoles of HCl is equal to the number of millimoles of NaOH. The concentration of the HCl solution is calculated by dividing that number of millimoles by the number of milliliters of the HCl solution.

■ **EXAMPLE 11.16**

An acid-base titration is performed as just described. Calculate the concentration of the HCl solution if the concentration of the NaOH solution is 2.000 M, the volume of HCl solution is 25.00 mL, the initial buret reading for the NaOH solution is 3.17 mL, and the final buret reading is 42.71 mL. (The lower numbers are toward the top of the buret.)

Solution

The volume of NaOH solution used is 42.71 mL − 3.17 mL = 39.54 mL. The number of millimoles of NaOH is

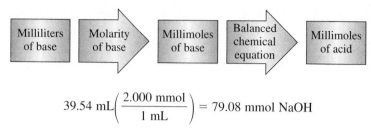

$$39.54 \text{ mL} \left(\frac{2.000 \text{ mmol}}{1 \text{ mL}} \right) = 79.08 \text{ mmol NaOH}$$

The number of millimoles of HCl is also 79.08 mmol because the titration was stopped when the numbers of millimoles of the two reactants were equal. The HCl concentration is therefore

$$\frac{79.08 \text{ mmol HCl}}{25.00 \text{ mL HCl}} = 3.163 \text{ M HCl}$$

Practice Problem 11.16 A titration is done to determine the concentration of NaOH in a certain solution. The concentration of the HCl solution used is 2.000 M, the volume of HCl solution is 25.00 mL, the initial buret reading for the NaOH solution is 2.83 mL, and the final buret reading is 47.19 mL. Calculate the molarity of the base. ▪

The indicator for an acid-base titration is an intensely colored dye, which is itself an acid or a base. It must be intensely colored in at least one of the two solutions so that its color is visible when it is present in a very low concentration. We should not use too much of the indicator because it, rather than the acid or base, would react with the other reagent; its quantity must be negligible, relative to the quantity of either reactant.

Why would we want to determine the concentration of an acid solution such as that of Example 11.16 when the acid gets converted to a salt in the process? That portion of acid used for the titration is indeed no longer useful, but the greater portion left in the stock bottle is the same concentration as that initially present in the flask. Thus, the titration allows us to determine the concentration of the major portion of the original solution.

▪ EXAMPLE 11.17

Calculate the concentration of H_2SO_4 solution that is completely neutralized in an acid-base titration if the concentration of NaOH solution is 2.000 M, the volume of H_2SO_4 solution is 25.00 mL, the initial buret reading for the NaOH solution is 2.77 mL, and the final buret reading is 40.91 mL.

Solution

The volume of NaOH solution used is 40.91 mL − 2.77 mL = 38.14 mL. The number of millimoles of NaOH is

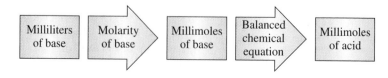

$$38.14 \text{ mL}\left(\frac{2.000 \text{ mmol NaOH}}{1 \text{ mL}}\right) = 76.28 \text{ mmol NaOH}$$

The balanced equation for this reaction is

$$H_2SO_4(aq) + 2 \text{ NaOH}(aq) \rightarrow Na_2SO_4(aq) + 2 H_2O(l)$$

The number of millimoles of H_2SO_4 is

$$76.28 \text{ mmol NaOH}\left(\frac{1 \text{ mmol } H_2SO_4}{2 \text{ mmol NaOH}}\right) = 38.14 \text{ mmol } H_2SO_4$$

The concentration of the H_2SO_4 solution is

$$\frac{38.14 \text{ mmol } H_2SO_4}{25.00 \text{ mL}} = 1.526 \text{ M } H_2SO_4$$

■ EXAMPLE 11.18

Two different indicators are used to show the two end points (one for each ionizable hydrogen atom) when H_2SO_4 is titrated with NaOH. In one laboratory, the labels fell off the indicator bottles. To determine which indicator was which, a student titrated 25.00 mL of 2.000 M H_2SO_4 with 31.79 mL of 1.573 M NaOH, using one of the indicators. Write the equation for the chemical reaction that occurred.

Solution

The numbers of millimoles of acid and base are calculated from the volumes and molarities given:

$$25.00 \text{ mL } H_2SO_4\left(\frac{2.000 \text{ mmol } H_2SO_4}{1 \text{ mL } H_2SO_4}\right) = 50.00 \text{ mmol } H_2SO_4$$

$$31.79 \text{ mL NaOH}\left(\frac{1.573 \text{ mmol NaOH}}{1 \text{ mL NaOH}}\right) = 50.01 \text{ mmol NaOH}$$

Since the numbers of millimoles of acid and base are equal, the acid and base react with each other in a 1:1 ratio. The balanced equation is therefore

$$NaOH(aq) + H_2SO_4(aq) \rightarrow NaHSO_4(aq) + H_2O(l)$$

Practice Problem 11.18 What volume of 2.573 M NaOH would be required to completely neutralize 25.00 mL of 2.000 M H_2SO_4? ■

■ EXAMPLE 11.19

Calculate the number of millimoles of Na_2CO_3 present in a sample if it takes 42.17 mL of 6.450 M HCl to convert the sample to NaCl, CO_2, and H_2O.

Solution

$$Na_2CO_3(s) + 2 \text{ HCl}(aq) \rightarrow 2 \text{ NaCl}(aq) + CO_2(g) + H_2O(l)$$

$$42.17 \text{ mL HCl}\left(\frac{6.450 \text{ mmol HCl}}{1 \text{ mL HCl}}\right)\left(\frac{1 \text{ mmol } Na_2CO_3}{2 \text{ mmol HCl}}\right) = 136.0 \text{ mmol } Na_2CO_3$$

Practice Problem 11.19 Calculate the number of millimoles of $NaHCO_3$ present in a sample if it takes 42.17 mL of 6.450 M HCl to convert the sample to NaCl, CO_2, and H_2O. ■

■ **EXAMPLE 11.20**

An unknown acid, represented as HA, with only one ionizable hydrogen atom per formula unit, is prepared in the laboratory. Calculate the molar mass of the acid if it takes 27.33 mL of 0.5000 M NaOH to neutralize a solution prepared by dissolving 1.373 g of the acid in water. The reaction may be represented as follows:

$$HA(aq) + NaOH(aq) \rightarrow NaA(aq) + H_2O(l)$$

Solution

The number of moles of base is calculated first:

$$27.33 \text{ mL NaOH}\left(\frac{1 \text{ L}}{1000 \text{ mL}}\right)\left(\frac{0.5000 \text{ mol NaOH}}{1 \text{ L NaOH}}\right) = 0.01366 \text{ mol NaOH}$$

The number of moles of acid is

$$0.01366 \text{ mol NaOH}\left(\frac{1 \text{ mol HA}}{1 \text{ mol NaOH}}\right) = 0.01366 \text{ mol HA}$$

The molar mass is the number of grams per mole:

$$\frac{1.373 \text{ g HA}}{0.01366 \text{ mol HA}} = 100.5 \text{ g/mol} \quad ■$$

■ ■ ■ ■ ■ ■ ■ **SUMMARY** ■ ■ ■ ■ ■ ■ ■

The concentration of a solute depends on the quantities of both the solute and the solvent or solution. Molarity is defined as the number of moles of solute per liter of solution. Molarity is calculated by dividing the number of moles of solute by the volume of the solution in liters, or alternatively, by dividing the number of millimoles of solute by the milliliters of solution. Since molarity is a ratio, it can be used as a conversion factor to change the volume of solution into the number of moles of solute, or vice versa.

If an aqueous solution is diluted with water, the number of moles of solute does not change, but the molarity does change. The final concentration of such a solution is calculated by dividing the number of moles of solute by the final volume. (The number of moles might have to be calculated from the initial volume and concentration.) If two solutions of the same solute are mixed, the total number of moles present in the final solution is the sum of the numbers of moles of the two original solutions. The molarities are *not* added.

The number of moles of a reactant involved in a reaction can be calculated from molarity and volume; that number of moles can then be used to calculate the number of moles of product. The number of moles of product can then be changed to a final molarity if a final volume is known. (Section 11.1)

The individual ions of an ionic compound may be regarded as separate solutes. The number of moles of each ion is calculated from the number of moles of the compound and the formula of the compound. If solutions of two compounds containing one ion in common are mixed, the number of moles of that ion is determined by adding the numbers of moles of the ion in the original solutions. In contrast, if solutions of ions that react with each other are mixed, the numbers of moles of the ions that react are subtracted from the original numbers of moles present, as in a problem involving a limiting quantity. The molarities of the ions in the final solution will be related to the numbers of moles of the ions remaining in that solution. (Section 11.2)

The experimental technique of titration is often used to determine the number of moles of a reactant in a given sample of an unknown, using a measured volume of a (standard) solution of known concentration. The color change of an indicator shows when the reaction has been completed. The concentration and volume of the standard solution give the number of moles of solute in the standard solution, and then the number of moles of the unknown substance may be calculated from the balanced chemical equation. If an unknown substance is dissolved in a measured volume of solution, its molarity can be calculated from its volume and calculated number of moles. (Section 11.3)

Items for Special Attention

■ Be sure to use mol as an abbreviation for mole—*not* M or m, which are used for other quantities related to moles. Otherwise, you may mix up quantities and concentrations.

■ Remember that the volume of *solution,* not the volume of solvent, is used in the definition of molarity.

■ Concentrations are *not* added when solutions are mixed.

■ In molarity problems involving two compounds, be sure to distinguish between those in which a reaction does not occur and those in which a reaction does occur.

Self-Tutorial Problems

11.1 Calculate the molarity of a solution of 2.00 mol of solute in

    (a) 1.00 L of solution   (b) 2.00 L of solution

    (c) 4.00 L of solution

11.2 What is the difference between (a) "dilute the solution to 3.00 L with water" and (b) "dilute the solution with 3.00 L of water"?

11.3 If 25.0 mL of a solution is poured from 100 mL of a 2.52 M sample, what is the concentration of the 25.0-mL portion?

11.4 What is the final concentration when 50.0 mL of 3.00 M sugar solution is added to 100.0 mL of 3.00 M sugar solution?

11.5 (a) If exactly one-thousandth of a 1.000-L sample of 4.000 M solution is poured into a small beaker, how many milliliters of solution are in the beaker?

    (b) How many millimoles of solute are in the beaker?

    (c) What is the concentration of the solution in the beaker?

    (d) What is the concentration in millimoles per milliliter in the beaker?

11.6 Which of the following combinations of solutions will result in a chemical reaction, which will result in a combi-

nation of the numbers of moles of a common ion, and which will result in a mere dilution?

    (a) $KOH(aq) + HClO_3(aq)$

    (b) $NaOH(aq) + NaClO_3(aq)$

    (c) $LiOH(aq) + KCl(aq)$

11.7 What is the final volume of solution if 1.5 L of solution is diluted (a) with 2.5 L of solvent or (b) to 2.5 L with solvent?

11.8 What is the concentration of each ion in the following solutions?

    (a) 1.0 M solution of KCl

    (b) 1.0 M solution of $MgCl_2$

    (c) 1.0 M solution of $AlCl_3$

    (d) 1.0 M solution of $KClO_3$

    (e) 1.0 M solution of $Mg(ClO_3)_2$

    (f) 1.0 M solution of $Al(ClO_3)_3$

11.9 (a) If 2 dozen couples get married at city hall on a certain weekend, how many brides are there? How many grooms?

    (b) What is the concentration of the cation in a 2.0 M solution of NaCl? What is the concentration of the anion?

■ ■ ■ PROBLEMS ■ ■ ■

11.1 Definition and Uses of Molarity

11.10 Calculate the molarity of a solution containing 2.00 mol of solute in 250 mL of solution.

11.11 Calculate the molarity of a solution containing 0.975 mol of solute in 750 mL of solution.

11.12 Calculate the molarity of 100 mL of a solution that contains 40.0 g of formaldehyde, CH_2O.

11.13 Calculate the molarity of 200 mL of a solution that contains 30.0 g of methyl alcohol, CH_3OH.

11.14 Calculate the molarity of a solution containing 125.0 mmol of solute in 400.0 mL of solution.

11.15 Calculate the molarity of a solution containing 0.1250 mol of solute in 500.0 mL of solution.

11.16 Calculate the number of moles of solute in 1.270 L of 1.755 M solution.

11.17 Calculate the number of grams of NaBr in 3.000 L of 4.000 M NaBr solution.

11.18 Calculate the number of moles of solute in 3.01 L of 2.22 M solution.

11.19 Calculate the number of grams of $MgCl_2$ in 3.00 L of 2.00 M $MgCl_2$ solution.

11.20 Calculate the volume of 1.27 M solution that contains 1.19 mol of solute.

11.21 Calculate the number of milliliters of 2.25 M NaCl solution that contains 7.14 g of NaCl.

11.22 What is the final concentration if 2.0 L of 1.5 M solution is diluted (a) with 3.0 L of solvent or (b) to 3.0 L with solvent?

11.23 Calculate the volume of 1.09 M solution that contains 8.29 mmol of solute.

11.24 Calculate the number of milliliters of 1.93 M NaCl solution that contains 6.28 g of NaCl.

11.25 Calculate the molarity of a solution prepared by diluting 25.0 mL of 1.15 M solution to 125.0 mL.

11.26 Calculate the volume of solution prepared by diluting 25.0 mL of 1.15 M solution to 0.800 M.

11.27 Calculate the volume of 4.50 M solution required to make 4.00 L of 1.75 M solution by dilution with water.

11.28 Calculate the molarity of a solution prepared by diluting 10.00 mL of 2.911 M solution to 50.00 mL.

11.29 Calculate the volume of solution prepared by diluting 2.102 L of 1.691 M solution to 1.123 M.

11.30 Calculate the volume of 5.00 M solution required to make 750 mL of 1.67 M solution by dilution with water.

11.2 Molarities of Ions

11.31 Calculate the concentration of each ion in each of the following solutions:

(a) 5.21 M NaCl   (b) 1.15 M K_3PO_4

(c) 1.79 M $(NH_4)_2SO_4$ (d) 0.750 M $Al_2(SO_4)_3$

(e) 0.200 M $NaClO_2$

11.32 Calculate the concentration of each ion in each of the following solutions:

(a) 6.00 M HCl   (b) 1.97 M $Cu(NO_3)_2$

(c) 0.875 M K_2SO_4 (d) 0.567 M $Al_2(SO_4)_3$

(e) 0.192 M NH_4ClO_3

11.33 Find the concentration of each type of ion in solution after 25.0 mL of 3.00 M $CoCl_2$ is diluted to 80.0 mL.

11.34 Find the concentration of each type of ion in solution after 10.0 mL of 3.00 M Na_2SO_4 is diluted to 25.0 mL.

11.35 In which of the following combinations of solutions will there be a chemical reaction? Which have ions in common? In which are the ions all different and unreactive?

(a) $KCl(aq) + CuCl_2(aq)$

(b) $NaCl(aq) + LiNO_3(aq)$

(c) $LiCl(aq) + AgNO_3(aq)$

(d) $Ba(OH)_2(aq) + H_2SO_4(aq)$

(e) $HCl(aq) + Na_2CO_3(aq)$

(f) $Na_2SO_4(aq) + KBr(aq)$

11.36 Calculate the *total* concentration of all the ions in each of the following solutions:

(a) 0.100 M NaCl   (b) 0.100 M $MgCl_2$

(c) 21.6 g of $(NH_4)_2SO_4$ in 252 mL of solution

11.37 If 0.250 mol of K_2SO_4 and 0.250 mol of Na_2SO_4 are dissolved in enough water to make 250 mL of solution, what is the concentration of each ion in the solution?

11.38 Calculate the concentration of each ion in 0.100 M $Hg(NO_3)_2$ solution.

11.39 What is the concentration of each type of ion in solution after 48.27 mL of 1.738 M $HClO_3$ is added to 50.00 mL of 0.5000 M NaOH? Assume that the final volume is the sum of the original volumes.

11.40 Calculate the concentration of each ion in solution after 272 mL of 1.42 M NaCl is mixed with 432 mL of 1.19 M $AlCl_3$ and the resulting solution is diluted to 1.00 L.

11.41 Calculate the concentration of each type of ion in solution after 50.0 mL of 2.00 M NaCl and 50.0 mL of 1.00 M NaBr are mixed. Assume that the final volume is 100 mL.

11.42 If 0.150 mol of $Al_2(SO_4)_3$ and 0.150 mol of Na_2SO_4 are dissolved in enough water to make 250 mL of solution, what is the concentration of each ion in the solution?

11.43 Calculate the concentration of each ion in 0.300 M $Fe_2(SO_4)_3$ solution.

11.44 Calculate the concentration of each ion in solution after 2.08 L of 3.38 M $MgCl_2$ is mixed with 1.50 L of 2.01 M $AlCl_3$ and the resulting solution is diluted to 5.00 L.

11.45 Calculate the concentration of each type of ion in solution after 50.0 mL of 2.00 M NaCl and 50.0 mL of 3.00 M Na_2SO_4 are mixed. Assume that the final volume is 100 mL.

11.46 What is the concentration of each type of ion in solution after 25.90 mL of 0.8230 M H_2SO_4 is added to 50.00 mL of 2.500 M NaOH? Assume that the final volume is the sum of the original volumes.

11.3 Titration

11.47 Calculate the concentration of an H_2SO_4 solution if 25.00 mL is completely neutralized by 46.19 mL of 1.500 M NaOH solution.

11.48 Calculate the concentration of a sulfuric acid solution if 25.00 mL is completely neutralized by 38.27 mL of 1.500 M sodium hydroxide solution.

11.49 Calculate the concentration of an H_2SO_4 solution if 25.00 mL is converted to $NaHSO_4$ by 41.73 mL of 3.000 M NaOH solution.

11.50 Calculate the concentration of a sulfuric acid solution if 25.00 mL is converted to sodium hydrogen sulfate by 38.77 mL of 2.500 M sodium hydroxide solution.

11.51 When 6.593 g of potassium hydrogen phthalate (symbolized as KHPh; molar mass = 204.2 g/mol) is titrated with KOH solution, it takes 41.99 mL of the base to achieve the end point. Calculate the concentration of the KOH solution.

$$KHPh(aq) + KOH(aq) \rightarrow K_2Ph(aq) + H_2O(l)$$

11.52 An antacid tablet contains 22.0 g of $NaHCO_3$. What volume of 4.47 M stomach acid (HCl) can this tablet neutralize?

11.53 An antacid tablet contains $NaHCO_3$. What mass of this compound is required to neutralize 0.875 L of 4.27 M stomach acid (HCl)?

11.54 How many millimoles of NaOH will react with 13.87 mL of 6.000 M HCl?

11.55 How many millimoles of AgNO₃ will react with 41.75 mL of 3.200 M LiCl?

11.56 How many millimoles of Ba(OH)₂ will react with 24.17 mL of 6.000 M HClO₃?

11.57 How many millimoles of AgNO₃ will react with 13.91 mL of 1.400 M CaCl₂?

■ ■ ■ GENERAL PROBLEMS ■ ■ ■

11.58 Calculate the concentration of each ion in solution after 1.91 L of 2.71 M NaCl is mixed with 2.27 L of 0.985 M AlCl₃ and then diluted to 5.00 L.

11.59 Calculate the concentration of each ion in solution after 1.79 L of 2.22 M NaOH is mixed with 2.19 L of 0.505 M H₂SO₄ and then diluted to 5.00 L.

11.60 Calculate the concentration of each ion in solution after 427 mL of 0.903 M NaCl is mixed with 215 mL of 1.31 M CuCl₂ and then diluted to 1.00 L.

11.61 Calculate the concentration of each ion in solution after 392 mL of 1.18 M NaOH is mixed with 141 mL of 0.800 M H₃PO₄ and then diluted to 800 mL.

11.62 (a) Calculate the concentration of each type of ion in 82.05 mL of a solution containing 40.50 mmol of NaCl plus 14.55 mmol of NaOH.

(b) Calculate the concentration of each type of ion in solution after 55.05 mL of 1.000 M NaOH is added to 27.00 mL of 1.500 M HCl. Assume that the volume of the final solution is the sum of the volumes of the two original solutions.

(c) Explain how parts (a) and (b) are related.

11.63 (a) Calculate the concentration of each type of ion in 75.90 mL of a solution containing 17.87 mmol of Na₂SO₄ plus 14.76 mmol of NaOH.

(b) Calculate the concentration of each type of ion in solution after 50.50 mL of 1.000 M NaOH is added to 25.40 mL of 0.7035 M H₂SO₄. Assume that the volume of the final solution is the sum of the volumes of the two original solutions.

(c) Explain how parts (a) and (b) are related.

11.64 Calculate the concentration of each type of ion in solution after 47.57 mL of 1.000 M HCl is added to 23.46 mL of 1.527 M NaOH. Use a net ionic equation in solving this problem. Assume that the final volume is equal to the sum of the volumes of the two original solutions.

11.65 When an alkali metal oxide is treated with water, it reacts with the water to form hydroxide ions. What concentration of hydroxide ions is present if 0.400 mol of solid Li₂O is treated with water and the final volume is 500 mL?

11.66 Calculate the concentration of H⁺ ion produced when H₂S is bubbled into 0.400 M Cu²⁺ solution, causing precipitation of all the copper(II) ion as CuS. Assume no change in the volume of the solution.

$$Cu^{2+}(aq) + H_2S(g) \rightarrow CuS(s) + 2\,H^+(aq)$$

11.67 When a lithium nitride is treated with water, it reacts with the water to form hydroxide ions and ammonia. What concentration of hydroxide ions is present if 0.250 mol of solid Li₃N is treated with water and diluted to 500 mL?

11.68 Calculate the concentration of CH₂O in a solution prepared by mixing 2.50 L of 1.28 M CH₂O and 1.70 L of 1.33 M CH₂O and diluting the mixture to 5.00 L with water.

11.69 Calculate the concentration of CH₂O in a solution prepared by mixing 55.3 mL of 3.84 M CH₂O and 27.8 mL of 0.922 M CH₂O and diluting the mixture to 100 mL with water.

11.70 Calculate the concentrations of acetate ion and acetic acid in solution after 100 mL of 2.00 M HC₂H₃O₂ and 100 mL of 1.00 M NaOH are mixed. Assume that the final volume is 200 mL and that the excess acetic acid yields no acetate ions to the final solution (since it is a *weak* acid).

$$OH^-(aq) + HC_2H_3O_2(aq) \rightarrow C_2H_3O_2^-(aq) + H_2O(l)$$

11.71 Calculate the concentrations of ammonium ion and ammonia in solution after 100 mL of 2.00 M NH₃ and 100 mL of 1.00 M HCl are mixed. Assume that the final volume is 200 mL and that the excess ammonia yields no ammonium ions to the final solution (since it is a *weak* base).

$$H^+(aq) + NH_3(aq) \rightarrow NH_4^+(aq)$$

11.72 Calculate the percentage of CaCO₃ in a 5.000-g sample of limestone if 19.79 mL of 4.000 M HCl is required to react completely with the CaCO₃. Assume that the rest of the limestone sample is inert.

11.73 In a certain experiment, 25.00 mL of 2.000 M H₃PO₄ was titrated to a certain end point with 30.95 mL of 3.231 M NaOH. Write the equation for the chemical reaction that occurred.

11.74 In a certain experiment, 25.00 mL of 2.000 M H₃PO₄ was titrated to a certain end point with 27.96 mL of 1.788 M NaOH. Write the equation for the chemical reaction that occurred.

11.75 What volume of 1.788 M NaOH would be required to completely neutralize the H₃PO₄ in Problem 11.74?

11.76 There are several acids in vinegar. Calculate the total concentration of acids in a 10.0-mL sample of vinegar if it takes 19.73 mL of 0.1000 M NaOH to neutralize the acids. Assume that each acid contains only one ionizable hydrogen atom per formula unit.

11.77 After a sample containing Na_2CO_3 and inert substances was treated with 27.16 mL of 6.000 M HCl, it took 2.471 mL of 1.000 M NaOH to neutralize the excess HCl. Calculate the mass of Na_2CO_3 in the sample.

11.78 After a 10.00-g sample containing Na_2CO_3 and inert substances was treated with 24.55 mL of 6.000 M HCl, it took 1.772 mL of 1.000 M NaOH to neutralize the excess HCl. Calculate the percent of Na_2CO_3 in the sample.

11.79 Calculate the molar mass of an unknown acid, HA, if a 8.153-g sample of the acid takes 43.18 mL of 4.000 M NaOH to neutralize it.

11.80 If 2.169 g of potassium hydrogen phthalate, an acid salt having one ionizable hydrogen atom and a molar mass of 204.2 g/mol, is used to neutralize 41.91 mL of NaOH solution, calculate the concentration of the base.

12 Gases

- 12.1 Gas Pressure

- 12.2 Boyle's Law

- 12.3 Charles' Law

- 12.4 The Combined Gas Law

- 12.5 The Ideal Gas Law

- 12.6 Gases in Chemical Reactions

- 12.7 Dalton's Law of Partial Pressures

- 12.8 Kinetic Molecular Theory of Gases

Key Terms (Key terms are defined in the Glossary.)

absolute temperature (12.3)
absolute zero (12.3)
atmosphere (12.1)
atmospheric pressure (12.1)
average kinetic energy (12.8)
barometer (12.1)
barometric pressure (12.1)
Boyle's law (12.2)
Charles' law (12.3)
combined gas law (12.4)
Dalton's law of partial pressures (12.7)
direct proportionality (12.3)

force (12.1)
gas (12.1)
Gay-Lussac's law of combining volumes (12.6)
ideal gas law (12.5)
inverse proportionality (12.2)
kelvin (12.3)
Kelvin scale (12.3)
kinetic molecular theory (12.8)
liquid (12.1)
partial pressure (12.7)
perfectly elastic collision (12.8)
phase (12.1)

pressure (12.1)
random motion (12.8)
solid (12.1)
standard atmosphere (12.1)
standard temperature and pressure (12.4)
state (12.1)
state of a system (12.2)
surroundings (12.2)
system (12.2)
torr (12.1)
vapor pressure (12.7)
volume ratio (12.6)

Symbols/Abbreviations

atm (atmosphere) (12.1)
K (kelvin) (12.3)
KE (kinetic energy) (12.8)
mm Hg (millimeters of mercury) (12.1)
n (moles of gas) (12.5)

N (Avogadro's number) (12.8)
P (pressure) (12.2)
Pa (pascal) (12.1)
R (ideal gas law constant) (12.5)
STP (standard temperature and pressure) (12.4)

t (Celsius temperature) (12.3)
T (absolute temperature) (12.3)
v (velocity of a molecule) (12.8)
V (volume) (12.2)

This book covers gases before liquids and solids because gases are easiest to understand and, historically, were understood first. Experimental results show that every sample of matter that exists as a gas, regardless of its composition, follows the laws presented in this chapter.

Section 12.1 introduces the concept of pressure and describes a simple way of measuring gas pressures, as well as the customary units used for pressure. Section 12.2 discusses Boyle's law, which describes the effect of the pressure of a gas on its volume. Section 12.3 examines the effect of temperature on volume and introduces a new temperature scale that makes the effect easy to understand. Section 12.4 covers the combined gas law, which describes the effect of changes in both temperature and pressure on the volume of a gas. The ideal gas law, introduced in Section 12.5, describes how to calculate the number of moles in a sample of gas from its temperature, volume, and pressure. The number of moles present can be used in related calculations—for example, to obtain the molar mass of the gas. Section 12.6 extends the concept of the number of moles of a gas to the stoichiometry of reactions in which at least one gas is involved. Dalton's law, presented in Section 12.7, allows the calculation of the pressure of an individual gas—for example, water vapor—in a mixture of gases. Section 12.8 presents the kinetic molecular theory of gases, the accepted explanation of why gases behave as they do, which is based on the behavior of their individual molecules.

12.1 Gas Pressure

OBJECTIVE

■ to measure gas pressures and convert them from one unit to another

Matter occurs in three **states,** or **phases:** solid, liquid, and gaseous. A **solid** has a definite shape and a definite volume. A **liquid** has a definite volume but assumes the shape of its container. A **gas** does not have a definite volume or shape; it expands to fill the entire volume of the container it occupies. The density of a gas is generally much lower than the density of the same substance in the solid or liquid state. A gas (like a liquid) exerts a pressure in all directions at any point within the gas.

Pressure is defined as force per unit area. A **force** is a push or a pull. The difference between force and pressure is important, both theoretically and practically. You might not object if someone pushed on your shoulder with an open palm with a force of 10 pounds. However, if someone pushed the point of a knife against your shoulder with a force of 10 pounds, you surely would object. The same force exerted over the tiny area of the point of a knife would cause serious damage!

All gases exert pressure. At any point, a gas exerts an equal pressure in all directions. Gases at rest exert a pressure equal to the pressure exerted on them. An easy method of measuring gas pressure is to use a simple **barometer,** as shown in Figure 12.1. There are two forces on the mercury in the tube: the force of gravity pulling down and the force due to air pressure pushing up. When these two forces balance each other, the mercury in the tube stops falling. The greater the air pressure, the higher the mercury stands in the tube above the level of mercury in the dish. A simple unit of gas pressure is millimeters of mercury (abbreviated mm Hg). The vertical height of the mercury column, corresponding to h in Figure 12.1, is a measure of pressure. A unit equal to the pressure necessary to support 1 mm Hg is the **torr,** named for Evangelista Torricelli (1608–1647), an Italian physicist who discovered the principle of the barometer. At sea level at

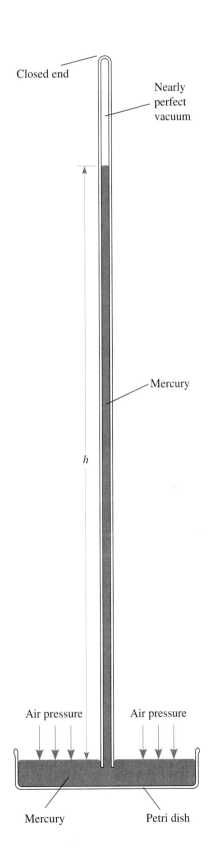

Figure 12.1 Simple Barometer (not drawn to scale)

A simple barometer is made with a long tube closed at one end. Fill the tube completely with mercury. Hold a finger over the open end, invert the tube and put the open end under the surface of mercury in a Petri dish, and hold the tube vertically. The mercury in the tube falls to a height, h, determined by the pressure of the air on the surface of the mercury in the Petri dish. There is essentially no gas pressure in the tube above the mercury. Caution: Mercury is toxic; use rubber gloves when handling it.

0°C on a "normal" day, the atmosphere can hold the mercury at a height of 760 mm; its pressure is 760 torr. Another unit of gas pressure, the **atmosphere** (atm), or **standard atmosphere,** is therefore defined as 760 torr:

$$1 \text{ atm} = 760 \text{ torr} = 760 \text{ mm Hg}$$

Note the difference between *1 atmosphere pressure* and **atmospheric pressure.** The former is a constant. The latter varies widely from place to place and also varies over time at the same place. Atmospheric pressure is often referred to as **barometric pressure.**

The SI unit of pressure is the pascal (Pa), which is such a small unit that the kilopascal (kPa) is used for atmospheric pressures under ordinary conditions. Chemists do not often use either the pascal or kilopascal. Since some modern reference books present pressure data in these units, however, you need to know how to convert them to atmospheres:

$$1.000 \text{ atm} = 101.3 \text{ kPa} = 1.013 \times 10^5 \text{ Pa}$$

12.2 Boyle's Law

Robert Boyle (1627–1691), an Irish physical scientist, discovered that the volume of a given sample of a gas at a constant temperature is *inversely proportional* to its pressure. This generalization, known as **Boyle's law,** applies approximately to any gas, no matter what its composition. (It does not apply to liquids or solids.)

Inverse proportionality is when one variable gets larger *by the same factor* as another gets smaller. For example, average speed and the time required to travel a certain distance are inversely proportional. The faster you go, the less time it takes you to complete the trip. Similarly, if the pressure on a given sample of gas at a given temperature is doubled (increased by a factor of 2), its volume is halved (decreased by a factor of 2).

Definitions of several terms will facilitate our discussion. The part of the world being investigated is called the **system;** the rest of the world is said to be the **surroundings.** The set of conditions (pressure, temperature, volume, phase, and so forth) that describe the system is called the **state of the system.** This use of the word *state* must be distinguished from that meaning "phase" (solid, liquid, or gas). For example, the state of a system about to undergo a chemical reaction includes the states (phases) of the reactants, and the state of the system after the reaction includes the states of the products.

Boyle might have observed the following data on volume and pressure for a given sample of gas at a given temperature, under four different sets of conditions (or four states):

State	Volume (L)	Pressure (atm)
1	8.00	1.00
2	4.00	2.00
3	2.00	4.00
4	1.00	8.00

Note that tabulating data is very helpful when two or more variables are being considered. The units are usually included in the column headings in such a table. The data in the table show that the product of the volume and the pressure is a constant. The table may be expanded to show this relationship:

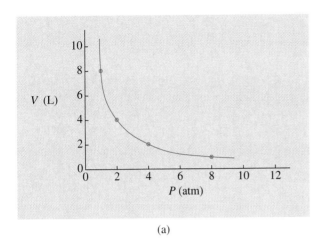

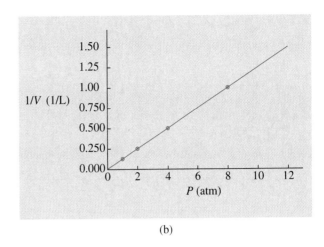

(a)   (b)

Figure 12.2 Graphs of Volume and Pressure Data

(a) Volume plotted against pressure results in a curved line. (b) When the reciprocal of volume is plotted against pressure, a straight line through the origin results. This indicates that $1/V$ is directly proportional to P and, therefore, that V is inversely proportional to P.

State	Volume (L)	Pressure (atm)	Volume $\times$ Pressure (L · atm)
1	8.00	1.00	8.00
2	4.00	2.00	8.00
3	2.00	4.00	8.00
4	1.00	8.00	8.00

Inverse proportionality may be expressed as a mathematical relationship in any of the following ways:

$$V \propto \frac{1}{P} \qquad V = \frac{k}{P} \qquad P = \frac{k}{V} \qquad PV = k$$

The proportionality sign ($\propto$) means that the value $1/P$ gets *bigger* when V gets *bigger*. Therefore, P must get smaller so that $1/P$ will get bigger. The k that appears in the equations is a constant—a given value for a specific sample of gas at a given temperature.

Another way to represent inverse proportionality is by graphing the data. If we place the values of P on the horizontal axis and the values of V on the vertical axis, plot the preceding tabulated values for P and V, and smoothly connect the points, we get a curve that can tell us what the volume will be at any intermediate pressure (Figure 12.2a). We can also plot $1/V$ versus P and get a straight line through the origin (Figure 12.2b).

State	V (L)	1/V (1/L)	P (atm)
1	8.00	0.125	1.00
2	4.00	0.250	2.00
3	2.00	0.500	4.00
4	1.00	1.00	8.00

■ **EXAMPLE 12.1**

The following data were obtained in a laboratory experiment:

P (atm)	V (L)
0.700	1.414
1.102	0.900
1.524	0.650
1.902	0.522
2.420	0.410

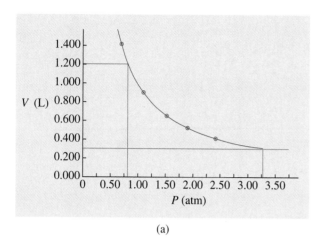

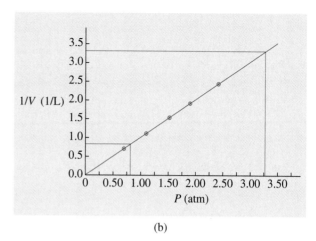

(a)   (b)

Figure 12.3 *Pressure-Volume Plots for Example 12.1*

(a) Plot of *V* versus *P*.
(b) Plot of 1/*V* versus *P*.

(a) Plot these data, and determine what pressure is required to get a volume of 1.200 L.

(b) Determine the pressure required to get a volume of 0.300 L.

(c) Plot *P* versus 1/*V*, and determine the two pressures again.

Solution

(a) The plot of the data is shown in Figure 12.3(a). With $V = 1.200$ L, the pressure is read as 0.82 atm.

(b) On Figure 12.3(a), the pressure is estimated to be 3.3 atm.

(c) The plot is shown in Figure 12.3(b). With $V = 0.300$ L, $1/V = 3.33/L$, and the pressure is read as 3.26 atm. With $V = 1.200$ L, $1/V = 0.833/L$, and the pressure is 0.82 atm. The linear plot of Figure 12.3(b) makes it easy to estimate data beyond the experimental points. Estimations beyond the experimental points are not as easy on the curve of Figure 12.3(a) because the curve's extent of bending is difficult to predict. ■

For a given sample of gas, $PV = k$. If we change that sample of gas from one pressure (P_1) to another (P_2) at constant temperature, the volume will change from V_1 to V_2. Both products of *P* and *V* must equal the same constant, *k*:

$$P_1V_1 = k = P_2V_2$$

Since both products are equal to the same constant, they must be equal to each other:

$$P_1V_1 = P_2V_2$$

We can use this expression to solve for any one of these variables when the other three are known.

■ **EXAMPLE 12.2**

A 2.00-L sample of gas has a pressure of 1.00 atm. Calculate the volume after its pressure is increased to 4.00 atm at constant temperature.

Solution

$$P_1V_1 = P_2V_2$$

In this type of problem, tabulating the given data is useful:

State	Pressure	Volume
1	1.00 atm	2.00 L
2	4.00 atm	V_2

Substitution of the values into the equation yields

$$(1.00 \text{ atm}) (2.00 \text{ L}) = (4.00 \text{ atm})V_2$$

$$V_2 = 0.500 \text{ L}$$

Note that multiplying the pressure by 4 causes the volume to be reduced to one-fourth.

Practice Problem 12.2   A sample of gas initially occupies 175 mL at 1.00 atm. Calculate the pressure if its volume is reduced to 125 mL at constant temperature. ■

The units of pressure and volume must be the same on each side of the equation $P_1V_1 = P_2V_2$. If they are not, one or more of the units must be converted.

■ EXAMPLE 12.3

A 6.00-L sample of gas has a pressure of 1.00 atm. Calculate the volume after its pressure is increased to 1140 torr at constant temperature.

Solution

Since the pressures are given in two different units, one of them must be changed.

State	Pressure		Volume
1	1.00 atm		6.00 L
2	1140 torr $\left(\dfrac{1 \text{ atm}}{760 \text{ torr}}\right) = 1.50$ atm		V_2

Now the problem can be solved as in Example 12.2:

$$P_1V_1 = P_2V_2$$

$$(1.00 \text{ atm}) (6.00 \text{ L}) = (1.50 \text{ atm}) \, V_2$$

$$V_2 = 4.00 \text{ L}$$

Alternatively, we could change the 1.00 atm to torr:

$$(760 \text{ torr}) (6.00 \text{ L}) = (1140 \text{ torr}) \, V_2$$

$$V_2 = 4.00 \text{ L}$$

Practice Problem 12.3   Calculate the volume of a sample of gas that is initially at 1.00 atm if its volume is changed to 450 mL as its pressure is changed to 717 torr at constant temperature. ■

◾ EXAMPLE 12.4

Calculate the pressure required to change a 1500-mL sample of gas initially at 1.00 atm to 4.18 L, at constant temperature.

Solution

$$P_1V_1 = P_2V_2$$

State	Pressure	Volume
1	1.00 atm	1500 mL = 1.50 L
2	P_2	4.18 L

$$(1.00 \text{ atm})(1.50 \text{ L}) = P_2(4.18 \text{ L})$$

$$P_2 = 0.359 \text{ atm}$$

The pressure must be lowered to 0.359 atm.

Practice Problem 12.4   Calculate the initial pressure of a 2.50-L sample of gas that has been changed at constant temperature to 1140 mL and 722 torr. ◾

12.3 Charles' Law

OBJECTIVE

◾ to calculate the volume or the temperature of a given sample of gas at a constant pressure

Figure 12.4   Dependence of Volume on Temperature at Constant Pressure

(a) Plot of the data given in Table 12.1. (b) Extension of the line in part (a) to absolute zero.

Section 12.2 discussed the effect of pressure on the volume of a gas at constant temperature. In 1787, 125 years after Boyle published the law that bears his name, J. A. C. Charles (1746–1823) discovered a law relating the volume of a given sample of gas to its **absolute temperature.** It took more than a century to discover this law because of the requirement that the temperature be absolute.

The volume of a sample of gas varies with the temperature, as shown in Table 12.1 and plotted in Figure 12.4(a) for a particular sample. While the volume changes with the Celsius temperature, the relationship is *not* a **direct proportionality.** That is, when the Celsius temperature doubles, the volume does not double, all other factors being held constant. On the graph, the plotted points form a straight line, but the line does *not* pass through the origin. For a direct pro-

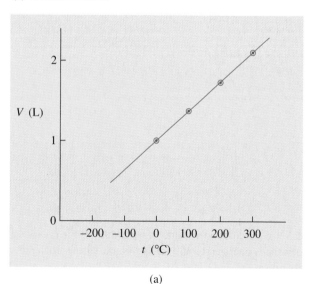

(a)

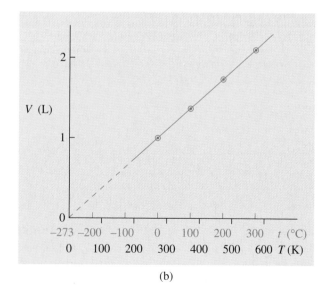

(b)

Table 12.1 Temperature and Volume Data for a Particular Sample of Gas at a Given Pressure

State	Temperature (°C)	Volume (L)
1	0	1.00
2	100	1.37
3	200	1.73
4	300	2.10

The Kelvin (absolute) scale must be used in all gas law problems involving temperature.

portionality to exist, the straight line must pass through the origin. If the straight line corresponding to the points in Table 12.1 is extended until the volume reaches 0 L, the Celsius temperature is −273°C (Figure 12.4b). Charles defined a new temperature scale in which the lowest possible temperature is 0° absolute, corresponding to −273°C. This temperature is called **absolute zero.** Each unit on this scale is defined as being the same size as a degree on the Celsius scale. Thus, each temperature on the absolute scale is 273 units greater than the same temperature on the Celsius scale (Figure 12.5). The symbol T is used for absolute temperature, and t is used for Celsius temperature:

$$T = t + 273$$

William Thomson (1824–1907), known as Lord Kelvin, first suggested using gas thermometers for measuring temperature. The absolute temperature scale is named after him. The **Kelvin scale** has units called **kelvins,** symbolized K, which are the same size as Celsius degrees. The word *degree* and the symbol for degrees (°) are not used with the Kelvin scale. The Kelvin (absolute) scale must be used in all gas law problems involving temperature.

■ EXAMPLE 12.5

Calculate the equivalent temperature on the other scale (Kelvin or Celsius) for each of the following:

(a) 100°C   (b) 300 K   (c) 0°C   (d) 0 K

Solution

(a) 100 + 273 = 373 K   (b) 300 − 273 = 27°C

(c) 273 K   (d) −273°C

Practice Problem 12.5   What is normal human body temperature, 98.6°F, on the Kelvin scale? (See Chapter 2, if necessary.) ■

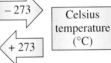

Figure 12.5 Comparison of Temperature Scales (Schematic)

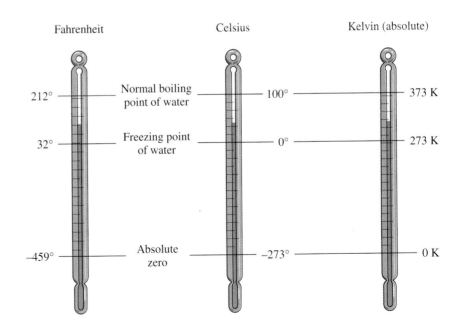

Charles' law states that the volume of a given sample of gas is directly proportional to its *absolute* temperature. In Table 12.2, the data of Table 12.1 are expanded to include the corresponding absolute temperatures. Note that dividing the volume by the corresponding absolute temperature for each state in Table 12.2 yields the same ratio for all the states. That means that the volume and the absolute temperature are directly proportional, which can be stated as an equation:

$$V = kT \quad \text{or} \quad \frac{V}{T} = k$$

The equation may also be written in terms of two states of a given sample of gas at constant pressure:

$$\frac{V_1}{T_1} = \frac{V_2}{T_2}$$

Note that T denotes *absolute* (*Kelvin*) temperature. Any unit of volume can be used, as long as the unit for both V_1 and V_2 is the same, but only one unit of temperature—kelvins—may be used.

■ **EXAMPLE 12.6**

Show that the data in Table 12.2 prove (a) that the Celsius temperature is *not* directly proportional to volume and (b) that the Kelvin temperature is directly proportional to volume.

Solution

(a) As the temperature 0°C is doubled (to 0°C), the volume does not change. The volume is *not* directly proportional to Celsius temperature.

(b) As the absolute temperature 273 K is increased to 373 K or 473 K, the volume increases to 373/273 = 1.37 or 473/273 = 1.73 times the original volume. The volume is directly proportional to *absolute* temperature. ■

As with the Boyle's law equation ($P_1V_1 = P_2V_2$), we can solve problems using the Charles' law equation:

$$\frac{V_1}{T_1} = \frac{V_2}{T_2}$$

Table 12.2   *Temperature and Volume Data
for a Particular Sample of Gas at a Given Pressure*

State	Temperature (°C)	Volume (L)	Temperature (K)	V/T (L/K)
1	0	1.00	273	0.00366
2	100	1.37	373	0.00367
3	200	1.73	473	0.00366
4	300	2.10	573	0.00366

In general, the values of three of these four variables are given, and the value for the remaining one is to be calculated. The units of temperature on both sides of this equation *must* be kelvins, however; they cannot merely be the same, as with the units of volume.

■ EXAMPLE 12.7

Calculate the temperature to which a 750-mL sample of gas at 0°C must be heated at constant pressure for the volume to change to 1.00 L.

Solution

The data are tabulated, with 750 mL converted to 0.750 L:

State	Temperature (°C)	Volume (L)	Temperature (K)
1	0	0.750	0 + 273 = 273 K
2	t_2	1.00	T_2

$$\frac{V_1}{T_1} = \frac{V_2}{T_2} \quad \text{or, inverting } both \text{ sides,} \quad \frac{T_1}{V_1} = \frac{T_2}{V_2}$$

$$\frac{273 \text{ K}}{0.750 \text{ L}} = \frac{T_2}{1.00 \text{ L}}$$

$$T_2 = 364 \text{ K} \qquad t_2 = 364 - 273 = 91°C$$

Practice Problem 12.7   Calculate the original temperature of a 1.78-L gas sample if it is expanded at constant pressure to 2.12 L at 50°C. ■

12.4 The Combined Gas Law

■ to calculate the volume, pressure, or temperature of a given sample of gas from other such data

Boyle's and Charles' laws may be merged into one law, called the **combined gas law,** expressed in equation form as follows:

$$\frac{PV}{T} = k \quad \text{or} \quad \frac{P_1V_1}{T_1} = \frac{P_2V_2}{T_2}$$

Most combined gas law problems involve solving for one of the variables in the second of these equations when the other five variables are given. Neither temperature nor pressure has to be held constant, but the law applies only to a given sample of gas and the temperatures must be in kelvins.

■ EXAMPLE 12.8

Calculate the volume of a sample of gas originally occupying 217 mL at 777 torr and 30°C after its temperature and pressure are changed to 60°C and 1.37 atm.

Solution

Again, tabulating the data is a good idea. The volume can be stated in milliliters in both states. The pressure can be stated in atmospheres in both. The temperature must be in kelvins in both states; it cannot merely be in the same unit in both states.

State	Pressure	Volume	Temperature
1	777 torr $\left(\dfrac{1 \text{ atm}}{760 \text{ torr}}\right) = 1.02$ atm	217 mL	$30 + 273 = 303$ K
2	1.37 atm	V_2	$60 + 273 = 333$ K

$$\frac{P_1 V_1}{T_1} = \frac{P_2 V_2}{T_2}$$

$$\frac{(1.02 \text{ atm})(217 \text{ mL})}{(303 \text{ K})} = \frac{(1.37 \text{ atm}) V_2}{(333 \text{ K})}$$

$$V_2 = 178 \text{ mL}$$

Practice Problem 12.8   Calculate the original volume of a sample of gas that is at 777 torr and 30°C before its volume, temperature, and pressure are changed to 217 mL, 60°C, and 1.37 atm. ■

A temperature of 0°C and a pressure of 1.00 atm constitute a set of conditions for a gas often called **standard temperature and pressure** (abbreviated STP).

■ **EXAMPLE 12.9**

A 3.25-L sample of gas originally at standard temperature and pressure is changed to 1.92 L at 1840 torr. Calculate its final temperature in degrees Celsius.

Solution

State	Pressure	Volume	Temperature
1	1.00 atm	3.25 L	$0 + 273 = 273$ K
2	1840 torr $\left(\dfrac{1 \text{ atm}}{760 \text{ torr}}\right) = 2.42$ atm	1.92 L	T_2

$$\frac{P_1 V_1}{T_1} = \frac{P_2 V_2}{T_2}$$

$$\frac{(1.00 \text{ atm})(3.25 \text{ L})}{(273 \text{ K})} = \frac{(2.42 \text{ atm})(1.92 \text{ L})}{T_2}$$

$$(0.0119)T_2 = 4.65 \text{ K}$$

$$T_2 = 390 \text{ K}$$

The problem requests the answer in Celsius:

$$390 - 273 = 117°C$$

Practice Problem 12.9   Calculate the volume at standard temperature and pressure of a sample of gas that has a volume of 251 mL at 55°C and 815 torr. ■

12.5 The Ideal Gas Law

The gas laws presented in the preceding sections apply to a given sample of gas. The combined gas law, written in the form

$$\frac{PV}{T} = k$$

may be rewritten as

$$\frac{PV}{T} = nR$$

where n represents the number of moles of gas molecules in the sample and R is another constant, valid for any sample of any gas. For the purpose of the gas laws, an atom of a monatomic gas, such as neon, is considered a molecule.

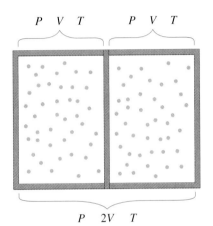

Figure 12.6 Combination of Two Identical Samples of Gas

Two samples of a gas that have the same volume, temperature, and pressure will have equal numbers of moles present. If the wall between the two samples were removed, the combined sample would have the same temperature and pressure as the original samples, but would have twice the volume and twice the number of moles. The volume of a gas under conditions of constant temperature and pressure is directly proportional to the number of moles present.

■ **EXAMPLE 12.10**

Consider a given sample of gas with pressure P and volume V at temperature T (Figure 12.6). A second sample of the same gas under the same conditions would have the same value of k, the constant in the combined gas law. What combined gas law constant would apply to the two samples of gas together?

Solution

Since the volume would be doubled if the gas samples were combined at the same temperature and pressure, the constant would also have to be doubled: It would equal $2k$. The more gas there is, the bigger the constant. Thus, separating the quantity of gas from the part of the constant that is not dependent on quantity is appropriate, which is what the ideal gas law does. The new constant is called R instead of k. The quantity of gas is given as n, the number of moles of gas. ■

The ideal gas law works approximately for all samples of any gas. The law works exactly only for a hypothetical "ideal" gas, from which it gets its name. The constant R has the value of 0.0821 L·atm/mol·K, no matter what gas or what size sample is under consideration. The value of R can be stated in other units, but the ideal gas law equation is easiest to use when pressure is in atmospheres and volume in liters. As usual, temperature must be expressed in kelvins. The equation expressing the **ideal gas law,** given previously as $PV/T = nR$, is usually rearranged to the form

$$PV = nRT$$

Note that R is the same for any sample of any gas, no matter what its composition.

■ **EXAMPLE 12.11**

Calculate the volume of 3.00 mol of carbon monoxide gas at 305 K and 1.00 atm.

Solution

$$PV = nRT$$

$$V = \frac{nRT}{P} = \frac{(3.00 \text{ mol})(0.0821 \text{ L·atm/mol·K})(305 \text{ K})}{1.00 \text{ atm}} = 75.1 \text{ L}$$

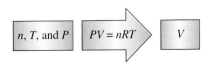

Practice Problem 12.11   Calculate the volume of 4.11 mol of methane gas, CH_4, at 309 K and 1.11 atm. ■

■ EXAMPLE 12.12

Calculate the volume of 1.75 mol of oxygen gas at 35°C and 748 torr.

Solution

Since the value of R given previously has units of liters, atmospheres, and kelvins, the data given here are converted to these units:

$$748 \text{ torr}\left(\frac{1 \text{ atm}}{760 \text{ torr}}\right) = 0.984 \text{ atm} \qquad 35 + 273 = 308 \text{ K}$$

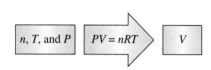

$$V = \frac{nRT}{P} = \frac{(1.75 \text{ mol})(0.0821 \text{ L} \cdot \text{atm/mol} \cdot \text{K})(308 \text{ K})}{0.984 \text{ atm}} = 45.0 \text{ L}$$

Practice Problem 12.12   Calculate the volume of 65.0 g of oxygen gas at 44°C and 678 torr. ■

A question that may have struck you by now is "How do I decide when to use the combined gas law and when to use the ideal gas law?" The answer depends on the problem, naturally. If moles are involved, the combined gas law cannot be used.

■ EXAMPLE 12.13

Decide which gas law should be used to solve each of the following:

(a) Calculate the final volume of a sample of gas that has an initial volume of 2.00 L at STP if the temperature and pressure are changed to 100°C and 785 torr.

(b) Calculate the volume of 2.00 mol of gas at 100°C and 785 torr.

Solution

(a) The combined gas law can be used.

(b) This problem involves moles and must be solved with the ideal gas law. ■

■ EXAMPLE 12.14

Calculate the volume of 125 g of water at 25°C and 1.00 atm.

Solution

Under these conditions, water is not a gas, and the ideal gas law cannot be used. The density of liquid water is 1.00 g/mL (Section 2.5), and thus the volume is 125 mL.

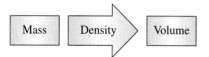

$$125 \text{ g}\left(\frac{1 \text{ mL}}{1.00 \text{ g}}\right) = 125 \text{ mL}$$

Not only the laws, but also when to use each one, must be learned. ■

■ EXAMPLE 12.15

Calculate the pressure of 12.5 g of nitrogen gas that occupies 17.0 L at 16°C.

Solution

The quantities given are converted to the units generally used with the ideal gas law equation. Note that nitrogen gas has the formula N_2, with a molar mass of 28.0 g/mol:

$$12.5 \text{ g N}_2\left(\frac{1 \text{ mol N}_2}{28.0 \text{ g N}_2}\right) = 0.446 \text{ mol N}_2 \qquad 16 + 273 = 289 \text{ K}$$

The pressure is then calculated from the ideal gas law equation:

$$P = \frac{nRT}{V} = \frac{(0.446 \text{ mol})(0.0821 \text{ L} \cdot \text{atm/mol} \cdot \text{K})(289 \text{ K})}{17.0 \text{ L}} = 0.622 \text{ atm}$$

Practice Problem 12.15 Calculate the pressure of 12.5 g of hydrogen gas that occupies 17.0 L at 16°C. Why does the answer to this problem differ from that for Example 12.15? ■

■ EXAMPLE 12.16

Calculate the number of moles of oxygen gas in a 40.0-L container at 31°C and 780 torr.

Solution

$$31 + 273 = 304 \text{ K} \qquad 780 \text{ torr}\left(\frac{1 \text{ atm}}{760 \text{ torr}}\right) = 1.03 \text{ atm}$$

$$n = \frac{PV}{RT} = \frac{(1.03 \text{ atm})(40.0 \text{ L})}{(0.0821 \text{ L} \cdot \text{atm/mol} \cdot \text{K})(304 \text{ K})} = 1.65 \text{ mol} \quad ■$$

12.6 Gases in Chemical Reactions

■ to calculate the numbers of moles involved in a chemical reaction using the properties of gases

Gases that are involved in chemical reactions obey the same laws of stoichiometry that apply to substances in any other state, as described in Chapters 8 and 10. Therefore, the ideal gas law can be used to calculate the quantities of gaseous substances involved in a reaction and then those results used to find the quantities of other substances. Figure 12.7 presents the conversions allowed by the ideal gas law, in addition to those originally shown in earlier figures.

■ EXAMPLE 12.17

How many liters of oxygen gas at 20°C and 2.00 atm can be prepared by thermal decomposition of 3.00 g of $KClO_3$?

Solution

The number of moles of oxygen produced in a chemical reaction can be calculated as shown in Chapter 10. Thus, the problem involves moles (even though the word *moles* is not explicitly stated), and so the ideal gas law is used.

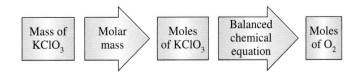

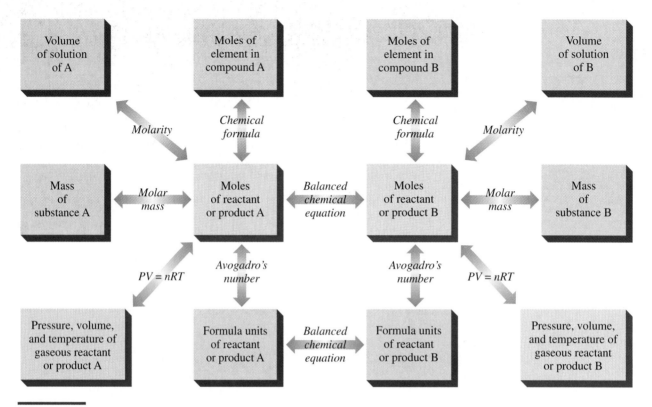

Figure 12.7 Mole Conversions, Including Application of the Ideal Gas Law to Determine the Number of Moles of a Gaseous Reactant or Product

$$3.00 \text{ g KClO}_3\left(\frac{1 \text{ mol KClO}_3}{122 \text{ g KClO}_3}\right) = 0.0246 \text{ mol KClO}_3$$

$$2 \text{ KClO}_3(s) \xrightarrow{\text{Heat}} 2 \text{ KCl}(s) + 3 \text{ O}_2(g)$$

The number of moles of O_2 produced is

$$0.0246 \text{ mol KClO}_3\left(\frac{3 \text{ mol O}_2}{2 \text{ mol KClO}_3}\right) = 0.0369 \text{ mol O}_2$$

Now you can use the ideal gas law equation:

$$V = \frac{nRT}{P} = \frac{(0.0369 \text{ mol})(0.0821 \text{ L} \cdot \text{atm/mol} \cdot \text{K})(293 \text{ K})}{2.00 \text{ atm}} = 0.444 \text{ L} \quad ■$$

■ **EXAMPLE 12.18**

A chemist decomposes 2.59 g of HgO in a sealed system. The oxygen produced has a pressure of 1.53 atm and a volume of 94.5 mL at 22°C. Calculate the value of R from these data.

Solution

$$2 \text{ HgO}(s) \xrightarrow{\text{Heat}} 2 \text{ Hg}(l) + \text{O}_2(g)$$

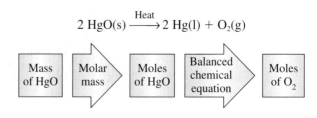

$$2.59 \text{ g HgO} \left(\frac{1 \text{ mol HgO}}{217 \text{ g HgO}} \right) \left(\frac{1 \text{ mol O}_2}{2 \text{ mol HgO}} \right) = 0.00597 \text{ mol O}_2$$

$$R = \frac{PV}{nT} = \frac{(1.53 \text{ atm})(0.0945 \text{ L})}{(0.00597 \text{ mol})(295 \text{ K})} = 0.0821 \text{ L} \cdot \text{atm/mol} \cdot \text{K}$$

Practice Problem 12.18   Calculate the volume of 1.00 mol of a gas at STP. ■

Volume Ratios in Chemical Reactions

For the special case where more than one gas is involved in a chemical reaction and all the gases are separate and measured at the same temperature and pressure, the **volume ratio** of the gases is equal to the mole ratio given in the balanced equation. This statement is known as **Gay-Lussac's law of combining volumes.** *This law is true only for separate gases and only when their temperatures and pressures are all equal.*

■ **EXAMPLE 12.19**

Ammonium carbonate decomposes when heated to yield carbon dioxide, ammonia, and water vapor. Calculate the ratio of the volume of ammonia to the volume of carbon dioxide, each at 500°C and 1.00 atm.

Solution

$$(\text{NH}_4)_2\text{CO}_3(s) \xrightarrow{\text{Heat}} 2 \text{ NH}_3(g) + \text{CO}_2(g) + \text{H}_2\text{O}(g)$$

The mole ratio of the gases, given in the balanced equation, is

$$2 \text{ mol NH}_3 : 1 \text{ mol CO}_2 : 1 \text{ mol H}_2\text{O}$$

Assume that the ammonia and carbon dioxide are separated and measured at the given temperature and pressure. The ratio of their volumes can be calculated as follows:

$$\frac{V_{\text{NH}_3}}{V_{\text{CO}_2}} = \frac{n_{\text{NH}_3}RT/P}{n_{\text{CO}_2}RT/P}$$

Since R is a constant and both T and P are the same for the two gases, this equation reduces to

$$\frac{V_{\text{NH}_3}}{V_{\text{CO}_2}} = \frac{n_{\text{NH}_3}}{n_{\text{CO}_2}}$$

The right side of this equation is the ratio of the numbers of moles—the ratio given by the balanced chemical equation. The left side of the equation is the ratio of the volumes, so the ratio given by the balanced chemical equation is equal to the volume ratio *under these conditions.* The ratio is 2 : 1. ■

■ **EXAMPLE 12.20**

If 2.00 L of H_2 and 1.00 L of O_2, both at standard temperature and pressure, are allowed to react, will the water vapor they form at 300°C and 1.00 atm occupy 2.00 L?

Solution

$$2 \, H_2(g) + O_2(g) \rightarrow 2 \, H_2O(g)$$

The volumes of H_2 and O_2 that react are in the ratio given in the balanced equation *because the two gases have the same temperature and pressure.* The volume of water vapor formed is *not* in that ratio, however, because its temperature is different. Its volume will be greater than 2 L. ■

Molar Masses and Molecular Formulas

The ideal gas law is another tool that we can use to determine the molar mass of a gaseous substance. To determine a molar mass, we need the mass of a given sample and also the number of moles in that sample. The ideal gas law can be used to determine the number of moles.

■ EXAMPLE 12.21

An 8.56-g sample of a pure gaseous substance occupies 4.55 L at 25°C and 1.44 atm. Calculate the molar mass of the gas.

Solution

You recognize immediately that given the volume, temperature, and pressure of a gaseous substance, you can calculate the number of moles:

$$n = \frac{PV}{RT} = \frac{(1.44 \text{ atm})(4.55 \text{ L})}{(0.0821 \text{ L} \cdot \text{atm/mol} \cdot \text{K})(298 \text{ K})} = 0.268 \text{ mol}$$

The molar mass is simply the mass divided by the number of moles:

$$\frac{8.56 \text{ g}}{0.268 \text{ mol}} = 31.9 \text{ g/mol}$$

Practice Problem 12.21   Calculate the molar mass of a gas if a 9.86-g sample occupies 2.50 L at 0°C and 1020 torr. ■

■ EXAMPLE 12.22

Calculate the volume that 6.92 g of oxygen gas occupies at standard temperature and pressure.

Solution

The molar mass of O_2 is used to calculate the number of moles of O_2 present:

$$6.92 \text{ g O}_2\left(\frac{1 \text{ mol O}_2}{32.0 \text{ g O}_2}\right) = 0.216 \text{ mol O}_2$$

The ideal gas law then allows you to calculate the volume under the stated conditions:

$$V = \frac{nRT}{P} = \frac{(0.216 \text{ mol})(0.0821 \text{ L} \cdot \text{atm/mol} \cdot \text{K})(273 \text{ K})}{1.00 \text{ atm}} = 4.84 \text{ L}$$

Practice Problem 12.22 Calculate the volume that 6.92 g of N_2 occupies at 0°C and 1.00 atm. Compare the answer with that of Example 12.22, and explain any difference. ■

With empirical formulas (Section 7.4) and molar masses, we can calculate molecular formulas (Section 7.5).

■ EXAMPLE 12.23

Calculate the molecular formula of a gaseous compound composed of 27.29% C and 72.71% O if 3.00 g of the compound occupies 1.35 L at 33°C and 1.27 atm pressure.

Solution

Even if you cannot see how to solve this problem completely at first glance, you can tell immediately that the empirical formula can be calculated from the percent composition.

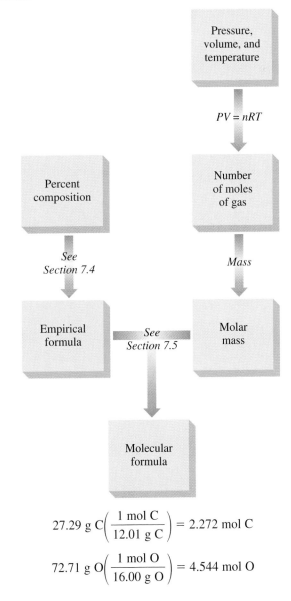

$$27.29 \text{ g C}\left(\frac{1 \text{ mol C}}{12.01 \text{ g C}}\right) = 2.272 \text{ mol C}$$

$$72.71 \text{ g O}\left(\frac{1 \text{ mol O}}{16.00 \text{ g O}}\right) = 4.544 \text{ mol O}$$

The mole ratio of oxygen to carbon is

$$\frac{4.544 \text{ mol O}}{2.272 \text{ mol C}} = \frac{2 \text{ mol O}}{1 \text{ mol C}}$$

so the empirical formula is CO_2.

You also know that you can calculate the number of moles of gas present from the pressure, volume, and temperature data:

$$n = \frac{PV}{RT} = \frac{(1.27 \text{ atm})(1.35 \text{ L})}{(0.0821 \text{ L} \cdot \text{atm/mol} \cdot \text{K})(306 \text{ K})} = 0.0682 \text{ mol}$$

The molar mass is the mass divided by the number of moles:

$$\frac{3.00 \text{ g}}{0.0682 \text{ mol}} = 44.0 \text{ g/mol}$$

The mass of a mole of empirical formula units is

$$12.0 \text{ g} + 2(16.0 \text{ g}) = 44.0 \text{ g}$$

so the empirical formula and the molecular formula are the same—CO_2.

Practice Problem 12.23   Calculate the molecular formula of a gaseous hydrocarbon of which 25.6 g occupies 14.6 L at 1090 torr and 25°C. The hydrocarbon consists of 79.9% carbon and the rest hydrogen. ■

12.7 Dalton's Law of Partial Pressures

OBJECTIVE

■ to calculate the properties of each gas in a mixture of gases

In a mixture of gases, all the components occupy the entire volume of the container, and all have the same temperature. The pressures of the individual gases, as well as the numbers of moles of the gases, vary.

A mixture of gases follows the same laws as a gas composed of only one substance. **Dalton's law of partial pressures** allows us to consider the properties of each component of a gas mixture. The law states that the sum of the individual pressures of the components of a gaseous mixture is equal to the total pressure of the mixture:

$$P_{\text{total}} = P_1 + P_2 + \cdots + P_n \qquad \text{(One pressure, } P_i \text{, for each gas)}$$

The pressures of the individual components of a mixture are called **partial pressures.** If the mixture is open to the atmosphere or if it is contained in a way that its pressure is equal to that of the atmosphere (Figure 12.8), the total pressure is equal to the barometric pressure.

■ **EXAMPLE 12.24**

A mixture of oxygen and nitrogen contains oxygen at a pressure of 323 torr and nitrogen at a pressure of 473 torr. What is the pressure of the mixture?

Solution

$$P_{\text{total}} = P_{N_2} + P_{O_2} = 473 \text{ torr} + 323 \text{ torr} = 796 \text{ torr}$$

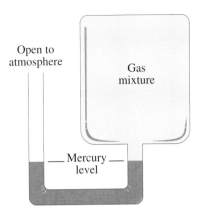

Figure 12.8 *Situation in Which the Total Pressure of a Mixture of Gases Equals Barometric Pressure*

Since the mercury levels are the same, the pressure of the gas mixture is equal to the pressure of the atmosphere—barometric pressure.

Practice Problem 12.24 A mixture of oxygen and nitrogen has a barometric pressure of 1.12 atm. If the pressure of the oxygen is 0.331 atm, what is the pressure of the nitrogen? ■

The ideal gas law applies to a mixture of gases as a whole and also to each of its components. That is,

$$P_{total}V = n_{total}RT \quad \text{and} \quad P_iV = n_iRT$$

where *i* stands for any component of the mixture.

■ EXAMPLE 12.25

Calculate the total number of moles in a 4.00-L sample of gas at 302 K, containing O_2 at 0.719 atm and N_2 at 0.289 atm. Also calculate the number of moles of O_2 present.

Solution

The total pressure of the gas mixture is

$$0.719 \text{ atm} + 0.289 \text{ atm} = 1.008 \text{ atm}$$

The total number of moles is

$$n_{total} = \frac{(1.008 \text{ atm})(4.00 \text{ L})}{(0.0821 \text{ L} \cdot \text{atm/mol} \cdot \text{K})(302 \text{ K})} = 0.163 \text{ mol}$$

The number of moles of O_2 is

$$n_{O2} = \frac{(0.719 \text{ atm})(4.00 \text{ L})}{(0.0821 \text{ L} \cdot \text{atm/mol} \cdot \text{K})(302 \text{ K})} = 0.116 \text{ mol } O_2$$

We can calculate the number of moles of oxygen another way. Since the oxygen and the gas mixture are both at the same temperature and have the same volume, the numbers of moles are proportional to the pressures:

$$\frac{P_{O2}V}{P_{total}V} = \frac{n_{O2}RT}{n_{total}RT}$$

$$\frac{P_{O2}}{P_{total}} = \frac{n_{O2}}{n_{total}}$$

$$n_{O2} = \frac{n_{total}P_{O2}}{P_{total}} = \frac{(0.163 \text{ mol})(0.719 \text{ atm})}{1.008 \text{ atm}} = 0.116 \text{ mol } O_2 \quad ■$$

Gases that are only slightly soluble in water are often collected over water (Figure 12.9). Gases in contact with liquid water will contain water vapor. The pressure of the water vapor in such a system does not vary as the pressure of an ordinary gas varies. For example, if we expand the gas, the pressure of the other components will drop (according to Boyle's law). However, more liquid water will then evaporate into the gas phase, and the pressure of the water vapor will remain constant (as long as there is liquid water left and once enough time has elapsed for the evaporation process to have taken place). If the gas mixture is compressed instead, the pressure of the water vapor does not increase, but some of the water vapor condenses into the liquid phase. At a given temperature, the

Table 12.3 Vapor
Pressure of Water
at Various Temperatures

Temperature (°C)	Pressure (torr)
−15	1.436
−10	2.149
−5	3.163
0	4.579
5	6.543
10	9.209
15	12.788
20	17.535
21	18.650
22	19.827
23	21.068
24	22.377
25	23.756
26	25.209
27	26.739
28	28.349
29	30.043
30	31.824
35	42.175
40	55.324
45	71.88
50	92.51
55	118.04
60	149.38
65	187.54
70	233.7
75	289.1
80	355.1
85	433.6
90	525.76
95	633.90
99	733.24
100	760.00
101	787.57
105	906.07
110	1074.56

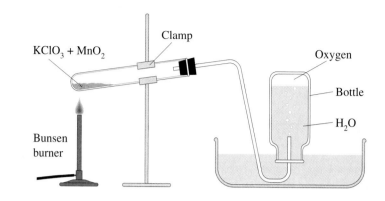

Figure 12.9   Collection of Oxygen over Water

A bottle filled with water is inverted in a pan of water. A test tube containing potassium chlorate and manganese(IV) oxide (the catalyst) is arranged so that the oxygen produced when the tube is heated displaces the water in the bottle. The mass lost from the test tube corresponds to the mass of oxygen produced. The oxygen in the bottle will have water vapor mixed with it.

pressure of water vapor in contact with liquid water is a constant, called the **vapor pressure** of water at that temperature. Vapor pressure is independent of the volume of the liquid and the volume of the gas mixture; it is also independent of the shape of the container and the surface area of the liquid. The *only* factor that determines the vapor pressure of pure water is *temperature*. The higher the temperature, the higher the vapor pressure of water is (Figure 12.10). The relationship is *not* a direct proportionality, however. Table 12.3 lists water vapor pressures at various temperatures. These values need not be memorized.

The law of partial pressures applies to mixtures containing water vapor, as well as to other gas mixtures. Problems involving water vapor in contact with liquid water can be solved by taking the vapor pressure from a table such as Table 12.3.

Figure 12.10   Variation
of the Vapor Pressure of Water
with Temperature

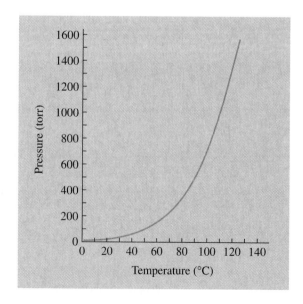

■ **EXAMPLE 12.26**

Oxygen gas is collected over water in an apparatus such as that shown in Figure 12.9 at a barometric pressure of 758 torr at 25°C.

(a) What is the pressure of the water vapor?

(b) What is the pressure of the oxygen gas?

Solution

(a) The water vapor pressure at 25°C is 24 torr, taken from Table 12.3.

(b) The pressure of the oxygen gas is

$$P_{O_2} = P_{total} - P_{H_2O} = 758 \text{ torr} - 24 \text{ torr} = 734 \text{ torr} \quad ■$$

■ **EXAMPLE 12.27**

What volume of oxygen, collected over water, will be obtained at 25°C and 760 torr barometric pressure from the thermal decomposition of 0.0200 mol of $KClO_3$?

Solution

$$2 \text{ KClO}_3(s) \xrightarrow{\text{Heat}} 2 \text{ KCl}(s) + 3 \text{ O}_2(g)$$

$$0.0200 \text{ mol KClO}_3 \left(\frac{3 \text{ mol O}_2}{2 \text{ mol KClO}_3} \right) = 0.0300 \text{ mol O}_2$$

The pressure of the *oxygen* is the barometric pressure minus the water vapor pressure (from Table 12.3):

$$P_{O_2} = 760 \text{ torr} - 24 \text{ torr} = 736 \text{ torr}$$

$$V = \frac{nRT}{P} = \frac{(0.0300 \text{ mol})(0.0821 \text{ L} \cdot \text{atm/mol} \cdot \text{K})(298 \text{ K})}{(736 \text{ torr})(1 \text{ atm/760 torr})}$$

$$= 0.758 \text{ L} = 758 \text{ mL} \quad ■$$

12.8 Kinetic Molecular Theory of Gases

■ to explain the behavior of gases using the kinetic molecular theory

When a wide variety of observations is grouped together in one generalization, that generalization is called a scientific law. The law of gravity is an example; it generalizes millions of observations that, when dropped, any object near the surface of the earth falls downward (toward the earth). When an explanation for a law is proposed, that explanation is called a hypothesis. When a hypothesis is accepted as true by the scientific community in general, the hypothesis becomes a theory. For example, Einstein's theory of relativity is an explanation of the law of gravity.

Several gas laws have been introduced in this chapter, but no explanation as to *why* those laws apply to all gases has been proposed. This section introduces the **kinetic molecular theory** of gases, which explains the gas laws and, when extended, also explains some properties of liquids and solids. Five postulates explain why gases behave as they do:

1. Gases are composed of small molecules that are in constant, *random motion.*

2. The volume of the molecules is insignificant compared to the volume occupied by the gas.

3. Forces between the molecules are negligible, except when the molecules collide with one another.

Figure 12.11 Lottery Machine
The molecules of a gas are somewhat like the balls in a lottery machine. The balls travel in (relatively)
straight paths until they hit the wall
or another ball. The collisions do
not produce much heat, so the total
kinetic energy of the balls does not
decrease much. Although the balls
"fill" the machine, their total volume
is much less than that of the machine, and there are no forces of
attraction between the balls.

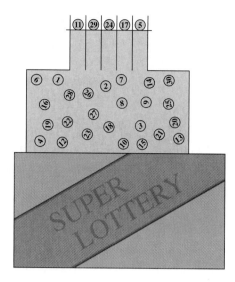

4. Molecular collisions are *perfectly elastic;* that is, no energy is lost when the molecules collide.

5. The average kinetic energy of the gas molecules is directly proportional to the absolute temperature of the gas:

$$KE = \frac{1.5RT}{N}$$

where R is the ideal gas law constant, T is the absolute temperature, and N is Avogadro's number. The value used for R is 8.31 J/mol · K, in units different from those used with the ideal gas law. Note that the kinetic energy depends only on temperature, and not on the composition of the gas.

The **random motion** referred to in postulate 1 means that the molecules travel in straight lines in any arbitrary direction until they hit other molecules or the walls of the container (Figure 12.11). A **perfectly elastic collision,** referred to in postulate 4, means that the molecules bounce off each other without any loss of total energy. No friction or energy loss of any kind occurs. (Bowling balls re-turning to the top of the alley often hit other bowling balls in almost perfectly elastic collisions.) The average kinetic energy of the molecules referred to in pos-tulate 5 does not mean that all the molecules travel with the same velocity at any one time or that one molecule travels with a constant velocity for any appreciable time. The **average kinetic energy** is the average of the kinetic energies of the individual molecules at any instant.

■ **EXAMPLE 12.28**

Calculate the volume occupied by 1.00 mol of H_2O at 100°C and 1.00 atm pressure when it is (a) in the vapor state and (b) in the liquid state (density = 0.958 g/mL).

Solution

(a) The ideal gas law (Section 12.5) yields the volume:

$$V = \frac{nRT}{P} = \frac{(1.00 \text{ mol})(0.0821 \text{ L} \cdot \text{atm/mol} \cdot \text{K})(373 \text{ K})}{1.00 \text{ atm}} = 30.6 \text{ L}$$

(b) The volume of liquid water is its mass divided by its density (Section 2.5):

$$1.00 \ mol\left(\frac{18.0 \ g \ H_2O}{1 \ mol \ H_2O}\right)\left(\frac{1 \ mL \ H_2O}{0.958 \ g \ H_2O}\right) = 18.8 \ mL$$

The same mole of H_2O occupies only 18.8 mL in the liquid state, but after it has been evaporated into the gas state, it occupies 30.6 L. As shown by the modern technique of electron diffraction, the molecules themselves do not expand. Since they do not expand, they cannot occupy more than 18.8 mL of the 30.6 L of the gas. Thus, most of the volume of the gas is composed of empty space! The volume occupied by the molecules is negligible (less than one-tenth of 1% of the total volume). ■

Why do gases—unlike liquids and solids—expand indefinitely to fill any container in which they are placed? The kinetic molecular theory explains this behavior by stating that the molecules move randomly. The molecules move until they hit other molecules or a wall of the container. The gas therefore is expanded to fully occupy the container. If the container is made bigger, the molecules simply travel farther until they hit a wall.

Where does gas pressure come from? The molecules' bombardment of the walls creates a continuous succession of forces on the walls. The total effect of those forces, divided by the area of the wall, is the gas pressure. Imagine a machine gun firing bullets very rapidly at a target. The target may be bent backward by the succession of impacts, whose effect is a continuing force on the target. Any force divided by an area is a pressure.

Why do gases mix with each other in any proportion? Gases are composed mostly of empty space anyway, and the molecules of one gas can interpenetrate the space occupied by another. Since there is no appreciable force between molecules, each molecule is unaffected by the presence of a molecule of a different substance in its vicinity.

How can Boyle's law be explained in terms of the kinetic molecular theory? For a simplified explanation, consider the cubic box pictured in Figure 12.12(a). Imagine one molecule of gas in the box, traveling between wall X and the wall opposite X. The molecule will strike wall X in a number of seconds, t, given by the equation

$$v = d/t \quad \text{or} \quad t = d/v$$

where d is distance and v is the velocity of the molecule. We now double the volume of the box by moving wall X twice as far from the wall opposite it (Figure 12.12b). We keep the kinetic energy of the molecule the same by keeping the temperature the same, so the molecule will have the same speed. It will take the

Figure 12.12 Effect of Doubling the Volume

(a) Cubic box. (b) Rectangular box with the same height and thickness but double the width—therefore, double the volume.

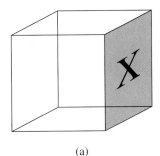

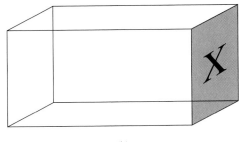

(a)   (b)

molecule twice as long to get to wall X each time because it has twice as far to go. In other words, the molecule will hit the wall half as often as it did in the original volume. Since the molecule is going as fast as it was originally, it will hit the wall just as hard each time. Since the number of collisions with the wall is halved, the force per unit time is halved, and the pressure is half of what it was before.

With many molecules in the box, if we ignore molecular interactions and assume that the molecules are moving perpendicularly to the walls, we find that the molecules hitting wall X and the wall opposite X do so half as often in the larger volume as in the smaller volume, resulting in half the pressure. The molecules moving perpendicularly to that path hit the other walls just as often and just as hard as they did before. But since the other walls have doubled in *area,* the *pressure* on those walls is halved—the same force divided by twice the area gives half the pressure. The net result is that doubling the volume of the box has halved the pressure, which is in accord with Boyle's law.

Charles' law is explained by noting that at higher temperature, molecules move faster and thus hit each other and the walls of the container more often and harder each time they hit. Each of these effects creates a greater force and thus a greater pressure. It can be shown quantitatively that the effect of an increase in absolute temperature is exactly what would be predicted by Charles' law.

■ ■ ■ ■ ■ ■ ■ **SUMMARY** ■ ■ ■ ■ ■ ■ ■

The physical properties of gases do not depend on the composition of the gas. Pressure is defined as force per unit area. The pressure of a gas may be measured with a simple barometer, and the usual units used for pressure are related to that apparatus. A torr is the pressure required to hold 1 mm of mercury vertically in the barometer, and a standard atmosphere is the pressure required to hold 760 mm Hg vertically in the barometer. (Section 12.1)

Boyle's law states that the volume of a given sample of gas at constant temperature is inversely proportional to its pressure. The law is often expressed in equation form as

$$P_1V_1 = P_2V_2 \quad \textit{(At constant temperature)}$$

If three of these quantities are known, the fourth may be calculated. In all gas law calculations, be sure to use the proper units for quantities. (Section 12.2)

Charles' law states that the volume of a given sample of gas at constant pressure is directly proportional to its *absolute* temperature. The Kelvin scale is used with all gas laws in which tempera-

ture is involved. A Kelvin (absolute) temperature is equal to the corresponding Celsius temperature plus 273. In equation form, Charles' law may be written as

$$\frac{V_1}{T_1} = \frac{V_2}{T_2} \quad \textit{(At constant pressure)}$$

If three of these quantities are known, the fourth can be calculated. (Section 12.3)

Boyle's and Charles' laws may be combined into one law related to a given sample of gas. This combined gas law may be expressed as:

$$\frac{P_1V_1}{T_1} = \frac{P_2V_2}{T_2}$$

If five of these quantities are given, the sixth may be calculated. Again, temperatures must be expressed in kelvins, and the units of pressure and of volume must be the same on both sides of the equation. Standard temperature and pressure, abbreviated STP, refers to 0°C and 1.00 atm. (Section 12.4)

The ideal gas law relates (at least approximately) the pressure, volume, number of moles, and temperature of any sample of any gas under any conditions: $PV = nRT$. The value of the constant R usually used is 0.0821 L·atm/mol·K. If any three of the four variables are given, the fourth may be calculated. Finding the number of moles of a gas is often especially useful. Since R is given in atmospheres and liters, these same units should be used for the pressure and volume, respectively. As always, the temperature must be in kelvins. (Section 12.5)

Gay-Lussac's law of combining volumes relates the volumes of gases involved in a reaction, all measured separately at the same temperature and pressure. The volume ratio under these conditions is equal to the mole ratio and, therefore, to the ratio of coefficients in the balanced chemical equation.

The molar mass of a gas may be calculated if the mass of a sample and the number of moles of the sample are both known. The ideal gas law may be used to determine the number of moles, from which the molar mass may be calculated. As introduced in Section 7.5, the molar mass, along with the empirical formula, may then be used to determine the molecular formula. (Section 12.6)

Dalton's law of partial pressures states that in a mixture of gases, the total pressure is equal to the sum of the pressures of the individual gases—the partial pressures. The partial pressures are directly proportional to the numbers of moles of the individual gases. The ideal gas law can be used for the mixture of gases as a whole or for any component in the mixture. Water vapor in a gas mixture, like any other gas in the mixture, obeys Dalton's law. However, the gaseous water will condense if the pressure of the water vapor is above a certain value, and if the mixture of gases is in contact with liquid water, the liquid water will evaporate when the pressure of the water vapor is below that value. The vapor pressure of water in contact with liquid water is thus a constant at any given temperature and depends only on temperature. (Section 12.7)

The kinetic molecular theory explains the behavior of gases in terms of characteristics of their molecules. It postulates that gases are made up of molecules that are in constant, random motion and whose sizes are insignificant compared to the total volume of the gas. Forces of attraction between the molecules are negligible, and when the molecules collide, the collisions are perfectly elastic. The average kinetic energy of the gas molecules is directly proportional to the absolute temperature:

$$KE = \frac{1.5RT}{N}$$

Application of these postulates explains the ability of gases to expand indefinitely, the existence of pressure, Boyle's and Charles' laws, and many other physical properties of gases. (Section 12.8)

Items for Special Attention

- Absolute (Kelvin) temperature must be used with all the gas laws in which temperature is a factor.

- The units must be the same on each side of the equation in all gas law calculations.

- In solving problems using Boyle's law, Charles' law, or the combined gas law, it does not make any difference which state is defined as the initial state and which is defined as the final state (see Problem 12.23).

- Note that the value of R in the units usually used with the ideal gas law equation has a zero to the right of the decimal point: 0.0821 L·atm/mol·K.

- All the gas laws are approximations for real gases. Only a hypothetical "ideal" gas follows the laws exactly.

- The volume ratios that correspond to mole ratios involve volumes of separate gas samples at the same temperature and pressure. In gas mixtures, the partial pressure ratios are equal to the mole ratios.

- In problems involving chemical reactions, be sure to use the ideal gas law only for substances that are gases.

- The gas laws deal with matter in the gaseous state. They have nothing to do with gasoline, a liquid.

- Molecules of a gas in a container are somewhat like the balls in a lottery machine. They are in motion and hit all the walls, and their volumes are much smaller than the volume of their container.

Self-Tutorial Problems

12.1 Solve the following equations for the unknown. (The integers are regarded as exact numbers, not measurements.)

(a) $\dfrac{1}{x} = \dfrac{(2)(7)}{(3)}$

(b) $\dfrac{1}{V} = \dfrac{(2)(7)}{(3)}$

(c) $\dfrac{x(27)}{(13)} = \dfrac{(7)(16)}{(2)}$

(d) $\dfrac{(2)(7)}{x} = \dfrac{(3)(10)}{(5)}$

(e) $\dfrac{(2.17)(7.71)}{x} = \dfrac{(3.21)(10.3)}{(5.15)}$

(f) $\dfrac{(7)(15)}{(3)} = \dfrac{x(17)}{(22)}$

(g) $\dfrac{(5)(19)}{(33)} = \dfrac{(2)(12)}{x}$

(h) $\dfrac{(5.17)(19.7)}{(33.3)} = \dfrac{(2.19)(12.5)}{x}$

12.2 How many significant digits are in the Kelvin scale temperature corresponding to 5°C?

12.3 On which temperature scale are there no negative temperatures?

12.4 Initially, a 3.00-L sample of gas is at 1.00 atm and 278 K. What is its final pressure if its final volume is 3000 mL and its final temperature is 5°C?

12.5 To which of the following substances, all at 0°C and 1.00 atm, do the gas laws apply?

CO N_2 H_2O Fe

12.6 To which of the following substances, all at 0°C and 1.00 atm, do the gas laws apply?

CO_2 O_2 $H_2O(g)$ $NH_3(aq)$

12.7 Show that the combined gas law equation simplifies to (a) the Boyle's law equation if the temperature is constant and (b) the Charles' law equation if the pressure is constant.

12.8 Starting with 2.00 L of a gas, what will be the final volume in each case?

(a) The volume is increased 3.00 L.

(b) The volume is increased to 3.00 L.

(c) The volume is increased by 3.00 L.

12.9 (a) What is the final volume of gas if the 2.00 L of gas is allowed to expand into the 3.00-L evacuated space, as shown in the accompanying figure?

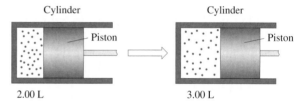

Initial   Final

2.00 L   3.00 L   2.00 L   3.00 L

(b) What is the final volume of the 2.00 L of gas in the cylinder if the piston is withdrawn, as shown in the accompanying figure?

Cylinder   Cylinder

— Piston   — Piston

2.00 L   3.00 L

12.10 If the Celsius temperature of a sample of gas is "doubled" from −10°C to −20°C at constant pressure, will its volume go up or down? Is Celsius temperature directly proportional to volume?

12.11 Compare the distance between molecules in a liquid and in a gas.

12.12 According to the kinetic molecular theory, what happens to gas molecules as a gas expands?

12.13 A tennis ball and a bowling ball both have the same energy—just enough to knock a certain box off a table. Which one is moving faster?

■ ■ ■ PROBLEMS ■ ■ ■

12.1 Gas Pressure

12.14 (a) What is the mass of a mercury column 1.00 cm² in cross-sectional area and 76.0 cm high? The density of mercury is 13.6 g/cm³.

(b) If the column in part (a) were 2.00 cm² in cross-sectional area, what would its mass be?

(c) Why does the column with smaller mass stand as high as the one with greater mass under the same atmospheric pressure?

12.15 What will happen if the gas in a cylinder has a greater pressure than the pressure of the piston on the gas? (The apparatus is shown in the figure for Problem 12.9[b].)

12.16 Calculate the pressure in torr corresponding to

    (a) 100.0 kPa

    (b) 1.00 Pa

    (c) 2.73×10^4 Pa

12.17 Change:

    (a) 4.00 atm to torr

    (b) 1.23 atm to mm Hg

    (c) 720 torr to atm

    (d) 920 torr to mm Hg

12.2 Boyle's Law

12.18 A sample of gas is compressed at constant temperature from 0.395 L at 1.14 atm to 197 mL.

    (a) Change the number of liters to milliliters.

    (b) Calculate the final pressure of the sample.

12.19 A sample of gas is compressed at constant temperature from 0.712 L at 2.11 atm to 557 mL. Calculate the final pressure of the sample.

12.20 Calculate the final pressure of a sample of gas that is changed at constant temperature to 1.17 L from 2.71 L at 781 torr.

12.21 Calculate the final pressure of a sample of gas that is changed at constant temperature to 2.19 L from 1.88 L at 1.41 atm.

12.22 (a) Calculate the final pressure of a sample of gas that is expanded to 2.26 L at constant temperature from 1.75 L at 770 torr.

    (b) Calculate the initial pressure of a sample of gas compressed at constant temperature to 1.75 L at 770 torr from 2.26 L.

12.23 (a) Calculate the final pressure of a sample of gas that is expanded to 29.5 L at constant temperature from 15.0 L at 681 torr.

    (b) Calculate the initial pressure of a sample of gas compressed at constant temperature to 15.0 L at 681 torr from 29.5 L.

12.24 Calculate the pressure required to change a 716-mL sample of gas at 777 torr and 25°C to 0.899 L at 25°C.

12.25 Calculate the pressure required to change a 213-mL sample of gas at 828 torr and 100°C to 1.27 L at 100°C.

12.26 (a) Using Figure 12.2(a), estimate the volume that the sample of gas would occupy at 6.00 atm and the given temperature.

    (b) Can you do the same estimation for a pressure of 12.0 atm? Explain.

12.27 Using Figure 12.2(b), estimate the values of $1/V$ and V at 12.00 atm.

12.28 Calculate the final pressure required to increase the volume of a 2.00-L sample of gas initially at 1.00 atm (a) by 4.00 L and (b) to 4.00 L.

12.3 Charles' Law

12.29 Plot the following data, and extrapolate to zero volume to estimate the value of absolute zero on the Celsius scale:

V (L)	t (°C)
1.23	10
1.45	60
1.66	110
1.88	160
2.10	210

12.30 Convert the temperatures of Table 12.1 to Fahrenheit, plot those versus the volumes, and determine the value of absolute zero on the Fahrenheit scale.

12.31 A sample of gas is heated at constant pressure from 1.0°C to 3.0°C.

    (a) Will the volume increase to three times its original volume?

    (b) If not, by what ratio will the volume increase?

12.32 A sample of gas is cooled at constant pressure from 2.0°C to 1.0°C.

    (a) Will the volume decrease to half its original volume?

    (b) If not, by what ratio will the volume decrease?

12.33 Calculate the final volume at 441 K of a 6.11-L sample of gas originally at 303 K if the pressure does not change.

12.34 Calculate the final volume at 617 K of a 7.17-L sample of gas originally at 381 K if the pressure does not change.

12.35 Calculate the final volume at 344°C of a 7.17-L sample of gas originally at 108°C if the pressure does not change.

12.36 Calculate the initial volume at 317 K of a sample of gas that is changed to 1.18 L by cooling to 302 K at constant pressure.

12.37 Calculate the final volume at 55°C of a 673-mL sample of gas that is originally at 20°C, assuming that the pressure does not change.

12.38 Calculate the initial volume at 70°C of a sample of gas that is changed to 1.71 L by cooling to 35°C at constant pressure.

12.4 The Combined Gas Law

12.39 Complete the following table:

	V_1	P_1	T_1	V_2	P_2	T_2
(a)	1.11 L	1.33 atm	303 K	3.00 L	1.71 atm	_____
(b)	6.17 L	762 torr	310 K	10.5 L	_____	310 K
(c)	1.79 L	6.11 atm	300 K	_____	6.11 atm	331 K
(d)	559 mL	777 torr	30°C	_____	2.00 atm	30°C
(e)	1.92 L	1.14 atm	288 K	1.18 L	_____	314 K
(f)	1.97 L	2.11 atm	_____	918 mL	3.09 atm	292 K
(g)	10.1 L	767 torr	25°C	4.10 L	1.10 atm	_____
(h)	0.973 L	_____	312 K	1.92 L	821 torr	270 K

12.40 Complete the following table:

	V_1	P_1	T_1	V_2	P_2	T_2
(a)	10.1 mL	1.20 atm	293 K	30.0 mL	1.20 atm	____
(b)	6.06 L	1.19 atm	308 K	4.11 L	____	317 K
(c)	127 mL	1.71 atm	300 K	____	1.50 atm	299 K
(d)	1.11 L	710 torr	25°C	____	4.01 atm	25°C
(e)	0.945 L	2.04 atm	273 K	1.25 L	____	300 K
(f)	255 mL	1.89 atm	____	1.02 L	6.18 atm	40°C
(g)	1.73 L	1.08 atm	22°C	1.77 L	2.11 atm	____
(h)	91.0 L	____	319 K	82.1 L	1.61 atm	42°C

12.41 Calculate the final pressure of a gas that is expanded from 725 mL at 30°C and 1.19 atm to 1.12 L at 43°C.

12.5 The Ideal Gas Law

12.42 Solve the problems in Example 12.13.

12.43 Solve the problem in Example 12.13(a), using the ideal gas law.

12.44 Calculate the value of R in (a) $L \cdot torr/mol \cdot K$ and (b) $mL \cdot atm/mol \cdot K$.

12.45 Determine the number of moles of gas in a volume of 2.33 L at 339 K and 780 torr.

12.46 Determine the volume of 0.114 mol of N_2 at 30°C and 781 torr.

12.47 Determine the number of moles of gas in a volume of 750 mL at 321 K and 795 torr.

12.48 Determine the volume of 0.200 mol of $NH_3(g)$ at 27°C and 1.13 atm.

12.49 Determine the pressure of 10.2 g of CO_2 gas in a volume of 17.7 L at 299 K.

12.50 Determine the number of moles of gas in a volume of 2.92 L at 319 K and 792 torr.

12.51 (a) Determine the temperature of a gas if 1.22 mol occupies 9.93 L at 1.01 atm.

(b) Is the gas more likely to be He or H_2O? Explain.

12.52 Determine the pressure of 0.0153 mol of CO_2 gas in a volume of 1.50 L at 285 K.

12.53 Determine the number of moles of gas in a volume of 1.33 L at 355 K and 1.12 atm.

12.54 Determine the temperature of a gas if 0.903 mol occupies 12.7 L at 921 torr.

12.6 Gases in Chemical Reactions

12.55 Explain why it is so important to write the correct formula for an elemental gas that exists as diatomic molecules.

12.56 How many moles of mercury(II) oxide are required to produce 750 mL of oxygen gas at 1.00 atm and 23°C?

12.57 What volume of hydrogen at 301 K and 1.00 atm can be produced by the reaction of 4.50 g of zinc with hydrochloric acid?

12.58 Determine the number of moles of aqueous sodium hydroxide required to react with 1.32 L of carbon dioxide gas at 325 K and 1.03 atm to form sodium carbonate.

12.59 How many moles of mercury(II) oxide are required to produce 5.00 L of oxygen gas at 1.00 atm and 17°C?

12.60 What ratio of volumes (all at 20°C and 1.00 atm) would be involved in the reaction represented by the following equation?

$$2\ O_2(g) + N_2(g) \rightarrow 2\ NO_2(g)$$

12.61 How many moles of aluminum metal are required to produce 1.24 L of hydrogen gas at 1.00 atm and 23°C by reaction with HCl?

12.62 What ratio of volumes (all at 18°C and 1.00 atm) would be involved in the reaction represented by the following equation?

$$4\ NH_3(g) + 5\ O_2(g) \rightarrow 4\ NO(g) + 6\ H_2O(g)$$

12.63 What volume of ammonia at 500 K and 2.00 atm can be produced by the reaction of 1.00 ton (1.00×10^6 g) of hydrogen gas with nitrogen gas?

12.64 Calculate the volume of CO_2 (measured after cooling to 25°C at 1.00 atm) that will be liberated by heating 2.75 g of $NaHCO_3$.

12.65 What volume of ammonia at 500 K and 8.00 atm can be produced by the reaction of 450 kg of hydrogen gas with nitrogen gas?

12.66 Calculate the volume of CO_2 (measured after cooling to 25°C at 1.00 atm) that will be liberated by heating 3.00 g of $Ba(HCO_3)_2$.

12.67 What mass of O_2 is required to react with H_2 to produce 16.0 L of $H_2O(g)$ at 122°C and 0.998 atm?

12.68 What mass of H_2 is required to react with O_2 to produce 1.17 L of $H_2O(g)$ at 100°C and 751 torr?

12.69 When 3.00 mol of H_2 reacts with 1.00 mol of N_2, 2.00 mol of NH_3 is formed.

(a) Write a balanced chemical equation for the reaction.

(b) Calculate the volumes of 3 mol of H_2, 1 mol of N_2, and 2 mol of NH_3 at STP.

(c) Compare the ratio of the volumes of the gases to the coefficients in the balanced equation. What can be stated about the volumes of gases involved in a reaction when they are *at the same temperature and pressure?*

12.70 Repeat Problem 12.69 at 25°C and 2.00 atm pressure. Are the results the same—that is, are the volume ratios the same as the mole ratios in the balanced chemical equation?

12.71 Repeat Problem 12.69 with any other given temperature and pressure. Are the results the same—that is, are the volume ratios the same as the mole ratios in the balanced chemical equation?

12.7 Dalton's Law of Partial Pressures

12.72 A mixture of gases exists in which the partial pressures of all the components are the same. What can be said about the numbers of moles of the components?

12.73 A gaseous mixture contains 1.00 mol of He, 1.00 mol of Ne, and 1.00 mol of Ar. The total pressure of the mixture is 1.50 atm. What is the partial pressure of each gas?

12.74 A gaseous mixture contains 0.773 mol of H_2, 0.271 mol of N_2, and 0.528 mol of Ne. The total pressure of the mixture is 2.13 atm. What is the partial pressure of each gas?

12.75 A gaseous mixture contains 0.616 mol of H_2, 0.173 mol of N_2, and 0.291 mol of Ne. The partial pressure of the hydrogen is 0.118 atm. What are the partial pressures of the nitrogen and the neon?

12.76 A gaseous mixture contains 0.600 mol of H_2, 0.200 mol of N_2, and 1.00 mol of Ne. The partial pressure of the nitrogen is 0.517 atm. What are the partial pressures of the hydrogen and the neon?

12.77 Oxygen gas is standing over water at a total pressure of 770 torr. The partial pressure of the oxygen is found to be 743 torr. Determine the temperature of the system.

12.78 Oxygen gas is standing over water at a total pressure of 770 torr at 27°C. Determine the partial pressure of the oxygen.

12.79 What volume will 2.00 g of hydrogen occupy when collected over water at 25°C and 1.00 atm barometric pressure?

12.80 What volume will 2.00 g of O_2 occupy when collected over water at 25°C and 1.00 atm barometric pressure?

12.81 What mass of $KClO_3$, must be decomposed to produce 1.00 L of O_2, collected over water, at 22°C and a barometric pressure of 771 torr?

12.82 What volume of O_2 can be collected over water at 25°C and 1.02 atm barometric pressure from thermal decomposition of 0.255 g of $KClO_3$?

12.83 What mass of $KClO_3$ must be decomposed to produce 551 mL of O_2, collected over water, at 25°C and a barometric pressure of 762 torr?

12.84 What volume of O_2 can be collected over water at 22°C and 1.00 atm barometric pressure from thermal decomposition of 0.717 g of $KClO_3$?

12.8 Kinetic Molecular Theory of Gases

12.85 Calculate the percentage of the 30.6 L occupied by the molecules of the 1.00 mol of water vapor at 1.00 atm and 100°C, as calculated in Example 12.28, assuming that the molecules of liquid water actually occupy all the 18.8-mL volume.

12.86 Use the kinetic molecular theory to explain why the number of moles of gas enclosed in a certain container at a certain temperature is directly proportional to the gas pressure.

12.87 Imagine a sheet of paper suspended by one edge and hanging vertically. What would happen if you shot a rapid succession of peas from a peashooter at the paper? Compare the impact of the peas to the impact of molecules of a gas bombarding a wall.

12.88 Use the kinetic molecular theory to explain why Dalton's law of partial pressures works.

■ ■ ■ GENERAL PROBLEMS ■ ■ ■

12.89 Which parts of Problem 12.39 may be solved with Boyle's law or with Charles' law?

12.90 Calculate the volume of a 1.00-L sample of gas after its pressure is halved and its absolute temperature is increased 20.0%.

12.91 A gaseous mixture contains 3.50 L of helium, some argon at 1.03 atm, and some neon at 27°C. For which gas can the number of moles be calculated?

12.92 A 10.0-L steel vessel that holds a sample of oxygen gas at 25°C and 1.00 atm develops a leak, and 6.00 g of oxygen escapes before the leak is repaired.

  (a) Calculate the initial number of moles of oxygen present.

  (b) Calculate the number of moles that escape.

  (c) Calculate the pressure of oxygen in the vessel, still at 25°C, after the leak is repaired.

12.93 A 10.0-L steel vessel that holds a sample of oxygen gas at 25°C and 1.00 atm develops a leak, and 6.00 g of oxygen escapes before the leak is repaired. After the leak is repaired, what is the pressure of oxygen in the vessel, if the temperature is still 25°C?

12.94 Determine the formula mass of a gaseous substance if 4.16 g occupies 1.38 L at 2.20 atm and 12°C.

12.95 Determine the formula of a gaseous substance if 4.20 g occupies 22.0 L at 2.20 atm and 10°C.

12.96 A 1.06-g sample of a gaseous hydrocarbon occupies 0.869 L at 23°C and 752 torr. The gas is composed of 79.89% C and 20.11% H. What is the molecular formula of the gas?

12.97 A 0.890-g sample of a gaseous hydrocarbon occupies 0.782 L at 23°C and 752 torr. The gas is composed of 85.63% C and 14.37% H. What is the molecular formula of the gas?

12.98 What are the volume and the pressure for helium, neon, and the mixture in the following experiments?

  (a) Samples of 0.200 mol of He and 0.100 mol of Ne are placed in identical 1.00-L containers at 25°C.

  (b) Samples of 0.200 mol of He and 0.100 mol of Ne are placed in a single 1.00-L container at 25°C.

12.99 (a) Using Boyle's law and Charles' law, show that the *pressure* of a gas is directly proportional to its absolute temperature at *constant volume*.

  (b) Repeat part (a), using the combined gas law instead.

12.100 A 10.0-L steel vessel that holds a sample of oxygen gas at 25°C and 748 torr develops a leak, and 5.55 g of oxygen escapes before the leak is repaired. After the leak is repaired, what is the pressure of oxygen in the vessel, if the temperature is still 25°C?

13 Atomic and Molecular Properties

- ■ 13.1 Atomic and Ionic Sizes
- ■ 13.2 Ionization Energy and Electron Affinity
- ■ 13.3 Electronegativity and Bond Polarity
- ■ 13.4 Molecular Shape
- ■ 13.5 Polar and Nonpolar Molecules
- ■ 13.6 Intermolecular Forces

■ **Key Terms** (*Key terms are defined in the Glossary.*)

angular molecule (13.4)
atomic size (13.1)
bent molecule (13.4)
central atom (13.4)
dipole (13.5)
dipole moment (13.5)
electron affinity (13.2)
electron group (13.4)
electronegativity (13.3)

hydrogen bonding (13.6)
intermolecular force (13.6)
ionic size (13.1)
ionization energy (13.2)
linear molecule (13.4)
molecular shape (13.4)
nonlinear molecule (13.4)
nonpolar bond (13.3)
nonpolar molecule (13.5)

polar bond (13.3)
polar molecule (13.5)
second ionization energy (13.2)
sublimation (13.6)
tetrahedral molecule (13.4)
third ionization energy (13.2)
trigonal planar molecule (13.4)
trigonal pyramidal molecule (13.4)
van der Waals force (13.6)

■ Symbol

δ (delta) (13.3)

This chapter examines many aspects of atomic and molecular properties. These properties affect the properties of the substances that the atoms and molecules constitute. We begin with the sizes of atoms and ions, especially the periodic variation of sizes (Section 13.1). We then discuss the effects the sizes have on the ability of the gaseous atom to lose or gain electrons—ionization energy and electron affinity, respectively (Section 13.2). From that more theoretical discussion of gaseous atoms and ions, we progress to electronegativity, a more practical, semiquantitative measure of the electron-attracting ability of atoms in molecules or ionic substances (Section 13.3). The atom's varying ability to attract electrons in molecules leads to polar and nonpolar covalent bonds (Section 13.3). Section 13.4 discusses molecular shapes. Depending on the molecular shapes, polar bonds can lead to dipolar molecules (Section 13.5). Intermolecular forces, the small forces between molecules that result from molecular polarity and other causes, are discussed in Section 13.6.

These discussions will allow us to consider the nature of the solid and liquid states in Chapter 14.

13.1 Atomic and Ionic Sizes

OBJECTIVE

■ to predict how changes in the numbers of protons, electrons, or both, affect the size of the resulting atom or ion, and to deduce the relative sizes of atoms from their positions in the periodic table

Before we discuss the sizes of atoms and ions, let us review briefly some basic principles:

1. Oppositely charged particles attract each other.
2. Like charged particles repel each other.
3. The closer the two charged bodies, the stronger the force of attraction or repulsion.
4. Electrons are negative and lie outside the nucleus, while protons are positive and lie within the nucleus.

The greater the number of electrons in an atom or ion, the larger the **atomic or ionic size.** The more electrons, the greater the electronic repulsion. With the same nuclear charge but greater electronic repulsion, the electrons occupy a larger volume of space.

■ **EXAMPLE 13.1**

Compare the sizes of Fe, Fe^{2+}, and Fe^{3+}.

Solution

In each case, the nucleus has 26 protons and, therefore, 26 positive charges. This quantity of positive charge can attract the 23 electrons in Fe^{3+} better than the 24 electrons in Fe^{2+} or the 26 electrons in Fe. Hence, Fe is the largest, and Fe^{3+} is the smallest.

Practice Problem 13.1   Compare the sizes of S and S^{2-}. ■

Understanding the size relationships is equally easy if the number of electrons is constant in several species and the number of protons changes. The greater the positive charge in the nucleus, the greater is the attraction for the set of electrons and, therefore, the smaller is the atom or ion.

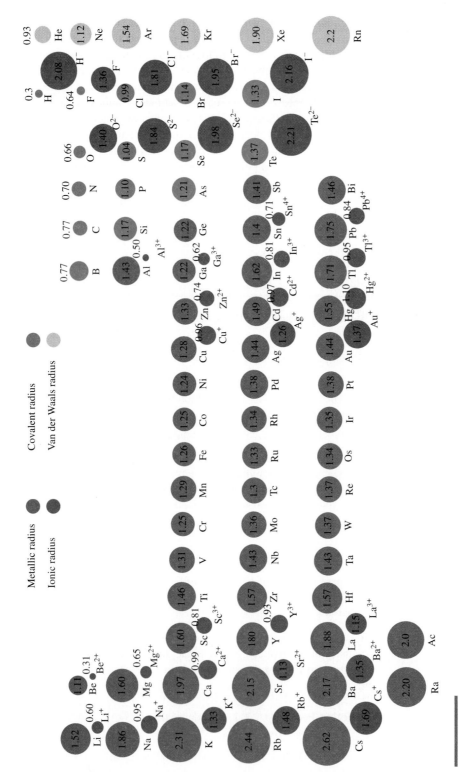

Figure 13.1 *Periodic Table Showing the Relative Sizes of Atoms and Ions (atomic and ionic radii in Ångström units; 1 Å = 10^{-10} m = 0.1 nm)*

The data include metallic radii, ionic radii, covalent radii, and van der Waals radii. While the various types of radii cannot be directly compared, the figure does illustrate the periodic trends in sizes.

■ EXAMPLE 13.2

Compare the sizes of F^-, Ne, and Na^+.

Solution

Each of these atoms or ions has 10 electrons. The greater positive charge in the Na^+ nucleus (11 protons) has a greater pull on the electrons than do the smaller charges in Ne (10 protons) and F^- (9 protons). Thus, Na^+ is the *smallest* and F^- is the *largest* of these three.

Practice Problem 13.2 Compare the sizes of S^{2-}, Ar, and Ca^{2+}. ■

Predicting size variations of neutral atoms of neighboring elements in the periodic table, in which both the number of protons and the number of neutrons change, is more difficult. Increasing the number of protons, which tends to make the atom or ion smaller, is more important than increasing the number of electrons, which tends to make the atom or ion larger, *except when the last electron starts a new shell of electrons.* That is, as we compare neighboring elements in a given period of the periodic table, the atoms get somewhat smaller as we go to the right. Starting a new shell of electrons (at each alkali metal) causes a large *increase* in size, so that in going down a group, the atoms get larger. Figure 13.1 shows a representation of the actual sizes of atoms and ions, and Figure 13.2 indicates the size variation of the neutral atoms.

■ EXAMPLE 13.3

Which atom in each of the following groups is largest?

(a) Na   Mg   Al   (b) Na   K   Rb   (c) Mg   Ca   Sr

Solution

(a) Na (The atoms get smaller as we go to the right.)

(b) Rb (The atoms get larger as we go down the group.)

(c) Sr (The atoms get larger as we go down every group, not only the alkali metal group.)

Figure 13.2 Periodic Variation of Atomic Size

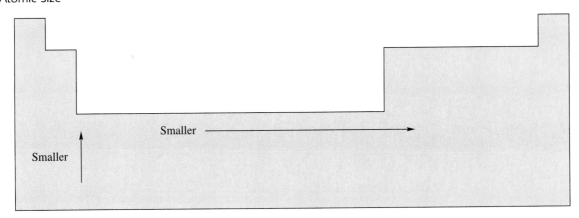

Practice Problem 13.3   Which atom in each of the following groups is largest?

(a) C   Si   Ge   (b) Ca   Zn   Br ■

13.2 Ionization Energy and Electron Affinity

OBJECTIVE

■ to deduce the relative ionization energies of different elements from their positions in the periodic table, and to distinguish (1) a combination of ionization energy and electron affinity of two elements from (2) the energy of reaction of the elements in their standard states

The **ionization energy** of an element is defined as the energy *required* to remove an electron from a *gaseous* atom to produce a *gaseous* cation. For example,

$$Na(g) \xrightarrow[\text{Ionization energy}]{} Na^+(g) + e^-$$

Note that this is *not* the energy for the more familiar reaction of solid sodium to produce sodium ions in a lattice or in solution.

The ionization energy of the elements varies periodically. In general, as an atom gets larger (and the outermost electron gets farther away from the nucleus), the energy needed to remove the electron decreases. Figure 13.3 shows ionization energies, and the periodic variation of ionization energies is represented in Figure 13.4.

■ **EXAMPLE 13.4**

Which atom in each of the following groups has the largest ionization energy?

(a) Na   Al   Cl   (b) Cl   Br   I

Solution

(a) Cl (The ionization energy gets larger as we go to the right.)

(b) Cl (The ionization energy gets smaller as we go down the group.)

Practice Problem 13.4   Which atom in each of the following groups has the largest ionization energy?

(a) Be   C   F   (b) N   P   As ■

Figure 13.3 First Ionization Energy As a Function of Atomic Number

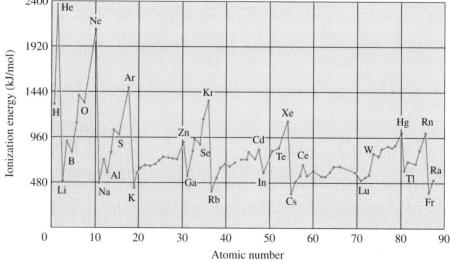

ENRICHMENT

While the general trend is for ionization energies to increase toward the right, some small exceptions exist. Many of these stem from the extra stability of a half-filled subshell of electrons.

The **second ionization energy** of an atom is the energy required to remove the second electron from the monopositive gaseous ion; the **third ionization energy** is the energy required to remove the third electron from the dipositive ion; and so on. For every element, the second ionization energy is higher than the first because it is harder to remove an electron (a negative particle) from a positively charged body than from the corresponding neutral one. The ionization energy increases dramatically as the octet of electrons is disrupted (see Table 13.1). The term *ionization energy* always means first ionization energy unless otherwise specified.

Table 13.1 Successive Ionization Energies for Some Representative Elements (kilojoules per mole)

	First	Second	Third
Na	495.8	4562	6912
Mg	737.7	1451	7733
Al	577.6	1817	2745

■ EXAMPLE 13.5

Predict which of the first three ionizations of calcium has the greatest *increase* in ionization energy from the preceding ionization.

Solution

The third ionization energy of calcium is much greater than the second because the octet of electrons is broken in ionizing that electron. (The actual values for the first three ionization energies are 589.8 kJ/mol, 1145 kJ/mol, and 4912 kJ/mol, respectively.)

Practice Problem 13.5   Predict which of the first three ionizations of lithium has the greatest *increase* in ionization energy from the preceding ionization. ■

Positive electron affinity values signify that energy is *liberated,* in contrast to ionization energy values and values of other variables to be introduced in Chapter 14.

Electron affinity is defined as the energy *liberated* when an electron is added to a *gaseous atom* to form a *gaseous anion*. For example,

$$Cl(g) + e^- \xrightarrow[\text{Electron affinity}]{} Cl^-(g)$$

Note that this energy is *not* the energy for the more familiar reaction of gaseous chlorine *molecules* to form chloride ions in a solid lattice or in solution.

Table 13.2 presents the electron affinities of a number of elements. The electron affinities of some elements are negative, meaning that energy is *required* to add an electron to the gaseous atoms. Periodic variation of electron affinity is

Figure 13.4 Periodic Variation of Ionization Energy

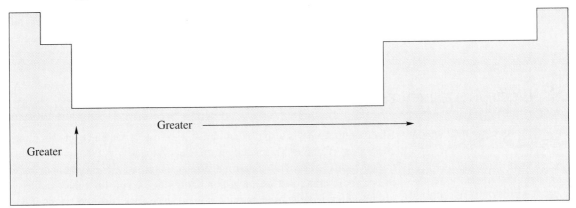

Greater

Greater

Table 13.2   Electron Affinities of Some Elements (kilojoules per mole)

H	72.6	K	48.2	Te	189.8
He	−21	Ca	−156	I	294.8
Li	59.7	Cr	64	Xe	−40
Be	−240	Ni	111	Cs	45.5
B	23	Cu	122.9	Ba	−52
C	122.3	Ga	36	Ta	80
N	0.0	Ge	116	W	50
O	141.1	As	77	Re	14
F	321.5	Se	194.5	Pt	204.9
Ne	−29	Br	323.9	Au	222.4
Na	52.8	Kr	−39	Tl	50
Mg	−230	Rb	46.8	Pb	101
Al	44	Sr	−168	Bi	101
Si	119	Mo	96	Po	173
P	74	Ag	125.5	At	270
S	200.0	In	34	Rn	−40
Cl	348	Sn	120	Fr	43.9
Ar	−35	Sb	101		

more difficult to rationalize than is the variation of ionization energy, but note that the halogens have the greatest electron affinities of all the elements.

■ EXAMPLE 13.6

Determine from Table 13.2 which periodic groups of elements generally require energy when an electron is added to their atoms.

Solution

The noble gases and alkaline earth metals—they are the only groups of elements with negative (first) electron affinities. ■

If we combine the ionization energy of sodium with the electron affinity of chlorine, we deduce that an overall energy input is required for the reaction

$$Na(g) + Cl(g) \rightarrow Na^+(g) + Cl^-(g)$$

because 496 kJ is *required* to remove the electrons from a mole of sodium atoms and only 348 kJ is *produced* when these electrons are added to a mole of chlorine atoms. This conclusion emphasizes that we are talking about *gaseous* atoms and ions. (When the energy to form a solid lattice of NaCl from the ions is taken into account [Chapter 14], the process of combining sodium and chlorine produces energy.)

13.3   Electronegativity and Bond Polarity

■ to use electronegativity differences to predict the bond types between atoms of the elements

Electronegativity is a semiquantitative measure of the electron-attracting ability of an atom (Chapter 5). The larger the electronegativity, the greater is the pull on the electrons. In general, electronegativity increases to the right and to the top of the periodic table, just as ionization energy does, but the lighter noble gases do not have defined electronegativity values, since they form no bonds. Electronegativity values for the main group elements are shown in Figure 5.1.

Table 13.3 Approximate Electronegativity Differences for Various Bond Types

Electronegativity Difference	Bond Type
1.2 or more	Ionic bond
0.3 to 1.0	Polar covalent bond
0.0 to 0.2	Nonpolar covalent bond

The charges on the atoms in a polar covalent bond are not full charges.

Electronegativity is useful in predicting the nature of the bond expected to form between two atoms (Table 13.3). If the electronegativity values are very different, an ionic bond is expected. If the electronegativity values are the same, a covalent bond will form with an equally shared pair of electrons. Such a bond is said to be **nonpolar.** For example, the bond in Cl_2 is nonpolar:

$$:\ddot{Cl}:\ddot{Cl}:$$

When electrons are shared between atoms of different elements, the electronegativities (the attractions of the atoms for the electrons) are generally not equal. Therefore, the electrons are not shared equally. (Since they are still shared, however, the bond is covalent.) The bond is said to be a **polar bond.** This unequal sharing of electrons may cause one end of the molecule to be *slightly* positive and the other end *slightly* negative. For example, in HCl, the bonding electrons are drawn *somewhat* more toward the chlorine atom than toward the hydrogen atom, yielding

$$^{\delta+}H \quad :\ddot{Cl}:^{\delta-}$$

These are not full ionic charges; there is about one-sixth of a single positive charge on the hydrogen atom in HCl and the same amount of negative charge on the chlorine atom. The *partial* nature of these charges is denoted by the lowercase Greek delta (δ). (Other polar bonds have partial charges of different magnitudes.) Polar bonds caused by unequal sharing of electrons can lead to polar molecules (see Section 13.5).

■ **EXAMPLE 13.7**

What type of bond would be formed between each of the following pairs of atoms?

(a) Mg and S   (b) C and Cl   (c) Br and Br

Solution

(a) The electronegativity difference is $2.5 - 1.2 = 1.3$, so the bond should be ionic.

(b) The electronegativity difference is $3.0 - 2.5 = 0.5$, so the bond should be polar covalent.

(c) The electronegativity difference is $2.8 - 2.8 = 0.0$, so the bond should be nonpolar covalent.

Practice Problem 13.7   What type of bond would be formed between each of the following pairs of atoms?

(a) P and Cl   (b) Ba and O ■

13.4 Molecular Shape

■ to predict the shape of a molecule or ion from its electron dot diagram

We discuss the shapes of molecules next, before we examine the effect of polar bonds on the properties of molecules as a whole. Both the bond polarity and the shape can have a distinct effect on the molecule's properties.

Molecular shape is defined by the locations of a molecule's *atoms,* not from the location of its electrons, because, experimentally, the scientist can determine the location of the atoms by X-ray diffraction and other studies. We can predict the shapes of molecules from their formulas by learning a few simple rules. The ability to make such predictions enables us to understand better the properties of the substances.

To predict the shape of a molecule from its formula, we first draw an electron dot diagram of the molecule (Section 5.4). We assume that the **electron groups** around the **central atom** are as far apart as possible while remaining attached to the central atom. The electron groups are ordinarily the electrons in the single, double, or triple bonds, or in the lone pairs on the central atom, which is the atom to which the other atoms are attracted. Then we bond the other atoms, using some or all of the electron groups, and finally describe the molecule by the locations of the atoms.

As a first example of the process, we consider the molecule BeH_2, experimentally known to be a linear molecule with two identical bonds. The electron dot diagram has only four electrons around the beryllium, two from the outermost shell of that atom and one each from the two hydrogen atoms:

$$:Be:$$

We place the electron pairs as far apart as possible (180°) while still being attached to the Be atom. Then we add the two hydrogen atoms to form the molecule:

$$H:Be:H$$

We have deduced that the molecule is linear, since all three atoms lie on a straight line.

Next, we consider BF_3. The B atom has three outermost electrons to contribute. The three fluorine atoms each contribute their one unpaired electron, which makes a total of six electrons available. These are distributed in three pairs symmetrically about the B atom:

We add the three fluorine atoms and again get a symmetrical molecule:

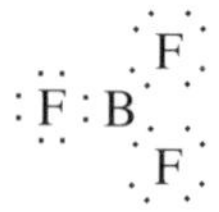

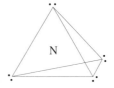

In this case, we call the molecular shape **trigonal planar,** since the triangle of fluorine atoms and the boron atom all lie in the same plane.

Our next example is CH_4, methane. The four outermost carbon electrons plus the four electrons from the hydrogen atoms total eight. These are distributed in four pairs as far apart as possible. In this case, the distribution is toward the corners of a tetrahedron (Figure 13.5a). Note that the electrons are not limited to a single plane. The addition of the hydrogen atoms produces a **tetrahedral molecule** (Figure 13.5b and c).

Ammonia, NH_3, is our next case. The five outermost electrons in the nitrogen atom, plus the three electrons from the hydrogen atoms, again make eight. Once more, these are distributed toward the corners of a tetrahedron:

This time, however, only three hydrogen atoms are to be attached. While the *electrons* are located toward the corners of a tetrahedron, the molecular shape is

Figure 13.5 Tetrahedral
Orientation of Electron Pairs
and Hydrogen Atoms

(a) Tetrahedral orientation of the
electron pairs around the carbon
atom. (b) A tetrahedral molecule is
produced when the four hydrogen
nuclei are attached. (c) The molecule
is usually shown as a ball-and-stick
model, where each stick represents
an electron pair.

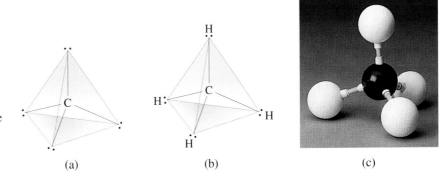

(a)   (b)   (c)

called **trigonal pyramidal,** not tetrahedral, because the *atoms* lie at the corners
of a triangular pyramid:

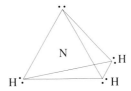

■ EXAMPLE 13.8

Deduce the electronic arrangement and the molecular shape of the water molecule.

Solution

The eight valence electrons, six from oxygen and two from hydrogen, are oriented toward the corners of a tetrahedron. Adding hydrogen atoms to any two of
those electron pairs results in a nonlinear molecule. The shape of water is described as **nonlinear, angular,** or **bent.**

Practice Problem 13.8   Determine the shapes of (a) PCl_3 and (b) SF_2. ■

> The electrons in each multiple
> bond are considered to be one
> group of electrons.

Compounds with multiple bonds are only slightly different. The electrons in each
multiple bond are considered to be one group of electrons.

■ EXAMPLE 13.9

Determine the shape of the carbon dioxide molecule.

Solution

The four electrons of carbon and the six from each oxygen atom make sixteen total. The electron dot diagram for CO_2 is

$$:\!\ddot{O}\!::\!C\!::\!\ddot{O}\!:$$

The two double bonds are considered two electron groups, and they arrange
themselves as far apart as possible: 180°. The addition of the oxygen atoms results in a **linear molecule:**

$$:\!\ddot{O}\!::\!C\!::\!\ddot{O}\!:$$

Table 13.4   Orientations of Electron Groups and Shapes of Molecules

Example	Number of Groups of Electrons	Number of Attached Atoms	Orientation of Electron Groups	Shape of Molecule
BeH_2	2	2	Linear	Linear
BF_3	3	3	Trigonal planar	Trigonal planar
CH_4	4	4	Tetrahedral	Tetrahedral
NH_3	4	3	Tetrahedral	Trigonal pyramidal
H_2O	4	2	Tetrahedral	Nonlinear
CO_2	2	2	Linear	Linear
H_2CO	3	3	Trigonal planar	Trigonal planar

Practice Problem 13.9   Determine the geometry of the F_2CO molecule.   ■

Table 13.4 presents several typical molecules with the orientations of their electron groups and their molecular shapes.

ENRICHMENT

If a molecule has two or more central atoms, the geometry around each is deduced separately. The orientation of one central atom to another is rather arbitrary if the central atoms are connected by single bonds.

■ EXAMPLE 13.10

Deduce the geometry of NH_2NH_2.

Solution

The orientation about each nitrogen atom is trigonal pyramidal, as deduced previously for ammonia. The hydrogen atoms on one nitrogen atom can be oriented at any angle with respect to those on the other, as shown in Figure 13.6.

Practice Problem 13.10   Deduce the geometry of NH_2OH. Draw the molecule in several orientations.   ■

13.5 Polar and Nonpolar Molecules

OBJECTIVE

■ to predict the polarity of a molecule by deduction from the orientations of its polar bonds

A molecule with its center of positive charge in a location different from its center of negative charge is said to be a **polar molecule** or to have a **dipole** or a **dipole moment.** One effect of such asymmetry of charge is intermolecular attraction (Section 13.6). Where do these dipoles come from? Do all molecules with polar bonds (Section 13.3) have dipoles?

Figure 13.6 Examples of Various Possible Orientations of One NH₂ Group with Respect to Another

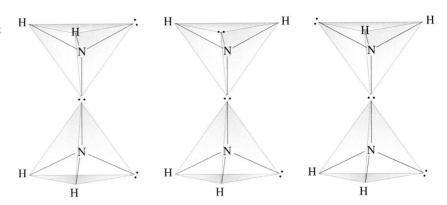

Care must be exercised in deciding whether molecules have dipoles. Every molecule with only one polar bond will be a polar molecule. In some molecules with more than one identical polar bond, the effect of one bond will exactly off-set the effect of the other(s), and the molecule as a whole will have no dipole.

In diatomic molecules, if the (only) bond is polar, the molecule as a whole has an asymmetric electronic distribution. For example, HCl has an unequally shared electron pair:

Partial positive charge   Partial negative charge

$$\delta+ \quad \delta-$$
$$H\ :\ddot{\underset{..}{Cl}}:$$

Electron pair closer to Cl

The charges in polar covalent bonds are *partial*. If there were a full charge on each atom, the bond would be ionic.

If the electrons are shared equally, the bond is nonpolar. For example, Cl_2 has a nonpolar bond.

In these cases of diatomic molecules, HCl and Cl_2, these molecules as a whole can be described as *dipolar* and *nondipolar,* respectively. For simplicity, many chemists refer to them as polar or nonpolar. When using these expressions, be careful to remember the difference between polar *bonds* and polar *molecules.*

When there are two or more bonds—that is, more than two atoms in the molecule—polar bonds might cancel out each other's effects, resulting in a **nonpolar molecule.** For example, in carbon dioxide, two polar bonds connect the carbon and oxygen atoms. However, these bonds lie exactly opposite one another (along a straight line), and the effect of one polar bond is canceled by the effect of the other, so the CO_2 molecule has no dipole; it is a nonpolar molecule.

$$\delta- \leftarrow \delta+ \rightarrow \delta-$$
$$:\underset{..}{O}::\ \ C\ \ ::\underset{..}{O}:$$

Although each oxygen atom attracts the electrons more than the carbon atom does (so two polar bonds result), the effect is somewhat like a tug-of-war between equally strong opponents. The electrons in the one bond are pulled toward the one oxygen atom exactly as much as the electrons in the other bond are pulled toward the other oxygen atom. The molecule as a whole has no net dipole.

Another way to look at the situation is that the center of the partial positive charge is on the carbon atom, and the center of the partial negative charges is halfway between the oxygen atoms. But the carbon atom is at a position halfway between the oxygen atoms, so the centers of positive and negative charge are at the same place, and there is no dipole. If the bonds were not identical or if they were at an angle other than 180° (other than in a straight line), a net dipole would result from two such polar bonds.

■ EXAMPLE 13.11

In water molecules, the hydrogen atoms lie at a 105° angle from each other. Does the water molecule have a dipole?

Solution

Since the two polar bonds do not exactly cancel each other, a polar molecule results.

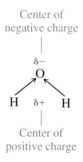

Center of
negative charge

Center of
positive charge

Practice Problem 13.11   State the bond type (polar or nonpolar) and the molecular polarity (dipole or no dipole) for the following substances:

(a) CS_2 (b) BF_3 (c) NF_3 (d) F_2 (e) ClF ■

Polar molecules align themselves in an external electric field. For example, HCl molecules have partial negative charges on the chlorine atoms and partial positive charges on the hydrogen atoms. In a field between two charged plates, HCl molecules tend to align themselves as shown in Figure 13.7.

Figure 13.7 Alignment of Dipolar Molecules in an Electric Field

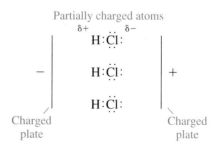

13.6 Intermolecular Forces

OBJECTIVE

■ to show how the three intermolecular forces affect the properties of substances

In addition to the chemical bonds that hold atoms or ions together, there are three **intermolecular forces** that tend to hold molecules to one another. Note that the atoms *within* molecules are held to each other by chemical bonds and that intermolecular forces are the forces *between* molecules. In general, intermolecular forces are much weaker than chemical bonds.

The following three molecules of hydrogen chloride illustrate the difference between chemical bonds and intermolecular forces:

$$\text{Chemical bonds}$$
$$H—Cl\cdots H—Cl\cdots H—Cl$$
$$\text{Intermolecular forces}$$

Within each molecule, the hydrogen atom is held to the chlorine atom by a covalent bond; the HCl molecules are attracted to each other by much weaker intermolecular forces.

The three intermolecular forces are dipolar attractions, van der Waals forces, and hydrogen bonding. The effects of dipoles (Section 13.5) are considered first, followed by discussions of van der Waals forces and hydrogen bonding.

Dipolar Attractions

As discussed in Section 13.5, when molecules are polar, partial positive and negative charges exist at different positions in molecules of the substance, and adjacent molecules tend to orient themselves so that the somewhat positively charged portion of one molecule is near the somewhat negatively charged portion of the next molecule (Figure 13.8). These portions of adjacent molecules will then exert an attractive force on each other. That is, the attractions of

Figure 13.8 Orientation of Dipoles

The molecule in (a) is rotated to the more favorable orientation in (b).

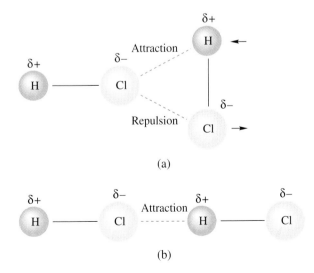

(a)

(b)

Figure 13.9 Origin
of van der Waals Forces

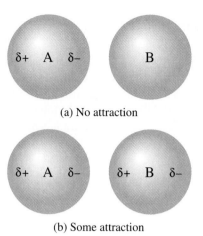

(a) No attraction

(b) Some attraction

positive charges to negative charges first tend to orient the molecules so that oppositely charged portions are adjacent to each other; the nearness of the oppositely charged portions of the molecules causes a net force of attraction between them. These attractions are an example of intermolecular forces known as dipolar attractions.

Van der Waals Forces

A molecule with no permanent dipole may have an unsymmetrical distribution of electrons for an instant. In that instant, the unsymmetrical charge distribution will tend to induce a similar charge dissymmetry in an adjacent molecule, creating an intermolecular force (Figure 13.9). Such a force is called a **van der Waals force,** after Johannes D. van der Waals (1837–1923), a Dutch physicist who worked on intermolecular forces. (Many chemists refer to van der Waals forces as *London forces* or *dispersion forces.*) Because the attraction between the molecules lasts for only an instant, this type of intermolecular force is weaker than dipolar attractions (all other factors being equal). The more electrons in the molecules of a substance, the greater is the van der Waals force. The greater the intermolecular force, the higher is the melting point, the boiling point, or the sublimation point of the substance, and the more energy is required to melt, vaporize, or sublime the substance. **Sublimation** is the process of changing directly from the solid state to the gas state.

■ EXAMPLE 13.12

Which of the following molecules is expected to have the larger van der Waals forces?

$$C_2H_6 \qquad C_8H_{18}$$

Solution

The larger molecule, C_8H_{18}, has more electrons and therefore should have greater van der Waals forces.

H:F: H:F: H:F: H:F:

Figure 13.10 Hydrogen Bonding in Hydrogen Fluoride

Each hydrogen atom is bonded to a fluorine atom and forms a hydrogen bond with another adjacent fluorine atom.

● Oxygen   ● Hydrogen

Figure 13.11 Hydrogen Bonding in Ice

Practice Problem 13.12 Which of the noble gases would be expected to exhibit the least van der Waals forces? ■

■ EXAMPLE 13.13

Which of the noble gases will boil at the highest temperature at 1 atm pressure?

Solution

The only intermolecular forces that occur in noble gases are the van der Waals forces. The more electrons present, the greater is the van der Waals force; the greater the van der Waals force, the higher is the normal boiling point. Thus, radon is expected to be the noble gas with the highest boiling temperature. ■

Hydrogen Bonding

Hydrogen bonding is the intermolecular force of attraction between a hydrogen atom in one molecule and a small, highly electronegative atom with an unshared pair of electrons in another molecule. The most electronegative atoms have a great affinity for electrons (Chapter 5). Fluorine, oxygen, nitrogen, and chlorine are the most electronegative elements, but a chlorine atom is too large to be very effective in hydrogen bonding. A bonded hydrogen atom has its bonding electrons oriented away from an adjacent molecule. That "outer" side of the hydrogen atom is rather positive and can attract an unshared pair of electrons, as shown in two molecules of hydrogen fluoride:

H:F: H:F:

The electrons bonding the hydrogen atom on the right are located to its right, which leaves the left side of that atom rather positive. It is attracted almost equally by the unshared electron pair of the fluorine atom to its left and the electron pair it shares with the fluorine atom to its right.

In fact, hydrogen atoms in liquid hydrogen fluoride migrate rather freely from one fluorine atom to another. They are attracted first by one, then by another, then by a third, and so on. As an overall result, these attractions tend to keep the fluorine atoms close together, resulting in a rather substantial intermolecular force. Consider the model of hydrogen fluoride in Figure 13.10. If the ends of the illustration are covered, it is hard to tell which pair of electrons each hydrogen atom is sharing covalently with a fluorine atom and which is involved in the hydrogen bonding.

■ ENRICHMENT

Hydrogen bonding is responsible for the low density of ice compared to liquid water (Figure 13.11). This difference is absolutely necessary for life as we know it. Because ice is less dense than water, it floats on top of water and acts as a blanket to protect the aquatic life below from the extremes of cold that occur above the ice. If it is pure, the water at the bottom never gets below 0°C (32°F) unless it freezes. Hydrogen bonding is also responsible for the helical structure of certain proteins in plants and animals (Figure 13.12).

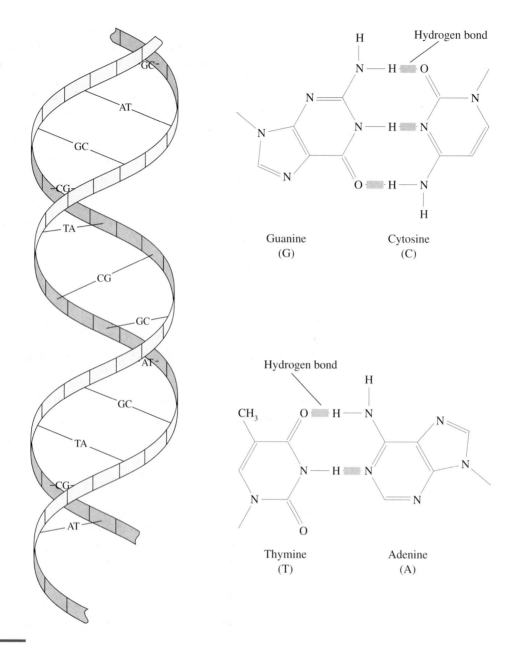

Figure 13.12 Hydrogen Bonding in the DNA Double Helix

The hydrogen bonding is indicated by the blue shading. The base groups detailed on the right are represented by their initials in the double helix on the left.

In summary, for hydrogen bonding to occur, the requirements are

1. A hydrogen atom

2. A nitrogen, oxygen, or fluorine atom

3. An unshared (lone) pair of electrons on the nitrogen, oxygen, or fluorine atom

■ **EXAMPLE 13.14**

Which of the following form hydrogen bonds?

(a) NH_3 (b) CH_4 (c) HI

Table 13.5 Forces of Attraction

Chemical bonds
  Covalent
  Ionic
  Metallic
Intermolecular forces
  van der Waals forces
  Dipolar attractions
  Hydrogen bonding

Solution

(a) NH_3 will form hydrogen bonds because it has (1) hydrogen atoms; (2) a small, highly electronegative atom; and (3) at least one free pair of electrons on that atom.

(b) CH_4 will not form hydrogen bonds because the carbon atom has no lone pair to bond with hydrogen atoms from other molecules.

(c) HI will not form hydrogen bonds because the iodine atom is too large and not electronegative enough.

Practice Problem 13.14   Which of the following substances are expected to form hydrogen bonds?

(a) CH_3OH (b) CH_3Br (c) H_2O ■

All the forces of attraction discussed so far are outlined in Table 13.5.

■ ■ ■ ■ ■ ■ ■ **SUMMARY** ■ ■ ■ ■ ■ ■ ■

The properties of atoms and molecules affect the properties of the substances they constitute. Ionic size is an important property, affecting the ability of the atom to gain or lose an electron. The size of an atom or ion increases as electrons are added, decreases with greater numbers of protons, and varies periodically if both electrons and protons are varied. (Section 13.1)

The ionization energy of an atom is the energy required to remove an electron from it; the electron affinity is the energy liberated when an electron is added to the atom. Both of these quantities refer to gaseous atoms producing gaseous ions. For some atoms, the electron affinity is negative—that is, energy is required to add an electron to the atom. (Section 13.2)

Electronegativity is a semiquantitative measure of the electron-attracting power of a bonded atom. The greater the electronegativity, the more the atom attracts electrons. Covalent bonding between atoms of different electronegativity yields polar bonds; that between atoms of the same electronegativity yields nonpolar bonds. (Section 13.3)

We can predict the shapes of many molecules from their electron dot diagrams and the fact that the electron groups on the central atom tend to get as far apart as possible while remaining attached to that atom. We then attach the outer atoms and describe the molecular shape by the position of the atoms. The electron groups and their attached atoms are not necessarily limited to a planar arrangement. (Section 13.4)

A molecule with more than one polar bond in which all the effects of the polar bonds do not cancel out has a dipole moment—an electrical dissymmetry that causes an intermolecular attraction between this molecule and other similar ones. This attraction is called a dipolar attraction, and it lowers the ability of the substance to exist in the gas phase. However, if the polar bonds in a molecule are oriented so that their effects are canceled out, as in carbon dioxide, then a molecule with no dipole results. (Section 13.5)

Intermolecular attractions include dipolar attractions, as well as van der Waals forces and hydrogen bonding. Van der Waals forces are similar to dipolar attractions but result from instantaneous dissymmetry of charge, which may disappear the next instant. The more electrons in the molecule, the greater is the van der Waals force. However, van der Waals forces tend to be lower in magnitude than dipolar attractions.

Hydrogen bonding is an intermolecular force between a hydrogen atom bonded to a fluorine, oxygen, or nitrogen atom and an unshared pair of electrons on another such atom in an adjacent molecule (or sometimes even the same molecule). The hydrogen atom may be bonded to one small electronegative atom and attracted to another at one instant and then bonded to the second and attracted to the first at the next instant. (Section 13.6)

Items for Special Attention

■ Be sure to distinguish between atomic size and ionic size.

■ Ionization energy and electronegativity vary in the same ways in the periodic table, but opposite to the way atomic size varies.

■ If a bond is either polar or nonpolar, it is covalent.

■ Adding a new shell of electrons considerably increases the size of the atom; breaking into a complete octet to ionize an electron requires a huge quantity of energy.

■ The term *ionization energy* means "first ionization energy" unless otherwise specified.

■ Any molecule with lone pair(s) on the central atom will have a different shape of the electron groups from the shape of the molecule (defined by the locations of the atoms).

■ Molecular geometry is actually determined experimentally. We can deduce the geometry of many molecules from their electron dot diagrams, with correct results in a great majority of cases.

■ Be careful to remember the difference between polar bonds and polar molecules. A molecule can have polar bonds without being a polar molecule.

■ Hydrogen bonding is the intermolecular force between molecules that have hydrogen atoms attached to very small, highly electronegative atoms with unshared pairs of electrons (nitrogen, oxygen, or fluorine atoms). It has nothing to do with the chemical bond between two hydrogen atoms in H_2 molecules.

Self-Tutorial Problems

13.1 Make a table of the number of protons, the number of electrons, and the relative sizes for each of the following sets:

(a) Li and Li^+ (b) F and F^-

(c) Ne and Na^+ (d) Ne and F^-

13.2 (a) In which periodic group does each atom have a new shell of electrons?

(b) In which periodic group does the size of each element increase markedly from the preceding atom in the periodic table?

13.3 (a) Which halogen atom is smallest?

(b) Which one has the smallest radius?

(c) What difference is there, if any, between these questions?

13.4 Would you expect a polyatomic anion (like ClO_4^-) to be larger or the same size as the ion of the central atom (for example, Cl^-)?

13.5 (a) Does any gaseous atom lose an electron spontaneously to form a gaseous ion?

(b) Is any ionization energy negative?

13.6 What is the difference between the bonding in the hydrogen molecule and hydrogen bonding?

13.7 State the difference between a polar bond and a polar molecule.

■ ■ ■ PROBLEMS ■ ■ ■

13.1 Atomic and Ionic Sizes

13.8 Which of the following is largest?

Ce   Gd   Lu

13.9 Which one in each of the following groups is largest?

(a) Cu Cu^+ Cu^{2+}

(b) Cr Cr^{2+} Cr^{3+}

(c) Cl Cl^-

13.10 Which one in each of the following groups is largest?

(a) Se^{2-} Br^- Kr Rb^+ Sr^{2+}

(b) Na^+ N^{3-} Li^+

13.11 Which second-period anion is smallest?

13.12 Without looking at the figures in this chapter, tell which element has (a) the largest atoms and (b) the smallest atoms, excluding hydrogen.

13.13 Select the largest atom in each of the following sets:

(a) Na   Mg   Al   (b) Na   K   Rb

(c) F   Cl   Br   I   (d) Na   S   Ar

(e) Fr   Ga   F   (f) O   F   S

13.14 Select the largest atom or ion in each of the following sets:

(a) Na^+ Ne F^- (b) Na^+ K^+ Rb^+

(c) F^- Cl^- Br^- I^- (d) Na^+ S^{2-} Ar

(e) Na^+ Na   (f) Br Br^-

13.15 Which anion in Figure 13.1, when compared to its parent anion, has had the greatest percentage increase in size in acquiring the extra electron(s)?

13.2 Ionization Energy and Electron Affinity

13.16 (a) What periodic group of elements has the lowest ionization energy?

(b) What main group elements are next lowest?

13.17 (a) Is hydrogen an alkali metal, as reflected by ionization energy?

(b) Is hydrogen an alkali metal, as reflected by size?

13.18 What periodic group of elements, on average, has the highest ionization energies?

13.19 (a) Without looking at the figures in this chapter, tell which element has the largest ionization energy.

(b) Which element has the smallest?

13.20 Select the atom with the largest ionization energy in each of the following sets:

(a) Na   Mg   Ar

(b) Na   K   Rb

(c) F   Cl   Br   I

(d) Na   S   Ar

(e) Fr   Ga   F

(f) O   F   S

13.21 Atoms of elements of which periodic group(s) do not produce energy as they gain electrons?

13.3 Electronegativity and Bond Polarity

13.22 Predict the bond type (ionic, polar covalent, nonpolar covalent) in each of the following:

(a) NaF   (b) PBr_3 (c) I_2

13.23 From the data of Figure 5.1, predict the bond type (ionic, polar covalent, nonpolar covalent) in each of the following:

(a) NCl_3 (b) SI_2

13.24 From the data of Figure 5.1, predict the bond type (ionic, polar covalent, nonpolar covalent) in each of the following:

(a) CS_2 (b) CI_4

13.25 From the data of Figure 5.1, predict the bond type (ionic, polar covalent, nonpolar covalent) in each of the following:

(a) ICl   (b) Cs_2O

13.26 From the data of Figure 5.1, predict the bond type (ionic, polar covalent, nonpolar covalent) in each of the following:

(a) CaF_2 (b) CCl_4

13.4 Molecular Shape

13.27 Deduce the shape of each of the following:

(a) CCl_4 (b) CH_4 (c) CH_2Cl_2

13.28 Determine the shape of (a) the BCl_3 molecule and (b) the SF_2 molecule.

13.29 Determine the shape of (a) the PCl_3 molecule and (b) the SeO_3 molecule.

13.5 Polar and Nonpolar Molecules

13.30 What is the difference between a nonpolar bond and a nonpolar molecule?

13.31 Can a molecule have polar bonds and not have a dipole?

13.32 Can a molecule have nonpolar bonds only and have a dipole?

13.33 Can a molecule have some nonpolar bonds and have a dipole?

13.34 (a) Will carbon dioxide molecules be oriented in an external electric field (see Figure 13.7)?

(b) Will water molecules?

13.35 (a) Would you expect either CH_4 or CCl_4 molecules to have dipoles?

(b) Would you expect CH_2Cl_2 molecules to have dipoles?

(c) Explain.

13.36 Give an example of a nonpolar molecule with:

(a) Two polar bonds

(b) Three polar bonds

(c) Four polar bonds

13.37 Consider the following rigid molecules. Does each have a dipole?

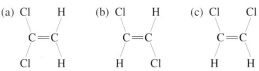

13.6 Intermolecular Forces

13.38 State whether each of the following compounds has only van der Waals forces, dipole moments but no hydrogen bonding, or hydrogen bonding:

(a) H_2O (b) H_2S (c) H_2

(d) SO_2 (e) CO_2

13.39 Explain how you can tell that the BF_3 molecule, which is triangular and planar, has no dipole. (See part [b] of Practice Problem 13.11.)

13.40 In which of the noble gases are van der Waals forces (a) least and (b) greatest?

■ ■ ■ GENERAL PROBLEMS ■ ■ ■

13.41 (a) Calculate the sum of the first *two* ionization energies of sodium and the first *three* ionization energies of aluminum. (Table 13.1)

(b) Would Na^{2+} or Al^{3+} be easier to obtain in the gas state?

(c) Which is a familiar ion?

13.42 Determine the shape of (a) the $POCl_3$ molecule and (b) the H_2SO_3 molecule.

13.43 Determine the shape of (a) the $COCl_2$ molecule and (b) the SiF_4 molecule.

13.44 Determine the shape of (a) the SO_2 molecule and (b) the SO_3 molecule.

13.45 Consider the size of H^+. Is it qualitatively any different from the size of every other cation? Would you expect such an ion to attract electrons well or poorly?

13.46 Which of the following is larger?

$$N^{3-} \qquad Li^+$$

13.47 (a) Predict the geometry of NO_2 (a nonoctet molecule with only seven electrons around the central nitrogen atom).

(b) Do you think that the angle in NO_2 is greater or less than that in NO_2^-, which has the extra electron on the nitrogen atom?

13.48 When a *gaseous* atom and ion are specified in Section 13.2, how close are their nearest neighbors? (See Section 12.8.) Are there apt to be any significant attractions between such atoms or ions?

13.49 The force of attraction between oppositely charged ions varies inversely as the square of the distance between them:

$$f = k/d^2$$

Ions in a solid might be at a distance about 2.0×10^{-10} m apart. If a pair of ions is separated to twice the distance in the solid, what percentage of the force of attraction remains? If they are separated to the average distance between gas molecules, how much force do you expect to remain?

13.50 Calculate the difference between the ionization energy of potassium and the electron affinity of fluorine. Deduce whether the transfer of an electron from one to the other in the gas phase is spontaneous.

13.51 Calculate the difference between the sum of the first two ionization energies of calcium and the sum of the first two electron affinities of oxygen. Deduce whether the transfer of two electrons from one to the other in the gas phase is spontaneous. The second electron affinity of oxygen is -816 kJ/mol.

13.52 Give an example of a polar molecule with four polar bonds.

13.53 Without looking at Figure 5.1, deduce the bond type (ionic, polar covalent, nonpolar covalent) in each of the following:

(a) OF_2 (b) SO_2 (c) SrF_2

13.54 (a) The AX_3 molecule, where A is one element and X is another element with a different electronegativity, has no dipole. What can you tell about its geometry?

(b) The ZX_3 molecule, where Z is one element and X is another element with a different electronegativity, has a finite dipole. What can you tell about its geometry?

14 Solids and Liquids, Energies of Physical and Chemical Changes

- 14.1 Nature of the Solid and Liquid States
- 14.2 Changes of Phase
- 14.3 Measurement of Energy Changes
- 14.4 Enthalpy Changes in Chemical Reactions

Key Terms *(Key terms are defined in the Glossary.)*

amorphous solid (14.1)
boiling point (14.2)
condensation (14.2)
crystalline solid (14.1)
distillation (14.2)
endothermic process (14.3)
enthalpy change (14.4)
enthalpy of formation (14.4)
enthalpy of fusion (14.3)
enthalpy of vaporization (14.3)
evaporation (14.2)
exothermic process (14.3)

fluidity (14.1)
freezing (14.2)
fusion (14.2)
heat (14.4)
heat of fusion (14.3)
heat of vaporization (14.3)
heating curve (14.3)
Hess's law (14.4)
ionic solid (14.1)
melting (14.2)
metallic solid (14.1)
molecular solid (14.1)

network solid (14.1)
normal boiling point (14.2)
phase change (14.2)
physical equilibrium (14.2)
specific heat (14.3)
standard state (14.4)
state function (14.4)
vapor (14.2)
vaporization (14.2)
vapor pressure (14.2)
work (14.4)

Symbols

Δ (change in) (14.3)
ΔE (change in energy) (14.4)

ΔH (change in enthalpy) (14.4)
ΔH_f (enthalpy of formation) (14.4)

q (heat) (14.4)
w (work) (14.4)

Table 14.1   Properties of the Particles of Solids, Liquids, and Gases

	Solids	**Liquids**	**Gases**
Strength of attractive forces	Strong	Moderately strong	Negligible
Components	Molecules or bonded atoms or ions	Molecules	Molecules
Distance between particles	Touching	Touching	Far apart
Permanence of position	Permanent	Variable	Variable

Gases are characterized by having little or no forces of attraction between their molecules (Section 12.8). In contrast, the forces between the particles of which liquids and solids are composed are significant. These forces range from ionic or covalent bonds, which are strong, to much weaker intermolecular forces. Intermolecular forces were discussed in Section 13.6. Substances that are liquid or gaseous at room temperature are composed of molecules. In contrast, solids may be composed of (1) macromolecules (atoms covalently bonded together in large networks), (2) ions bonded together, (3) molecules, or (4) atoms bonded together by loosely held valence electrons. The particles that make up solids are held more or less stationary in their positions; the molecules that make up liquids and gases are much more free to move about. In gases, the distance between molecules is very large compared to the size of the molecules themselves; in liquids and solids, the particles are essentially touching each other. These characteristics are summarized in Table 14.1.

Section 14.1 examines the liquid and solid states. Section 14.2 discusses changes of phase and concepts related to systems with two phases in contact, such as vapor pressure. In Section 14.3, the measurement of quantities of heat added to or removed from a system is introduced, and the energies involved in phase changes are calculated. Section 14.4 deals with the enthalpies involved in chemical reactions.

14.1 Nature of the Solid and Liquid States

The Solid State

Solids are classified as **crystalline solids** or **amorphous solids.** Crystalline solids, such as an ice cube, have a definite melting point. Amorphous solids, such as a chocolate bar, get softer and softer as the temperature is raised. The structures of crystalline solids feature regularly repeating arrangements of the constituent particles. The structure of amorphous solids is not regular, but something like that of liquids; sometimes, amorphous solids are called "supercooled liquids."

Crystalline solids may be classified as (1) **ionic solids,** in which the repeating units are ions; (2) **network solids** (or macromolecular solids), in which covalently bonded atoms are the repeating units; (3) **molecular solids,** in which individual molecules are the repeating units; and (4) **metallic solids,** in which individual metal atoms are held together by their loosely held valence electrons.

■ EXAMPLE 14.1

(a) What is the difference between a stack of concrete blocks and a concrete wall?

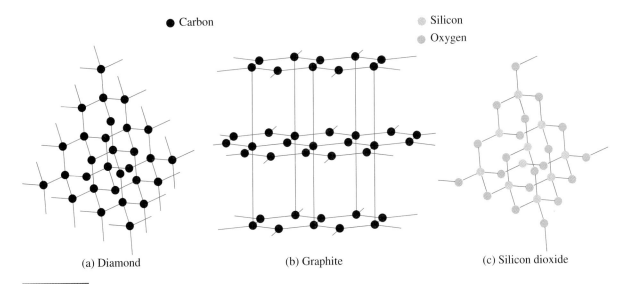

● Carbon

◉ Silicon
◉ Oxygen

(a) Diamond   (b) Graphite   (c) Silicon dioxide

Figure 14.1   Tiny Portions of Structures of Three Macromolecular Solids
The structures extend almost indefinitely in all directions, left and right, up and down, in and out.

(b) Since the atoms in both network (macromolecular) solids and molecular solids are covalently bonded, what is the difference in their structures?

Solution

(a) The concrete in each block in a stack is held firmly to the rest of that block, but the blocks are not held tightly to each other. The concrete wall is held together tightly from one end to the other.

(b) In molecular solids, a few to hundreds of atoms are bonded into each molecule, as for example in I_2, $B_{10}H_{14}$, or S_4N_4. These individual molecules are held to each other by intermolecular forces. In macromolecular solids, covalent bonds connect all the atoms—thousands or even millions of them—in a network (Figure 14.1).

Practice Problem 14.1   What type of solid is described by each of the following statements?

(a) Five atoms are bonded in each molecule.

(b) About 500 000 atoms are bonded to each other with covalent bonds.

(c) About 500 million ions are bonded to each other. ■

The regularity of the units is characteristic of crystalline solids. The sodium chloride structure (also called the rock salt structure) is an example of an ionic lattice (Figure 14.3). Diamond, graphite (pencil lead), and silica (sand) are examples of network lattices (Figure 14.1). Molecular solids include such solids as ice and naphthalene (mothballs), $C_{10}H_8$. Most elemental metals are examples of metallic solids. Regardless of their specific classifications, all crystalline solids are characterized by regularly repeating arrangements of their component particles.

Carbon exists in a newly discovered form, with 60 atoms covalently bonded in the shape of a ball (Figure 14.2).

■ **EXAMPLE 14.2**

(a) Is the 60-carbon molecule a macromolecular form of carbon?

(b) Would you expect its melting temperature to be as high as that of graphite?

Solution

(a) No, although a large molecule, it is not macromolecular.

(b) Its melting point is much lower.

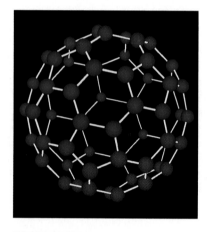

Figure 14.2 The
60-Carbon Molecule

The 60-carbon molecule in the shape of a geodesic dome is named after Buckminster Fuller. Its nickname "Buckyball" is catchier than its full name—Buckminster Fullerene.

■ **EXAMPLE 14.3**

How many pairs of ions are included in a tiny cube of rock salt, if 150 000 sodium ions occur on each edge in the three perpendicular directions?

Solution

The cube contains $(1.5 \times 10^5)^3 = 3.4 \times 10^{15}$ sodium ions. There are also 3.4×10^{15} chloride ions in this tiny particle of NaCl. (Remember that 1.00 mol, or 58.5 g, of NaCl includes 6.02×10^{23} NaCl ion pairs, almost 200 million times as many as in this tiny cube.)

An ionic substance is composed of individual positive and negative ions, which can form a three-dimensional lattice. The attraction of one ion is not limited to any single other ion but extends to all the oppositely charged ions around it. Thus, the attractive forces throughout the lattice are strong, and, therefore, the structure is hard to break up. Ionic substances are solids at room temperature.

Practice Problem 14.3   How many iodine molecules and how many iodine atoms are included in a cube of solid I_2 with 150 000 molecules on each edge? ■

Figure 14.3 Sodium
Chloride Structure

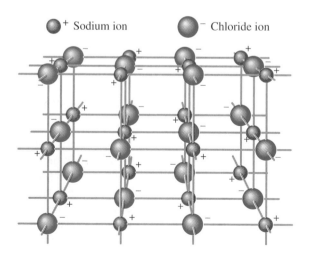

○+ Sodium ion   ○− Chloride ion

Table 14.2 Types of Crystalline Solids

Type	Units	Relative Melting and Boiling Points
Ionic	Ions	High to very high
Network	Atoms covalently bonded throughout	Very high
Molecular	Small molecules	Very low to moderate
Metallic	Metal atoms bonded by valence electrons	Low to very high

Molecular solids are relatively easy to melt or sublime, but network or ionic solids have very high melting or sublimation points. In general, the weaker the force holding the particles together, the easier it is to get them apart. The strong forces (chemical bonds) that hold network or ionic solids together take a great deal of energy to overcome in order to get a substance into the liquid or the gaseous state.

In macromolecules such as diamond and silica, covalent bonds link millions of atoms into a giant molecule, rather than linking only a few atoms into each of many small molecules. Since the atoms are held together by strong forces, the solid is difficult to "get apart." Diamond and graphite are still solids at temperatures up to 3500°C. Ionic compounds, made up of oppositely charged ions with ionic bonds extending throughout the solid, also have high melting points (typically at least 500°C). In contrast, solids composed of molecules, which are held to each other by (weak) intermolecular forces (Section 13.6), are much more easily melted or sublimed.

The four types of crystalline solids are summarized in Table 14.2.

■ EXAMPLE 14.4

State the type of bonding or force holding together each of the following solids, and indicate whether the melting point is low or high:

(a) MgO   (b) Diamond   (c) Ice

Solution

(a) Ionic bonding, high melting point

(b) (Network) covalent bonding, high melting point

(c) Hydrogen bonding (a type of intermolecular force), low melting point

Practice Problem 14.4 Three solids have the following melting points: 100°C, 1000°C, and 3000°C. Which one is molecular? ■

The Liquid State

Solids have strong forces holding the particles in their proper positions. Only molecular solids (and some metals) can be melted at temperatures anywhere near room temperature. The energy involved in disrupting the covalent or ionic bonds of other types of solids makes the melting points of such solids very high, up to hundreds or even thousands of degrees Celsius (Table 14.3). The weaker intermolecular forces involved in molecular solids allow them to be melted at temperatures below about 200°C. Substances that exist as liquids at these relatively

Table 14.3   *Melting Points of Various Types of Solids*

Solid	Formula	Melting Point (°C)
Molecular		
Benzoic acid	$HC_6H_5CO_2$	122.4
Carbon tetrabromide	CBr_4	90.1
Carbon tetrachloride	CCl_4	−23
Carbon tetrafluoride	CF_4	−184
Helium	He	−272.2 (at 26 atm)
Naphthalene	$C_{10}H_8$	80.22
Tetrasulfur tetranitride	S_4N_4	179 (sublimes)
Water	H_2O	0.00
Ionic		
Aluminum oxide	Al_2O_3	2045
Magnesium oxide	MgO	2800
Sodium chloride	NaCl	801
Macromolecular		
Diamond	C	>3550
Graphite	C	3652 (sublimes)
Quartz	SiO_2	1610
Metallic		
Iron	Fe	1535
Mercury	Hg	−38.87
Sodium	Na	97.8
Tungsten	W	3410

low temperatures are composed of molecules, uncombined atoms (noble gases), or weakly bonded atoms (metals such as mercury).

The molecules of a liquid are in contact with one another, as in molecular solids, but the regularity of their arrangement does not extend very far. You might think of liquids as being like a crowd of commuters in a railroad station and solids being more like an army of soldiers on parade. The molecules in a liquid are not arranged in any regular pattern, although some may form pairs or small groups. Molecules in a liquid are able to slip and slide past one another, giving liquids their **fluidity.** In contrast, molecules (or other particles) of solids cannot move past one another, giving solids their unyielding structures.

A material is a solid because the energy available to its particles is not sufficient to overcome the relatively strong attractive forces holding the individual particles together. The molecules of liquids exhibit attractive forces strong enough to hold them in close proximity to one another, but not to hold them relatively motionless. Gases, of course, have very weak intermolecular forces relative to their energies, and their molecules are almost completely independent of one another.

14.2   Changes of Phase

Any process that results in a change of state (or phase) for a sample of matter is called a **phase change.** When a solid changes to a liquid as a result of a rise in temperature, the process is called **melting,** or **fusion.** The process of changing a liquid to a gas is called **evaporation,** or **vaporization.** A gas in contact with its liquid is often called a **vapor.** Changing a solid directly into a gas is called subli-

Figure 14.4 Terminology of Phase Changes

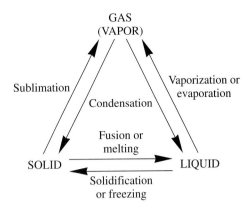

mation. Changing a liquid to a solid is called **freezing,** and changing a gas to either a solid or a liquid is called **condensation.** These states and changes are summarized in Figure 14.4.

As an example of these phase changes, we will investigate the vaporization process in some detail. It takes energy to cause a liquid to evaporate. We can put a pan of water on a stove and add heat to it to cause rapid vaporization. Coming out of a swimming pool on a hot, breezy day makes you aware that energy is used to evaporate a liquid. The breeze causes evaporation, which uses energy, thereby lowering your temperature. You grab for a towel to stop the evaporation and resulting cooling.

A process called **distillation** is often used to purify liquids. The liquid is heated to vaporize it, and then the vapor is cooled to condense it back to a liquid in a different place (Figure 14.5). Impurities that are less easily vaporized are left behind in the original container, and those that are more easily vaporized distill and are discarded before the desired product distills.

On a molecular level, the change from a liquid to a gas is accompanied by a separation of the molecules—from close proximity to far apart. Only the most energetic of the liquid molecules have sufficient energy to overcome the attractive forces and go into the gas phase. Thus, the process of vaporization leaves behind the less energetic molecules. As you learned in Section 12.8, the absolute

Figure 14.5 Distillation Apparatus

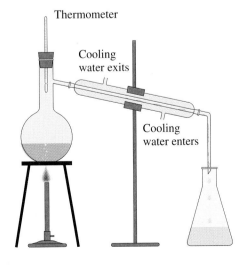

Figure 14.6 Water Boiling at Room Temperature at Reduced Pressure

temperature is directly proportional to the average kinetic energy of gas molecules. The same is also true for liquid (and solid) molecules. Thus, when the highest-energy molecules go into the gas phase, the molecules left in the liquid phase have less energy, on average, and therefore, the liquid has a lower temperature unless energy is added from the surroundings. In addition, if the gas molecules do work, such as pushing back the atmosphere, additional energy is required. If no energy is added from the surroundings, this energy must come from the molecules themselves. That is, the gas molecules also lose energy, and therefore, the gas is cooled.

In a closed system at constant temperature, as liquid molecules escape into the gas phase, the pressure of the vapor builds up. More and more gas molecules are present, and they can condense back to liquid. Thus, the rate of condensation increases. When the rate of condensation matches the rate of evaporation, a condition called **physical equilibrium** is attained. The rate of evaporation of liquid molecules is equal to the rate of condensation of gas molecules. Two opposite processes are occurring at equal rates, and no net change is taking place. Specifically, the pressure of the vapor is constant under these conditions and is called the **vapor pressure** of the substance at that temperature.

If the temperature of the system is increased, the number of liquid molecules with sufficient energy to escape into the gas phase will increase, but the number of gas molecules available to condense will not immediately be affected. Therefore, the rate of evaporation will exceed the rate of condensation until the number of molecules in the gas phase builds up, and a new and higher vapor pressure results. (If all the liquid evaporates, no equilibrium is possible.)

The vapor pressure of a pure liquid is determined solely by the temperature. (Section 12.7 showed that the vapor pressure of water is determined by its temperature alone.) If the surface area for evaporation is larger, so also is the surface area for condensation. If there is more volume for the gas to occupy, more molecules will be able to evaporate before the rate of condensation equals the rate of evaporation, but it will take that many more molecules to build up to the same vapor pressure. The volume and shape occupied by the liquid also have no effect on the vapor pressure of a pure liquid.

Boiling occurs when the vapor pressure of the liquid equals the pressure of the surroundings. At that point, bubbles appear within the liquid itself. The **boiling point** of a liquid is the temperature at which the vapor pressure of the liquid equals the pressure on the system. The **normal boiling point** is the boiling point at a pressure of 1.00 atm. For example, when liquid water at 100°C and 1.00 atm is heated, it boils. Liquids can boil at temperatures other than their normal boiling points if the pressure on them is different from 1.00 atm. For example, water can boil at 25°C if the pressure above it is 24 torr (Figure 14.6). In contrast, pressure cookers are used to allow water to boil at a higher temperature, so that food will cook faster.

14.3 Measurement of Energy Changes

O B J E C T I V E

■ to calculate the energy required to change a substance from one temperature to another, or from one phase to another, or both

When we add energy to a sample of matter, we generally expect the sample to warm up. The sample does warm up, unless it happens to be a pure substance at its melting, sublimation, or boiling point. The amount of heat required to warm a sample of matter is given by the equation

$$\text{Heat} = mc\Delta t$$

Table 14.4 Specific Heat Capacities of Selected Substances

Substance	Formula	Specific Heat Capacity $(J/g \cdot C°)$
Covalent molecules		
Carbon dioxide	CO_2	0.852
Carbon monoxide	CO	1.04
Hydrogen	H_2	14.4
Nitrogen	N_2	1.04
Oxygen	O_2	0.922
Water, gaseous	$H_2O(g)$	2.042
Water, liquid	$H_2O(l)$	4.184
Water, solid	$H_2O(s)$	2.089
Metals		
Aluminum	Al	0.90
Chromium	Cr	0.45
Cobalt	Co	0.46
Copper	Cu	0.385
Gold	Au	0.129
Iron	Fe	0.442
Lead	Pb	0.13
Magnesium	Mg	1.0
Silver	Ag	0.24
Tin	Sn	0.22
Zinc	Zn	0.388

where m is the mass, c is the specific heat capacity, usually called the **specific heat,** and Δt is the *change* in temperature in either degrees Celsius or kelvins. The Greek capital delta (Δ) means "change in." The change in temperature is defined as the *final* temperature minus the *initial* temperature.

Specific heats of some substances are given in Table 14.4.

■ EXAMPLE 14.5

(a) Calculate the daily room rate per person for a hotel room if the cost for two people for three nights is $660.

(b) Calculate the specific heat of water if 41.84 J is required to raise the temperature of 5.000 g of water by 2.000°C.

Solution

(a) Cost = (number of people)(rate)(number of nights)

660 dollars = (2 people)(rate)(3 nights)

$$\text{Rate} = \frac{110 \text{ dollars}}{\text{person} \cdot \text{night}}$$

(b) $c = \dfrac{\text{heat}}{m\Delta t} = \dfrac{41.84 \text{ J}}{(5.000 \text{ g})(2.000°C)} = 4.184 \text{ J/g} \cdot °C$

Note that the unit for specific heat, like that for the room rate in part (a), has two different units in its denominator (g and °C).

Practice Problem 14.5   Calculate the specific heat of ice if 50.1 J is required to raise the temperature of 4.00 g of ice by 6.00°C. ■

■ **EXAMPLE 14.6**

(a) Calculate the amount of money that will be charged to house 17 students for three nights in a guest house at a rate of 16 dollars per student per night.

(b) Calculate the quantity of energy required to raise the temperature of 17.0 g of water by 16.0°C. The specific heat of water is 4.184 J/g·°C.

Solution

(a) $\text{Cost} = 17 \text{ students} \left(\dfrac{16 \text{ dollars}}{\text{student} \cdot \text{night}} \right) 3 \text{ nights} = 816 \text{ dollars}$

(b) $\text{Heat} = 17.0 \text{ g} \left(\dfrac{4.184 \text{ J}}{\text{g} \cdot {}^\circ\text{C}} \right) 16.0{}^\circ\text{C} = 1140 \text{ J} = 1.14 \text{ kJ}$ ■

■ **EXAMPLE 14.7**

Calculate the number of joules required to heat 8.17 g of water from 19.0°C to 34.1°C.

Solution

The equation to use is

$$\text{Heat} = mc\Delta t$$

The mass of the water is given in the problem. The *change* in temperature is the final temperature minus the initial temperature:

$$\Delta t = t_2 - t_1 = 34.1{}^\circ\text{C} - 19.0{}^\circ\text{C} = 15.1{}^\circ\text{C}$$

The specific heat of the water, as calculated in Example 14.5 and listed in Table 14.4, is 4.184 J/g·°C. Thus, the heat required is

$$\text{Heat} = (8.17 \text{ g})(4.184 \text{ J/g} \cdot {}^\circ\text{C})(15.1{}^\circ\text{C}) = 516 \text{ J}$$

Practice Problem 14.7   Calculate the number of joules required to heat 8.17 g of iron from 19.0°C to 34.1°C. Does heating the iron take more energy than heating the same mass of water? ■

■ **EXAMPLE 14.8**

Calculate the final temperature after 1445 J of energy is added to 10.16 g of water at 40.00°C.

Solution

The equation is rearranged to give the *change in temperature:*

$$\Delta t = \frac{\text{heat}}{mc} = \frac{1445 \text{ J}}{(10.16 \text{ g})(4.184 \text{ J/g} \cdot {}^\circ\text{C})} = 33.99{}^\circ\text{C}$$

The final temperature is 33.99°C greater than the initial temperature. That is, since $\Delta t = t_2 - t_1$,

$$t_2 = \Delta t + t_1 = 33.99{}^\circ\text{C} + 40.00{}^\circ\text{C} = 73.99{}^\circ\text{C}$$ ■

Energy *added to a system* is defined as *positive,* and energy *released from a system* is defined as *negative.* A process in which energy is added to a system is said to be an **endothermic process.** A process in which energy is released from a system is said to be an **exothermic process.**

■ EXAMPLE 14.9

Calculate the final temperature after 1445 J of energy is removed from 10.16 g of water at 40.00°C.

Solution

In this case, since energy is removed from the system, you use −1445 J in the equation:

$$\Delta t = \frac{\text{heat}}{mc} = \frac{-1445 \text{ J}}{(10.16 \text{ g})(4.184 \text{ J/g} \cdot °\text{C})} = -33.99°\text{C}$$

The final temperature is

$$t_2 = \Delta t + t_1 = -33.99°\text{C} + 40.00°\text{C} = 6.01°\text{C}$$

That the temperature is lowered when energy is removed from the system should not be surprising. ■

When heat is transferred from one object to another, with no energy gained from or lost to anything else (the surroundings), the total change in energy is the sum of the changes in energy of the two objects. For example, if a piece of hot metal is placed in a sample of cold water, the water is warmed and the metal is cooled. Both the metal and the water will reach the same temperature. The heat transferred to or from the surroundings is zero. Thus,

$$\text{Heat} = 0 = (m_{\text{water}})(c_{\text{water}})(\Delta t_{\text{water}}) + (m_{\text{metal}})(c_{\text{metal}})(\Delta t_{\text{metal}})$$

This relationship may be used to find the specific heat of a metal.

■ EXAMPLE 14.10

A 123-g sample of a metal at 90.0°C is placed in 145 g of water at 15.0°C, and the final temperature of the system is 28.1°C. Calculate the specific heat of the metal. Which of the metals in Table 14.4 is it?

Solution

$$\text{Heat} = 0 = (m_{\text{water}})(c_{\text{water}})(\Delta t_{\text{water}}) + (m_{\text{metal}})(c_{\text{metal}})(\Delta t_{\text{metal}})$$

The change in the temperature of the water is 28.1°C − 15.0°C = 13.1°C. The change in the temperature of the metal is 28.1°C − 90.0°C = −61.9°C. Note that both the water and the metal wind up at 28.1°C. Note also that the change in temperature of the metal is negative; it cooled down. Substituting into the equation yields

$$0 = (145 \text{ g})(4.184 \text{ J/g} \cdot °\text{C})(13.1°\text{C}) + (123 \text{ g})(c_{\text{metal}})(-61.9°\text{C})$$
$$c_{\text{metal}} = 1.04 \text{ J/g} \cdot °\text{C}$$

The metal in the table with the heat capacity closest to 1.04 J/g·°C is magnesium.

Practice Problem 14.10   A 53.2-g sample of a metal at 80.0°C is placed in 101 g of water at 25.0°C, and the final temperature of the system is 30.6°C. Calculate the specific heat of the metal. Which of the metals in Table 14.4 is it? ■

■ **EXAMPLE 14.11**

Calculate the final temperature after 25.5 g of a metal ($c = 0.975$ J/g·°C) at 75.0°C is placed in 125 g of water at 21.3°C.

Solution

$$\text{Heat} = 0 = (m_{\text{water}})(c_{\text{water}})(\Delta t_{\text{water}}) + (m_{\text{metal}})(c_{\text{metal}})(\Delta t_{\text{metal}})$$

Let t_f represent the final temperature of both the metal and the water.

$$0 = (m_{\text{water}})(c_{\text{water}})(\Delta t_{\text{water}}) + (m_{\text{metal}})(c_{\text{metal}})(\Delta t_{\text{metal}})$$

Substituting the known quantities gives

$$0 = (125 \text{ g})(4.184 \text{ J/g·°C})(t_f - 21.3°C) + (25.5 \text{ g})(0.975 \text{ J/g·°C})(t_f - 75.0°C)$$

$$0 = (125)(4.184)(t_f - 21.3°C) + (25.5)(0.975)(t_f - 75.0°C)$$

The equation is simplified algebraically:

$$0 = 523t_f - 11\,100°C + 24.9t_f - 1860°C$$
$$548t_f = 13\,000°C$$
$$t_f = 23.7°C$$

> The algebraic simplification is based on the equality $wx(y - z) = wxy - wxz$.

Practice Problem 14.11   Calculate the final temperature after 10.9 g of a metal ($c = 0.385$ J/g·°C) at 80.3°C is placed in 222 g of water at 21.1°C. ■

When energy is added to or removed from a pure substance and that substance *changes phase* as a result, the *temperature does not change*. That is, when a pure substance melts, solidifies, sublimes, evaporates, or condenses, it does so at a given temperature (under constant pressure). For example, when energy is added to liquid water at 100°C and 1.00 atm, the water boils. However, as long as some liquid water remains, the water and vapor are still at 100°C. The additional energy goes into potential energy as the molecules go from their relative closeness in the liquid state to their well-separated positions in the gas phase. This energy is called the **enthalpy of vaporization,** or the **heat of vaporization.**

The amount of heat associated with the phase change of a pure substance depends on the nature of the substance and the quantity of substance in the sample. For example, the **enthalpy of fusion,** or **heat of fusion,** of water—the heat necessary to change water from solid to liquid—is 335 J/g. The unit for this value includes no degree, since the temperature does not change when solid water melts. In general, the heat term is named according to the name of the process the substance undergoes. For example:

Process	*Name of Heat Term*	*More Formal Name*
Fusion	Heat of fusion	Enthalpy of fusion
Vaporization	Heat of vaporization	Enthalpy of vaporization
Sublimation	Heat of sublimation	Enthalpy of sublimation

Some selected values of heats of phase change are presented in Table 14.5.

Table 14.5 Selected Heats of Phase Change

Substance	Phase Change (at 1 atm)	Temperature	Heat Involved (J/g)
Ammonia	Boiling	−33°C	1380
Carbon dioxide	Sublimation	−90°C	3680
Sulfur dioxide	Boiling	−10°C	402
Water	Melting	0°C	335
Water	Boiling	100°C	2260

■ EXAMPLE 14.12

Calculate the energy required to melt 27.8 g of ice at 0°C.

Solution

The heat of fusion of water at 0°C is 335 J/g (from Table 14.5).

$$\text{Heat} = (27.8 \text{ g})(335 \text{ J/g}) = 9.31 \times 10^3 \text{ J}$$

Practice Problem 14.12 Calculate the heat required to vaporize 27.8 g of water at 100°C. Compare this quantity of heat to that required to melt the same mass of ice (Example 14.12). ■

When we add heat to a substance that is not at a temperature at which a phase change will occur, its temperature rises. Suppose that the temperature rises to the temperature of a phase change. Then, the phase change will occur at constant temperature until it is complete, after which the temperature will start to rise again. For example, if we add heat to ice at −20°C, the ice will warm to 0°C. At that temperature, melting begins, and the temperature stays at 0°C until all the ice has melted. Then the temperature of the liquid water rises as we add even more heat. After the temperature of the liquid reaches the boiling point, the addition of more heat causes it to boil. Further heating increases the temperature of the vapor. We can draw a graph that shows what happens as we heat a pure substance. Such a graph is called a **heating curve** (Figure 14.7).

Figure 14.7 Heating Curve for Water

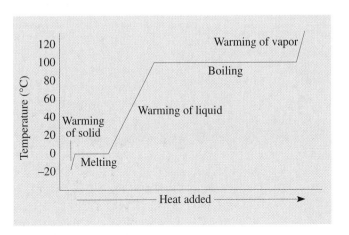

How do we calculate the energy required to raise the temperature of a substance and cause it to change phase? We do separate calculations for each step and then sum the results.

■ **EXAMPLE 14.13**

Calculate the energy required to change a 15.0-g sample of water from 85.0°C to steam at 113.0°C and 1.00 atm. Use data from Tables 14.4 and 14.5.

Solution

The water in the liquid phase will be heated to 100.0°C. It will then boil, producing water vapor at 100°C. Finally, the water vapor produced will be heated to 113.0°C. You do one calculation for each of these three steps. For heating the liquid water:

$$\text{Heat} = mc\Delta t = (15.0 \text{ g})(4.184 \text{ J/g} \cdot °C)(100.0°C - 85.0°C) = 941 \text{ J}$$

To vaporize the water:

$$\text{Heat} = mc_{vap} = (15.0 \text{ g})(2260 \text{ J/g}) = 3.39 \times 10^4 \text{ J} = 33\ 900 \text{ J}$$

For heating the water vapor:

$$\text{Heat} = mc\Delta t = (15.0 \text{ g})(2.042 \text{ J/g} \cdot °C)(113.0°C - 100.0°C) = 398 \text{ J}$$

You sum the energies:

$$\text{Total heat} = 941 \text{ J} + 33\ 900 \text{ J} + 398 \text{ J} = 35\ 200 \text{ J} = 3.52 \times 10^4 \text{ J}$$

Note that you cannot combine the calculations for the two heating steps, since liquid water and water vapor have different heat capacities.

Practice Problem 14.13   Calculate the quantity of energy that is removed when 15.0 g of water at 10.0°C is changed to ice at −5.0°C. ■

14.4   Enthalpy Changes in Chemical Reactions

OBJECTIVE

■ to calculate enthalpies of many reactions from known data for other reactions and to recognize the difference between energy and enthalpy

Heat, symbolized q, is different from other kinds of energy. All other kinds of energy can be converted entirely to heat, but heat cannot be converted entirely to any other form of energy. We classify all other kinds of energy as **work,** symbolized w. Thus, the change in energy of a system, ΔE, is merely the heat added to the system plus the work added to the system:

$$\Delta E = q + w$$

By convention, energy, heat, and work added to a system are regarded as positive, while that removed from a system is regarded as negative. You must exercise great care in energy calculations to get the sign right.

■ **EXAMPLE 14.14**

(a) A certain combustion process gives off 1000 kJ of heat. What is the sign of q?

(b) All of the heat is used to warm a sample of water. What is the sign of q for the warming process?

(c) What is the value of q for the overall process?

Solution

(a) Since heat is given off by the reaction, q is negative.

(b) Since the heat is absorbed by the water, this q is positive.

(c) The overall q is zero, since no heat was added to or taken from the surroundings. All the heat from the reaction went into the warming process. ■

We cannot measure directly the heat produced by a chemical reaction; we measure it by determining its effect on something. For example, the heat produced by burning 100 g of natural gas might be measured by determining the increase in temperature of a certain mass of water warmed by the heat. In such an experiment, we must try to ensure that *all* of the heat produced by the combustion reaction is used to warm the water and that none escapes to the surroundings.

■ **EXAMPLE 14.15**

Burning 1.00 mol of CH_4 warms 2500 g of water by 84.3°C. Calculate the heat produced in the combustion process.

$$CH_4(g) + 2\ O_2(g) \rightarrow CO_2(g) + 2\ H_2O(l)$$

Solution

The heat added to the water is

$$\text{Heat} = mc\Delta t = (2500\ \text{g})(4.184\ \text{J/g·°C})(84.3°C) = 882\ 000\ \text{J}$$
$$= 882\ \text{kJ}$$

That is the magnitude of the heat produced by the reaction. Since the reaction *produced* the heat, $q = -882$ kJ.

Practice Problem 14.15 Burning 0.0500 mol of C_2H_6 warms 250 g of water by 73.7°C. Calculate the heat produced in the combustion process. ■

Once the value for a particular reaction has been determined, it can be used to calculate values for other experiments. For example, if we burn the same gas under the same conditions, we will get the same quantity of heat. Twice as much gas will yield twice as much heat, and so on. Moreover, we can find out information about other reactions.

Enthalpy Change

Enthalpy change, ΔH, is equal to the heat involved in a process when the process is done under a constant pressure and involves no work except perhaps expansion (or contraction) against the atmosphere. When these conditions are not met, ΔH is a more fundamental quantity than heat. For example, ΔH is a **state function,** which means that its value is independent of the path in going from the initial state to the final state. (Heat, in contrast, does depend on the path, except when $q = \Delta H$.) The ΔH value is important in both theoretical and practical terms in chemistry.

The ΔH value of a certain process conventionally is called the enthalpy of that process. For example, the enthalpy change of the vaporization process is called the enthalpy of vaporization, $\Delta H_{\text{vaporization}}$. The enthalpy change for the fusion (the melting) process is ΔH_{fusion}. The enthalpy of combustion of a substance is, by definition, the enthalpy change accompanying the burning of the substance *in excess oxygen*. Essentially, however, enthalpy change is a single function that can be applied to many different processes, and the name used as a subscript to ΔH is the name of the particular process.

Enthalpy of Formation

One particularly useful type of process for us to consider is the formation of a substance in its standard state from its elements in their standard states. **Standard state** is the state in which the substance is most stable. For example, the standard state of elemental oxygen is $O_2(g)$, not $O(g)$ or $O_3(l)$. Thus, the equation for the formation reaction of water is

$$2\,H_2(g) + O_2(g) \rightarrow 2\,H_2O(l)$$

The enthalpy change for this reaction can be called the **enthalpy of formation** of water, $\Delta H_f(H_2O)$. Table 14.6 shows standard enthalpies of formation. Table 14.7 is a separate table for unbranched hydrocarbons.

Values of ΔH for many reactions can be determined experimentally; for other reactions, measuring ΔH is impossible because the process is too slow or for some other reason. Values of ΔH_f are useful because we can calculate the ΔH for

> Be sure to distinguish between ΔH_f and ΔH for a reaction.

> ΔH_f of an element in its standard state is zero by definition.

Table 14.6   Standard Enthalpies of Formation at 298 K

	State	ΔH_f(kJ/mol)		State	ΔH_f(kJ/mol)
AgCl	s	−127.0	HCl	g	−92.13
AgBr	s	−99.6	HF	g	−269
AgI	s	−62.3	HI	g	25.9
Al$_2$O$_3$	s	−1670	H$_2$S	g	−20.15
Au(OH)$_3$	s	−418	HgO	s	−90.8
B$_2$H$_6$	g	31	KCl	s	−435.9
BF$_3$	g	−1110	NF$_3$	g	−127
B$_2$O$_3$	s	−1277	NH$_3$	g	−46.19
BaO	s	−558.6	N$_2$H$_4$	g	50.42
BaCO$_3$	s	−1217	NH$_4$Cl	s	−315.4
BaSO$_4$	s	−1465	N$_2$O	g	81.55
CO	g	−110.5	NO	g	90.37
CO$_2$	g	−393.5	NO$_2$	g	33.8
CaO	s	−635.5	NaCl	s	−411.0
CaCO$_3$	s	−1207	NaHCO$_3$	s	−710.4
CaCl$_2$	s	−795.0	Na$_2$CO$_3$	s	−1430.1
Ca(OH)$_2$	s	−986.6	PH$_3$	g	23
Cr$_2$O$_3$	s	−1141	P$_2$O$_5$	s	−1531
CuO	s	−157	SO$_2$	g	−296.9
Cu$_2$S	s	−79.5	SO$_3$	g	−395.2
Fe$_2$O$_3$	s	−822.2	SiO$_2$	s	−878.2
H$_2$O	l	−285.9	SiCl$_4$	g	−609.6
H$_2$O	g	−241.8	SiF$_4$	g	−1511.6
H$_2$SO$_4$	l	−811.3	WO$_3$	s	−843.1
HBr	g	−36.2	ZnO	s	−348
			ZnS	s	−203

Table 14.7 Enthalpies of Formation at 25°C of Unbranched Hydrocarbons

Compound	ΔH_f (kJ/mol)
CH_4	−74.5
C_2H_6	−83.7
C_3H_8	−105
C_4H_{10}	−126
C_5H_{12}	−146
C_6H_{14}	−167
C_7H_{16}	−187
C_8H_{18}	−208

any reaction by subtracting the sum of the ΔH_f values of the reactants of the reaction from the sum of the ΔH_f values of the products of the reaction:

$$\Delta H = \text{Sum of } \Delta H_f(\text{products}) - \text{Sum of } \Delta H_f(\text{reactants})$$

The ΔH_f of an element in its standard state is zero because converting an element in its standard state to the element in its standard state does not involve any change. That "reaction" will have a ΔH of zero.

■ **EXAMPLE 14.16**

Calculate ΔH for the complete combustion of 1.00 mol of propane gas, C_3H_8, which is sold in tanks to heat mobile homes and houses not connected to a central gas supply.

Solution

$$C_3H_8(g) + 5\ O_2(g) \rightarrow 3\ CO_2(g) + 4\ H_2O(l)$$

Values of ΔH_f are obtained from Tables 14.6 and 14.7. For the products,

$$\Delta H_f = 3\ \text{mol CO}_2\left(\frac{-393\ \text{kJ}}{1\ \text{mol CO}_2}\right) + 4\ \text{mol H}_2\text{O}\left(\frac{-286\ \text{kJ}}{1\ \text{mol H}_2\text{O}}\right) = -2320\ \text{kJ}$$

For the reactants,

$$\Delta H_f = 1\ \text{mol C}_3\text{H}_8\left(\frac{-105\ \text{kJ}}{1\ \text{mol C}_3\text{O}_8}\right) + 5\ \text{mol O}_2\left(\frac{0\ \text{kJ}}{1\ \text{mol O}_2}\right) = -105\ \text{kJ}$$

$$\Delta H = \Delta H_f(\text{prod}) - \Delta H_f(\text{react}) = (-2320\ \text{kJ}) - (-105\ \text{kJ}) = -2220\ \text{kJ}$$

(Note that the value of ΔH_f of O_2 is not in Table 14.6 because we know that ΔH_f of every uncombined element is zero.)

> Be especially careful about the signs when doing these calculations!

Practice Problem 14.16 Calculate ΔH for the complete combustion of 1.00 mol of methane gas, CH_4, the principal component of natural gas. ■

We can multiply the number of kilojoules per mole by the number of moles undergoing reaction to get the number of kilojoules for any given quantity of reactant or product involved. Thus, for Example 14.16, if 2.00 mol of C_3H_8 had been involved, the enthalpy change would have been

$$2.00\ \text{mol}\left(\frac{-2220\ \text{kJ}}{1\ \text{mol}}\right) = -4440\ \text{kJ}$$

■ **EXAMPLE 14.17**

Calculate the enthalpy change for the combustion of 1.00 mol of C_3H_8 to give CO and H_2O. Use the following equation:

$$2\ C_3H_8(g) + 7\ O_2(g) \rightarrow 6\ CO(g) + 8\ H_2O(l)$$

Solution

For the products,

$$\Delta H_f = 6\ \text{mol CO}\left(\frac{-110\ \text{kJ}}{1\ \text{mol CO}}\right) + 8\ \text{mol H}_2\text{O}\left(\frac{-286\ \text{kJ}}{1\ \text{mol H}_2\text{O}}\right) = -2950\ \text{kJ}$$

For the reactants,

$$\Delta H_f = 2 \text{ mol } C_3H_8 \left(\frac{-105 \text{ kJ}}{1 \text{ mol } C_3H_8} \right) + 7 \text{ mol } O_2 \left(\frac{0 \text{ kJ}}{1 \text{ mol } O_2} \right) = -210 \text{ kJ}$$

For this equation,

$$\Delta H = \Delta H_f(\text{prod}) - \Delta H_f(\text{react}) = (-2950 \text{ kJ}) - (-210 \text{ kJ}) = -2740 \text{ kJ}$$

However, the equation involves 2 mol C_3H_8, and the problem requires 1.00 mol.

$$1.00 \text{ mol} \left(\frac{-2740 \text{ kJ}}{2 \text{ mol}} \right) = -1370 \text{ kJ}$$

Practice Problem 14.17   Repeat Example 14.17, using the following equation, and compare the results:

$$C_3H_8(g) + \tfrac{7}{2} O_2(g) \rightarrow 3 \text{ CO}(g) + 4 \text{ H}_2O(l) \quad ■$$

Hess's Law

Another way to calculate values of ΔH for reactions involves manipulating equations for other reactions with known ΔH values. When chemical equations are added to yield a different chemical equation, the corresponding ΔH values are added to get the ΔH for the desired equation. This principle is called **Hess's law.** For example, we can calculate the ΔH for the reaction of carbon with oxygen gas to yield carbon dioxide from the values for the reaction of carbon with oxygen to yield carbon monoxide and that of carbon monoxide plus oxygen to yield carbon dioxide:

Desired: $C(s) + O_2(g) \rightarrow CO_2(g)$

		ΔH **(kJ)**
Given:	$C(s) + \tfrac{1}{2} O_2(g) \rightarrow CO(g)$	-110
	$CO(g) + \tfrac{1}{2} O_2(g) \rightarrow CO_2(g)$	-283

Adding the two chemical equations results in the desired equation:

$$C(s) + CO(g) + O_2(g) \rightarrow CO_2(g) + CO(g)$$

or, eliminating CO from both sides,

$$C(s) + O_2(g) \rightarrow CO_2(g)$$

Therefore, adding these two ΔH values will give the ΔH desired:

$$\Delta H = (-110 \text{ kJ}) + (-283 \text{ kJ}) = -393 \text{ kJ}$$

Note that we have not used enthalpies of formation explicitly in this process. Enthalpies of *any* type of reaction will do.

■ **EXAMPLE 14.18**

Calculate the enthalpy change of the general reaction

$$Z + 2\,Y \rightarrow X + W$$

from the following data:

	ΔH **(kJ)**
$Z + 2\,Y \rightarrow V + U$	-14.7
$V + U \rightarrow X + W$	-55.4

Solution

Adding the two equations will result in the elimination of V and U and yield the desired equation. We then merely add the corresponding ΔH values:

$$\Delta H = (-14.7 \text{ kJ}) + (-55.4 \text{ kJ}) = -70.1 \text{ kJ}$$

Practice Problem 14.18 Calculate the enthalpy change of the general reaction

$$Z + 2\,Y \longrightarrow X + W$$

from the following data:

	ΔH (kJ)
$Z + 2\,Y \longrightarrow V + U$	-14.7
$V + U \longrightarrow Q + 2\,R$	$+50.2$
$Q + 2\,R \longrightarrow X + W$	-105.6 ■

Using Hess's law is easy if the equations given are in a form where simply adding them yields the equation desired. Sometimes, however, to get the desired equation, we must multiply or divide a given equation by a small integer, and then multiply or divide the corresponding ΔH value by that same integer. Sometimes the given equation must be turned around, whereupon the sign of the corresponding ΔH must be changed. Sometimes, both of these processes are necessary. You can decide which of these steps to take by looking at the desired equation to see where you want each reactant and product, and how many moles of each you want.

■ **EXAMPLE 14.19**

Calculate ΔH for the reaction

$$C_2H_4(g) + H_2(g) \longrightarrow C_2H_6(g)$$

from the following data:

	ΔH (kJ)
(A) $C_2H_4(g) + 3\,O_2(g) \longrightarrow 2\,CO_2(g) + 2\,H_2O(l)$	-1387
(B) $2\,H_2(g) + O_2(g) \longrightarrow 2\,H_2O(l)$	-572
(C) $2\,C_2H_6(g) + 7\,O_2(g) \longrightarrow 4\,CO_2(g) + 6\,H_2O(l)$	-3082

Solution

Since we have 1 mol of C_2H_4 on the left of the desired equation, and the only place it appears in the equations given is on the left of equation A, we leave equation A alone. Since 1 mol of H_2 is on the left in the equation desired, we must divide equation B by 2, leaving the H_2 on the left. The ΔH value is also divided by 2:

(B) $H_2(g) + \frac{1}{2} O_2(g) \longrightarrow H_2O(l)$ $\Delta H = -286$ kJ

Since we need only 1 mol of C_2H_6, and it is on the right in the equation desired, we reverse equation C (changing the sign of ΔH) and also divide it by 2:

(C) $2\,CO_2(g) + 3\,H_2O(l) \longrightarrow C_2H_6(g) + \frac{7}{2} O_2(g)$ $\Delta H = +1541$ kJ

Adding these three equations results in the equation desired; adding the corresponding ΔH values yields the requested ΔH:

	ΔH (kJ)
(A) $C_2H_4(g) + 3\ O_2(g) \rightarrow 2\ CO_2(g) + 2\ H_2O(l)$	-1387
(B) $H_2(g) + \frac{1}{2} O_2(g) \rightarrow H_2O(l)$	-286
(C) $2\ CO_2(g) + 3\ H_2O(l) \rightarrow C_2H_6(g) + \frac{7}{2} O_2(g)$	1541

The 2 mol CO_2, 3 mol H_2O and 3.5 mol O_2 all cancel out, leaving

$$C_2H_4(g) + H_2(g) \rightarrow C_2H_6(g)$$

The sum of the ΔH values is -132 kJ. ■

■ ■ ■ ■ ■ ■ ■ **SUMMARY** ■ ■ ■ ■ ■ ■ ■

Compared to the molecules of gases, the particles that make up liquids and solids are (1) closer together, (2) more strongly held together, and (3) for solids, not necessarily molecules. The forces that hold solids together include chemical bonds and intermolecular forces—dipole moments, van der Waal forces, and hydrogen bonding. In general, the stronger the intermolecular attractions are, the higher the boiling point and melting point of the substance and the more energy it takes to get the substance to melt or vaporize. Intermolecular forces are not as strong as chemical bonds, such as the ionic bonds that hold the ions in sodium chloride together or the covalent bonds that hold the atoms in diamond together.

A crystalline solid is characterized by a regularly repeating arrangement of the particles that make it up. The particles may be ions, covalently bonded atoms (in macromolecules), small molecules held to each other by intermolecular forces, or metal atoms held to each other by metallic bonding. In contrast, liquids at room temperature are composed of molecules. The molecules are not regularly arranged over long distances (on a molecular level) but change their relative positions rather easily, giving a liquid its fluidity. (Section 14.1)

Substances can change phase—from solid to liquid or from liquid to gas, for example. It takes energy to change a substance from solid to liquid, solid to gas, or liquid to gas, but energy is released in the reverse processes. On a molecular level, energy is needed to overcome the high attractive forces between molecules in the solid state to get to the somewhat lower attractions in the liquid state or the negligible attractions in the gas state.

A liquid exerts a certain vapor pressure as a given fraction of its molecules escape from the liquid surface into the gas phase. The relative rates of escape of molecules from the liquid phase and of their return to it determine how many molecules per unit volume will be in the gas phase, and these rates are determined by the nature of the liquid and its temperature. Vapor pressure increases with increasing temperature (but not in direct proportion) for any substance in the liquid state. Thus, the vapor pressure of a pure liquid is governed by temperature alone. Boiling occurs when the vapor pressure equals the pressure of the surrounding atmosphere, and the normal boiling point is the boiling point when the pressure is equal to 1.00 atm. (Section 14.2)

When energy is added to a pure substance, either its temperature will rise or it will change phase (unless it undergoes a chemical reaction). Alternatively, both a temperature rise and a phase change may occur (one at a time). The quantity of energy required to raise the temperature of a substance is given by the equation

$$\text{Heat} = mc\Delta t$$

where m is the mass, c is the specific heat capacity, and Δt is the change in temperature. Any one of these four quantities may be calculated if the other three are known. Note that the change in temperature involves an initial and a final temperature, either of which may be unknown.

When a pure substance changes phase, it does so at constant temperature. The quantity of energy involved is simply the mass times the heat required per gram (called the enthalpy of fusion, enthalpy of vaporization, or enthalpy of sublimation, for example). When energy is added to a system, the sign on the quantity of energy is plus; when energy is removed from the system, the sign is minus. (Section 14.3)

The enthalpy change (ΔH) for any reaction is equal to the sum of the enthalpies of formation of the products minus the sum of the enthalpies of formation of the reactants. Values of ΔH can also be obtained from other reactions using Hess's law, which involves adding the chemical equations and the corresponding ΔH values. (Section 14.4)

Items for Special Attention

■ In *normal boiling point*, the word *normal* means "at 1 atm."

■ For a solid or liquid to change to a gas, the forces holding the particles together must be overcome. The stronger the forces holding the solid together, the more energy is required to melt or sublime it. Also, the stronger the forces holding the solid together, the higher is the sublimation point or melting point.

■ Heat capacities have units with degrees Celsius (or kelvins) in the denominator, but heats of phase change do not because the temperature does not change when a pure substance changes phase.

■ Be especially careful about the signs and the units when doing the calculations in this chapter.

■ Be careful to distinguish between ΔH_f and ΔH for a reaction.

Self-Tutorial Problems

14.1 Explain how you know that gases at room temperature are not composed of ions.

14.2 Which of the phases is characterized by having strong intermolecular forces among molecules that are not permanently located in or near one site?

14.3 List the types of forces that hold particles of various types of solids together, and give an example of a solid that is held together by each type.

14.4 If 47.6 g of a metal at 70.2°C is placed in 125 g of water at 17.9°C and it warms the water to 22.4°C, what is the final temperature of the metal?

14.5 Identify the initial temperature, the final temperature, and the change in temperature for each of the following cases:

(a) Water is warmed from 10.0°C to 24.1°C.

(b) Water is warmed 24.1°C from 10.0°C.

(c) Water is warmed by 24.1°C to 35.7°C.

(d) Water is warmed 24.1°C to 35.7°C.

(e) Water is warmed 35.7°C from 24.1°C.

(f) Water is changed from 24.1°C to 10.0°C.

14.6 If it takes 5440 J of energy to warm 50.0 g of water from 22.0°C to 48.8°C, how much energy is required to cool 50.0 g of water from 48.8°C to 22.0°C?

14.7 Calculate the final temperature if the initial temperature of a system is 20.0°C and the temperature change is (a) 10.0°C and (b) −10.0°C.

14.8 (a) What is the difference between vaporization of a liquid to a gas and condensation of a gas to a liquid?

(b) Which process is endothermic?

(c) What is the difference between the heat of vaporization and the heat of condensation?

14.9 What types of calculations would you have to make to find the energy required to change 150 g of ice at −12°C to water at 50°C?

14.10 The specific heat of ice is 2.09 J/g·°C, and that of water is 4.184 J/g·°C. The heat of fusion of ice is 335 J/g.

(a) Calculate the heat required to warm 50.0 g of ice from −10.0°C to 0.0°C.

(b) Calculate the heat required to melt the ice.

(c) Calculate the heat required to warm the resulting water to 25.0°C.

(d) Calculate the total heat required for all three processes.

14.11 (a) What is the difference between fusion of a solid and solidification of a liquid?

(b) Which process is endothermic?

(c) What is the difference between the heat of fusion and the heat of solidification?

■ ■ ■ PROBLEMS ■ ■ ■

14.1 Nature of the Solid and Liquid States

14.12 Carbon dioxide, CO_2, sublimes at $-78°C$ and silicon dioxide, SiO_2, melts at 1713°C. What type of solid is each of these?

14.13 Would you expect Br_2 or ICl, which have the same total number of electrons and approximately the same molar mass, to have a higher normal boiling point? Explain.

14.14 Which of the following substances do you expect to have the lowest melting point?

$$NaCl \qquad PCl_3 \qquad SiO_2 \text{ (a network solid)}$$

14.15 Diphosphorus trioxide, P_2O_3 melts at 23.8°C, and aluminum oxide, Al_2O_3 melts at 2045°C. What type of solid is each of these?

14.16 Would you expect SO_2 or SO_3 to have a higher normal boiling point? Explain.

14.17 Which of the following substances do you expect to have the lowest melting point?

$$He \qquad Ne \qquad Ar \qquad Kr$$

14.18 Explain why, under ordinary conditions, elemental bromine exists as a liquid, iodine exists as a solid, and fluorine and chlorine both exist as gases.

14.19 Explain why solid $NaCl$ and solid CCl_4 melt at such widely different temperatures: 801°C and $-23°C$, respectively.

14.20 Suppose you could look at one atom of a substance in a certain state and when you looked along a certain line from that atom, you saw many atoms of the same element occurring at regular intervals. Each such atom had the same other atoms surrounding it as the original atom. In what state is the substance?

14.21 Which has the higher melting point—an ionic compound in which the ions have single charges or an analogous ionic compound in which each ion has a double charge?

14.22 Which of the following compounds is highest melting?

$$SCl_2 \qquad H_2O \qquad MgCl_2$$

14.2 Changes of Phase

14.23 Suppose all the liquid water in a closed container evaporates. Is it certain that the pressure of the water vapor in the container is equal to the vapor pressure of water?

14.24 (a) Determine from the graph of Figure 14.7 what will happen first to liquid water at 100°C as heat is removed from it.

(b) Repeat part (a) for water at 0°C.

14.25 Which of the following will increase the vapor pressure of water?

(a) Increasing the volume of the liquid

(b) Increasing the volume occupied by the vapor

(c) Increasing the surface area of the liquid

(d) Increasing the temperature

(e) Increasing the pressure of dry air over the liquid

14.26 (a) Determine from the graph of Figure 14.7 what will happen first to liquid water at 100°C as heat is added to it.

(b) Repeat part (a) for water at 0°C.

14.27 A sealed rectangular box is half-filled with water and has water vapor in the upper half at a pressure equal to the vapor pressure of water at the temperature. Will the pressure change if the box is placed horizontally, thereby changing the surface area of the water? (See the accompanying illustrations.)

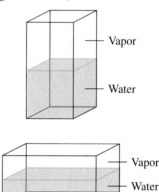

14.3 Measurement of Energy Changes

14.28 (a) Calculate the heat required to warm 20.0 g of water from 22.2°C to 53.3°C.

(b) Calculate the heat required to cool 20.0 g of water from 53.3°C to 22.2°C.

14.29 (a) Calculate the heat required to warm 10.0 g of ice from $-19.0°C$ to $-5.0°C$.

(b) Calculate the heat required to cool 10.0 g of ice from $-5.0°C$ to $-19.0°C$.

14.30 Calculate the final temperature after 273 J of heat is added to 25.0 g of zinc at 22.7°C.

14.31 How much energy is required to raise the temperature of 49.8 g of zinc by 14.7°C?

14.32 Calculate the final temperature after 155 J of heat is added to 40.0 g of aluminum at 22.7°C.

14.33 How much energy is required to raise the temperature of 150 g of cobalt by 10.6°C?

14.34 How much energy is required to raise the temperature of 150 g of cobalt from 18.7°C to 29.3°C?

14.35 Calculate the specific heat of a metal if 80.5 g of the metal at 75.0°C warms 150 g of water at 20.0°C to 23.0°C. Which metal in Table 14.4 is this metal most likely to be?

14.36 Calculate the heat required to change 10.0 g of liquid water at 92.0°C to water vapor at 115°C.

14.37 Calculate the specific heat of a metal if 102 g of the metal at 77.7°C warms 250 g of water at 18.2°C to 20.7°C. Which metal in Table 14.4 is this metal most likely to be?

14.38 Calculate the heat required to change 25.0 g of ice at −11.2°C to liquid water at 20.7°C.

14.39 How much energy is required to raise the temperature of 121.0 g of iron from 27.0°C to 79.2°C?

14.40 Calculate the final temperature after 39.1 g of a metal ($c = 0.0650$ J/g·°C) at 65.7°C is placed in 102 g of water at 19.7°C.

14.41 A certain reaction emits 1.00 kJ of heat. What temperature change will 100.0 g of water undergo if the heat from the reaction is added to it?

14.4 Enthalpy Changes in Chemical Reactions

14.42 The value of ΔH for the reaction $A + B \rightarrow C$ is −10.0 kJ. What is the value of ΔH for each of the following reactions?

(a) $2 A + 2 B \rightarrow 2 C$ (b) $C \rightarrow A + B$

14.43 Write an equation to which each of the following corresponds:

(a) $\Delta H_{fusion}(H_2O)$

(b) $\Delta H_f(HCl)$

(c) $\Delta H_{combustion}(CH_4)$

(d) $\Delta H_{vaporization}(H_2O)$

(e) $\Delta H_f(CO)$

(f) $\Delta H_{combustion}(CO)$

14.44 The value of $\Delta H_{combustion}$ (C_2H_4) is −1410 kJ. Calculate its ΔH_f.

14.45 Calculate ΔH for the reaction of 1.00 mol of C_8H_{18} with limited O_2 to form CO and H_2O.

14.46 Calculate ΔH for the reaction of 155 g of CuO according to the equation

$3 CuO(s) + CH_4(g) \rightarrow CO(g) + 2 H_2O(l) + 3 Cu(s)$

14.47 Which of the following processes represents the reaction corresponding to the enthalpy of formation of AgCl? Explain.

(a) $Ag^+(aq) + Cl^-(aq) \rightarrow AgCl(s)$

(b) $Ag(s) + \frac{1}{2} Cl_2(g) \rightarrow AgCl(s)$

(c) $AgCl(s) \rightarrow Ag(s) + \frac{1}{2} Cl_2(g)$

(d) $Ag(s) + AuCl(s) \rightarrow Au(s) + AgCl(s)$

14.48 Given the following information:

$A + B \rightarrow C + D$ $\Delta H° = -10.0$ kJ
$C + D \rightarrow E$ $\Delta H° = 15.0$ kJ

calculate $\Delta H°$ for each of the following reactions:

(a) $C + D \rightarrow A + B$

(b) $2 C + 2 D \rightarrow 2 A + 2 B$

(c) $A + B \rightarrow E$

14.49 Calculate the enthalpy change at 25°C for the reaction of 50.0 g of CO with oxygen gas to form CO_2.

14.50 Which of the following equations have enthalpy changes equal to

(a) $\Delta H_f°(CO_2)$ (b) $\Delta H_{comb}°(C)$

(c) $\Delta H_{comb}°(CO)$ (d) $\Delta H_f°(CO)$

(i) $C + O_2 \rightarrow CO_2$

(ii) $C + \frac{1}{2} O_2 \rightarrow CO$

(iii) $CO + \frac{1}{2} O_2 \rightarrow CO_2$

14.51 Given the following reactions with their enthalpy changes at 25°C:

$N_2(g) + 2 O_2(g) \rightarrow 2 NO_2(g)$ $\Delta H° = 67.70$ kJ
$N_2(g) + 2 O_2(g) \rightarrow N_2O_4(g)$ $\Delta H° = 9.67$ kJ

calculate the enthalpy of reaction of NO_2 to form N_2O_4. Is N_2O_4 apt to be stable with respect to NO_2 at 25°C? Explain your answer.

14.52 From the following data, calculate the enthalpy of hydrogenation of C_2H_2 to C_2H_6:

$C_2H_2 + 2 H_2 \rightarrow C_2H_6$

	ΔH (kJ)
$C_2H_2 + 2\frac{1}{2} O_2 \rightarrow 2 CO_2 + H_2O$	−1305
$C_2H_6 + 3\frac{1}{2} O_2 \rightarrow 2 CO_2 + 3 H_2O$	−1541
$H_2 + \frac{1}{2} O_2 \rightarrow H_2O$	−286

14.53 From the following data, calculate the enthalpy of hydrogenation of C_2H_2 to C_2H_4:

$C_2H_2 + H_2 \rightarrow C_2H_4$

	ΔH (kJ)
$C_2H_2 + 2\frac{1}{2} O_2 \rightarrow 2 CO_2 + H_2O$	−1305
$C_2H_4 + 3 O_2 \rightarrow 2 CO_2 + 2 H_2O$	−1387
$H_2 + \frac{1}{2} O_2 \rightarrow H_2O$	−286

14.54 What quantity of heat is yielded to the surroundings when 0.100 mol of C_8H_{18} is completely burned at constant pressure in 1.25 mol of oxygen gas at 25°C, yielding CO_2 and H_2O at 25°C?

14.55 Calculate the enthalpy of formation of C_2H_4 from the following data:

	ΔH (kJ)
$C_2H_4 + 3 O_2 \rightarrow 2 CO_2 + 2 H_2O$	−1387
$H_2 + \frac{1}{2} O_2 \rightarrow H_2O$	−285.9
$C + O_2 \rightarrow CO_2$	−393.5

14.56 Determine the ΔH for the reaction of 96.0 g of SO_3 with BaO at 25°C. Values of ΔH_f are −559 kJ/mol for BaO, −395 kJ/mol for SO_3, and −1465 kJ/mol for $BaSO_4$.

14.57 From the following data, calculate the enthalpy of hydrogenation of C_2H_2 to C_2H_6:

$C_2H_2 + 2 H_2 \rightarrow C_2H_6$

	ΔH (kJ)
$2 C_2H_2 + 5 O_2 \rightarrow 4 CO_2 + 2 H_2O$	−2610
$2 C_2H_6 + 7 O_2 \rightarrow 4 CO_2 + 6 H_2O$	−3082
$2 H_2 + O_2 \rightarrow 2 H_2O$	−572

14.58 Given the following enthalpies of combustion, calculate the enthalpy change for the reaction

$$3\ C_2H_2 \rightarrow C_6H_6$$

C_2H_2 -1305 kJ/mol

C_6H_6 -3273 kJ/mol

14.59 Consider the following reaction:

$$3\ C_2H_2 \rightarrow C_6H_6$$

Given the following enthalpies of combustion:

C_2H_2 -1305 kJ/mol

C_6H_6 -3273 kJ/mol

Calculate the enthalpy change for:

(a) The reaction of 95.0 g of C_2H_2

(b) The production of 95.0 g of C_6H_6

(c) Compare and explain the results of (a) and (b).

(d) The reaction of 1.50 mol of C_2H_2

(e) The production of 1.50 mol of C_6H_6

(f) Compare and explain the results of (d) and (e).

■ ■ ■ GENERAL PROBLEMS ■ ■ ■

14.60 The molar heat capacity of a substance is the number of joules required to raise the temperature of a mole of the substance by 1°C. (Note the difference between this use of the word *molar* and that in Chapter 11.) Calculate the molar heat capacity of (a) water and (b) iron.

14.61 Lithium fluoride and calcium oxide have the same crystal structure. Which one do you think melts at the higher temperature?

14.62 Barium oxide and sodium chloride have the same crystal structure. Which one do you think melts at the higher temperature?

14.63 Explain why a carbon dioxide fire extinguisher sprays "snowlike" solid CO_2 when discharged, even though it is at room temperature before being used.

14.64 Label the portions of the heating curve in Figure 14.7 to show the areas in which the energy input goes into kinetic energy of the molecules and the areas in which the energy goes into potential energy.

14.65 A certain amount of steam at 100°C is bubbled into 100.0 g of water at 19.0°C. When the process is ended, there is 104.0 g of water at 42.9°C.

(a) What mass of steam is used in this experiment?

(b) How much energy is supplied to the cold water?

(c) As it cooled, how much energy did the hot water (formed by condensation of the steam at 100°C) supply to the cold water?

(d) How much energy did the condensation process supply to the cold water?

(e) What is the heat of vaporization of water per gram at 100°C?

14.66 A certain amount of steam at 100°C is bubbled into 100.0 g of water at 19.0°C. When the process is ended, there is 104.0 g of water at 42.9°C. What is the heat of vaporization of water per gram at 100°C?

14.67 A certain amount of ice at 0°C is placed in 50.0 g of water at 50.7°C. When the process is ended, there is 58.0 g of water at 32.6°C.

(a) What mass of ice is used in this experiment?

(b) How much energy is provided by the warm water?

(c) As it warmed, how much energy did the cold water (formed by melting of the ice at 0°C) accept from the hot water?

(d) How much energy did the melting process accept from the water?

(e) What is the heat of fusion of ice per gram at 0°C?

14.68 Explain why both HF and HI boil higher than HCl.

14.69 Compare problems 14.52 and 14.57, and explain the results.

14.70 If 1000 J of energy is added to a solution, the solution is warmed.

(a) Does it make any difference if the added energy is electrical, chemical, or heat energy?

(b) If a chemical reaction supplies energy to the solution, does the reaction have a positive, negative, or zero ΔH?

(c) Is adding heat to the solution a positive, negative, or zero ΔH?

(d) Does the overall process have a positive, negative, or zero ΔH?

14.71 A certain amount of ice at 0°C is placed in 50.0 g of water at 50.7°C. When the process is ended, there is 58.0 g of water at 32.6°C. What is the heat of fusion of water per gram at 0°C?

14.72 Calculate the molar heat capacity (see Problem 14.60) of each element in Table 14.4. What can you generalize about the molar heat capacities of gaseous elements and of metallic elements?

14.73 A certain solid does not have regularly repeating atoms, ions, or molecules in any direction in its structure. What can you say about its melting point?

14.74 It takes energy to melt a pure substance, but the substance has the same temperature after being melted as before. Since the kinetic energy of the substance is directly proportional to the absolute temperature, both solid and fluid have the same kinetic energy. What happened to the energy that was added?

14.75 It takes energy to boil a pure liquid substance, but the substance has the same temperature after being boiled as before. Since the kinetic energy of the substance is directly proportional to the absolute temperature, both liquid and vapor have the same kinetic energy. What happened to the energy that was used to change the liquid to vapor?

14.76 Water molecules are linked by hydrogen bonds, but H_2S, H_2Se, and H_2Te molecules are not. Plot the normal boiling points of these three compounds (see the data that follow) versus the atomic number of the central atom, and extend the best straight line through them to atomic number 8. Predict what the normal boiling point of water would be in the absence of hydrogen bonding.

Substance	Normal Boiling Point (°C)
H_2S	−60.7
H_2Se	−30
H_2Te	−2

14.77 Ammonia molecules are linked by hydrogen bonds, but PH_3, AsH_3, and SbH_3 molecules are not. Plot the normal boiling points of these compounds (see the data that follow) versus the atomic number of the central atom, and extend the best straight line through them to atomic number 7. Predict what the normal boiling point of ammonia would be in the absence of hydrogen bonding.

Substance	Normal Boiling Point (°C)
NH_3	−33.35
PH_3	−87.7
AsH_3	−55
SbH_3	−17

14.78 Relative humidity is defined as 100 times the ratio of the humidity at a given temperature to the maximum humidity that the air can have at that temperature (its vapor pressure):

Relative humidity

$$= \frac{\text{Actual pressure of water vapor}}{\text{Vapor pressure}} \times 100\%$$

On a certain day, the relative humidity is 100%. If the temperature rises with no change in the concentration of water vapor in the air, what happens to the relative humidity?

14.79 Calculate the energy required to convert 153 g of water at 85.0°C and 1.00 atm to water vapor at 121.0°C and 1.00 atm.

14.80 What is the final temperature of a 22.2-g sample of ice that is initially at −15°C after 3021 J of energy is added to it. (See Tables 14.4 and 14.5 for specific heats and heats of phase change.)

14.81 What is the final temperature of a 103-g sample of water that is initially at 15°C after 15 210 J of energy is removed from it. (See Tables 14.4 and 14.5 for specific heats and heats of phase change.)

14.82 Calculate the energy required to convert 10.0 g of water at 98.7°C and 1.00 atm to water vapor at 111.1°C and 1.00 atm.

14.83 Calculate the enthalpy of combustion of 1.00 mol of CO at 125°C. The value at 25°C is −283 kJ/mol. **Hint:** Use Hess's law, including equations for heating or cooling the reactants and products of the combustion reaction.

14.84 (a) Calculate the final temperature after 5.52 kJ of energy is added to 200.0 mL of 0.500 M NaCl solution (density 1.01 g/mL, $c = 4.10$ J/g·°C) at 25.0°C.

(b) Calculate the number of kilojoules of heat produced when 0.100 mol of NaOH(aq) is allowed to react with 0.100 mol of HCl(aq).

$$\Delta H_{\text{neutralization}} = -55.2 \text{ kJ/mol of water formed}$$

(c) Calculate the final temperature when 100.0 mL of 1.000 M HCl at 25.0°C reacts with 100.0 mL of 1.000 M NaOH at 25.0°C.

14.85 Calculate the final temperature when 100.0 mL of 1.000 M HCl at 25.0°C reacts with 100.0 mL of 1.000 M NaOH at 25.0°C (density of final solution = 1.01 g/mL, $c = 4.10$ J/g·°C).

$$\Delta H_{\text{neutralization}} = -55.2 \text{ kJ/mol of water formed}$$

14.86 Calculate the heat produced when 1.000 gallon of octane, C_8H_{18}, reacts with oxygen to form carbon monoxide and water at 25°C. (The density of octane is 0.7025 g/mL; 1.000 gallon = 3.785 L. The complete combustion of octane yields 5450 kJ/mol, and the combustion of CO yields 283.0 kJ/mol.)

14.87 When 12.0 g of carbon reacted with oxygen to form CO and CO_2 at 25°C and constant pressure, 313.8 kJ of heat was liberated, and no carbon remained. Calculate the mass of oxygen that reacted.

15 Solutions

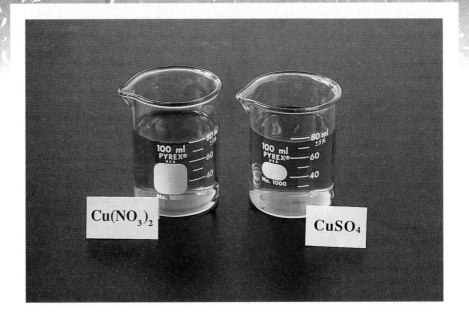

- 15.1 The Solution Process
- 15.2 Saturated, Unsaturated, and Supersaturated Solutions
- 15.3 Molality
- 15.4 Mole Fraction
- 15.5 Colligative Properties

Key Terms (Key terms are defined in the Glossary.)

boiling-point elevation (15.5)
colligative properties (15.5)
freezing-point depression (15.5)
ideal solution (15.5)
molal (15.3)
molality (15.3)

mole fraction (15.4)
nonvolatile (15.5)
osmotic pressure (15.5)
Raoult's law (15.5)
saturated solution (15.2)

solubility (15.2)
supersaturated solution (15.2)
unsaturated solution (15.2)
vapor-pressure lowering (15.5)
volatile (15.5)

Symbols

k_b (boiling-point elevation constant) (15.5)
k_f (freezing-point depression constant) (15.5)
m (molal) (15.3)

m (molality) (15.3)
$P°$ (vapor pressure of pure substance) (15.5)
ΔP (vapor-pressure lowering) (15.5)
π (osmotic pressure) (15.5)

Δt_b (boiling-point elevation) (15.5)
Δt_f (freezing-point depression) (15.5)
X_A (mole fraction of component A) (15.4)

The concept of solutions was introduced in Chapter 1, and molarity was considered in Chapter 11. Molarity is defined in terms of volume of solution, and volume can change with temperature, even if the mass of solvent and of solute does not change. When dealing with solution concentrations in systems at varying temperatures, we need concentration units that do not vary with temperature. Two such units are introduced here.

This chapter first considers the nature of the solution process (Section 15.1) and the quantity of solute that can dissolve in a given quantity of solvent (Section 15.2). Covered next are the temperature-independent measures of concentration—molality (Section 15.3) and mole fraction (Section 15.4). Finally, colligative properties of solutions are considered in Section 15.5. Normality, another unit of concentration, is discussed in Chapter 16.

15.1 The Solution Process

OBJECTIVE

■ to explain, using the energetics of the solution process, how to predict which types of solvents are most likely to dissolve a given solute

Solutions of two substances may be any of the types shown in Table 15.1. Most solutions discussed in this book are solutions of a solid, liquid, or gas in a liquid (very often water).

Substances dissolve in one another because the solution they form is more stable than a heterogeneous mixture of the individual substances. That statement implies that the attractions between the solvent molecules and the solute particles (atoms, molecules, or ions) are at least as great as the attractive forces between the particles of the solute or the molecules of the solvent. For an ionic solid to dissolve in a liquid, attractions between the solvent molecules and the ions of the solid must overcome the large attractions between the oppositely charged ions. The solvent molecules generally must have dipoles to be able to exert significant attractions on the charged ions. Thus, solutes with polar molecules are able to dissolve some ionic substances, but those with nonpolar molecules cannot. In contrast, nonpolar solutes are more apt to dissolve in nonpolar solvents. The solvent-solvent interactions cannot be too strong, otherwise the solvent-solute interactions will not be able to overcome them. (That is, if a solute is to dissolve in a polar solvent, considerable forces of attraction between the solvent molecules must be overcome. The solute-solvent attractions must be strong enough to do that. Nonpolar solutes do not have

Table 15.1 Types of Solutions Containing Two Substances

Type	Example
Solid in solid	Brass (zinc in copper)
Solid in liquid	Sugar water
Solid in gas	Moth balls in nitrogen (or in air)
Liquid in solid	Mercury in silver (dental amalgam)
Liquid in liquid	Martini (no olive)
Liquid in gas	Water vapor in nitrogen (or in air)
Gas in solid	Hydrogen in platinum
Gas in liquid	Club soda
Gas in gas	Oxygen in nitrogen (a mixture like air)

Figure 15.1 Dissolving Action of Water on an Ionic Solid

All of the ions at the surface of the solid are affected by water molecules. However, the ones at the edges and corners are affected more, since there are more water molecules in the vicinity and fewer ions of opposite charge. The ions in solution have the most water molecules around them.

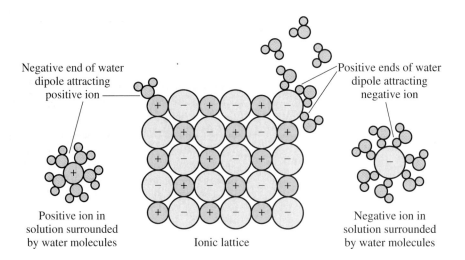

Negative end of water dipole attracting positive ion

Positive ends of water dipole attracting negative ion

Positive ion in solution surrounded by water molecules

Ionic lattice

Negative ion in solution surrounded by water molecules

sufficient attractions for polar solvents to do the job.) Thus, a general rule emerges: "Like dissolves like." That is, polar solvents are more likely to dissolve ionic and polar solutes, and nonpolar solvents are more likely to dissolve nonpolar solutes.

When sodium chloride dissolves in water, the H_2O molecules orient their dipoles around the Na^+ and Cl^- ions so that their oppositely charged ends are adjacent to each ion (Figure 15.1). Each sodium or chloride ion in solution is surrounded by many water molecules, lessening the attractions between the ions. Silver chloride, AgCl, does not dissolve in water. Evidently, the ion-dipole attractions are not sufficient to overcome the ion-ion attractions of this solid lattice. The nonpolar solvent benzene, C_6H_6, cannot dissolve either of these ionic compounds.

■ **EXAMPLE 15.1**

Which solvent—liquid ammonia (NH_3) or octane (C_8H_{18})—is more likely to dissolve each of the following solutes?   (a) AgCl   (b) C_6H_6 (c) H_2O

Solution

(a) AgCl is more likely to dissolve in the polar solvent, NH_3.

(b) C_6H_6 is more likely to dissolve in the nonpolar solvent, C_8H_{18}.

(c) H_2O is more likely to dissolve in NH_3 because both substances are polar and capable of hydrogen bonding.

Practice Problem 15.1   Which type of solvent—polar or nonpolar—is more likely to dissolve NH_2OH? ■

15.2 Saturated, Unsaturated, and Supersaturated Solutions

O B J E C T I V E

■ to determine if more or less solute can be held in a given solution from its concentration, the solubility of the solute at various temperatures, and the temperature

Most solutes will dissolve only to a certain extent in a given solvent at a certain temperature. For example, only 34.7 g of potassium chloride, KCl, will dissolve in 100 g of water at 20°C. If we add more than 34.7 g of KCl to the water at that temperature, the excess quantity of KCl will not dissolve. We state that the **solubility** of KCl in water at 20°C is 34.7 g per 100 g of water.

A solution containing as much solute as it can stably hold at a given temperature is said to be a **saturated solution.** If the solution contains less solute, it is

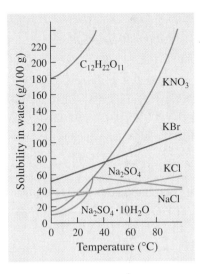

Figure 15.2 Variation of Solubility with Temperature

said to be an **unsaturated solution.** Thus, a solution containing 30.0 g of KCl in 100 g of water at 20°C is unsaturated.

A change in temperature affects the solubility of a solute in a given solvent (Figure 15.2). Most solid solutes get more soluble in liquid solvents as the temperature is raised, but gases dissolved in liquids get less soluble. If we have a saturated solution of such a solid solute at a high temperature and we lower the solution's temperature, we expect that the excess solute will crystallize from the solution at the lower temperature. For example, at 100°C, the solubility of KCl in water is 56.7 g per 100 g of water. If a solution containing 56.7 g of KCl in 100 g of water at 100°C is cooled to 20°C, only 34.7 g of the KCl will continue to be soluble, and 22.0 g of KCl will crystallize from the solution. (This type of process, called *recrystallization,* is often used to purify solid substances.)

When we cool saturated solutions of certain solutes, such as sodium acetate, the solute often has difficulty forming the first crystal, resulting in an unstable solution. More solute is in the solvent than is stable at the lower temperature. This type of solution is called a **supersaturated solution.** If a supersaturated solution is shaken or the inner surface of the container is scratched with a glass rod, the excess solute may crystallize out. A more certain way to get a supersaturated solution to crystallize is to add to it a tiny crystal of the solid solute, on which the excess solute can crystallize (Figure 15.3).

■ EXAMPLE 15.2

The number of grams of sodium acetate that will dissolve in 100 g of water is 119 g at 0°C and 170 g at 100°C.

(a) If 170 g of sodium acetate is placed in 100 g of water at 0°C, how much will dissolve?

(b) Is the resulting system a mixture or a solution?

(c) Ignoring any excess solid solute that may be present, is the solution saturated, unsaturated, or supersaturated?

(d) If the system is then raised to 100°C, is it a mixture or a solution?

(e) Is the solution saturated, unsaturated, or supersaturated?

(f) If the system is then carefully cooled back to 0°C, and no crystals appear, is the system a solution or a mixture?

(g) Is the solution saturated, unsaturated, or supersaturated?

Figure 15.3 Crystallization of a Solute from a Supersaturated Solution

Solution

(a) Only 119 g of sodium acetate will dissolve in 100 g of water at 0°C.

(b) The system is a mixture of the solution and the excess solid.

(c) The solution part is saturated; it holds as much solute as it stably can hold at 0°C.

(d) When the system is heated to 100°C, all 170 g of solute will dissolve, and the system is a solution.

(e) The solution is saturated; it holds 170 g in 100 g of water, which is the maximum it can hold stably *at this temperature.*

(f) Since no solute crystallizes, the system is still a solution.

(g) The solution is supersaturated, since the 170 g of solute that it holds is more than the 119 g that would be stable at this temperature. If a small crystal of sodium acetate is added, the excess sodium acetate will crystallize out (see Figure 15.3).

Practice Problem 15.2

(a) What would your answer to part (e) of Example 15.2 be if only 150 g of sodium acetate were used?

(b) What would your answer to part (d) be if 190 g of sodium acetate were used? ▪

15.3 Molality

▪ to do calculations using a temperature-independent measure of concentration

Molality is the number of moles of solute per kilogram of solvent.

The concept of molarity was defined and used in Chapter 11. Molarity is based on the volume of the solution, which may change with temperature. That is, if the solution is heated, its volume increases and its molarity decreases, even if the solution's components do not change. Since chemists sometimes need to measure concentrations in experiments in which different temperatures must be used, they developed another way to express concentrations. **Molality** is the number of moles of solute per kilogram of solvent:

$$\text{Molality} = \frac{\text{moles of solute}}{\text{kilograms of solvent}}$$

Molality is symbolized *m*. The unit of molality is **molal,** and the symbol for molal is m. The great similarities in name and meaning between molarity and molality can be confusing. However, the concepts differ in two ways: For molality, (1) *mass* is used, not volume; and (2) the *solvent* is measured, not the solution. Be sure to use M (capital) to represent molar, and m (lowercase) to represent molal.

▪ EXAMPLE 15.3

Calculate the molality of a solution prepared by dissolving 17.5 g of NaCl in 559 g of water.

Solution

The number of moles of solute and the number of kilograms of solvent are calculated first:

$$17.5 \text{ g NaCl}\left(\frac{1 \text{ mol NaCl}}{58.5 \text{ g NaCl}}\right) = 0.299 \text{ mol NaCl}$$

$$559 \text{ g H}_2\text{O}\left(\frac{1 \text{ kg H}_2\text{O}}{1000 \text{ g H}_2\text{O}}\right) = 0.559 \text{ kg H}_2\text{O}$$

The number of moles of NaCl is then divided by the number of kilograms of H_2O to obtain the molality:

$$m = \frac{0.299 \text{ mol NaCl}}{0.559 \text{ kg } H_2O} = 0.535 \text{ m}$$

Practice Problem 15.3 Calculate the molality of a solution prepared by dissolving 77.3 g of $C_6H_{12}O_6$ in 481 mL of water (density = 1.00 g/mL). ■

■ EXAMPLE 15.4

The solubility of LiCl in water at 0°C is 63.7 g per 100 g of water. Express this solubility as a molality.

Solution

$$\frac{(63.7 \text{ g LiCl})(1 \text{ mol LiCl}/42.4 \text{ g LiCl})}{(100 \text{ g } H_2O)(1 \text{ kg}/1000 \text{ g})} = \frac{1.50 \text{ mol LiCl}}{0.100 \text{ kg } H_2O} = \frac{15.0 \text{ mol LiCl}}{1 \text{ kg } H_2O}$$
$$= 15.0 \text{ m LiCl} ■$$

Many problems involving molality are similar to those involving molarity.

■ EXAMPLE 15.5

Calculate the number of moles of solute in a 1.75 m solution containing 0.750 kg of solvent.

Solution

Since molality is a ratio, it may be used as a factor:

$$0.750 \text{ kg} \left(\frac{1.75 \text{ mol solute}}{1 \text{ kg solvent}} \right) = 1.31 \text{ mol solute}$$

Practice Problem 15.5 Calculate the mass of $HClO_4$ needed to make a 3.00 m solution in 981 g of water. ■

■ EXAMPLE 15.6

Calculate the mass of solvent required to make a 0.480 m solution that contains 0.684 mol of solute.

Solution

The inverse of the ratio expressing the molality may also be used as a factor:

$$0.684 \text{ mol solute} \left(\frac{1 \text{ kg solvent}}{0.480 \text{ mol solute}} \right) = 1.42 \text{ kg solvent}$$

Practice Problem 15.6 Calculate the number of grams of water that must be mixed with 51.9 g of acetaldehyde, C_2H_4O, to make 6.92 m C_2H_4O. ■

■ EXAMPLE 15.7

What is the molality of a solution prepared by mixing a 2.00 m solution of KCl containing 3.50 mol of solute with a 4.00 m solution of KCl containing 2.50 mol of solute?

Solution

The total number of moles of KCl is 6.00 mol. The number of kilograms of solvent must be calculated for each solution and then added to get the number of kilograms of solvent in the final solution:

$$3.50 \text{ mol KCl}\left(\frac{1 \text{ kg solvent}}{2.00 \text{ mol KCl}}\right) = 1.75 \text{ kg solvent}$$

$$2.50 \text{ mol KCl}\left(\frac{1 \text{ kg solvent}}{4.00 \text{ mol KCl}}\right) = 0.625 \text{ kg solvent}$$

The total mass of solvent is 2.38 kg, so the molality is

$$m = \frac{6.00 \text{ mol KCl}}{2.38 \text{ kg solvent}} = 2.52 \text{ m}$$

Practice Problem 15.7 Calculate the molality of a certain solute if a 1.50 m solution of the solute is combined with a 3.00 m solution of the solute. The first solution contains 2.50 kg of solvent, and the second solution contains 750 g of the same solvent. ■

15.4 Mole Fraction

OBJECTIVE

■ to do calculations using another temperature-independent measure of concentration, in which the difference between solute and solvent is immaterial

The **mole fraction** of a component of a solution is simply the number of moles of the component divided by the total number of moles in the solution. The mole fraction of component A is symbolized by X_A. A mole fraction is similar to a percentage in that it represents a part of a whole. The sum of the mole fractions of all the components of a solution is equal to 1, just as the sum of the percentages of all components of a whole must be 100%. With mole fraction, solute and solvent are not differentiated; every component is treated the same. A mole fraction has no units because it is obtained by dividing moles by moles.

■ EXAMPLE 15.8

Calculate the mole fraction of methyl alcohol, CH_3OH, in a solution composed of 0.200 mol of methyl alcohol and 0.300 mol of ethyl alcohol, C_2H_5OH.

Solution

The mole fraction is

$$X_{CH3OH} = \frac{0.200 \text{ mol}}{0.200 \text{ mol} + 0.300 \text{ mol}} = 0.400$$

Practice Problem 15.8 Calculate the mole fraction of C_2H_4O and of C_2H_5OH in a solution containing equal masses of both. (**Hint:** Assume 100 g of each.) ■

■ EXAMPLE 15.9

Show that the sum of the mole fractions of the components of the solution of Example 15.8 is equal to 1.

Solution

The mole fraction of the ethyl alcohol is

$$X_{C_2H_5OH} = \frac{0.300 \text{ mol}}{0.200 \text{ mol} + 0.300 \text{ mol}} = 0.600$$

Adding this mole fraction to that of methyl alcohol (obtained in Example 15.8) gives

$$0.400 + 0.600 = 1.000$$

Practice Problem 15.9   Which, if either, of the two substances in Example 15.8 is regarded as the solute? ■

15.5 Colligative Properties

OBJECTIVE

■ to do calculations using some of the properties of the solution that depend on the concentrations of the dissolved particles, rather than their nature

Certain physical properties of solutions depend on the concentrations of the particles (molecules or ions) dissolved in the solution, rather than the nature of these solute particles. In this section, we will consider four such **colligative properties.** The word *colligative* means "depending on the number of particles."

Vapor-Pressure Lowering

The vapor pressure of a pure substance is dependent only on the nature of the substance and the temperature (Chapter 14). If we add a solute to the substance, its vapor pressure is lowered because the molecules of the substance cannot evaporate from the surface as rapidly as they could in the absence of the solute. The **vapor-pressure lowering** depends on the *concentration* of the solute particles, not on their nature. **Raoult's law** states that the vapor pressure of a component of an ideal solution is equal to the mole fraction of the component times its vapor pressure when it is a pure substance. That is, for component Z:

$$P_Z = X_Z P^\circ_Z$$

where P_Z is the vapor pressure of the component in the solution and P°_Z is its vapor pressure when it is a pure substance. Many solutions follow Raoult's law approximately. An **ideal solution** is defined as a solution that follows Raoult's law exactly.

■ **EXAMPLE 15.10**

The vapor pressure of methyl alcohol at 25°C is 105 torr. Calculate the vapor pressure of methyl alcohol at 25°C if 0.300 mol of glucose is dissolved in 1.00 mol of the alcohol.

Solution

The mole fraction of the alcohol is

$$X_{alcohol} = \frac{1.00 \text{ mol}}{1.30 \text{ mol}} = 0.769$$

The vapor pressure of the alcohol in the solution is

$$P_{\text{alcohol}} = X_{\text{alcohol}}P^{\circ}_{\text{alcohol}} = (0.769)(105 \text{ torr}) = 80.7 \text{ torr}$$

Practice Problem 15.10 Calculate the vapor pressure of benzene in a solution in which the mole fraction of benzene is 0.750. The vapor pressure of pure benzene is 96.0 torr. ■

■ **EXAMPLE 15.11**

Calculate the vapor-pressure *lowering* of the alcohol in Example 15.10.

Solution

The lowering of the vapor pressure may be defined as the vapor pressure of the pure substance minus its vapor pressure in the solution. It is symbolized ΔP:

$$\Delta P = P^{\circ} - P = 105 \text{ torr} - 80.7 \text{ torr} = 24 \text{ torr} \quad ■$$

Note that it is vapor-pressure lowering, *not* vapor pressure, that is directly proportional to the mole fraction of solute. For a solution of a **nonvolatile** (nonevaporating) solute in a **volatile** (easily evaporated) solvent, the greater the number of moles of *solute* present, the smaller is the mole fraction of the solvent. The smaller the mole fraction of the solvent, the lower is the vapor pressure of the solvent. Since the solute does not evaporate, the lower the vapor pressure of the solvent, the lower is the vapor pressure of the solution and the larger is the vapor-pressure lowering. That is, the smaller the value of P, the larger is the value of ΔP.

Freezing-Point Depression

Figure 15.4 Product for Common Use of Freezing-Point Depression

Calcium chloride is often used to melt ice on sidewalks. It is less harmful to lawns than is sodium chloride.

Addition of a solute to a solvent will cause the freezing point of the solvent to go down. The solute particles interfere with the process of making the orderly array of molecules characteristic of a crystalline solid. Antifreeze is added to car radiators to lower the freezing point of the water to prevent the coolant in the radiator from freezing in the winter. A salt is spread on icy sidewalks to make the freezing point of the water lower than the external temperature and thus melt the ice (Figure 15.4). In general, the more solute added per kilogram of solvent, the *greater* is the freezing-point depression and the *lower* is the freezing point. That is, the **freezing-point depression**, Δt_{f}, is directly proportional to the molality of the solute particles; the proportionality constant is symbolized as k_{f}:

$$\Delta t_{\text{f}} = k_{\text{f}}m$$

where

$$\Delta t_{\text{f}} = t_{\text{solvent}} - t_{\text{solution}}$$

The freezing point of a solution is always lower than that of the pure solvent, and the freezing-point depression may be defined as negative ($\Delta t_{\text{f}} = t_{\text{solution}} - t_{\text{sovlent}}$). Since the value of Δt_{f} is referred to as a depression, however, its value is often given without a minus sign. That practice will be followed in this book.

The value of k_{f} depends on the nature of the *solvent* and is independent of the solute used (as long as the solute does not ionize). The nature of the solute does not affect the freezing-point depression. Some k_{f} values and freezing points of

Table 15.2 Some Freezing-Point Depression Data

Solvent	Formula	Freezing Point (°C)	k_f(°C/m)
Benzene	C_6H_6	5.5	5.12
Naphthalene	$C_{10}H_8$	80.22	6.85
Phenol	C_6H_5OH	43	7.40
Water	H_2O	0.00	1.86

substances frequently used as solvents are given in Table 15.2. Note that each value of k_f in the table is for that particular solvent and that the solute dissolved in it does not make any difference in the value. (However, for ionic solutes, the total molality of cations *and* anions must be considered.)

■ **EXAMPLE 15.12**

Calculate the freezing point of a solution containing 0.300 mol of a nonionic solute in 2.00 kg of benzene.

Solution

The freezing-point depression is proportional to the molality. The molality is

$$m = \frac{0.300 \text{ mol}}{2.00 \text{ kg}} = 0.150 \text{ m}$$

The value of k_f is found in Table 15.2. Then,

$$\Delta t_f = k_f m = (5.12°C/m)(0.150 \text{ m}) = 0.768°C$$

The freezing point of the solution is equal to the freezing point of the solvent, pure benzene, minus the freezing-point depression (Δt_f):

$$t_{solution} = t_{solvent} - \Delta t_f = 5.5°C - 0.768°C = 4.7°C$$

Note that you must differentiate between the freezing point and the freezing-point depression, as well as between the freezing point of the solvent and the freezing point of the solution.

Practice Problem 15.12   Calculate the freezing point of a 0.200 m solution of anthracene, $C_{14}H_{10}$, in naphthalene. ■

The molar mass of a solute may be determined from the number of grams of solute dissolved in a given mass of solvent and the freezing-point depression for the solution. The freezing-point depression yields the number of moles of solute per kilogram of solvent, and the mass data yield the number of grams per kilogram of solvent. Dividing the number of grams per kilogram by the number of moles per kilogram yields the number of grams per mole—the molar mass.

■ **EXAMPLE 15.13**

A solution containing 3.00 g of nonionic solute in 80.0 g of water freezes at $-1.22°C$. Calculate the molar mass of the solute.

Solution

The molality of the solution is determined from the freezing-point depression and the value of k_f for water. The freezing-point depression for this solution is 1.22°C.

$$m = \frac{\Delta t}{k_f} = \frac{1.22°C}{1.86°C/m} = 0.656 \text{ m}$$

The number of moles of solute in 1.00 kg of solvent is therefore 0.656 mol. The mass of solute per 1.00 kg of solvent is calculated next:

$$1.00 \text{ kg H}_2\text{O}\left(\frac{3.00 \text{ g solute}}{0.0800 \text{ kg H}_2\text{O}}\right) = 37.5 \text{ g solute}$$

The molar mass is the ratio of the number of grams in a kilogram to the number of moles in a kilogram:

$$\text{Molar mass} = \frac{37.5 \text{ g}}{0.656 \text{ mol}} = 57.2 \text{ g/mol} \quad ■$$

Boiling-Point Elevation

The presence of a nonvolatile solute causes the boiling point of a solution to be raised in comparison to the boiling point of the pure solvent. The antifreeze added to water in car radiators causes the liquid in the radiator to boil at a temperature higher than 100°C, which is important in summertime to avoid engine overheating and radiator boilover. Like the freezing-point depression, the **boiling-point elevation** is directly proportional to the molality of the solute particles. In this case, however, the effect is positive; the boiling point is *raised*. Thus, the presence of a solute makes the liquid range of a solution longer, extending it at both the freezing-point end and the boiling-point end. That is, the solution is in the liquid state over a wider range of temperatures than is the pure solvent (Figure 15.5). The equation that relates boiling-point elevation to molality is the same as the one for freezing-point depression, except that the constant is k_b. Even for a given solvent, the values of k_f and k_b are different. Some boiling-point elevation data are given in Table 15.3.

■ **EXAMPLE 15.14**

Calculate the boiling-point elevation for a 0.200 m solution of glucose, $C_6H_{12}O_6$, in water. Glucose is nonvolatile and nonionic.

Figure 15.5 Liquid Ranges of a Solvent and Its Solution

The freezing point of the solution is lower than that of the solvent, and the boiling point of the solution is higher than that of the solvent. The liquid range has been extended on both ends.

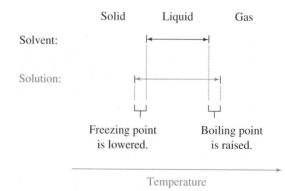

Table 15.3 *Some Boiling-Point Elevation Data*

Solvent	Formula	Normal Boiling Point (°C)	k_b(°C/m)
Benzene	C_6H_6	80.1	2.53
Carbon tetrachloride	CCl_4	−22.99	5.03
Chloroform	$CHCl_3$	61.2	3.63
Water	H_2O	100.0	0.512

Solution

The boiling-point elevation constant for water is 0.512°C/m (Table 15.3). Thus,

$$\Delta t_b = k_b m = (0.512°C/m)(0.200\ m) = 0.102°C$$

Practice Problem 15.14   Calculate the boiling-point elevation for a 0.200 m solution of ethylene glycol, $C_2H_4(OH)_2$, in water. Ethylene glycol is the main constituent of permanent antifreeze. ■

■ **EXAMPLE 15.15**

Calculate the boiling point of a 0.200 m solution of glucose in water at 1.00 atm.

Solution

The boiling-point elevation was calculated in Example 15.14. The boiling point is therefore

$$t_{solution} = t_{solvent} + \Delta t_b = 100.00°C + 0.102°C = 100.10°C$$

Note the difference between boiling point and boiling-point elevation.

Practice Problem 15.15   Calculate the boiling point of 0.200 m solution of naphthalene (a nonvolatile solute) in benzene. ■

Osmotic Pressure

When a solution is separated from its solvent by a semipermeable membrane, the solvent molecules can pass through the membrane but the solute particles cannot. The word *semipermeable* in this context means "porous to some molecules and not to others." The rate of passage of solvent molecules from pure solvent to solution is greater than the rate of their passage in the other direction because the presence of the solute particles reduces the number of solvent molecules at the membrane. Therefore, if all other factors are equal, the liquid will rise on the solution side as more of the solvent passes through the membrane to that side than in the other direction (Figure 15.6). The more concentrated the solute, the higher the liquid level rises on the solution side. The effect can be measured by seeing how much external pressure it takes to keep the liquid levels even on the two sides. That pressure is equal to the **osmotic pressure** of the solution. The osmotic pressure is governed by an equation analogous to the ideal gas law:

$$\pi V = nRT$$

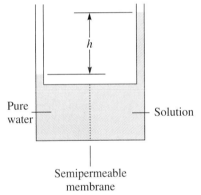

Figure 15.6 Osmotic Pressure
Because water molecules can pass through the semipermeable membrane from the pure water side to the solution side more easily than in the other direction, the level of pure water falls and that of the solution rises. The difference in the heights of the levels, *h*, corresponds to the osmotic pressure of the solution.

A process called *reverse osmosis* is used to remove salts from seawater to make drinking water for human consumption. If a pressure greater than the osmotic pressure is applied to the solution side of an apparatus such as shown in Figure 15.6, water is forced from the solution (the seawater) to the pure solvent (water) side. This process is used industrially for water purification.

where π is the osmotic pressure, V is the volume of the solution, n is the number of moles of solute, R is the gas constant, and T is the absolute temperature. If $0.0821 \ L \cdot atm/mol \cdot K$ is used for the value of R and the volume is given in liters, then π is in atmospheres. This equation can be rearranged to relate the osmotic pressure to *molarity*—the number of moles of solute per liter of solution:

$$\pi = \frac{nRT}{V} = \left(\frac{n}{v}\right)RT = MRT \qquad \text{Thus,} \qquad M = \frac{n}{V} = \frac{\pi}{RT}$$

■ **EXAMPLE 15.16**

Calculate the osmotic pressure of a 0.200 M solution of glucose, $C_6H_{12}O_6$, in water at 27°C.

Solution

$$\pi = \frac{nRT}{V} = \left(\frac{n}{V}\right)RT$$

$$= (0.200 \ M)(0.0821 \ L \cdot atm/mol \cdot K)(300 \ K) = 4.93 \ atm \ ■$$

Example 15.16 shows that osmotic pressure is rather high, even for a dilute solution.

Osmotic pressure is extremely important in biological systems. For example, in trees, it helps in moving liquids from the root systems to the tops. Water returns from human tissue to blood capillaries because of the greater concentration of solutes in the blood. The use of saline solution when replacing blood in accident victims must be carefully controlled so that the fluid's osmotic pressure is the same as that of blood.

■ ■ ■ ■ ■ ■ ■ SUMMARY ■ ■ ■ ■ ■ ■ ■

Substances dissolve in solvents to form solutions because the attractions between solute particles and solvent particles are at least as strong as those holding the solute or solvent particles together. In general, nonpolar solutes tend to dissolve in nonpolar solvents, and polar and ionic solutes tend to dissolve in polar solvents. (Section 15.1)

Most solutes have a certain limit to their solubility in a given solvent at a given temperature. A solution at that concentration is said to be saturated. If less of the solute is dissolved in the solvent at that temperature, the solution is unsaturated. If more is dissolved, the solution is supersaturated. A change in temperature changes the solubility of a solute in a given solvent. Raising the temperature makes most solids more soluble in liquids but makes gases less soluble in liquids. Supersaturated solutions are not stable; the excess solute is likely to crystallize out of solution if the solution is shaken, if the inside of the container is scratched, or especially if a crystal of the solute is added. Supersaturated solutions are typically prepared by dissolving a large quantity of solute in a given quantity of solvent at a high temperature, then carefully cooling the resulting solution. If the solute has difficulty crystallizing, a supersaturated solution may result. (Section 15.2)

Molality (m) is a temperature-independent measure of concentration, defined as the number of moles of solute per kilogram of solvent. It differs from molarity (M) in that it is based on a *mass of solvent*, rather than a volume of solution. Like molarity, molality can be used as a factor to solve problems. However, molality is also used in problems involving colligative properties. (Section 15.3)

Another temperature-independent measure of concentration is mole fraction. It does not distinguish between solute and solvent. The mole fraction is defined as the number of moles of one component, say component A, divided by the total number of moles in the solution:

$$X_A = \frac{\text{number of moles of A}}{\text{total number of moles}}$$

(Section 15.4)

The presence of a solute affects some of the physical properties of a solution, but the identity of the solute makes little difference in the colligative properties. Vapor-pressure lowering, freezing-point depression, boiling-point elevation, and osmotic pressure are four such properties.

The vapor pressure of a solvent in the presence of a nonvolatile solute is given by Raoult's law:

$$P_A = X_A P_A^\circ$$

where P_A is the vapor pressure of the solvent in the solution, X_A is the mole fraction of A, and P_A° is the vapor pressure of the solvent when pure. Since X_A must be lower than 1 in a solution of A, P_A must be lower than P_A°; a vapor-pressure lowering has resulted because of the presence of the solute.

The freezing point of a solvent is lowered by the presence of a solute, and the boiling point of a solvent is raised by the presence of a solute. The freezing-point depression and the boiling-point elevation are both directly proportional to the molality of the solute particles. Determining the molality therefore allows you to calculate the freezing point or boiling point of a solution; conversely, determining the freezing point or boiling point allows you to calculate the molality. With the molality and other data, you can calculate the number of moles and the molar mass, if required.

Osmotic pressure is another property due to dissolved substances. The presence of solute particles lowers the ability of solvent molecules to pass through a semipermeable membrane. Osmotic pressure is very important in biological systems, and an application of the theory behind osmotic pressure allows purification of seawater. The osmotic pressure of a solution, π, is proportional to the molarity (the number of moles per liter):

$$\pi V = nRT$$

or

$$\pi = \left(\frac{n}{V}\right)RT$$

where R is the ideal gas law constant. Osmotic pressure measurement is a sensitive means of determining the number of moles of solute in a given volume of solution and, together with other data, the molar mass. (Section 15.5)

Items for Special Attention

■ Molarity and molality are very similar; be careful not to confuse them. Also, be sure to use M to represent molar and m to represent molal.

■ Osmotic pressure is a colligative property.

■ Be sure to remember the difference between *vapor pressure* and *vapor-pressure lowering,* between *freezing point* and *freezing-point depression,* and between *boiling point* and *boiling-point elevation.*

Self-Tutorial Problems

15.1 A 6.00-g sample of a solid substance is placed in 50.0 g of water at 20°C, and all of the solid dissolves. Then another 3.00 g of the substance is added at 20°C, and all of it dissolves.

   (a) Is the first solution saturated, unsaturated, or supersaturated?

   (b) What can you tell about the saturation, unsaturation, or supersaturation of the final solution?

15.2 List as many differences as you can between molarity and molality.

15.3 List as many similarities and differences as you can between mole fraction and percent composition by mass.

15.4 In a solution of CH_2O in water, the mole fraction of H_2O is equal to 0.850. Calculate the mole fraction of CH_2O.

15.5 Calculate the freezing-point depression of a solution of each of the following in water:

   (a) 0.100 m formaldehyde (CH_2O)

   (b) 0.100 m methyl alcohol (CH_3OH)

   (c) 0.100 m glucose ($C_6H_{12}O_6$)

15.6 In a solution of CH_3OH and C_2H_5OH in water, the mole fraction of H_2O is equal to 0.650 and that of C_2H_5OH is 0.100. Calculate the mole fraction of CH_3OH.

15.7 Consider a 0.15 m solution and a 0.30 m solution of CH_3OH in water.

   (a) Which solution has the greater freezing-point depression?

   (b) Which one has the higher freezing point?

15.8 Consider a 0.10 m solution and a 0.20 m solution of $C_6H_{12}O_6$, a nonvolatile solute, in water.

   (a) Which solution has the greater boiling-point elevation?

   (b) Which one has the higher boiling point?

15.9 Which measure of concentration is used with (a) vapor-pressure lowering, (b) freezing-point depression, (c) boiling-point elevation, and (d) osmotic pressure?

15.10 Using Figure 15.2, determine whether KNO_3 or KBr is more soluble in water at (a) 20°C and (b) 80°C.

■ ■ ■ PROBLEMS ■ ■ ■

15.1 The Solution Process

15.11 Explain why CH_3OH is soluble in water, but C_6H_6 is not.

15.12 Explain why CH_3OH is soluble in CH_2O but C_6H_6 is not.

15.13 Explain why $C_{10}H_8$ is soluble in C_6H_6, but NaCl is not.

15.14 (a) When water dissolves solid sodium chloride, which end, if either, of a water molecule's dipole is expected to be nearer to an Na^+ ion?

   (b) When water dissolves solid sodium chloride, which end, if either, of a water molecule's dipole is expected to be nearer to a Cl^- ion?

   (c) When water dissolves liquid formaldehyde, which end, if either, of a water molecule's dipole is expected to be nearer to the oxygen atom of the CH_2O molecule?

15.15 Glucose, $C_6H_{12}O_6$, has several covalently bonded —OH groups in its molecule. Explain why glucose is so soluble in water.

15.2 Saturated, Unsaturated, and Supersaturated Solutions

15.16 From Figure 15.2, determine the solubility of KNO_3 in water at 60°C.

15.17 Which solute in Figure 15.2 has the solubility that is least temperature-dependent?

15.18 From Figure 15.2, determine whether each of the following solutions is saturated, unsaturated, or supersaturated:

   (a) 53 g KBr in 100 g of water at 50°C

   (b) 53 g KBr in 100 g of water at 0°C

   (c) 20 g KBr in 100 g of water at 30°C

15.19 From Figure 15.2, determine whether a solution of 35.0 g of NaCl in 100 g of water at 25°C is saturated, unsaturated, or supersaturated.

15.20 Which compound in Figure 15.2 changes solubility most with increasing temperature?

15.21 If you place 3 teaspoons of table sugar in a glass of iced tea, it does not all dissolve right away. Examine Figure 15.2 and conclude if the sugar ($C_{12}H_{22}O_{11}$) is not very soluble or not soluble very quickly.

15.22 A tiny crystal of a solid substance is added to an aqueous solution of the same substance. What would you expect to see if the original solution was (a) supersaturated, (b) unsaturated, or (c) saturated?

15.23 Explain how to prepare 100 mL of a saturated solution of a given salt in water.

15.3 Molality

15.24 Calculate the molality of a solution of 7.12 g of $C_{12}H_{22}O_{11}$ in 10.0 g of water.

15.25 Calculate the molality of a solution containing 0.2083 mol of CsCl and 100.0 g of water.

15.26 Calculate the number of moles of solute in a 2.00 m solution that was made with 50.0 g of water.

15.27 Calculate the mass of solvent in a 4.00 m solution containing 7.00 mol of solute.

15.28 Calculate the molality of a solution of 22.73 g of NaCl in 100.0 g of water.

15.29 Calculate the molality of chloride ions in a solution of 12.7 g of $MgCl_2$ in 200 g of water.

15.30 Calculate the molality of 108.7 mL of a solution containing 0.2083 mol of CsCl and 100.0 g of water.

15.31 Calculate the number of moles of $CuCl_2$ needed to make a solution with 500 g of water that is 0.800 m in Cl^-.

15.32 Calculate the mass of solvent in a 2.50 m solution containing 0.851 mol of solute.

15.4 Mole Fraction

15.33 Calculate the mole fraction of water in each of the following solutions:

(a) 10.0 mol of C_2H_5OH, 10.0 mol of CH_2O, and 10.0 mol of H_2O

(b) 10.0 g of C_2H_5OH, 10.0 g of CH_2O, and 10.0 g of H_2O

15.34 Calculate the mole fraction of ammonia in a 3.00 m aqueous solution of ammonia. (**Hint:** Assume that 1.00 kg of water is present.)

15.35 Calculate the mole fraction of water in each of the following solutions:

(a) 1.50 mol H_2O and 10.0 g C_2H_5OH

(b) 73.6 g CH_2O and 44.4 g H_2O

15.36 Calculate the molality of alcohol in water if the mole fraction of alcohol is 0.150. (**Hint:** Assume 1.00 mol total in the solution.)

15.37 Calculate the mole fraction of NaCl in a solution containing 5.35 g NaCl and 74.5 g $CaCl_2$ in 500.0 g of water.

15.5 Colligative Properties

15.38 Calculate the vapor pressure of benzene, C_6H_6, at 25°C in an ideal solution containing 2.50 mol of benzene and 1.75 mol of toluene. The vapor pressure of pure benzene is 96.0 torr.

15.39 Calculate the vapor pressure of toluene, C_7H_8, at 25°C in an ideal solution containing 1.25 mol of benzene and 2.10 mol of toluene. The vapor pressure of pure toluene is 27.0 torr.

15.40 (a) The temperature of a sample of pure water is −2°C. Is the sample in the liquid or solid state?

(b) If the freezing point of a sample of water is depressed to −4°C by the addition of salt, in what state will the sample be at −2°C?

(c) Explain why salt is used on icy streets and sidewalks.

15.41 The freezing point of pure ethylene glycol, $C_2H_4(OH)_2$, which is used in permanent antifreeze, is −11.5°C. Would a solution of 5 g of water in 100 g of ethylene glycol freeze below −11.5°C, at −11.5°C, or above −11.5°C? Explain.

15.42 Calculate the freezing-point depression of a 0.150 m solution of (a) CH_3OH in water, (b) C_2H_5OH in water, and (c) C_3H_7OH in water.

15.43 Sucrose, $C_{12}H_{22}O_{11}$, is nonvolatile and nonionic. For a 0.750 m solution of sucrose in water, calculate:

(a) The freezing point of the solution

(b) The boiling point of the solution at 1.00 atm

15.44 Glucose, $C_6H_{12}O_6$, is nonvolatile and nonionic. For a 1.75 m solution of glucose in water, calculate:

(a) The freezing point of the solution

(b) The boiling point of the solution at 1.00 atm

15.45 Calculate the freezing point of (a) 0.100 m benzene in naphthalene and (b) 0.100 m naphthalene in benzene.

15.46 Calculate the freezing point of (a) 0.150 m benzene in naphthalene and (b) 0.150 m naphthalene in benzene.

15.47 An aqueous solution freezes at −0.457°C. Calculate the molality of the nonionic solute.

15.48 An aqueous solution freezes at −1.50°C. Calculate the molality of the nonionic solute.

15.49 A solution of a nonionic solute in benzene freezes at 3.00°C. Calculate the molality.

15.50 A solution of a nonionic solute in naphthalene freezes at 78.17°C. Calculate the molality.

15.51 Calculate the molar mass of a nonionic solute if a solution containing 13.00 g in 250.0 g of water freezes at −1.22°C.

15.52 Calculate the molar mass of a nonionic solute if a solution containing 2.00 g in 110.0 g of benzene freezes at 4.1°C.

15.53 Calculate the osmotic pressure of a 0.250 M solution of glucose in water at 25°C.

15.54 A 5.00-g sample of a nonionic solute is dissolved in 20.0 g of water. The solution freezes at −1.20°C. Calculate the molar mass of the solute.

15.55 Calculate the osmotic pressure in torr of a 0.0111 M solution of sucrose (table sugar) in water at 25°C.

15.56 A 22.7-g sample of a nonionic solute is dissolved in 210 g of water. The solution freezes at −1.27°C. Calculate the molar mass of the solute.

15.57 At 25°C, the vapor pressure of pure benzene, C_6H_6, is 100 torr, and that of pure ethyl alcohol, C_2H_5OH, is 44.0 torr. Assuming ideal behavior, calculate the vapor pressure at 25°C of a solution that contains 25.0 g of each.

15.58 At 25°C, the vapor pressure of methanol, CH_3OH, is 96.0 torr. What is the mole fraction of CH_3OH in a solution in which the partial pressure of CH_3OH is 26.5 torr at 25°C?

■ ■ ■ GENERAL PROBLEMS ■ ■ ■

15.59 Calculate the total pressure of an ideal solution containing 1.75 mol of benzene and 1.25 mol of toluene. The vapor pressures of the pure components are 105 torr and 34.0 torr, respectively.

15.60 Calculate the total pressure of an ideal solution containing 0.251 mol of benzene and 0.125 mol of toluene. The vapor pressures of the pure components are 105 torr and 34.0 torr, respectively.

15.61 Calculate the mole fraction of CH_3OH in a solution containing equal masses of CH_3OH and CH_2O.

15.62 Calculate the mole fraction of C_2H_5OH in a solution containing equal masses of C_2H_5OH and C_3H_7OH.

15.63 Calculate the freezing point of an aqueous solution that boils at 102.1°C at 1.00 atm.

15.64 A concentrated solution of ammonia in water is 12 M at 25°C and 1.0 atm. Calculate the ratio of moles of ammonia per liter of this solution to moles per liter of ammonia gas at 25°C and 1.00 atm.

15.65 The boiling point at 1.00 atm of a solution containing 14.19 g of a nonionic solute and 100.0 g of water is 101.22°C. Calculate the molar mass of the solute.

15.66 The boiling point at 1.00 atm of a solution containing 0.625 g of a nonionic solute and 5.00 g of water is 100.15°C. Calculate the molar mass of the solute.

15.67 Calculate the number of moles of Na^+ ions in a 0.200 m solution of Na_2SO_4 containing 0.650 kg of water.

15.68 The percent composition of a nonionic solute is 40.0% C, 6.7% H, and 53.3% O. If a solution containing 2.90 g of this substance in 106 g of water freezes at −0.41°C, calculate the molecular formula of the substance.

15.69 The percent composition of a solute is 93.75% C and 6.25% H. If a solution containing 2.15 g of this substance in 10.0 g of benzene freezes at −2.90°C, calculate the molecular formula of the substance.

15.70 The molality of a solution can be calculated from its molarity if the density of the solution is known. In such a calculation, you must keep track not only of the units involved but of the materials to which the units apply. Calculate the molality of a 0.324 M $CaCl_2$ solution that has a density of 1.029 g/mL.

15.71 The vapor pressure of pure benzene C_6H_6, at 50°C is 268 torr. How many moles of nonvolatile solute per mole of benzene is needed to prepare a solution having a vapor pressure of 177.5 torr at 50°C?

15.72 The density of 10.0% by mass KCl solution in water is 1.06 g/mL. Calculate the molarity, molality, and mole fraction of KCl in this solution.

15.73 The molality of a solution can be calculated from its molarity if the density of the solution is known. In such a calculation, you must keep track not only of the units involved but of the materials to which the units apply. Calculate the molality of a 0.259 M $Pb(NO_3)_2$ solution that has a density of 1.073 g/mL.

15.74 Calculate the molality of methyl alcohol, CH_3OH, in an aqueous solution in which the mole fraction of methyl alcohol is 0.123.

15.75 Calculate the molality of methyl alcohol, CH_3OH, in ethyl alcohol, C_2H_5OH, in which the mole fraction of methyl alcohol is 0.200.

15.76 A solution is described as 1.35×10^2 m CH_3OH in water. Calculate the molality of water in CH_3OH.

15.77 A solution is described as 72.0 m CH_3OH in ethyl alcohol, C_2H_5OH. Calculate the molality of ethyl alcohol in CH_3OH.

15.78 A solution contains 22.7 g of C_2H_5OH and 75.0 g of H_2O.

(a) Calculate the molality of C_2H_5OH in the water.

(b) Calculate the molality of H_2O in the alcohol.

(c) Calculate the mole fraction of each component.

(d) What does identifying the solvent have to do with determining the mole fractions?

15.79 A solution contains 50.0 g of CH_2O and 42.7 g of CH_3OH.

(a) Calculate the molality of CH_2O in CH_3OH.

(b) Calculate the molality of CH_3OH in CH_2O.

(c) Calculate the mole fraction of each component.

(d) What does identifying the solvent have to do with determining the mole fractions?

15.80 Which compound in Figure 15.2 could be purified best by dissolving it in hot water and cooling the solution?

15.81 For an ideal solution of one nonvolatile solute in a solvent, show that

$$\Delta P_{solvent} = X_{solute} P^\circ_{solvent}$$

16 Oxidation Numbers

- 16.1 Assigning Oxidation Numbers
- 16.2 Using Oxidation Numbers in Naming Compounds
- 16.3 Periodic Variation of Oxidation Numbers
- 16.4 Balancing Oxidation-Reduction Equations
- 16.5 Equivalents and Normality (*Optional*)

■ Key Terms (*Key terms are defined in the Glossary.*)

"control" of shared electrons (16.1)
disproportionation (16.4)
equivalent (16.5)
equivalent mass (16.5)
half-reaction (16.4)
half-reaction method (16.4)

normal (16.5)
normality (16.5)
oxidation (16.4)
oxidation number (16.1)
oxidation-reduction reaction (16.4)

oxidation state (16.1)
oxidizing agent (16.4)
reducing agent (16.4)
reduction (16.4)
Stock system (16.2)

■ Symbols/Abbreviations

e⁻ (electron) (16.4)
equiv (equivalent) (16.5)

N (normal) (16.5)
N (normality) (16.5)

In Chapter 5, you learned to write formulas for ionic compounds from the charges on the ions and to recognize the ions from the formulas of the compounds. For example, you know that calcium chloride is $CaCl_2$ and that $FeCl_3$ contains Fe^{3+} ions. We cannot make comparable deductions for covalent compounds because they have no ions; there are no charges to balance. To make similar predictions for species with covalent bonds, we need to use the concept of oxidation number, also called oxidation state. A system with some arbitrary rules allows us to predict formulas for covalent compounds from the positions of the elements in the periodic table and also to balance complicated equations for oxidation-reduction reactions.

Section 16.1 introduces the concept of oxidation number and how to calculate the oxidation number of an element from the formula of the compound or ion of which it is a part. Section 16.2 describes how to use the oxidation numbers to name compounds, formalizing and extending the rules given in Chapter 6. Section 16.3 shows how to predict possible oxidation numbers from the position of the element in the periodic table and how to use these oxidation numbers to write probable formulas for covalent compounds. Section 16.4 presents a systematic method for balancing equations in which oxidation numbers change. In Section 16.5, oxidation-reduction and acid-base equations are used to define a new chemical unit of quantity and a new concentration unit.

16.1 Assigning Oxidation Numbers

The **oxidation number,** also known as the **oxidation state,** of an atom in a compound (or in a free element or polyatomic ion) is defined as the number of electrons that an uncombined atom of the element has minus the number that is assigned to the atom in the compound (or element or ion):

$$\text{Oxidation number} = \text{Number of electrons in uncombined atom} - \\ \text{Number of electrons in atom in compound}$$

Alternatively, we can use the numbers of *valence* electrons in the free atom and in the atom in the compound to calculate the oxidation number.

For compounds of monatomic ions, the number of electrons assigned to each ion is easy to calculate, and the oxidation number turns out to be equal to the charge on the ion. For example, in magnesium chloride, the magnesium ion has 10 electrons remaining after transferring 2 electrons to the chlorine atoms. Since an uncombined magnesium atom has 12 electrons, the oxidation number of magnesium in this compound is $12 - 10 = +2$. Each chlorine atom started out with 17 electrons and has gained 1 electron to form the chloride ion in the compound, so the oxidation number of chlorine is $17 - 18 = -1$. Alternatively, the number of valence electrons in magnesium went from 2 to 0, for a $+2$ oxidation number. The number in chlorine went from 7 to 8, for a -1 oxidation number. In each case, the oxidation number is equal to the charge on the ion.

However, we do not really need oxidation numbers when working with compounds of monatomic ions; we can use the charges to write formulas, and we can predict the charges from the periodic table or the formulas. For working with compounds with covalent bonds and polyatomic ions (which also contain covalent bonds), however, we need oxidation numbers. Oxidation numbers for such species must be assigned by allocating the shared electrons in the covalent bonds to one or the other of the atoms, or equally to both if the atoms are the same. We arbitrarily allocate all the electrons in the bond to the more electronegative of the atoms involved (see Figure 5.1). In most cases, that will be the atom farther to the right or farther toward the top of the periodic table. The electrons are assigned to the more nonmetallic of the elements involved in the bond. For example, in SCl_2, the bonds are covalent, so there are no charges. Since sulfur lies closer to the metal area of the periodic table, each chlorine atom is assigned **"control" of the shared electrons.** This control is not physically real; if the chlorine atoms really had complete control of the electrons, they would be chloride ions. This compound is covalent. However, the concept of control is useful for the purposes of assigning oxidation numbers.

We can assign oxidation numbers from the electron dot diagram of a compound, such as SCl:

$$:\!\ddot{C}l\!: \; :\!\ddot{S}\!: \; :\!\ddot{C}l\!:$$

The sulfur atom, when uncombined, has 6 valence electrons. In the compound, the shared electrons are "controlled" by the chlorine atoms, so the sulfur atom retains "control" of only the unshared pairs of electrons. It "controls" only 4 electrons. Its oxidation state is $6 - 4 = +2$. The uncombined chlorine atom has 7 valence electrons; but in the compound, each chlorine atom "controls" 8, so its oxidation number is $7 - 8 = -1$.

	Sulfur	*Chlorine*
Number of valence electrons of free atom	6	7
minus Number of electrons "controlled" in compound	-4	-8
Oxidation number	$+2$	-1

Note that an oxidation number, like the charge on an ion, is assigned to *each atom* in a compound. Thus, *each* chlorine atom in SCl_2 has an oxidation number of -1.

In most cases, assigning oxidation numbers from formulas by drawing electron dot diagrams is more time-consuming than necessary. The set of rules that follows speeds the assignment of most oxidation numbers. If these rules do not work for assigning oxidation numbers in a particular case, the electron dot method may be used.

Rules for Assigning Oxidation Numbers

1. The total of the oxidation numbers of all the atoms (not simply all the elements) is equal to the charge on the atom, molecule, or ion. Thus, for any compound, the sum of the oxidation numbers totals zero because the net charge on any compound is zero. The sum of the oxidation numbers of the atoms of any polyatomic ion is equal to the charge on that ion.

2. The oxidation number of any uncombined element is zero. Since the element is not combined with any other element, there is either one atom alone or a group of identical atoms. Since the atoms are alike, the electrons are shared equally, and the number "controlled" is equal to the number in the free atom. For example, the oxidation number of potassium as a free element and that of bromine in Br_2 are determined as follows:

$$K\cdot \qquad :\!\ddot{B}r\!:\!\ddot{B}r\!:$$

	Potassium as Free Element	*Bromine in Br_2*
Number of valence electrons of free atom	1	7
minus Number of electrons "controlled"	-1	-7
Oxidation number	0	0

Thus, this rule is actually a corollary to rule 1; the oxidation number is equal to the charge on the species—zero in these cases.

3. The oxidation number of a monatomic ion is equal to the charge on the ion. This rule is also a corollary of rule 1. If the sum of all the oxidation numbers is equal to the charge and there is only one oxidation number because there is only one atom, then that oxidation number must equal the charge on the monatomic ion.

4. The oxidation number of any alkali metal atom *in any of its compounds* is always equal to $+1$.

5. The oxidation number of any alkaline earth metal atom *in any of its compounds* is always equal to $+2$.

6. The oxidation number of hydrogen in its compounds is equal to $+1$, except in its binary compounds with an alkali metal, an alkaline earth metal, or aluminum, when it is -1.

7. The oxidation number of oxygen in its compounds is generally -2.

8. The oxidation number of a halogen atom in its compounds is -1, except for chlorine, bromine, or iodine when combined with oxygen or another halogen that is nearer the top of the periodic table.

The examples that follow illustrate these rules.

■ **EXAMPLE 16.1**

Determine the oxidation numbers of phosphorus and chlorine in phosphorus trichloride, PCl_3.

Solution

In this compound, the oxidation number of each chlorine atom is -1 (rule 8). To make the total of all the oxidation numbers for PCl_3 equal zero, the phosphorus must have an oxidation number of $+3$. To do this calculation systematically, let x equal the oxidation number of phosphorus in PCl_3. Then

$$x + 3(-1) = 0$$

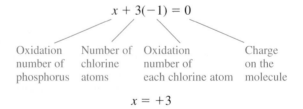

| Oxidation number of phosphorus | Number of chlorine atoms | Oxidation number of each chlorine atom | Charge on the molecule |

$$x = +3$$

Practice Problem 16.1   Determine the oxidation number of sulfur in SO_2. ■

ENRICHMENT

In peroxides, the oxidation number of oxygen is -1, and in superoxides, it is $-\frac{1}{2}$. In oxygen's rarely encountered compounds with fluorine—OF_2 and O_2F_2—the oxygen atoms have oxidation states of $+2$ and $+1$, respectively. The only common superoxides are those of hydrogen and the alkali metals from potassium through francium. Peroxides occur with hydrogen, sodium, barium, and chromium, and in these compounds, each of these elements exists in its highest oxidation state. Therefore, the oxidation number of oxygen is -2 in most of its compounds.

■ EXAMPLE 16.2

What is the oxidation number of Mg?

Solution

Zero. (This is elemental magnesium.) ■

■ EXAMPLE 16.3

What is the oxidation number of nitrogen in the nitrate ion, NO_3^-?

Solution

The oxidation number of oxygen in most of its compounds is -2 (rule 7). The oxidation numbers of the three oxygen atoms plus that of the nitrogen atom must total -1, the charge on the ion:

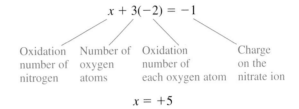

$$x + 3(-2) = -1$$

| Oxidation number of nitrogen | Number of oxygen atoms | Oxidation number of each oxygen atom | Charge on the nitrate ion |

$$x = +5$$

Practice Problem 16.3   What is the oxidation number of sulfur in the sulfate ion, SO_4^{2-}? ■

■ EXAMPLE 16.4

Calculate the oxidation number of chromium in the dichromate ion, $Cr_2O_7^{2-}$.

Solution

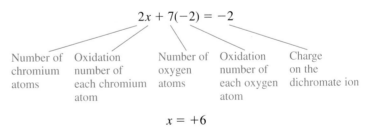

$$2x + 7(-2) = -2$$

| Number of chromium atoms | Oxidation number of each chromium atom | Number of oxygen atoms | Oxidation number of each oxygen atom | Charge on the dichromate ion |

$$x = +6$$

Practice Problem 16.4   Calculate the oxidation number of sulfur in $S_2O_3^{2-}$, the thiosulfate ion. ■

■ EXAMPLE 16.5

What is the oxidation number of hydrogen in (a) H_2O and (b) NaH?

Solution

(a) Oxygen has an oxidation number of -2 in most of its compounds (rule 7), so that of hydrogen in water must be positive; it is $+1$ (rule 6).

(b) The oxidation number of sodium in its compounds is always +1 (rule 4), so that of hydrogen in NaH must be negative; it is −1 (rule 6).

Practice Problem 16.5 What is the oxidation number of hydrogen in (a) HCl and (b) BeH_2? ■

Occasionally, oxidation numbers are fractional. For example, the oxidation number of manganese in Mn_3O_4 is $+2\frac{2}{3}$. This does *not* mean that electrons have been split; the oxidation number simply gives the average of the numbers of electrons still controlled by the three manganese atoms.

■ **EXAMPLE 16.6**

Calculate the oxidation number of nitrogen in (a) hydrazoic acid, HN_3, and (b) ammonia, NH_3.

Solution

The hydrogen atoms in each compound have the oxidation number +1. Nitrogen must have an oxidation number such that the overall sum of oxidation numbers for each compound is zero:

(a) $-\frac{1}{3}$ (b) −3 ■

16.2 Using Oxidation Numbers in Naming Compounds

OBJECTIVE

- to use oxidation numbers in naming compounds

In Chapter 6, you learned how to name cations. In the **Stock system,** the charges on monatomic ions are used to distinguish between different ions of the same element. For example, Cu^+ and Cu^{2+} are named copper(I) ion and copper(II) ion, respectively. The Roman numeral actually represents the oxidation number, not the charge on the ion. Of course, for monatomic ions, the charge is equal to the oxidation number, and thus we can use the charge to determine which Roman numeral to use. By using oxidation numbers, however, we can extend our compound-naming ability to include compounds other than those of monatomic ions. For example, Hg_2^{2+} is called the mercury(I) ion because the *oxidation number of each mercury atom* is +1.

The Stock system, which indicates the oxidation numbers in the name, can be extended to covalent compounds, such as binary nonmetal-nonmetal compounds. For example, PCl_3 was called phosphorus trichloride (Chapter 6), and chemists still commonly use that name. However, the compound can also be named phosphorus(III) chloride because the *oxidation number* of the phosphorus is +3. The Stock system may become more widely used by chemists to name covalent compounds in the future.

■ **EXAMPLE 16.7**

Name each of the following compounds in two ways:

(a) SO_3 (b) SF_4

Solution

(a) Sulfur trioxide *or* sulfur(VI) oxide

(b) Sulfur tetrafluoride *or* sulfur(IV) fluoride

Practice Problem 16.7   Write formulas for (a) phosphorus(V) bromide and (b) nitrogen(III) oxide. ■

■ **EXAMPLE 16.8**

What are the Stock system names for (a) sulfur dichloride and (b) sulfur dioxide?

Solution

(a) Sulfur(II) chloride   (b) sulfur(IV) oxide

Practice Problem 16.8   What is the Stock system name for carbon monoxide? ■

The prefixes and suffixes used in Chapter 6 can also be interpreted in terms of oxidation numbers, rather than numbers of oxygen atoms. For example, the ending *-ous* corresponds to the oxyacid in which the central atom has the lower oxidation state in each case, and the ending *-ic* corresponds to the higher oxidation state. For example phosphorous acid has phosphorus in the $+3$ oxidation state; phosphoric acid has phosphorus in the $+5$ oxidation state.

16.3   Periodic Variation of Oxidation Numbers

OBJECTIVE
■ to predict possible oxidation numbers of an element from its position in the periodic table and to use those numbers to predict possible formulas for compounds

In Section 16.1, you learned how to determine oxidation numbers of atoms of elements from the formulas of their ions or molecules. This section shows the opposite—how to write formulas for compounds based on knowledge of the possible oxidation numbers of the atoms of the elements. Predicting possible oxidation numbers is easy, but learning which are the most important oxidation numbers of the most familiar elements takes a great deal of experience.

Predicting Oxidation Numbers

For most of the elements, the *maximum* oxidation number is equal to the classical group number of the element. That is, the maximum oxidation number of lithium is $+1$ because lithium is in group IA, and the maximum oxidation number of chlorine is $+7$ because chlorine is in group VIIA. Exceptions to this rule are shown in Figure 16.1. Several elements, such as fluorine and most elements in group VIII, have maximum oxidation numbers less than their group number. Several of the elements of two groups—the coinage metals and the noble gases—have maximum oxidation numbers greater than their classical group number. Copper and gold have maximum oxidation numbers of $+2$ and $+3$, respectively, which are greater than the number of their group (IB). (Each of these elements also has an oxidation number of $+1$, but that is not the *maximum*.) Atoms of the noble gases that form compounds have positive oxidation numbers, all of which are greater than their group number, 0. Despite these exceptions, this simple rule allows us to predict at least one oxidation number for more than 90 elements.

Every element has an oxidation number of zero when it is uncombined (rule 2, Section 16.1). This second simple rule allows us to predict about 100 more oxidation numbers. (Some *compounds* have atoms with oxidation numbers of zero. For example, the carbon atom in CH_2F_2 has an oxidation number of zero.)

Figure 16.1 Exceptions to Maximum Oxidation Number Rule

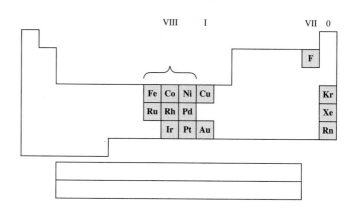

☐ Elements with oxidation numbers that exceed the group number.

☐ Elements that do not have oxidation numbers as high as the group number.

The *minimum* oxidation number of the metallic elements and the noble gases is generally zero and that for hydrogen is -1. The minimum oxidation number of most other nonmetallic elements is equal to the classical group number of the element minus 8.

Minimum Oxidation Number

For metals and noble gases:	0
For hydrogen:	-1
For other nonmetals:	Group number -8

■ **EXAMPLE 16.9**

What is the minimum oxidation number of sulfur in any of its compounds?

Solution

Minimum oxidation number = Group number $-8 = 6 - 8 = -2$

Practice Problem 16.9 What is the minimum oxidation number of potassium? ■

■ **EXAMPLE 16.10**

What are three possible oxidation states of phosphorus?

Solution

The maximum oxidation number of phosphorus is $+5$, equal to its group number. Phosphorus also has the oxidation state 0 when it is not combined and a minimum oxidation state of $5 - 8 = -3$. (Phosphorus can also have oxidation states $+3$ and $+1$, not covered by the rules given so far.)

Practice Problem 16.10 What are three possible oxidation states of copper? ■

The rules given previously do not predict all of the oxidation numbers of the elements in their compounds, and there is no certainty that the oxidation numbers

Figure 16.2 Periodic Variation of Oxidation Number

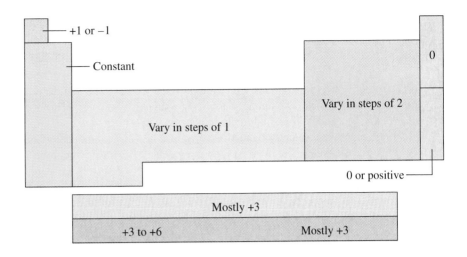

predicted by these rules will be those of the element in its most stable or most important compounds. Other oxidation states can be predicted by the location of the element in the periodic table. In their compounds, the main group metals to the left of the transition group elements in the periodic table have constant oxidation states. The main group elements to the right of the transition group elements have oxidation numbers that usually vary in steps of 2. The transition group elements have oxidation numbers that may vary in steps of 1, and only a few of these elements have oxidation numbers equal to +1. The oxidation numbers of the lanthanides are mostly +3. The elements in the first half of the actinide series have oxidation numbers that increase from +3 to +6, but the elements of the last half of that series have oxidation numbers equal to +3. These generalizations are illustrated in Figure 16.2.

■ **EXAMPLE 16.11**

What are the possible positive oxidation numbers of sulfur?

Solution

The maximum oxidation number of sulfur, equal to its group number, is +6. Since the oxidation numbers vary mostly in steps of 2 in this portion of the periodic table, you can predict (correctly) that sulfur also has +4 and +2 as oxidation numbers. Examples of compounds in which sulfur has these three oxidation numbers are SF_6, SF_4, and SF_2.

Practice Problem 16.11 What are four positive oxidation numbers of chlorine? ■

The rules discussed so far allow us to make "educated guesses" about possible compounds formed by elements. However, not all compounds predicted by application of the rules actually exist. Moreover, some elements have more oxidation numbers than the rules predict. Nitrogen, for example, exhibits every integral oxidation number from −3 to +5, as well as a fractional oxidation number, $-\frac{1}{3}$, in HN_3. The halogens, except for fluorine, also exhibit most of the integral oxidation numbers from −1 to +7. Detailed study of the chemistry of the elements and their compounds is necessary to know which compounds actually exist.

ENRICHMENT

Metals with high oxidation numbers tend to act somewhat like nonmetals. For example, many transition metals form oxoanions, such as permanganate ion, chromate ion, and dichromate ion, in which the metal is covalently bonded to oxygen. The ability to form covalent bonds to oxygen is evidence of these metals' more covalent nature. (In their low oxidation states, most metals typically exist in ionic compounds as monatomic cations.) Titanium(IV) chloride is an example of a compound in which the metal has a high oxidation number. It is a liquid at room temperature, and the liquid does not conduct electricity well. These characteristics are typical of covalent compounds. Since this is a binary covalent compound, it may be named titanium tetrachloride, using the prefix in Table 6.1, but modern practice prefers the Stock system name. It is easy enough to write the formula for the compound from either name, and either name is acceptable.

Writing Formulas for Covalent Compounds

Knowledge of possible oxidation numbers for the elements allow us to predict formulas for covalent compounds, just as we did for ionic compounds in Chapter 5. Atoms of elements with positive oxidation numbers and atoms of elements with negative oxidation numbers are combined in a formula so that the total of the oxidation numbers of all the atoms equals zero. The resulting formula may be a compound of the elements.

■ EXAMPLE 16.12

Determine the formula of a compound of carbon and chlorine.

Solution

Chlorine in its compounds with elements other than oxygen and fluorine exhibits a negative oxidation number, -1 (rule 8, Section 16.1). Thus, the oxidation number of carbon in this compound must be positive. The maximum oxidation number of carbon in its compounds is $+4$ because it is in group IV. Thus, the formula is CCl_4. The sum of the four (-1) oxidation numbers for the chlorine atoms and the one $(+4)$ oxidation number for the carbon atom is zero, the charge on the compound. ■

16.4 Balancing Oxidation-Reduction Equations

OBJECTIVE

■ to use a systematic method to balance equations in which oxidation numbers change

When an atom increases its oxidation number during a reaction, it is said to have been oxidized. When an atom decreases its oxidation number during a reaction, it is said to have been reduced. **Oxidation** is an increase in oxidation number as a result of the loss of electrons or of "control" of electrons. **Reduction** is a decrease in oxidation number due to the gain of electrons or of "control" of electrons. A chemical species serves as an **oxidizing agent** by causing the oxidation of another species. When it acts as an oxidizing agent, one or more of its elements is reduced. A chemical species serves as a **reducing agent** by causing the reduction of another species. When it acts as a reducing agent, one or more of its elements is oxidized.

A reaction in which an oxidation and a reduction occur is called an **oxidation-reduction reaction,** or *redox reaction* for short. Every redox reaction must have an oxidation and a reduction. You can never have one of these processes without the other. A gain of electrons is not possible without a loss of electrons, and vice versa. For example, the reaction of MnO_4^- with Fe^{2+} is an oxidation-reduction reaction:

$$8\ H^+(aq) + MnO_4^-(aq) + 5\ Fe^{2+}(aq) \rightarrow Mn^{2+}(aq) + 4\ H_2O(l) + 5\ Fe^{3+}(aq)$$

The iron is oxidized, and the manganese is reduced. The iron(II) ion is the reducing agent, and the permanganate ion is the oxidizing agent.

■ **EXAMPLE 16.13**

A dish towel used to dry dishes can be classified as a drying agent, and the dishes can be classified as wetting agents.

(a) What happens to the drying agent in the process of drying the dishes? What happens to the wetting agent? What happens to the water?

(b) When an oxidation-reduction reaction occurs, what happens to the reducing agent, the oxidizing agent, and the electrons?

Solution

(a) The dish towel is the drying agent; it gets wet. The dishes are the wetting agents; they get dry. The water is transferred from the wetting agent to the drying agent.

(b) The electrons in an oxidation-reduction reaction act like the water in part (a): They are transferred from the reducing agent to the oxidizing agent. The reducing agent is oxidized; the oxidizing agent is reduced. ■

We have balanced a number of simple oxidation-reduction equations, starting in Chapter 8. Most combination, decomposition, substitution, and combustion reactions are oxidation-reduction reactions. However, many oxidation-reduction reactions are much more complicated than the ones we have already considered, and we must use a systematic method for balancing equations for them. Unfortunately, many different systematic methods are used, and each chemistry instructor seems to have his or her favorite method. Most instructors will accept any valid method that a student understands, however. The method outlined here is a standard method that should be acceptable.

All methods of balancing oxidation-reduction equations are based on the overall gain of oxidation numbers in a reaction being the same as the overall loss of oxidation numbers in the reaction (because the same number of electrons must be gained as lost).

The first step in any method of balancing oxidation-reduction equations is to identify the element that is oxidized and the one that is reduced. Since the change in oxidation number is equal to a change in the number of electrons "controlled," and the electrons must be "controlled" by some atom, the total gain in oxidation number is equal to the total loss in oxidation number. The oxidation half of a reaction may be written in one equation, and the reduction half in another. Neither **half-reaction** can be carried out without the other, but they can be done in different locations if they are connected in such a way that a complete electrical circuit is made. The **half-reaction method** is illustrated by balancing the equation for the reaction of aluminum metal with dilute nitric acid to produce ammonium ion, aluminum ion, and water:

$$H^+(aq) + NO_3^-(aq) + Al(s) \rightarrow Al^{3+}(aq) + NH_4^+(aq) + H_2O(l)$$

First, we consider the reduction half-reaction:

Step 1: Determine the oxidation number of each element in the reaction. Identify the element that is reduced, and balance the number of atoms of that element, if necessary. In this example, the nitrogen atom is reduced from an oxidation state of $+5$ to -3:

$$NO_3^-(aq) \rightarrow NH_4^+(aq)$$

Now that the element whose oxidation number changes has been balanced, no element will change oxidation number in steps 2 through 4.

Step 2: Balance any other element present that is not oxygen, hydrogen, or the element whose oxidation number changed. This example has no other elements besides those listed.

Step 3: Balance the oxygen atoms by adding water to the side deficient in oxygen:

$$NO_3^-(aq) \rightarrow NH_4^+(aq) + 3 H_2O(l)$$

Step 4: Balance the hydrogen atoms by adding hydrogen ions (H^+) to the side deficient in hydrogen:

$$10 H^+(aq) + NO_3^-(aq) \rightarrow NH_4^+(aq) + 3 H_2O(l)$$

Step 5: Balance the charge by adding electrons (e^-) to the more positive, or less negative, side:

$$8 e^- + 10 H^+(aq) + NO_3^-(aq) \rightarrow NH_4^+(aq) + 3 H_2O(l)$$
(Balanced half-reaction)

Repeat steps 1 through 5 to balance the oxidation half-reaction.

Step 1: Balance the atoms of the element that is oxidized:

$$Al(s) \rightarrow Al^{3+}(aq)$$

Steps 2–4: No other elements need to be balanced.

Step 5: Balance the charge:

$$Al(s) \rightarrow Al^{3+}(aq) + 3 e^- \qquad \textit{(Balanced half-reaction)}$$

Combine the balanced oxidation half-reaction and the balanced reduction half-reaction after checking that each has the same number of electrons involved.

Step 6: The oxidation half-reaction for this example has three electrons involved, and the reduction half-reaction has eight electrons involved. To make the numbers of electrons equal, we multiply the former by 8 and the latter by 3:

$$8 \times [\qquad\qquad Al(s) \rightarrow Al^{3+}(aq) + 3 e^- \quad]$$
$$3 \times [8 e^- + 10 H^+(aq) + NO_3^-(aq) \rightarrow NH_4^+(aq) + 3 H_2O(l)]$$

or

$$8 Al(s) \rightarrow 8 Al^{3+}(aq) + 24 e^-$$
$$24 e^- + 30 H^+(aq) + 3 NO_3^-(aq) \rightarrow 3 NH_4^+(aq) + 9 H_2O(l)$$

Adding the two half-reactions gives

$$30\ H^+(aq) + 3\ NO_3^-(aq) + 8\ Al(s) \rightarrow 8\ Al^{3+}(aq) + 3\ NH_4^+(aq) + 9\ H_2O(l)$$

■ EXAMPLE 16.14

Balance the equation for the reaction of dichromate ion with iron(II) ion to produce chromium(III) ion, iron(III) ion, and other products.

Solution

$$Cr_2O_7^{2-}(aq) + Fe^{2+}(aq) \rightarrow Cr^{3+}(aq) + Fe^{3+}(aq) + ?$$

The oxidation half-reaction is easy to determine:

$$Fe^{2+}(aq) \rightarrow Fe^{3+}(aq) + e^- \qquad \textit{(Balanced half-reaction)}$$

The reduction half-reaction is worked out as follows:

Step 1: $Cr_2O_7^{2-}(aq) \rightarrow Cr^{3+}(aq)$

$Cr_2O_7^{2-}(aq) \rightarrow 2\ Cr^{3+}(aq)$

(The chromium is balanced first.)

Step 2: No change.

Step 3: $Cr_2O_7^{2-}(aq) \rightarrow 2\ Cr^{3+}(aq) + 7\ H_2O(l)$

Step 4: $14\ H^+(aq) + Cr_2O_7^{2-}(aq) \rightarrow 2\ Cr^{3+}(aq) + 7\ H_2O(l)$

Step 5: $6\ e^- + 14\ H^+(aq) + Cr_2O_7^{2-}(aq) \rightarrow 2\ Cr^{3+}(aq) + 7\ H_2O(l)$
$$\textit{(Balanced half-reaction)}$$

Step 6: Combine the two half-reactions, after making the number of electrons in each the same by multiplying the oxidation half-reaction by 6:

$$6\ Fe^{2+}(aq) \rightarrow 6\ Fe^{3+}(aq) + 6\ e^-$$

The complete balanced equation is

$$6\ Fe^{2+}(aq) + 14\ H^+(aq) + Cr_2O_7^{2-}(aq) \rightarrow 2\ Cr^{3+}(aq) + 7\ H_2O(l) + 6\ Fe^{3+}(aq)$$

Practice Problem 16.14   Complete and balance the equation for the reaction of the chromate ion with iodide ion in acid solution to give chromium(III) ion and iodine, as well as other products:

$$CrO_4^{2-}(aq) + I^-(aq) \rightarrow Cr^{3+}(aq) + I_2(aq) + ?\ ■$$

■ EXAMPLE 16.15

Balance the equation for the oxidation of H_2O_2 by MnO_4^- to give oxygen gas (O_2) and manganese(II) ion (Mn^{2+}) in acid solution.

Solution

$$H_2O_2(aq) + MnO_4^-(aq) \rightarrow O_2(g) + Mn^{2+}(aq) + ?$$

Step 1: $H_2O_2(aq) \rightarrow O_2(g)$

Steps 2 and 3: No change.

Step 4: $H_2O_2(aq) \rightarrow O_2(g) + 2\ H^+(aq)$

Step 5: $H_2O_2(aq) \rightarrow O_2(g) + 2\ H^+(aq) + 2\ e^- \qquad \textit{(Balanced half-reaction)}$

Step 1: $MnO_4^-(aq) \rightarrow Mn^{2+}(aq)$

Step 2:   No other elements are present besides manganese and oxygen.

Step 3:   Balance the oxygen atoms:

$$MnO_4^-(aq) \rightarrow Mn^{2+}(aq) + 4\ H_2O(l)$$

Step 4:   Balance the hydrogen atoms:

$$8\ H^+(aq) + MnO_4^-(aq) \rightarrow Mn^{2+}(aq) + 4\ H_2O(l)$$

Step 5:   Balance the charge:

$$5\ e^- + 8\ H^+(aq) + MnO_4^-(aq) \rightarrow Mn^{2+}(aq) + 4\ H_2O(l)$$
                                    (Balanced half-reaction)

Step 6:   Multiply the oxidation half-reaction by 5 to get 10 electrons:

$$5\ H_2O_2(aq) \rightarrow 5\ O_2(g) + 10\ H^+(aq) + 10\ e^-$$

Multiply the reduction half-reaction by 2 to get 10 electrons:

$$10\ e^- + 16\ H^+(aq) + 2\ MnO_4^-(aq) \rightarrow 2\ Mn^{2+}(aq) + 8\ H_2O(l)$$

Combine the two half-reactions:

$$16\ H^+(aq) + 5\ H_2O_2(aq) + 2\ MnO_4^-(aq) \rightarrow$$
$$2\ Mn^{2+}(aq) + 8\ H_2O(l) + 5\ O_2(g) + 10\ H^+(aq)$$

Eliminating 10 of the H^+ ions, which appear on both sides, yields the complete balanced equation:

$$6\ H^+(aq) + 5\ H_2O_2(aq) + 2\ MnO_4^-(aq) \rightarrow$$
$$2\ Mn^{2+}(aq) + 8\ H_2O(l) + 5\ O_2(g)$$

Practice Problem 16.15   Complete and balance the equation for the reaction of bromine with permanganate ion in acid solution to give bromate ion and manganese(II) ion, as well as other products.

$$Br_2(aq) + MnO_4^-(aq) \rightarrow BrO_3^-(aq) + Mn^{2+}(aq) + ?\ ■$$

Sometimes, a single reactant can act as both oxidizing agent and reducing agent in the same reaction. Such a reactant is said to undergo auto–oxidation-reduction, or **disproportionation.**

■ **EXAMPLE 16.16**

When Cl_2 gas disproportionates in concentrated NaOH solution, $NaClO_3$ is produced. What other chlorine-containing compound is produced? Write a balanced chemical equation for the reaction.

Solution

Since the chlorine is oxidized to +5 in the sodium chlorate, it must be reduced in the other compound. The only oxidation number lower than the zero for chlorine in Cl_2 is −1, and the compound is NaCl. The balanced equation is

$$3\ Cl_2(g) + 6\ NaOH(aq) \rightarrow 5\ NaCl(aq) + NaClO_3(aq) + 3\ H_2O(l)$$

Practice Problem 16.16   If left standing, H_2O_2 decomposes. What is the oxidizing agent, and what is the reducing agent? Write a balanced equation for the reaction. ■

16.5   Equivalents and Normality (*Optional*)

OBJECTIVE

■ to calculate concentrations using a unit associated with the number of moles of electrons or the number of moles of hydrogen ions or hydroxide ions involved in a substance's reactions

Describing the concentration of a substance in terms of the quantity of electrons or hydrogen ions that it will react with or produce is sometimes desirable. For example, the ability of an ionic substance in solution to carry an electrical current is related not only to the concentration of the substance but also to the charge on each of its ions. The concentration of charge is a useful concept in some advanced areas of study.

The **equivalent** (abbreviated equiv) is defined as the quantity of a substance that will react with or produce

1. One mole of electrons in an oxidation-reduction reaction, or
2. One mole of H^+ or OH^- ions in an acid-base reaction.

Note that the equivalent is defined in terms of a particular reaction. A given number of moles of a substance may be different numbers of equivalents in different reactions.

■ EXAMPLE 16.17

Determine the number of equivalents per mole of H_3PO_4 for each of the following reactions:

(a) $H_3PO_4(aq) + 3\,NaOH(aq) \rightarrow Na_3PO_4(aq) + 3\,H_2O(l)$
(b) $H_3PO_4(aq) + 2\,NaOH(aq) \rightarrow Na_2HPO_4(aq) + 2\,H_2O(l)$
(c) $H_3PO_4(aq) + 8\,H^+(aq) + 8\,e^- \rightarrow PH_3(aq) + 4\,H_2O(l)$

Solution

(a) This is an acid-base reaction. Each mole of H_3PO_4 reacts with 3 mol of OH^- ions, so each mole of H_3PO_4 is 3 equivalents.
(b) This is an acid-base reaction. Each mole of H_3PO_4 reacts with 2 mol of OH^- ions, so each mole of H_3PO_4 is 2 equivalents.
(c) This is an oxidation-reduction reaction. Each mole of H_3PO_4 reacts with 8 mol of electrons, so each mole of H_3PO_4 is 8 equivalents.

Practice Problem 16.17   Determine the number of equivalents per mole of H_2SO_4 for each of the following reactions:

(a) $H_2SO_4(aq) + 2\,NaOH(aq) \rightarrow Na_2SO_4(aq) + 2\,H_2O(l)$
(b) $H_2SO_4(aq) + NaOH(aq) \rightarrow NaHSO_4(aq) + H_2O(l)$
(c) $H_2SO_4(aq) + 8\,H^+(aq) + 8\,e^- \rightarrow H_2S(g) + 4\,H_2O(l)$ ■

If a problem involving equivalents of a substance does not mention the reaction that the substance will undergo, assume that a complete acid-base reaction will occur.

▪ EXAMPLE 16.18

What is the number of equivalents in 2.00 mol of H_2SO_4?

Solution

For this problem, assume that the reaction is the acid-base reaction of Practice Problem 16.17(a)—complete neutralization of the acid.

$$2.00 \text{ mol } H_2SO_4 \left(\frac{2 \text{ equiv } H_2SO_4}{1 \text{ mol } H_2SO_4} \right) = 4.00 \text{ equiv } H_2SO_4 \quad ▪$$

One of the main ways in which equivalents are useful is that one equivalent of any substance in a given reaction will react with one equivalent of any other substance in the reaction. The balanced chemical equation has already been used to determine the number of equivalents, so further calculations to find equivalents of other substances are not necessary. This generality will be very useful in the next subsection.

▪ EXAMPLE 16.19

In a certain chemical reaction, how many equivalents of NaOH will react with 2.25 equivalents of H_3PO_4?

Solution

The given quantity of H_3PO_4 reacts with 2.25 equivalents of NaOH. You do not know how many hydrogen atoms per molecule have been neutralized, but the number of equivalents has already been calculated, using the correct equation.

Normality

The **normality** of a solution is the number of equivalents of solute per liter of solution. Normality is symbolized N. The unit of normality is **normal,** symbolized N.

▪ EXAMPLE 16.20

Calculate the normality of a solution of 3.61 equivalents of H_3PO_4 in 2.51 L of solution.

Solution

$$\text{Normality} = \frac{3.61 \text{ equiv}}{2.51 \text{ L}} = 1.44 \text{ N} \quad ▪$$

▪ EXAMPLE 16.21

Calculate the normality of 1.72 M H_3PO_4 that is to be neutralized to Na_3PO_4.

Solution

$$H_3PO_4(aq) + 3 \text{ NaOH}(aq) \rightarrow Na_3PO_4(aq) + 3 \text{ H}_2O(l)$$

The number of equivalents per mole is 3; the normality is

$$1.72 \text{ M} = \left(\frac{1.72 \text{ mol}}{1 \text{ L}}\right)\left(\frac{3 \text{ equiv}}{1 \text{ mol}}\right) = \frac{5.16 \text{ equiv}}{1 \text{ L}} = 5.16 \text{ N}$$

Practice Problem 16.21   Calculate the normality of 1.72 M H_3PO_4, which is to be neutralized to NaH_2PO_4. ■

■ EXAMPLE 16.22

Calculate the normality of a solution that is prepared with 112 g of H_3PO_4 in 555 mL of solution and is to be titrated to Na_2HPO_4.

Solution

$$112 \text{ g } H_3PO_4\left(\frac{1 \text{ mol } H_3PO_4}{98.0 \text{ g } H_3PO_4}\right) = 1.14 \text{ mol } H_3PO_4$$

There are 2 equivalents per mole, as shown by the balanced chemical equation:

$$H_3PO_4(aq) + 2 \text{ NaOH}(aq) \rightarrow Na_2HPO_4(aq) + 2 \text{ H}_2O(l)$$

$$1.14 \text{ mol } H_3PO_4\left(\frac{2 \text{ equiv } H_3PO_4}{1 \text{ mol } H_3PO_4}\right) = 2.28 \text{ equiv } H_3PO_4$$

The normality is

$$N = \frac{2.28 \text{ equiv}}{0.555 \text{ L}} = 4.11 \text{ N}$$

Practice Problem 16.22   Calculate the normality of a solution that is prepared with 90.0 g of NaH_2PO_4 in 555 mL of solution and is to be titrated to Na_3PO_4. ■

The number of equivalents per liter (the normality) times the number of liters of a solution is equal to the number of equivalents of the solute. Since the numbers of equivalents of two substances in the same reaction are equal, the products of normalities (N) and volumes (V) are also equal:

$$N_1V_1 = N_2V_2$$

> Caution: This equation works only for normalities. In general, the analogous equation does not work for molarities (unless the substances happen to react in a 1:1 mole ratio).

The equation relating the normality and volume of one substance to that of another is easy for technicians to learn and use under the direction of a professional chemist, who will ensure that the normalities are calculated from the proper balanced equation.

■ EXAMPLE 16.23

Calculate the volume of 3.000 N NaOH required to react with 50.00 mL of 4.500 N H_2SO_4.

Solution

$$N_1V_1 = N_2V_2$$

$$(3.000 \text{ N})V_1 = (4.500 \text{ N})(0.0500 \text{ L})$$

$$V_1 = 0.07500 \text{ L} = 75.00 \text{ mL}$$

Practice Problem 16.23   Calculate the normality of an NaOH solution if 42.8 mL is required to react with 21.3 mL of 4.50 N H_3PO_4. ■

Equivalent Mass

The **equivalent mass,** or equivalent weight, of a substance is defined as the number of grams per equivalent of the substance. Equivalent mass is useful for characterizing a newly prepared or discovered substance whose formula has not yet been established, as well as for other calculations.

■ **EXAMPLE 16.24**

Calculate the equivalent mass of a newly prepared acid if 2.50 g of the acid requires 42.7 mL of 0.400 N NaOH to be completely neutralized. Can you tell if each molecule of the acid contains one, two, or three ionizable hydrogen atoms?

Solution

The number of equivalents of acid is equal to the number of equivalents of base, which is

Number of equivalents = (0.0427 L)(0.400 equiv/L) = 0.0171 equiv

Dividing this number of equivalents into the mass yields the equivalent mass:

$$\text{Equivalent mass} = \frac{2.50 \text{ g}}{0.0171 \text{ equiv}} = 146 \text{ g/equiv}$$

The information given is insufficient to determine how many ionizable hydrogen atoms there are per molecule of acid. (See Problem 16.81 at the end of the chapter.) ■

Figure 16.3 summarizes the conversions that can be carried out using the number of equivalents per mole.

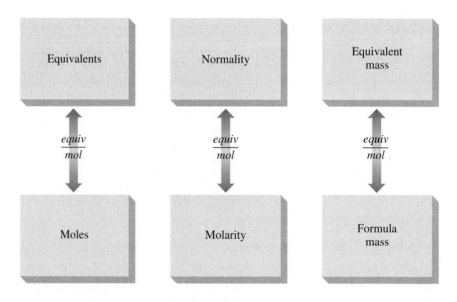

Figure 16.3 Using Equivalents Per Mole As a Conversion Factor

■ ■ ■ ■ ■ ■ ■ SUMMARY ■ ■ ■ ■ ■ ■ ■

Oxidation numbers (also called oxidation states) are used as a sort of bookkeeping method for keeping track of the electrons in compounds or polyatomic ions that have covalent bonds. (For monatomic ions, the charge on the ions works just as well.) Oxidation number is defined as the number of electrons in a free atom minus the number "controlled" by that atom in the compound. The "control" of electrons in a covalent bond is assigned to the more electronegative atom of the bond. Eight simple rules can be used to determine the oxidation number of an element from the formula of its compound or ion. (Section 16.1)

Oxidation numbers are used in the Stock system for naming compounds. Positive oxidation numbers are denoted as Roman numerals in parentheses in the *names* of the compounds; the numbers of atoms or ions can be deduced from the oxidation numbers. (In contrast, the subscripts in formulas give the numbers of atoms or ions, from which the oxidation numbers may be deduced.) (Section 16.2)

The oxidation numbers of the elements are periodic properties, and common oxidation numbers can often be deduced from the classical periodic group number of an element. The possible oxidation numbers can be used to write possible formulas for compounds. (Section 16.3)

Oxidation is defined as a gain in oxidation number, caused by a loss of electrons or of "control" of electrons. Reduction is defined as a loss in oxidation number, caused by a gain of electrons or of "control" of electrons. Complicated oxidation-reduction equations must be balanced according to some systematic method, since they are too complex to be balanced by inspection. Although neither can take place alone, the oxidation and the reduction can occur in different locations if suitable electrical connections are provided. In the half-reaction method, the equation for the half-reaction involving oxidation and that for the half-reaction involving reduction are balanced separately; then the two are combined. Each may be multiplied by a small integer if necessary to balance the numbers of electrons involved.

Some reactants can be both oxidized and reduced in the same reaction. They are said to undergo disproportionation. (Section 16.4)

An equivalent is defined as the quantity of substance that reacts with or produces 1 mole of electrons in an oxidation-reduction reaction or 1 mole of hydrogen ions or hydroxide ions in an acid-base reaction. Some integral number of equivalents $(1, 2, 3, \ldots)$ constitutes a mole of the substance in a given reaction. Normality is the number of equivalents of solute per liter of solution. Equivalent mass is the number of grams per equivalent of a substance. (Section 16.5)

Items for Special Attention

- Shared electrons are counted for each of the bonded atoms when writing electron dot diagrams, but only for the more electronegative atom when determining oxidation numbers.

- Oxidation numbers are usually but not always integral.

- The oxidation numbers predicted by Figure 16.2 will not be correct for all elements in all compounds.

- Be sure to start the half-reaction method by balancing the number of atoms of the element oxidized or reduced (before proceeding to step 2). The other elements involved cannot change oxidation number after the element oxidized or reduced has been balanced (in step 1). For example, do not add O_2 to balance oxygen atoms in step 3.

- Do not confuse oxidation numbers with charges when balancing oxidation-reduction equations. Use Roman numerals for positive oxidation numbers and Arabic numbers for charges. (If you need to denote negative oxidation numbers, you might use Arabic numerals below the formula and circle them, so that you do not get them mixed up with charges. The Romans did not have negative numbers.)

- Balanced equations for half-reactions *always* include electrons; balanced equations for overall reactions *never* include electrons.

- Always check a completed equation for either a half-reaction or an overall reaction to make sure that each element is balanced and also that the net charge is balanced.

- The number of equivalents of a substance is always a whole number $(1, 2, 3, \ldots)$ times the number of moles of that substance.

Self-Tutorial Problems

16.1 Write the formula for:

(a) An ionic compound whose cation (M) has a 1+ charge and whose anion (A) has a 2− charge

(b) A compound with element Y in a +1 oxidation state and element Z in a −2 oxidation state

16.2 Determine the oxidation number of each atom in (a) Na_2S and (b) PCl_5.

16.3 Determine the oxidation number of each atom in (a) Cu_2O and (b) CO_2.

16.4 What is the *sum* of the oxidation numbers of all the atoms in each of the following?

(a) $Cr_2O_7^{2-}$ (b) CH_3OH (c) ClO_3^- (d) $AlCl_3$

16.5 What is the *sum* of the oxidation numbers of all the atoms in each of the following?

(a) CN^- (b) $CH_3NH_3^+$ (c) MnO_2 (d) $H_2PO_4^-$

16.6 What is the oxidation number of the element in each of the following ions?   (a) Ba^{2+} (b) O_2^{2-} (c) S_2^{2-}

16.7 What is the oxidation number of sodium on each side of the following equation?

$$2\,Na(s) + S(s) \xrightarrow{\text{Heat}} Na_2S(s)$$

16.8 Why isn't H_2 used instead of H^+ to balance hydrogen atoms in step 4 of the half-reaction method?

16.9 How many electrons should be added to an equation for a half-reaction in which the left side has a net charge of 5− and the right side has a net charge of 1+? To which side should they be added?

16.10 (a) Which of the elementary types of reactions discussed in Chapter 8 is generally *not* an oxidation-reduction reaction?

(b) Do *all* oxidation-reduction equations have to be balanced by a systematic method?

■ ■ ■ PROBLEMS ■ ■ ■

16.1 Assigning Oxidation Numbers

16.11 Draw an electron dot diagram for $CSCl_2$, in which the carbon is the central atom. Determine the oxidation number of each element, and compare with the oxidation numbers in $COCl_2$.

16.12 Draw an electron dot diagram for CS_2. Determine the oxidation number of each element, and compare with the oxidation numbers of carbon and oxygen in carbon dioxide.

16.13 Determine the oxidation number of chlorine in each of the following:

(a) ClO_3^- (b) ClO_2^- (c) $NaClO$ (d) HCl

16.14 Determine the oxidation number of sulfur in each of the following:

(a) H_2SO_3 (b) Na_2SO_3 (c) SO_3^{2-} (d) SO_3

16.15 Determine the oxidation number of carbon in each of the following compounds:

(a) CH_4 (b) CH_3Cl (c) CH_2Cl_2
(d) $CHCl_3$ (e) CCl_4

16.16 Determine the oxidation number of carbon in each of the following:

(a) CO_2 (b) CO_3^{2-} (c) CH_3OH
(d) CH_2O (e) $HCHO_2$

16.17 What is the oxidation number of (a) chromium in CrO_4^{2-} and (b) sulfur in SO_4^{2-}?

16.18 What is the oxidation number of (a) manganese in MnO_4^- and (b) chlorine in ClO_4^-?

16.19 Determine the oxidation number of silicon in (a) SiF_4, (b) SiO_2, and (c) Na_2SiO_3.

16.20 Determine the oxidation number of chlorine in each of the following compounds:

(a) ClO (b) Cl_2O (c) ClO_2
(d) ClO_3 (e) Cl_2O_5 (f) Cl_2O_7

16.21 What is the oxidation number of nitrogen in each of the following compounds?

(a) N_2H_4 (b) NH_3 (c) N_2O
(d) N_2 (e) HN_3 (f) NH_2OH
(g) N_2O_3 (h) NO (i) NO_2
(j) N_2O_5 (k) N_2O_4

16.22 Determine the oxidation number of chlorine in each of the following ions:

(a) ClO^- (b) ClO_2^- (c) ClO_3^- (d) ClO_4^-

16.23 What is the oxidation number of phosphorus in (a) phosphorous acid and (b) phosphoric acid?

16.24 What is the oxidation number of nitrogen in (a) nitric acid and (b) the nitrate ion?

16.25 What is the oxidation number of iron in each of the following compounds?

(a) FeO (b) Fe_2O_3 (c) Fe_3O_4

16.26 Find the oxidation number of each element in each substance in the following reactions:

(a) $MgO(s) + CO_2(g) \rightarrow MgCO_3(s)$

(b) $2\,Mg(s) + O_2(g) \rightarrow 2\,MgO(s)$

16.27 Find the oxidation number of hydrogen in each of the following compounds:

(a) HCl   (b) LiH   (c) CaH_2 (d) $LiAlH_4$

16.28 In the reaction of SO_3^{2-} with an oxygen atom to form SO_4^{2-}, explain why the sulfur is oxidized. Does the sulfur atom actually lose any electrons?

16.2 Using Oxidation Numbers in Naming Compounds

16.29 Write the formulas for (a) copper(I) oxide, (b) copper(II) oxide, and (c) mercury(I) chloride.

16.30 Write the formulas for (a) carbon(IV) sulfide, (b) carbon(II) oxide, and (c) phosphorus(V) bromide.

16.31 Name each of the following compounds, using the prefixes of Table 6.1:

(a) Vanadium(V) chloride   (b) Carbon(IV) fluoride

(c) Phosphorus(V) oxide

16.32 Name each of the following compounds, using the prefixes of Table 6.1:

(a) Tin(IV) chloride   (b) Sulfur(IV) fluoride

(c) Phosphorus(III) oxide

16.33 Name (a) PbS, (b) CrF_3, and (c) $Fe(NO_3)_3$.

16.34 Name (a) $Hg(NO_3)_2$, (b) $Co(H_2PO_4)_2$, and (c) $Cu(NO_3)_2$.

16.35 Name each of the following compounds according to the Stock system:

(a) Dinitrogen trioxide   (b) Chlorine pentafluoride

(c) Nitrogen trichloride

16.3 Periodic Variation of Oxidation Numbers

16.36 What are the maximum and minimum oxidation numbers of hydrogen?

16.37 What is the maximum oxidation number of fluorine?

16.38 The maximum oxidation number for any element is +8, as for osmium in OsO_4; the maximum charge on a monatomic cation is 4+, as for Ce^{4+}. Explain the difference.

16.39 What is the minimum oxidation number of (a) nickel, (b) nitrogen, and (c) argon?

16.40 Predict the most probable positive oxidation numbers for phosphorus other than +5. Write the formula for a phosphorus-containing acid as an example of each.

16.41 In what ways are group IB elements and group VIII elements exceptions to the rule that the maximum oxidation number is usually equal to the group number?

16.42 In what ways are the noble gases and fluorine exceptions to the rule that the maximum oxidation number is usually equal to the group number?

16.43 What is the minimum oxidation number of (a) sulfur, (b) calcium, and (c) lead?

16.44 Predict the most probable positive oxidation number(s) for carbon other than +4.

16.45 Predict the formulas for three compounds of fluorine and sulfur.

16.46 In which oxoanions of Table 6.4 do elements have the highest oxidation numbers?

16.47 For each of the elements combined with oxygen in Table 6.4, in which ion does the element have the lowest oxidation number?

16.48 Predict the formulas for four compounds of fluorine with iodine.

16.49 Which suffix for an oxoacid (-ous or -ic) signifies the higher oxidation number for the central atom?

16.50 What can be said about the oxidation numbers of the group VA elements in oxoanions whose names have the suffix -ate?

16.51 Consider the bromide ion and the oxoanions of bromine. Which suffix is used for the ion in which bromine has the lowest oxidation number, and which is used for the ion in which bromine has the highest?

16.4 Balancing Oxidation-Reduction Equations

16.52 What is the oxidizing agent, the reducing agent, the element oxidized, and the element reduced in each of the following equations?

(a) $2 CuO(s) \xrightarrow{\text{Heat}} 2 Cu(s) + O_2(g)$

(b) $Mg(s) + PbCl_2(s) \xrightarrow{\text{Heat}} MgCl_2(s) + Pb(s)$

(c) $8 H_2SO_4(aq) + 2 KMnO_4(aq) + 10 KCl(aq) \rightarrow$
$5 Cl_2(aq) + 2 MnSO_4(aq) + 6 K_2SO_4(aq) + 8 H_2O(l)$

(d) $Fe(s) + 2 FeCl_3(aq) \rightarrow 3 FeCl_2(aq)$

16.53 What is the oxidizing agent, the reducing agent, the element oxidized, and the element reduced in each of the following equations?

(a) $2 H_2O(l) \xrightarrow{\text{Electricity}} 2 H_2(g) + O_2(g)$

(b) $Ca(s) + SnCl_2(s) \xrightarrow{\text{Heat}} CaCl_2(s) + Sn(s)$

(c) $2 KI(aq) + 2 FeCl_3(aq) \rightarrow$
$2 FeCl_2(aq) + I_2(aq) + 2 KCl(aq)$

16.54 What is the oxidizing agent, the reducing agent, the element oxidized, and the element reduced in each of the following equations? Which equation is easier to work with?

(a) $4 HNO_3(aq) + Cu(s) \rightarrow$
$Cu(NO_3)_2(aq) + 2 NO_2(g) + 2 H_2O(l)$

(b) $4 H^+(aq) + 2 NO_3^-(aq) + Cu(s) \rightarrow$
$Cu^{2+}(aq) + 2 NO_2(g) + 2 H_2O(l)$

16.55 Balance the following oxidation-reduction equation:

$$Pb^{4+}(aq) + C_2O_4^{2-}(aq) \rightarrow CO_2(g) + Pb^{2+}(aq)$$

16.56 Complete and balance each of the following oxidation-reduction equations:

(a) $Cu(s) + H^+(aq) + NO_3^-(aq) \rightarrow$
$$Cu^{2+}(aq) + NO(g) + ?$$

(b) $H_2O_2(aq) + Co^{3+}(aq) \rightarrow Co^{2+}(aq) + H^+(aq) + O_2(g)$

(c) $BiO_3^-(aq) + Mn^{2+}(aq) \rightarrow$
$$MnO_4^-(aq) + Bi^{3+}(aq) + ?$$

(d) $H_2SO_4(concentrated) + Al(s) \rightarrow$
$$H_2S(g) + Al^{3+}(aq) + ?$$

(e) $MnO_4^-(aq) + Pb^{2+}(aq) \rightarrow Mn^{2+}(aq) + Pb^{4+}(aq) + ?$

16.57 Balance the following oxidation-reduction equation:
$$Br^-(aq) + Ce^{4+}(aq) \rightarrow Br_2(aq) + Ce^{3+}(aq)$$

16.58 Balance the following oxidation-reduction equation:
$$I^-(aq) + Cu^{2+}(aq) \rightarrow CuI(s) + I_2(aq)$$

16.59 Complete and balance each of the following oxidation-reduction equations:

(a) $Zn(s) + H^+(aq) + NO_3^-(aq) \rightarrow$
$$Zn^{2+}(aq) + NH_4^+(aq) + ?$$

(b) $H_2SO_4(aq) + Zn(s) \rightarrow H_2(g) + ZnSO_4(aq)$

(c) $MnO_4^-(aq) + Cl^-(aq) \rightarrow Mn^{2+}(aq) + Cl_2(g) + ?$

(d) $H_2O_2(aq) + CrO_4^{2-}(aq) + H^+(aq) \rightarrow$
$$Cr^{3+}(aq) + O_2(g) + ?$$

16.60 Complete and balance each of the following equations:

(a) $HAsO_2(s) + H_2O(l) + Ce^{4+}(aq) \rightarrow$
$$Ce^{3+}(aq) + H_3AsO_4(aq) + H^+(aq)$$

(b) $Pb^{4+}(aq) + Br^-(aq) + H_2O(l) \rightarrow$
$$Pb^{2+}(aq) + BrO_3^-(aq) + H^+(aq)$$

(c) $MnO_2(s) + V^{2+}(aq) \rightarrow VO_2^+(aq) + Mn^{2+}(aq)$

(d) $Br_2(aq) + C_2O_4^{2-}(aq) \rightarrow Br^-(aq) + CO_2(g)$

16.5 Equivalents and Normality

16.61 Calculate the number of equivalents per mole of HCl for each of the following reactions:

(a) $HCl(aq) + NaOH(aq) \rightarrow NaCl(aq) + H_2O(l)$

(b) $2 HCl(aq) + Ba(OH)_2(aq) \rightarrow BaCl_2(aq) + 2 H_2O(l)$

(c) $10 HCl(aq) + 6 H^+(aq) + 2 MnO_4^-(aq) \rightarrow$
$$5 Cl_2(g) + 2 Mn^{2+}(aq) + 8 H_2O(l)$$

16.62 Calculate the number of equivalents per mole of H_2SO_4 for each of the following reactions:

(a) $H_2SO_4(aq) + NaOH(aq) \rightarrow NaHSO_4(aq) + H_2O(l)$

(b) $H_2SO_4(aq) + Ba(OH)_2(aq) \rightarrow BaSO_4(s) + 2 H_2O(l)$

(c) $5 H_2SO_4(aq) + 4 Zn(s) \rightarrow$
$$4 ZnSO_4(aq) + H_2S(g) + 4 H_2O(l)$$

16.63 Calculate the number of equivalents of H_2SO_4 in 5.00 mol of H_2SO_4 undergoing complete neutralization.

16.64 Calculate the number of equivalents of H_2SO_4 in 5.00 mol of H_2SO_4 undergoing partial neutralization to $NaHSO_4$.

16.65 Calculate the normality of an acid if 25.00 mL of the acid is neutralized by 38.35 mL of 0.2193 N (a) NaOH or (b) $Ba(OH)_2$.

16.66 Calculate the normality of an acid if 25.00 mL of the acid is neutralized by (a) 43.77 mL of 0.4113 N NaOH or (b) 0.1571 equiv of $La(OH)_3$.

16.67 Calculate the equivalent mass of an unknown base if 2.135 g of the base requires 31.97 mL of 1.500 N HCl for neutralization.

16.68 Calculate the equivalent mass of an unknown solid acid if 10.01 g of the acid requires 41.73 mL of 4.000 N KOH for neutralization.

16.69 Calculate the number of milliequivalents of solid acid that is neutralized by 31.49 mL of 3.127 N NaOH.

16.70 Calculate the normality of a base if 25.00 mL of 0.1000 N HCl is neutralized by 44.16 mL of the base.

16.71 Calculate the number of milliequivalents of solid acid that is neutralized by 39.47 mL of 4.000 N KOH.

16.72 Calculate the normality of a base if 25.00 mL of 0.5000 N HCl is neutralized by 42.72 mL of the base.

16.73 Calculate the equivalent mass of H_2SeO_4 that undergoes complete neutralization:
$$H_2SeO_4(aq) + 2 NaOH(aq) \rightarrow Na_2SeO_4(aq) + 2 H_2O(l)$$

■ ■ ■ GENERAL PROBLEMS ■ ■ ■

16.74 Give an example of each type of compound referred to in Problem 16.1.

16.75 Draw an electron dot diagram for CN^-. Determine the oxidation number of each element, and compare with the oxidation numbers of carbon and oxygen in carbon monoxide, which has the same number of electrons. Why is there a difference?

16.76 Consider the names of the oxyanions in Table 6.4. The ending -ate is used for ions with "more oxygen." How could that rule be restated in terms of oxidation numbers?

16.77 What is the oxidation number of hydrogen in (a) hydrogen ion and (b) hydride ion?

16.78 The compound $SnCl_4$ is mainly covalent. Name it in two ways.

16.79 Explain the reason that we could use charges instead of oxidation numbers for naming monatomic cations in Chapter 6.

16.80 Explain why a balanced equation for a half-reaction must always include at least one type of ion.

16.81 (a) When 4.54 g of a monoprotic acid (an acid that contains one ionizable hydrogen atom per molecule), denoted by HX, is titrated with 1.000 M NaOH, it takes 42.73 mL of the base to neutralize the acid. Calculate the molar mass of the acid.

(b) When 4.54 g of a diprotic acid (an acid that contains two ionizable hydrogen atoms per molecule), denoted by H_2X, is titrated with 1.000 M NaOH, it takes 42.73 mL of the base to neutralize the acid. Calculate the molar mass of the acid.

(c) From the mass of the acid and the volume and concentration of the base alone, can you tell if the acid is monoprotic or diprotic?

16.82 Calculate the equivalent mass of both acids in Problem 16.81.

16.83 Determine the oxidation number of the oxygen in each of the following compounds:

(a) Na_2O_2 (b) KO_2 (c) PbO_2

16.84 Based on equations in this chapter, select the good oxidizing agents and the good reducing agents from the following list:

$KMnO_4$ KI HNO_3 Zn $K_2Cr_2O_7$

16.85 Name the following compounds:

(a) BaO_2 (b) RbO_2 (c) SnO_2

16.86 Explain why hydrogen peroxide can act sometimes as an oxidizing agent and sometimes as a reducing agent.

16.87 Consider the following equation:

$$14 \, KMnO_4(s) + 5 \, C_3H_8O_3(l) \xrightarrow[\text{temperature}]{\text{High}}$$
$$7 \, K_2CO_3(s) + 14 \, MnO(s) + 8 \, CO_2(g) + 20 \, H_2O(g)$$

(a) What is the change in oxidation number for manganese?

(b) What is (are) the change(s) in oxidation number for carbon?

(c) Why wasn't a net ionic equation written for this reaction?

16.88 (a) If 2.46 g of an unknown acid is completely neutralized with 41.73 mL of 2.500 N NaOH, what is the number of equivalents of acid?

(b) How many moles of acid are involved if the acid is monoprotic?

(c) How many moles of acid are involved if the acid is diprotic?

16.89 Calculate the change in oxidation state of chromium during the following reaction:

$$2 \, H^+(aq) + 2 \, CrO_4^{2-}(aq) \rightarrow Cr_2O_7^{2-}(aq) + H_2O(1)$$

16.90 Complete and balance each of the following equations, in which more than one element is oxidized. (**Hint:** Choose any reasonable oxidation numbers for the elements in the second ion in each reaction.)

(a) $Cr_2O_7^{2-}(aq) + CN^-(aq) \rightarrow$
$$Cr^{3+}(aq) + CO_2(g) + NO_2(g) + ?$$

(b) $Cr_2O_7^{2-}(aq) + CNO^-(aq) \rightarrow$
$$Cr^{3+}(aq) + CO_2(g) + NO_2(g) + ?$$

(c) $Cr_2O_7^{2-}(aq) + CNS^-(aq) \rightarrow$
$$Cr^{3+}(aq) + CO_2(g) + NO_2(g) + SO_4^{2-}(aq) + ?$$

16.91 Complete and balance each of the following equations:

(a) $Cl_2(g) + NaOH(aq) \rightarrow NaCl(aq) + NaClO_3(aq) + ?$

(b) $Cl_2(g) + NaOH(aq) \rightarrow NaCl(aq) + NaClO(aq) + ?$

(c) $P_4(s) + NaOH(aq) \rightarrow PH_3(g) + NaH_2PO_3(aq) + ?$

17 Reaction Rates and Chemical Equilibrium

- 17.1 Rates of Reaction
- 17.2 The Condition of Equilibrium
- 17.3 LeChâtelier's Principle
- 17.4 Equilibrium Constants

■ **Key Terms** (*Key terms are defined in the Glossary.*)

catalyst (17.1)
enzyme (17.1)
equilibrium (17.2)
equilibrium constant (17.4)

equilibrium constant expression (17.4)
LeChâtelier's principle (17.3)
proceed to the left (17.2)
proceed to the right (17.2)

rate (17.1)
shift (17.3)
state of subdivision (17.1)
stress (17.3)

■ **Symbols**

$\rightleftharpoons$ (equilibrium) (17.2)
[A] (molarity of A) (17.4)
K (equilibrium constant) (17.4)

You know that (pure) liquid water at 0°C will freeze if heat is removed from it and that ice at 0°C will melt if heat is added to it. However, you may not know that if no heat is added or removed from a mixture of ice and liquid water at 0°C, both the melting and freezing processes still continue. In this case, the rate of melting and the rate of freezing are the same, so the mass of the ice and the mass of the liquid water do not change. (The shape of the ice might change, however, as some melting and freezing take place.) When exactly opposite physical processes or chemical reactions occur together at the *same rate,* a condition of *equilibrium* is said to be established. When two opposite processes occur at equal rates, nothing *appears* to be happening. However, both processes continue at the rates dictated by the conditions. If the conditions change, a change in the system may be observed.

Since chemical equilibrium involves rates of reactions, this chapter first investigates the factors that affect the rate of a reaction (Section 17.1). The molecular basis of chemical equilibrium and some of its terminology are then presented in Section 17.2. LeChâtelier's principle, discussed in Section 17.3, explains qualitatively what happens to a system at equilibrium when a change is imposed on the system. Section 17.4 presents the equilibrium constant, which allows us to obtain quantitative results for systems at equilibrium.

17.1   Rates of Reaction

OBJECTIVE

■ to predict how changes in the conditions affect the rate of a chemical reaction

Any of six factors can affect the **rate** of a chemical reaction: (1) the nature of the reactants, (2) the temperature, (3) the presence of a catalyst, (4) the concentration of a reactant in solution, (5) the pressure of a gaseous reactant, and (6) the state of subdivision of a solid reactant.

Some reactions, such as a dynamite explosion, inherently tend to proceed rapidly. Others, such as the reaction of solid limestone with carbon dioxide in water to form calcium hydrogen carbonate, tend to proceed slowly. It may take centuries for underground streams to form caverns by reaction with limestone (Figure 8.10). For a given reaction, chemists have no control over the nature of the reactants.

Raising the temperature generally increases the rate of a chemical reaction. A rule of thumb is that a 10°C rise in temperature approximately doubles the rate of a reaction. The actual increase for any given reaction will not follow this rule exactly, of course, but many reactions follow it approximately.

A **catalyst** increases the rate of a chemical reaction (Chapter 8). The thousands of different catalysts work in many different ways. For example, **enzymes** are catalysts that accelerate reactions in living things. A given reaction may proceed at the same rate without the enzyme but only at a much higher temperature. For example, glucose, $C_6H_{12}O_6$, reacts with oxygen to form carbon dioxide and water. Without any enzymes and at body temperature, glucose does not react with oxygen at all. In the body, many enzymes cause rapid conversion of glucose. In the absence of such enzymes, the system has to be heated to get glucose to react with oxygen at any reasonable rate.

In general, the higher the concentration of a reactant in a solution, the higher is the rate of reaction. Increases in the concentration of a reactant involved in one reaction will cause different rate increases than the same increases in concentration of a reactant involved in another reaction. In general, however, an increase in concentration causes an increase in rate. Conversely, a decrease in concentration generally causes a decrease in rate.

The pressure of a gaseous reactant is a measure of its concentration. At a given temperature, pressure is proportional to the number of moles per liter, as given by the ideal gas law (Chapter 12):

$$P = \left(\frac{n}{V}\right)RT$$

Thus, this effect is a corollary of the concentration effect. In general, the higher the pressure, the higher is the rate of reaction.

The **state of subdivision** of a solid is important to the rate of its reaction. When a solid reacts with a liquid or a gas, contact between the two reactants occurs only at the surface of the solid. The more surface area there is per unit mass, the faster the reaction can occur. Thus, finely divided solids tend to react more rapidly than the same solids in particles of larger sizes. An alarming illustration of this effect is a coal dust explosion in a mine. The finely divided coal, set off by a spark, can react almost instantaneously with oxygen. Coal in lump form burns slowly and quietly and must be ignited with kindling to get it started in the first place.

17.2 The Condition of Equilibrium

OBJECTIVE

■ to predict the behavior of an equilibrium system once it has been formed

Suppose we introduce some nitrogen and hydrogen with a catalyst (such as platinum, [Pt]) into a vessel and increase the temperature of the system to 500°C (Figure 17.1). The nitrogen begins to react with the hydrogen to form ammonia, according to the equation

$$3\ H_2(g) + N_2(g) \xrightarrow{Pt} 2\ NH_3(g)$$

The rate of the reaction is rapid at first but slows down as some reactants are used up. Since the pressures of the nitrogen and the hydrogen both decrease, the rate at which the combination occurs also decreases.

Suppose we introduce some ammonia into a similar vessel with the same catalyst and increase the temperature of the system to 500°C (Figure 17.1). A decomposition occurs, according to the equation

$$2\ NH_3(g) \xrightarrow{Pt} 3\ H_2(g) + N_2(g)$$

Figure 17.1 *Approaching Equilibrium from Two Directions*

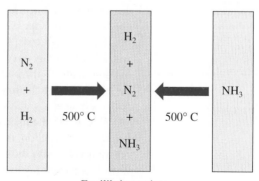

Equilibrium mixture

Figure 17.2 Rates of Forward and Reverse Reactions

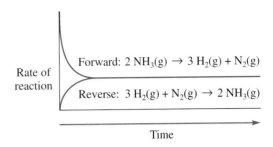

The initial rate of this reaction is lowered as some of the ammonia is used up. However, as hydrogen and nitrogen are produced, they begin to react to form ammonia. They react slowly at first, since little of either is present to react. As their pressures increase as a result of the ammonia decomposition, their rate of combination increases. So in this vessel, we have a decreasing rate of decomposition of ammonia and an increasing rate of combination of hydrogen and nitrogen. When these two rates become equal (Figure 17.2), a state of **equilibrium** will exist. Equilibrium is a *dynamic* state; both reactions continue to occur. Since the rate of combination equals the rate of decomposition, no *net* reaction occurs. The decomposition reaction does not go to completion (Section 10.5). Neither reaction stops, but the effect of the two opposing reactions will make the decomposition reaction *appear* to stop. No matter how long this system exists, as long as we do not change the conditions on the system, none of the pressures of any of the gases will change. (A similar situation is reached in the vessel in which the hydrogen and nitrogen were placed.)

For efficiency, we can write both the forward and reverse reactions of an equilibrium system in one equation, using a double arrow to denote that the reaction goes in both directions. For example:

$$2\ NH_3(g) \overset{Pt}{\rightleftharpoons} 3\ H_2(g) + N_2(g)$$

If you start with the substances on the left, you produce the substances on the right. If you start with the substances on the right, you produce the substances on the left. Because either set of substances produces the other, either set can be placed on the left. The substances written on the left side are conventionally

ITEM OF INTEREST

The equilibrium system for the combination of hydrogen and nitrogen is economically very important:

$$3\ H_2(g) + N_2(g) \xrightarrow{\text{Catalyst}} 2\ NH_3(g)$$

Nitrogen gas is very difficult to get to react to form compounds, yet compounds of nitrogen are necessary for preparing fertilizers and explosives, among many other useful products. The German chemist Fritz Haber (1868–1934) devised the *Haber process* to "fix" nitrogen (convert it to a compound). The process was first used to produce explosives during World War I, when Germany was cut off from imports of natural nitrates by a shipping blockade.

called the reactants, and those written on the right side are called the products. When the substances on the left undergo reaction to produce those on the right, the reaction is said to **proceed to the right.** When the substances on the right undergo reaction to produce those on the left, the reaction is said to **proceed to the left.**

17.3 LeChâtelier's Principle

OBJECTIVE

- to predict the effect of a change of conditions on a system initially at equilibrium

The addition of a catalyst affects the rates of both forward and reverse reactions equally and thus does not cause any shift in the equilibrium.

When the conditions that affect the rate of a chemical reaction are changed, the rates of a forward and a reverse reaction may be affected differently. If these rates become different, more reactants or products are produced. The direction of the shift of an equilibrium can be predicted qualitatively using LeChâtelier's principle. **LeChâtelier's principle** states that if a *stress* is applied to a system *at equilibrium,* the equilibrium will tend to *shift* in a direction to relieve that stress.

A **stress** is a change of conditions, such as a change in the concentration or pressure of one or more of the reactants or products or a change in the system's temperature. A **shift** is a net reaction of reactants to form products or of products to form reactants. If no further stress is applied, a new equilibrium will be established, with different concentrations of reactants and products from those in the original system. Some stresses cause no shift. The addition of a catalyst affects the rates of both forward and reverse reactions equally and thus does not cause any shift in the equilibrium.

■ EXAMPLE 17.1

What will be the effect of adding ammonia to the following system at equilibrium?

$$2\ NH_3(g) \rightleftharpoons 3\ H_2(g) + N_2(g)$$

Solution

The addition of ammonia is a stress, causing an increase in the ammonia pressure. The equilibrium can shift to the right to use up some of the added ammonia, tending to relieve the stress. After the shift, more nitrogen and hydrogen are present. Also, less ammonia is present than there would have been if no shift had occurred, but more than was present in the original equilibrium mixture. The system reaches a new equilibrium.

Practice Problem 17.1 What will be the effect of adding nitrogen to the following system at equilibrium?

$$2\ NH_3(g) \rightleftharpoons 3\ H_2(g) + N_2(g)$$ ■

■ EXAMPLE 17.2

Explain in terms of reaction rates why the addition of ammonia causes the equilibrium of Example 17.1 to shift to the right.

Solution

Before the additional ammonia is added, the system is at equilibrium, which means that the rate of ammonia formation and the rate of ammonia decomposition are equal. The addition of ammonia increases the ammonia pressure, which increases the rate of ammonia decomposition. At the instant when ammonia is added, the rate of combination of nitrogen and hydrogen is not affected. Thus, for some period of time, the rate of decomposition is greater than the rate of combination. The result is a net decomposition reaction until the pressures of the hydrogen and the nitrogen build up, the combination reaction thus speeds up, and a new equilibrium is established. ■

■ EXAMPLE 17.3

What will be the effect of the removal of ammonia from the following system at equilibrium?

$$2 \, NH_3(g) \rightleftharpoons 3 \, H_2(g) + N_2(g)$$

Solution

The stress is the removal of ammonia, causing a decrease in the ammonia pressure. The equilibrium will shift to the left to replace some of the ammonia, tending to relieve the stress. After the shift, less nitrogen and hydrogen are present. Also, more ammonia is present than there would have been if no shift had occurred, but less than was present in the original equilibrium mixture. The system reaches a new equilibrium.

Practice Problem 17.3 What will be the effect of the removal of nitrogen from the following system at equilibrium at high temperature?

$$2 \, NO_2(g) \rightleftharpoons 2 \, O_2(g) + N_2(g)$$ ■

■ EXAMPLE 17.4

What will be the effect of adding heat to the following system at equilibrium?

$$2 \, NH_3(g) + heat \rightleftharpoons 3 \, H_2(g) + N_2(g)$$

Solution

The stress is the addition of heat, causing an increase in temperature. The equilibrium can shift to the right to use up some of the added heat, tending to relieve the stress. After the shift, more nitrogen and hydrogen but less ammonia are present. The system reaches a new equilibrium at a higher temperature. ■

■ EXAMPLE 17.5

What will be the effect of a temperature increase on the following system at equilibrium?

$$2 \, NH_3(g) + heat \rightleftharpoons 3 \, H_2(g) + N_2(g)$$

Solution

The temperature is increased by adding energy, usually in the form of heat. Some of the added heat can be removed by a shift of the equilibrium to the right, so more nitrogen and hydrogen are produced and some ammonia is used up. (This problem is a rewording of that presented in Example 17.4.) ■

■ EXAMPLE 17.6

What effect will lowering the temperature have on the following equilibrium system?

$$3 H_2(g) + N_2(g) \rightleftharpoons 2 NH_3(g) + heat$$

Solution

The temperature will be lowered by removal of heat. The equilibrium will shift to the right to replace some of the heat removed. Note that this equation is the reverse of the equation in Example 17.5. The effect here is opposite to that caused by increasing the temperature in that example, even though both equilibria shifted to the right. Because the reactions are written differently, a shift to the right in that case produced more hydrogen and nitrogen and in this case produces more ammonia. ■

A change in the partial pressure of a gaseous reactant or product will cause a shift in an equilibrium system. Note that a change in the *total pressure* does not necessarily shift the equilibrium, but a change in all the partial pressures may do so. If all the partial pressures are changed at once—for example, by a change in the volume of the container—the equilibrium will shift to reduce the stress if possible. For example, consider the reaction of ammonia to produce nitrogen and hydrogen:

$$2 NH_3(g) \rightleftharpoons 3 H_2(g) + N_2(g)$$

If this system is at equilibrium in a container of a certain volume and the volume of the container is reduced, all the partial pressures are increased. The equilibrium will shift to reduce the total pressure by making fewer moles; that is, it will shift to the left. (Reaction of every 4 mol of the elements produces only 2 mol of ammonia.) Making fewer moles reduces the stress of the increased pressure.

■ EXAMPLE 17.7

Predict how reducing the volume of the container will affect the equilibrium state of the reaction of nitrogen and oxygen to make nitrogen monoxide, NO:

$$N_2(g) + O_2(g) \rightleftharpoons 2 NO(g)$$

Solution

Since there are 2 mol of gas on each side of this equation, neither a shift to the left nor a shift to the right will change the total pressure. The reduction in volume causes an increase in pressure but will not shift the equilibrium. ■

■ EXAMPLE 17.8

Predict how reducing the volume of the container will affect the equilibrium state of the reaction of solid carbon and oxygen gas to make carbon monoxide gas:

$$2 C(s) + O_2(g) \rightleftharpoons 2 CO(g)$$

Solution

The reduction in volume causes an increase in partial pressure of each of the *gases*. The carbon is a solid and has no partial pressure. Therefore, the equilibrium will shift to the left to reduce the number of moles of gas at equilibrium. ■

17.4 Equilibrium Constants

■ to use equilibrium constants to determine the concentrations of reactants and products in a system at equilibrium

LeChâtelier's principle (Section 17.3) allows us to make qualitative predictions about the effects of changes of conditions on an equilibrium system but does not allow quantitative calculations. However, at equilibrium at a given temperature, a certain ratio of concentration terms is a constant for all solutes and gases involved in the reaction. (Solids and pure liquids are not included in the ratio.) For the general reaction

$$a\,A + b\,B \rightleftharpoons c\,C + d\,D$$

the following ratio is a constant at equilibrium at a given temperature:

$$K = \frac{[C]^c[D]^d}{[A]^a[B]^b}$$

The square brackets mean "the molarity of" the substance they enclose, and the constant K is called the **equilibrium constant.** The entire equation is known as the **equilibrium constant expression.** No matter what the initial concentrations of reactants or products, the ratio of the concentrations *at equilibrium* will be equal to the constant K. The value of K depends only on the specific chemical equation and on the temperature. It does not depend on any of the other factors that can affect the rate of a reaction. For example, if different quantities of the same reactants and products are introduced into different reaction vessels, they will react with each other until, at equilibrium, the same ratio of concentrations (each raised to the appropriate power) is established.

The value of K for a reaction gives quantitative information about the extent of the reaction. That is, a large value of K (about 10^4 or larger) means that, at equilibrium, the reaction proceeds almost completely to the right; a small value of K (less than 10^{-4} or so) means that, at equilibrium, the reactants have not reacted much at all (or the products have reacted almost completely). If concentrations of the reactants and products at equilibrium are known, a value for K can be calculated. If a value for K and some concentration data are known, other concentrations at equilibrium can be calculated.

Several points concerning the equilibrium constant expression must be emphasized:

1. *Equilibrium* concentrations are used.
2. The concentrations (raised to the appropriate powers) are multiplied and divided, not added or subtracted.
3. The exponents of the concentration terms are equal to the coefficients in the balanced chemical equation.
4. The concentrations of the substances on the right side of the equation appear in the numerator of the equilibrium constant expression, and those of the substances on the left side appear in the denominator.
5. No terms related to solids or pure liquids, including water in dilute aqueous solutions, are included in the equilibrium constant expression.

■ **EXAMPLE 17.9**

Write an equilibrium constant expression for the reaction of nitrogen and hydrogen to form ammonia at 500°C.

Solution

$$3 H_2(g) + N_2(g) \rightleftharpoons 2 NH_3(g)$$

$$K = \frac{[NH_3]^2}{[H_2]^3[N_2]}$$

Practice Problem 17.9   Write an equilibrium constant expression for the following reaction at 500°C:

$$2 NO(g) \rightleftharpoons N_2(g) + O_2(g) \quad ■$$

■ **EXAMPLE 17.10**

Write an equilibrium constant expression for the reaction of ammonia to form nitrogen and hydrogen at 500°C.

Solution

$$2 NH_3(g) \rightleftharpoons 3 H_2(g) + N_2(g)$$

$$K = \frac{[H_2]^3[N_2]}{[NH_3]^2}$$

This expression is the reciprocal of that obtained in Example 17.9 because the chemical reaction is written in the reverse direction.

Practice Problem 17.10   Write an equilibrium constant expression for the following reaction:

$$N_2O(g) \rightleftharpoons N_2(g) + \tfrac{1}{2} O_2(g) \quad ■$$

Finding Values of Equilibrium Constants

> The values used in the equilibrium constant expression are *concentrations* at equilibrium.

The values used in the equilibrium constant expression are *concentrations*, not numbers of moles. Moreover, they are concentrations of the reactants and products *at equilibrium*.

■ **EXAMPLE 17.11**

For the following reaction at high temperature,

$$CO_2(g) + H_2(g) \rightleftharpoons CO(g) + H_2O(g)$$

the concentrations of $CO_2(g)$ and $H_2(g)$ in a particular equilibrium mixture are each 0.80 mol/L, and the concentrations of $CO(g)$ and $H_2O(g)$ are each 1.00 mol/L. Calculate the value of the equilibrium constant at this temperature.

Solution

The equilibrium constant expression for this reaction is

$$K = \frac{[CO][H_2O]}{[CO_2][H_2]}$$

Substituting the equilibrium concentrations yields the value of the equilibrium constant:

$$K = \frac{[CO][H_2O]}{[CO_2][H_2]} = \frac{(1.00 \text{ mol/L})^2}{(0.80 \text{ mol/L})^2} = 1.6 \quad \blacksquare$$

If the equilibrium concentrations are not given in the problem, they can usually be calculated from data given in the problem.

■ **EXAMPLE 17.12**

For the reaction

$$CO_2(g) + H_2(g) \rightleftharpoons CO(g) + H_2O(g)$$

2.000 mol of carbon dioxide gas and 1.000 mol of hydrogen are placed in a 1.000-L flask and allowed to come to equilibrium at 1000°C. At that point, the concentration of water vapor is found to be 0.723 mol/L. Calculate the value of the equilibrium constant.

Solution

Two initial concentrations and one equilibrium concentration are given in this problem. You can tabulate the values given and then work out all the equilibrium concentrations. The units are all moles per liter.

	$CO_2(g)$	+	$H_2(g)$	$\rightleftharpoons$	$CO(g)$	+	$H_2O(g)$
Initial concentrations	2.000		1.000		0.000		0.000
Change due to reaction							
Equilibrium concentrations							0.723

You see immediately that the 0.723 mol/L of water vapor must have been produced by the reaction, and you deduce from the balanced equation that 0.723 mol/L of carbon monoxide has also been produced and that 0.723 mol/L of each reactant has been used up.

	$CO_2(g)$	+	$H_2(g)$	$\rightleftharpoons$	$CO(g)$	+	$H_2O(g)$
Initial concentrations	2.000		1.000		0.000		0.000
Change due to reaction	−0.723		−0.723		+0.723		+0.723
Equilibrium concentrations							0.723

With the initial concentrations and the changes brought about by the reaction, you can calculate the rest of the equilibrium concentrations:

	$CO_2(g)$	+	$H_2(g)$	$\rightleftharpoons$	$CO(g)$	+	$H_2O(g)$
Initial concentrations	2.000		1.000		0.000		0.000
Change due to reaction	−0.723		−0.723		+0.723		+0.723
Equilibrium concentrations	1.277		0.277		0.723		0.723

Finally, put these *equilibrium* concentrations in the equilibrium constant expression and solve:

$$K = \frac{[CO][H_2O]}{[CO_2][H_2]} = \frac{(0.723 \text{ mol/L})^2}{(1.277 \text{ mol/L})(0.277 \text{ mol/L})} = 1.48 \quad \blacksquare$$

■ **EXAMPLE 17.13**

If 0.400 mol of SO_2 and 0.209 mol of O_2 are placed in a 1.00-L vessel and allowed to come to equilibrium at a high temperature, 0.075 mol of SO_2 remains. Calculate the value of the equilibrium constant for the reaction

$$2 SO_2(g) + O_2(g) \rightleftharpoons 2 SO_3(g)$$

Solution

Tabulate the concentrations given and the concentrations that can be deduced from them:

	$2 SO_2(g)$	$+$	$O_2(g)$	$\rightleftharpoons$	$2 SO_3(g)$
Initial concentrations	0.400		0.209		0.000
Change due to reaction	−0.325		−0.162		+0.325
Equilibrium concentrations	0.075		0.047		0.325

Solve for *K*:

$$K = \frac{[SO_3]^2}{[SO_2]^2[O_2]} = \frac{(0.325)^2}{(0.075)^2(0.047)} = 400 = 4.0 \times 10^2$$

Note that the *changes in concentration* are in the same ratio as the coefficients in the balanced chemical equation but that neither the initial concentrations nor the equilibrium concentrations are in that ratio.

Practice Problem 17.13   If 0.400 mol of SO_3 and 0.010 mol of O_2 are placed in a 1.00-L vessel at high temperature, 0.325 mol of SO_3 remains at equilibrium. Calculate the value of the equilibrium constant for the reaction

$$2 SO_3(g) \rightleftharpoons 2 SO_2(g) + O_2(g)$$

Compare this value to that found in Example 17.13, and also compare the equilibrium concentrations. ■

Calculations Using Equilibrium Constants

If the value of the equilibrium constant is given, we may solve for the equilibrium concentrations in terms of initial concentrations. Algebraic variables, such as *x* or *y*, are used for the equilibrium concentration terms.

■ **EXAMPLE 17.14**

Calculate the concentration of CO_2 at equilibrium if 1.50 mol of CO and 1.50 mol of H_2O are introduced into a 1.00-L flask at 1000°C and allowed to come to equilibrium. The value of the equilibrium constant is 0.63.

$$CO(g) + H_2O(g) \rightleftharpoons CO_2(g) + H_2(g)$$

Solution

The equilibrium constant expression is

$$K = \frac{[CO_2][H_2]}{[CO][H_2O]} = 0.63$$

Let x equal the equilibrium concentration of CO_2. Then,

	$CO(g)$	+	$H_2O(g)$	$\rightleftharpoons$	$CO_2(g)$	+	$H_2(g)$
Initial concentrations	1.50		1.50		0.00		0.00
Change due to reaction	$-x$		$-x$		$+x$		$+x$
Equilibrium concentrations	$1.50 - x$		$1.50 - x$		x		x

Substituting these values into the equilibrium constant expression yields

$$K = \frac{[CO_2][H_2]}{[CO][H_2O]} = \frac{x^2}{(1.50 - x)^2} = 0.63$$

Taking the square root of both sides yields

$$\frac{x}{(1.50 - x)} = 0.79$$

$$x = 0.79(1.50 - x) = 1.2 - 0.79x$$

$$1.79x = 1.2$$

$$x = 0.67$$

The concentration of each product is thus 0.67 mol/L, and that of each reactant is

$$1.50 \text{ mol/L} - 0.67 \text{ mol/L} = 0.83 \text{ mol/L}$$

To check, put these values into the equilibrium constant expression, and solve for K:

$$K = \frac{[CO_2][H_2]}{[CO][H_2O]} = \frac{(0.67)^2}{(0.83)^2} = 0.65$$

The value is close enough to the original value given for K, allowing for rounding errors. ■

■ **EXAMPLE 17.15**

At 525°C, $K = 3.35 \times 10^{-3}$ for the reaction

$$CaCO_3(s) \rightleftharpoons CaO(s) + CO_2(g)$$

Calculate the concentration of carbon dioxide at equilibrium.

Solution

Since substances that are solids do not have terms in the equilibrium constant expression,

$$K = [CO_2] = 3.35 \times 10^{-3}$$

Thus, the concentration of CO_2 is 3.35×10^{-3} mol/L. ■

If the equilibrium constant is very large, we expect very little of at least one of the reactants to remain unreacted. In contrast, if the equilibrium constant is very small, we expect very little of the products to be produced. We can use these facts to approximate some solutions to problems involving equilibrium constants because a small value *added to or subtracted from* a larger quantity has a negligible effect. In this book, approximations that cause an error of no greater than 5–10% are acceptable.

■ **EXAMPLE 17.16**

Consider the general equation

$$2\,A(g) + B(g) \rightleftharpoons C(g) \qquad K = 1.0 \times 10^{-8}$$

If 0.25 mol of A and 0.20 mol of B are placed in a 1.0-L vessel and allowed to come to equilibrium, what will the equilibrium concentration of C be?

Solution

	$2\,A(g)$	$+$	$B(g)$	$\rightleftharpoons$	$C(g)$
Initial concentrations	0.25		0.20		0.00
Change due to reaction	$-2x$		$-x$		$+x$
Equilibrium concentrations	$0.25 - 2x$		$0.20 - x$		x

$$K = \frac{[C]}{[A]^2[B]} = \frac{x}{(0.25 - 2x)^2(0.20 - x)}$$

Since the value of K is so low, you expect x to have a very small value. You can thus ignore x and $2x$ when subtracted from the larger values 0.25 and 0.20, which gives

$$K = \frac{[C]}{[A]^2[B]} = \frac{x}{(0.25)^2(0.20)}$$

$$\frac{x}{0.012} = 1.0 \times 10^{-8}$$

$$x = 0.012 \times 10^{-8} = 1.2 \times 10^{-10}$$

This small value of x, or even twice this value, $2x$, is certainly insignificant when subtracted from 0.25 or 0.20.

Substitute the final concentration values into the equilibrium constant expression to check the approximation:

$$K = \frac{[C]}{[A]^2[B]} = \frac{1.2 \times 10^{-10}}{(0.25)^2(0.20)} = 9.6 \times 10^{-9}$$

Since this value of K is close to the value given in the statement of the problem, the approximation is allowable.

Practice Problem 17.16   Consider the general equation

$$2\,A(g) + B(g) \rightleftharpoons C(g) + 2\,D(g) \qquad K = 1.0 \times 10^{-8}$$

If 0.25 mol of A and 0.20 mol of B are placed in a 1.0-L vessel and allowed to come to equilibrium, what will the equilibrium concentration of C be? Explain why this value is so different from that found in Example 17.16. ■

■ **EXAMPLE 17.17**

The following reaction occurs at 400°C:

$$2\,HI(g) \rightleftharpoons H_2(g) + I_2(g) \qquad K = 0.0200$$

If 0.0800 mol of H_2 and 0.0800 mol of I_2 are placed in a 1.00-L vessel and allowed to come to equilibrium, can an approximation be used in the calculation of the concentration of each substance at equilibrium? What is the concentration of gaseous iodine at equilibrium?

Solution

With this equation, the "products" react to form the "reactant," but that makes no difference to the solution method.

$$2 \text{ HI(g)} \quad \rightleftharpoons \quad H_2(g) \quad + \quad I_2(g)$$

Initial concentrations	0.00	0.0800	0.0800
Change due to reaction	$2x$	$-x$	$-x$
Equilibrium concentrations	$2x$	$0.0800 - x$	$0.0800 - x$

$$K = \frac{[H_2][I_2]}{[HI]^2} = \frac{(0.0800 - x)^2}{(2x)^2} = 0.0200$$

You approximate by assuming that x is insignificant when subtracted from 0.0800:

$$K = \frac{(0.0800)^2}{(2x)^2} = 0.0200$$

$$x^2 = 0.0800$$

$$x = 0.283$$

This value of x is certainly *not* insignificant relative to 0.0800. It exceeds 0.0800, and it cannot be neglected. A more exact solution method is required. Begin by taking the square root of each side of the original equilibrium constant expression:

Take the square root:
$$\frac{(0.0800 - x)^2}{(2x)^2} = 0.0200$$

$$\frac{0.0800 - x}{2x} = 0.141$$

$$x = 0.0624$$

$$[I_2] = 0.0800 - x = 0.0176 \text{ mol/L}$$

Checking:

$$\frac{(0.0176)^2}{4(0.0624)^2} = 0.0199$$

This method is accurate. ■

■ ■ ■ ■ ■ ■ **SUMMARY** ■ ■ ■ ■ ■ ■

Six factors affect the rate of a chemical reaction: (1) the nature of the reactants, (2) the temperature, (3) the presence of a catalyst, (4) the concentration of a solute, (5) the pressure of a gas, and (6) the state of subdivision of a solid. The last two of these may be considered as effects of concentration. (Section 17.1)

Two exactly opposite processes occurring in the same place at the same time at the same rate constitute a state called equilibrium. Although no reaction appears to be occurring in a mixture at equilibrium because the effects of the opposite processes cancel each other out, each process continues. We say that equilibrium is a dynamic state. Both reactions can be represented in one chemical equation, using a double arrow to indicate an equilibrium. (Section 17.2)

LeChâtelier's principle governs the response of a system initially at equilibrium to a stress placed upon it. The equilibrium tends to shift to reduce the stress. For example, an equilibrium will shift to reduce the concentration of a substance that has been added to the system, to use up heat in a system whose temperature has been raised (heat has been added), or to produce fewer moles of gas (thus lowering the pressure) because of an increase in the total pressure. However, addition of a catalyst causes no shift in an equilibrium. LeChâtelier's principle is qualitative; it indicates the direction of an equilibrium shift but not the extent. (Section 17.3)

Quantitative calculations can be made for systems at equilibrium using the equilibrium constant expression. For the general reaction

$$a \, A + b \, B \rightleftharpoons c \, C + d \, D$$

the following ratio is a constant:

$$K = \frac{[C]^c [D]^d}{[A]^a [B]^b}$$

The concentrations of the substances on the right side of the chemical equation are placed in the numerator and divided by the concentrations of the substances on the left side, all raised to their appropriate powers (equal to the coefficients in the balanced equation).

Given equilibrium concentrations, the value of K can be calculated by simply substituting those concentrations in the equilibrium constant expression. A somewhat more difficult problem gives initial concentrations of some reactants and one equilibrium concentration and requires determination of the equilibrium concentrations of the other substances and then substitution into the equilibrium constant expression to solve for K. You can apply the concepts used to solve problems involving limiting quantities (Chapter 10) to find the equilibrium concentrations and use a tabulation method to make the solution process easier. The balanced chemical equation governs the *changes* in concentrations. If initial concentrations of some reactants and the value of K are given, algebraic variables (such as x) are used for the equilibrium concentrations. You substitute these quantities in the equilibrium constant expression and solve for x, the unknown equilibrium concentration. Approximations that involve neglecting small concentrations when they are added to or subtracted from larger ones are often helpful. (Section 17.4)

Items for Special Attention

■ The concentrations and not numbers of moles are the factors in equilibrium constant expressions.

■ In equilibrium constant expressions, the concentration terms are multiplied and divided, not added or subtracted. The concentrations in the numerator are those of the substances on the right side of the chemical reaction, and those in the denominator are for the substances on the left side. The exponents in the equilibrium constant expression correspond to the coefficients in the balanced chemical equation.

■ The balanced chemical equation governs the concentrations produced and used up by the reaction (the second row of a tabulation such as those used in the examples in Section 17.4). We can use concentrations for this purpose because the concentration ratio of the substances in a chemical reaction is the same as the mole ratio. The reactants or products all have the same volume because they are all in the same container or solution.

Self-Tutorial Problems

17.1 Which of the following statements is a correct interpretation of this balanced chemical equation?

$$2 \, CO(g) + O_2(g) \rightleftharpoons 2 \, CO_2(g)$$

(a) You can put only 2 mol of CO and 1 mol of O_2 into a vessel.

(b) You can add to a vessel only 2 mol of CO for every 1 mol of O_2 added.

(c) No matter how much of each reactant is added to a vessel, only 2 mol of CO and 1 mol of O_2 will react.

(d) When 2 mol of CO and 1 mol of O_2 are placed in a vessel, they will react to give 2 mol of CO_2.

(e) CO and O_2 react in a $2:1$ mole ratio.

17.2 (a) What is the rate of combination of nitrogen and hydrogen at the point where ammonia is first introduced into an empty reaction vessel at 500°C?

(b) What happens to this rate of combination as time passes?

17.3 Explain why the concentration ratios in addition to mole ratios are governed by the balanced chemical equation for a gaseous equilibrium system.

17.4 Write an equilibrium constant expression for each of the following reactions:

(a) $H_2(g) + I_2(g) \rightleftharpoons HI(g) + HI(g)$

(b) $H_2(g) + I_2(g) \rightleftharpoons 2 HI(g)$

17.5 Write an equilibrium constant expression for each of the following reactions:

(a) $N_2(g) + 3 H_2(g) \rightleftharpoons 2 NH_3(g)$

(b) $N_2(g) + 3 H_2(g) \rightleftharpoons 2 NH_3(g) + heat$

17.6 What is the relationship between the equilibrium constant values for the two equations in each of the following sets?

(a) $CO(g) + \frac{1}{2} O_2(g) \rightleftharpoons CO_2(g)$

$2 CO(g) + O_2(g) \rightleftharpoons 2 CO_2(g)$

(b) $2 NO_2(g) \rightleftharpoons N_2O_4(l)$

$N_2O_4(l) \rightleftharpoons 2 NO_2(g)$

17.7 Write an equilibrium constant expression for each of the following reactions:

(a) $CO(g) + H_2O(g) \rightleftharpoons CO_2(g) + H_2(g)$

(b) $CO(g) + H_2O(g) \rightleftharpoons heat + CO_2(g) + H_2(g)$

■ ■ ■ PROBLEMS ■ ■ ■

17.1 Rates of Reaction

17.8 (a) At which of the following temperatures will the combination reaction of the Haber process go fastest?

25°C   200°C   500°C

(b) At which of these temperatures will the decomposition reaction go fastest?

17.9 A certain reaction produces 0.0100 mol/L of product per minute at 25°C. How could you get that same reaction to produce 0.0200 mol/L of product per minute without changing the contents of the reaction vessel?

17.10 How can scouts on a camping trip get a log to burn faster?

17.11 A certain reaction produces 3.0 mol/L of product per minute at 25°C. How could you get that same reaction to produce 6.0 mol/L of product per minute without changing the nature or concentration of the reactants or the temperature?

17.12 (a) Calculate the surface area of a cube that is 1.00 cm on each edge.

(b) How much will the surface area increase if the cube is cut in half, as shown in the accompanying figure?

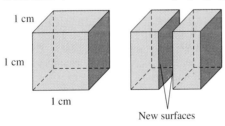

1 cm

1 cm

1 cm

New surfaces

(c) How much will the surface area increase if the cube is cut in half three times—vertically front to back, vertically side to side, and horizontally?

17.2 The Condition of Equilibrium

17.13 (a) At a certain point in a reaction leading to the equilibrium system

$$2 CO(g) + O_2(g) \rightleftharpoons 2 CO_2(g)$$

the rate of decomposition of CO_2 is 9.72×10^{-4} mol/L·s. What is the rate of combination of CO with O_2 if at this point the system is at equilibrium?

(b) What is the rate of combination of CO with O_2 if this point is at the start of this experiment?

17.14 (a) At a certain point in a reaction leading to the equilibrium system

$$2 SO_2(g) + O_2(g) \rightleftharpoons 2 SO_3(g)$$

the rate of decomposition of SO_3 is 2.61×10^{-3} mol/L·s, and the rate of production of SO_3 is 3.17×10^{-3} mol/L·s. When the reaction gets to equilibrium, will there be more or less SO_3 present than there is at this point?

(b) Was SO_3 or SO_2 plus O_2 the starting material in this experiment?

17.15 At a certain point in a reaction leading to the equilibrium system

$$2 CO(g) + O_2(g) \rightleftharpoons 2 CO_2(g)$$

the rate of combination of CO with O_2 is 4.00×10^{-4} mol/L·s. What is the rate of decomposition of CO_2 at this point?

17.3 LeChâtelier's Principle

17.16 What is the effect of raising the temperature on the following system at equilibrium?

$$2 NO(g) + O_2(g) \rightleftharpoons 2 NO_2(g) + heat$$

17.17 What is the effect of raising the temperature on the following system at equilibrium?

$$2 SO_2(g) + O_2(g) \rightleftharpoons 2 SO_3(g) + heat$$

17.18 What is the effect, if any, of decreasing the total volume on each of the following systems at equilibrium?

(a) $2 CO(g) + O_2(g) \rightleftharpoons 2 CO_2(g)$

(b) $2 C(s) + O_2(g) \rightleftharpoons 2 CO(g)$

(c) $H_2(g) + CO_2(g) \rightleftharpoons CO(g) + H_2O(g)$

17.19 What shift, if any, will occur in the following system at equilibrium as a result of each change?

$$2\,CO(g) + O_2(g) \rightleftharpoons 2\,CO_2(g) + heat$$

(a) Adding CO

(b) Removing O_2

(c) Adding O_2

(d) Removing CO_2

(e) Adding a catalyst

(f) Decreasing the volume

(g) Increasing the temperature

17.20 What shift, if any, will occur in the following system at equilibrium as a result of each change?

$$4\,NH_3(g) + 5\,O_2(g) \rightleftharpoons 4\,NO(g) + 6\,H_2O(g) + heat$$

(a) Adding O_2

(b) Removing NO

(c) Adding NH_3

(d) Adding $H_2O(g)$

(e) Increasing the temperature

(f) Decreasing the volume

(g) Adding a catalyst

17.21 State the direction of the shift of equilibrium (if any) in each of the following systems that will be caused by the indicated change:

(a) $4\,NH_3(g) + 5\,O_2(g) \rightleftharpoons 4\,NO(g) + 6\,H_2O(g)$

Adding O_2

(b) $2\,Cl_2(g) + O_2(g) \rightleftharpoons 2\,Cl_2O(g)$

Removing O_2

(c) $2\,NO_2(g) \rightleftharpoons N_2O_4(l) + heat$

Decreasing the total volume

(d) $CH_4(g) + H_2O(g) \overset{\text{Heat}}{\rightleftharpoons} CO(g) + 3\,H_2(g)$

Adding a catalyst

(e) $CaCO_3(s) + H_2O(l) + CO_2(g) \rightleftharpoons Ca(HCO_3)_2(aq)$

Removing CO_2

(f) $Heat + CO_2(g) + H_2(g) \rightleftharpoons CO(g) + H_2O(g)$

Raising the temperature

17.22 State the direction of the shift of equilibrium (if any) in each of the following systems that will be caused by the indicated change:

(a) $CCl_4(l) + \frac{1}{2}\,O_2(g) \rightleftharpoons COCl_2(g) + Cl_2(g)$

Adding O_2

(b) $PCl_5(g) \rightleftharpoons PCl_3(g) + Cl_2(g)$

Removing Cl_2

(c) $2\,HI(g) \rightleftharpoons H_2(g) + I_2(g)$

Increasing the volume

(d) $S_8(s) + 8\,Cl_2(g) \rightleftharpoons 8\,SCl_2(g)$

Adding SCl_2

(e) $2\,P_4(s) + 12\,Cl_2(g) \rightleftharpoons 8\,PCl_3(g) + heat$

Lowering the temperature

(f) $2\,NO_2(g) \rightleftharpoons N_2O_4(l) + heat$

Heating the system

17.4 Equilibrium Constants

17.23 Write an equilibrium constant expression for each of the following reactions:

(a) $SO_2(g) + Cl_2(g) \rightleftharpoons SO_2Cl_2(g)$

(b) $PF_3(g) + F_2(g) \rightleftharpoons PF_5(g)$

17.24 Write an equilibrium constant expression for the following reaction:

$$CO(g) + \frac{1}{2}\,O_2(g) \rightleftharpoons CO_2(g)$$

17.25 At a certain temperature, the value of the equilibrium constant for the following reaction is 0.010:

$$NO_2(g) \rightleftharpoons \frac{1}{2}\,N_2(g) + O_2(g)$$

Calculate the value of the equilibrium constant for each of the following reactions:

(a) $2\,NO_2(g) \rightleftharpoons N_2(g) + 2\,O_2(g)$

(b) $\frac{1}{2}\,N_2(g) + O_2(g) \rightleftharpoons NO_2(g)$

(c) $N_2(g) + 2\,O_2(g) \rightleftharpoons 2\,NO_2(g)$

17.26 At a certain temperature, the value of the equilibrium constant for the following reaction is 1.40:

$$\frac{1}{2}\,Br_2(g) + \frac{1}{2}\,Cl_2(g) \rightleftharpoons BrCl(g)$$

Calculate the value of the equilibrium constant for each of the following reactions:

(a) $Br_2(g) + Cl_2(g) \rightleftharpoons 2\,BrCl(g)$

(b) $2\,BrCl(g) \rightleftharpoons Br_2(g) + Cl_2(g)$

(c) $BrCl(g) \rightleftharpoons \frac{1}{2}\,Br_2(g) + \frac{1}{2}\,Cl_2(g)$

17.27 Calculate the value of the equilibrium constant for the reaction $2\,A + 3\,B \rightleftharpoons C + D$ if 2.00 mol of A and 3.00 mol of B are introduced into a 2.00-L reaction vessel and allowed to come to equilibrium, at which point 0.10 mol/L of A remains.

17.28 Calculate the value of the equilibrium constant for the reaction $R + 2\,T \rightleftharpoons X + Z$ if 1.00 mol of R and 2.00 mol of T are introduced into a 2.00-L reaction vessel and allowed to come to equilibrium, at which point 0.30 mol/L of R remains.

17.29 If 4.00 mol of T is placed in a 1.00-L vessel and allowed to come to equilibrium, what will be the equilibrium concentration of R? Solve this problem with each of the following equations, and compare the answers:

(a) $2\,T \rightleftharpoons R + Z \qquad K = 1.0 \times 10^{-8}$

(b) $T \rightleftharpoons \frac{1}{2}\,R + \frac{1}{2}\,Z \qquad K = 1.0 \times 10^{-4}$

17.30 At 700°C, $K = 9.00$ for the following reaction, which is one of the reactions in the commercial production of sulfuric acid:

$$SO_2(g) + NO_2(g) \rightleftharpoons SO_3(g) + NO(g)$$

If $[NO] = 0.300$ mol/L, $[NO_2] = 0.025$ mol/L, and $[SO_2] = 0.500$ mol/L at equilibrium, what is the concentration of SO_3?

17.31 Calculate the equilibrium concentration of C for the following reaction if 2.00 mol of A and 2.00 mol of B are introduced into a 2.00-L reaction vessel and allowed to come to equilibrium:

$$2 A + 2 B \rightleftharpoons C + D \qquad K = 2.50 \times 10^{-9}$$

17.32 For the following reaction at 600°C, $K = 0.410$:

$$CO_2(g) + H_2(g) \rightleftharpoons CO(g) + H_2O(g)$$

Calculate the concentration at equilibrium of CO if 0.500 mol of CO_2 and 0.500 mol of H_2 are placed in a 1.00-L container and allowed to come to equilibrium at 600°C.

17.33 Calculate the value of K for the following equilibrium at high temperature (where all the substances involved are gases) if 0.200 mol of Br_2 and 0.400 mol of Cl_2 are placed in a 2.00-L vessel, and at equilibrium, there is 0.045 mol/L of Br_2 remaining:

$$Br_2(g) + Cl_2(g) \rightleftharpoons 2 BrCl(g)$$

■ ■ ■ GENERAL PROBLEMS ■ ■ ■

17.34 (a) At a certain point in a reaction leading to the equilibrium system

$$2 SO_2(g) + O_2(g) \rightleftharpoons 2 SO_3(g)$$

the rate of decomposition of SO_3 is 6.92×10^{-3} mol/L · s, and the rate of reaction of O_2 is 4.07×10^{-3} mol/L · s. When the reaction gets to equilibrium, will there be more or less SO_3 present than there is at this point?

(b) Was SO_3 or SO_2 plus O_2 the starting material in this experiment?

17.35 Calculate the total surface area after a cube 1.00 cm on each edge has been divided into cubes 1.00 mm on each edge. Compare this surface area to that of the 1-cm³ cube of Problem 17.12(a).

17.36 What is the effect on the temperature of adding nitrogen to the following system at equilibrium?

$$\text{Heat} + 2 NH_3(g) \rightleftharpoons 3 H_2(g) + N_2(g)$$

17.37 Calculate the equilibrium concentration of each species after 0.600 mol of NO is heated to 2000 K in a 1.000-L vessel, at which point it partially decomposes to its elements. (**Hint:** Let x equal the equilibrium concentration of NO.)

$$N_2(g) + O_2(g) \rightleftharpoons 2 NO(g) \qquad K = 4.00 \times 10^{-4}$$

17.38 Calculate the equilibrium concentration of each species after 1.000 mol of NO_2 is heated to a high temperature in a 1.000-L vessel, at which point it partially decomposes to its elements.

$$N_2(g) + 2 O_2(g) \rightleftharpoons 2 NO_2(g) \qquad K = 5.00 \times 10^{-6}$$

17.39 (a) Will the following equilibrium proceed to the right most completely at a temperature of 25°C, 200°C, or 500°C?

$$N_2(g) + 3 H_2(g) \rightleftharpoons 2 NH_3(g) + \text{heat}$$

(b) Explain why a temperature of 500°C is used in the industrial production of ammonia (the Haber process). (**Hint:** See Problem 17.8.)

17.40 Predict, if possible, the direction of the shift in the equilibrium of the following system as a result of each pair of changes:

$$\text{heat} + CO(g) + H_2O(g) \rightleftharpoons CO_2(g) + H_2(g)$$

(a) Adding H_2 and decreasing the temperature

(b) Adding CO_2 and raising the temperature

(c) Adding a catalyst and removing CO_2

(d) Adding CO and CO_2

17.41 Predict, if possible, the direction of the shift in the equilibrium of the following system as a result of each pair of changes:

$$2 SO_2(g) + O_2(g) \rightleftharpoons 2 SO_3(g) + \text{heat}$$

(a) Adding O_2 and decreasing the temperature

(b) Adding SO_3 and raising the temperature

(c) Adding a catalyst and removing SO_3

(d) Adding SO_2 and SO_3

17.42 What effect, if any, is predicted by LeChâtelier's principle if the temperature is increased for a mixture of nitrogen and oxygen only in a 1:2 mole ratio?

$$N_2(g) + 2 O_2(g) \rightleftharpoons 2 NO_2(g)$$

17.43 A 1.00-L mixture of SO_2, O_2, and SO_3 at 100°C is heated to 500°C at constant pressure. Explain why the general gas law cannot be used to calculate the new volume of the mixture.

17.44 At a certain point in a reaction leading to the equilibrium system

$$2 NH_3(g) \rightleftharpoons 3 H_2(g) + N_2(g)$$

the rate of decomposition of NH_3 is 2.00×10^{-3} mol/L · s, and the rate of reaction of H_2 is 4.50×10^{-3} mol/L · s. In which direction is the reaction proceeding?

17.45 At a certain point in a reaction leading to the equilibrium system

$$2 NH_3(g) \rightleftharpoons 3 H_2(g) + N_2(g)$$

the rate of decomposition of NH_3 is 2.00×10^{-3} mol/L · s, and the rate of reaction of N_2 is 2.00×10^{-3} mol/L · s. In which direction is the reaction proceeding?

17.46 An equilibrium mixture of N_2, H_2, and NH_3 contains isotopically pure ^1H. That is, there is no deuterium (^2H, also denoted by D) in the system. A mixture of D_2, N_2, and ND_3 in the same ratio as the equilibrium mixture is added to that mixture. What hydrogen-containing substances will be found in the equilibrium mixture after the passage of sufficient time? Explain.

17.47 If 0.200 mol of Cl_2 and 0.400 mol of NO_2 are placed in a 1.00-L vessel and allowed to come to equilibrium at a certain temperature, the equilibrium concentration of NO_2Cl is 0.153 mol/L. What is the value of K for the following equilibrium at that temperature?

$$2\ NO_2Cl(g) \rightleftharpoons 2\ NO_2(g) + Cl_2(g)$$

17.48 If 0.100 mol of Cl_2 and 0.200 mol of NO_2 are placed in a 1.00-L vessel and allowed to come to equilibrium at a certain temperature, the equilibrium concentration of NO_2Cl is 0.054 mol/L. What is the value of K for the following equilibrium at that temperature?

$$2\ NO_2(g) + Cl_2(g) \rightleftharpoons 2\ NO_2Cl(g)$$

18 Acid-Base Theory

- 18.1 The Brønsted Theory
- 18.2 Ionization Constants
- 18.3 Autoionization of Water
- 18.4 Buffer Solutions

Key Terms (Key terms are defined in the Glossary.)

acid dissociation constant (18.2)
acid strength (18.1)
amphiprotic (18.1)
autoionization (18.3)
base dissociation constant (18.2)
base strength (18.1)
Brønsted acid (18.1)
Brønsted base (18.1)

Brønsted theory (18.1)
buffer solution (18.4)
conjugate acid (18.1)
conjugate acid-base pair (18.1)
conjugate base (18.1)
feeble acid (18.1)
feeble base (18.1)

hydronium ion (18.1)
ionization constant (18.2)
pH (18.3)
proton (18.1)
proton acceptor (18.1)
proton donor (18.1)
water ionization constant (18.3)

Symbols

K_a (acid dissociation constant) (18.2)
K_b (base dissociation constant) (18.2)

K_i (ionization constant) (18.2)
K_w (water ionization constant) (18.3)

In Chapter 8, you learned the Arrhenius definition of acids and bases—that an acid is a substance that can increase the concentration of H^+ ions in water and a base is a substance that can increase the concentration of OH^- ions in water. In Chapter 17, you learned about equilibrium systems. This chapter extends both of these concepts in discussing acid-base equilibria in aqueous solutions, which are extremely important to biological as well as chemical processes.

Section 18.1 discusses the Brønsted theory of acids and bases, which extends the concepts of acid and base beyond aqueous solutions and also explains the acidic or basic nature of solutions of salts. Ionization constants, the equilibrium constants for the reactions of weak acids or bases with water, are introduced in Section 18.2. The concept of the ionization of covalent compounds is extended to water itself in Section 18.3, which also covers pH, a scale of acidity and basicity. Section 18.4 describes buffer solutions, which resist change in their acidity or basicity even when strong acid or base is added. Both the preparation and action of such solutions are explained.

18.1 The Brønsted Theory

OBJECTIVE

■ to extend the definitions of acids and bases to explain why many salts give acidic or basic solutions in water

An acid is a proton donor; a base is a proton acceptor.

In 1923, J. Brønsted (1879–1947) and T. M. Lowry (1874–1936) independently defined acids and bases in a different way from the Arrhenius definitions. The resulting theory is sometimes called the Brønsted-Lowry theory, but more often is referred to as just the Brønsted theory. The **Brønsted theory** extends the definitions of acid and base in a way that explains more than the Arrhenius definitions can explain. According to this theory, a **Brønsted acid** is a **proton donor,** and a **Brønsted base** is a **proton acceptor.**

In this context, a **proton** is defined as the nucleus of a hydrogen atom. It has nothing to do with the protons in a carbon atom, a sodium atom, or any other atom. The nucleus of a normal hydrogen atom (^1H) consists of one proton, but the Brønsted definition of a proton includes the nucleus of heavy hydrogen (^2H). The proton is not a stable chemical species. Because it has no electrons and is a tiny particle, it has an extremely high density of positive charge. It is capable of sharing (negatively charged) electrons with any species that has a lone pair of electrons. In an aqueous solution, it shares a pair of electrons with a water molecule, forming the **hydronium ion** (H_3O^+):

$$H^+ \quad + \quad :\overset{..}{\underset{H}{O}}:H \quad \rightarrow \quad H:\overset{..}{\underset{H}{O}}:H^+$$

Hydronium ion

The atoms in the hydronium ion have the required numbers of electrons to be stable (Chapter 5), and the ion is chemically stable (in the absence of bases and active metals).

Acids are proton donors. When hydrogen fluoride gas dissolves in water, it reacts with the water, as follows:

$$HF(g) + H_2O(l) \rightarrow H_3O^+(aq) + F^-(aq)$$

The HF molecule donates its proton (H^+) to the H_2O molecule, so the hydrogen fluoride is acting as a Brønsted acid. However, the water molecule accepts that proton, so in this reaction, the water molecule is acting as a Brønsted base. Moreover, the reaction is an equilibrium reaction because it proceeds to the left as well.

In the reverse reaction, the hydronium ion donates the proton to the fluoride ion, and these ions are thus an acid and a base, respectively:

$$HF(g) + H_2O(l) \rightleftharpoons H_3O^+(aq) + F^-(aq)$$

    Acid   Base   Acid   Base

The acid on the left (HF) and the base on the right (F^-) are related to each other; they differ by one proton (H^+). They are called a **conjugate acid-base pair.** Similarly, the base on the left (H_2O) and the acid on the right (H_3O^+) are related in the same way and are also a conjugate pair. Every Brønsted acid has a **conjugate base,** and every Brønsted base has a **conjugate acid.**

The reaction of ammonia with water extends these concepts:

$$H_2O(l) + NH_3(aq) \rightleftharpoons NH_4^+(aq) + OH^-(aq)$$

    Acid   Base   Acid   Base

In the forward reaction, the ammonia molecule accepts a proton from the water molecule and thus is a base. The water molecule donates the proton, so in this reaction, water acts as an acid. In the reverse reaction, the ammonium ion acts as an acid, and the hydroxide ion acts as a base.

Thus, water can act as an acid, and it also can act as a base, even though we are not used to thinking of it as being either acidic or basic. It acts as an acid in the presence of a stronger base and as a base in the presence of a stronger acid. Chemists say that water is **amphiprotic.**

■ **EXAMPLE 18.1**

(a) Classify each species in the following reactions as an acid or a base, and connect each conjugate pair:

$$H_3PO_4(aq) + OH^-(aq) \rightleftharpoons H_2PO_4^-(aq) + H_2O(l)$$

$$H_2PO_4^-(aq) + OH^-(aq) \rightleftharpoons HPO_4^{2-}(aq) + H_2O(l)$$

$$HPO_4^{2-}(aq) + OH^-(aq) \rightleftharpoons PO_4^{3-}(aq) + H_2O(l)$$

(b) Identify the species in part (a) that are Brønsted acids in one reaction and Brønsted bases in another.

Solution

(a)
$$H_3PO_4(aq) + OH^-(aq) \rightleftharpoons H_2PO_4^-(aq) + H_2O(l)$$

      Acid   Base   Base   Acid

$$H_2PO_4^-(aq) + OH^-(aq) \rightleftharpoons HPO_4^{2-}(aq) + H_2O(l)$$

      Acid   Base   Base   Acid

$$\text{HPO}_4{}^{2-}(\text{aq}) \ + \ \text{OH}^-(\text{aq}) \ \rightleftharpoons \ \text{PO}_4{}^{3-}(\text{aq}) \ + \ \text{H}_2\text{O}(\text{l})$$

Acid   Base   Base   Acid

(b) The ion $\text{H}_2\text{PO}_4{}^-$ is a base in the first reaction and an acid in the second. The ion $\text{HPO}_4{}^{2-}$ is a base in the second reaction and an acid in the third.

Practice Problem 18.1   Rewrite the equations of Example 18.1 as total equations, using NaOH as the base. Do all the acids have hydrogen written first? ■

An aqueous solution is acidic if and only if the hydronium ion concentration is greater than the hydroxide ion concentration, and such a solution is basic if and only if the hydroxide ion concentration is greater than the hydronium ion concentration. The solution is neutral if the concentrations of these two ions are equal.

$[\text{H}_3\text{O}^+] > [\text{OH}^-]$	Acidic
$[\text{H}_3\text{O}^+] = [\text{OH}^-]$	Neutral
$[\text{H}_3\text{O}^+[< [\text{OH}^-]$	Basic

You learned in Chapter 8 that some acids are strong and others are weak. Its **acid strength** determines how much any acid reacts with *water* to form hydronium ions. The equilibrium for the reaction of a strong acid with water proceeds far to the right. Essentially, 100% of a strong molecular acid reacts with water to form hydronium ions. The equilibrium for ionization of a weak acid in water proceeds very little to the right. (Remember that both strong and weak acids react extensively with Arrhenius bases.) Since different acids ionize to different extents, it is not surprising that their conjugate bases also have different tendencies to react with hydronium ions or with water. That is, *the stronger the conjugate acid, the weaker is its conjugate base.*

Hydrogen chloride is a strong acid, meaning that it reacts with water essentially 100%. The chloride ion formed has virtually no attraction for hydronium ion at room temperature. Chloride ion can be said to be a **feeble base.** Less strong acids, such as acetic acid, ionize to the extent of a few percent. Their conjugate bases are stronger than those of strong acids. Some acids, such as boric acid, ionize to a very limited extent; an acid like this may be called a **feeble acid.** Its conjugate base reacts almost 100% with hydronium ion to form the molecular acid. In fact, conjugates of feeble acids react extensively even with water to form the acid. In short, the stronger the acid or base, the weaker is its conjugate. The relative strengths of conjugate acid-base pairs are summarized in Table 18.1.

Since chloride ion (Cl^-) has no tendency to react with hydronium ion, how much should we expect it to react with water? (Hydronium ion is certainly a stronger acid than water; it produces water as its conjugate base when it reacts with another base.) Since chloride ion does not react with hydronium ion, it certainly has no tendency to react with water; it is a feeble base.

■ EXAMPLE 18.2

Acetic acid is a weak acid. How strong is its conjugate base?

Solution

Table 18.1 shows that the conjugate of a weak acid is a weak base. The acetate ion is weakly basic.

Practice Problem 18.2   How strong a base is the nitrate ion? ■

Table 18.1   *Relative Strengths of Acids and Bases and Their Conjugates*

Conjugate Acid	Conjugate Base
Strong	Feeble
Weak	Weak
Feeble	Strong

Figure 18.1 Basicity of Sodium Dihydrogen Borate Solution

A solution of NaH_2BO_3 in water turns phenolphthalein red because of the extensive reaction of $H_2BO_3^-$ with water.

■ **EXAMPLE 18.3**

Boric acid is a feeble acid. (It is so weak that it is used in solution to bathe infected eyes.) How strong a base is its conjugate—the dihydrogen borate ion, $H_2BO_3^-$?

Solution

Since the acid is feeble, its conjugate base is strong. The $H_2BO_3^-$ ion reacts extensively with water to form H_3BO_3 and OH^- (Figure 18.1):

$$H_2BO_3^-(aq) + H_2O(l) \rightleftharpoons H_3BO_3(aq) + OH^-(aq) \quad ■$$

Molecular bases behave analogously to molecular acids. **Base strength** determines the extent to which a molecular base reacts with water to form ions. Also, the stronger the base, the weaker is its conjugate acid.

Aqueous solutions of salts often test acidic or basic. We can explain this in terms of the strengths of Brønsted acids and bases. For example, an aqueous solution of sodium acetate tests basic to litmus paper. The solutes are the feeble acid Na^+ and the weak base $C_2H_3O_2^-$. Since the base present is a *stronger* base (not a strong base) than the acid present is an acid, the solution tests basic:

$$C_2H_3O_2^-(aq) + H_2O(l) \rightleftharpoons HC_2H_3O_2(aq) + OH^-(aq)$$

$$Na^+(aq) + H_2O(l) \longrightarrow nr$$

The presence of excess OH^- (relative to the quantity of H_3O^+) in the solution is what makes the solution basic. ■

■ **EXAMPLE 18.4**

A solution of ammonium chloride tests acidic to litmus paper. Explain why such a solution is acidic.

Solution

The solution contains a feeble base (chloride ion) and a weak acid (ammonium ion). The acid is stronger than the base, so the solution is somewhat acidic:

$$NH_4^+(aq) + H_2O(l) \rightleftharpoons NH_3(aq) + H_3O^+(aq)$$

$$Cl^-(aq) + H_2O(l) \longrightarrow nr$$

The hydronium ion concentration is greater than the hydroxide ion concentration, so the solution is somewhat acidic. ■

The Brønsted theory extends the concepts of acid and base beyond reactions in aqueous solution. For example, it describes the reaction of ammonia gas with hydrogen fluoride gas to form solid ammonium fluoride:

$$NH_3(g) + HF(g) \rightleftharpoons NH_4^+(s) + F^-(s)$$

Here, hydrogen fluoride donates its proton to the lone pair of electrons on the ammonia molecule and is therefore a Brønsted acid. The ammonia molecule is a base, since it accepts that proton.

18.2   Ionization Constants

■ to use equilibrium constants for acid-base reactions

Since the reactions of Brønsted acids and bases with water are equilibria, we can write equilibrium constant expressions for them. For example, for the reaction of acetic acid with water, the equilibrium constant expression is as follows:

$$HC_2H_3O_2(aq) + H_2O(l) \rightleftharpoons H_3O^+(aq) + C_2H_3O_2^-(aq)$$

$$K_a = \frac{[H_3O^+][C_2H_3O_2^-]}{[HC_2H_3O_2]}$$

The concentration of water is not included explicitly in this expression (it is incorporated into the value of K_a) because in dilute aqueous solution, the concentration of water is a constant, almost equal to the concentration of water when pure (55.5 M). The constant K_a is called the **acid dissociation constant.** The subscript a indicates that the constant is for the ionization of an *acid*.

Similarly, an equilibrium constant expression can be written for the reaction of a weak base with water:

$$NH_3(aq) + H_2O(l) \rightleftharpoons NH_4^+(aq) + OH^-(aq)$$

$$K_b = \frac{[NH_4^+][OH^-]}{[NH_3]}$$

The constant K_b is called the **base dissociation constant.** The subscript b indicates that the constant is for the ionization of a *base*.

Either K_a or K_b may be represented by K_i, called an **ionization constant** because ions are formed in these reactions.

Selected values of K_a and K_b are presented in Table 18.2. Although the reactions are equilibria and can be written with either set of species on the left, these equilibrium constant values represent the equations with the molecular acid or base on the *left*.

■ EXAMPLE 18.5

Calculate the hydronium ion concentration of 0.391 M HCl.

Table 18.2   Selected K_a and K_b Values

Acids	K_a	Bases	K_b
Acetic acid, $HC_2H_3O_2$	1.8×10^{-5}	Ammonia, NH_3	1.8×10^{-5}
Formic acid, $HCHO_2$	1.7×10^{-4}	Methylamine, CH_3NH_2	4.4×10^{-4}
Hydrofluoric acid, HF	6.7×10^{-4}	Pyridine, C_5H_5N	2.3×10^{-9}
Hydrosulfuric acid*, H_2S	1.0×10^{-7}		
Hydrogen sulfide ion*, HS^-	1.0×10^{-14}		
Nitrous acid, HNO_2	4.5×10^{-4}		
Phenol, HOC_6H_5	1.3×10^{-10}		

*The constants for H_2S and HS^- are referred to as K_1 and K_2 for H_2S, respectively.

Solution

Hydrochloric acid is a strong acid and in aqueous solution is 100% ionized to H_3O^+ and Cl^-. The hydronium ion concentration is therefore 0.391 M. (The chloride ion concentration is also 0.391 M.) ■

■ EXAMPLE 18.6

When 0.100 mol of an unknown acid, represented as HX, is dissolved in enough water to make 1.00 L of solution, the hydronium ion concentration is found to be 3.00×10^{-3} M. Calculate the value of K_a for HX.

$$HX(aq) + H_2O(l) \rightleftharpoons H_3O^+(aq) + X^-(aq)$$

Solution

The initial concentrations, changes in concentration, and equilibrium concentrations can be tabulated as in Chapter 17:

	HX(aq)	+	H₂O(l)	⇌	H₃O⁺(aq)	+	X⁻(aq)
Initial	0.100				0		0
Change	−0.00300				+0.00300		+0.00300
Equilibrium	0.097				0.00300		0.00300

The value of K_a can then be calculated, using the equilibrium concentrations:

$$K_a = \frac{[H_3O^+][X^-]}{[HX]} = \frac{(0.00300)^2}{(0.097)} = 9.3 \times 10^{-5} \quad ■$$

■ EXAMPLE 18.7

Calculate the hydronium ion concentration and the acetate ion concentration that result when 0.120 mol of acetic acid is dissolved in enough water to make 1.00 L of solution.

Solution

The initial concentration of acetic acid (before ionization) is 0.120 M. After it ionizes, its concentration and those of its products can be tabulated:

	HC₂H₃O₂(aq)	+	H₂O(l)	⇌	H₃O⁺(aq)	+	C₂H₃O₂⁻(aq)
Initial	0.120				0.00		0.00
Change	−x				+x		+x
Equilibrium	0.120 − x				x		x

The equilibrium constant expression and the value of K_a from Table 18.2 can be used to solve for x:

$$K_a = \frac{[H_3O^+][C_2H_3O_2^-]}{[HC_2H_3O_2]}$$

$$= \frac{x^2}{(0.120 - x)} = 1.8 \times 10^{-5}$$

Neglecting x when it is subtracted from 0.120 yields

$$x^2 = 2.2 \times 10^{-6}$$

Taking the square root of both sides of the equation yields

$$x = 1.5 \times 10^{-3}$$

(To get a square root on a calculator, press the $\boxed{\sqrt{x}}$ key or the $\boxed{\text{INV}}$ or $\boxed{\text{2ndF}}$ key plus another key. See Appendix 1.) Since x represents both the hydronium ion concentration and the acetate ion concentration, both of these are equal to 1.5×10^{-3} M. Finally, 1.5×10^{-3} M is about 1% of 0.120 M, so the approximation is acceptable.

Practice Problem 18.7   Calculate the hydroxide ion concentration that results when 0.120 mol of ammonia is dissolved in enough water to make 1.00 L of solution. ■

18.3 Autoionization of Water

<table>
<tr><td>

O B J E C T I V E

■ to calculate the effects of the ionization of water

</td><td>

You learned in Section 18.1 that water can act as either an acid or a base. In fact, water can act as both an acid and a base *in the same reaction:*

$$H_2O(l) + H_2O(l) \rightleftharpoons H_3O^+(aq) + OH^-(aq)$$

This reaction, which is an example of **autoionization,** does not proceed to the right to a very large extent, since water is such a weak acid and base. To write an equilibrium constant expression for this reaction, we again consider the concentration of water itself to be a constant, which we incorporate into the value of K. We get another type of constant, called the **water ionization constant** and denoted K_w:

$$K_w = [H_3O^+][OH^-]$$

</td></tr>
</table>

$$K_w = [H_3O^+][OH^-]$$
$$= 1.0 \times 10^{-14}$$

Note that this equilibrium constant expression has no denominator and also that the corresponding chemical equation is written with water as the reactant (on the left). The value of K_w for dilute aqueous solutions at 25°C is 1.0×10^{-14}. Since the product $[H_3O^+][OH^-]$ is a constant, the concentration of hydronium ion in any dilute aqueous solution can be calculated from the concentration of hydroxide ion, and vice versa.

■ **EXAMPLE 18.8**

Calculate the hydronium ion concentration of a solution in which the hydroxide ion concentration is 5.0×10^{-6} M.

Solution

$$K_w = [H_3O^+][OH^-] = [H_3O^+](5.0 \times 10^{-6}) = 1.0 \times 10^{-14}$$

$$[H_3O^+] = \frac{1.0 \times 10^{-14}}{5.0 \times 10^{-6}} = 2.0 \times 10^{-9} \text{ M} \quad ■$$

Note that both hydronium ions and hydroxide ions are present in every dilute aqueous solution. A solution is basic if the hydroxide ion concentration is greater than the hydronium ion concentration, neutral if these two concentrations are equal, and acidic if the hydronium ion concentration is greater than the hydroxide ion concentration:

$$[H_3O^+] > [OH^-] \quad \text{Acidic}$$

$$[H_3O^+] = [OH^-] \quad \text{Neutral}$$

$$[H_3O^+] < [OH^-] \quad \text{Basic}$$

■ EXAMPLE 18.9

Calculate the hydroxide ion concentration in (a) a 0.010 M NaOH solution and (b) a 0.010 M HCl solution.

Solution

(a) NaOH is a strong, soluble base, so 0.010 M NaOH consists of 0.010 M Na^+ and 0.010 M OH^-. The hydroxide ion concentration is 0.010 M.

(b) HCl is a strong acid, so the hydronium ion concentration is 0.010 M. To find the hydroxide ion concentration, use the K_w expression:

$$K_w = [H_3O^+][OH^-] = (0.010 \text{ M})[OH^-] = 1.0 \times 10^{-14}$$

Thus,

$$[OH^-] = 1.0 \times 10^{-12} \text{ M}$$

Practice Problem 18.9 Calculate the hydronium ion concentration of (a) 0.0010 M HCl and (b) 0.0010 M NaOH. ■

To enable scientists to express the acidity of a solution without using exponential notation, a biologist invented the pH scale. The **pH** of a solution is defined as the negative of the logarithm of the *hydronium ion* concentration:

$$pH = -\log [H_3O^+]$$

The pH values for some common items are presented in Table 18.3.

To get the pH of a solution on an electronic calculator, key in the hydronium ion concentration, press the ⬚log key, and then press the ⬚+/− key. To get the hydronium ion concentration from the pH, enter the pH value, press the ⬚+/− key, press the ⬚INV or ⬚2ndF key, and then press the ⬚log key. (See Appendix 1.)

Table 18.3 Approximate pH Values for Some Common Items

Item	pH
Gastric juice	1–2
Soft drinks	3
Vinegar	3
Orange juice	3.5
Tomatoes	4
Rainwater	6
Pure water	7
Human blood	7.3
Seawater	8.5
Baking soda, aqueous	8.5
Ammonia, aqueous	11–12
Washing soda, aqueous	12

■ EXAMPLE 18.10

Calculate the pH of a solution that has a hydronium ion concentration of 4.6×10^{-6} M.

Solution

$$pH = -\log [H_3O^+] = -\log (4.6 \times 10^{-6}) = 5.34$$

Practice Problem 18.10 Calculate the hydronium ion concentration of a solution with a pH of 7.930. ■

The significant digits in pH values are the *decimal place digits.* The integer digit(s) indicate(s) only the magnitude of the hydronium ion concentration. For example, if the coefficient of the hydronium ion concentration has two significant digits, use two decimal places in the corresponding pH value.

■ EXAMPLE 18.11

Consider three solutions with the following hydronium ion concentrations:

$$2.89 \times 10^{-1} \text{ M} \qquad 2.89 \times 10^{-7} \text{ M} \qquad 2.89 \times 10^{-13} \text{ M}$$

(a) How many significant digits are in each concentration?
(b) What negative logarithm value is shown on an electronic calculator for each?
(c) Which digits are the significant digits in each calculator value?

Solution

(a) There are three significant digits in each concentration.
(b) The values obtained on a calculator are 0.539102157, 6.539102157, and 12.539102157, respectively.
(c) The first three digits *after the decimal point* (539) are the significant digits in each calculator value. The integer digits tell only to which power of 10 the base is raised. The pH values should be reported as 0.539, 6.539, and 12.539, respectively.

Practice Problem 18.11   Calculate the pH of a solution for which $[H_3O^+]$ is 2.9×10^{-3} M. ■

■ EXAMPLE 18.12

Calculate the hydronium ion concentration of a solution with pH = 3.192.

Solution

$$[H_3O^+] = 6.43 \times 10^{-4} \text{ M}$$

On the calculator, enter 3.192, press the $\boxed{+/-}$ key (to get −3.192), and press $\boxed{\text{INV}}$ or $\boxed{\text{2ndF}}$ and then $\boxed{\text{log}}$ to get the answer. Use three significant digits because there are three digits after the decimal point in the value given for the pH.

Practice Problem 18.12   Calculate the hydronium ion concentration of a solution with pH = 9.19. ■

The pH is less than 7 for acidic solutions, equal to 7 for neutral solutions, and greater than 7 for basic solutions.

$[H_3O^+] > [OH^-]$	Acidic	pH < 7
$[H_3O^+] = [OH^-]$	Neutral	pH = 7
$[H_3O^+] < [OH^-]$	Basic	pH > 7

■ EXAMPLE 18.13

Calculate the pH of a solution that is 0.0070 M in NaOH.

Solution

The pH is based on the *hydronium ion* concentration. The hydroxide ion concentration is 0.0070 M, so you first calculate the hydronium ion concentration, using the K_w expression:

$$K_w = [H_3O^+][OH^-] = 1.0 \times 10^{-14}$$

$$[H_3O^+](0.0070) = 1.0 \times 10^{-14}$$

$$[H_3O^+] = 1.4 \times 10^{-12} \text{ M}$$

Converting to a pH value gives

$$pH = 11.85$$

Practice Problem 18.13   Calculate the pH of a 0.300 M NaCl solution.   ■

■ EXAMPLE 18.14

Calculate the pH of a solution that is 0.00400 M in NH_3.

Solution

The pH is based on the *hydronium ion* concentration. Since you are given the ammonia concentration, you first use it to calculate the hydroxide ion concentration. The initial concentration of ammonia (before reaction) is 0.00400 M. After it reacts with water, its concentration and those of its products are as given in the following table:

	$NH_3(aq)$	$+$	$H_2O(l)$	$\rightleftharpoons$	$NH_4^+(aq)$	$+$	$OH^-(aq)$
Initial	0.00400				0.00		0.00
Change	$-x$				$+x$		$+x$
Equilibrium	$0.00400 - x$				x		x

The equilibrium constant expression and the value of K_b from Table 18.2 can be used to solve for x:

$$K_b = \frac{[OH^-][NH_4^+]}{[NH_3]} = \frac{x^2}{0.00400 - x} = 1.8 \times 10^{-5}$$

Neglecting x when subtracted from 0.00400 yields

$$x^2 = 7.2 \times 10^{-8}$$

Taking the square root of both sides of the equation yields

$$x = 2.7 \times 10^{-4}$$

The hydroxide ion concentration is 2.7×10^{-4} M. As a check, you can see that 2.7×10^{-4} M is 7% of 0.00400 M, so the approximation is barely acceptable.

You can now use the K_w expression to calculate the hydronium ion concentration:

$$K_w = [H_3O^+][OH^-] = 1.0 \times 10^{-14}$$

$$[H_3O^+](2.7 \times 10^{-4}) = 1.0 \times 10^{-14}$$

Thus,

$$[H_3O^+] = 3.7 \times 10^{-11}\ M$$

and

$$pH = -\log (3.7 \times 10^{-11}) = 10.43$$

Practice Problem 18.14 Calculate the pH of a 0.300 M $HC_2H_3O_2$ solution. ■

18.4 Buffer Solutions

OBJECTIVE

- to make buffer solutions and to calculate their resistance to changes in pH

An aqueous solution of a weak acid or a weak base contains two substances that react with each other to some extent—the acid or base and the water. When another substance is added to this solution, it can affect the original pair of reactants without necessarily reacting directly with either. It may, as predicted by LeChâtelier's principle, suppress the reaction of the original two reactants. For example, if sodium acetate, $NaC_2H_3O_2$, is added to an aqueous solution of acetic acid, $HC_2H_3O_2$, it does not react directly with the acid or with the water. Instead, the acetate ion represses the ionization of the acetic acid. The equilibrium for the acetic acid solution is

$$HC_2H_3O_2(aq) + H_2O(l) \rightleftharpoons C_2H_3O_2^-(aq) + H_3O^+(aq)$$

When $NaC_2H_3O_2$ is added to this solution, the Na^+ ion does not affect this equilibrium because it is not involved in the reaction. The acetate ion, $C_2H_3O_2^-$, appears on the right side of the equation. Addition of a species that appears on the right shifts the equilibrium to the left.

Suppose we make a solution containing both a weak acid and its conjugate base (as a soluble salt). Such a solution is called a **buffer solution.** We can make a buffer solution by dissolving acetic acid and sodium acetate in water, for example. For acetic acid, $HC_2H_3O_2$, the conjugate base is the acetate ion, $C_2H_3O_2^-$. (Remember that soluble salts, strong acids, and soluble hydroxides are completely ionized in solution.) Thus, in the same solution both acetic acid and its conjugate base, the acetate ions are present. The acidity of this solution is less than that of a solution of acetic acid alone because the acetate ion represses the ionization of the acetic acid. Furthermore, the buffer solution has the special property of resisting change in its pH even when some (reasonable quantity of) strong acid or strong base is added. Let us see how this property arises. The following equation gives the equilibrium of the acid and water, with the relative concentration of each species in the buffer solution written below it:

$HC_2H_3O_2(aq)$	+	$H_2O(l)$	$\rightleftharpoons$	$C_2H_3O_2^-(aq)$	+	$H_3O^+(aq)$
Large concentration		Very large concentration		Large concentration		Tiny concentration

When H_3O^+ is added to the buffer solution, this equilibrium will shift to the left, using up most of the added H_3O^+. The pH will not drop as much as it would if the equilibrium were not present. If OH^- is added to the system instead of H_3O^+, the OH^- reacts with the H_3O^+ present. However, in response to the stress of lowering the concentration of H_3O^+, the equilibrium shifts to the right to produce more H_3O^+. The pH is not increased as much as it would be if the equilibrium were not present.

Figure 18.2 A Product That Uses the Principle of Buffering

Buffered aspirin lessens the chance that the acid of plain aspirin will upset the stomach.

A buffer solution has an excess of all the reactants and products, except for H_3O^+. Thus, when we try to change the concentration of H_3O^+ by adding strong acid or base, the equilibrium shifts, in accordance with LeChâtelier's principle, to resist that change. The pH changes very little. (We will do calculations to show quantitatively how much the pH changes in such systems later in this section.)

To summarize the important facts about buffer solutions:

1. They are prepared by combining, in aqueous solution, a weak acid and its conjugate base or a weak base and its conjugate acid.

2. They resist changes in their pH values, even if a limited quantity of strong acid or base is added.

3. Buffer solutions of molecular acids and their conjugates accomplish that stability of pH by shifting their equilibria to use up added H_3O^+ or to replace H_3O^+ that has reacted with added OH^-.

The principle of buffering is used in a commercial pain reliever to reduce the acidity of plain aspirin (Figure 18.2).

■ **EXAMPLE 18.15**

Explain how a weak base and its conjugate acid form a buffer solution.

Solution

The equilibrium of a weak base and water has OH^- as a product. If strong acid is added to the system, the strong acid reacts with the OH^-. However, more OH^- is produced as the equilibrium shifts to the right in accordance with LeChâtelier's principle, so the pH does not change much. If a strong base is added to the system, the equilibrium shifts to the left to use up some of the added OH^-, and again, the pH is fairly stable.

■ **EXAMPLE 18.16**

Explain why a solution of a strong acid and its conjugate base does not act as a buffer solution.

Solution

A strong acid in solution is completely ionized. The equilibrium cannot shift to the right to replace H_3O^+ because it is already as far to the right as it can go. Also, it cannot shift left, since, by definition, a strong acid in solution is completely ionized.

Consider a solution of HCl and NaCl, for example:

$$HCl(aq) + H_2O(l) \rightarrow H_3O^+(aq) + Cl^-(aq)$$

Since HCl is a strong acid, this "equilibrium" proceeds completely to the right. If OH^- ions are added, no more HCl can react to replace the H_3O^+ ions used up because all of the HCl has already reacted. If H_3O^+ ions are added, the equilibrium will not shift to the left because HCl is a strong acid.

Practice Problem 18.16   Explain why a solution of a strong base and its conjugate acid do not form a buffer solution. ■

The same procedures we have used for calculations involving weak acid and weak base equilibria can be used for calculations concerning buffer solutions. The major difference is that *nonzero* concentrations of both the weak acid or base and its conjugate are initially present.

■ **EXAMPLE 18.17**

Calculate the pH of a solution of 0.120 mol of $HC_2H_3O_2$ and 0.140 mol of $NaC_2H_3O_2$ in 1.00 L of solution.

Solution

The acid ionizes according to the equation

$$HC_2H_3O_2(aq) + H_2O(l) \rightleftharpoons C_2H_3O_2^-(aq) + H_3O^+(aq)$$

Sodium acetate is a salt, so initially, there is a concentration of 0.140 M $C_2H_3O_2^-$. The acetate ion affects the position of the acid equilibrium.

	$HC_2H_3O_2(aq)$	+	$H_2O(l)$	$\rightleftharpoons$	$C_2H_3O_2^-(aq)$	+	$H_3O^+(aq)$
Initial	0.120				0.140		0
Change	$-x$				$+x$		$+x$
Equilibrium	$0.120 - x$				$0.140 + x$		x

$$K_a = \frac{[C_2H_3O_2^-][H_3O^+]}{[HC_2H_3O_2]}$$

Neglecting x when added to or subtracted from larger quantities and using the value of K_a in Table 18.2 yields:

$$K_a = \frac{(0.140)(x)}{0.120} = 1.8 \times 10^{-5}$$

$$x = 1.5 \times 10^{-5}$$

The approximations made by neglecting x when added to 0.140 M and subtracted from 0.120 M are valid. Since $[H_3O^+]$ is 1.5×10^{-5} M,

$$pH = 4.82$$

Practice Problem 18.17 Calculate the pH of a solution containing 0.130 M NH_3 and 0.145 M NH_4Cl. ■

If strong acid or strong base is added to a buffer solution, a net reaction takes place. To calculate the pH of such solutions, we first assume that the strong acid or base reacts as completely as possible, given that some reactant is present in limiting quantity, as discussed in Chapter 10. Then we concentrate on the equilibrium calculation.

■ **EXAMPLE 18.18**

Calculate the pH after 0.0100 mol of NaOH is added to 1.00 L of a solution containing 0.120 mol of $HC_2H_3O_2$ and 0.140 mol of $NaC_2H_3O_2$. Assume that the volume of the solution does not change.

Solution

When a base is added to an acid, they react to form a salt (Chapter 8). Here, the base (NaOH) is in limiting quantity, since a greater number of moles of the acid are present, and the acid and base react in a 1:1 ratio:

$$HC_2H_3O_2(aq) \quad + \quad NaOH(s) \quad \rightarrow \quad NaC_2H_3O_2(aq) \quad + \quad H_2O(l)$$

Initial concentrations	0.120	0.0100	0.140	Excess
Change due to acid-base reaction	−0.0100	−0.0100	+0.0100	+0.0100
Concentrations before equilibrium	0.110	0.0000	0.150	Excess

Right after the NaOH is added, 0.0100 mol of OH^- reacts with 0.0100 mol of $HC_2H_3O_2$, leaving 0.110 mol of $HC_2H_3O_2$ and 0.150 mol of $C_2H_3O_2^-$ in the solution. Thus, these are the initial concentrations *for the equilibrium reaction:*

$$HC_2H_3O_2(aq) \quad + \quad H_2O(l) \quad \rightleftharpoons \quad C_2H_3O_2^-(aq) \quad + \quad H_3O^+(aq)$$

Initial	0.110		0.150	0
Change	−x		+x	+x
Equilibrium	0.110 − x		0.150 + x	x

$$K_a = \frac{[C_2H_3O_2^-][H_3O^+]}{[HC_2H_3O_2]} = \frac{(0.150)(x)}{0.110} = 1.8 \times 10^{-5}$$

$$x = 1.3 \times 10^{-5}$$

The hydronium ion concentration is 1.3×10^{-5} M, and

$$pH = 4.89$$

Comparing the pH of the solution in Example 18.17 to this pH, which is for the same solution after 0.0100 mol of strong base has been added to it, you can see that only a very small change has taken place—from 4.82 to 4.88. If 0.0100 mol of NaOH were added to a solution of a strong acid with pH 4.82 (which was not buffered), the pH would have risen to 12.00.

■ EXAMPLE 18.19

Calculate the pH of 1.00 L of a solution originally containing 1.5×10^{-5} M HCl (which has the same hydronium ion concentration as the buffer solution of Example 18.17 and the original buffer solution of Example 18.18) after 0.0100 mol of NaOH has been added to it.

Solution

After neutralization of all the HCl present, there is essentially 0.0100 mol of NaOH left. In 1.00 L of solution, the hydroxide ion concentration is 0.0100 M, and the pH is 12.000. Note the difference between the effect of the base on this solution and its effect on the buffered solution of Example 18.18. ■

■ ■ ■ ■ ■ ■ ■ SUMMARY ■ ■ ■ ■ ■ ■ ■

The Brønsted theory of acids and bases extends the definition of acids and bases, which allows an explanation of why most salts dissolved in water do not form neutral solutions. Acids are defined as proton donors, and bases are proton acceptors. An excess of H_3O^+ ions over OH^- ions makes an aqueous solution acidic, and an excess of OH^- ions over H_3O^+ ions makes the solution basic. Neutral solutions have equal concentrations of these two ions. Some substances, most notably water and also acid salts such as $NaHCO_3$, can act as either acids or bases, depending on what other species is present. According to the Brønsted theory, an acid reacts with a base to produce a conjugate base of the original acid and a conjugate acid of the original base. The stronger the acid, the weaker is its conjugate base. Strong acids, for example, have feeble conjugate bases. Solutions of salts in water often test acidic or basic because one of their ions reacts with the water more than the other does. (Section 18.1)

For reactions of weak acids or weak bases with water, the specialized equilibrium constant is denoted K_a or K_b, respectively. Neither K_a nor K_b includes the concentration of water in its equilibrium constant expression. If you are given initial concentrations and one equilibrium concentration, you can calculate the equilibrium constant. If you are given initial concentrations and the value of the equilibrium constant, you can calculate the equilibrium concentrations. (Section 18.2)

Water can react with itself to a very limited extent, and every aqueous solution has at least *some* concentration of hydronium ion and of hydroxide ion, enough to satisfy the equation

$$K_w = [H_3O^+][OH^-] = 1.0 \times 10^{-14}$$

Given a hydronium ion concentration, you can calculate the hydroxide ion concentration using this equation, and vice versa. The pH, defined as $-\log [H_3O^+]$, is a convenient way of expressing the acidity of a solution. (Section 18.3)

A buffer solution resists changes in its pH due to addition of small quantities of strong acid or strong base. A buffer solution may be formed by combining a weak acid and its conjugate base or a weak base and its conjugate acid. The resistance of the buffer solution to pH changes is based on LeChâtelier's principle. For example, if a small quantity of H_3O^+ ions is added to a solution containing the conjugate base of a weak acid, the base will react with most of the added H_3O^+ ions to form more conjugate weak acid, and the pH will not change much. If a small quantity of OH^- ions is added to a solution containing an un-ionized weak acid, it will react with the acid, producing more conjugate base. The added OH^- ions are no longer present to increase the pH. Given a quantity of strong acid or base that is added to a buffer solution, you first calculate how much of a reactant that is present will react completely with the added quantity, giving consideration to the limiting quantity (Chapter 10). You next deduce the new concentrations of the weak acid and its conjugate and then proceed with the equilibrium calculation. (Section 18.4)

Items for Special Attention

■ Concentrations of strong acids or bases are stated without reference to their ionizations. For example, 1 M HCl contains *no* HCl molecules, but 1 M H_3O^+ and 1 M Cl^-.

■ Weak acids react extensively with bases. It is their incomplete reactions with *water* that differentiate them from strong acids.

■ Values of K_a and K_b given in tables are for reactions with the un-ionized molecules on the left. However, weak acids and bases can be products of the reactions of their ions. For example:

$$H_3O^+(aq) + C_2H_3O_2^-(aq) \rightleftharpoons HC_2H_3O_2(aq) + H_2O(l)$$

■ The pH is the negative logarithm of the *hydronium ion* concentration, not the hydroxide ion concentration, the acetate ion concentration, or any other concentration.

■ Not every substance added to a buffer solution will react with the weak acid or base; some substances simply affect the position of the equilibrium already present. For example, sodium fluoride added to a solution containing HF and F^- reacts with neither of these but shifts the equilibrium toward the side on which HF is present by partially reacting with some of the H_3O^+ ions present.

Self-Tutorial Problems

18.1 Which of the following are weak acids, and which are strong acids?

(a) HF   (b) $HC_2H_3O_2$ (c) $HClO_3$

(d) $HClO_2$ (e) HCl

18.2 Which of the following are weak acids, and which are strong acids?

(a) H_2SO_4 (b) H_3PO_4 (c) HNO_3

(d) HBr   (e) H_2S

18.3 What ions or molecules are present in solution after 0.100 mol of each of the following is dissolved in enough water to make 1.00 L of solution?

(a) NaCl   (b) NH_3 (c) K_2SO_4

(d) $(NH_4)_2SO_4$

18.4 What ions or molecules are present in solution after 0.100 mol of each of the following is dissolved in enough water to make 1.00 L of solution?

(a) HNO_2 (b) $NaNO_3$

18.5 Among the following, identify the compounds that exist essentially as ions in solution and those that do not:

(a) KBr   (b) $HClO_2$ (c) H_2SO_3

(d) $Ca(HCO_3)_2$ (e) NH_4ClO_3 (f) CH_3OH

18.6 Label each of the species in the following equilibrium system as either an acid or base in the Brønsted sense:

$$HF(aq) + H_2O(l) \rightleftharpoons F^-(aq) + H_3O^+(aq)$$

18.7 Label each of the species in the following equilibrium system as either an acid or base in the Brønsted sense:

$$NH_3(aq) + H_2O(l) \rightleftharpoons NH_4^+(aq) + OH^-(aq)$$

18.8 Calculate the pH of each of the following solutions:

(a) 0.300 M H_3O^+ (b) 0.300 M HCl

18.9 Calculate the pH of a solution having each of the following hydronium ion concentrations:

(a) 9.63×10^{-4} M   (b) 2.28×10^{-9} M

(c) 5.08×10^{-4} M   (d) 6.97×10^{-7} M

(e) 1.00 M

18.10 Calculate the pH of a solution having each of the following hydronium ion concentrations:

(a) 7.77×10^{-7} M

(b) 9.03×10^{-11} M

(c) 2.24×10^{-2} M

(d) 5.00×10^{-12} M

(e) 10.0 M

18.11 Calculate the hydronium ion concentration of a solution having each of the following pH values:

(a) 12.207   (b) 6.109   (c) 12.110

(d) 6.998   (e) 2.901

18.12 Calculate the pH of a 6.49×10^{-2} M KNO_3 solution.

18.13 Classify the following solutions as acidic, basic, or neutral:

(a) pH = 7.00

(b) $[H_3O^+] = 1.0 \times 10^{-9}$ M

(c) pH = 0.00

(d) $[H_3O^+] = 1.0 \times 10^{-6}$ M

(e) pH = 8.00

(f) $[H_3O^+] = 1.0 \times 10^{-14}$

18.14 Calculate the acetic acid concentration and the acetate ion concentration in each of the following solutions:

(a) 0.200 mol of $HC_2H_3O_2$ plus 0.130 mol of $NaC_2H_3O_2$ in 1.00 L of solution

(b) 0.330 mol of $HC_2H_3O_2$ plus 0.130 mol of NaOH in 1.00 L of solution

18.15 List six singly charged cations that are feeble acids.

18.16 Give examples of compounds that could provide the ions to make the following reaction proceed to the right (reverse of the ionization of chlorous acid):

$$ClO_2^-(aq) + H_3O^+(aq) \rightleftharpoons HClO_2(aq) + H_2O(l)$$

18.17 A salt in Arrhenius terminology is composed of two types of ions. Describe them in terms of the Brønsted theory.

■ ■ ■ PROBLEMS ■ ■ ■

18.1 The Brønsted Theory

18.18 What is the conjugate of each of the following?

(a) $HClO_2$ (b) CH_3NH_2 (a weak base like NH_3)

(c) H_3PO_4

18.19 What is the conjugate of each of the following?

(a) HNO_3 (b) H_2S (c) $HClO_4$

18.20 Write the formula for a molecule that is the conjugate of each of the following:

(a) H_3O^+ (b) ClO_2^- (c) NH_4^+ (d) OH^-

18.21 Write the formula for the conjugate of each of the following:

(a) S^{2-} (b) CO_3^{2-} (c) OH^- (d) NO_2^-

18.22 Label each of the species in the following equations as an acid or a base, and connect the conjugate pairs:

$$H_2SO_4(aq) + OH^-(aq) \rightleftharpoons HSO_4^-(aq) + H_2O(l)$$
$$HSO_4^-(aq) + OH^-(aq) \rightleftharpoons SO_4^{2-}(aq) + H_2O(l)$$

18.23 Label each of the species in the following equations as an acid or a base, and connect the conjugate pairs:

$$H_3PO_4(aq) + OH^-(aq) \rightleftharpoons H_2PO_4^-(aq) + H_2O(l)$$
$$H_2PO_4^-(aq) + OH^-(aq) \rightleftharpoons HPO_4^{2-}(aq) + H_2O(l)$$
$$HPO_4^{2-}(aq) + OH^-(aq) \rightleftharpoons PO_4^{3-}(aq) + H_2O(l)$$

18.24 Which of the following 0.100 M solutions are acidic, which are basic, and which are neutral?

(a) $(NH_4)_2SO_4$ (b) NaF   (c) $NaClO_2$

(d) KNO_3 (e) $KC_2H_3O_2$

18.25 Which of the following 0.100 M solutions are acidic, which are basic, and which are neutral?

(a) K_2CO_3 (b) $NaHSO_4$ (c) $BaCl_2$

(d) KNO_2 (e) MgS

18.26 Write formulas for the conjugate acid and the conjugate base of each of the following:

(a) HSO_4^- (b) H_2O

18.27 Write formulas for the conjugate acid and the conjugate base of each of the following:

(a) $H_2PO_4^-$ (b) HPO_4^{2-}

18.28 Label each of the species in the following equations as an acid or a base, and connect the conjugate pairs:

$$HClO_2(aq) + H_2O(l) \rightleftharpoons ClO_2^-(aq) + H_3O^+(aq)$$
$$C_5H_5N(aq) + H_2O(l) \rightleftharpoons C_5H_5NH^+(aq) + OH^-(aq)$$

18.29 Label each of the species in the following equations as an acid or a base, and connect the conjugate pairs:

$$S^{2-}(aq) + H_2O(l) \rightleftharpoons HS^-(aq) + OH^-(aq)$$
$$NH_4^+(aq) + H_2O(l) \rightleftharpoons NH_3(aq) + H_3O^+(aq)$$

18.2 Ionization Constants

18.30 A 0.100 M solution of a weak acid, represented as HA, has a hydronium ion concentration of 4.79×10^{-5} M. Calculate the value of K_a for HA. The equation for the ionization of this acid is

$$HA(aq) + H_2O(l) \rightleftharpoons H_3O^+(aq) + A^-(aq)$$

18.31 A 0.200 M solution of a weak base, represented as B, has a hydroxide ion concentration of 5.19×10^{-6} M. Calculate the value of K_b for B. The equation for the ionization of this base is

$$B(aq) + H_2O(l) \rightleftharpoons BH^+(aq) + OH^-(aq)$$

18.32 A solution is described as a 0.200 M solution of $NaClO_2$. What are the actual solutes? What is the meaning of 0.200 M?

18.33 A solution is described as a 0.300 M solution of CaF_2. What are the actual solutes? What is the meaning of 0.300 M?

18.34 Calculate the hydronium ion concentration of 0.250 M formic acid, $HCHO_2$. Formic acid ionizes according to the equation

$$HCHO_2(aq) + H_2O(l) \rightleftharpoons CHO_2^-(aq) + H_3O^+(aq)$$

The acid dissociation constant is 1.7×10^{-4}.

18.35 Which base in Table 18.2 has the lowest hydroxide ion concentration in 0.100 M solution?

18.36 Calculate the hydronium ion concentration of 0.200 M phenol, HOC_6H_5. Phenol ionizes according to the equation

$$HOC_6H_5(aq) + H_2O(l) \rightleftharpoons OC_6H_5^-(aq) + H_3O^+(aq)$$

The acid dissociation constant is 1.3×10^{-10}.

18.37 Calculate the percent ionization of methyl amine, CH_3NH_2, in (a) a 0.333 M solution, (b) a 0.100 M solution, and (c) a 0.0333 M solution. Tabulate the initial concentration of methyl amine, the hydroxide ion concentration, and the percent ionization for these cases.

18.3 Autoionization of Water

18.38 Calculate the pH of each of the following solutions:

(a) 0.100 M HNO_3 (b) 0.100 M KOH

(c) 0.100 M KNO_3

18.39 Calculate the pH of each of the following solutions:

(a) 0.0100 M H_2SO_4 (b) 0.200 M NaOH

(c) 0.400 M NaCl

18.40 Calculate the hydronium ion concentration of a solution having each of the following pH values:

(a) 1.000   (b) 13.000   (c) 7.000

(d) 0.000   (e) 14.000

18.41 Calculate the pH of each of the following solutions:

(a) 0.0173 M $HClO_3$ (b) 0.00424 M KOH

(c) 1.33×10^{-4} M $Ba(OH)_2$

18.42 Calculate the pH of each of the following solutions:

(a) 1.12×10^{-3} M HI   (b) 2.99×10^{-4} M $Ba(OH)_2$

18.43 Calculate the hydronium ion concentration of a solution having each of the following pH values:

(a) 7.127   (b) 13.093   (c) 10.10

18.44 What is the hydronium ion concentration of a solution having each of the following pH values?

(a) 1.033   (b) 0.921   (c) 9.191

18.45 Calculate the pH of a solution having each of the following hydroxide ion concentrations:

(a) 6.39×10^{-3} M   (b) 7.28×10^{-4} M

(c) 2.18×10^{-7} M

18.46 A 0.100 M solution of a weak acid, represented as HA, has a pH of 4.93. Calculate the value of K_a for HA.

18.47 Calculate $[H_3O^+]$ and pH for a 0.321 M solution of NH_3.

18.48 A 0.100 M solution of a weak acid, represented as HA, has a pH of 2.29. Calculate the value of K_a for HA.

18.49 A 0.200 M solution of a weak base, represented as B, has a pH of 12.79. Calculate the value of K_b for B.

18.50 Calculate the hydroxide ion concentration of a solution having each of the following pH values:

(a) 6.127

(b) 11.210

(c) 2.912

18.51 What is the hydroxide ion concentration of a solution having each of the following pH values?

(a) 6.111

(b) 9.21

(c) 2.441

18.52 Calculate the pH of a solution having each of the following hydroxide ion concentrations:

(a) 3.42×10^{-3} M

(b) 4.53×10^{-3} M

(c) 7.19×10^{-8} M

18.53 A 0.100 M solution of a weak acid, represented as HA, has a pH of 3.89. Calculate the value of K_a for HA.

18.54 Calculate $[H_3O^+]$ and pH for a 0.413 M solution of CH_3NH_2.

18.4 Buffer Solutions

18.55 A 0.130 M solution of a weak base, represented as B, is also 0.140 M in BHCl and has a pH of 12.15. Calculate the value of K_b for B.

18.56 A 0.170 M solution of a weak base, represented as B, is also 0.130 M in $BHNO_3$ and has a pH of 10.74. Calculate the value of K_b for B.

18.57 A 0.213 M solution of a weak acid, represented as HA, is also 0.311 M in NaA and has a pH of 4.27. Calculate the value of K_a for HA.

18.58 A 0.172 M solution of a weak acid, represented as HA, is also 0.218 M in NaA and has a pH of 3.991. Calculate the value of K_a for HA.

18.59 Calculate the hydronium ion concentration of 1.00 L of a solution made from each of the following combinations:

(a) 0.130 mol of formic acid, $HCHO_2$, and 0.120 mol of sodium formate, $NaCHO_2$

(b) 0.250 mol of formic acid and 0.120 mol of NaOH

18.60 Calculate the hydronium ion concentration of 1.00 L of a solution prepared with each of the following combinations:

(a) 0.100 mol of NH_3 and 0.170 mol of NH_4NO_3

(b) 0.290 mol of NH_3 and 0.150 mol of HNO_3

18.61 Calculate the pH of each solution in Problem 18.14.

■ ■ ■ GENERAL PROBLEMS ■ ■ ■

18.62 Calculate the acetate ion concentration in 1.00 L of a solution containing 0.120 mol of acetic acid and 0.140 mol of HCl.

18.63 Calculate the ammonium ion concentration in 1.00 L of a solution containing 0.130 mol of NH_3 and 0.130 mol of NaOH.

18.64 A solution containing 0.200 mol of phenol, HOC_6H_5, and 0.200 mol of its conjugate base, $OC_6H_5^-$ (the phenolate ion), has a hydronium ion concentration 1.3×10^{-10} M. The solution is basic. Explain how a solution with *any* acid in it can be basic.

18.65 Other than adding a weak acid to its conjugate base, what are two ways in which a buffer solution involving the same equilibrium system can be made?

18.66 Which of the following will form a buffer solution when added to water?

(a) 0.100 mol $HC_2H_3O_2$ plus 0.100 mol $C_2H_3O_2^-$

(b) 0.200 mol $HC_2H_3O_2$ plus 0.100 mol NaOH

(c) 0.200 mol $HC_2H_3O_2$ plus 0.200 mol NaOH

(d) 0.200 mol $NaC_2H_3O_2$ plus 0.100 mol HCl

(e) 0.200 mol $NaC_2H_3O_2$ plus 0.200 mol HCl

(f) 0.100 mol NH_3 plus 0.100 mol NH_4Cl

18.67 (a) A 1.00-L solution contains 0.200 mol of $HC_2H_3O_2$ and 0.200 mol of $NaC_2H_3O_2$. Calculate the hydronium ion concentration of the solution.

(b) The solution is diluted to 2.00 L. Calculate the new hydronium ion concentration.

18.68 What is the pH of a solution in which 0.125 M $HCHO_2$ is exactly half-neutralized with NaOH during a titration?

18.69 Analogously to pH, the pOH can be defined as

$$pOH = -\log[OH^-]$$

Show that pH + pOH = 14.00 in all dilute aqueous solutions.

18.70 Which of the following will form a buffer solution when added to water?

(a) 0.100 mol $HC_2H_3O_2$ plus 0.100 mol HCl

(b) 0.200 mol NH_3 plus 0.100 mol NaOH

(c) 0.200 mol NaCl plus 0.200 mol NaOH

(d) 0.200 mol NaCl plus 0.100 mol HCl

(e) 0.200 mol NaOH plus 0.200 mol HCl

(f) 0.100 mol NH_3 plus 0.100 mol NH_4NO_3

(g) 0.100 mol HCl plus 0.200 mol NaOH

19 Organic Chemistry

- 19.1 Hydrocarbons
- 19.2 Isomerism
- 19.3 Some Other Classes of Organic Compounds
- 19.4 Polymers
- 19.5 Foods

■ Key Terms *(Key terms are defined in the Glossary.)*

alcohol (19.3)
aldehyde (19.3)
alkane (19.1)
alkene (19.1)
alkyne (19.1)
amide (19.3)
amine (19.3)
aromatic hydrocarbon (19.1)
benzene (19.1)
carbohydrate (19.5)
disaccharide (19.5)

ester (19.3)
ether (19.3)
fat (19.5)
functional group (19.3)
hydrocarbon (19.1)
isomers (19.2)
ketone (19.3)
line formula (19.2)
monomer (19.4)
monosaccharide (19.5)
organic acid (19.3)

organic halide (19.3)
polymer (19.4)
polysaccharide (19.5)
protein (19.4)
radical (19.3)
saturated hydrocarbon (19.1)
soap (19.5)
sugar (19.5)
total bond order (19.3)
unsaturated hydrocarbon (19.1)

■ Symbols/Suffixes

⬡ (benzene) (19.1)
R (radical) (19.3)
X (halogen) (19.3)
-al (19.3)

-ane (19.1)
-ate (19.3)
-ene (19.1)
-oic acid (19.3)

-ol (19.3)
-one (19.3)
-yl (19.3)
-yne (19.1)

Organic chemistry can be defined as the chemistry of carbon and its compounds. However, this definition is too broad because many compounds of carbon, such as carbon dioxide and carbon tetrachloride, are regarded as inorganic. Until almost 200 years ago, organic chemistry was defined as the chemistry of compounds derived from living things. In 1828, Friedrich Wöhler (1800–1882) did a laboratory experiment that converted ammonium cyanate (an inorganic compound) to urea (an organic compound), which showed that the "life force" was not necessary to make organic compounds:

$$NH_4NCO(aq) \xrightarrow{\text{Heat}} NH_2CNH_2(aq)$$
$$\overset{\|}{O}$$

Ammonium cyanate   Urea

(Note that structural formulas for organic compounds are often written without the dots representing unshared electrons.) Perhaps the best definition of an organic compound is that it contains at least one carbon-carbon or carbon-hydrogen bond. However, this definition excludes urea and thiourea, so it is not perfect.

Hydrocarbons, the simplest class of organic compounds, are described in Section 19.1. The four major series of hydrocarbons are presented, and the most modern system for naming them, as well as the basic geometry of the simplest molecules, is introduced. Section 19.2 examines isomers, different compounds that have the same molecular formula. The discussion in this section is limited to isomerism of hydrocarbons, but the phenomenon exists in all classes of organic compounds. Organic compounds containing elements in addition to carbon and hydrogen are covered in Section 19.3. With the inclusion of oxygen, nitrogen, and the halogens, nine additional classes of organic compounds are formed. Section 19.4 briefly describes polymers, whose huge molecules are formed from simpler molecules by ordinary types of reactions such as those already described. A brief look at the organic chemistry of foods is presented in Section 19.5.

19.1 Hadrocarbons

19.1 Hydrocarbons

OBJECTIVE

■ to write general formulas for the different series of hydrocarbons, to write formulas for the individual members of the series, and to describe a few of their properties

Hydrocarbons are compounds containing only carbon and hydrogen. Not only are hydrocarbons important in themselves, but they are the foundation of all the other classes of organic compounds. You must learn the names, formulas, and some simple reactions of the hydrocarbons. The number of such compounds is almost uncountable because carbon atoms can bond to other carbon atoms in seemingly limitless numbers. (An 8-carbon chain is shown in Figure 19.1. Compounds with extremely long chains are discussed in Section 19.4.) Moreover, the atoms can bond in different ways (see Section 19.2), which greatly increases the number

Figure 19.1 Octane, an 8-Carbon Unbranched Hydrocarbon

This computer-generated image of an octane molecule shows that its 8 carbon atoms (blue) form a continuous chain with a total of 18 hydrogen atoms (white) attached. (Image from HyperChem™ software, copyright© Autodesk, Inc.)

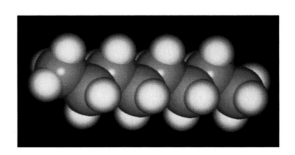

of possible compounds. Study of the chemistry of such a huge collection of compounds is kept manageable by dividing the hydrocarbons into four main series: alkanes, alkenes, alkynes, and aromatic hydrocarbons.

The Alkanes

The **alkanes** are the most fundamental class of hydrocarbons. Their molecular formulas may be represented as C_nH_{2n+2}, where n represents some integer. For example, if $n = 2$, the formula is C_2H_6, and if $n = 4$, the formula is C_4H_{10}. Characterized by having only *single* covalent bonds within their molecules, these compounds are said to be **saturated hydrocarbons,** since their carbon backbones cannot hold any more hydrogen atoms. Because they are relatively inert, alkanes are also known as *paraffins* (from the Latin, meaning "against reaction"). The names and formulas of the first 10 unbranched alkanes are given in Table 19.1. Each of the alkanes has a name that ends in *-ane*. The beginnings of the names of the fifth through tenth alkanes are Greek or Latin prefixes for the numbers that correspond to the numbers of carbon atoms in these molecules. For example, *oct*ane has *eight* carbon atoms in each of its molecules. You should learn the names of the first 10 alkanes and the associated numbers of carbon atoms in their molecules, since the names of the other classes of hydrocarbons and other types of organic compounds are based on these names.

At high temperatures, the alkanes react with excess oxygen to yield water and carbon dioxide or with limited oxygen to yield water and carbon monoxide:

$$CH_4(g) + 2\ O_2(g, \text{excess}) \xrightarrow{\text{High temperature}} CO_2(g) + 2\ H_2O(g)$$

$$2\ CH_4(g) + 3\ O_2(g, \text{limited supply}) \xrightarrow{\text{High temperature}} 2\ CO(g) + 4\ H_2O(g)$$

Alkanes are widely used as fuels: Methane is familiar as natural gas, propane is used as bottled gas and as a fuel for welding torches, butane is used as cigarette lighter fluid, and octane is in gasoline. The alkanes also react with ele-

Table 19.1   *The First Ten Unbranched Alkanes*

Name	Molecular Formula	Line Formula
Methane	CH_4	CH_4
Ethane	C_2H_6	CH_3CH_3
Propane	C_3H_8	$CH_3CH_2CH_3$
Butane	C_4H_{10}	$CH_3CH_2CH_2CH_3$
Pentane	C_5H_{12}	$CH_3CH_2CH_2CH_2CH_3$
Hexane	C_6H_{14}	$CH_3CH_2CH_2CH_2CH_2CH_3$
Heptane	C_7H_{16}	$CH_3CH_2CH_2CH_2CH_2CH_2CH_3$
Octane	C_8H_{18}	$CH_3CH_2CH_2CH_2CH_2CH_2CH_2CH_3$
Nonane	C_9H_{20}	$CH_3CH_2CH_2CH_2CH_2CH_2CH_2CH_2CH_3$
Decane	$C_{10}H_{22}$	$CH_3CH_2CH_2CH_2CH_2CH_2CH_2CH_2CH_2CH_3$

ITEM OF INTEREST

Ignorance is not bliss! Burning charcoal or hydrocarbons in a limited supply of oxygen produces carbon monoxide, a deadly, poisonous gas. Internal combustion engines also produce carbon monoxide. *Don't* run automobiles in closed spaces such as garages. *Don't* heat your RV or station wagon with a charcoal or propane heater with all the windows closed or nearly closed. *Don't* weather-strip the doors and windows of your home too well, especially if you use gas for cooking and/or heating. *Do* check your furnace for proper functioning periodically and keep your chimney in good repair and unblocked. Failure to follow these rules has cost too many human lives.

mental halogens at high temperatures to produce halogenated hydrocarbons. For example:

$$CH_4(g) + Cl_2(g, \text{ limited supply}) \xrightarrow{\text{High temperature}} CH_3Cl(g) + HCl(g)$$

$$CH_4(g) + 4\ Cl_2(g, \text{ excess}) \xrightarrow{\text{High temperature}} CCl_4(l) + 4\ HCl(g)$$

The reactions of all the alkanes are similar; they differ in degree, rather than kind. That fact makes the study of the alkanes the study of a single class of compounds, rather than of the millions of individual compounds that make up the class.

■ EXAMPLE 19.1

Write equations for the reaction of ethane with (a) excess oxygen and (b) chlorine in limited supply.

Solution

(a) $2\ C_2H_6(g) + 7\ O_2(g) \rightarrow 4\ CO_2(g) + 6\ H_2O(g)$

(b) $C_2H_6(g) + Cl_2(g) \rightarrow C_2H_5Cl(g) + HCl(g)$

Practice Problem 19.1   Write equations for the reaction of ethane with (a) excess chlorine and (b) oxygen in limited supply. ■

The Alkenes

An **alkene** is a hydrocarbon that has one carbon-to-carbon double bond in its structure. The general molecular formula for this class of hydrocarbons is C_nH_{2n}. The two electrons involved as the second pair of the double bond are not shared with hydrogen atoms, so each molecule of an alkene has two fewer hydrogen atoms than that of the corresponding alkane. Alkenes are therefore said to be **unsaturated hydrocarbons.** The alkenes are named systematically, using the name of the alkane having the same number of carbon atoms and changing the ending to *-ene*. The location of the double bond is indicated when necessary. The names and formulas of the first four unbranched alkenes are presented in Table 19.2. Ethylene (ethene) and propylene (propene) are the raw materials used in the manufacture of the common plastics polyethylene and polypropylene.

Table 19.2 The First Four Unbranched Alkenes

Systematic Name	Common Name	Molecular Formula	Line Formula	Structural Formula
Ethene	Ethylene	C_2H_4	$CH_2{=}CH_2$	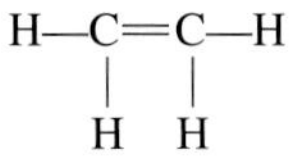
Propene	Propylene	C_3H_6	$CH_3CH{=}CH_2$	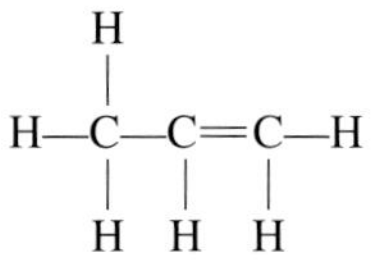
1-Butene		C_4H_8	$CH_2{=}CHCH_2CH_3$	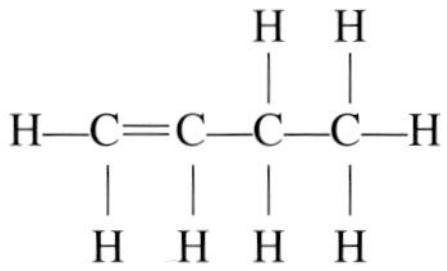
2-Butene		C_4H_8	$CH_3CH{=}CHCH_3$	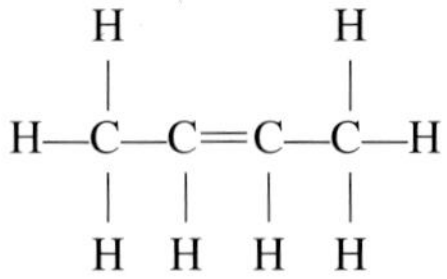

The systematic naming of simple organic compounds is based on the longest continuous chain of carbon atoms in the molecule. A branch, functional group (Section 19.3), or multiple bond involving a carbon atom in such a chain is given an "address," or position on the carbon chain, which is the number of that carbon atom counting from the *nearer end* of the chain. A one-carbon branch, CH_3—, is called *methyl.* For example, consider the following formulas, shown for clarity without the hydrogen atoms on the longest continuous chain:

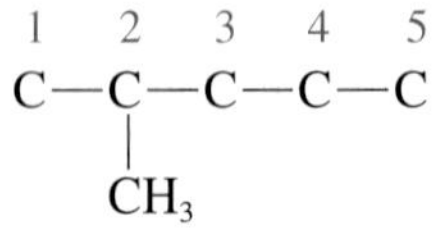
Compound A

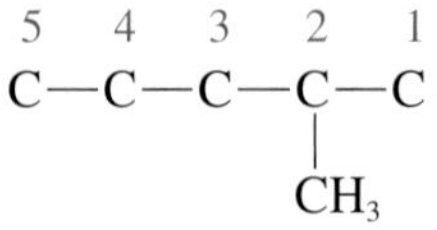
Compound B

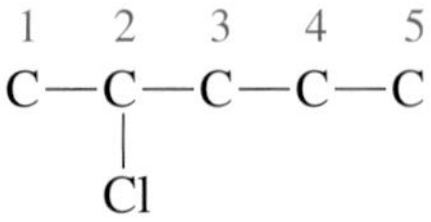
Compound C

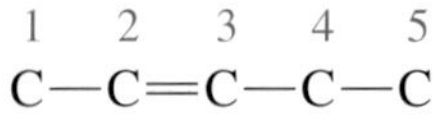
Compound D

In compound A, the longest continuous chain of carbon atoms is five, and there are no multiple bonds, so this compound is named as a *pentane* (Table 19.1). The carbon atom to which the methyl group (CH_3—) is attached is identified with the number 2, since that carbon is the second from the nearer end of the chain. The name is 2-methylpentane. Note that the number 2 is an *address* and does not mean two methyl groups.

The formula for compound B represents the same compound as compound A but written in the reverse direction. We start numbering the carbon chain from the end nearer the methyl group, and we get the same name—2-methylpentane.

We name organic compounds containing chlorine in an analogous manner; compound C is 2-chloropentane (see p. 418).

A hydrocarbon with a double bond is an alkene, named with the suffix *-ene.* Of the *two* carbon atoms involved in the double bond, we use the address of the carbon closer to the nearer end of the longest chain in the name. Compound D is 2-pentene.

When no ambiguity would result from the lack of an address, none is given. For example, the following compound is called simply methylpropane:

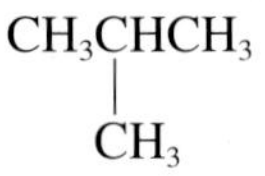

No address is needed for the methyl group, since there is only one carbon atom where a branch may occur. If the methyl group is attached to either carbon atom at the end of the chain, a longer continuous chain is formed:

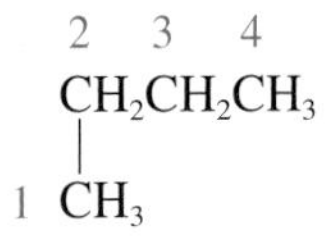

$$
\begin{array}{c}
\overset{2}{}\ \ \overset{3}{}\ \ \overset{4}{} \\
CH_2CH_2CH_3 \\
| \\
1\ \ CH_3
\end{array}
$$

This compound is butane.

■ EXAMPLE 19.2

Name each of the following alkenes:

(a) $CH_3CH\!=\!CHCH_2CH_3$ (b) $CH_3CH_2CH_2CH\!=\!CH_2$ (c) $CH_2\!=\!CHCH_3$

Solution

(a) 2-pentene   (b) 1-pentene   (c) propene

Practice Problem 19.2 Write line formulas for (a) 1-heptene and (b) 3-heptene. ■

■ EXAMPLE 19.3

Explain why no alkene has only one carbon atom.

Solution

By definition, an alkene is a hydrocarbon with a carbon-to-carbon double bond. Therefore, it must have at least two carbon atoms. ■

The alkenes are more reactive than the alkanes. For example, alkene molecules undergo addition reactions in which the atoms from halogen molecules or from hydrogen halides bond to the carbon atoms involved in the double bond:

$$CH_2\!=\!CH_2(g) + Br_2(l) \rightarrow CH_2BrCH_2Br(l)$$

$$CH_2\!=\!CH_2(g) + HBr(g) \rightarrow CH_3CH_2Br(l)$$

These reactions can take place at room temperature, in contrast to the high temperatures needed for an alkane to react with a halogen molecule.

The Alkynes

An **alkyne** is a hydrocarbon with one carbon-to-carbon triple bond per molecule. The series has molecules with the general molecular formula C_nH_{2n-2}. The first alkyne, called ethyne and also known as acetylene, has the formula C_2H_2 or $HC\!\equiv\!CH$, also written $CH\!\equiv\!CH$. Acetylene is used as a fuel in high-temperature welding torches. The alkynes are named similarly to the alkanes and alkenes, but with the ending -*yne*.

Like the alkenes, the alkynes are unsaturated and more reactive than the alkanes. Alkynes can add two molecules of halogen across the triple bond:

$$CH\!\equiv\!CH(g) + 2\ Br_2(l) \rightarrow CHBr_2CHBr_2(l)$$

■ **EXAMPLE 19.4**

Name $CH_3CH_2C{\equiv}CH$.

Solution

The name is 1-butyne. (The carbon atom involved in the triple bond that is nearer the end of the chain is the *first* from the *right.*)

Practice Problem 19.4 Write the line formula for (a) 2-pentyne and (b) 3-hexyne. ■

The Aromatic Hydrocarbons

A class of hydrocarbons whose molecules contain a ring of six carbon atoms, each with only one hydrogen atom (or other group) attached, has special properties. The class is known as the **aromatic hydrocarbons.** The simplest member is **benzene,** and all other members have at least one benzene ring included in their structures (Figure 19.2). A special kind of bonding in the ring gives the aromatic hydrocarbons more stability than is expected for similar compounds with ordinary double bonds. For example, benzene reacts with a halogen molecule by substitution, in a manner that is more like the reaction of an alkane than of an alkene. Iron is used as a catalyst in the substitution reaction:

$$C_6H_6(l) + Br_2(l) \xrightarrow[\text{Heat}]{\text{Fe}} C_6H_5Br(l) + HBr(g)$$

■ **EXAMPLE 19.5**

Draw the structure of bromobenzene, C_6H_5Br, an aromatic organic halide. (Organic halides are defined in Section 19.3.)

Solution

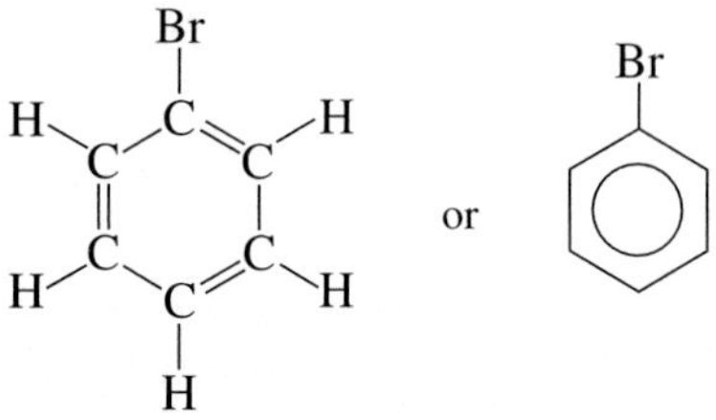

Figure 19.2 *Several Representations of Benzene*

(a) Note that the double bonds can be in different locations. (b) The letters representing the carbon and hydrogen atoms are often omitted. (c) The circle represents the possible locations of the double bonds averaged over time.

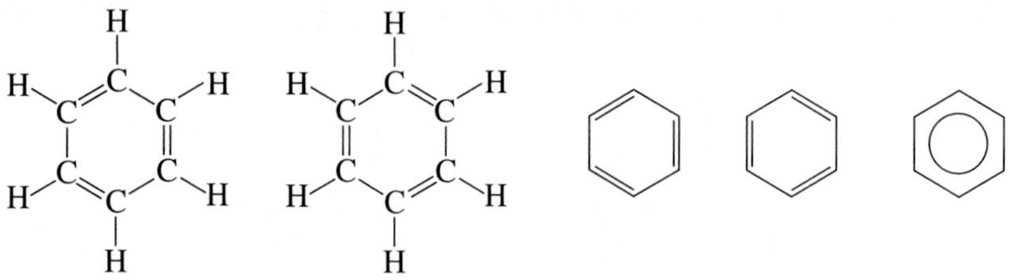

Bromobenzene

(a)   (b)   (c)

Practice Problem 19.5 How many hydrogen atoms are in dibromo-benzene? ■

Compounds such as trinitrotoluene, known familiarly as TNT, are derived from aromatic hydrocarbons:

CH₃

O₂N ⧵C⧸ C ⧸NO₂

Trinitrotoluene (TNT)

19.2 Isomerism

OBJECTIVE

■ to write formulas for isomers, and to distinguish formulas for isomers from formulas that represent a single compound written in different ways

When a carbon atom is involved in four single bonds, those bonds are oriented toward the corners of a tetrahedron (Figure 19.3). The angle between any two of the bonds is 109.5°. A chain of such atoms in a hydrocarbon can assume a zigzag shape. Organic molecules are usually represented in one of six different ways (Figure 19.4). Models of the molecule are easiest to visualize but hardest to produce. A ball-and-stick model (Figure 19.4a) shows the bond angles best, but a space-filling model (Figure 19.4b) is more realistic and it better shows the relative sizes of the atoms and the distances between them. Chemists could draw two-dimensional pictures of the two types of models, of course. To save time and

(a)   (b)   (c)

Figure 19.3 Tetrahedral Nature of the Carbon Atom

(a) A tetrahedron has four triangular sides (including the base). (b) The carbon atom of CH₄ is typical of carbon atoms in alkanes. (c) The tetrahedral nature of the carbon atom is apparent in that its bonds with four hydrogen atoms point toward the corners of a tetrahedron. (Model courtesy of Molecular Models Co.)

Figure 19.4 Representations of Butane

(a) Ball-and-stick model, with black carbon atoms and white hydrogen atoms. (Model courtesy of Molecular Models Co.) (b) Space-filling model, with blue carbon atoms and white hydrogen atoms. (Image from HyperChem™ software, copyright© Autodesk, Inc.) (c) Structural formula showing actual bond angles. (d) "Two-dimensional" structural formula. (e) Line formula. (f) Molecular formula.

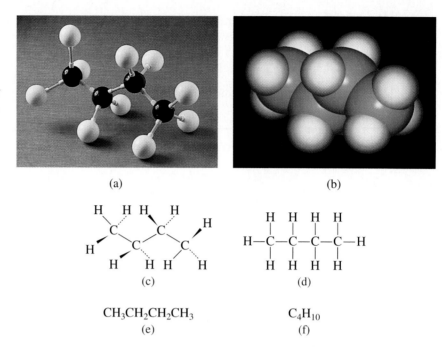

(a)   (b)

(c)   (d)

$CH_3CH_2CH_2CH_3$ C_4H_{10}

(e)   (f)

trouble, however, they often draw structural formulas. The actual geometry of the molecule may be indicated in a structural formula (Figure 19.4c), but usually it is not (Figure 19.4d). Figure 19.4(d) is similar to an electron dot diagram, with the shared pairs of electrons represented by lines, rather than pairs of dots, and the unshared pairs generally not shown. In a **line formula,** so called because it is written on a single line, groups of atoms that are connected in the molecule are written together (Figure 19.4e). This representation is easier to write than a structural formula, but it requires the chemist to remember more about the bonding of the atoms in the molecule. Finally, a molecular formula (Figure 19.4f) is the easiest to write but gives far less information. A molecular formula may even represent more than one compound.

In an alkane molecule, one part of the molecule can rotate about one (or more) of the single bonds to assume different shapes (Figure 19.5). Such different orientations are different forms of the same molecule.

Figure 19.5 Two Different Orientations of Butane

(Model courtesy of Molecular Models Co.)

■ EXAMPLE 19.6

Write structural formulas that show the two different orientations for butane illustrated in Figure 19.5.

Solution

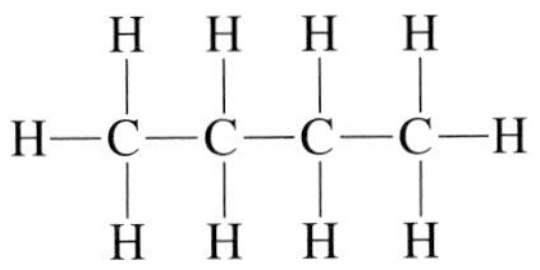

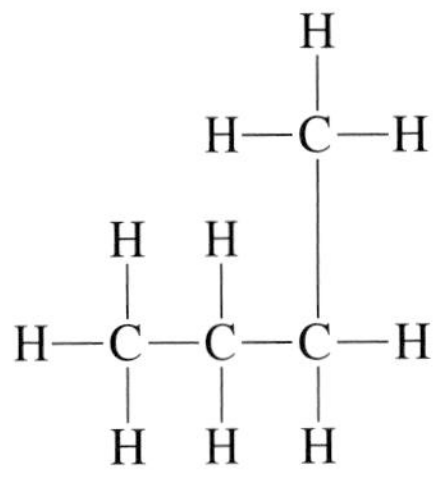

```
                                              H
                                              |
                                          H—C—H
                                              |
      H   H   H   H                   H   H   |
      |   |   |   |                   |   |   |
  H—C—C—C—C—H               H—C—C—C—H
      |   |   |   |                   |   |   |
      H   H   H   H                   H   H   H
```

These structural formulas look less like the same molecule than do the models in Figure 19.5. Be aware, however, that they are representations of the same molecule.

Practice Problem 19.6 Write another structural formula for butane that shows both ends of the molecule rotated from the zigzag shape in Figure 19.4(c). ■

When more than three carbon atoms are involved, the possibility exists of having two different hydrocarbons with the same molecular formula but different structures. Such compounds are structural **isomers** of each other. Each isomer is a separate compound; isomers are related only by having the same molecular formula. (Of course, their properties are similar, since they are in the same class of hydrocarbons.) For example, the molecular formula C_4H_{10} corresponds to the following two compounds, which are isomers of each other:

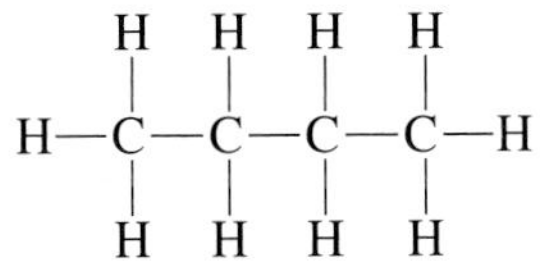

```
      H   H   H   H                   H   H   H
      |   |   |   |                   |   |   |
  H—C—C—C—C—H               H—C—C—C—H
      |   |   |   |                   |   |   |
      H   H   H   H                   H   |   H
                                          |
                                      H—C—H
                                          |
                                          H
           Butane                   Methylpropane
```

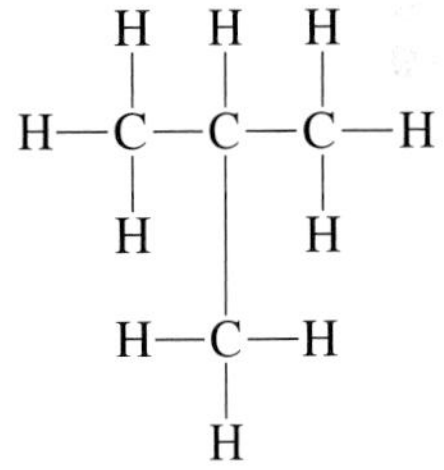

Compounds with branches on the longest continuous chain of carbon atoms are named as derivatives of that chain. The branches are named using the stem of the alkane having the same number of carbon atoms with the ending changed to *-yl*. Thus, for the branched chain compound whose structural formula was just illustrated, the side chain CH_3— is called methyl, and the three-carbon continuous chain is called propane. Two, three, or four identical side chains are named using the prefix *di-, tri,* or *tetra-,* respectively.

■ EXAMPLE 19.7

Does the structural formula of methylpropane shown previously represent another orientation of the molecule in Example 19.6, or is it another molecule?

Solution

It is another molecule. No amount of rotation about bonds can move the last carbon atom of the four-carbon chain of butane from the third to the second carbon atom in the molecule.

Practice Problem 19.7 A student incorrectly named two compounds (a) 1-methylpropane and (b) 3-methylbutane. Write structural formulas corresponding to these names, and give the correct names for the compounds. ■

■ EXAMPLE 19.8

Write line formulas for (a) butane and (b) methylpropane.

Solution

(a) $CH_3CH_2CH_2CH_3$ (b) $CH_3CH(CH_3)_2$

Practice Problem 19.8 Write a line formula for 2,4-dimethylhexane. ■

■ EXAMPLE 19.9

Write structural formulas for three compounds with the molecular formula C_5H_{12}. Name these compounds.

Solution

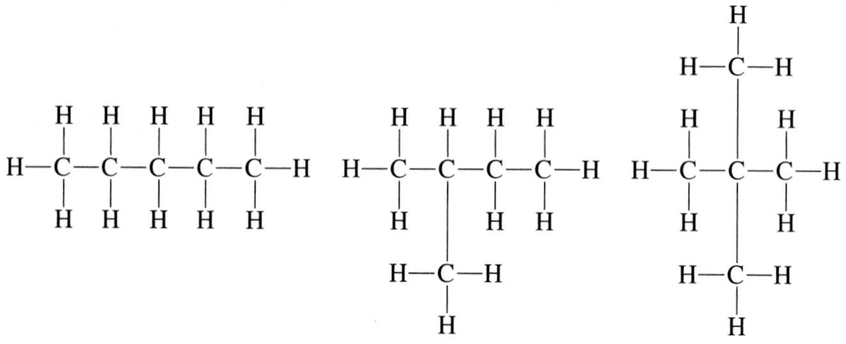

Pentane   Methylbutane   Dimethylpropane

Practice Problem 19.9 Write structural formulas for the two compounds with the molecular formula C_3H_7Cl. Name these compounds. ■

In unsaturated hydrocarbons, isomerism can occur because of the position of the double or triple bond.

■ EXAMPLE 19.10

Draw structural formulas for two compounds with the molecular formula C_4H_8 and with the four carbon atoms connected in a continuous chain. Name them.

Solution

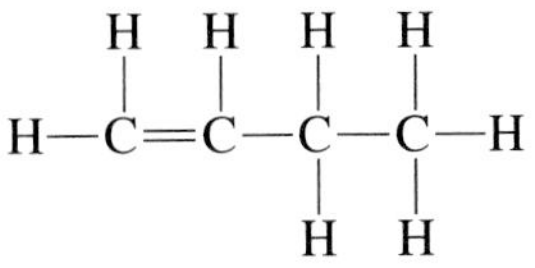

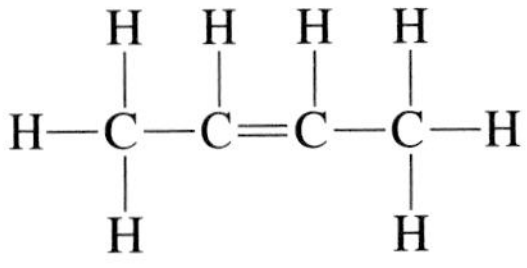

$$H-\overset{\overset{\displaystyle H}{|}}{C}=\overset{\overset{\displaystyle H}{|}}{C}-\overset{\overset{\displaystyle H}{|}}{\underset{\underset{\displaystyle H}{|}}{C}}-\overset{\overset{\displaystyle H}{|}}{\underset{\underset{\displaystyle H}{|}}{C}}-H \qquad H-\overset{\overset{\displaystyle H}{|}}{\underset{\underset{\displaystyle H}{|}}{C}}-\overset{\overset{\displaystyle H}{|}}{C}=\overset{\overset{\displaystyle H}{|}}{C}-\overset{\overset{\displaystyle H}{|}}{\underset{\underset{\displaystyle H}{|}}{C}}-H$$

These two compounds are called 1-butene and 2-butene, respectively. ■

■ EXAMPLE 19.11

Draw a structural formula for another isomer with the molecular formula C_4H_8, besides the two shown in Example 19.10.

Solution

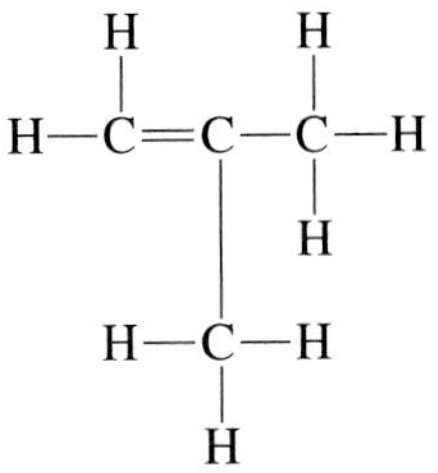

$$H-\overset{\overset{\displaystyle H}{|}}{C}=\overset{\overset{\displaystyle H}{|}}{C}-\overset{\overset{\displaystyle H}{|}}{\underset{\underset{\displaystyle H}{|}}{C}}-H$$

$$H-\overset{\overset{\displaystyle H}{|}}{\underset{\underset{\displaystyle H}{|}}{C}}-H$$

Methylpropene ■

ENRICHMENT

Petroleum is a mixture of many compounds, mainly hydrocarbons containing varying numbers of carbon atoms per molecule. Hydrocarbons with more than 12 carbon atoms per molecule are oily, greasy, or waxy substances. To produce smaller molecules, which are characteristic of hydrocarbons used as fuels, oil companies *crack* the petroleum by heating it in the absence of air. The heating causes some of the carbon-carbon bonds to break and results in the production of a range of smaller molecules. Cracking in the presence of certain catalysts optimizes the yield of those isomers that are useful as fuels. The cracking process is followed by distillation, in which various types of pretroleum products are separated from each other by means of their different boiling points. Two isomers of octane, C_8H_{18}, are shown in Figure 19.6.

Figure 19.6 Two Isomers of Octane

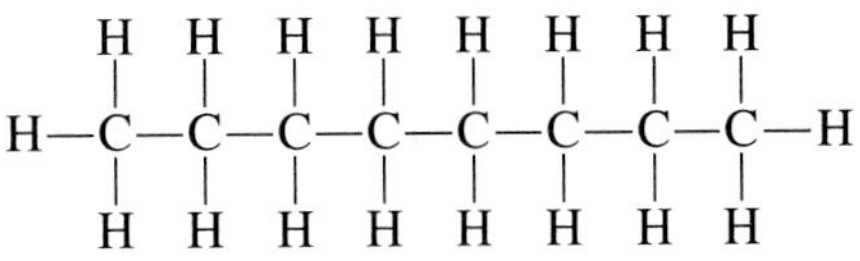

Octane

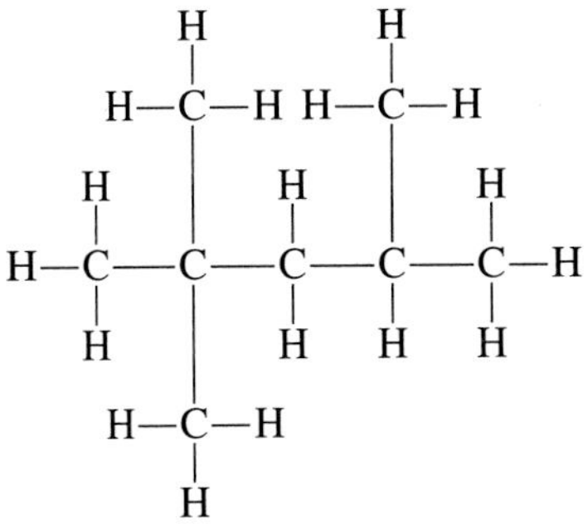

2,2,4-Trimethylpentane

Isomers exist in all classes of organic compounds, some of which are described in the next section.

19.3 Some Other Classes of Organic Compounds

▪ to write general formulas for several classes of organic compounds, to identify their functional groups, and to write equations for some of their reactions

A wide variety of organic compounds contain at least one other element in addition to carbon and hydrogen. The other elements most commonly found in organic compounds are the nonmetals oxygen, nitrogen, fluorine, chlorine, bromine, iodine, sulfur, and phosphorus. The **total bond order** of an atom is its number of bonding electron pairs. Just as a hydrogen atom always shares one electron pair to form one covalent bond in organic compounds and a carbon atom shares four electron pairs (in four single bonds, two double bonds, two single bonds and one double bond, *or* one single bond and one triple bond), these other elements also have characteristic total bond orders (Table 19.3). Knowing these total bond orders is essential to interpreting line formulas correctly.

Since the hydrocarbon portion of an organic molecule is relatively inert, like an alkane or aromatic hydrocarbon, the atoms of the other elements form the centers of reaction. The hydrocarbon portion is termed a **radical** and is often represented by the symbol R. The radical is essentially the parent hydrocarbon with a hydrogen atom removed. The reactive portion of the molecule, containing the other element(s), is called the **functional group.** Radicals are named just like side chains (Section 19.2): The name of the parent hydrocarbon has its ending changed to *-yl.* (A radical may be a hydrogen atom in some classes containing carbon atoms in the functional group.)

▪ **EXAMPLE 19.12**

What are the names of the radicals formed when a hydrogen atom is removed from one end of each of the first four unbranched alkanes? (See Table 19.1 if necessary.)

Solution

The names are methyl, ethyl, propyl, and butyl, respectively.

Practice Problem 19.12   What are the names of the radicals formed from the unbranched alkanes with five and six carbon atoms? ▪

Table 19.3 Total Bond Orders of Atoms in Organic Molecules

Atom	Total Bond Order
C	4
H	1
O	2
N	3
S	2
P	3
X*	1

*X refers to any halogen atom—-F, Cl, Br, or I.

The subsections that follow briefly describe nine different classes of organic compounds: organic halides, alcohols, ethers, aldehydes, ketones, acids, esters, amines, and amides.

Organic Halides

Organic halide is a general term that refers to any fluoride, chloride, bromide, or iodide of a hydrocarbon. The symbol X is often used to represent any one of the halogen atoms—F, Cl, Br, or I. Organic halides may be produced by the reaction of the elemental halogen and a hydrocarbon or by the reaction of a hydrogen halide and an unsaturated hydrocarbon, among other methods:

$$X_2 + CH_4 \rightarrow CH_3X + HX \qquad (X = F, Cl, Br, or I)$$

$$HX + CH_2{=}CH_2 \rightarrow CH_3CH_2X$$

Freons are organic halides in which one to three fluorine atoms are substituted for the chlorine atoms of carbon tetrachloride. An example is Freon 12, CCl_2F_2. Such compounds are used as refrigerants. They are relatively inert, boil in a suitable temperature range, and have a relatively high heat of vaporization. The Freon is allowed to evaporate in the coils in the inner part of the refrigerator; the evaporation process cools. The Freon is then pumped as a gas to coils outside the food compartment, where it is compressed back to the liquid state. The liquefaction gives off heat, which is discharged into the room. The whole process is repeated over and over. (The refrigeration process is an example of heat being moved from a cold place to a warmer place, but the process is not spontaneous. Energy is required to accomplish this change.)

The escape of Freons into the atmosphere is thought to contribute to the destruction of the ozone layer of the earth's atmosphere. For this reason, Freons are no longer used in aerosol cans for shaving cream, hair spray, and other products, and they are being phased out of refrigeration systems.

Reaction of excess halogen with methane can result in replacement of more than one hydrogen atom:

$$2 X_2 + CH_4 \rightarrow CH_2X_2 + 2 HX$$

$$3 X_2 + CH_4 \rightarrow CHX_3 + 3 HX$$

$$4 X_2 + CH_4 \rightarrow CX_4 + 4 HX$$

Chloroform, $CHCl_3$, and carbon tetrachloride, CCl_4, are familiar products of the chlorination of methane. Halogenated hydrocarbons are intermediates in the production of many other types of organic compounds.

DDT (Figure 19.7) is short for DichloroDiphenylTrichloroethane, which is systematically named 1,1-di(4-chlorophenyl)-2,2,2-trichloroethane. It is a chlorinated hydrocarbon that has had widespread success in controlling mosquitoes and preventing thousands of deaths from malaria. The insecticide is not easily decomposed in the environment, however, and finds its way into the food chain of higher animals, with harmful results. For this reason, its use has been banned in the United States, and substitute insecticides are being used.

Alcohols and Ethers

If a hydrogen atom on a hydrocarbon is replaced by a functional group consisting of covalently bonded oxygen and hydrogen (—OH), an **alcohol** is formed. The systematic name of each alcohol is formed from the name of the parent hydrocarbon by changing the ending to -*ol*. For example, the compound CH_3CH_2OH is ethan*ol*. An older nomenclature system used the name of the radical (the name of the parent hydrocarbon with the -*yl* ending) plus the word *alcohol*. Thus, ethanol is also known as ethyl alcohol and, since it is the most familiar alcohol, even simply as alcohol. The formulas of organic compounds can be written in the reverse direction, so ethanol can be represented by $HOCH_2CH_3$, as well as by CH_3CH_2OH.

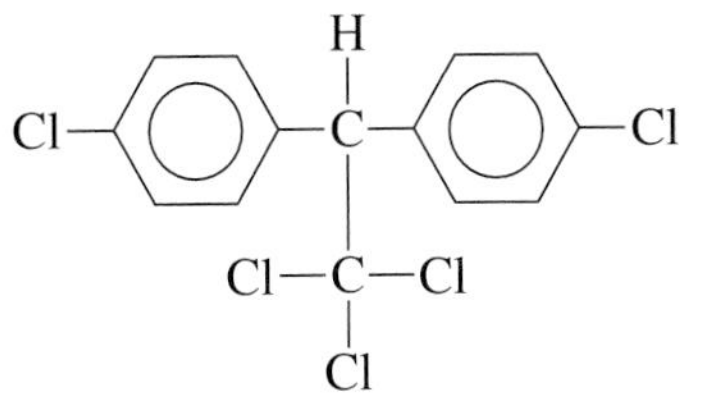

The reaction of ethane, CH_3CH_3, to produce ethanol is not easy to accomplish directly; the previously mentioned replacement of a hydrogen is a mental

Figure 19.7 Structural Formula for DDT

Caution: The physiological properties of ethanol are different. Even ethanol must be used with extreme caution, but the other alcohols are deadly poisonous if ingested.

replacement. The CH_3CH_2— part of the molecule is the radical, and the —OH part is the functional group. Since the radical is not very reactive and since radicals of similar hydrocarbons have very similar properties, the characteristic properties of an alcohol come from its functional group. For example, methanol, CH_3OH, and 1-propanol, $CH_3CH_2CH_2OH$, have chemical properties very similar to those of ethanol. Moreover, most alcohols behave rather similarly. Thus, we can write general reactions of all alcohols:

1. They react with very active metals, such as sodium, to produce hydrogen and the corresponding organic salt. For example:

$$2\ CH_3CH_2OH(l) + 2\ Na(s) \longrightarrow 2\ CH_3CH_2ONa(s) + H_2(g)$$

2. They react when heated to produce ethers:

$$\underset{\text{Ethanol}}{CH_3CH_2OH(l)} + \underset{\text{Ethanol}}{HOCH_2CH_3(l)} \longrightarrow \underset{\text{Diethyl ether}}{CH_3CH_2OCH_2CH_3(l)} + H_2O(l)$$

3. They react with organic acids, as described later in this section.

Several alcohols have commercial importance. Ethanol is used in intoxicating beverages. A dialcohol, ethylene glycol, $HOCH_2CH_2OH$, is the major ingredient in permanent antifreeze for cars. 2-Propanol, $CH_3CHOHCH_3$, is commonly known as rubbing alcohol. Methanol, CH_3OH, is used as a portable source of heat (Sterno), among many other uses.

Ethers are formed by the reaction of alcohols, as just shown in item 2 of the previous list. If two different alcohol molecules combine, a mixed ether is formed:

$$\underset{\text{Methanol}}{CH_3OH(l)} + \underset{\text{Ethanol}}{HOCH_2CH_3(l)} \longrightarrow \underset{\text{Methyl ethyl ether}}{CH_3OCH_2CH_3(l)} + H_2O(l)$$

Ethers are named using the radical names derived from their parent hydrocarbons. Their names end with the word *ether*. Diethyl ether was used as an anesthetic until compounds with fewer undesirable side effects were found. It is still widely known as ether, just as ethanol is known as alcohol. Ethers and alcohols can be isomers of each other, since both classes have the general molecular formula $C_nH_{2n+2}O$.

■ **EXAMPLE 19.13**

Write structural formulas for three ethers that are isomers of 1-butanol.

Solution

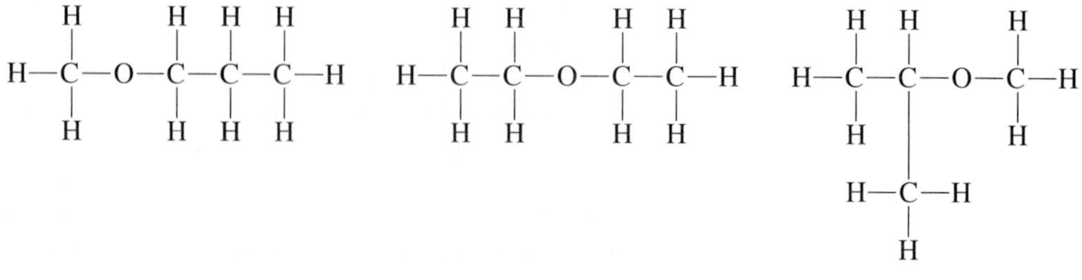

Practice Problem 19.13 Write line formulas for four alcohols that are isomers of diethyl ether. ■

Aldehydes and Ketones

Both aldehydes and ketones have a carbonyl group as the functional group:

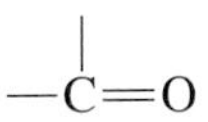

$$-\overset{\displaystyle |}{C}=O$$

Aldehydes have the carbonyl group on the end of a carbon chain, and **ketones** have it somewhere other than the end. Since the end carbon atom also has a hydrogen atom attached, the formal functional group for aldehydes is written:

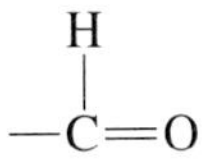

$$\overset{\displaystyle H}{\underset{\displaystyle }{-\overset{\displaystyle |}{C}=O}}$$

The systematic names of **al**dehydes end in *-al;* the systematic names of ket**one**s end in *-one.* The simplest ketone, propanone, known familiarly as acetone, has three carbon atoms. Because of its solvent properties, acetone is used in nail polish remover. The simplest aldehyde, methanal, also known as formaldehyde, is used as a preservative for biological specimens.

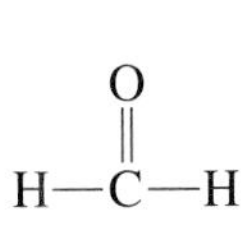

$$\underset{\text{Methanal (formaldehyde)}}{HCHO \quad \text{or} \quad H-\overset{\displaystyle O}{\overset{\displaystyle \|}{C}}-H} \qquad \underset{\text{Propanone (acetone)}}{CH_3COCH_3 \quad \text{or} \quad CH_3-\overset{\displaystyle O}{\overset{\displaystyle \|}{C}}-CH_3}$$

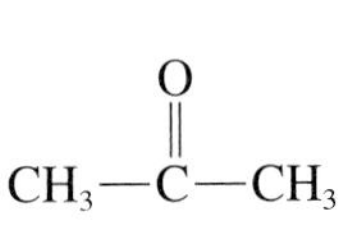

Aldehydes and ketones are produced by the mild oxidation of alcohols. If the —OH group of the alcohol is on the end carbon atom, an aldehyde is produced; if that functional group is on a carbon atom that is connected to two other carbon atoms, a ketone is produced. Aldehydes and ketones can be isomers of each other, since both have the general molecular formula $C_nH_{2n}O$.

■ **EXAMPLE 19.14**

Draw structural formulas for three ketones that are isomers of pentanal. Name these compounds.

Solution

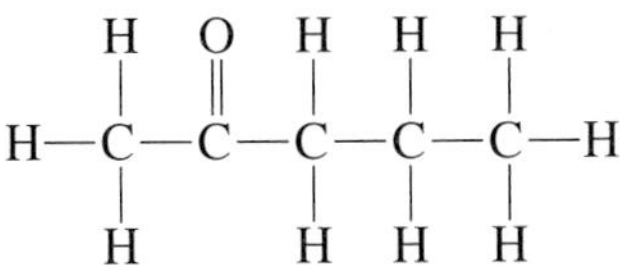

2-Pentanone 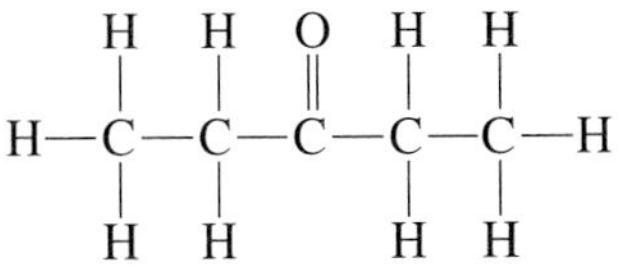   3-Pentanone 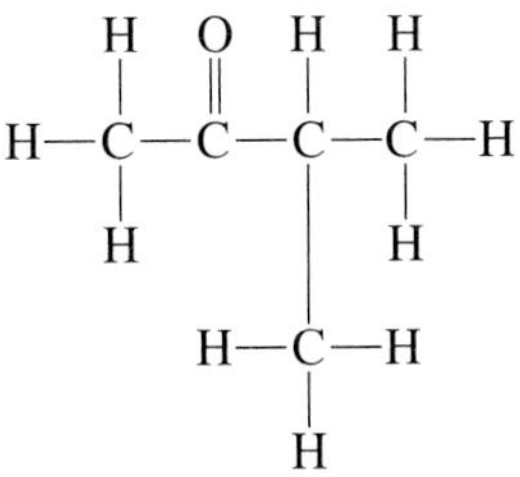   Methylbutanone

Practice Problem 19.14   Why are numbers unnecessary in the last isomer shown in Example 19.14? ■

Organic Acids and Esters

An **organic acid** has the following functional group:

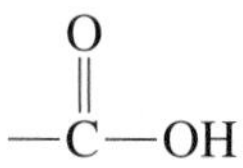

$$\begin{matrix} & O \\ & \| \\ -C & -OH \end{matrix}$$

The systematic name of an organic acid ends in *-oic acid.* The hydrogen is not necessarily written first in formulas for organic acids. While the ionizable hydrogen atom of such an acid looks as if it is part of an —OH group, the —OH group is not a hydroxide ion, and the compound is not a base. The —OH group is covalently bonded to a carbon atom that is also doubly bonded to an oxygen atom. As a rule, organic acids are weak; the hydrogen atom ionizes to a slight extent in water. Perhaps the most familiar organic acid is acetic acid, denoted $HC_2H_3O_2$ earlier in this book; here are its structural and line formulas:

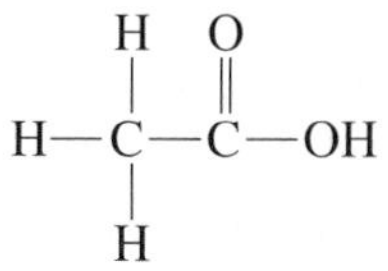

Molecules with covalently bonded —OH groups are not bases.

$$\begin{matrix} H & O \\ | & \| \\ H-C-C & -OH \\ | \\ H \end{matrix} \qquad CH_3COOH$$

Acetic acid is the most abundant acid in vinegar. Its equilibrium reaction with water may be written as follows:

$$CH_3COOH(aq) + H_2O(l) \rightleftharpoons CH_3COO^-(aq) + H_3O^+(aq)$$

Acetic acid reacts with NaOH to produce sodium acetate, CH_3COONa.

Organic acids also react with alcohols to produce **esters** plus water:

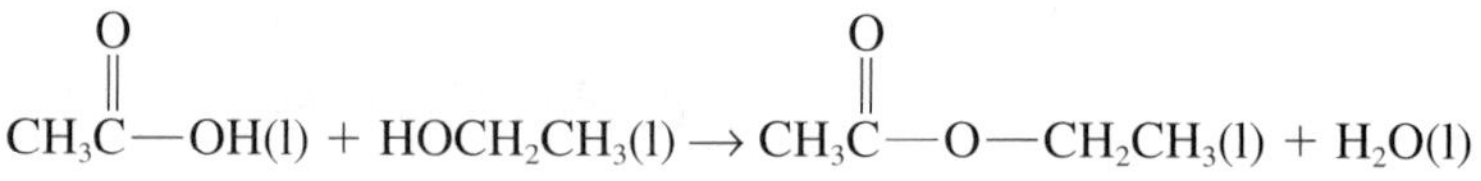

$$\begin{matrix} O & & & O \\ \| & & & \| \\ CH_3C-OH(l) + HOCH_2CH_3(l) \rightarrow CH_3C & -O-CH_2CH_3(l) + H_2O(l) \end{matrix}$$

or

$$CH_3COOH(l) + HOCH_2CH_3(l) \rightarrow CH_3COOCH_2CH_3(l) + H_2O(l)$$
$$\text{Acetic acid} \qquad \text{Ethanol} \qquad \text{Ethyl acetate} \qquad \text{Water}$$

The functional group of the ester is related to that of the acid group, with the hydrogen group replaced by a carbon atom:

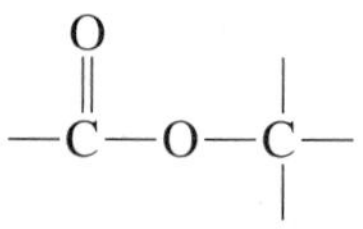

$$\begin{matrix} & O & & \\ & \| & & | \\ -C & -O-C & - \\ & & & | \end{matrix}$$

The name of an ester is a combination of the name of the hydrocarbon radical of the alcohol plus the name of the parent acid with its ending changed to *-ate.* The ester of *ethyl* alcohol plus acetic acid is thus *ethyl acetate.* Ethyl acetate is used to provide an odor to artificial fruits, as a solvent for lacquers and varnishes, in the manufacture of photographic films, and for many other purposes. Many esters occur naturally; their sweet odors are responsible for the pleasant fragrances of flowers and fruits. Esters, especially those containing more than one ester linkage in a molecule, are components of fats (Section 19.5) and many polymers (Section 19.4).

■ **EXAMPLE 19.15**

Write an equation for the reaction of 1-propanol with acetic acid, and name the products.

Solution

$$CH_3COOH(l) + HOCH_2CH_2CH_3(l) \rightarrow CH_3COOCH_2CH_2CH_3(l) + H_2O(l)$$

  Acetic acid   1-Propanol   Propyl acetate   Water

Practice Problem 19.15   Write an equation for the reaction of ethanol with propanoic acid, and name the products. ■

Amines and Amides

Amines may be regarded as organic derivatives of ammonia. One or more of the hydrogen atoms of ammonia may be replaced with a radical to produce an **amine.** If one H atom is replaced, a *primary* amine is produced; if two are replaced, a *secondary* amine is produced; if three are replaced, a *tertiary* amine is produced:

  CH_3NH_2 $(CH_3)_2NH$ $(CH_3)_3N$
  Methyl amine   Dimethyl amine   Trimethyl amine
  A primary amine   A secondary amine   A tertiary amine

Like ammonia, the amines as a class act as weak bases in water solution (Table 18.2):

$$CH_3NH_2(aq) + HCl(aq) \rightleftharpoons CH_3NH_3^+(aq) + Cl^-(aq)$$

When amines are not in water solution, they can react with organic acids to produce another class of organic compounds. **Amides** are formed by the reaction of organic acids with ammonia or amines:

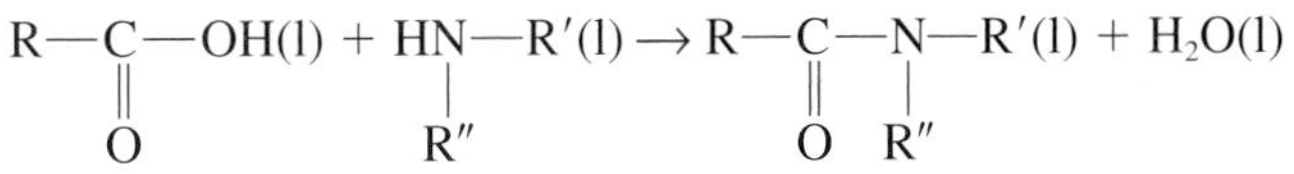

$$R{-}\underset{\underset{O}{\|}}{C}{-}OH(l) + H\underset{\underset{R''}{|}}{N}{-}R'(l) \rightarrow R{-}\underset{\underset{O}{\|}}{C}{-}\underset{\underset{R''}{|}}{N}{-}R'(l) + H_2O(l)$$

  Acid   Amine   Amide

(Here R, R′, and R″ represent radicals that may be the same or different from one another; any or all of them may be hydrogen atoms instead.) The reaction is reversible if the reaction conditions are changed. The functional group that characterizes an amide is

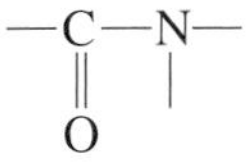

$$-\underset{\underset{O}{\|}}{C}{-}\underset{|}{N}{-}$$

The classes of organic compounds introduced in this section are summarized in Table 19.4.

In addition to the most modern systematic nomenclature system, several older nomenclature systems are still in use in organic chemistry. The simplest compounds are more often known by their older names than their most modern names, especially in commerce. For example, ethanol is often called ethyl alcohol or just alcohol, and methanol is called methyl alcohol. Even older names, grain alcohol and wood alcohol, respectively, are still used for these two

Table 19.4　Classes of Organic Compounds

Class	Functional Group	General Formula*	Ending for Name	Example
Organic halides	—X	RX		CH_3Cl, chloromethane (methyl chloride)
Alcohols	—OH	ROH	-ol	CH_3CH_2OH, ethanol (ethyl alcohol)
Ethers	—O—	ROR′	ether	CH_3OCH_3, dimethyl ether (methyl ether)
Aldehydes	—C=O with H	RCHO	-al	CH_3CHO, ethanal (acetaldehyde)
Ketones	—C=O with —	RC=O with R′	-one	CH_3COCH_3, propanone (acetone)
Organic acids	—C—OH, ‖O	RCOH, ‖O	-oic acid	CH_3C—OH, ethanoic acid (acetic acid), ‖O
Esters	—C—O—, ‖O	RCOR′, ‖O	-ate	CH_3C—OCH_3, methyl ethanoate (methyl acetate), ‖O
Amines	—NH_2	RNH_2	amine	CH_3NH_2, aminomethane (methyl amine)
Amides	—C—N—, ‖O	RC—NR′, ‖O R″	amide	CH_3C—$NHCH_3$, methyl ethanamide (methyl acetamide), ‖O

*The radicals designated R, R′, and R″ can be the same as or different from each other.

compounds. Similarly, methanal is more generally called formaldehyde, methanoic acid is frequently referred to as formic acid, and methylpropane is often called isobutane.

19.4 Polymers

OBJECTIVE

■ to write equations for the formation of macromolecules called polymers, and to describe some of the properties of the compounds they comprise

Polymers are familiar to the general public in the form of plastics and synthetic fibers, such as polyethylene, Teflon, nylon, and polyester. When two organic molecules that each contain one functional group react with each other, they usually form a new molecule whose size is about that of the two reacting molecules combined. However, an organic molecule can have more than one functional group. In such a molecule, each functional group acts more or less independently of the other(s). When such molecules react, a huge molecule can result. For example, consider the reaction of $NH_2(CH_2)_6NH_2$ with $HOCO(CH_2)_4COOH$, each of which has two functional groups. Note that only two molecules of each reactant are pictured in the following equation, representing a large number (n) of each:

$$NH_2(CH_2)_6NH_2 \quad HO-\underset{O}{\overset{\|}{C}}-(CH_2)_4-\underset{O}{\overset{\|}{C}}-OH \quad NH_2(CH_2)_6NH_2 \quad HO-\underset{O}{\overset{\|}{C}}-(CH_2)_4-\underset{O}{\overset{\|}{C}}-OH \quad \rightarrow$$

$$\sim\!\!\sim NH(CH_2)_6NHCO(CH_2)_4CONH(CH_2)_6NHCO(CH_2)_4CO\!\sim\!\!\sim \quad + \quad 2n\ H_2O$$

When the —NH_2 group of the first molecule reacts with the —COOH group of the second molecule, an amide linkage (Section 19.3) is formed, and the resulting molecule is about twice the size of each of the reacting molecules. That molecule still has the two functional groups on its two ends and is capable of further reaction. When the —COOH group at the right reacts with the adjacent —NH_2 group, a still larger molecule is formed. This sort of reaction can continue until the supply of reactant molecules, called **monomers,** is exhausted (or until a ring is formed). The large molecule that is the product of such a reaction is called a **polymer** (from the Greek *poly,* meaning "many," and *mer,* meaning "parts"). A polymer can have a molecular mass of hundreds of thousands of atomic mass units (amu) or even more. The polymer formed in the preceding reaction is called nylon.

A similar reaction takes place when an amino acid (an organic molecule with both an amine and an acid group) reacts with itself:

$$NH_2CH_2COOH \quad NH_2CH_2COOH \quad NH_2CH_2COOH \quad NH_2CH_2COOH \quad \rightarrow$$

$$\sim\!\!\sim NHCH_2CO\!-\!NHCH_2CO\!-\!NHCH_2CO\!-\!NHCH_2CO\!\sim\!\!\sim \quad + \quad n\,H_2O$$

A **protein** is a product of the polymerization of amino acids. Twenty-one different natural amino acids, each with a different R group in the general formula, are components of human proteins.

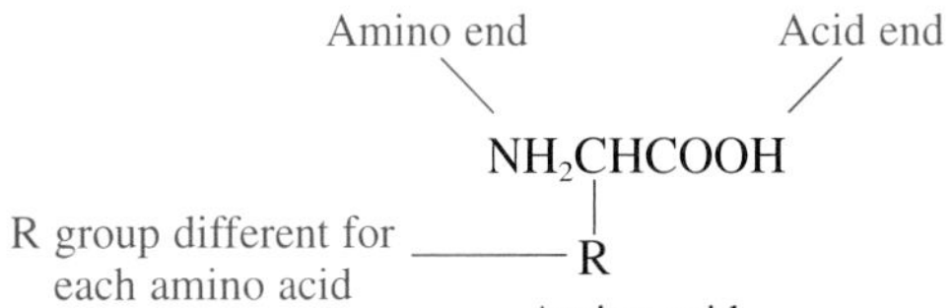

A polymerization reaction can also take place between molecules containing two organic acid groups and molecules containing two alcohol groups; a polyester, widely used in clothing manufacture, is formed.

The reactions described so far are examples of condensation polymerization. In such reactions, small molecules (in these cases, water molecules) are eliminated from the reacting functional groups. Another type of polymerization is addition polymerization, illustrated by the polymerization of ethylene:

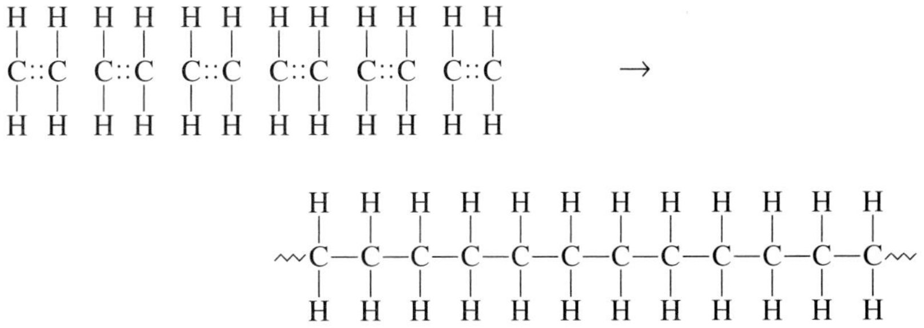

In this polymerization process, no small molecules are produced. The monomer molecules add together by shifting electrons from their double bonds to form new carbon-carbon single bonds.

Table 19.5   Commercial Polymers Related to Ethylene

Name and Formula of Monomer	Polymer	Use
Ethylene $CH_2{=}CH_2$	Polyethylene	Wrapping film
Propylene $CH_3CH{=}CH_2$	Polypropylene	Wrapping film
Vinyl chloride $CH_2{=}CHCl$	Polyvinylchloride (PVC)	Raincoats, bottles
Styrene $C_6H_5CH{=}CH_2$	Polystyrene	Molded plastic insulation
Tetrafluoroethylene $CF_2{=}CF_2$	Teflon	Chemical- and heat-resistant plastics

■ **EXAMPLE 19.16**

Show how shifting the electrons of three C_2H_4 molecules causes reaction leading to polymerization.

Solution

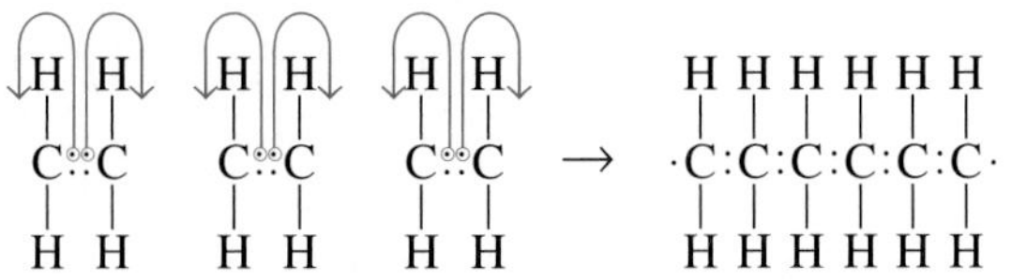

This process continues until a polymer is formed. ■

Several commercial polymers are related to polyethylene (Table 19.5). For example, fluorinated polymers, such as Teflon, are familiar commercial products. Having fluorine atoms in place of all the hydrogen atoms makes the polymer non-flammable and much more resistant to oxidation and thermal decomposition.

19.5 Foods

OBJECTIVE

■ to apply the principles of organic chemistry to some of the foods we eat

Fats

Fats are triesters of the trialcohol glycerine:

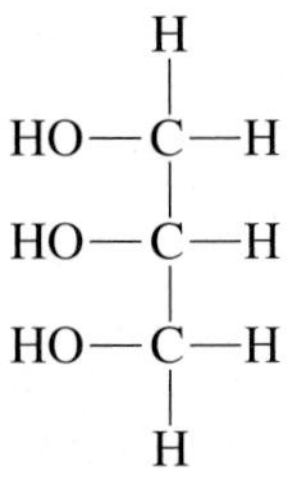

Glycerine, or glycerol
A trialcohol

The radicals of acid molecules that react with glycerine to make animal fats have long chains of carbon atoms. Perhaps most typical of these so-called fatty acids is stearic acid, $CH_3(CH_2)_{16}COOH$. The reaction of three stearic acid molecules with a molecule of glycerine produces a fat:

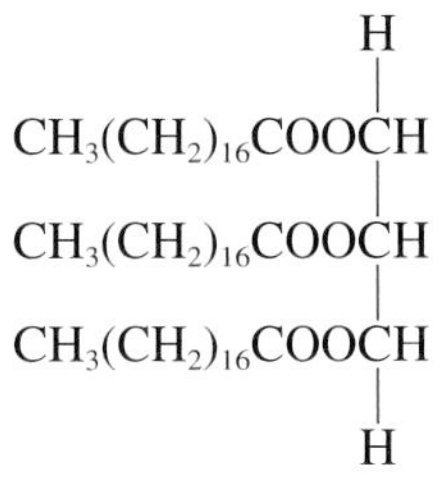

$$
\begin{array}{c}
H \\
| \\
CH_3(CH_2)_{16}COOCH \\
| \\
CH_3(CH_2)_{16}COOCH \\
| \\
CH_3(CH_2)_{16}COOCH \\
| \\
H
\end{array}
$$

An animal fat
A triester

This fat can be broken down into glycerine and the sodium salt of the acid (sodium stearate) by treating it with NaOH.

Sodium stearate is a **soap.** It has a long-chain hydrocarbon-like end, which dissolves well in greases and oils, and it has an ionic end, which dissolves well in water:

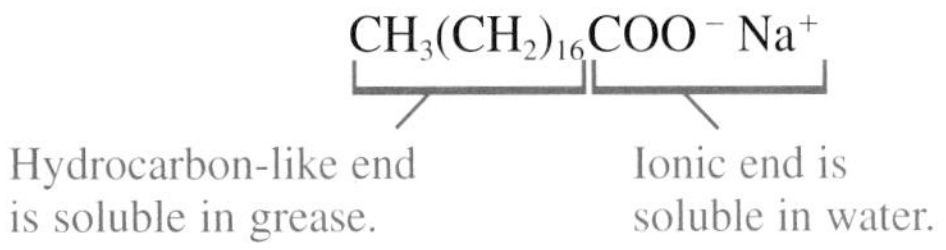

$$CH_3(CH_2)_{16}COO^-\,Na^+$$

Hydrocarbon-like end
is soluble in grease.

Ionic end is
soluble in water.

Such a molecule can cause particles of grease or oil to mix somewhat with water and to be washed from a dirty article. Synthetic detergents are similar to soaps in that they have an ionic end and a large hydrocarbon-like end.

If triesters are made with unsaturated fatty acids—ones with double bonds in their hydrocarbon-like parts—oils are produced. If only single bonds are in the fatty acids, fats are produced. Since consumers seem to prefer solid fats (such as butter) to liquid oils, some food manufacturers hydrogenate the unsaturated oils to reduce the number of double bonds in the molecules and produce a more solid product:

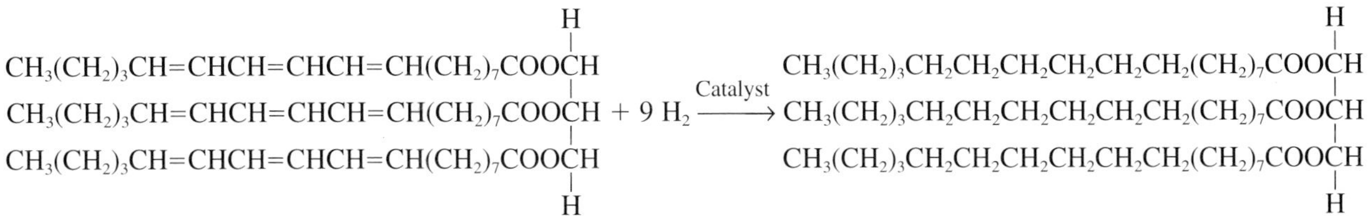

$$
\begin{array}{c}
H \\
| \\
CH_3(CH_2)_3CH{=}CHCH{=}CHCH{=}CH(CH_2)_7COOCH \\
| \\
CH_3(CH_2)_3CH{=}CHCH{=}CHCH{=}CH(CH_2)_7COOCH + 9\,H_2 \\
| \\
CH_3(CH_2)_3CH{=}CHCH{=}CHCH{=}CH(CH_2)_7COOCH \\
| \\
H
\end{array}
\quad\xrightarrow{\text{Catalyst}}\quad
\begin{array}{c}
H \\
| \\
CH_3(CH_2)_3CH_2CH_2CH_2CH_2CH_2CH_2(CH_2)_7COOCH \\
| \\
CH_3(CH_2)_3CH_2CH_2CH_2CH_2CH_2CH_2(CH_2)_7COOCH \\
| \\
CH_3(CH_2)_3CH_2CH_2CH_2CH_2CH_2CH_2(CH_2)_7COOCH \\
| \\
H
\end{array}
$$

Unsaturated

Saturated

Unsaturated fats seem to be more healthful than saturated ones, especially in reducing the buildup of deposits in the arteries. Do not expect to reap the benefits of unsaturated fats, however, if the product you eat is fully hydrogenated.

Carbohydrates

Sugars and starches are two classes of **carbohydrates,** compounds containing carbon, hydrogen, and oxygen in which the hydrogen-to-oxygen ratio is $2:1$. Despite the *hydrate* in the name, a carbohydrate molecule contains no water. **Sugars** are compounds whose molecules have many alcohol groups, plus an aldehyde or ketone group.

Glucose and fructose, two simple sugars, have the following structures:

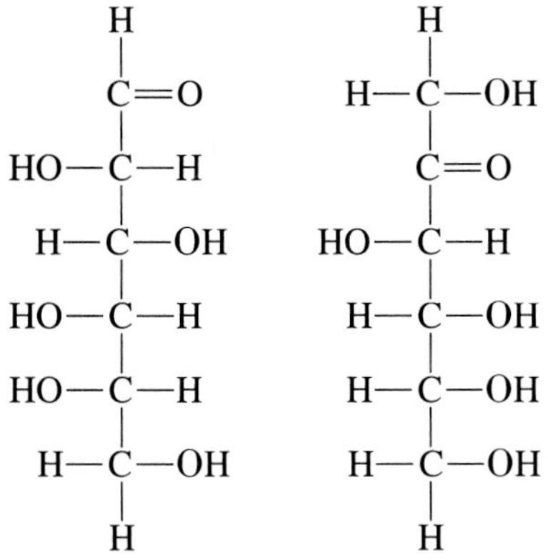

Glucose Fructose

These sugars are highly water-soluble because of the great similarity between their —OH groups and water molecules.

Molecules of two of these simple sugars—also called **monosaccharides**—can react by eliminating one water molecule to produce a double molecule. Sucrose, ordinary table sugar, is an example of such a double molecule, formed from a glucose molecule and a fructose molecule; it is a **disaccharide.** Its molecular formula is $C_{12}H_{22}O_{11}$. The two simple sugars that make it up both have the formula $C_6H_{12}O_6$, and twice that formula yields $C_{12}H_{24}O_{12}$. Elimination of one water molecule, H_2O, yields the correct molecular formula for sucrose (Figure 19.8). Starches and cellulose are polymers of monosaccharides linked together. They are **polysaccharides.**

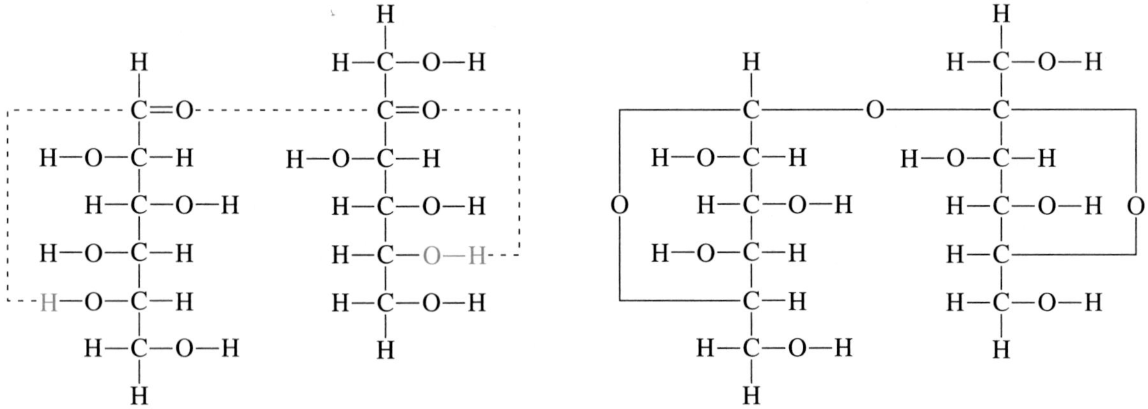

Glucose Fructose Glucose unit Fructose unit

(a) Sucrose

(b)

Figure 19.8 Relationship of Glucose and Fructose to Sucrose

(a) In the molecules of glucose and fructose, the two hydrogen atoms and the oxygen atom that will be eliminated as water are shown in color. The dashed lines represent the bonds that will be formed. (b) In sucrose, the subunits actually have a cyclic shape; the bonds are not as long as shown here and do not bend.

■ ■ ■ ■ ■ ■ SUMMARY ■ ■ ■ ■ ■ ■

Organic compounds (except urea and thiourea) are compounds containing at least one carbon-carbon or carbon-hydrogen bond. Hydrocarbons are compounds containing only carbon and hydrogen. The hydrocarbons to learn first are the alkanes, which have single bonds only. Their systematic names end in *-ane,* and their general molecular formula is C_nH_{2n+2}. The systematic names of most other organic compounds are based on the names of the alkanes. The alkanes are relatively inert, but they do react with oxygen to form water and carbon monoxide or carbon dioxide, or with a halogen to form a hydrogen halide and the halogenated alkane.

The unsaturated hydrocarbons include the alkenes, which contain one double bond per molecule, and the alkynes, which contain one triple bond per molecule. Their systematic names begin as the names of the corresponding alkanes do, but they end with *-ene* or *-yne,* respectively. The location of the multiple bond may have to be specified in the name by including the address of the multiple-bonded carbon atom that is closer to the end of the chain. The alkenes and alkynes are more active than the alkanes; for example, they react with halogens to form halogenated hydrocarbons under less severe conditions than are required for the reactions of alkanes with halogens.

Aromatic hydrocarbons are compounds containing a benzene ring. The ring structure and special bonding in benzene give the aromatic hydrocarbons added stability. (Section 19.1)

Molecules with the same molecular formula but different structures are called isomers of each other. They are different compounds, with different properties. Structural isomers exist because of different points of attachment of groups on a chain of carbon atoms, because of different locations of multiple bonds in the chain, or because of other differences in arrangement. Determining whether two structures represent the same molecule or are isomers of each other may seem difficult at first, because two-dimensional representations do not accurately portray the three-dimensional molecules. (Section 19.2)

In addition to the hydrocarbons are nine other important classes of organic compounds for you to learn: organic halides, alcohols, ethers, aldehydes, ketones, organic acids, esters, amines, and amides. The hydrocarbon-like part of such molecules is called the radical; the other part is called the functional group. The names of radicals are derived from the names of the corresponding alkanes, with the ending changed to *-yl.* The functional groups are named as shown in Table 19.4; all of those characteristic functional groups and their names must be learned. Isomerism exists within all classes of organic compounds and between certain classes as well. For example, most alcohols have isomers among the ethers. (Section 19.3)

Each functional group in any organic molecule can react more or less independently of any others in the molecule. Therefore, many molecules with two or more functional groups can react in sequences to form very large molecules, called polymers. Nylon, polyethylene, and protein are familiar examples of polymers. (Section 19.4)

The foods we consume are mainly organic compounds. Proteins are polymers of amino acids. Fats are triesters of glycerine, a trialcohol. When fats are treated with a strong base, soaps and glycerine result. Unsaturated fats contain double bonds in the hydrocarbon-like part of the ester; saturated fats have only single bonds there. Carbohydrates are compounds of carbon, hydrogen, and oxygen, in which hydrogen and oxygen generally occur in a $2:1$ mole ratio of atoms. Despite the name, water molecules are not present in carbohydrates. Molecules of the simplest carbohydrates contain alcohol and either aldehyde or ketone functional groups. Carbohydrates include sugars, starches, and cellulose; the last two are polymers of simpler carbohydrates. (Section 19.5)

Items for Special Attention

■ Hydrocarbons are compounds containing only carbon and hydrogen. Carbohydrates are compounds of carbon, hydrogen, and oxygen.

■ The formulas $CH_3CH=CH_2$ and $CH_2=CHCH_3$ both represent the same compound, propene. They are simply written in the opposite order.

■ Hydrogen is not necessarily written first in the formula for an organic acid, nor is an organic compound whose formula has hydrogen written first necessarily an acid.

■ In a line formula, an oxygen atom doubly bonded to a carbon atom is usually written to the right of the carbon atom.

Self-Tutorial Problems

19.1 How many carbon atoms are in (a) pentane, (b) 1-pentene, (c) 1-pentyne, and (d) a pentyl radical?

19.2 Write the line formula for each of the following compounds in the reverse order.

(a) CH_3CH_2OH　　　(b) CH_3CHO

(c) CH_3COOH　　　(d) CH_3NH_2

19.3 We can picture methanol, CH_3OH, as a water molecule with one hydrogen atom replaced by a radical (CH_3—). How can each of the following be similarly described?

(a) CH_3NH_2　　　(b) CH_3OCH_3

(c) CH_3NHCH_3　　　(d) $ClCH_2CH_3$

19.4 How many hydrogen atoms are present in an alkane with 40 carbon atoms?

19.5 Explain why no location numbers are needed in the names (a) propanone and (b) propanal.

19.6 What is the nonorganic product of the reaction of (a) an alcohol with another alcohol, (b) an alcohol with an organic acid, and (c) an organic acid with an amine?

19.7 In which classes of Table 19.4 must the radical R be *different* from a hydrogen atom?

■ ■ ■ **PROBLEMS** ■ ■ ■

19.1 Hydrocarbons

19.8 Draw a structural formula for: (a) $C(CH_3)_4$ and (b) $CH_3(CH_2)_5CH_3$.

19.9 How many carbon atoms are in (a) 2-methylheptane, (b) 3-ethylheptane, and (c) 2,3-dimethylheptane?

19.10 How many hydrogen atoms are in each of the compounds in Problem 19.9?

19.11 Identify each of the following as an alkane, alkene, or alkyne:

(a) C_2H_4　　　(b) C_2H_2　　　(c) C_2H_6

19.12 How many carbon atoms are in (a) 2,4-dimethylhexane, (b) 3-ethyl-4-propyl-octane, and (c) 2,2,3,4-tetramethyl-hexane?

19.13 How many hydrogen atoms are in each of the compounds in Problem 19.12?

19.14 Identify each of the following as an alkane, alkene, or alkyne:

(a) $C_{20}H_{40}$　　　(b) $C_{30}H_{58}$　　　(c) $C_{40}H_{82}$

19.15 Write a line formula for each of the following compounds:

(a)

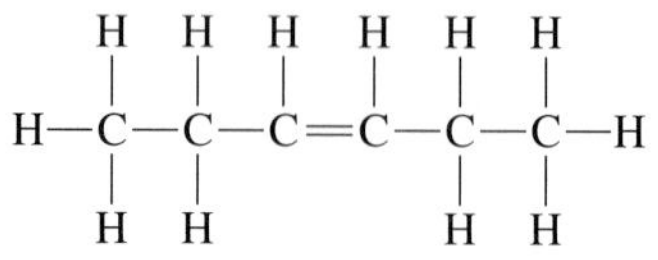

(b)

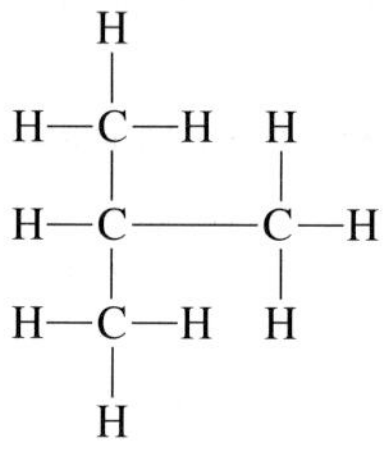

(c)

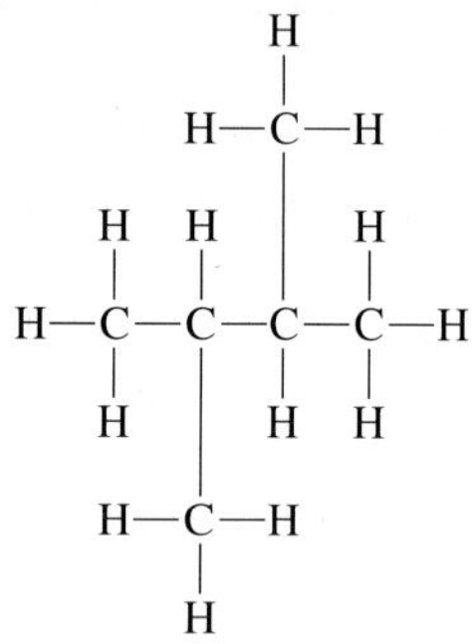

(d)

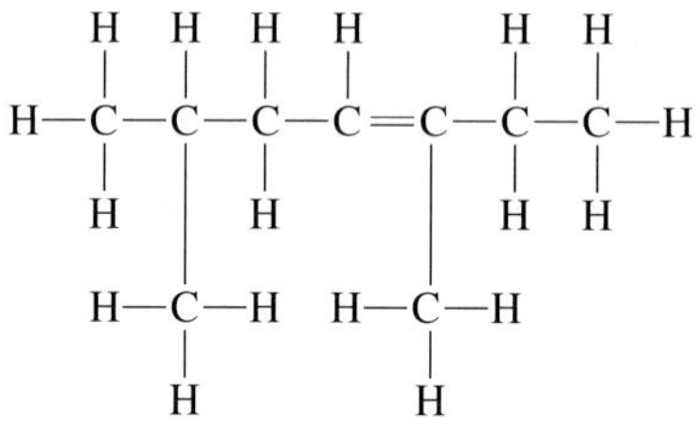

19.16 Write a line formula for each of the following compounds:

(a)

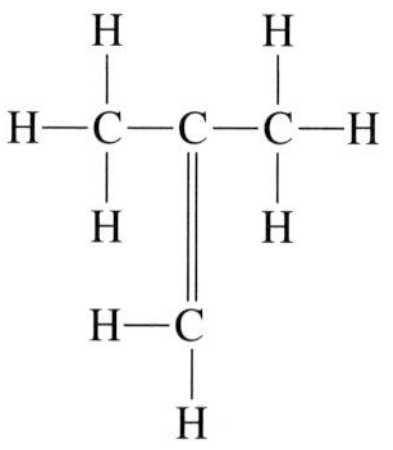

(b)

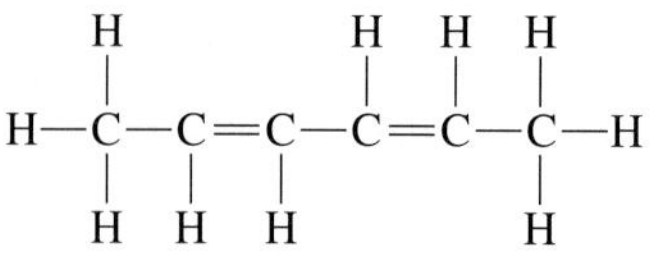

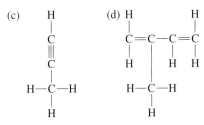

19.2 Isomerism

19.17 Draw structural formulas for all isomers of C_6H_{14}. Name each one.

19.18 Draw structural formulas for five structural isomers of C_5H_{10}. Name each one.

19.19 Write a formula for (a) 2-methylhexane and (b) 3-methylhexane. (c) Explain why 4-methylhexane does not exist.

19.20 Write a formula for (a) 2-methyl-1-pentene, (b) 3-methyl-1-pentene, and (c) 4-methyl-1-pentene. (d) Explain why 4-ethyl-1-pentene does not exist.

19.21 Draw structural formulas for (a) all isomers of C_4H_8 and (b) all unbranched isomers of C_8H_{16}.

19.22 Draw structural formulas for (a) all four isomers of C_4H_6 and (b) all unbranched isomers of C_5H_8.

19.23 Write a structural formula for the isomer of C_7H_{16} that has an ethyl branch.

19.3 Some Other Classes of Organic Compounds

19.24 Draw structural formulas for all isomers of C_3H_7Br.

19.25 Draw structural formulas for all isomers of C_3H_8O.

19.26 Write the line formula of the simplest member of each of the following classes:

(a) Alkyne   (b) Alkene

(c) Alkane   (d) Organic acid

(e) Alkyl chloride   (f) Alcohol   (g) Ketone

19.27 Name the compounds you wrote formulas for in Problem 19.26.

19.28 Give the systematic name for (a) ethyl alcohol, (b) formaldehyde, (c) acetone, and (d) formic acid.

19.29 Identify the class of each of the following compounds:

(a) CH_3NH_2 (b) CH_3OH (c) HCHO

(d) HCOOH   (e) CH_3OCH_3 (f) CH_3COCH_3

19.30 Write a structural formula for (a) 1-bromopentane and (b) 2-bromopentane. (c) Explain why there is no compound named 5-bromopentane.

19.31 Give the common name for each of the following:

(a) Methanol   (b) Methanal   (c) Methanoic acid

19.32 Write the structural formula for an isomer of each of the following:

(a) CH_3OCH_3 (b) CH_3NHCH_3

(c) $HCOOCH_3$ (d) CH_3COCH_3

19.33 Identify the class of compound of both isomers for each part of Problem 19.32.

19.34 Name both compounds in each part of Problem 19.32.

19.35 Write the structural formula for one isomer of each of the following. The isomer should be in a different class of organic compound.

(a) $CH_3CH_2CH_2OCH_3$ (b) $CH_3CHOHCH_3$

(c) $CH_3COOCH_2CH_3$ (d) $CH_3CH_2COCH_2CH_3$

19.36 Identify the class of compound of both isomers for each part of Problem 19.35.

19.37 Name both compounds in each part of Problem 19.35.

19.38 Write the line formula of an isomer of each of the following:

(a) $CH_3CH_2OCH_2CH_3$ (b) $CH_3N(CH_3)_2$

(c) $CH_3CONHCH_3$ (d) $(CH_3)_2CHCOOH$

19.39 What is the class of the compound whose line formula is given in each part of Problem 19.38?

19.40 What is the class of each isomer whose formula you wrote as an answer in Problem 19.38?

19.41 For which pair of groups discussed in the subsections of Section 19.3 is isomerism between a member of one class and a member of the other impossible?

19.42 Name the compounds whose line formulas are given in Problem 19.38.

19.43 Name the compounds whose line formulas you wrote as answers in Problem 19.38.

19.44 Draw structural formulas for all possible isomers of $C_5H_{11}Br$.

19.45 Give the systematic name for (a) alcohol, (b) acetic acid, (c) acetone, (d) methyl chloride, and (e) ether.

19.4 Polymers

19.46 A styrene molecule can be considered to be an ethylene molecule with one hydrogen replaced by a phenyl radical (C_6H_5—). Using R for C_6H_5, draw a representation of polystyrene like that of polyethylene shown in Section 19.4.

19.47 A propylene molecule can be considered to be an ethylene molecule with one hydrogen replaced by a methyl radical. Using R for CH_3, draw a representation of polypropylene like that of polyethylene shown in Section 19.4.

19.48 Do you expect polymers to be gaseous at room temperature and atmospheric pressure? Explain.

19.49 Name all the plastic materials that existed when your great-grandparents were your age.

19.5 Foods

19.50 Explain the difference between an amino acid and an amide.

19.51 Draw a structural formula for each of the following:

(a) The simplest amino acid

(b) A soap

(c) Glycerine (glycerol)

19.52 What elements are contained in each of the following?

(a) Proteins

(b) Sugars

(c) Fats

■ ■ ■ **GENERAL PROBLEMS** ■ ■ ■

19.53 (a) Calculate the oxidation number of the carbon atoms in each of the following molecules:

CH_3CH_2OH CH_3OCH_3 CH_3CH_2CHO

CH_3COCH_3 $HCOOCH_3$ CH_3COOH

(b) Deduce a generalization about oxidation numbers and isomerism.

19.54 Calculate the oxidation number of the carbon atom in each of the following:

(a) $HCOOH$ (b) $HCONH_2$

(c) CH_3OH (d) $HCHO$

19.55 Complete the following table by writing the molecular formula for each compound. If no compound exists for a particular place in the table, write "None."

Number of Carbon Atoms	Alkane	Alkene	Alkyne	Alkyl Radical
1	——	——	——	——
2	——	——	——	——
3	——	——	——	——
4	——	——	——	——
5	——	——	——	——

19.56 Which food groups can be considered polymers?

19.57 In dilute aqueous solution, the amino end of an amino acid molecule acts as a base, and the acid end acts as an acid. Write a line formula for an amino acid in neutral solution after the two ends have reacted with each other.

19.58 Write a formula for a portion of a protein formed solely from glycine monomers. Glycine is NH_2CH_2COOH.

19.59 Write a formula for a portion of a protein formed from monomers represented by $NH_2CHRCOOH$.

19.60 Using molecular formulas, write an equation for the combination of glucose and fructose (each $C_6H_{12}O_6$) into sucrose ($C_{12}H_{22}O_{11}$).

19.61 Explain how the following compound can act as a detergent, making grease and water mix:

$$CH_3(CH_2)_nCH_2SO_3^- \ K^+$$

19.62 (a) Deduce the numbering system for the benzene rings in DDT (Figure 19.7) from the systematic name given.

(b) Draw a structural formula for di(2-chlorophenyl) methane.

19.63 TNT (Section 19.1) is more precisely known as 2,4,6-trinitrotoluene. Define the numbering system for the benzene ring of toluene.

19.64 What functional group is common to both nylon and proteins?

19.65 Explain why no addresses (positional numbers) need to be used in the names (a) butanal and (b) butanone.

19.66 Draw a structural formula for (a) 1-propanol and (b) 2-propanol. (c) Explain why there is no 3-propanol.

19.67 Which classes of organic compound do *not* need positional numbers for their functional groups in their names?

19.68 Draw formulas for all structural isomers of the monobromo compound produced by treating a limited quantity of Br_2 with (a) 2-methylpentane and (b) 3-methylpentane. (c) Why are fewer isomers produced for one of these than for the other?

19.69 Would you be willing to pay extra to get the health benefits of a fully hydrogenated unsaturated fat compared to a saturated fat? Explain.

20 Nuclear Reactions

- 20.1 Natural Radioactivity
- 20.2 Half-Life
- 20.3 Nuclear Fission
- 20.4 Nuclear Fusion

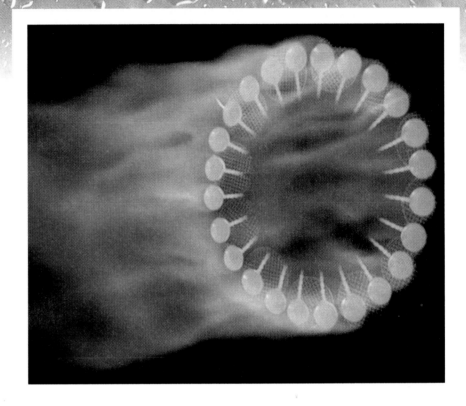

Key Terms (Key terms are defined in the Glossary.)

alpha particle (20.1)
atom smasher (20.3)
atomic bomb (20.3)
beta particle (20.1)
chain reaction (20.3)
control rod (20.3)
critical mass (20.3)
daughter isotope (20.1)
deuteron (20.4)
disintegration (20.1)

electromagnetic radiation (20.1)
event (20.1)
gamma particle (20.1)
Geiger counter (20.1)
half-life (20.2)
hydrogen bomb (20.4)
isotopes (20.1)
nuclear fission (20.3)
nuclear fusion (20.4)
nuclear radiation (20.1)

parent isotope (20.1)
positron (20.3)
radioactive dating (20.2)
radioactive decay (20.1)
radioactive series (20.1)
radioactivity (20.1)
tracer (20.1)
transmutation (20.3)
tritium (20.4)

Symbols

α (alpha particle) (20.1)
β (beta particle) (20.1)
$_{+}^{0}\beta$ (positron) (20.4)

d (deuteron, with symbol d) (20.4)
γ (gamma particle) (20.1)
n (neutron) (20.3)

N (number of nuclei) (20.2)
N_o (original number of nuclei) (20.2)
$t_{1/2}$ (half-life) (20.2)

None of the reactions or processes studied in previous chapters affected the nucleus of an atom. No atom changed from one element to another. This chapter considers the effects of nuclear change. In most cases, such changes cause a change from one element to another. They include the natural radioactivity of certain isotopes (Chapter 3), as well as the artificial nuclear reactions discovered during the twentieth century. Nuclear reactions differ from ordinary chemical reactions in the following ways:

Nuclear Reactions	*Chemical Reactions*
1. Atomic numbers may change.	1. Atomic numbers do not change.
2. Isotopes of an element have different properties.	2. Isotopes of a given element behave almost identically.
3. The total quantity of matter in the reaction changes in a small but significant way; some matter is converted to energy.	3. The total quantity of matter in the reaction does not change significantly.
4. Individual atoms are usually used in calculations.	4. Mole quantities are usually used in calculations.

Natural radioactivity, the spontaneous disintegration of unstable nuclei, is described in Section 20.1. The subatomic particles involved, the balancing of nuclear equations, and a practical use of such reactions are also discussed. Section 20.2 examines half-life, the time it takes for half of any particular sample of a given radioactive isotope to disintegrate spontaneously. Artificially induced nuclear reactions are introduced in Section 20.3. Many such reactions produce two large product nuclei from each target nucleus, a process called nuclear fission. The subatomic products of some of these reactions cause further reactions, producing a chain reaction. Nuclear fusion, the formation of larger nuclei from smaller ones, is discussed in Section 20.4.

20.1 Natural Radioactivity

OBJECTIVE

■ to write equations describing how certain isotopes disintegrate naturally and spontaneously, and to describe how these processes occur

The subatomic particles that are the major components of the atom were introduced in Chapter 3. Protons, neutrons, and electrons have the properties presented in Table 20.1. The atomic number of an atom is the number of protons in its nucleus, and the mass number of an atom is the number of protons plus neutrons in the atom's nucleus. **Isotopes** are atoms having the same number of protons (the same atomic number) and different numbers of neutrons (and therefore different mass numbers).

Table 20.1   Properties of Subatomic Particles

	Charge (*e*)*	Mass (amu)†	Location in the Atom
Proton	1+	1.0073	In the nucleus
Neutron	0	1.0087	In the nucleus
Electron	1−	0.000549	Outside the nucleus

*The charges given are relative charges, based on the charge on the electron, *e*, as the fundamental unit of charge (1 *e* = 1.60×10^{-19} coulomb).
†The masses are given in atomic mass units (amu), described in Section 3.4.

The symbol for an isotope stands for the *nucleus* of that isotope. Recall that the mass number of an isotope is written as a left superscript on the symbol of the element and that the atomic number may appear as a left subscript. For example, carbon-12 may be represented as either ^{12}C or $^{12}_6C$. The symbol of the element tells the atomic number, but for balancing nuclear equations, writing the subscripts explicitly is more convenient.

Some isotopes of certain elements have nuclei that disintegrate (break down) spontaneously, with the emission of one of three types of **nuclear radiation**—alpha particles, beta particles, and gamma particles—as well as a considerable quantity of energy. The properties of these particles are presented in Table 20.2. **Alpha particles** are identical to helium-4 nuclei. **Beta particles** are electrons with very high energies. **Gamma particles** are identical to very high-energy electromagnetic radiation (a form of light, but with much higher energy than visible light). Gamma radiation was originally referred to as a ray, but since electromagnetic radiation is now known to have the characteristics of particles as well as of waves, a gamma ray can also be thought of as a stream of gamma particles (called photons). Similarly, a stream of alpha or beta particles is often called an alpha ray or a beta ray, respectively.

The process involving spontaneous emission of particles from nuclei is called **radioactivity,** or **radioactive decay.** The isotopes that undergo such **disintegration** are said to be radioactive. Each disintegration is called an **event.** Since the nucleus of the atom is undergoing the change, the chemical environment of the atom makes little difference to the process of radioactive decay. That is, uranium metal or uranium in any one of its compounds undergoes nuclear disintegration in the same manner and at the same rate.

Superscripts and subscripts may be used with the Greek letters that represent alpha, beta, and gamma particles. For example, an alpha particle can be represented by the Greek letter alpha (α) or the symbol ^4He. In either case, the subscript can be used or omitted. With the Greek letter, the superscript is also optional. In short, the alpha particle can be represented as α, $^4\alpha$, $^4_2\alpha$, ^4He, or ^4_2He. In this book, we will usually use ^4_2He.

Nuclear radiation must be distinguished from electromagnetic radiation. **Electromagnetic radiation** is light in its various forms, including visible light, ultraviolet light, infrared light, microwaves, X rays, and gamma rays. Radiation emitted from the nuclei of atoms can be dangerous, as can electromagnetic radiation. Except for gamma rays, however, the two forms of radiation are not the same. Gamma rays consist of a stream of high-energy light particles.

During a nuclear disintegration, the total of the charges and the total of the mass numbers of the particles involved do not change, but the quantity of matter does change as some matter is transformed into energy. The isotope undergoing

Table 20.2 Products of Natural Radioactivity

Particle*	Symbol	Charge	Mass Number	Identity
Alpha	α	2+	4	Helium nucleus
Beta	β	1−	0	Electron
Gamma	γ	0	0	Photon of light

*Sometimes, a stream of any of these types of particles is called a ray, as in gamma ray.

decay is called the **parent isotope,** and the isotope produced (along with a small particle from Table 20.2) is called the **daughter isotope.** For example, the natural decay of ^{238}U produces ^{234}Th and an alpha particle. The ^{238}U is the parent isotope, and the ^{234}Th is the daughter isotope. In an equation written to represent this process, the subscripts on the left and right sides of the arrow total to the same number, and the superscripts also have the same totals:

$$\text{Superscripts:} \quad 238 \quad = 234 \quad + 4$$
$$^{238}_{92}U \longrightarrow {}^{234}_{90}Th + {}^{4}_{2}He$$
$$\text{Subscripts:} \quad 92 \quad = 90 \quad + 2$$

The superscripts represent the mass numbers, and the (optional) subscripts represent the atomic numbers or charges. Knowing that the superscripts and subscripts must balance allows us to deduce one species involved in a reaction if all others are given.

■ EXAMPLE 20.1

In addition to ^{234}U, what is the other product of disintegration of ^{234}Pa?

Solution

Since the charges must balance, you must write the subscripts for each isotope. You determine the atomic numbers from the periodic table and start to write an equation:

$$^{234}_{91}Pa \longrightarrow {}^{234}_{92}U + ?$$

You see that the superscript of the unknown product must be 0 for the superscripts to balance:

$$^{234}_{91}Pa \longrightarrow {}^{234}_{92}U + {}^{0}?$$

The subscript of the unknown product must be -1 for the charges to balance:

$$^{234}_{91}Pa \longrightarrow {}^{234}_{92}U + {}^{0}_{-1}?$$

The particle in Table 20.2 that has a single negative charge and a zero mass number is a beta particle (electron), so you can complete the equation:

$$^{234}_{91}Pa \longrightarrow {}^{234}_{92}U + {}^{0}_{-1}\beta$$

Practice Problem 20.1 Determine what isotope is produced along with a beta particle by the decomposition of ^{214}Pb. ■

The emission of a gamma particle does not change the atomic number or the mass number of the parent isotope because the gamma particle has zero charge and zero mass number. For example, the emission of a gamma particle from $^{119}_{50}Sn$ yields a lower-energy form of the *same isotope:*

$$^{119}_{50}Sn \longrightarrow {}^{119}_{50}Sn + {}^{0}_{0}\gamma$$

■ EXAMPLE 20.2

Show that the emission of a gamma particle does not change the atomic number or the mass number of the parent isotope.

Solution

Since the superscript and the subscript on the symbol for the gamma particle are both zero, the superscript and the subscript of the daughter isotope must be the same as those of the parent isotope for the mass numbers and charges to balance. ■

■ EXAMPLE 20.3

Consider the following equation:

$$^{230}_{90}\text{Th} \longrightarrow {}^{226}_{88}\text{Ra} + {}^{4}_{2}\text{He}$$

Does this equation refer to the nuclei of these elements or to atoms of the elements as a whole?

Solution

Nuclear equations are written to describe nuclear changes. However, since the correct number of electrons is available for the products, the equations can also be regarded as describing the reactions of complete atoms, as well as their nuclei.

Practice Problem 20.3   How many electrons are associated with the atoms on each side of the equation in Example 20.3? ■

Radioactive Series

The daughters of most radioactive isotopes with very high atomic numbers will themselves disintegrate, and a whole **radioactive series** takes place. For example, ^{238}U loses eight alpha particles and six beta particles (as well as some gamma particles) as it successively disintegrates into stable ^{206}Pb. The reactions involving the alpha and beta particles are shown in Figure 20.1(a). Gamma particle emissions are not shown in the figure because the emission of a gamma particle does not change the atomic number or the mass number of the isotope (only its energy). A graph of the mass number versus the atomic number for these isotopes is presented in Figure 20.1(b). One of the isotopes in this series can disintegrate in either of two ways.

The mass number changes by 4 when an alpha particle is emitted but does not change at all when a beta or a gamma particle is emitted. Therefore, the mass numbers of all the isotopes in a given series differ from each other by some integral multiple (0, 1, 2, . . .) of 4. Because of this fact, there are four different series. One of them has mass numbers evenly divisible by 4; it is called the $4n$ series. In another series, all of the mass numbers exceed a multiple of 4 by 1; it is called the $4n + 1$ series. Similarly, the other two series are the $4n + 2$ and the $4n + 3$ series. The disintegrations in the $4n$, $4n + 1$, and $4n + 3$ series are shown in Figure 20.2.

■ EXAMPLE 20.4

Without looking at any of the figures, determine the designation of the series that starts with ^{237}Np.

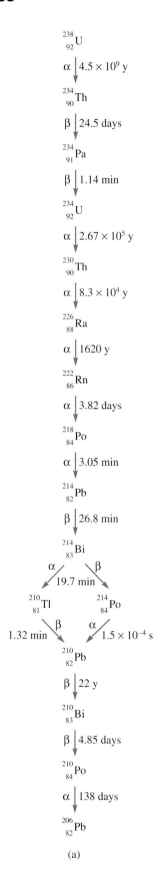

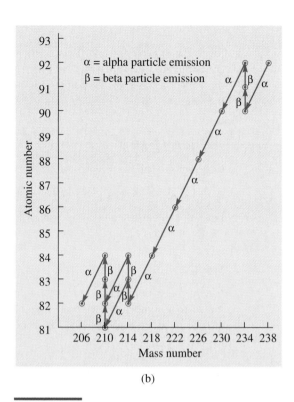

(b)

Figure 20.1 Disintegration Series for Uranium-238

(a) The diagram shows the particle emitted and the half-life (Section 20.2) for each event. (b) Atomic number is plotted versus mass number for each isotope in the series diagrammed in part (a).

Solution

Since ^{237}Np has a mass number that is 1 greater than an even multiple of 4, the series is the $4n + 1$ series:

$$237 = 4(59) + 1$$

Practice Problem 20.4   What series involves ^{232}Th? ■

■ **EXAMPLE 20.5**

Radium was discovered by Marie Curie (1867–1934) in pitchblende, a uranium ore. Radium is produced from a series of disintegrations, starting with ^{238}U and each producing an alpha particle or a beta particle (and possibly a gamma particle). Without looking at any figure or table, but using the mass number of ^{226}Ra, deduce how many alpha particles have been emitted from ^{238}U to produce this isotope of radium.

Solution

The emission of an alpha particle lowers the mass number of the daughter isotope by 4; the emission of a beta particle or a gamma particle produces no loss in mass number. Since the mass number of ^{226}Ra is 12 less than that of ^{238}U, three alpha particles must have been lost.

Figure 20.2 Other
Disintegration Series

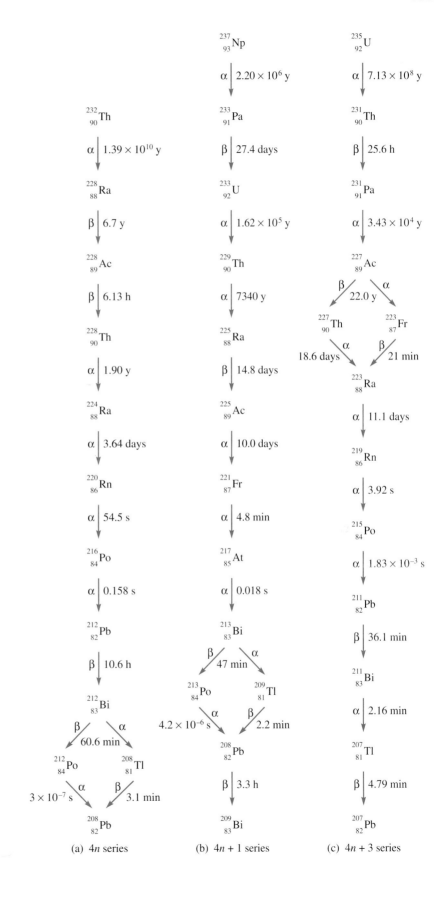

(a) 4n series   (b) 4n + 1 series   (c) 4n + 3 series

The general public has a great deal of fear of nuclear energy and nuclear reactions of any type. The concept of radioactivity is unfamiliar to many people and therefore frightens them. For example, a suburban New York legislator introduced a bill in the county legislature that would have banned transportation of *any amount* of *any radioactive material* through the county. The legislator withdrew the bill when he was informed that a tiny percentage of the carbon in *every human being* is radioactive ^{14}C. If his bill had passed, no one would have been allowed to use the county highways.

Practice Problem 20.5   How many alpha particles are emitted from ^{235}U in its disintegration to ^{207}Pb? ■

■ EXAMPLE 20.6

How many beta particles are emitted in the disintegration of ^{238}U to ^{226}Ra?

Solution

The emission of the three alpha particles (see Example 20.5) lowers the atomic number by 6; however, radium's atomic number is only 4 units lower than uranium's. Therefore, two beta particles must also have been emitted, each event raising the atomic number by 1.

Practice Problem 20.6   How many beta particles are emitted from $^{235}_{92}U$ in its disintegration to $^{207}_{82}Pb$? ■

Tracers

A **Geiger counter** is a device that can detect radioactive disintegrations (Figure 20.3). The high-energy particles emitted from a radioactive sample cause the gas in a Geiger counter's tube to be ionized and to carry an electric current for a short period of time. The "blip" of the Geiger counter signals a radioactive event. Other devices, such as scintillation counters and film badges, are also used to detect and measure disintegrations.

Radioactive isotopes have the same chemical properties as the nonradioactive isotopes of the same element. Since they undergo the same chemical reactions, radioactive atoms are often used as **tracers** to determine what ordinary atoms are

Figure 20.3 Simplified Diagram of a Geiger Counter

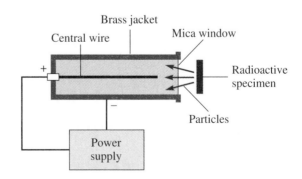

doing. For example, to detect problems in the human thyroid gland, physicians often prescribe iodine that includes a tiny fraction of ^{131}I, a radioactive isotope of iodine. The body should utilize all the iodine in the thyroid gland. With a Geiger counter, the physician can follow the path of the radioactive isotope. If the radioactive iodine is not absorbed by the thyroid, then the regular iodine has not been absorbed either, and the physician has confirmed that a certain problem exists.

20.2 Half-Life

■ to predict the time in which a given fraction of a naturally radioactive sample will disintegrate

Radioactive isotopes have widely different stabilities. They disintegrate in times ranging from fractions of a second to billions of years. The atoms of a given isotope do not disintegrate all at once; they undergo their particle emissions in a pattern that is statistically predictable. While scientists cannot tell when any particular atom of an isotope will disintegrate, they can predict what fraction of the atoms in any given sample of the isotope will disintegrate during a given period of time. Life insurance companies work on the same principle; they cannot predict accurately whether any particular individual will die in a given year, but they can tell what fraction of a particular age group (in a given state of health, etc.) will die within the year. Their statistics are quite accurate (and have to be if the companies are to survive), even though they are based on only a few million people. The statistics for radioactive decay are based on billions of billions of atoms and thus are very precise indeed.

The number of radioactive disintegrations that will occur in a given unit of time in a given sample depends not only on the relative instability of the isotope but also on the number of atoms in the sample. As a sample of an isotope disintegrates, the number of atoms of the isotope gets smaller. As the number of atoms gets smaller, the number of disintegrations per second gets smaller. The time it takes for a sample of an isotope to get to half of its original number of atoms is called the **half-life** of the isotope. The size of the original sample does not affect the length of the half-life. Bigger samples will disintegrate faster, and half of a larger sample will disintegrate in the same time as half of a smaller sample. For example, a 2.0-kg sample of an isotope, which we can designate A, will have twice as many disintegrations per second as a 1.0-kg sample of the same isotope, which we can designate B. In the time it takes 1.0 kg of sample A to disintegrate, only 0.50 kg of sample B disintegrates because it is disintegrating only half as fast:

$$\text{Sample A} \quad \begin{matrix} 2.0 \text{ kg} \\ \downarrow \\ 1.0 \text{ kg} \end{matrix} \quad \} \quad 1 \text{ half-life} \quad \{ \quad \begin{matrix} 1.0 \text{ kg} \\ \downarrow \\ 0.50 \text{ kg} \end{matrix} \quad \text{Sample B}$$

After that first half-life, sample A will contain 1.0 kg of the original isotope (plus other products) and that 1.0 kg will disintegrate only as fast as sample B did *originally.*

■ **EXAMPLE 20.7**

A quiz show gives away 50% of its current jackpot to each contestant who answers a question correctly. Show how much is left in the jackpot after four contestants answer correctly if the jackpot starts at (a) $24,000 and (b) $12,000.

Solution

(a) $24 000 $\xrightarrow[\substack{\text{to first} \\ \text{contestant}}]{\substack{\text{Give} \\ \text{away} \\ \$12\,000}}$ $12 000 $\xrightarrow[\substack{\text{to second} \\ \text{contestant}}]{\substack{\text{Give} \\ \text{away} \\ \$6000}}$ $6000 $\xrightarrow[\substack{\text{to third} \\ \text{contestant}}]{\substack{\text{Give} \\ \text{away} \\ \$3000}}$ $3000 $\xrightarrow[\substack{\text{to fourth} \\ \text{contestant}}]{\substack{\text{Give} \\ \text{away} \\ \$1500}}$ $1500

(b) $12 000 $\xrightarrow[\substack{\text{to first} \\ \text{contestant}}]{\substack{\text{Give} \\ \text{away} \\ \$6000}}$ $6000 $\xrightarrow[\substack{\text{to second} \\ \text{contestant}}]{\substack{\text{Give} \\ \text{away} \\ \$3000}}$ $3000 $\xrightarrow[\substack{\text{to third} \\ \text{contestant}}]{\substack{\text{Give} \\ \text{away} \\ \$1500}}$ $1500 $\xrightarrow[\substack{\text{to fourth} \\ \text{contestant}}]{\substack{\text{Give} \\ \text{away} \\ \$750}}$ $750

You can see that the value after the first contestant in part (a) is the same as the value at the beginning of part (b). At each stage, the jackpot for part (a) and the amount given away are twice as great as for part (b). The same principle works in the case of natural radioactivity. ■

Half-lives of some typical isotopes are listed in Table 20.3.

■ EXAMPLE 20.8

From the data in Table 20.3, calculate the time required for a 1.00-kg sample of ^{140}La to be reduced to 0.250 kg of ^{140}La.

Solution

The half-life of ^{140}La is 40 hours. That means that in 40 hours, half of the 1.00-kg sample will have disintegrated; 0.500 kg of ^{140}La will remain. After another 40 hours, half of *that sample* of ^{140}La—that is, 0.250 kg of ^{140}La—will remain. It takes two half-lives to get to one-quarter of the original sample size, so the time required is 80 hours.

Table 20.3 Half-Lives
of Some Radioactive Isotopes

Isotope	Half-life	Particle Emitted
^{187}Re	7×10^{10} years = 70 billion years	Beta
^{232}Th	1.41×10^{10} years = 14.1 billion years	Alpha
^{238}U	4.51×10^{9} years = 4.51 billion years	Alpha
^{235}U	7.1×10^{8} years = 710 million years	Alpha
^{14}C	5.73×10^{3} years = 5730 years	Beta
^{90}Sr	28.1 years	Beta
^{3}H	12.3 years	Beta
^{228}Th	1.90 years	Alpha
^{131}I	8.0 days	Beta
^{140}La	40 hours	Beta
^{94}Kr	1.4 seconds	Beta
^{216}Po	0.158 second	Alpha

Practice Problem 20.8 From the data in Table 20.3, calculate the time required for three-quarters of a 1.00-kg sample of ^{140}La to disintegrate. What is the difference between this problem and the one in Example 20.8? ■

■ **EXAMPLE 20.9**

(a) If 0.660 g of a certain radioactive isotope disintegrates to 0.0825 g in 10.0 minutes (600 seconds), what is the half-life of the isotope?

(b) How long would it take for 1.54 g of the isotope to disintegrate to 0.385 g?

Solution

(a) The fraction of the original sample remaining is

$$\frac{0.0825 \text{ g}}{0.660 \text{ g}} = 0.125 = \frac{1}{8}$$

The isotope in the sample has disintegrated to one-eighth of its original mass, so one-eighth of its original number of atoms must remain. Three half-lives must have elapsed:

$$1 \underset{1}{\rightarrow} \frac{1}{2} \underset{2}{\rightarrow} \frac{1}{4} \underset{3}{\rightarrow} \frac{1}{8}$$

The 600 seconds represent three half-lives; therefore, one half-life is 200 seconds.

(b)
$$\frac{0.385 \text{ g}}{1.54 \text{ g}} = 0.250 = \frac{1}{4}$$

The isotope in the sample has disintegrated to one-fourth of its original size, which takes two half-lives. The time necessary is thus

$$2(200 \text{ seconds}) = 400 \text{ seconds} \quad ■$$

Note that although the mass of the particular isotope drops to one-half of its original value in one half-life, the total mass of the sample is reduced only slightly. In addition to the mass of the parent isotope that remains, the masses of the daughter isotope and the alpha or beta particles must be included in the total mass. For example, if 1.00 mol of ^{119}In is allowed to disintegrate for 2.1 minutes (one half-life), 0.500 mol of ^{119}In remains:

$$^{119}_{49}\text{In} \rightarrow ^{119}_{50}\text{Sn} + ^{0}_{-1}\beta$$

However, 0.500 mol of ^{119}Sn and 0.500 mol of electrons are formed in the process, and the total mass of the remaining ^{119}In, the ^{119}Sn, and the electrons is only slightly less than the original mass of the ^{119}In.

For times that are not an integral multiple of the half-life, the following equation can be used to calculate disintegration times (t) or half-lives ($t_{1/2}$):

$$\log\left(\frac{N_0}{N}\right) = \left(\frac{0.301}{t_{1/2}}\right)t$$

The ratio N_0/N is the original number of atoms of the isotope (N_0) divided by the number of atoms (N) at time t. That ratio is equal to the mole ratio and also to the mass ratio *of the same isotope,* and so those quantities may be used if they are given in a problem. To calculate a disintegration time on your scientific calculator, given a half-life and the numbers of atoms (or masses) of an isotope at the start and at the end of the period of time, first rearrange the previous equation to

$$t = \left(\frac{t_{1/2}}{0.301}\right) \log \left(\frac{N_\text{o}}{N}\right)$$

Step 1: Enter the original number of atoms (N_o) into your calculator:

Step 2: Divide by the number of atoms (N) at time t and press the equals key, $\boxed{=}$:

$$\frac{N_\text{o}}{N}$$

Step 3: Press the $\boxed{\log}$ key:

$$\log \left(\frac{N_\text{o}}{N}\right)$$

Step 4: Multiply that result by the half-life ($t_{1/2}$):

$$t_{1/2} \log \left(\frac{N_\text{o}}{N}\right)$$

Step 5: Divide by 0.301 and press the equals key:

$$\left(\frac{t_{1/2}}{0.301}\right) \log \left(\frac{N_\text{o}}{N}\right) \quad ■$$

■ EXAMPLE 20.10

Calculate the time it will take for 3.76×10^{23} atoms of an isotope to disintegrate to a point at which 4.92×10^{22} atoms of the isotope remain. The half-life of the isotope is 3.31 years.

Solution

$$\log \left(\frac{N_\text{o}}{N}\right) = \left(\frac{0.301}{t_{1/2}}\right)t$$

$$t = \left(\frac{t_{1/2}}{0.301}\right) \log \left(\frac{N_\text{o}}{N}\right)$$

Steps 1, 2, and 3:

$$\log \left(\frac{3.76 \times 10^{23}}{4.92 \times 10^{22}}\right) = 0.883$$

Steps 4 and 5:

$$t = \frac{(3.31 \text{ years})(0.883)}{0.301} = 9.71 \text{ years}$$

This is a reasonable time. Since the number of atoms of the isotope has been reduced to less than one-fourth of the original number but not to one-eighth, the time elapsed should be between two and three half-lives.

Practice Problem 20.10 Calculate the time it takes for a sample of an isotope to be reduced to one-fourth of its original mass if the half-life is 12.3 seconds. ■

■ ■ ■ ■ ■ ■ ■

Isotopes with a half-life of intermediate length are the most dangerous to human beings. Isotopes of long half-life, such as uranium (4.5 billion years), do not undergo many nuclear reactions in a short period of time. Isotopes of short half-life are not around long enough to do major damage. Although the isotopes of intermediate half-life are dangerous if not handled with great care, even they can be beneficial. For example, ^{60}Co, with a half-life of 5.26 years, has been used in nuclear medicine in a treatment called *radiation therapy* because its radiation kills a greater percentage of cancer cells than of ordinary cells.

■ **EXAMPLE 20.11**

Calculate the half-life of an isotope if a sample was reduced to 30.0% of its original number of atoms in 6.93 days.

Solution

Here, the ratio of atoms, N_o/N, is equal to 100/30.0:

$$\log\left(\frac{100}{30.0}\right) = \left(\frac{0.301}{t_{1/2}}\right)t$$

The left side of this equation can be evaluated first, using a scientific calculator:

$$\log\left(\frac{100}{30.0}\right) = 0.523$$

Substituting this value yields

$$0.523 = \left(\frac{0.301}{t_{1/2}}\right)t$$

Rearranging gives

$$t_{1/2} = \frac{0.301t}{0.523} = \frac{(0.301)(6.93\text{ days})}{0.523} = 3.99\text{ days}$$

This answer is reasonable, since less than two half-lives is required.

Practice Problem 20.11 Calculate the half-life of an isotope if a sample was reduced to 30.0% its original *mass* of the isotope in 6.93 days. ■

■ **EXAMPLE 20.12**

(a) Cobalt-62 has a half-life of 1.62 minutes. Starting with 1.00 mol of ^{62}Co atoms, calculate the number of atoms of ^{62}Co that will remain after 1.00 hour.

(b) Discuss how many atoms would remain after 3.00 hours.

Solution

(a) $\log\left(\dfrac{N_o}{N}\right) = \left(\dfrac{0.301}{t_{1/2}}\right)t = \left(\dfrac{0.301}{1.62\text{ minutes}}\right)(1.00\text{ hour})(60\text{ minutes/hour})$

$= 11.1$

Taking the antilogarithm gives

$$\frac{N_o}{N} = 1 \times 10^{11} = \frac{6.02 \times 10^{23}}{N}$$

Solving for N yields

$$N = 6 \times 10^{12} \text{ atoms}$$

Of every mole of atoms (6.02 thousand billion billion atoms), 6×10^{12} will remain after 1.00 hour.

(b) A similar calculation with a time of 180 minutes shows that 2×10^{-10} atom would remain after 3.00 hours, but a fraction of an atom cannot exist. During the second or third hour, the number of atoms of the isotope becomes so small that the statistics involved break down. The isotope virtually disappears. ■

The slowest step in one of the disintegration series generally determines the time in which half of the original parent is converted to the final, stable product. For example, if we add up the half-lives of all the isotopes in Figure 20.1, we get a sum not significantly different from the half-life of ^{238}U, which is 4.5×10^9 years.

Archaeologists and geologists apply the principles of half-life measurement to determine the age of samples of interest to them. The method they use is called **radioactive dating.** For example, a geologist might be interested in the age of a certain rock formation. If the rock contained uranium-238 but no lead when it solidified from a molten state, the geologist can tell how long ago the rock solidified. Uranium-238 disintegrates in a series of events to lead-206, which is stable. Eight alpha particles and six beta particles are emitted in the process. The atoms of lead-206 currently in the rock were formed by the disintegration of the uranium-238 that was originally present. Thus, N_o is equal to the number of atoms of lead plus the number of atoms of uranium now present, and N is the number of atoms of uranium. These values can be used in the equation we have been discussing to calculate the age of the rock. A representative sample of the rock is analyzed to discover the numbers of atoms.

■ EXAMPLE 20.13

A geologist analyzes a sample of rock and finds that it contains 3.16×10^{16} atoms of ^{206}Pb for every 7.19×10^{16} atoms of ^{238}U. Calculate the age of the rock, assuming that no lead was present when the rock solidified.

Solution

$$N_o = 3.16 \times 10^{16} + 7.19 \times 10^{16} = 10.35 \times 10^{16} \text{ atoms}$$

$$N = 7.19 \times 10^{16} \text{ atoms}$$

$$\log\left(\frac{N_o}{N}\right) = \log\left(\frac{10.35 \times 10^{16}}{7.19 \times 10^{16}}\right) = 0.158$$

$$t = \frac{(0.158)(4.51 \times 10^9 \text{ years})}{(0.301)} = 2.37 \times 10^9 \text{ years}$$

The rock is 2.37 billion years old.

Practice Problem 20.13   Calculate the age of a sample of rock containing 9.07 $\times 10^{18}$ atoms of ^{40}K for every 5.11×10^{17} atoms of ^{40}Ar, its stable daughter isotope. Assume that no argon was in the rock when it solidified and that no argon escaped from the rock since its solidification. The half-life of ^{40}K is 1.3×10^9 years. ■

■ EXAMPLE 20.14

Calculate the age of a sample of rock containing 22.6 g of ^{206}Pb for every 597 g of ^{238}U.

Solution

Although the ratio of masses of *a given isotope* is equal to the ratio of numbers of atoms (since each atom of a given isotope has the same mass at time zero as at time t), the ratio of the masses of different isotopes (^{206}Pb and ^{238}U, for example) is *not* equal to the ratio of atoms. You must change each mass to moles of atoms. To three significant digits, the mass of a mole of each of the isotopes involved in this problem is equal to its mass number (in grams):

$$22.6 \text{ g } {}^{206}\text{Pb}\left(\frac{1 \text{ mol } {}^{206}\text{Pb}}{206 \text{ g } {}^{206}\text{Pb}}\right) = 0.110 \text{ mol } {}^{206}\text{Pb}$$

$$597 \text{ g } {}^{238}\text{U}\left(\frac{1 \text{ mol } {}^{238}\text{U}}{238 \text{ g } {}^{238}\text{U}}\right) = 2.51 \text{ mol } {}^{238}\text{U}$$

The number of moles of ^{238}U originally present, n_o, is the sum of these numbers of moles:

$$n_o = 0.110 \text{ mol} + 2.51 \text{ mol} = 2.62 \text{ mol}$$

$$\log\left(\frac{n_o}{n}\right) = \log\left(\frac{2.62 \text{ mol}}{2.51 \text{ mol}}\right) = 0.019$$

$$t = \frac{(0.019)(4.51 \times 10^9 \text{ years})}{0.301} = 2.8 \times 10^8 \text{ years}$$

Practice Problem 20.14   Calculate the age of a sample of rock containing 7.11 g of ^{40}K and 7.11 g of ^{40}Ar, its daughter isotope. Assume that no argon was in the rock when it solidified and that no argon escaped from the rock since its solidification. The half-life of ^{40}K is 1.3×10^9 years. ■

Carbon-14 is created continuously in the upper atmosphere by bombardment of ^{14}N atoms by cosmic rays:

$$^{14}_{7}\text{N} + {}^{1}_{0}\text{n} \rightarrow {}^{14}_{6}\text{C} + {}^{1}_{1}\text{H}$$

The carbon finds its way into carbon dioxide and eventually into living things through the process of photosynthesis. An archaeologist can date a sample of wood by analyzing the amount of ^{14}C in it because after a tree has died, it no longer takes in ^{14}C in carbon dioxide.

■ EXAMPLE 20.15

Calculate the age of a piece of wood that is disintegrating at a rate of 9.29 counts per minute per gram of carbon. A modern sample of carbon disintegrates at a rate

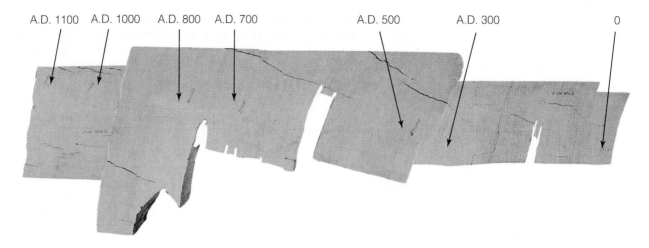

A.D. 1100   A.D. 1000   A.D. 800   A.D. 700   A.D. 500   A.D. 300   0

Figure 20.4 Cross Section of a Tree

The age of a tree can be determined by counting the rings, each of which is formed in a year's growth.

of 15.3 dis/min · gram. Assume that the ratio of ^{14}C to ^{12}C in the atmosphere was the same as it is today.

Solution

$$\log\left(\frac{15.3 \text{ dis/min} \cdot \text{gram}}{9.29 \text{ dis/min} \cdot \text{gram}}\right) = \left(\frac{0.301}{5730 \text{ years}}\right)t$$

$$t = 4120 \text{ years}$$

The ^{14}C in the wood disintegrated through less than one half-life, decreasing from 15.3 dis/min · gram to 9.29 dis/min · gram.

Practice Problem 20.15   At what rate would a 4000-year-old piece of wood disintegrate? ■

Archaeologists can test the assumption that the ratio of ^{14}C to ^{12}C was the same in ancient times as it is now by dating a sample of wood of known age. The age of a particular sample of wood from the cross section of a tree (Figure 20.4) can be determined by counting the annual rings, and then a portion of the wood from an inner ring is tested.

20.3 Nuclear Fission

■ to predict the products of artificially induced reactions in which nuclei break down into smaller nuclei and subatomic particles

Scientists have learned how to make isotopes undergo nuclear reactions. Artificial radioactivity is induced by bombardment of certain nuclei with subatomic particles (or atoms), which are produced either by other nuclear reactions or in machines referred to as **atom smashers.** For example, the first artificially induced nuclear reaction was produced by Ernest Rutherford (1871–1937) in 1919:

$$^{14}N + {}^4He \longrightarrow {}^{17}O + {}^1H$$

Nitrogen-14 was bombarded by alpha particles (helium nuclei), producing oxygen-17 and protons (hydrogen nuclei). In 1934, Irène Joliot-Curie (1897–1956), the daughter of Marie Curie, produced an isotope of phosphorus by bombarding aluminum-27 with alpha particles from polonium:

$$^{27}Al + {}^4He \longrightarrow {}^{30}P + {}^1n$$

Table 20.4   *Small Particles Involved in Artificial Radioactivity*

Particle	Symbol	Charge	Mass Number	Identity
Neutron	n	0	1	Uncharged nuclear particle
Proton	p	1+	1	Hydrogen nucleus
Deuteron	d	1+	2	Nucleus of ^{2}H
Positron	$^0_{+}\beta$	1+	0	Positively charged electron

Several small particles, in addition to those involved in natural radioactivity, are involved in artificial nuclear reactions. Some of these additional particles are listed in Table 20.4. They are used as projectiles to bombard nuclei and/or are produced along with other products of such reactions.

When an isotope of large mass number undergoes a nuclear reaction to produce two isotopes of much lower mass numbers plus some subatomic particles, the isotope is said to have undergone **nuclear fission.**

After such artificial **transmutations** (changes of one element into another), the product nucleus often disintegrates spontaneously in further nuclear reactions. For example, the ^{30}P produced in the reaction described previously disintegrates with a half-life of 2.50 minutes, producing ^{30}Si:

$$^{30}\text{P} \longrightarrow {}^{30}\text{Si} + {}^{0}_{+}\beta$$

The positive particle produced along with the silicon isotope is a **positron,** a particle identical to an electron except that it has a positive charge. When a positron collides with an electron, both are annihilated and a great deal of energy is produced.

Nuclear equations for artificial reactions can be balanced by making the superscripts on the left and right sides sum to the same quantity, and also the subscripts, just as we did for natural radioactive reactions.

Chain Reactions

One of the projectiles most often used in modern times to initiate nuclear reactions is the neutron. To be effective, a neutron does not need as high an energy as an alpha particle or a proton because it is uncharged and can penetrate a nucleus more easily than a positively charged particle can. Some nuclear reactions initiated with neutrons produce more neutrons. For example, the bombardment of ^{235}U with a neutron produces two large nuclei plus two or three new neutrons. The following are examples of the many possible reactions:

$$^{235}\text{U} + {}^{1}\text{n} \longrightarrow {}^{140}\text{Ba} + {}^{94}\text{Kr} + 2\ {}^{1}\text{n}$$

$$^{235}\text{U} + {}^{1}\text{n} \longrightarrow {}^{90}\text{Sr} + {}^{143}\text{Xe} + 3\ {}^{1}\text{n}$$

When a neutron enters a sample of ^{235}U, it may collide with the nucleus of one of the atoms, producing a reaction in which two or three new neutrons are produced. Each of these may react with another nucleus, producing more reactions and an increased number of neutrons (Figure 20.5). Such reactions, all started by a single neutron, can continue until the entire sample of ^{235}U has reacted. The sequence of reactions is called a **chain reaction** and is the source of energy by which nuclear power plants operate. The **atomic bomb,** which should more accurately be called the nuclear bomb, also uses a chain reaction.

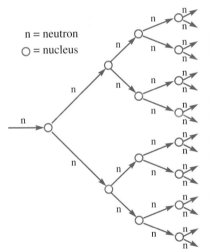

n = neutron
O = nucleus

Figure 20.5 Schematic Representation of a Chain Reaction

In this representation, each collision of a neutron with a nucleus causes a reaction that produces two new neutrons (plus other products and energy). The new neutrons can cause further reactions.

ITEM OF INTEREST

Nuclear energy supplies a significant percentage of the commercial electricity in the United States and an even greater percentage in Europe and Japan, where the cost of conventional fuels is greater. As conventional fuels are used up, nuclear energy may become even more important. Nuclear energy can be very dangerous if used incorrectly but also very beneficial if handled correctly. (The same statement can be made about electricity and automobiles, of course.) Nuclear power plants can cause problems, but conventional plants utilizing fossil fuels can also be troublesome. Conventional plants produce air pollution. Production of fossil fuels causes pollution of a different type; strip-mining of coal and oil-well fires are only two problems associated with the production of fuels for conventional power plants. Moreover, the world's supply of fossil fuels is rapidly being depleted. Solar energy and wind energy are two sources of energy that produce little pollution and pose little danger, but they also have their limitations. The problem of how to dispose of nuclear wastes—the products of nuclear reactions, which take thousands of years to disintegrate—is one that today's scientists and/or future scientists must solve.

■ **EXAMPLE 20.16**

How many neutrons are produced at the end if the following reaction is run through ten cycles and every neutron produced in the first nine cycles causes another such reaction? How many neutrons does the eleventh cycle produce?

$$^{235}U + {}^{1}n \longrightarrow {}^{140}Ba + {}^{94}Kr + 2 \, {}^{1}n$$

Solution

The first reaction produces 2^1 neutrons; those two, in the second cycle, produce $2^2 = 4$ neutrons; and so forth in each cycle, until in the tenth cycle, $2^{10} = 1024$ neutrons are produced. The eleventh cycle produces 2048 neutrons.

Practice Problem 20.16   How many neutrons are produced at the end if the following reaction is run through ten cycles and every neutron produced in the first nine cycles causes another such reaction?

$$^{235}U + {}^{1}n \longrightarrow {}^{90}Sr + {}^{143}Xe + 3 \, {}^{1}n \quad ■$$

To control such a reaction, the mass of uranium may be kept small so that many of the neutrons that are produced escape from the sample without causing other nuclear reactions. The smallest mass in which a chain reaction can continue is called the **critical mass.** Another way to control such a reaction is to use rods of some material that absorbs neutrons without causing a reaction (Figure 20.6). Such **control rods** are part of nuclear reactors in which uranium is used as "fuel" to produce electricity commercially. When the reaction is to be slowed down, the control rods are inserted farther into the sample of fuel. When more energy from the reaction is desired, the rods are withdrawn somewhat. The control rods are operated automatically.

Figure 20.6 Simplified Diagram of a Nuclear Reactor

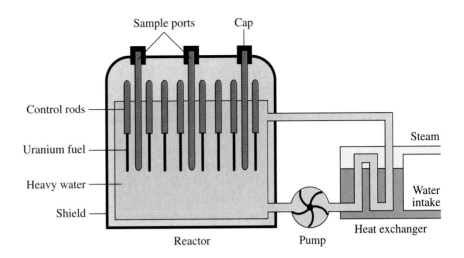

Energetics of Nuclear Reactions

In the process of changing a heavy nucleus into two lighter nuclei plus some small particles, a great deal of energy is emitted. Some matter is converted to energy. The mass of this portion of matter is related to the energy by Einstein's famous equation

$$E = mc^2$$

where m is the decrease in mass of the matter, E is the energy produced, and c is the velocity of light in a vacuum, a constant equal to 3.00×10^8 m/s. The c^2 in the equation is such a large number that conversion of a small quantity of matter produces a large quantity of energy. This is the source of the energy of nuclear bombs and nuclear power plants.

■ **EXAMPLE 20.17**

How many joules of energy (1 J $= 1$ kg $\cdot$ m²/s²) is produced by conversion of 1.00 g of matter to energy?

Solution

$$E = mc^2 = 1.00 \text{ g}\left(\frac{1 \text{ kg}}{1000 \text{ g}}\right)\left(\frac{3.00 \times 10^8 \text{ m}}{\text{s}}\right)^2$$

$$= 9.00 \times 10^{13} \text{ kg} \cdot \text{m}^2/\text{s}^2 = 9.00 \times 10^{13} \text{ J}$$

$$= 9.00 \times 10^{10} \text{ kJ} = 90.0 \text{ billion kJ}$$

Contrast that quantity of energy with the amount that can be produced by burning 1.00 g of carbon, which is 32.8 kJ. ■

20.4 Nuclear Fusion

In contrast to nuclear fission reactions, in which large nuclei are broken into smaller nuclei, **nuclear fusion** involves the combination of small nuclei into larger ones. In nuclear fusion reactions, tremendous quantities of energy are released. The atomic bomb uses a fission reaction as its source of energy; the so-called **hydrogen bomb** uses a fusion reaction as its source of energy. In this fusion reaction, deuterium (^{2}H) nuclei, also called **deuterons,** with symbol **d** are combined into nuclei of **tritium** (^{3}H) and then into nuclei of an isotope of helium. This reaction is less "dirty" than a fission reaction in that the products of fusion are not radioactive. However, fusion reactions require an extremely high temperature to get them started, so a small atomic bomb is used as the energy source to trigger a hydrogen bomb.

The energy of the sun and stars comes from nuclear fusion reactions, which have the overall effect of transforming hydrogen nuclei to alpha particles (helium nuclei). The temperature of the particular star determines the mechanism by which this transformation takes place. The sun, a moderately small star, is thought to be powered by the following sequence of reactions:

$$^1H + {}^1H \longrightarrow {}^2H + {}^0_+\beta$$

$$^1H + {}^2H \longrightarrow {}^3He + {}^0_0\gamma$$

$$^3He + {}^3He \longrightarrow {}^4He + 2\,{}^1_1H$$

Each of these reactions emits energy, as well as the products listed. The positron released in the first reaction can react with an electron to yield even more energy. The net reaction is the conversion of four protons into an alpha particle plus a great deal of energy. Hotter stars convert protons to alpha particles via another series of reactions (see Problem 20.27 at the end of the chapter).

■ EXAMPLE 20.18

When a positron and an electron meet, the matter in both is totally converted to energy. (They annihilate each other.) Assuming that the mass of each of these particles is 9.10×10^{-31} kg, calculate the number of joules of energy released in each annihilation reaction (1 J = 1 kg · m^2/s^2).

Solution

The total mass of a positron and an electron involved in the annihilation reaction is 9.1×10^{-31} kg + 9.1×10^{-31} kg = 1.82×10^{-30} kg. The energy is given by Einstein's equation:

$$E = mc^2 = (1.82 \times 10^{-30} \text{ kg})(3.00 \times 10^8 \text{ m/s})^2$$
$$= 1.54 \times 10^{-31} \text{ J}$$

ITEM OF INTEREST

In 1989, the scientific community was startled by the announcement of two chemists that they had succeeded in causing a fusion reaction to occur near room temperature. This *cold fusion* would have allowed the population of the earth to be supplied with almost limitless energy without the radioactivity associated with the operation of ordinary nuclear power plants. The effect on the scientific and economic communities was profound. Unfortunately, so far, the results reported by the scientists have not been repeated or confirmed, and cold fusion is still a dream.

■ ■ ■ ■ ■ ■ ■ **SUMMARY** ■ ■ ■ ■ ■ ■ ■

Nuclear reactions, unlike ordinary chemical reactions, usually change one element into one or more other elements. The energy produced in nuclear reactions is much greater than that involved in any chemical reaction.

Natural radioactivity (on earth) occurs when radioactive nuclei disintegrate spontaneously, yielding alpha, beta, and gamma particles (rays) plus daughter isotopes almost as massive as the parent isotopes. A nuclear equation can be balanced by making both the total of the mass numbers and the total of the charges the same on the two sides of the equation. When the products of a nuclear reaction are radioactive, a series of disintegrations may occur. The isotopes involved, including the original parent isotope and all daughter isotopes, constitute a radioactive series. Each member of the series differs in mass number from the other members of the series by some integral multiple of 4 (0, 4, 8, . . .). Four such series occur naturally.

Radioactive isotopes react chemically just like the nonradioactive isotopes of the same element. Because of this, physicians can add a little radioactive isotope to a sample of an element, which is then ingested or injected into the body. They can then determine the element's location in the body by detecting the particles that the radioactive isotope emits. The radioactive isotope is called a tracer, and this technique is used extensively in medicine and other fields. (Section 20.1)

All the atoms of a radioactive isotope in a sample do not disintegrate at once, but in accordance with the laws of statistics. The rate of disintegration is proportional to the number of atoms of the radioactive isotope present. The time it takes for half of a given number of atoms of the isotope to disintegrate spontaneously is called the half-life of the isotope. The half-life does *not* depend on the sample size. The following equation, which governs radioactive disintegration, can be used to calculate the half-life from data on the numbers of atoms or to calculate the number of atoms remaining from knowledge of the half-life:

$$\log\left(\frac{N_\text{o}}{N}\right) = \frac{(0.301)t}{t_{1/2}}$$

Here, N_o is the original number of atoms of the isotope, N is the number at time t, and $t_{1/2}$ is the half-life. Archaeological and geological samples may be dated by half-life measurement. (Section 20.2)

Bombardment of certain nuclei with small particles, such as alpha particles or neutrons, can lead to artificial nuclear reactions. The splitting of heavy atoms is one such process, called nuclear fission. Two fairly massive products plus some small particles are apt to result from splitting one large nucleus with a projectile particle. For example:

$$^{235}\text{U} + {}^{1}\text{n} \longrightarrow {}^{140}\text{Ba} + {}^{94}\text{Kr} + 2\,{}^{1}\text{n}$$

Chain reactions result when more neutrons are produced by a nuclear reaction than are used up in one step. For example, for every neutron that causes a nuclear reaction of ^{235}U, two or three new neutrons are produced. If each of these product neutrons, in turn, causes another nuclear reaction, many more neutrons will be produced, creating a chain reaction. If some of the product neutrons escape from the sample or are absorbed by other nuclei, the overall reaction can be controlled. The operation of commercial nuclear reactors is based on this principle. In nuclear reactions, a certain quantity of matter is converted to energy. The mass of the matter converted is related to the energy produced by Einstein's equation, $E = mc^2$. (Section 20.3)

In contrast to fission reactions, some other nuclear reactions cause very small nuclei to fuse into larger ones. An example of such nuclear fusion is one of the reactions thought to power the sun:

$$^{1}\text{H} + {}^{1}\text{H} \longrightarrow {}^{2}\text{H} + {}^{0}_{+}\beta$$

(Section 20.4)

Items for Special Attention

■ In the statement of a problem involving a half-life, be careful to distinguish between the number of atoms *remaining* and the number of atoms that *have disintegrated*.

■ Recall that the mass number of an isotope is generally not given in the periodic table.

■ The m in Einstein's equation $E = mc^2$ is the *change in mass* of the matter undergoing a nuclear reaction, not the total mass.

Self-Tutorial Problems

20.1 What is the difference in α, $^4\alpha$, $^4_2\alpha$, and $^4_2$He?

20.2 What is the difference between the symbols Co and ^{60}Co?

20.3 In a nuclear equation, the subscripts are often omitted, but the superscripts always appear. If you want to write a nuclear equation, where can you look to find the subscripts? Why *can't* you look there for the superscripts?

20.4 If half of a radioactive sample disintegrates in 10.0 minutes, why doesn't it all disintegrate in 20.0 minutes?

20.5 Substitute the correct symbol for X for each of the following particles (without looking at Tables 20.1, 20.2, and 20.4 if possible):

(a) $^0_0$X   (b) $^2_1$X   (c) $^3_1$X   (d) $^0_{-1}$X

(e) $^0_{+1}$X   (f) $^4_2$X   (g) $^1_0$X   (h) $^1_1$X

20.6 In the equation that allows a disintegration time to be calculated from a half-life, what do the symbols N_0 and N stand for? What do t and $t_{1/2}$ stand for?

20.7 State whether each of the following nuclear reactions is spontaneous, a fission, or a fusion:

(a) $^1_1$H + $^2_1$H $\rightarrow$ $^3_2$He

(b) ^{214}Pb $\rightarrow$ $_{-1}\beta$ + ^{214}Bi

(c) ^{1}n + $^{235}_{92}$U $\rightarrow$ $^{140}_{56}$Ba + $^{94}_{36}$Kr + 2 ^{1}n

(d) ^{27}Al + ^{4}He $\rightarrow$ ^{30}P + ^{1}n

■ ■ ■ PROBLEMS ■ ■ ■

20.1 Natural Radioactivity

20.8 Complete each of the following nuclear equations:

(a) $^{217}_{85}$At $\rightarrow$ $^4_2\alpha$ + ?

(b) $^{227}_{89}$Ac $\rightarrow$ $^0_{-1}\beta$ + ?

(c) $^{119}_{50}$Sn $\rightarrow$ $^0_0\gamma$ + ?

20.9 Complete each of the following nuclear equations:

(a) $^{225}_{88}$Ra $\rightarrow$ $^{225}_{89}$Ac + ?

(b) $^{221}_{87}$Fr $\rightarrow$ $^{217}_{85}$At + ?

(c) $^{211}_{83}$Bi $\rightarrow$ $^{207}_{81}$Tl + ?

(d) $^{212}_{83}$Bi $\rightarrow$ $^{208}_{81}$Tl + ?

20.10 Complete each of the following nuclear equations:

(a) ^{228}Th $\rightarrow$ ^{224}Ra + ?

(b) ^{220}Rn $\rightarrow$ ^{216}Po + ?

(c) ^{207}Tl $\rightarrow$ ^{207}Pb + ?

20.11 Complete each of the following nuclear equations:

(a) ^{233}Pa $\rightarrow$ ^{233}U + ?

(b) ^{221}Fr $\rightarrow$ ^{217}At + ?

(c) ^{213}Bi $\rightarrow$? + $^4\alpha$

(d) ^{213}Bi $\rightarrow$ ^{213}Po + ?

20.12 Radon-222 is a noble gas with a high density that is a product of the decay of uranium-238. Since radon is chemically inert, explain how it can cause "poisoning" in the basements of homes in certain parts of the United States.

20.13 Iodine-131 is used for medical treatment of thyroid problems. It decays by emission of a beta particle. Write a nuclear equation for this reaction.

20.2 Half-Life

20.14 A certain isotope has a half-life of 6.80 years. How much of a 60-mg sample of this isotope will remain after 13.6 years?

20.15 A certain isotope has a half-life of 2.68 seconds. How much of a 7.00-g sample of this isotope will remain after 8.04 seconds?

20.16 Calculate the half-life of each of the following isotopes:

(a) A 6.21-g sample of the isotope disintegrates to 0.881 g in 6.42 days.

(b) Of a 6.21-g sample of the isotope, 0.881 g disintegrates in 6.42 days.

20.17 Calculate the half-life of each of the following isotopes:

(a) A 1.73-g sample of the isotope disintegrates to 903 mg in 6.75 minutes.

(b) Of a 7.95-g sample of the isotope, 2.47 g disintegrates in 6.27 days.

20.18 Create a graph for the disintegration series that starts with ^{235}U like that for ^{238}U in Figure 20.1(b).

20.19 Show that $N_o/N = m_o/m$, where m represents the mass of the isotope at time t, and m_o represents the mass of the isotope at time zero.

20.20 Show that $N_o/N = n_o/n$, where n represents the number of moles of the isotope at time t, and n_o represents the number of moles of the isotope at time zero.

20.21 Calculate the age of a sample of rock containing 8.96×10^{21} atoms of ^{40}K for every 6.44×10^{22} atoms of ^{40}Ar, its stable daughter isotope. Assume that no argon was in the rock when it solidified and that no argon escaped from the rock since its solidification. The half-life of ^{40}K is 1.3×10^9 years.

20.22 Calculate the age of a piece of wood that is disintegrating at a rate of 9.81 counts per minute per gram of carbon. A modern sample of carbon disintegrates at a rate of 15.3 dis/min · gram. Assume that the ratio of ^{14}C to ^{12}C in the atmosphere was the same as it is today.

20.3 Nuclear Fission

20.23 Complete each of the following nuclear equations:

(a) $^{235}_{92}U + ^{1}_{0}n \rightarrow ? + ^{90}_{38}Sr + 3\,^{1}_{0}n$

(b) $^{27}_{13}Al + ^{4}_{2}\alpha \rightarrow ^{1}_{0}n + ?$

(c) $^{14}_{7}N + ? \rightarrow ^{17}_{8}O + ^{1}_{1}H$

(d) $^{235}_{92}U + ? \rightarrow ^{140}_{56}Ba + ^{94}_{36}Kr + 2\,^{1}_{0}n$

20.24 Complete each of the following nuclear equations:

(a) $^{235}U + ^{1}n \rightarrow ^{95}Br + ? + 2\,^{1}n$

(b) $^{238}U + ^{1}n \rightarrow ^{239}U + ?$

(c) $^{239}Pu + ? \rightarrow ? + ^{235}U$

20.25 Complete each of the following nuclear equations:

(a) $^{9}Be + ^{4}He \rightarrow ^{12}C + ?$

(b) $^{238}U + ^{1}n \rightarrow ^{239}U + ?$

(c) $^{239}Pu + ^{1}n \rightarrow ?$

20.26 Complete each of the following nuclear equations:

(a) $^{237}U + ^{2}H \rightarrow ^{238}Np + ?$

(b) $^{7}Li + ^{1}H \rightarrow ^{4}He + ?$

20.4 Nuclear Fusion

20.27 Complete each of the following nuclear equations, the sequence of which is thought to be the source of the energy of some stars:

(a) $^{12}C + ^{1}H \rightarrow ?$

(b) $^{13}N \rightarrow ^{13}C + ?$

(c) $^{13}C + ^{1}H \rightarrow ?$

(d) $^{14}N + ^{1}H \rightarrow ?$

(e) $^{15}O \rightarrow ^{15}N + ?$

(f) $^{15}N + ^{1}H \rightarrow ^{12}C + ?$

20.28 Add the equations you wrote to answer Problem 20.27, and cancel the species appearing on both sides. What is the net reaction?

20.29 Complete each of the following nuclear equations:

(a) $^{2}H + ^{2}H \rightarrow ^{3}H + ?$

(b) $^{2}H + ^{2}H \rightarrow ^{3}He + ?$

20.30 If a neutron escapes from a sample undergoing a chain reaction, it may disintegrate into a proton plus an electron. Calculate the energy released during this process. The masses are as follows: 1.00728 amu for a proton, 0.00054858 amu for an electron, and 1.00867 amu for a neutron.

20.31 Complete each of the following nuclear equations:

(a) $^{1}H + ^{1}H \rightarrow ^{2}H + ?$

(b) $^{1}H + ^{2}H \rightarrow ^{3}He + ?$

(c) $^{2}H + ^{3}H \rightarrow ^{4}He + ?$

(d) $^{10}B + ^{4}He \rightarrow ^{13}C + ?$

(e) $^{8}Be \rightarrow ^{8}Li + ?$

(f) $2\,^{3}He \rightarrow 2\,^{1}H + ?$

(g) $^{3}He + ^{2}H \rightarrow ^{1}H + ?$

■ ■ ■ GENERAL PROBLEMS ■ ■ ■

20.32 Calculate the time required for 90.0% of a sample with a 7.50-minute half-life to disintegrate.

20.33 One-eighth of the radioactive atoms of a certain sample are present 24.0 minutes after the original measurement. How many will be present after another 8.00 minutes?

20.34 Which of the following reactions represent natural radioactive decay, and which are artificially induced?

(a) $^{214}_{83}Bi \rightarrow ^{210}_{81}Tl + ^{4}_{2}\alpha$

(b) $^{238}_{92}U + ^{1}_{0}n \rightarrow ^{239}_{92}U + ^{0}_{0}\gamma$

(c) $^{14}_{7}N + ^{4}_{2}\alpha \rightarrow ^{17}_{8}O + ^{1}_{1}H$

20.35 A certain rock contains 4.79×10^{-3} g of ^{238}U and 2.93×10^{-4} g of ^{206}Pb. Assuming that no lead was in the rock when it solidified and that no uranium escaped, calculate the age of the rock. **Note:** To three significant digits, the masses of the atoms are equal to their mass numbers.

20.36 It takes 7.34 hours for 10.0% of a certain isotope to disintegrate. Calculate its half-life.

20.37 Does boiling water in which a dangerous radioactive isotope is dissolved make the water safe to drink?

20.38 Look up the number of people killed in the Chernobyl nuclear accident. Look up the number of people killed in the United States in accidents involving drunken drivers in 1989. Look up the number of people who die in one year worldwide as a result of cigarette smoking. Comment on the relative risk of mortality from nuclear accidents as opposed to cigarette smoking and drunken driving.

20.39 What is the mass of a gamma particle having an energy of 1.00×10^{-11} J?

20.40 What decrease in the mass of the matter accompanies the chemical combustion of 1.00 mol of carbon in oxygen to form carbon dioxide if 393 kJ of energy is produced?

20.41 From the information in Table 20.3, does the half-life seem to depend on the type of particle emitted?

20.42 Calculate the number of joules emitted by the conversion of 1.00 amu of matter to energy.

20.43 (a) What fraction of the neutrons generated in the chain reaction illustrated in Figure 20.5 must be absorbed by control rods or must escape from the sample to keep the reaction from accelerating?

(b) What fraction would have to be absorbed or escape if the following reaction were the only one occurring?

$$^{1}\text{n} + {}^{235}\text{U} \longrightarrow {}^{90}\text{Sr} + {}^{143}\text{Xe} + 3\ {}^{1}\text{n}$$

20.44 Complete each of the following nuclear equations:

(a) $^{67}\text{Cu} + {}^{1}\text{n} \longrightarrow 2\ {}^{1}\text{n} + ?$

(b) $^{2}\text{H} + {}^{12}\text{C} \longrightarrow {}^{4}\text{He} + ?$

(c) $^{1}\text{n} + {}^{6}\text{Li} \longrightarrow {}^{4}\text{He} + ?$

Appendix 1 Scientific Calculations

This appendix introduces two mathematics topics important for chemistry students: (1) scientific algebra and (2) electronic calculator mathematics. The scientific algebra section (Section A.1) presents the relationships between scientific algebra and ordinary algebra. The two topics are much more similar than different; however, since you already know ordinary algebra, the differences are emphasized here. The calculator math section (Section A.2) discusses points with which students most often have trouble. This section is not intended to replace the instruction booklet that comes with a calculator, but to emphasize the points in that booklet that are most important to science students.

For more practice with the concepts in this appendix, you might recalculate the answers to some of the examples in the text.

A.1 Scientific Algebra

Designation of Variables

We all know how to solve an algebraic equation such as

$$3x + 23 = 50$$

We first isolate the term containing the unknown $(3x)$ by addition or subtraction on each side of the equation of any terms not containing the unknown. In this case, we subtract 23 from each side:

$$3x + 23 - 23 = 50 - 23$$
$$3x = 27$$

We then isolate the variable by multiplication or division. In this case, we divide by 3:

$$\frac{3x}{3} = \frac{27}{3}$$
$$x = 9$$

If values are given for some variables—for example, if $y = 50$ and $z = 10$ are given for the equation

$$x = y/z$$

we simply substitute the given values and solve:

$$x = 50/10 = 5$$

457

In chemistry and other sciences, such equations are used continually. Since density is defined as mass divided by volume, we could use the equation

$$x = y/z$$

with y representing the mass and z representing the volume, to solve for x, the density. However, we find it easier to use letters (or combinations of letters) that remind us of the quantities they represent. Thus, we write

$$d = m/V$$

with m representing the mass, V representing the volume, and d representing the density. In this way, we do not have to keep looking at the statement of the problem to see what x represents. However, we solve this equation in exactly the same way as the equation in x, y, and z.

Chemists need to represent so many different kinds of quantities that the same letter may have to represent more than one quantity. The necessity for duplication is lessened in the following ways:

Method	*Example*
Using *italic* letters for variables and roman (regular) letters for units	m for mass and m for meter
Using capital and lowercase (small) letters to mean different things	V for volume and v for velocity
Using subscripts to differentiate values of the same type	P_1 for the first pressure, P_2 for the second, and so on
Using Greek letters	π (pi) for osmotic pressure
Using combinations of letters	MM for molar mass

Each such symbol is handled just like an ordinary algebraic variable, such as x or y.

■ **EXAMPLE A.1**

Solve each of the following equations for the first variable, assuming that the second and third variables are equal to 24 and 3, respectively. For example, in part (a), M is the first variable, n is the second, and V is the third. So n is set equal to 24 and V is set equal to 3, allowing you to solve for M.

(a) $M = n/V$

(b) $n = m/\text{MM}$ (where MM is a single variable)

(c) $v = \lambda \nu$ (λ and ν are the Greek letters lambda and nu.)

(d) $P_1 V_1 = P_2(2.00)$ (P_1 and P_2 represent different pressures.)

Solution

(a) $M = 8$

(b) $n = 24/3 = 8$

(c) $v = (24)(3) = 72$

(d) $P_1 = (3)(2.00)/24 = 0.25$ ■

■ **EXAMPLE A.2**

Solve the following equation, in which each letter stands for a different quantity, for n:

$$PV = nRT$$

Solution

Dividing both sides of the equation by R and T yields

$$n = \frac{PV}{RT} \quad ■$$

■ **EXAMPLE A.3**

Solve the following equation for F in terms of C:

$$\frac{C}{F - 32.0} = \frac{5}{9}$$

Solution

Inverting each side of the equation yields

$$\frac{F - 32.0}{C} = \frac{9}{5}$$

Simplifying gives

$$F = \tfrac{9}{5}C + 32.0 \quad ■$$

■ **EXAMPLE A.4**

Using the equation in Example A.3, find the value of F if $C = 18.0$.

Solution

$$F = \tfrac{9}{5}C + 32.0 = \tfrac{9}{5}(18.0) + 32.0 = 64.4 \quad ■$$

■ **EXAMPLE A.5**

Solve the following equation for t:

$$90.0(4.184)(t - 27.0) + 20.0(0.300)(t - 82.0) = 0$$

Solution

$$377(t - 27.0) + 6.00(t - 82.0) = 0$$
$$377t - 10\,200 + 6.00t - 492 = 0$$
$$383t - 10\,700 = 0$$
$$383t = 10\,700$$
$$t = 27.9$$

Some of these steps could have been combined for a more efficient solution process. ■

Units

Perhaps the biggest difference between ordinary algebra and scientific algebra is that scientific measurements (and most other measurements) are always

Figure A.1 *Addition and Multiplication of Lengths*

(a) When two (or more) lengths are added, the result is a length, and the unit is a unit of length, such as yard. (b) When two lengths are multiplied, the result is an area, and the unit is the square of the unit of length, such as square yards.

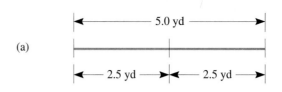

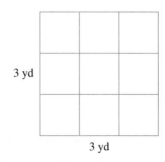

expressed *with units.* Like variables, units have standard symbols. The units are part of the measurements and can often help you determine what operation to perform.

Units are often multiplied or divided, but never added or subtracted. (The associated quantities may be added or subtracted, but the units are not.) For example, if we add the lengths of two ropes, each of which measures 2.50 yards (Figure A.1a), the final answer includes just the unit yards (abbreviated yd). Two units of distance are multiplied to get area, and three units of distance are multiplied to get the volume of a rectangular solid (such as a box). For example, to get the area of a carpet, we multiply its length in yards by its width in yards. The result has the unit square yards (Figure A.1b):

$$\text{yard} \times \text{yard} = \text{yard}^2$$

Be careful to distinguish between similarly worded phrases, such as "3.00 yards, squared" and "3.00 square yards" (Figure A.2).

■ **EXAMPLE A.6**

What is the unit of the volume of a cubic box whose edge measures 4.00 ft?

Solution

A cube has the same length along each of its edges, so the volume is

$$V = (4.00\ \text{ft})^3 = 64.0\ \text{ft}^3$$

Figure A.2 *An Important Difference in Wording*

Knowing the difference between such phrases as "3 yards, squared" and "3 square yards" is important. Multiplying 3 yards by 3 yards gives 3 yards, squared, which is equivalent to 9 square yards, or nine blocks with sides 1 yard long, as you can see. In contrast, 3 square yards is three blocks, each having sides measuring 1 yard.

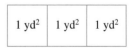

The unit is

$$ft \times ft \times ft = ft^3$$

Practice Problem A.6 What is the unit of the price of gasoline? ■

The unit of a quantity may be treated like an algebraic variable. For example, how many liters of soda are purchased if someone buys a 1.00-L bottle of soda plus two 2.00-L bottles of soda?

$$1.00 \text{ L} + 2(2.00 \text{ L}) = 5.00 \text{ L}$$

The same result would have been obtained if L were an algebraic variable instead of a unit. In dollars, how much will the 5.00 L of soda cost if the average price is 95 cents per liter?

$$5.00 \cancel{L} \left(\frac{95 \text{ cents}}{1 \cancel{L}} \right) = 475 \cancel{\text{ cents}} \left(\frac{1 \text{ dollar}}{100 \cancel{\text{ cents}}} \right) = 4.75 \text{ dollars}$$

■ **EXAMPLE A.7**

What is the unit of the price of ground meat at the supermarket?

Solution

The price is given in dollars per pound. ■

If two or more quantities representing the same type of measurement—for example, a distance—are multiplied, they are usually expressed in the same unit. For example, to calculate the area of a rug that is 6.0 feet wide and 4.0 yards long, we express the length and the width in the same unit before they are multiplied. The width in yards is

$$6.0 \text{ ft} \left(\frac{1 \text{ yd}}{3 \text{ ft}} \right) = 2.0 \text{ yd}$$

The area is

$$(4.0 \text{ yd})(2.0 \text{ yd}) = 8.0 \text{ yd}^2$$

If we had multiplied the original measurements without first converting one to the unit of the other, we would have obtained an incomprehensible set of units:

$$(6.0 \text{ ft})(4.0 \text{ yd}) = 24 \text{ ft} \cdot \text{yd}$$

■ **EXAMPLE A.8**

What is the cost of 8.00 ounces of hamburger if the store charges $2.95 per pound?

Solution

Do not simply multiply:

$$8.00 \text{ ounces} \left(\frac{2.95 \text{ dollars}}{1 \text{ pound}} \right) = \frac{23.60 \text{ ounce} \cdot \text{dollar}}{\text{pound}}$$

Instead, first convert one of the quantities to a unit that matches that of the other quantity:

$$8.00 \text{ ounces}\left(\frac{1 \text{ pound}}{16 \text{ ounces}}\right) = 0.500 \text{ pound}$$

$$0.500 \text{ pound}\left(\frac{2.95 \text{ dollars}}{1 \text{ pound}}\right) = 1.48 \text{ dollars}$$

The same principles apply to the metric units used in science. ■

■ **EXAMPLE A.9**

A car accelerates from 0.0 miles/hour to 50.0 miles/hour in 10.0 seconds. What is the acceleration of the car? (Acceleration is the change in velocity per unit of time.)

Solution

The change in velocity is

$$50.0 \text{ miles/hour} - 0.0 \text{ miles/hour} = 50.0 \text{ miles/hour}$$

$$\frac{50.0 \text{ miles/hour}}{10.0 \text{ seconds}} = \frac{5.00 \text{ miles/hour}}{\text{second}}$$

The acceleration is 5.00 miles per hour per second. This is an example of one of the few times when two different units are used for the same quantity (time) in one value (the acceleration).

Practice Problem A.9   A sprinter accelerates from 0.00 meters/second to 6.00 meters/second in 0.200 second. What is his acceleration? ■

Quadratic Equations

A quadratic equation is an equation of the form

$$ax^2 + bx + c = 0$$

The solution is

$$x = \frac{-b \pm \sqrt{b^2 - 4ac}}{2a}$$

This equation giving the values of x is known as the quadratic formula. Two answers are given by this equation (depending on whether the plus or minus sign is used), but often, only one of them has any physical significance.

■ **EXAMPLE A.10**

Determine the values of a, b, and c in each of the following equations (after it is put in the form $ax^2 + bx + c = 0$). Then calculate two values for x in each case.

(a) $x^2 - 4x + 3 = 0$ (b) $x^2 + 4x = 1$

Solution

(a) Here $a = 1$, $b = -4$, and $c = 3$.

$$x = \frac{-(-4) \pm \sqrt{(-4)^2 - 4(1)(3)}}{2(1)}$$

Using the plus sign before the square root yields

$$x = \frac{+4 + \sqrt{16 - 12}}{2} = 3$$

Using the minus sign before the square root yields

$$x = \frac{4 - \sqrt{4}}{2} = 1$$

The two values for x are 3 and 1. Check:

$$(3)^2 - 4(3) + 3 = 0$$

$$(1)^2 - 4(1) + 3 = 0$$

(b) First, rearrange the equation into the form

$$ax^2 + bx + c = 0$$

In this case, subtracting 1 from each side yields

$$x^2 + 4x - 1 = 0$$

Thus, $a = 1$, $b = 4$, and $c = -1$. The two values of x are

$$x = \frac{-4 + \sqrt{16 + 4}}{2} = \frac{-4 + \sqrt{20}}{2} = -2 + \sqrt{5} = 0.236$$

and

$$x = \frac{-4 - \sqrt{20}}{2} = -4.236 \quad \blacksquare$$

Conversion to Integral Ratios

It is sometimes necessary to convert a ratio of decimal fraction numbers to integral ratios (Chapters 3 and 7). (Note that you cannot round off a number more than about 1%.) The steps necessary to perform this operation follow, with an example given at the right:

Step	*Example*
1. Divide the larger of the numbers by the smaller to get a new ratio of equal value with a denominator equal to 1.	$1.481/1.111 = 1.333/1$
2. Recognize the *fractional* part of the numerator of the new fraction as a common fraction. (The whole-number part does not matter.) Table A.1 lists decimal fractions and their equivalents.	0.333 is the decimal equivalent of 1/3

Table A.1　Certain Decimal Fractions and Their Common Fraction Equivalents

Decimal Part of Number	Common Fraction Equivalent	Multiply by	Example
0.5	1/2	2	$\dfrac{1.5 \times 2}{1 \times 2} = \dfrac{3}{2}$
0.333	1/3	3	$\dfrac{1.333 \times 3}{1 \times 3} = \dfrac{4}{3}$
0.667	2/3	3	$\dfrac{4.667 \times 3}{1 \times 3} = \dfrac{14}{3}$
0.25	1/4	4	$\dfrac{2.25 \times 4}{1 \times 4} = \dfrac{9}{4}$
0.75	3/4	4	$\dfrac{1.75 \times 4}{1 \times 4} = \dfrac{7}{4}$
0.20	1/5	5	$\dfrac{1.2 \times 5}{1 \times 5} = \dfrac{6}{5}$

3. Multiply both numerator and denominator of the calculated ratio in step 1 by the denominator of the common fraction of step 2.

$$\frac{1.333 \times 3}{1 \times 3} = \frac{4}{3}$$

A.2　Calculator Mathematics

A chemistry student needs to have and know how to operate a scientific calculator capable of handling exponential numbers. A huge variety of features is available on calculators, but any calculator with exponential capability should be sufficient for this and other introductory chemistry courses. Once you have obtained such a calculator, practice doing calculations with it so that you will not have to think about how to use the calculator while you should be thinking about how to solve the chemistry problems.

Some calculators have almost twice as many functions as function keys. Each key stands for two different operations, one typically labeled *on* the key and the other *above* the key. To use the second function (above the key), you press a special key first and then the function key. This special key is labeled INV (for inverse) on some calculators, 2ndF (for second function) on others, and something else on still others. The second function key and other special keys are identified for three calculators in Figure A.3. On a few calculators, the MODE key gives several keys a third function. Most scientific calculators have a memory in which you can store a value, using the STO key, and later recall the value, using the RCL key.

Some of the examples in this section involve very simple calculations. The idea is to learn to use the calculator for operations that you can easily do in your head; if you make a mistake with the calculator, you will recognize it immediately. After you practice with simple calculations, you will have the confidence to do more complicated types.

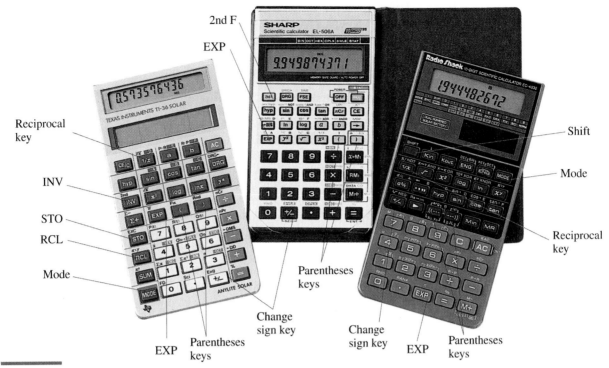

2nd F

EXP

Reciprocal
key

INV

STO

RCL

Mode

Shift

Mode

Reciprocal
key

Parentheses
keys

EXP   Parentheses
keys

Change
sign key

Change
sign key   EXP   Parentheses
keys

Figure A.3 Typical
Electronic Calculators

Precedence Rules

In algebra, when more than one operation is indicated in a calculation, the operations are done in a prescribed order. The order in which they are performed is called the precedence, or priority, of the operations. The order of the common algebraic operations is given in Table A.2. If operations having the same precedence are used, they are performed as they appear from left to right (except for exponentiation and unary minus, which are done right to left). For example, if a calculation involves a multiplication and an addition, the multiplication should be done first, since it has a higher precedence. In each of the following calculations, the multiplication should be done before the addition:

$$y = 2 + 3 \times 4$$
$$y = 3 \times 4 + 2$$

Table A.2 Order of Precedence of Common Operations

	Calculator	**Algebra**
Highest	Parentheses	Parentheses
	Exponentiation (root) or unary minus*	Exponentiation (root) or unary minus*
	Multiplication or division	Multiplication
		Division
Lowest	Addition or subtraction	Addition or subtraction

*Unary minus makes a single value negative, such as in the number -2.

The answer in each case is 14. Try each of these calculations on your calculator to make sure that it does the operations in the correct order automatically.

If parentheses are used in an equation, they indicate that all calculations within the parentheses are to be done *before* the result is used for the rest of the calculations. For example,

$$y = (2 + 3) \times 4$$

means that the addition (within the parentheses) is to be done first, before the other operation (multiplication). *The parentheses override the normal precedence rules.* We might say that parentheses have the highest precedence.

When you are using a calculator, some operation may be waiting for its turn to be done. For example, when $2 + 3 \times 4$ is being entered, the addition will not be done when the multiplication key is pressed. It will await the final equals key, when first the multiplication and then the addition will be carried out. If you want the addition to be done first, you may press the equals key right after entering the three. If you want the calculator to do operations in an order different from that determined by the precedence rules, you may insert parentheses (if they are provided on your calculator; see Figure A.3) or press the equals $\boxed{=}$ key to finalize all calculations so far before you continue with others.

▪ EXAMPLE A.11

What result will be shown on the calculator for each of the following sequences of keystrokes?

(a) $\boxed{5}\ \boxed{\times}\ \boxed{3}\ \boxed{+}\ \boxed{4}\ \boxed{=}$

(b) $\boxed{(}\ \boxed{5}\ \boxed{\times}\ \boxed{3}\ \boxed{)}\ \boxed{+}\ \boxed{4}\ \boxed{=}$

(c) $\boxed{5}\ \boxed{\times}\ \boxed{(}\ \boxed{3}\ \boxed{+}\ \boxed{4}\ \boxed{)}\ \boxed{=}$

(d) $\boxed{(}\ \boxed{5}\ \boxed{\times}\ \boxed{3}\ \boxed{+}\ \boxed{4}\ \boxed{)}\ \boxed{=}$

Solution

(a) 19   (b) 19   (c) 35   (d) 19

▪ EXAMPLE A.12

What result will be shown on the calculator for each of the following sequences of keystrokes?

(a) $\boxed{5}\ \boxed{+}\ \boxed{3}\ \boxed{\times}\ \boxed{4}\ \boxed{=}$

(b) $\boxed{5}\ \boxed{+}\ \boxed{3}\ \boxed{=}\ \boxed{\times}\ \boxed{4}\ \boxed{=}$

Solution

(a) 17   (b) 32 ▪

The rules of precedence followed by the calculator are exactly the same as those for algebra and arithmetic, except that in algebra and arithmetic, multiplication is done before division (multiplication has a higher precedence than division). Algebraically, *ab/cd* means that the product of *a* and *b* is divided by the product of c and d. On the calculator, since multiplication and division have equal precedence, if the keys are pressed in the order

$$a \;\boxed{\times}\; b \;\boxed{\div}\; c \;\boxed{\times}\; d \;\boxed{=}$$

the quotient ab/c is *multiplied* by d. To get the correct algebraic result for ab/cd, the *division* key should be pressed before the value of d is entered, or parentheses should be used around the denominator.

■ **EXAMPLE A.13**

What result will be obtained from pressing the following keys?

$$\boxed{1}\;\boxed{2}\;\boxed{\div}\;\boxed{3}\;\boxed{\times}\;\boxed{5}\;\boxed{=}$$

Solution

The result is 20; first, 12 is divided by 3, and then the result, 4, is multiplied by 5. If you want 12 divided by both 3 and 5, use either of the following keystroke sequences:

$$\boxed{1}\;\boxed{2}\;\boxed{\div}\;\boxed{3}\;\boxed{\div}\;\boxed{5}\;\boxed{=}$$

$$\boxed{1}\;\boxed{2}\;\boxed{\div}\;\boxed{(}\;\boxed{3}\;\boxed{\times}\;\boxed{5}\;\boxed{)}\;\boxed{=} \quad ■$$

■ **EXAMPLE A.14**

Solve:

(a) $\left(\dfrac{3.6}{6.0}\right) \times 3.0$ (b) $\dfrac{3.6}{6.0 \times 3.0}$

Solution

(a) The answer is 1.8. The keystrokes are

$$\boxed{3}\;\boxed{.}\;\boxed{6}\;\boxed{\div}\;\boxed{6}\;\boxed{.}\;\boxed{0}\;\boxed{\times}\;\boxed{3}\;\boxed{.}\;\boxed{0}\;\boxed{=}$$

(b) The answer is 0.20. The keystrokes used in part (a) do not carry out the calculations required for this part. The correct keystroke sequence is

$$\boxed{3}\;\boxed{.}\;\boxed{6}\;\boxed{\div}\;\boxed{6}\;\boxed{.}\;\boxed{0}\;\boxed{\div}\;\boxed{3}\;\boxed{.}\;\boxed{0}\;\boxed{=}$$

or

$$\boxed{3}\;\boxed{.}\;\boxed{6}\;\boxed{\div}\;\boxed{(}\;\boxed{6}\;\boxed{.}\;\boxed{0}\;\boxed{\times}\;\boxed{3}\;\boxed{.}\;\boxed{0}\;\boxed{)}\;\boxed{=}$$

Either *divide* by *each number* in the denominator, or use parentheses around the two numbers so that their product will be divided into the numerator. ■

Division

In algebra, division is represented in any of the following ways, all of which mean the same thing:

$$\frac{a}{b} \qquad a/b \qquad a \div b$$

Note that any *operation* in the numerator or in the denominator of a built-up fraction (a fraction written on two lines), no matter what its precedence, is done *before* the division indicated by the fraction bar. For example, to simplify the expression

$$\frac{a + b}{c - d}$$

the sum $(a + b)$ is divided by the difference $(c - d)$. The addition and subtraction, despite being lower in precedence, are done before the division. In the other forms of representation, this expression is written as

$$(a + b)/(c - d) \qquad \text{or} \qquad (a + b) \div (c - d)$$

where the parentheses are required to signify that the addition and subtraction are to be done first. A calculator has only one operation key for division, $\boxed{\div}$. When using your calculator, be careful to indicate what is divided by what when more than two variables are involved.

▪ **EXAMPLE A.15**

Write the sequence of keystrokes required to perform each of the following calculations, where *a, b,* and *c* represent numbers to be entered:

(a) $\dfrac{a + b}{c + d}$ (b) $a + b/c + d$ (c) $a + b/(c + d)$

Solution

(a) $\boxed{(}\ a\ \boxed{+}\ b\ \boxed{)}\ \boxed{\div}\ \boxed{(}\ c\ \boxed{+}\ d\ \boxed{)}\ \boxed{=}$

(b) $a\ \boxed{+}\ b\ \boxed{\div}\ c\ \boxed{+}\ d\ \boxed{=}$

(c) $a\ \boxed{+}\ b\ \boxed{\div}\ \boxed{(}\ c\ \boxed{+}\ d\ \boxed{)}\ \boxed{=}$ ▪

The Change Sign Key

If we want to enter a negative number, the number is entered first, and then its sign is changed with the change sign $\boxed{+/-}$ key (see Figure A.3), not the subtraction $\boxed{-}$ key.

▪ **EXAMPLE A.16**

Write the sequence of keystrokes required to calculate the product of 4 and -2.

Solution

$$\boxed{4}\ \boxed{\times}\ \boxed{2}\ \boxed{+/-}\ \boxed{=}$$

The change sign key converts 2 to -2 before the two numbers are multiplied. ▪

▪ **EXAMPLE A.17**

(a) What result will be displayed after the following sequence of keystrokes?

$$\boxed{5}\ \boxed{+/-}\ \boxed{x^2}\ \boxed{=}$$

(b) What keystrokes are needed to calculate the value of -5^2?

Solution

(a) The $\boxed{+/-}$ key changes 5 to -5 (minus 5). The squaring key $\boxed{x^2}$ squares the quantity in the display (-5), yielding $+25$.

(b) For the algebraic quantity -5^2, the operations are done right to left; that is, the squaring is done first, and the final answer is -25. To make the operations on the calculator follow the algebraic rules, enter 5, press $\boxed{x^2}$, and then press $\boxed{+/-}$. ■

Exponential Numbers

When displaying a number in exponential notation, most calculators show the coefficient followed by a space or a minus sign and then two digits giving the value of the exponent. For example, the following numbers represent 1.23×10^4 and 1.23×10^{-4}, respectively:

$$1.23 \quad 04$$
$$1.23 {-} 04$$

$$\underbrace{\text{Coefficient}} \quad \underbrace{\text{Exponent}}$$

Note that the base (10) is not shown explicitly on most calculators. If it is shown, interpreting the values of numbers in exponential notation is slightly easier.

To enter a number in exponential form, press the keys corresponding to the coefficient, then press either the $\boxed{\text{EE}}$ key or the $\boxed{\text{EXP}}$ key (you will have one or the other on your calculator), and finally press the keys corresponding to the exponent. For example, to enter 6×10^3 into the calculator, press

$$\boxed{6}\ \boxed{\text{EE}}\ \boxed{3} \quad \text{or} \quad \boxed{6}\ \boxed{\text{EXP}}\ \boxed{3}$$

Do not press the times $\boxed{\times}$ *key or the* $\boxed{1}$ *and* $\boxed{0}$ *keys when entering an exponential number!* The $\boxed{\text{EE}}$ or $\boxed{\text{EXP}}$ key stands for "times 10 to the power." For simplicity, we will use EXP to mean either EXP or EE from this point on.

If the *coefficient* is negative, press the $\boxed{+/-}$ key *before* the $\boxed{\text{EXP}}$ key. If the *exponent* is negative, press the $\boxed{+/-}$ key *after* the $\boxed{\text{EXP}}$ key.

> Caution: Do not press the $\boxed{\times}$ key or the $\boxed{1}$ and $\boxed{0}$ keys when entering an exponential number!

■ **EXAMPLE A.18**

What keys should be pressed to enter the number 6.11×10^{-7}?

Solution

$$\boxed{6}\ \boxed{.}\ \boxed{1}\ \boxed{1}\ \boxed{\text{EXP}}\ \boxed{7}\ \boxed{+/-}$$

or

$$\boxed{6}\ \boxed{.}\ \boxed{1}\ \boxed{1}\ \boxed{\text{EXP}}\ \boxed{+/-}\ \boxed{7}$$

Practice Problem A.18 What keys should be pressed to enter the number -6.11×10^{-7}? ■

■ **EXAMPLE A.19**

What value will be displayed if you enter the following sequence of keystrokes?

$$\boxed{6}\ \boxed{.}\ \boxed{1}\ \boxed{1}\ \boxed{\times}\ \boxed{1}\ \boxed{0}\ \boxed{\text{EXP}}\ \boxed{7}\ \boxed{=}$$

Solution

These keystrokes perform the calculation

$$6.11 \times (10 \times 10^7) =$$

The resulting value will be 6.11×10^8 (which might be displayed in floating point format as 611 000 000). These keystrokes instruct the calculator to multiply 6.11 times 10×10^7, which yields a value 10 times larger than was intended if you wanted to enter 6.11×10^7. ■

■ **EXAMPLE A.20**

To get the value of the quotient

$$\frac{3 \times 10^5}{3 \times 10^5}$$

a student presses the following sequence of keys:

$$\boxed{3}\; \boxed{\times}\; \boxed{1}\; \boxed{0}\; \boxed{\text{EXP}}\; \boxed{5}\; \boxed{\div}\; \boxed{3}\; \boxed{\times}\; \boxed{1}\; \boxed{0}\; \boxed{\text{EXP}}\; \boxed{5}\; \boxed{=}$$

What value is displayed on the calculator as a result?

Solution

The result is 1×10^{12}. The calculator divides 3×10^6 by 3, then multiplies that answer by 10 times 10^5. (See the precedence rules in Table A.1.) This answer is wrong because any number divided by itself should give an answer of 1. (You should always check to see if your answer is reasonable.) ■

■ **EXAMPLE A.21**

What keystrokes should the student have used to get the correct result in Example A.20?

Solution

$$\boxed{3}\; \boxed{\text{EXP}}\; \boxed{5}\; \boxed{\div}\; \boxed{3}\; \boxed{\text{EXP}}\; \boxed{5}\; \boxed{=}$$

The precedence rules are not invoked for this sequence of keystrokes, since only one operation, division, is done. ■

Some calculators display answers in decimal notation unless they are programmed to display them in scientific notation. If a number is too large to fit on the display, such a calculator will use scientific notation automatically. To get a display in scientific notation for a reasonably sized decimal number, press the $\boxed{\text{SCI}}$ key or an equivalent key, if available. (See your instruction booklet.) If automatic conversion is not available on your calculator, you can multiply the decimal value by 1×10^{10} (if the number is greater than 1) or 1×10^{-10} (if the number is less than 1), and mentally subtract or add 10 to the resulting exponent.

The Reciprocal Key

The reciprocal of a number is 1 divided by the number. It has the same number of significant digits as the number itself. For example, the reciprocal of 5.00 is 0.200. A number times its reciprocal is equal to 1.

On the calculator, the reciprocal $\boxed{1/x}$ key (see Figure A.3) takes the reciprocal of whatever value is in the display. To get the reciprocal of *a/b,* enter the value of *a,* press the division $\boxed{\div}$ key, enter the value of *b,* press the equals $\boxed{=}$ key, and finally press the reciprocal $\boxed{1/x}$ key. Alternatively, enter the value of *b,* press the division key, enter the value of *a,* and press the equals key. The reciprocal of *a/b* is *b/a.*

The reciprocal key is especially useful if you have a calculated value in the display that you want to use as a denominator. For example, if you want to calculate *a/(b + c)* and you already have the value of *b + c* in the display, you divide by *a,* press the equals key, and then press the reciprocal key to get the answer. Alternatively, with the value of *b + c* in the display, you can press the reciprocal key and then *multiply* that value by *a.*

$$\frac{a}{b+c} = \left(\frac{1}{b+c}\right)a$$

■ **EXAMPLE A.22**

The value equal to 6.19 + 3.33 = 9.12 is in the display of your calculator. What keystrokes should you use to calculate the value of the following expression?

$$\frac{8.28}{6.19 + 3.33}$$

Solution

$$\boxed{1/x}\ \boxed{\times}\ \boxed{8}\ \boxed{.}\ \boxed{2}\ \boxed{8}\ \boxed{=}$$

or

$$\boxed{\div}\ \boxed{8}\ \boxed{.}\ \boxed{2}\ \boxed{8}\ \boxed{=}\ \boxed{1/x}$$

The display should read 0.8697. . . . ■

Logarithms and Antilogarithms

Determining the logarithm of a number or what number a certain value is the logarithm of (the antilogarithm of the number) is sometimes necessary. The logarithm $\boxed{\log}$ key takes the *common logarithm of the value in the display.* The $\boxed{\text{INV}}$ key or the $\boxed{\text{2ndF}}$ key, followed by the $\boxed{\log}$ key, takes the *antilogarithm of the value in the display.* That is, this sequence gives the number whose logarithm was in the display.

■ **EXAMPLE A.23**

What sequence of keystrokes is required to determine (a) the logarithm of 2.46 and (b) the antilogarithm of 2.46?

Solution

(a) $\boxed{2}\ \boxed{.}\ \boxed{4}\ \boxed{6}\ \boxed{\log}$

(b) $\boxed{2}\ \boxed{.}\ \boxed{4}\ \boxed{6}\ \boxed{\text{2ndF}}\ \boxed{\log}$ or $\boxed{2}\ \boxed{.}\ \boxed{4}\ \boxed{6}\ \boxed{\text{INV}}\ \boxed{\log}$ ■

▪ EXAMPLE A.24

What sequence of keystrokes is required to determine (a) the logarithm of (5.1/5.43) and (b) the antilogarithm of (5.1/5.43)?

Solution

(a) $\boxed{5}$ $\boxed{.}$ $\boxed{1}$ $\boxed{\div}$ $\boxed{5}$ $\boxed{.}$ $\boxed{4}$ $\boxed{3}$ $\boxed{=}$ $\boxed{\log}$

(b) $\boxed{5}$ $\boxed{.}$ $\boxed{1}$ $\boxed{\div}$ $\boxed{5}$ $\boxed{.}$ $\boxed{4}$ $\boxed{3}$ $\boxed{=}$ $\boxed{2ndF}$ $\boxed{\log}$

In part (a), be sure to press $\boxed{=}$ before $\boxed{\log}$ so that you take the logarithm of the quotient, not that of the denominator. In part (b), press $\boxed{=}$ before $\boxed{2ndF}$ (or $\boxed{INV}$). ▪

▪ EXAMPLE A.25

Write the sequence of keystrokes required to solve for *x,* and calculate the value of *x* in each case:

(a) $x = \log (9.63/2.44)$ (b) $x = 9.63/(\log 2.44)$

Solution

(a) $\boxed{9}$ $\boxed{.}$ $\boxed{6}$ $\boxed{3}$ $\boxed{\div}$ $\boxed{2}$ $\boxed{.}$ $\boxed{4}$ $\boxed{4}$ $\boxed{=}$ $\boxed{\log}$
     $x = 0.596$

(b) $\boxed{9}$ $\boxed{.}$ $\boxed{6}$ $\boxed{3}$ $\boxed{\div}$ $\boxed{2}$ $\boxed{.}$ $\boxed{4}$ $\boxed{4}$ $\boxed{\log}$ $\boxed{=}$
     $x = 24.9$

Note the similarity of the keystroke sequences but the great difference in answers. ▪

Significant Figures

An electronic calculator gives its answers with as many digits as are available on the display unless the last digits are zeros to the right of the decimal point. The calculator has no regard for the rules of significant figures (see Section 2.3). *You must apply the rules when reporting the answer.* For example, the reciprocal of 3.00 is really 0.333, but the calculator displays something like 0.333333333. Similarly, dividing 6.69 by 2.23 should yield 3.00, but the calculator displays 3. *You must report only the three significant threes in the first example and must add the two significant zeros in the second example.*

Items for Special Attention

- Variables, constants, and units are represented with standard symbols in scientific mathematics (for example, *d* for density). Be sure to learn and use the standard symbols.

- Units may be treated like algebraic quantities.

- Be sure to distinguish between similar symbols for variables and units. Variables are often printed in *italics.* For example, mass is symbolized by *m,* and meter is represented by m. Capitalization can be crucial: *v* represents velocity, and *V* stands for volume.

- Algebraic and calculator operations must be done in the proper order (according to the rules of precedence). (See Table A.1.) Note that multiplication and division are treated somewhat differently on the calculator than in algebra.

- Operations of equal precedence are done left to right except for exponentiation and unary minus, which are done right to left.

Self-Tutorial Problems

A.1 Simplify each of the following:

(a) $2x = 26$ (b) $\dfrac{15x}{3y}$

(c) $\dfrac{1}{x} = 5.00$ (d) $12\left(\dfrac{x}{42}\right)$

(e) $25(x - 72) = 0$ (f) $(x + 7) - (2y + 3)$

A.2 Simplify each of the following, if possible:

(a) $\dfrac{V_1 T_2}{V_1}$ (b) $\dfrac{V_1 T_2}{V_2}$ (c) $\dfrac{P_1}{3T_2} = \dfrac{P_2}{T_2}$

A.3 Using your calculator, find the value of x from

$$x = (a + b) - (c + d)$$

where $a = -5$, $b = -6$, $c = -7$, and $d = -8$. Repeat this calculation twice, once using the parentheses keys and once without using them.

A.4 Using your calculator, find the value of x from

$$x = (a + b) - (c + d)$$

where $a = -9$, $b = 7$, $c = -5$, and $d = 4$.

A.5 What is the difference between (3.0 cm^3) and $(3.0 \text{ cm})^3$?

A.6 In each of the following sets, are the expressions equal?

(a) $\dfrac{a}{b + c}$ $\left(\dfrac{1}{b + c}\right)a$

(b) $\dfrac{a}{bc}$ $\dfrac{1}{b} \times \dfrac{a}{c}$ $\dfrac{a}{b} \times \dfrac{1}{c}$

(c) $\dfrac{a}{b + c}$ $\dfrac{a}{b} + \dfrac{a}{c}$ $\dfrac{a}{b} + \dfrac{1}{c}$

A.7 Evaluate each of the following pairs of expressions, using $a = 4$ and $b = 2$. In each case, determine if the two expressions are equal.

(a) $\dfrac{1}{a} + \dfrac{1}{b}$ $\dfrac{1}{a + b}$

(b) $\dfrac{1}{a} - \dfrac{1}{b}$ $\dfrac{1}{a - b}$

(c) $\dfrac{1}{a} \times \dfrac{1}{b}$ $\dfrac{1}{ab}$

A.8 Are the following expressions equal to each other?

(a) $a(b + c) - d(e - f)$ $ab + ac - de - df$

(b) $(a + b)(c - d) - (e + f)(g - h)$
$$ac + bc - ad - bd - eg + eh - fg + fh$$

A.9 What units are obtained when a mass in grams is divided by a volume in (a) milliliters (mL), (b) cubic centimeters (cm^3), and (c) cubic meters (m^3)?

A.10 What units are obtained when a volume in liters is divided into a mass in (a) kilograms, (b) grams, and (c) milligrams?

A.11 Simplify each of the following:

(a) $9.00 \text{ dollars}\left(\dfrac{1 \text{ pound}}{2.50 \text{ dollars}}\right)$

(b) $150.0 \text{ miles}\left(\dfrac{1 \text{ gallon}}{27.2 \text{ miles}}\right)$

A.12 Note the difference between $(1.23)^2$ and 1.23×10^2. Which calculator operation key should be used for entering each of these?

A.13 If you have a value for the ratio x/y in the display of your calculator, what key(s) do you press to get the value y/x?

A.14 Calculate the reciprocal of 4.0×10^{-9} m.

A.15 Perform the following calculations in your head. Check the results on your calculator:

(a) $4 + 3 \times 7$ (b) $4 \times 3 + 7$ (c) $4 \times (3 - 7)$

A.16 Perform the following calculations in your head. Check the results on your calculator:

(a) $3 \times (-7)$ (b) $2 \times 3 + (-7)$ (c) $-2 \times (3 - 7)$

A.17 Write the exponential number corresponding to each of the following displays on a scientific calculator:

(a) 1.23 06 (b) 1.23−06

(c) −1.23 06 (d) −1.23−06

(e) 1 11

A.18 What value will be obtained if 3 is entered on a calculator and then the square $\boxed{x^2}$ key is pressed *twice*?

A.19 Write the display on a scientific calculator corresponding to each of the following exponential numbers:

(a) 7.71×10^3 (b) 7.7×10^{-3}

(c) -7.71×10^3 (d) -7.71×10^{-3}

(e) 2.2×10^{22}

■ ■ ■ PROBLEMS ■ ■ ■

A.1 Scientific Algebra

A.20 (a) Solve $PV = nRT$ for R.

(b) Solve $M = n/V$ for n.

(c) Solve $d = m/V$ for V.

(d) Solve $\dfrac{P_1}{T_1} = \dfrac{P_2}{T_2}$ for P_2.

(e) Solve $\pi V = nRT$ for T.

A.21 (a) Solve $PV = nRT$ for V.

(b) Solve $M = n/V$ for V.

(c) Solve $d = m/V$ for m.

(d) Solve $\dfrac{P_1}{T_1} = \dfrac{P_2}{T_2}$ for T_1.

(e) Solve $\pi V = nRT$ for n.

(f) Solve $P_1V_1 = P_2V_2$ for V_1.

(g) Solve $E = h\nu$ for h.

(h) Solve $\Delta t = k_b m$ for m.

A.22 What *unit* does the answer have in each of the following:

(a) 1 cm is added to 1 cm

(b) 1 cm is multiplied by 1 cm

(c) 1 cm is multiplied by 1 cm²

(d) 1 cm³ is divided by 1 cm²

A.23 What *unit* does the answer have in each of the following:

(a) 1 g is divided by 1 cm³

(b) 1 g is divided by 1 cm and by 2 cm²

A.24 What unit does the result have when each of the following expressions is simplified?

(a) $7.00 \text{ feet}\left(\dfrac{12 \text{ inches}}{1 \text{ foot}}\right)\left(\dfrac{4.95 \text{ dollars}}{1 \text{ inch}}\right)$

(b) $(16.0 \text{ cm}^3)/(3.00 \text{ cm})$

(c) $(7.71 \text{ cm}^2) \times (3.37 \text{ cm})$

(d) $(7.71 \text{ cm}^2)/(3.37 \text{ cm})$

A.25 Assuming that the units of density are g/mL, what are the units of the reciprocal of density?

A.26 The specific gravity of a solid or liquid is defined as the density of the substance divided by the density of water at 4°C. What can you say about the unit used with specific gravity?

A.27 Solve each of the following equations for the indicated variable, given the other values listed.

Equation	Solve for	Given
(a) $PV = nRT$	P	$n = 3.00$ mol
		$R = 0.0821$ L · atm/mol · K
		$T = 295$ K
		$V = 6.00$ L
(b) $M = n/V$	V	$M = 0.953$ mol/L
		$n = 0.317$ mol
(c) $d = m/V$	V	$d = 4.73$ g/mL
		$m = 122$ g
(d) $\dfrac{P_1}{T_1} = \dfrac{P_2}{T_2}$	T_1	$T_2 = 273$ K
		$P_1 = 750$ torr
		$P_2 = 760$ torr

(e) $\pi V = nRT$ π $V = 3.00$ L

$n = 0.0100$ mol

$T = 293$ K

$R = 0.0821$ L · atm/mol · K

(f) $P_1V_1 = P_2V_2$ P_2 $V_1 = 425$ mL

$V_2 = 722$ mL

$P_1 = 0.989$ atm

(g) $E = h\nu$ ν $E = 4.45 \times 10^{-19}$ J

$h = 6.63 \times 10^{-34}$ J · s

(h) $\Delta t = k_b m$ k_b $\Delta t = 0.300$°C

$m = 0.400$ m

A.28 Simplify:

$$33.00 \text{ dollars}\left(\frac{3 \text{ pairs of socks}}{11.00 \text{ dollars}}\right)\left(\frac{2 \text{ socks}}{1 \text{ pair of socks}}\right) =$$

A.2 Calculator Mathematics

A.29 Put parentheses around two successive variables in each expression so that the value of the expression is not changed. Use the precedence rules for algebra. For example, $a \times b + c$ is the same as $(a \times b) + c$.

(a) $a - b - c$ (b) $a/b + c$

(c) a/bc (d) ab/c

A.30 Write the sequence of keystrokes required to do each of the following calculations on your calculator:

(a) 2×3^2 (b) $(2 \times 3)^2$ (c) -2^2 (d) 2^3

A.31 Determine the value of each of the following expressions:

(a) $\dfrac{(825)(1/760)(2.11)}{(0.0821)(295)}$ (b) $\dfrac{(0.0821)(298)}{(1.07)(1.23)}$

(c) $\dfrac{(3.00 \times 10^8)(6.63 \times 10^{-34})}{6.91 \times 10^{-10}}$

(d) $\dfrac{(2.11)(8.25 \times 10^2)/(7.60 \times 10^2)}{(8.21 \times 10^{-2})(2.95 \times 10^2)}$

A.32 Which part of Problem A.29 has a different answer if the precedence rules for calculators are used?

A.33 Solve each of the following equations for x:

(a) $\log x = 6.18$ (b) $\log (x/2.00) = 6.18$

(c) $\log (2.00/x) = 6.18$

A.34 Report the result for each of the following expressions in scientific notation:

(a) $(4 \times 10^5)(3 \times 10^6)$

(b) $\dfrac{9 \times 10^3}{3 \times 10^4}$

(c) $\dfrac{3.3 \times 10^{-2}}{2.2 \times 10^2}$

(d) $(3.5 \times 10^7)(8.0 \times 10^{-3})$

(e) $\dfrac{-4.0 \times 10^7}{8.0 \times 10^{-2}}$

(f) $\dfrac{4.0 \times 10^{-7}}{-8.0 \times 10^{-2}}$

(g) $\dfrac{(1.0 \times 10^4)(8.0 \times 10^2)}{2.0 \times 10^5}$

A.35 Report the result for each of the following expressions to the proper number of significant digits:

(a) $3.98 - 7.77 \times 3.28$

(b) $1.93/(11.0)(90.2)$

(c) $222/19.32 - 44.6$

(d) $(73.2)^3$

(e) $(-3.59)(77.2)(66.3)$

(f) $-12.5 + (-44.9) \times 12.6$

(g) $-13.6 \times (-22.8)/66.8$

(h) $2.00^3 \times 3.00^2$

(i) $13.9/(-33.6 + 19.5)$

(j) $73.66/83.9 - 44.4$

A.36 Report the result for each of the following expressions to the proper number of significant digits:

(a) $18.3 - 9.21$

(b) 18.3×9.21

(c) $-18.3 \div 9.21$

(d) $(1.83 \times 10^1) - (9.21 \times 10^0)$

(e) $(1.83 \times 10^{11}) - (9.21 \times 10^{10})$

A.37 Report the result for each of the following expressions to the proper number of significant digits:

(a) $(6.97 \times 10^4) \times (2.11 \times 10^{-3})$

(b) $(8.42 \times 10^2)/(9.01 \times 10^{-7})$

(c) $(3.45 \times 10^6)/(1.89 \times 10^{-7})$

(d) $(2.46 \times 10^5)^4$

(e) $(-1.81 \times 10^{-1})(118.4)$

(f) $(1.00 \times 10^{17}) - (1.00 \times 10^{15})$

A.38 Perform each of the following calculations, and report each answer to the proper number of significant digits:

(a) 1.48×6.87

(b) $1.93/273$

(c) $8.14/29.13$

(d) $(6.69)^2$

(e) $(-3.59)(33.8)$

(f) $-29.4 + (-4.94)$

(g) $-12 \times (-22.8)$

(h) 3.00^4

(i) $27.5/(-28.9)$

(j) $112.1 - (-66.273)$

A.39 Perform each of the following calculations, and report each answer to the proper number of significant digits:

(a) $(7.15 \times 10^4) + (3.52 \times 10^3)$

(b) $(2.00)/(4.00)$

(c) $7.20 + 0.909$

(d) $-8.137 + (-2.1)$

(e) $17.1921 - 17.1909$

A.40 To calculate 4.00×4.125, one student pressed the following keys:

$\boxed{4}\ \boxed{.}\ \boxed{0}\ \boxed{0}\ \boxed{\times}\ \boxed{4}\ \boxed{.}\ \boxed{1}\ \boxed{2}\ \boxed{5}\ \boxed{=}$

Another student pressed the following keys:

$\boxed{4}\ \boxed{\times}\ \boxed{4}\ \boxed{.}\ \boxed{1}\ \boxed{2}\ \boxed{5}\ \boxed{=}$

Both students got the same answer: 16.5. Explain why.

A.41 Calculate the result for each of the following expressions to the proper number of significant digits:

(a) $(6.023 \times 10^{23})(1.602 \times 10^{-19})$

(b) $(96\ 490)/(1.602 \times 10^{-19})$

■ ■ ■ GENERAL PROBLEMS ■ ■ ■

A.42 Report the result for each of the following expressions to the proper number of significant digits:

(a) $1.98 \times 10^3 - 1.03 \times 10^4$

(b) $\dfrac{7.33 \times 10^{16}}{1.98 \times 10^3 - 1.03 \times 10^4}$

(c) $9.62 \times 10^7 + 6.02 \times 10^2$

(d) $9.62 \times 10^{-7} + 6.02 \times 10^{-2}$

(e) $4.184(100)(73.0 - 72.1)$

(f) $6.842/41.052$

(g) $17.0137 \div 17.0127$

A.43 State how you would obtain a value for the expression 4^{x^2}, given that $x = 3$.

A.44 Solve for t:

$$t = \left(\frac{133 \text{ years}}{0.301}\right) \log\left(\frac{72.3 \text{ g}}{8.11 \text{ g}}\right)$$

A.45 Solve for m:

$$\log\left(\frac{10.0 \text{ g}}{m}\right) = \frac{(0.301)(7.65 \text{ years})}{12.18 \text{ years}}$$

A.46 Determine the values of a, b, and c in each of the following quadratic equations after you put the equation in the form $ax^2 + bx + c = 0$. Then calculate two values for each x.

(a) $x^2 + 2x + 1 = 0$

(b) $x^2 - 2x + 1 = 0$

(c) $x^2 - 4 = 0$

(d) $x^2 - 4x + 4 = 0$

(e) $4x^2 + 4x + 1 = 0$ (f) $x^2 + 2x = -1$

(g) $x^2 = 2x - 1$ (h) $x^2 = 4$

(i) $x^2 + 4 = 4x$ (j) $4x^2 = -(1 + 4x)$

A.47 Determine the values of a, b, and c in each of the following quadratic equations after you put the equation in the form $ax^2 + bx + c = 0$. Then calculate two values for each x.

(a) $x^2 + (-4.5 \times 10^{-3})x - (3.8 \times 10^{-5}) = 0$

(b) $\dfrac{x^2}{0.100 - x} = 2.23 \times 10^{-2}$

A.48 What answer is displayed on the calculator after each of the following sequences of key strokes?

(a) $\boxed{(}$ $\boxed{1}$ $\boxed{.}$ $\boxed{3}$ $\boxed{\text{EXP}}$ $\boxed{+/-}$ $\boxed{1}$ $\boxed{1}$ $\boxed{-}$
$\boxed{7}$ $\boxed{.}$ $\boxed{2}$ $\boxed{\text{EXP}}$ $\boxed{+/-}$ $\boxed{1}$ $\boxed{0}$ $\boxed{)}$ $\boxed{\div}$ $\boxed{3}$

(b) $\boxed{1}$ $\boxed{.}$ $\boxed{3}$ $\boxed{\text{EXP}}$ $\boxed{+/-}$ $\boxed{1}$ $\boxed{1}$ $\boxed{-}$
$\boxed{7}$ $\boxed{.}$ $\boxed{2}$ $\boxed{\text{EXP}}$ $\boxed{+/-}$ $\boxed{1}$ $\boxed{0}$ $\boxed{\div}$ $\boxed{3}$

(c) $\boxed{3}$ $\boxed{.}$ $\boxed{3}$ $\boxed{\text{EXP}}$ $\boxed{8}$ $\boxed{\log}$

(d) $\boxed{5}$ $\boxed{1/x}$

(e) $\boxed{2}$ $\boxed{+}$ $\boxed{5}$ $\boxed{=}$ $\boxed{1/x}$ $\boxed{\times}$ $\boxed{1}$ $\boxed{4}$ $\boxed{=}$

(f) $\boxed{1}$ $\boxed{1}$ $\boxed{\text{INV}}$ $\boxed{\log}$
(or $\boxed{\text{2ndF}}$)

(g) $\boxed{2}$ $\boxed{0}$ $\boxed{\log}$

(h) $\boxed{2}$ $\boxed{x^2}$ $\boxed{\text{STO}}$ $\boxed{x^2}$ $\boxed{\times}$ $\boxed{\text{RCL}}$ $\boxed{=}$

A.49 Determine the value of x in each of the following:

(a) $x = \log 13.5$

(b) $\log x = 13.5$

(c) $x = \sqrt[3]{91.2}$

(d) $x = (14.2)^5$

(e) $x = \sqrt{1.06 \times 10^{18}}$

Appendix 2

Tables of Symbols, Abbreviations, and Prefixes and Suffixes

Symbols for Variables and Constants

Symbol	Meaning	Unit(s)	Chapter Reference
A	Area	m^2	2
c	Specific heat capacity	$J/g \cdot {}^{\circ}C$	14
C	Celsius temperature	${}^{\circ}C$	2
d	Density	kg/m^3	2
e	Charge on the electron	C	3
E	Energy	J	1
ΔE	Change in energy	J	14
F	Fahrenheit temperature	${}^{\circ}F$	2
H	Enthalpy	J	14
ΔH	Change in enthalpy	J	14
ΔH_f	Enthalpy of formation	J	14
K	Kelvin temperature	K	2
K	Equilibrium constant	Vary	17
K_a	Acid equilibrium constant	mol/L	18
K_b	Base equilibrium constant	mol/L	18
K_w	Water equilibrium constant	mol^2/L^2	18
k	Proportionality constant	Vary	2
k_b	Boiling-point constant	${}^{\circ}C/m$	15
k_f	Freezing-point constant	${}^{\circ}C/m$	15
KE	Kinetic energy	J	12
l	Length	m	2
m	Mass	g	2
m	Molality	mol/kg	15
M	Molarity	mol/L	11
MM	Molar mass	g/mol	7
n	Number of moles	mol	12
N	Avogadro's number	None	7
N	Normality	equiv/L	16
N	Number of atoms	None	20
N_o	Original number of atoms	None	20
P	Pressure	atm or torr	12
π	Osmotic pressure	atm or torr	15
pH	$-\log [H_3O^+]$	None	18
q	Heat	J	14
r	Radius	m	2
R	Gas constant	$L \cdot atm/mol \cdot K$	12
STP	Standard temperature and pressure	$0{}^{\circ}C$ and 1 atm	12
t	Temperature, in Celsius	${}^{\circ}C$	1, 12

continued

Symbols for Variables and Constants—cont'd

Symbol	Meaning	Unit(s)	Chapter Reference
T	Absolute temperature	K	1, 12
Δt	Change in temperature	°C	14
v	Velocity	m/s	12
V	Volume	m^3 or L	2
w	Work	J	14
X_A	Mole fraction of A	None	15

Symbols or Abbreviations for Units*

Symbol or Abbreviation	Unit	Use	Chapter Reference
amu	Atomic mass unit	Mass of atoms, etc.	3
atm	Atmosphere	Pressure	12
C	Coulomb	Electric charge	3
°C	Degree Celsius	Temperature	2
equiv	Equivalent	Quantity of matter	16
°F	Degree Fahrenheit	Temperature	2
g	Gram	Mass	2
h	Hour	Time	2
J	Joule	Energy	2
K	Kelvin	Temperature	2, 12
L	Liter	Volume	2
m	Meter	Distance (length)	2
m	Molal	Molality	15
M	Molar	Molarity	11
mol	Mole	Quantity of matter	7
N	Normal	Normality	16
s	Second	Time	2
torr	Torr	Pressure	12

*See Table 2.1 for metric prefixes.

Symbols for Subatomic Particles

Symbol	Name of Particle	Chapter Reference
p	Proton	3
e	Electron	3
n	Neutron	3
d	Deuteron	20
α	Alpha (helium nucleus)	20
$_-\beta$	Beta (electron)	20
$_+\beta$	Positron	20
γ	Gamma (photon)	20

Other Greek Letter Symbols

Symbol	Name	Meaning	Chapter Reference
δ	Delta	Partial charge	13
Δ	Delta	Change in	14
μ	Mu	Micro	2
π	Pi	Osmotic pressure	15

Prefixes and Suffixes

Prefix	Meaning	Chapter Reference
bi-	A half-neutralized acid salt of an acid with two ionizable hydrogen atoms	6
centi-	One-hundredth	2
di-	Two	6
hexa-	Six	6
hydro-	No oxygen (in an acid)	6
hypo-	Fewer oxygen atoms (in oxoanion or oxoacid)	6
kilo-	One thousand	2
micro-	One-millionth	2
milli-	One-thousandth	2
mono-	One	6
penta-	Five	6
per-	More oxygen atoms	6
tetra-	Four	6
tri-	Three	6

Suffix	Use	Chapter Reference
-al	Name ending for aldehydes	19
-ane	Name ending for alkanes	19
-ate	Name ending for anions with certain numbers of oxygen atoms	6
-ate	Name ending for esters	19
-ene	Name ending for alkenes	19
-ic	Name ending for cation having the larger positive charge	6
-ic acid	Name ending for oxoacid with more oxygen atoms than corresponding acid named with -ous	6
-ide	Name ending for monatomic anions and also cyanide and hydroxide	5
-ide	Name ending for binary nonmetal-nonmetal compounds	6

continued

Prefixes and Suffixes—cont'd

Suffix	Use	Chapter Reference
-ite	Name ending for oxoanions with fewer oxygen atoms than corresponding ions named with -ate	6
-oic acid	Name ending for organic acids	19
-ol	Name ending for alcohols	19
-one	Name ending for ketones	19
-ous	Name ending for cation having the lower positive charge	6
-ous acid	Name ending for oxoacid with fewer oxygen atoms than the corresponding acid named with -ic	6
-yl	Name ending for hydrocarbon radical	19
-yne	Name ending for alkynes	19

Equation	Name	Chapter Reference
$V = l \times h \times w$	Volume of a rectangular solid	2
$d = m/V$	Density	2
$C = \frac{5}{9}(F - 32)$	Temperature conversion, Celsius and Fahrenheit	2
$K = C + 273$	Temperature conversion, Kelvin and Celsius	2
$KE = \frac{1}{2}mv^2$	Kinetic energy	2
$A = p + n$	Mass number	3
$Z = p$	Atomic number	3
$M = \dfrac{\text{moles of solute}}{\text{liters of solution}}$	Molarity	11
$P_1V_1 = P_2V_2$	Boyle's law	12
$\dfrac{V_1}{T_1} = \dfrac{V_2}{T_2}$	Charles' law	12
$\dfrac{P_1V_1}{T_1} = \dfrac{P_2V_2}{T_2}$	Combined gas law	12
$PV = nRT$	Ideal gas law	12
$P_{\text{total}} = P_1 + P_2 + \ldots$	Dalton's law	12
$KE = 1.5RT/N$	Kinetic energy of an average molecule	12
$\text{Heat} = mc\Delta t$	Heat capacity equation	14
$\Delta E = q + w$	Change in energy	14
$\Delta H = \Delta H_f(\text{products}) - \Delta H_f(\text{reactants})$	Enthalpy change from enthalpy of formation	14
$m = \dfrac{\text{moles of solute}}{\text{kilograms of solvent}}$	Molality	15
$X_A = \dfrac{\text{moles of A}}{\text{total moles in solution}}$	Mole fraction of A	15
$P_Z = X_Z P_Z^\circ$	Vapor pressure of Z	15
$\Delta P = P^\circ - P$	Vapor-pressure lowering	15
$\Delta t_f = k_f m$	Freezing-point depression	15
$\Delta t_b = k_b m$	Boiling-point elevation	15
$\pi = \dfrac{nRT}{V}$	Osmotic pressure	15
$N = \dfrac{\text{equivalents of solute}}{\text{liters of solution}}$	Normality	16
$K = \dfrac{[C]^c[D]^d}{[A]^a[B]^b}$	Equilibrium constant expression	17
$K_a = \dfrac{[H_3O^+][A^-]}{[HA]}$	Acid dissociation constant expression	18
$K_b = \dfrac{[BH^+][OH^-]}{[B]}$	Base dissociation constant expression	18

continued

Equation	Name	Chapter Reference
$K_{\mathrm{w}} = [\mathrm{H_3O^+}][\mathrm{OH^-}]$	Water ionization constant expression	18
$\mathrm{pH} = -\log\,[\mathrm{H_3O^+}]$	Definition of pH	18
$\log\left(\dfrac{N_\mathrm{o}}{N}\right) = \left(\dfrac{0.301}{t_{1/2}}\right)t$	Half-life equation	20
$E = mc^2$	Einstein's equation	20

1 Basic Concepts

PP1.2 Homogeneous

PP1.3 (a) False. (Elements are also substances.)
(b) True. (However, the elements may be combined into compounds.)
(c) False.
(d) False. (Substances are also homogeneous.)

PP1.4 You cannot tell.

PP1.5 No; it is a compound (a chemical combination of the shiny substance—a metal—and something else).

PP1.6 (a) Kinetic energy is being converted to electrical energy.
(b) Electrical energy is being converted to mechanical energy, which is being converted to potential energy.

PP1.7 Cesium (Cs) begins the sixth period, radon (Rn) ends it, and there are 32 elements in the period, including the first series of inner transition elements.

PP1.8 Calcium (Ca) is more likely to be similar to magnesium (Mg), since they are in the same group.

PP1.9 28

PP1.10 Period 7

PP1.11 (a) Carbon is a nonmetal in any of its forms.
(b) Copper is a metal.

PP1.12 Lithium (Li) is the first alkali metal. (Hydrogen is not a metal.)

2 Measurement

PP2.1 $456 \text{ seconds}\left(\dfrac{1 \text{ minute}}{60 \text{ seconds}}\right) = 7.60 \text{ minutes}$

PP2.2 $7.00 \text{ miles}\left(\dfrac{1 \text{ hour}}{35.0 \text{ miles}}\right) = 0.200 \text{ hour (12.0 minutes)}$

PP2.4 $400 \text{ people}\left(\dfrac{53 \text{ men}}{100 \text{ people}}\right) = 212 \text{ men}$

PP2.5 $3 \text{ weeks}\left(\dfrac{7 \text{ days}}{1 \text{ week}}\right)\left(\dfrac{24 \text{ hours}}{1 \text{ day}}\right)\left(\dfrac{60 \text{ minutes}}{1 \text{ hour}}\right)\left(\dfrac{60 \text{ seconds}}{1 \text{ minute}}\right)$
$= 1\,814\,400 \text{ seconds}$

PP2.6 $3.10 \text{ yd}^3\left(\dfrac{27 \text{ ft}^3}{1 \text{ yd}^3}\right) = 83.7 \text{ ft}^3$

PP2.7 $\dfrac{100.0 \text{ yd}}{9.40 \text{ s}}\left(\dfrac{1 \text{ mile}}{1760 \text{ yd}}\right)\left(\dfrac{3600 \text{ s}}{1 \text{ hr}}\right) = \dfrac{21.8 \text{ miles}}{1 \text{ hr}}$
$= 21.8 \text{ miles per hour}$

PP2.8 (a) $29.5 \text{ m}\left(\dfrac{1 \text{ cm}}{0.01 \text{ m}}\right) = 2950 \text{ cm}$

(b) $29.5 \text{ ft}\left(\dfrac{12 \text{ inches}}{1 \text{ ft}}\right) = 354 \text{ inches}$

(c) The English system conversion requires a calculator (or pencil and paper).

PP2.9 $3.59 \text{ mL}\left(\dfrac{0.001 \text{ L}}{1 \text{ mL}}\right) = 0.00359 \text{ L}$

PP2.10 (a) $200x + 35x = 235x$

(b) $35.00y\left(\dfrac{x}{0.01y}\right) = 3500x$
$200x + 3500x = 3700x$

PP2.11 $(2.50x)(3.00x) = 7.50x^2$
The x here and the unit in the example are treated the same algebraically.

PP2.12 $(100x)/(20z^3) = 5.0x/z^3$

PP2.13 $3.00 \text{ m}\left(\dfrac{1 \text{ mm}}{0.001 \text{ m}}\right) = 3000 \text{ mm}$

$10.00 \text{ m}\left(\dfrac{1 \text{ mm}}{0.001 \text{ m}}\right) = 10\,000 \text{ mm}$

PP2.14 $73.2 \text{ cm}\left(\dfrac{0.01 \text{ m}}{1 \text{ cm}}\right)\left(\dfrac{1 \text{ mm}}{0.001 \text{ m}}\right) = 732 \text{ mm}$

PP2.15 $1 \text{ mg}\left(\dfrac{0.001 \text{ g}}{1 \text{ mg}}\right) = 0.001 \text{ g}$

PP2.16 $172 \text{ lb}\left(\dfrac{1 \text{ kg}}{2.20 \text{ lb}}\right) = 78.2 \text{ kg}$

PP2.17 1000 cm^3

PP2.18 (a) Yes   (b) No   (c) It is more precise.

PP2.20 Two significant digits, in meters or in centimeters:
(2.4 m = 240 cm   The zero is not significant.)

PP2.21 (a) 20.0 cm   (b) 101.20 cm   (c) 3.002 cm
(d) 4000 cm
???

PP2.24 85.6 cm

PP2.25 (a) 9.8 cm^3 (b) 0.27 g/cm^3

PP2.26 $5.4321 \text{ g}\left(\dfrac{1 \text{ cg}}{0.01 \text{ g}}\right) = 543.21 \text{ cg}$

PP2.27 (a) -2 g (b) -1 g (c) -2 g (d) -2 g

PP2.28 (a) 1.27 m   (b) 1.28 m   (c) 1.28 m

PP2.29 $277\,000$ km^2

PP2.30 $\left(\dfrac{2.00 \times 10^{11} \text{ dollars}}{2500 \text{ years}}\right)\left(\dfrac{1 \text{ year}}{365 \text{ days}}\right)\left(\dfrac{1 \text{ day}}{24 \text{ hours}}\right) \times$

$\left(\dfrac{1 \text{ hour}}{3600 \text{ seconds}}\right) = 2.54$ dollars/second

PP2.31 2×10^{10}

PP2.32 Only the number in part (c) is given in scientific notation.

PP2.33 (a) 2.4×10^3 g   (b) 2.40×10^3 g   (c) 2.400×10^3 g

PP2.35 (a) 2.02×10^5 (b) 4.000×10^5
(c) $6.70 \times 10^0 = 6.70$

PP2.36 (a) 4.5×10^{15} (b) 1.9×10^7 (c) 6.4×10^{13}

PP2.37 $5.0 \times 10^2 = 500$

PP2.38 5.65

PP2.39 (a) 1.01×10^{-8} (b) 2.000×10^{-1}
(c) 3.00×10^{-1}

PP2.40 2.5×10^{-3}

PP2.41 (a) -5.55×10^4 (b) -9.13×10^5

PP2.42 9.0×10^{-6} cm^2

PP2.43 $\sqrt[3]{8.00 \times 10^9} = \sqrt[3]{8.00} \times \sqrt[3]{10^9} = 2.00 \times 10^3$

PP2.45 $\dfrac{113 \text{ g}}{(5.00 \text{ cm})(2.00 \text{ cm})(1.00 \text{ cm})} = 11.3$ g/cm^3

PP2.46 668 g Hg$\left(\dfrac{1 \text{ mL}}{13.6 \text{ g Hg}}\right) = 49.1$ mL

PP2.47 1.000 L Al$\left(\dfrac{1000 \text{ mL}}{1 \text{ L}}\right)\left(\dfrac{2.702 \text{ g}}{1 \text{ mL}}\right) = 2702$ g

PP2.48 The density of the solid bar is

$\dfrac{566 \text{ g}}{(9.00 \text{ cm})(4.00 \text{ cm})(2.00 \text{ cm})} = 7.86$ g/cm^3

Since the density of the object is greater than that of water, it will not float.

PP2.49 Iron. (See Table 2.5.)

PP2.51 The coffee is cooled, and the milk is warmed.

3 Atoms and Atomic Masses

PP3.1 Anywhere between 27.29% and 42.9% C, depending on how much of each compound is in the mixture.

PP3.2 100 g C$\left(\dfrac{100.0 \text{ g CO}_2}{27.29 \text{ g C}}\right) = 366$ g CO$_2$

PP3.3 Per gram of oxygen, there are

$\dfrac{69.60 \text{ g Mn}}{30.40 \text{ g O}} = \dfrac{2.289 \text{ g Mn}}{1 \text{ g O}}$ $\dfrac{77.45 \text{ g Mn}}{22.55 \text{ g O}} = \dfrac{3.435 \text{ g Mn}}{1 \text{ g O}}$

The ratio of grams of manganese in the two compounds per gram of oxygen is

$\dfrac{3.435 \text{ g}}{2.289 \text{ g}} = \dfrac{1.501}{1 \text{ g}}$

Multiplying numerator and denominator by 2 yields a whole-number ratio:

$\dfrac{1.501 \text{ g} \times 2}{1 \text{ g} \times 2} = \dfrac{3 \text{ g}}{2 \text{ g}}$

PP3.4 They are not isotopes of the same element. Since $p + n$ is the same for both, if n is different, p must also be different. The atoms are $^{127}_{53}$I and $^{127}_{52}$Te.

PP3.5 $^{127}_{52}$Te has 52 electrons, and $^{127}_{53}$I has 53.

PP3.6 $0.5725(120.9038 \text{ amu}) + 0.4275(122.9041 \text{ amu}) = 121.76$ amu

PP3.7 $\dfrac{6300 \text{ lb}/3x}{1400 \text{ lb}/x} = \dfrac{1.5}{1} = 1.5 : 1$

PP3.8 No elements are out of order according to atomic number.

4 Electronic Configuration of the Atom

PP4.1 The change in energy is the energy of the final state minus the energy of the initial state. Since the electron is falling to a lower energy state, light is emitted.
(a) $(-5.445 \times 10^{-19} \text{ J}) - (-2.420 \times 10^{-19} \text{ J})$
$= -3.025 \times 10^{-19}$ J
(b) $(-2.178 \times 10^{-18} \text{ J}) - (-2.420 \times 10^{-19} \text{ J})$
$= -1.936 \times 10^{-18}$ J

PP4.2 The number is equal to n.   (a) Four   (b) One
(The values are listed in Example 4.2.)

PP4.3 The number is $2\ell + 1$.   (a) Seven   (b) One
(The values are listed in Example 4.3.)

PP4.4 With $n = 4$, the maximum possible value for ℓ is 3. The lowest m value is therefore -3.

PP4.6 Four

$n =$	1	1	2	2
$\ell =$	0	0	0	0
$m =$	0	0	0	0
$s =$	$-\frac{1}{2}$	$+\frac{1}{2}$	$-\frac{1}{2}$	$+\frac{1}{2}$

PP4.8 (a) < (e) < (c) < (b) < (d)
(The $n + \ell$ values are 5, 5, 6, 6, and 7, respectively.)

PP4.9 Two. (There is a maximum of two in any $\ell = 0$ subshell.)

PP4.11 Carbon (atomic number $6 = $ sum of superscripts)

PP4.12 Nickel (atomic number $28 = $ sum of superscripts)

PP4.13 (a) Zero   (b) One

PP4.14 They all have $ns^2\,np^6$ outermost electronic configurations except for helium:

He $1s^2$

Ne $1s^2\,2s^2\,2p^6$

Ar $1s^2\,2s^2\,2p^6\,3s^2\,3p^6$

Kr $1s^2\,2s^2\,2p^6\,3s^2\,3p^6\,4s^2\,3d^{10}\,4p^6$

Xe $1s^2\,2s^2\,2p^6\,3s^2\,3p^6\,4s^2\,3d^{10}\,4p^6\,5s^2\,4d^{10}\,5p^6$

Rn $1s^2\,2s^2\,2p^6\,3s^2\,3p^6\,4s^2\,3d^{10}\,4p^6\,5s^2\,4d^{10}\,5p^6\,6s^2\,5d^{10}\,4f^{14}\,6p^6$

PP4.15 (a) Sn $1s^2\,2s^2\,2p^6\,3s^2\,3p^6\,4s^2\,3d^{10}\,4p^6\,5s^2\,4d^{10}\,5p^2$
(b) Ga $1s^2\,2s^2\,2p^6\,3s^2\,3p^6\,4s^2\,3d^{10}\,4p^1$
(c) Se $1s^2\,2s^2\,2p^6\,3s^2\,3p^6\,4s^2\,3d^{10}\,4p^4$

PP4.16 $[\text{Xe}]\,6s^2\,5d^{10}\,4f^{14}\,6p^3$

5 Chemical Bonding

PP5.1 (a) $CH_3(CH_2)_3CH_3$ (b) C_5H_{12}

PP5.2 (a) 1 Co, 3 N, 6 O   (b) 1 Co, 1 C, 3 O

PP5.3 $1s^2\,2s^2\,2p^6$

PP5.4 (a) Sn^{2+} $[\text{Kr}]\,5s^2\,4d^{10}\,5p^0$ or $[\text{Kr}]\,5s^2\,4d^{10}$
(b) Sn^{4+} $[\text{Kr}]\,5s^0\,4d^{10}\,5p^0$ or $[\text{Kr}]\,4d^{10}$

PP5.5 Cr^{2+} $1s^2\,2s^2\,2p^6\,3s^2\,3p^6\,4s^0\,3d^4$
Cr^{3+} $1s^2\,2s^2\,2p^6\,3s^2\,3p^6\,4s^0\,3d^3$

PP5.6 Li$_3$N

PP5.7 (a) CuF$_2$ (b) Co$_3$P$_2$

PP5.8 Ca^{2+} and P^{3-}

PP5.9 (a) Ag$^+$ (b) Mn^{2+}

PP5.10 (a) $:\!\ddot{\text{C}}\text{l}\!\cdot$ $:\!\ddot{\text{C}}\text{l}\!:$ $\cdot\ddot{\text{C}}\text{l}\!:$ $:\!\ddot{\text{C}}\text{l}\!:$

(b) $\cdot\dot{\ddot{\text{S}}}\cdot$ $\cdot\ddot{\text{S}}\!:$ $:\!\dot{\ddot{\text{S}}}\cdot$ $:\!\ddot{\text{S}}\!:$ $\cdot\ddot{\text{S}}\!:$ $:\!\ddot{\text{S}}\!:$

PP5.11

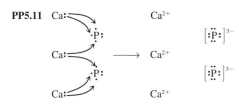

PP5.12 Single bond

:F:F:

Lone pairs

PP5.13 (a) Molecules (b) Ions (c) Uncombined atoms
(d) Ions

PP5.14 H:Ö:Ö:H

PP5.15 :Cl:C:Cl:
.. .. ..
:Ö:

PP5.16 H H H
H:C:C:C:H
.. .. ..
H H H

The carbon atoms must be the central atoms, because hydrogen atoms can have only two electrons each.

PP5.17 $\begin{bmatrix} & :\ddot{O}: & \\ :\ddot{O}:P:\ddot{O}: & \\ & :\ddot{O}: & \end{bmatrix}^{3-}$

PP5.18 $\begin{bmatrix} :\ddot{O}::N:\ddot{O}: \\ :\ddot{O}: \end{bmatrix}^-$ or $\begin{bmatrix} :\ddot{O}:N::O: \\ :\ddot{O}: \end{bmatrix}^-$ or $\begin{bmatrix} :\ddot{O}:N:\ddot{O}: \\ :\ddot{O}: \end{bmatrix}^-$

PP5.19 H:Ö:N::O: or H:Ö:N:Ö:
:Ö: .. ..
:Ö:

PP5.20 Na^+ and $Cr_2O_7^{2-}$

6 Nomenclature

PP6.1 (a) I (b) Xe (c) O

PP6.2 (a) Iodine pentafluoride (b) Iodine trichloride
(c) Sulfur tetrafluoride (d) Phosphorus pentachloride
(e) Arsenic tribromide (f) Selenium dioxide

PP6.3 (a) S_4F_4 (b) N_2O_3

PP6.5 The anion is MoO_4^{2-}. You know that the charge is 2− because it takes two (1+) ammonium ions to balance each anion.

PP6.6 (a) Boron trichloride (b) Chromium(III) chloride
(c) Aluminum chloride

PP6.7 (a) NiO (b) NiO_2

PP6.8 Manganese(III) ion

PP6.9 (a) IO_2^- (b) IO_4^-

PP6.10 (a) NO_3^- (b) SO_3^{2-} (c) PO_4^{3-} (d) CO_3^{2-}

PP6.11 Silicate ion (analogous to carbonate ion)

PP6.12 Magnesium sulfate

PP6.13 (a) Chromium(III) acetate (b) Calcium chlorate

PP6.14 (a) TiF_3 (b) $Al_2(SO_4)_3$

PP6.15 The two hydrogen atoms represented first in the formula are the only ionizable ones.

PP6.16 (a) Bromic acid (b) Nitrous acid (c) Nitric acid

PP6.17 (a) $HBrO_2$ (b) $HBrO_3$

PP6.18 (a) $NaHSO_3$ (b) $Mg(HSO_3)_2$

7 Formula Calculations

PP7.1 F_2 and IF have molecular masses; they are composed of molecules because they consist of nonmetals only. (LiF is ionic.)

PP7.2 (a) 40.1 amu + 4(12.0 amu) + 6(1.0 amu) + 4(16.0 amu)
= 158.1 amu

(b) 58.9 amu + 3(35.5 amu) + 9(16.0 amu) = 309.4 amu

PP7.3
3 N	3×14.01 amu	= 42.03 amu
12 H	12×1.008 amu	= 12.10 amu
P	1×30.97 amu	= 30.97 amu
4 O	4×16.00 amu	= 64.00 amu
	Total	= 149.10 amu

$$\%N = \left(\frac{42.03 \text{ amu}}{149.10 \text{ amu}}\right) \times 100\% = 28.19\% \text{ N}$$

$$\%H = \left(\frac{12.10 \text{ amu}}{149.10 \text{ amu}}\right) \times 100\% = 8.115\% \text{ H}$$

$$\%P = \left(\frac{30.97 \text{ amu}}{149.10 \text{ amu}}\right) \times 100\% = 20.77\% \text{ P}$$

$$\%O = \left(\frac{64.00 \text{ amu}}{149.10 \text{ amu}}\right) \times 100\% = 42.92\% \text{ O}$$

Total = 100.00%

PP7.4 $\left(\dfrac{4.00 \text{ inches}}{1 \text{ box}}\right)\left(\dfrac{12 \text{ boxes}}{1 \text{ dozen}}\right)\left(\dfrac{1 \text{ dozen}}{4.00 \text{ feet}}\right) = \dfrac{12.0 \text{ inches}}{1 \text{ foot}}$

PP7.5 100.09 g/mol

PP7.6 (a) $17.0 \text{ g } C_3H_6\left(\dfrac{1 \text{ mol } C_3H_6}{42.0 \text{ g } C_3H_6}\right) = 0.405 \text{ mol } C_3H_6$

(b) 6.95×10^{24} molecules $C_3H_6 \times$
$\left(\dfrac{1 \text{ mol } C_3H_6}{6.02 \times 10^{23} \text{ molecules } C_3H_6}\right) = 11.5 \text{ mol } C_3H_6$

PP7.7 3.71×10^{22} molecules $CCl_4\left(\dfrac{1 \text{ mol } CCl_4}{6.02 \times 10^{23} \text{ molecules } CCl_4}\right) \times$
$\left(\dfrac{154 \text{ g } CCl_4}{1 \text{ mol } CCl_4}\right) = 9.49 \text{ g } CCl_4$

PP7.8 9.161×10^{21} atoms $O\left(\dfrac{1 \text{ mol } O}{6.023 \times 10^{23} \text{ atoms } O}\right) \times$
$\left(\dfrac{1 \text{ mol } Mg(HCO_3)_2}{6 \text{ mol } O}\right) = 2.535 \times 10^{-3} \text{ mol } Mg(HCO_3)_2$

PP7.9 (a) $50.0 \text{ g } (NH_4)_3PO_4\left(\dfrac{1 \text{ mol } (NH_4)_3PO_4}{149 \text{ g } (NH_4)_3PO_4}\right)\left(\dfrac{1 \text{ mol } P}{1 \text{ mol } (NH_4)_3PO_4}\right) \times$
$\left(\dfrac{31.0 \text{ g P}}{1 \text{ mol } P}\right) = 10.4 \text{ g } P$

(b) $50.0 \text{ g } P\left(\dfrac{1 \text{ mol } P}{31.0 \text{ g } P}\right)\left(\dfrac{1 \text{ mol } (NH_4)_3PO_4}{1 \text{ mol } P}\right) \times$
$\left(\dfrac{149 \text{ g } (NH_4)_3PO_4}{1 \text{ mol } (NH_4)_3PO_4}\right) = 240 \text{ g } (NH_4)_3PO_4$

PP7.10
6 C	= 72.06 g C
12 H	= 12.10 g H
6 O	= 96.00 g O
Total	= 180.16 g

$$\left(\frac{72.06 \text{ g C}}{180.16 \text{ g total}}\right) \times 100\% = 40.00\% \text{ C}$$

$$\left(\frac{12.10 \text{ g H}}{180.16 \text{ g total}}\right) \times 100\% = 6.716\% \text{ H}$$

$$\left(\frac{96.00 \text{ g O}}{180.16 \text{ g total}}\right) \times 100\% = 53.29\% \text{ O}$$

Total = 100.01%

PP7.11 **(a)** AB   **(b)** A_2B_3 (multiply the given ratio by 2)
     (c) A_3B_4 (multiply the given ratio by 3)
     (d) A_4B_5 (multiply the given ratio by 4)
     (e) A_3B_5 (multiply the given ratio by 3)
     (f) A_4B_7 (multiply the given ratio by 4)

PP7.12 $90.50 \text{ g C} \left(\dfrac{1 \text{ mol C}}{12.01 \text{ g C}} \right) = 7.535 \text{ mol C}$

$9.50 \text{ g H} \left(\dfrac{1 \text{ mol H}}{1.008 \text{ g H}} \right) = 9.42 \text{ mol H}$

$\dfrac{9.42 \text{ mol H}}{7.535 \text{ mol C}} = \dfrac{1.25 \text{ mol H}}{1 \text{ mol C}} = \dfrac{5 \text{ mol H}}{4 \text{ mol C}}$

The empirical formula is C_4H_5.

PP7.13 $35.88 \text{ g Na} \left(\dfrac{1 \text{ mol Na}}{22.99 \text{ g Na}} \right) = 1.561 \text{ mol Na}$

$50.05 \text{ g S} \left(\dfrac{1 \text{ mol S}}{32.06 \text{ g S}} \right) = 1.561 \text{ mol S}$

$37.46 \text{ g O} \left(\dfrac{1 \text{ mol O}}{16.00 \text{ g O}} \right) = 2.341 \text{ mol O}$

$\dfrac{1.561 \text{ mol Na}}{1.561} = 1.000 \text{ mol Na}$

$\dfrac{1.561 \text{ mol S}}{1.561} = 1.000 \text{ mol S}$

$\dfrac{2.341 \text{ mol O}}{1.561} = 1.500 \text{ mol O}$

The empirical formula is $Na_2S_2O_3$.

PP7.14 A hydrocarbon contains hydrogen and carbon only. The percentage of hydrogen is

$100.00\% \text{ total} - 85.63\% \text{ C} = 14.37\% \text{ H}$

The numbers of moles are

$85.63 \text{ g C} \left(\dfrac{1 \text{ mol C}}{12.01 \text{ g C}} \right) = 7.130 \text{ mol C}$

$14.37 \text{ g H} \left(\dfrac{1 \text{ mol H}}{1.008 \text{ g H}} \right) = 14.26 \text{ mol H}$

There is a 1:2 mole ratio, and the empirical formula is CH_2. The empirical formula mass is 14.03 g/mol of empirical formula units:

$12.01 \text{ g/mol} + 2.016 \text{ g/mol} = 14.03 \text{ g/mol}$

The number of moles of empirical formula units per mole is

$\dfrac{98.2 \text{ g/mol}}{14.03 \text{ g/mol empirical formula units}}$

$= \dfrac{7 \text{ mol empirical formula units}}{1 \text{ mol}}$

The molecular formula is therefore C_7H_{14}.

8 Chemical Reactions

PP8.1 $Ba(OH)_2 + HBr \longrightarrow BaBr_2 + H_2O$
$1\, Ba(OH)_2 + ?\, HBr \longrightarrow ?\, BaBr_2 + ?\, H_2O$
$1\, Ba(OH)_2 + ?\, HBr \longrightarrow 1\, BaBr_2 + ?\, H_2O$
$1\, Ba(OH)_2 + 2\, HBr \longrightarrow 1\, BaBr_2 + ?\, H_2O$
$1\, Ba(OH)_2 + 2\, HBr \longrightarrow 1\, BaBr_2 + 2\, H_2O$
$Ba(OH)_2 + 2\, HBr \longrightarrow BaBr_2 + 2\, H_2O$

PP8.2 $CH_2O + O_2 \longrightarrow CO + H_2O$
$1\, CH_2O + ?\, O_2 \longrightarrow ?\, CO + ?\, H_2O$
$1\, CH_2O + ?\, O_2 \longrightarrow 1\, CO + 1\, H_2O$
$1\, CH_2O + 0.5\, O_2 \longrightarrow 1\, CO + 1\, H_2O$
$2\, CH_2O + O_2 \longrightarrow 2\, CO + 2\, H_2O$

PP8.3 $CCl_4 + O_2 \longrightarrow COCl_2 + Cl_2$
$?\, CCl_4 + ?\, O_2 \longrightarrow 1\, COCl_2 + ?\, Cl_2$
$1\, CCl_4 + \frac{1}{2}\, O_2 \longrightarrow 1\, COCl_2 + ?\, Cl_2$
$1\, CCl_4 + \frac{1}{2}\, O_2 \longrightarrow 1\, COCl_2 + 1\, Cl_2$
$2\, CCl_4 + 1\, O_2 \longrightarrow 2\, COCl_2 + 2\, Cl_2$
$2\, CCl_4 + O_2 \longrightarrow 2\, COCl_2 + 2\, Cl_2$

PP8.4 $Na_2S + FeCl_2 \longrightarrow FeS + 2\, NaCl$

PP8.5 $Mg(s) + N_2(g) \longrightarrow Mg_3N_2(s)$ (Unbalanced)
$3\, Mg(s) + N_2(g) \longrightarrow Mg_3N_2(s)$ (Balanced)

PP8.6 No reaction (The two metals do not combine chemically.)

PP8.7 **(a)** $2\, Al(s) + 3\, Pb(NO_3)_2(aq) \longrightarrow 2\, Al(NO_3)_3(aq) + 3\, Pb(s)$
     (b) $2\, HCl(aq) + Ca(s) \longrightarrow CaCl_2(aq) + H_2(g)$

PP8.8 **(a)** $Ba(OH)_2(aq) + 2\, HNO_3(aq) \longrightarrow Ba(NO_3)_2(aq) + 2\, H_2O(l)$
     (b) $KCl(aq) + AgC_2H_3O_2(aq) \longrightarrow AgCl(s) + KC_2H_3O_2(aq)$

PP8.9 **(a)** $(NH_4)_2CO_3(aq) + 2\, HCl(aq) \longrightarrow$
$2\, NH_4Cl(aq) + H_2O(l) + CO_2(g)$
     (b) $(NH_4)_2CO_3(aq) + 2\, NaOH(aq) \longrightarrow$
$2\, NH_3(aq) + Na_2CO_3(aq) + 2\, H_2O(l)$

PP8.10 Covalent compounds are formed in each case:
     (a) $Li_2O(s) + H_2SO_4(aq) \longrightarrow Li_2SO_4(aq) + H_2O(l)$
     (b) $LiHCO_3(aq) + HCl(aq) \longrightarrow LiCl(aq) + H_2O(l) + CO_2(g)$

PP8.13 A mixture of CO and CO_2 will be produced, along with water.

PP8.15 $C_{12}H_{22}O_{11}(s) + 6\, O_2(g) \longrightarrow 12\, CO(g) + 11\, H_2O(l)$

PP8.16 The pink color disappears.

PP8.17 -1

9 Net Ionic Equations

PP9.2 $Ag^+(aq) + Cl^-(aq) \longrightarrow AgCl(s)$

PP9.3 **(a)–(d)** $H^+(aq) + OH^-(aq) \longrightarrow H_2O(l)$

PP9.4 $NH_3(aq) + H^+(aq) \longrightarrow NH_4^+(aq)$

PP9.5 Of many possible examples, two are
$Br_2(aq) + 2\, KI(aq) \longrightarrow I_2(aq) + 2\, KBr(aq)$
$Br_2(aq) + 2\, NaI(aq) \longrightarrow I_2(aq) + 2\, NaBr(aq)$

PP9.6 $Cu(s) + Cu^{2+}(aq) + 2\, Br^-(aq) \longrightarrow 2\, CuBr(s)$
(There are no spectator ions in this reaction.)

10 Stoichiometry

PP10.1 $\dfrac{3 \text{ mol HCl}}{1 \text{ mol La(OH)}_3} \quad \dfrac{3 \text{ mol HCl}}{1 \text{ mol LaCl}_3} \quad \dfrac{3 \text{ mol HCl}}{3 \text{ mol H}_2O}$

$\dfrac{1 \text{ mol La(OH)}_3}{3 \text{ mol HCl}} \quad \dfrac{1 \text{ mol La(OH)}_3}{1 \text{ mol LaCl}_3} \quad \dfrac{1 \text{ mol La(OH)}_3}{3 \text{ mol H}_2O}$

$\dfrac{1 \text{ mol LaCl}_3}{3 \text{ mol HCl}} \quad \dfrac{1 \text{ mol LaCl}_3}{1 \text{ mol La(OH)}_3} \quad \dfrac{1 \text{ mol LaCl}_3}{3 \text{ mol H}_2O}$

$\dfrac{3 \text{ mol H}_2O}{3 \text{ mol HCl}} \quad \dfrac{3 \text{ mol H}_2O}{1 \text{ mol La(OH)}_3} \quad \dfrac{3 \text{ mol H}_2O}{1 \text{ mol LaCl}_3}$

PP10.2 $2.27 \text{ mol S} \left(\dfrac{1 \text{ mol Na}_2S}{1 \text{ mol S}} \right) = 2.27 \text{ mol Na}_2S$

PP10.3 $H_3PO_4(aq) + 3\, NaOH(aq) \longrightarrow Na_3PO_4(aq) + 3\, H_2O(l)$

$2.74 \text{ mol Na}_3PO_4 \left(\dfrac{1 \text{ mol H}_3PO_4}{1 \text{ mol Na}_3PO_4} \right) = 2.74 \text{ mol H}_3PO_4$

PP10.4 $3\, H_2(g) + N_2(g) \longrightarrow 2\, NH_3(g)$

Notice the difference in wording between this problem and Example 10.4. Here, the number of moles of reactant *remaining* is given. The number of moles of hydrogen that reacted is

0.300 mol H_2 present $-$ 0.096 mol H_2 left
$$= 0.204 \text{ mol } H_2 \text{ reacted}$$

The number of moles of ammonia is

$$0.204 \text{ mol } H_2\left(\frac{2 \text{ mol } NH_3}{3 \text{ mol } H_2}\right) = 0.136 \text{ mol } NH_3 \text{ produced}$$

PP10.5 $3.40 \text{ mol } Cl_2\left(\frac{2 \text{ mol NaCl}}{1 \text{ mol } Cl_2}\right)\left(\frac{58.5 \text{ g NaCl}}{1 \text{ mol NaCl}}\right) = 398 \text{ g NaCl}$

PP10.6 $227 \text{ g NaCl}\left(\frac{1 \text{ mol NaCl}}{58.5 \text{ g NaCl}}\right)\left(\frac{2 \text{ mol Na}}{2 \text{ mol NaCl}}\right)\left(\frac{23.0 \text{ g Na}}{1 \text{ mol } N_a}\right)$
$$= 89.7 \text{ g Na}$$

PP10.7 $Cl_2(g) + 2 Na(s) \rightarrow 2 NaCl(s)$

$$10.9 \text{ g NaCl}\left(\frac{1 \text{ mol NaCl}}{58.5 \text{ g NaCl}}\right)\left(\frac{2 \text{ mol Na}}{2 \text{ mol NaCl}}\right)\left(\frac{23.0 \text{ g Na}}{1 \text{ mol Na}}\right)$$
$$= 89.2 \text{ g Na}$$

PP10.8 $CuS(s) + O_2(g) \xrightarrow{\text{Heat}} Cu(s) + SO_2(g)$

$$2.75 \times 10^6 \text{ g CuS}\left(\frac{1 \text{ mol CuS}}{95.5 \text{ g CuS}}\right)\left(\frac{1 \text{ mol Cu}}{1 \text{ mol CuS}}\right)\left(\frac{63.5 \text{ g Cu}}{1 \text{ mol Cu}}\right)$$
$$= 1.83 \times 10^6 \text{ g Cu} = 1.83 \text{ metric tons Cu}$$

PP10.10 $CH_4(g) + 2 O_2(g) \rightarrow CO_2(g) + 2 H_2O(l)$

$$173 \text{ g CH}_4\left(\frac{1 \text{ mol CH}_4}{16.0 \text{ g CH}_4}\right)\left(\frac{2 \text{ mol } H_2O}{1 \text{ mol CH}_4}\right)\left(\frac{18.0 \text{ g } H_2O}{1 \text{ mol } H_2O}\right) \times$$
$$\left(\frac{1 \text{ mL } H_2O}{1.00 \text{ g } H_2O}\right) = 389 \text{ mL } H_2O$$

PP10.11 $N_2(g) + 3 H_2(g) \rightarrow 2 NH_3(g)$

$$0.170 \text{ mol } NH_3\left(\frac{3 \text{ mol } H_2}{2 \text{ mol } NH_3}\right)\left(\frac{6.02 \times 10^{23} \ H_2}{1 \text{ mol } H_2}\right) =$$
$$1.54 \times 10^{23} \ H_2 \text{ molecules}$$

PP10.12 $2 C_3H_6(g) + 9 O_2(g) \rightarrow 6 CO_2(g) + 6 H_2O(l)$

$$7.18 \times 10^{23} \text{ molecules } C_3H_6\left(\frac{1 \text{ mol } C_3H_6}{6.02 \times 10^{23} \text{ molecules } C_3H_6}\right) \times$$
$$\left(\frac{6 \text{ mol } H_2O}{2 \text{ mol } C_3H_6}\right)\left(\frac{18.0 \text{ g } H_2O}{1 \text{ mol } H_2O}\right) = 64.4 \text{ g } H_2O$$

PP10.13 $2 Na(s) + ZnCl_2(s) \xrightarrow{\text{Heat}} 2 NaCl(s) + Zn(s)$

$$121 \text{ g Zn}\left(\frac{1 \text{ mol Zn}}{65.4 \text{ g Zn}}\right)\left(\frac{2 \text{ mol Na}}{1 \text{ mol Zn}}\right)\left(\frac{6.02 \times 10^{23} \text{ atoms Na}}{1 \text{ mol Na}}\right)$$
$$= 2.23 \times 10^{24} \text{ atoms Na}$$

PP10.14 $3 NH_3(aq) + H_3PO_4(aq) \rightarrow (NH_4)_3PO_4(aq)$

$$3.00 \text{ kg } H_3PO_4\left(\frac{1000 \text{ g}}{1 \text{ kg}}\right)\left(\frac{1 \text{ mol } H_3PO_4}{98.0 \text{ g } H_3PO_4}\right) = 30.6 \text{ mol } H_3PO_4$$

$$30.6 \text{ mol } H_3PO_4\left(\frac{1 \text{ mol } (NH_4)_3PO_4}{1 \text{ mol } H_3PO_4}\right)\left(\frac{3 \text{ mol N}}{1 \text{ mol } (NH_4)_3PO_4}\right) \times$$
$$\left(\frac{14.0 \text{ g N}}{1 \text{ mol N}}\right) = 1290 \text{ g N}$$

PP10.15 $100.00 \text{ dollars}\left(\frac{1 \text{ pound}}{8.00 \text{ dollars}}\right) = 12.5 \text{ pounds}$

The dollars are in limiting quantity, so only 12.5 pounds of nuts can be purchased.

PP10.16 (a) $4.16 \text{ mol HCl}\left(\frac{1 \text{ mol } H_2}{2 \text{ mol HCl}}\right) = 2.08 \text{ mol } H_2$

(b) $2.08 \text{ mol Zn}\left(\frac{1 \text{ mol } H_2}{1 \text{ mol Zn}}\right) = 2.08 \text{ mol } H_2$

(c) $1.00 \text{ mol Zn}\left(\frac{1 \text{ mol } H_2}{1 \text{ mol Zn}}\right) = 1.00 \text{ mol } H_2$

PP10.17 $2 Na(s) + S(s) \rightarrow Na_2S(s)$

(a) *Present* *Required*

$$\frac{12.9 \text{ mol Na}}{7.80 \text{ mol S}} = \frac{1.65 \text{ mol Na}}{1 \text{ mol S}} \qquad \frac{2 \text{ mol Na}}{1 \text{ mol S}}$$

Sodium is in limiting quantity; sulfur is in excess.

(b) $12.9 \text{ mol Na}\left(\frac{1 \text{ mol } Na_2S}{2 \text{ mol Na}}\right) = 6.45 \text{ mol } Na_2S$

The tabulation confirms this answer:

	2 Na(s)	+ S(s)	→ Na$_2$S(s)
Present initially	12.9 mol	7.80 mol	0.00 mol
Change due to reaction	−12.9 mol	−6.45 mol	+6.45 mol
Present finally	0.0 mol	1.35 mol	6.45 mol

PP10.18 $Ba(OH)_2(aq) + 2 HCl(aq) \rightarrow BaCl_2(aq) + 2 H_2O(l)$

Present *Required*

$$\frac{4.50 \text{ mol HCl}}{2.50 \text{ mol } Ba(OH)_2} = \frac{1.80 \text{ mol HCl}}{1 \text{ mol } Ba(OH)_2} \qquad \frac{2 \text{ mol HCl}}{1 \text{ mol } Ba(OH)_2}$$

HCl is limiting.

$$4.50 \text{ mol HCl}\left(\frac{1 \text{ mol } BaCl_2}{2 \text{ mol HCl}}\right) = 2.25 \text{ mol } BaCl_2$$

PP10.19 (a) $2 NaI(aq) + Pb(NO_3)_2(aq) \rightarrow PbI_2(s) + 2 NaNO_3(aq)$

Present *Required*

$$\frac{4.00 \text{ mol NaI}}{2.00 \text{ mol } Pb(NO_3)_2} = \frac{2.00 \text{ mol NaI}}{1 \text{ mol } Pb(NO_3)_2} \qquad \frac{2 \text{ mol NaI}}{1 \text{ mol } Pb(NO_3)_2}$$

Neither reactant is in excess, and either may be used:

$$2.00 \text{ mol } Pb(NO_3)_2\left(\frac{1 \text{ mol } PbI_2}{1 \text{ mol } Pb(NO_3)_2}\right) = 2.00 \text{ mol } PbI_2$$

or

$$4.00 \text{ mol NaI}\left(\frac{1 \text{ mol } PbI_2}{2 \text{ mol NaI}}\right) = 2.00 \text{ mol } PbI_2$$

(b)

Present *Required*

$$\frac{2.00 \text{ mol NaI}}{4.00 \text{ mol } Pb(NO_3)_2} = \frac{0.500 \text{ mol NaI}}{1 \text{ mol } Pb(NO_3)_2} \qquad \frac{2 \text{ mol NaI}}{1 \text{ mol } Pb(NO_3)_2}$$

$Pb(NO_3)_2$ is in excess.

$$2.00 \text{ mol NaI}\left(\frac{1 \text{ mol } PbI_2}{2 \text{ mol NaI}}\right) = 1.00 \text{ mol } PbI_2$$

PP10.20 $3 Zn(s) + 2 H_3PO_4(aq) \rightarrow 3 H_2(g) + Zn_3(PO_4)_2(s)$

Present *Required*

$$\frac{2.50 \text{ mol Zn}}{2.25 \text{ mol } H_3PO_4} = \frac{1.11 \text{ mol Zn}}{1 \text{ mol } H_3PO_4} \qquad \frac{3 \text{ mol Zn}}{2 \text{ mol } H_3PO_4}$$
$$= \frac{1.50 \text{ mol Zn}}{1 \text{ mol } H_3PO_4}$$

Zinc is in limiting quantity.

$$2.50 \text{ mol Zn}\left(\frac{2 \text{ mol } H_3PO_4}{3 \text{ mol Zn}}\right) = 1.67 \text{ mol } H_3PO_4 \text{ reacts}$$

$2.25 \text{ mol } H_3PO_4$ present $-$ 1.67 mol H_3PO_4 reacts
$$= 0.58 \text{ mol } H_3PO_4 \text{ excess}$$

Tabulating, all in moles, shows the same result:

	3 Zn(s)	+ 2 H$_3$PO$_4$(aq)	→ 3 H$_2$(g)	+ Zn$_3$(PO$_4$)$_2$(s)
Present initially	2.50	2.25	0.00	0.000
Change	−2.50	−1.67	+2.50	+0.833
Present finally	0.00	0.58	2.50	0.833

PP10.21 $HCl(aq) + NaHCO_3(aq) \rightarrow NaCl(aq) + CO_2(g) + H_2O(l)$

$$26.0 \text{ g NaHCO}_3\left(\frac{1 \text{ mol NaHCO}_3}{84.0 \text{ g NaHCO}_3}\right) = 0.310 \text{ mol NaHCO}_3$$

$$15.3 \text{ g HCl}\left(\frac{1 \text{ mol HCl}}{36.5 \text{ g HCl}}\right) = 0.419 \text{ mol HCl}$$

Present		*Required*
$\dfrac{0.419 \text{ mol HCl}}{0.310 \text{ mol NaHCO}_3} =$	$\dfrac{1.35 \text{ mol HCl}}{1 \text{ mol NaHCO}_3}$	$\dfrac{1 \text{ mol HCl}}{1 \text{ mol NaHCO}_3}$

HCl is in excess.

$$0.310 \text{ mol NaHCO}_3\left(\frac{1 \text{ mol NaCl}}{1 \text{ mol NaHCO}_3}\right)\left(\frac{58.5 \text{ g NaCl}}{1 \text{ mol NaCl}}\right)$$
$$= 18.1 \text{ g NaCl}$$

PP10.22 $H_2SO_4(aq) + 2 KOH(aq) \rightarrow K_2SO_4(aq) + 2 H_2O(l)$

$$42.7 \text{ g KOH}\left(\frac{1 \text{ mol KOH}}{56.1 \text{ g KOH}}\right) = 0.761 \text{ mol KOH}$$

$$21.1 \text{ g H}_2\text{SO}_4\left(\frac{1 \text{ mol H}_2\text{SO}_4}{98.0 \text{ g H}_2\text{SO}_4}\right) = 0.215 \text{ mol H}_2\text{SO}_4$$

Present		*Required*
$\dfrac{0.761 \text{ mol KOH}}{0.215 \text{ mol H}_2\text{SO}_4} =$	$\dfrac{3.54 \text{ mol KOH}}{1 \text{ mol H}_2\text{SO}_4}$	$\dfrac{2 \text{ mol KOH}}{1 \text{ mol H}_2\text{SO}_4}$

KOH is in excess.

$$0.215 \text{ mol H}_2\text{SO}_4\left(\frac{1 \text{ mol K}_2\text{SO}_4}{1 \text{ mol H}_2\text{SO}_4}\right)\left(\frac{174 \text{ g K}_2\text{SO}_4}{1 \text{ mol K}_2\text{SO}_4}\right)$$
$$= 37.4 \text{ g K}_2\text{SO}_4$$

PP10.24 $3.45 \text{ g PCl}_3\left(\dfrac{1 \text{ mol PCl}_3}{138 \text{ g PCl}_3}\right) = 0.0250 \text{ mol PCl}_3$

$$1.80 \text{ g Cl}_2\left(\frac{1 \text{ mol Cl}_2}{70.9 \text{ g Cl}_2}\right) = 0.0254 \text{ mol Cl}_2$$

Present		*Required*
$\dfrac{0.0254 \text{ mol Cl}_2}{0.0250 \text{ mol PCl}_3} =$	$\dfrac{1.02 \text{ mol Cl}_2}{1 \text{ mol PCl}_3}$	$\dfrac{1 \text{ mol Cl}_2}{1 \text{ mol PCl}_3}$

PCl_3 is present in limiting quantity, so the answer is based on the PCl_3 present, just as in Example 10.24. The answer is 99.6%.

PP10.26 $Ag^+(aq) + Cl^-(aq) \rightarrow AgCl(s)$

$$212 \text{ g AgCl}\left(\frac{1 \text{ mol AgCl}}{143 \text{ g AgCl}}\right)\left(\frac{1 \text{ mol Ag}^+}{1 \text{ mol AgCl}}\right) = 1.48 \text{ mol Ag}^+$$

$$1.48 \text{ mol Ag}^+\left(\frac{1 \text{ mol AgClO}_3}{1 \text{ mol Ag}^+}\right)\left(\frac{191 \text{ g AgClO}_3}{1 \text{ mol AgClO}_3}\right)$$
$$= 283 \text{ g AgClO}_3$$

11 Molarity

PP11.3 Be sure to choose the correct values to calculate the molarity, which is moles of solute divided by liters of solution:

$$\frac{2.50 \text{ mol solute}}{4.00 \text{ L solution}} = 0.625 \text{ M}$$

PP11.4 $15.2 \text{ g KBr}\left(\dfrac{1 \text{ mol KBr}}{119 \text{ g KBr}}\right) = 0.128 \text{ mol KBr}$

$$\frac{0.128 \text{ mol KBr}}{2.17 \text{ L}} = 0.0590 \text{ M KBr}$$

PP11.5 $58.5 \text{ mg NaCl}\left(\dfrac{1 \text{ mmol NaCl}}{58.5 \text{ mg NaCl}}\right) = 1.00 \text{ mmol NaCl}$

$$\frac{1.00 \text{ mmol NaCl}}{1.50 \text{ mL}} = 0.667 \text{ M NaCl}$$

PP11.6 $25.00 \text{ mL}\left(\dfrac{3.000 \text{ mmol}}{1 \text{ mL}}\right) = 75.00 \text{ mmol NaClO}_3$

PP11.8 $150 \text{ mL}\left(\dfrac{2.72 \text{ mmol}}{1 \text{ mL}}\right) = 408 \text{ mmol}$

$$175 \text{ mL}\left(\frac{1.78 \text{ mmol}}{1 \text{ mL}}\right) = 312 \text{ mmol}$$
$$\text{Total} = 720 \text{ mmol}$$

$$\frac{720 \text{ mmol}}{600 \text{ mL}} = 1.20 \text{ M}$$

PP11.10 $Ba(OH)_2(aq) + 2 HNO_3(aq) \rightarrow Ba(NO_3)_2(aq) + 2 H_2O(l)$

$$31.70 \text{ mL HNO}_3\left(\frac{0.2112 \text{ mmol HNO}_3}{1 \text{ mL HNO}_3}\right) = 6.695 \text{ mmol HNO}_3$$

$$6.695 \text{ mmol HNO}_3\left(\frac{1 \text{ mmol Ba(OH)}_2}{2 \text{ mmol HNO}_3}\right) = 3.348 \text{ mmol Ba(OH)}_2$$

$$3.348 \text{ mmol Ba(OH)}_2\left(\frac{1 \text{ mL}}{0.2710 \text{ mmol Ba(OH)}_2}\right) = 12.35 \text{ mL}$$

PP11.11 $HCl(aq) + NaOH(aq) \rightarrow NaCl(aq) + H_2O(l)$

$$22.0 \text{ g HCl}\left(\frac{1 \text{ mol HCl}}{36.5 \text{ g HCl}}\right) = 0.603 \text{ mol HCl}$$

$$0.553 \text{ L NaOH}\left(\frac{3.00 \text{ mol NaOH}}{1 \text{ L NaOH}}\right) = 1.66 \text{ mol NaOH}$$

HCl is in limiting quantity.

$$0.603 \text{ mol HCl}\left(\frac{1 \text{ mol NaCl}}{1 \text{ mol HCl}}\right)\left(\frac{58.5 \text{ g NaCl}}{1 \text{ mol NaCl}}\right)$$
$$= 35.3 \text{ g NaCl}$$

PP11.12 The concentration of A must be very small, since the concentration of B cannot be greater than about 10^1 M.

PP11.13 (a) $\dfrac{1.12 \text{ mol Al(NO}_3)_3}{3.00 \text{ L}} = 0.373 \text{ M Al(NO}_3)_3$

(b) $1.12 \text{ mol Al(NO}_3)_3\left(\dfrac{1 \text{ mol Al}^{3+}}{1 \text{ mol Al(NO}_3)_3}\right) = 1.12 \text{ mol Al}^{3+}$

$$1.12 \text{ mol Al(NO}_3)_3\left(\frac{3 \text{ mol NO}_3^-}{1 \text{ mol Al(NO}_3)_3}\right) = 3.36 \text{ mol NO}_3^-$$

$$\frac{1.12 \text{ mol Al}^{3+}}{3.00 \text{ L}} = 0.373 \text{ M Al}^{3+}$$

$$\frac{3.36 \text{ mol NO}_3^-}{3.00 \text{ L}} = 1.12 \text{ M NO}_3^-$$

PP11.14 $2.00 \text{ L}\left(\dfrac{1.72 \text{ mol (NH}_4)_3\text{PO}_4}{1 \text{ L}}\right) = 3.44 \text{ mol (NH}_4)_3\text{PO}_4$

$$2.50 \text{ L}\left(\frac{0.985 \text{ mol NH}_4\text{Cl}}{1 \text{ L}}\right) = 2.46 \text{ mol NH}_4\text{Cl}$$

3.44 mol $(NH_4)_3PO_4$ consists of
10.3 mol NH_4^+ and 3.44 mol PO_4^{3-}

2.46 mol NH_4Cl consists of
2.46 mol NH_4^+ and 2.46 mol Cl^-

The total number of moles of NH_4^+ is 12.8 mol, and the concentrations are

$$\frac{12.8 \text{ mol } NH_4^+}{5.00 \text{ L}} = 2.56 \text{ M } NH_4^+$$

$$\frac{3.44 \text{ mol } PO_4^{3-}}{5.00 \text{ L}} = 0.688 \text{ M } PO_4^{3-}$$

$$\frac{2.46 \text{ mol } Cl^-}{5.00 \text{ L}} = 0.492 \text{ M } Cl^-$$

PP11.15 $H^+(aq) + OH^-(aq) \rightarrow H_2O(l)$

$$2.00 \text{ L}\left(\frac{3.23 \text{ mol } H^+}{1 \text{ L}}\right) = 6.46 \text{ mol } H^+$$

$$2.50 \text{ L}\left(\frac{0.975 \text{ mol } OH^-}{1 \text{ L}}\right) = 2.44 \text{ mol } OH^-$$

The OH^- is in limiting quantity, so 2.44 mol OH^- reacts with 2.44 mol H^+, leaving no OH^- and 6.46 mol − 2.44 mol = 4.02 mol H^+. The concentrations are

$$\frac{6.46 \text{ mol } Cl^-}{5.00 \text{ L}} = 1.29 \text{ M } Cl^-$$

$$\frac{4.02 \text{ mol } H^+}{5.00 \text{ L}} = 0.804 \text{ M } H^+$$

$$\frac{2.44 \text{ mol } Na^+}{5.00 \text{ L}} = 0.488 \text{ M } Na^+$$

PP11.16 $V_{NaOH} = 47.19 \text{ mL} - 2.83 \text{ mL} = 44.36 \text{ mL}$
$HCl(aq) + NaOH(aq) \rightarrow NaCl(aq) + H_2O(l)$

$$25.00 \text{ mL HCl}\left(\frac{2.000 \text{ mmol HCl}}{1 \text{ mL HCl}}\right) = 50.00 \text{ mmol HCl}$$

$$50.00 \text{ mmol HCl}\left(\frac{1 \text{ mmol NaOH}}{1 \text{ mmol HCl}}\right) = 50.00 \text{ mmol NaOH}$$

$$\frac{50.00 \text{ mmol NaOH}}{44.36 \text{ mL NaOH}} = 1.127 \text{ M NaOH}$$

PP11.18 $H_2SO_4(aq) + 2 NaOH(aq) \rightarrow Na_2SO_4(aq) + 2 H_2O(l)$

$$25.00 \text{ mL}\left(\frac{2.000 \text{ mmol } H_2SO_4}{1 \text{ mL}}\right) = 50.00 \text{ mmol } H_2SO_4$$

$$50.00 \text{ mmol } H_2SO_4\left(\frac{2 \text{ mmol NaOH}}{1 \text{ mmol } H_2SO_4}\right) = 100.0 \text{ mmol NaOH}$$

$$100.0 \text{ mmol NaOH}\left(\frac{1 \text{ mL NaOH}}{2.573 \text{ mmol NaOH}}\right) = 38.87 \text{ mL NaOH}$$

PP11.19 $NaHCO_3(aq) + HCl(aq) \rightarrow NaCl(aq) + CO_2(g) + H_2O(l)$

$$42.17 \text{ mL HCl}\left(\frac{6.450 \text{ mmol HCl}}{1 \text{ mL HCl}}\right) = 272.0 \text{ mmol HCl}$$

$$272.0 \text{ mmol HCl}\left(\frac{1 \text{ mmol } NaHCO_3}{1 \text{ mmol HCl}}\right)$$
$$= 272.0 \text{ mmol } NaHCO_3$$

12 Gases

PP12.2
$$P_1V_1 = P_2V_2$$
$$(1.00 \text{ atm})(175 \text{ mL}) = P_2(125 \text{ mL})$$
$$P_2 = 1.40 \text{ atm}$$

PP12.3

State	P	V
1	1.00 atm	V_1
2	717 torr	450 mL

$$P_1V_1 = P_2V_2$$

$$(1.00 \text{ atm}) V_1 = (717 \text{ torr})\left(\frac{1 \text{ atm}}{760 \text{ torr}}\right)(450 \text{ mL})$$
$$V_1 = 425 \text{ mL}$$

PP12.4 $P_1V_1 = P_2V_2$

$$P_1(2.50 \text{ L}) = (722 \text{ torr})(1140 \text{ mL})\left(\frac{1 \text{ L}}{1000 \text{ mL}}\right)$$

$$P_1 = 329 \text{ torr}$$

PP12.5 $C = \frac{5}{9}(F - 32.0) = \frac{5}{9}(98.6 - 32.0) = \frac{5}{9}(66.6) = 37.0°C$
$T = 37.0 + 273 = 310 \text{ K}$

PP12.7 $\dfrac{T_1}{V_1} = \dfrac{T_2}{V_2}$

$$\frac{T_1}{1.78 \text{ L}} = \frac{(273 + 50) \text{ K}}{2.12 \text{ L}}$$
$$T_1 = 271 \text{ K}$$

PP12.8
$$\frac{P_1V_1}{T_1} = \frac{P_2V_2}{T_2}$$

$$\frac{(777 \text{ torr}) [(1 \text{ atm})/(760 \text{ torr})]V_1}{} = \frac{(1.37 \text{ atm}) (217 \text{ mL})}{(333 \text{ K})}$$
$$V_1 = 265 \text{ mL}$$

PP12.9
$$\frac{P_1V_1}{T_1} = \frac{P_2V_2}{T_2}$$

$$\frac{(815 \text{ torr}) (251 \text{ mL})}{(328 \text{ K})} = \frac{(760 \text{ torr})V_2}{(273 \text{ K})}$$
$$V_2 = 224 \text{ mL}$$

PP12.11 $V = \dfrac{nRT}{P} = \dfrac{(4.11 \text{ mol}) (0.0821 \text{ L} \cdot \text{atm/mol} \cdot \text{K}) (309 \text{ K})}{1.11 \text{ atm}}$
$$= 93.9 \text{ L}$$

PP12.12 $65.0 \text{ g } O_2\left(\dfrac{1 \text{ mol } O_2}{32.0 \text{ g } O_2}\right) = 2.03 \text{ mol } O_2$

$$678 \text{ torr}\left(\frac{1 \text{ atm}}{760 \text{ torr}}\right) = 0.892 \text{ atm}$$

$$44 + 273 = 317 \text{ K}$$

$$V = \frac{nRT}{P} = \frac{(2.03 \text{ mol}) (0.821 \text{ L} \cdot \text{atm/mol} \cdot \text{K}) (317 \text{ K})}{0.892 \text{ atm}}$$
$$= 59.2 \text{ L}$$

PP12.15 $12.5 \text{ g } H_2\left(\dfrac{1 \text{ mol } H_2}{2.016 \text{ g } H_2}\right) = 6.20 \text{ mol } H_2$

$$P = \frac{nRT}{V} = \frac{(6.20 \text{ mol}) (0.0821 \text{ L} \cdot \text{atm/mol} \cdot \text{K}) (289 \text{ K})}{17.0 \text{ L}}$$
$$= 8.65 \text{ atm}$$

The answer differs because 12.5 g of H_2 is a much greater number of moles of H_2 than 12.5 g of N_2 is of N_2.

PP12.18 $V = \dfrac{nRT}{P} = \dfrac{(1.00 \text{ mol}) (0.0821 \text{ L} \cdot \text{atm/mol} \cdot \text{K}) (273 \text{ K})}{1.00 \text{ atm}}$
$$= 22.4 \text{ L}$$

PP12.21 $n = \dfrac{PV}{RT} = \dfrac{(1.342 \text{ atm}) (2.50 \text{ L})}{(0.0821 \text{ L} \cdot \text{atm/mol} \cdot \text{K}) (273 \text{ K})} = 0.150 \text{ mol}$

$$\text{Molar mass} = \frac{9.86 \text{ g}}{0.150 \text{ mol}} = 65.7 \text{ g/mol}$$

PP12.22 $6.92 \text{ g } N_2\left(\dfrac{1 \text{ mol } N_2}{28.0 \text{ g } N_2}\right) = 0.247 \text{ mol } N_2$

$$V = \frac{nRT}{P}$$

$$= \frac{(0.247 \text{ mol}) (0.0821 \text{ L} \cdot \text{atm/mol} \cdot \text{K}) (273 \text{ K})}{1.00 \text{ atm}}$$

$$= 5.54 \text{ L}$$

Although the masses of the gases are the same, the numbers of moles are different because the gases have different molar masses. Since the numbers of moles are different, the volumes are different.

PP12.23 $n = \dfrac{PV}{RT} = \dfrac{(1.43 \text{ atm}) (14.6 \text{ L})}{(0.0821 \text{ L} \cdot \text{atm/mol} \cdot \text{K}) (298 \text{ K})} = 0.853 \text{ mol}$

$$\text{Molar mass} = \frac{25.6 \text{ g}}{0.853 \text{ mol}} = 30.0 \text{ g/mol}$$

$$79.9 \text{ g C} \left(\frac{1 \text{ mol C}}{12.0 \text{ g C}} \right) = 6.66 \text{ mol C}$$

$$20.1 \text{ g H} \left(\frac{1 \text{ mol H}}{1.008 \text{ g H}} \right) = 19.9 \text{ mol H}$$

$$\frac{19.9 \text{ mol H}}{6.66 \text{ mol C}} = \frac{2.99 \text{ mol H}}{1 \text{ mol C}}$$

The empirical formula is CH_3, with an empirical formula mass of 15.0 g/mol of empirical formula units:

$$\frac{30.0 \text{ g/mol}}{15.0 \text{ g/mol empirical formula units}}$$

$$= \frac{2 \text{ mol empirical formula units}}{1 \text{ mol}}$$

The molecular formula is C_2H_6.

PP12.24 $P_{N_2} = P_{\text{total}} - P_{O_2} = 1.12 \text{ atm} - 0.331 \text{ atm} = 0.79 \text{ atm}$

13 Atomic and Molecular Properties

PP13.1 S^{2-} is much bigger because it has added two electrons, but the nucleus has not changed. The greater interelectronic repulsion causes an increase in size (from 1.04×10^{-10} m to 1.84×10^{-10} m).

PP13.2 $Ca^{2+} < Ar < S^{2-}$
All three species have 18 electrons. Ca^{2+} has the most positive nucleus (20 protons) and, therefore, the smallest size; S^{2-} has the least positive nucleus (16 protons) and, therefore, the largest size.

PP13.3 (a) Ge (b) Ca

PP13.4 (a) F (b) N

PP13.5 The second ionization, which breaks the filled inner shell of electrons

PP13.7 (a) Polar covalent (b) Ionic

PP13.8 (a) Trigonal pyramidal (b) Angular

PP13.9 Trigonal planar (The four electrons of the double bond are counted as one group; the three groups repel each other to about 120° angles.)

PP13.10 The atoms bonded to the nitrogen atom are trigonally pyramidal; those around the oxygen atom are angular. Examples of different orientations show free rotation about the N—O bond:

PP13.11

	Bond Type	Molecular Polarity
(a) CS_2	Polar	No dipole
(b) BF_3	Polar	No dipole
(c) NF_3	Polar	Dipole
(d) F_2	Nonpolar	No dipole
(e) ClF	Polar	Dipole

The molecules of CS_2 and BF_3 are symmetric, so the effects of the polar bonds cancel out. In NF_3, the polar bonds do not balance one another. In diatomic molecules, such as F_2 and ClF, a polar bond always results in a polar molecule, and a nonpolar bond yields a nonpolar molecule.

PP13.12 Helium, which has the least mass and the fewest electrons

PP13.14 (a) CH_3OH forms hydrogen bonds.
(b) CH_3Br does not form hydrogen bonds.
(c) H_2O forms hydrogen bonds.

14 Solids and Liquids, Energies of Physical and Chemical Changes

PP14.1 (a) Molecular (b) Macromolecular (c) Ionic

PP14.3 $(1.5 \times 10^5)^3 = 3.4 \times 10^{15} \text{ I}_2$ molecules
This number of molecules contains 6.8×10^{15} I atoms.

PP14.4 Since molecular solids have relatively weak intermolecular forces, the molecular solid must be the one with melting point of 100°C.

PP14.5 $\text{Heat} = mc\Delta t$

$$50.1 \text{ J} = (4.00 \text{ g})(c)(6.00°C)$$

$$c = \frac{50.1}{(4.00 \text{ g})(6.00°C)} = 2.09 \text{ J/g} \cdot °C$$

PP14.7 $\text{Heat} = (8.17 \text{ g})(0.442 \text{ J/g} \cdot °C)(15.1°C) = 54.5 \text{ J}$

Heating the iron takes much less energy since its heat capacity is so much lower.

PP14.10 $0 = m_{\text{metal}} c_{\text{metal}} \Delta t_{\text{metal}} + m_{\text{water}} c_{\text{water}} \Delta t_{\text{water}}$

$$(53.2 \text{ g})(c_{\text{metal}})(30.6°C - 80.0°C)$$
$$= -(101 \text{ g})(4.184 \text{ J/g} \cdot °C)(30.6°C - 25.0°C)$$
$$c_{\text{metal}}(53.2 \text{ g})(-49.4°C) = -2.4 \times 10^3 \text{ J}$$
$$c_{\text{metal}} = 0.90 \text{ J/g} \cdot °C$$

The metal is aluminum.

PP14.11 $0 = m_{\text{metal}} c_{\text{metal}} \Delta t_{\text{metal}} + m_{\text{water}} c_{\text{water}} \Delta t_{\text{water}}$

$$0 = (10.9 \text{ g})(0.385 \text{ J/g} \cdot °C)(t_f - 80.3°C)$$
$$+ (222 \text{ g})(4.184 \text{ J/g} \cdot °C)(t_f - 21.1°C)$$
$$4.20t_f - 337°C + 929t_f - 19\,600°C = 0$$
$$933t_f = 19\,900°C$$
$$t_f = 21.3°C$$

PP14.12 $27.8 \text{ g} (2260 \text{ J/g}) = 6.28 \times 10^4 \text{ J}$

The heat required to vaporize water is much greater than that required to melt the same mass of ice.

PP14.13 For cooling the liquid water to 0°C:

$$\text{Heat} = (15.0 \text{ g})(4.184 \text{ J/g} \cdot °C)(0.0°C - 10.0°C)$$
$$= -628 \text{ J}$$

To freeze the water:

$$\text{Heat} = 15.0 \text{ g} (-335 \text{ J/g}) = -5020 \text{ J}$$

For cooling the ice:

$$\text{Heat} = (15.0 \text{ g})(2.089 \text{ J/g} \cdot °C)(-5.0°C - 0.0°C)$$
$$= -157 \text{ J}$$

Total heat $= -628 \text{ J} + (-5020 \text{ J}) + (-157 \text{ J})$
$$= -5800 \text{ J}$$

In each process, the minus sign indicates that heat is being removed.

PP14.15 Heat = $(250 \text{ g})(4.184 \text{ J/g} \cdot °\text{C})(73.7°\text{C}) = 77\,100 \text{ J} = 77.1 \text{ kJ}$

Since 77.1 kJ was added to the water, $\Delta H = -77.1 \text{ kJ}$ for the reaction.

PP14.16 $CH_4(g) + 2 O_2(g) \rightarrow CO_2(g) + 2 H_2O(l)$

$$\Delta H = \Delta H_f(CO_2) + 2 \Delta H_f(H_2O) - \Delta H_f(CH_4)$$
$$= (-393.5 \text{ kJ}) + 2(-285.9 \text{ kJ}) - (-74.5 \text{ kJ}) = -890.8 \text{ kJ}$$

PP14.17 For the products:

$$\Delta H_f = 3 \text{ mol CO}\left(\frac{-110 \text{ kJ}}{1 \text{ mol CO}}\right) + 4 \text{ mol H}_2\text{O}\left(\frac{-286 \text{ kJ}}{1 \text{ mol H}_2\text{O}}\right)$$
$$= -1474 \text{ kJ}$$

For the reactants:

$$\Delta H_f = 1 \text{ mol C}_3\text{H}_8\left(\frac{-105 \text{ kJ}}{1 \text{ mol C}_3\text{H}_8}\right) = -105 \text{ kJ}$$

$$\Delta H = \Delta H_f(\text{products}) - \Delta H_f(\text{reactants}) = (-1474 \text{ kJ}) - (-105 \text{ kJ})$$
$$= -1369 \text{ kJ}$$

For 1.00 mol, the answer is the same within limits of significant digits.

PP14.18 Adding the three equations given yields the desired equation, so we merely add the three ΔH values, yielding -70.1 kJ.

15 Solutions

PP15.1 A polar solvent would be more likely to dissolve NH_2OH because it is polar. (It is especially soluble in a solvent capable of hydrogen bonding.)

PP15.2 (a) The solution would be unsaturated in that case because it would be holding less solute than is stable at 100°C.

(b) Not all of the 190 g of solute would dissolve, and a heterogeneous mixture would result.

PP15.3 $77.3 \text{ g C}_6\text{H}_{12}\text{O}_6\left(\frac{1 \text{ mol C}_6\text{H}_{12}\text{O}_6}{180 \text{ g C}_6\text{H}_{12}\text{O}_6}\right) = 0.429 \text{ mol C}_6\text{H}_{12}\text{O}_6$

$481 \text{ mL H}_2\text{O}\left(\frac{1.00 \text{ g H}_2\text{O}}{1 \text{ mL H}_2\text{O}}\right)\left(\frac{1 \text{ kg}}{1000 \text{ g}}\right) = 0.481 \text{ kg H}_2\text{O}$

$\dfrac{0.429 \text{ mol}}{0.481 \text{ kg}} = 0.892 \text{ m}$

PP15.5 $981 \text{ g H}_2\text{O}\left(\frac{1 \text{ kg H}_2\text{O}}{1000 \text{ g H}_2\text{O}}\right)\left(\frac{3.00 \text{ mol HClO}_4}{1 \text{ kg H}_2\text{O}}\right) \times$

$\left(\frac{100.5 \text{ g HClO}_4}{1 \text{ mol HClO}_4}\right) = 296 \text{ g HClO}_4$

PP15.6 $51.9 \text{ g C}_2\text{H}_4\text{O}\left(\frac{1 \text{ mol C}_2\text{H}_4\text{O}}{44.0 \text{ g C}_2\text{H}_4\text{O}}\right)\left(\frac{1 \text{ kg H}_2\text{O}}{6.92 \text{ mol C}_2\text{H}_4\text{O}}\right) \times$

$\left(\frac{1000 \text{ g H}_2\text{O}}{1 \text{ kg H}_2\text{O}}\right) = 170 \text{ g H}_2\text{O}$

PP15.7 $2.50 \text{ kg}\left(\frac{1.50 \text{ mol}}{1 \text{ kg}}\right) = 3.75 \text{ mol solute}$

$0.750 \text{ kg}\left(\frac{3.00 \text{ mol}}{1 \text{ kg}}\right) = 2.25 \text{ mol solute}$

Total = 6.00 mol solute

Total mass of solvent = $2.50 \text{ kg} + 0.750 \text{ kg} = 3.25 \text{ kg}$

Molality = $\dfrac{6.00 \text{ mol}}{3.25 \text{ kg}} = 1.85 \text{ m}$

PP15.8 $100 \text{ g C}_2\text{H}_4\text{O}\left(\frac{1 \text{ mol C}_2\text{H}_4\text{O}}{44.0 \text{ g C}_2\text{H}_4\text{O}}\right) = 2.27 \text{ mol C}_2\text{H}_4\text{O}$

$100 \text{ g C}_2\text{H}_5\text{OH}\left(\frac{1 \text{ mol C}_2\text{H}_5\text{OH}}{46.0 \text{ g C}_2\text{H}_5\text{OH}}\right) = 2.17 \text{ mol C}_2\text{H}_5\text{OH}$

$X_{\text{C}_2\text{H}_4\text{O}} = \dfrac{2.27 \text{ mol}}{2.27 \text{ mol} + 2.17 \text{ mol}} = 0.511$

$X_{\text{C}_2\text{H}_5\text{OH}} = \dfrac{2.17 \text{ mol}}{2.27 \text{ mol} + 2.17 \text{ mol}} = 0.489$

Check: $0.511 + 0.489 = 1.000$

PP15.9 Neither substance is necessarily regarded as the solute or the solvent.

PP15.10 $P_{\text{benzene}} = X_{\text{benzene}}P°_{\text{benzene}} = (0.750)(96.0 \text{ torr}) = 72.0 \text{ torr}$

PP15.12 You get the value of k_f for naphthalene from Table 15.2:

$\Delta t_f = k_f m = (6.85°\text{C/m})(0.200 \text{ m}) = 1.37°\text{C}$

The freezing point is lowered from 80.22°C by 1.37°C, so the freezing point of the solution is

$80.22°\text{C} - 1.37°\text{C} = 78.85°\text{C}$

PP15.14 Ethylene glycol is nonvolatile and nonionic. You get the value of k_b for water from Table 15.3:

$\Delta t_b = k_b m = (0.512°\text{C/m})(0.200 \text{ m}) = 0.102°\text{C}$

$t_b = 100.00°\text{C} + 0.102°\text{C} = 100.10°\text{C}$

PP15.15 $\Delta t_b = k_b m = (2.53°\text{C/m})(0.200 \text{ m}) = 0.506°\text{C}$

$t_b = 80.1°\text{C} + 0.506°\text{C} = 80.6°\text{C}$

16 Oxidation Numbers

PP16.1 Let x represent the oxidation number of sulfur:

$x + 2(-2) = 0$

$x = +4$

PP16.3 $x + 4(-2) = -2$

$x = +6$

PP16.4 $2x + 3(-2) = -2$

$x = +2$

PP16.5 (a) $+1$ (as in all hydrogen compounds except metallic hydrides)

(b) -1 (This compound is a metallic hydride.)

PP16.7 (a) PBr_5 (b) N_2O_3

PP16.8 Carbon(II) oxide (This name is practically never used for CO.)

PP16.9 Zero (Metals do not normally have negative oxidation numbers.)

PP16.10 $+2, +1, 0$

PP16.11 $+7, +5, +3, +1$ (in steps of 2)

PP16.14 $\quad CrO_4^{2-}(aq) + I^-(aq) \rightarrow Cr^{3+}(aq) + I_2(aq) + ?$

Step 1: $\quad CrO_4^{2-}(aq) \rightarrow Cr^{3+}(aq)$

Step 2: No change.

Step 3: $\quad CrO_4^{2-}(aq) \rightarrow Cr^{3+}(aq) + 4 H_2O(l)$

Step 4: $8 H^+(aq) + CrO_4^{2-}(aq) \rightarrow Cr^{3+}(aq) + 4 H_2O(l)$

Step 5: $3 e^- + 8 H^+(aq) + CrO_4^{2-}(aq) \rightarrow$
$$Cr^{3+}(aq) + 4 H_2O(l)$$

Step 1: $\quad 2 I^-(aq) \rightarrow I_2(aq)$

Steps 2–4: No change.

Step 5: $\quad 2 I^-(aq) \rightarrow I_2(aq) + 2 e^-$

$6 I^-(aq) + 16 H^+(aq) + 2 CrO_4^{2-}(aq) \rightarrow$
$$2 Cr^{3+}(aq) + 8 H_2O(l) + 3 I_2(aq)$$

PP16.15 $Br_2(aq) + MnO_4^-(aq) \rightarrow BrO_3^-(aq) + Mn^{2+}(aq) + ?$

Oxidation half-reaction:

$$Br_2(aq) \rightarrow 2 BrO_3^-(aq)$$

$$6 H_2O(l) + Br_2(aq) \rightarrow 2 BrO_3^-(aq)$$

$$6 H_2O(l) + Br_2(aq) \rightarrow 2 BrO_3^-(aq) + 12 H^+(aq)$$
$$6 H_2O(l) + Br_2(aq) \rightarrow 2 BrO_3^-(aq) + 12 H^+(aq) + 10 e^-$$

Reduction half-reaction:

$$MnO_4^-(aq) \rightarrow Mn^{2+}(aq)$$
$$MnO_4^-(aq) \rightarrow Mn^{2+}(aq) + 4 H_2O(l)$$
$$8 H^+(aq) + MnO_4^-(aq) \rightarrow Mn^{2+}(aq) + 4 H_2O(l)$$
$$5 e^- + 8 H^+(aq) + MnO_4^-(aq) \rightarrow Mn^{2+}(aq) + 4 H_2O(l)$$

Complete balanced equation:

$$4 H^+(aq) + 2 MnO_4^-(aq) + Br_2(aq) \rightarrow$$
$$2 Mn^{2+}(aq) + 2 H_2O(l) + 2 BrO_3^-(aq)$$

PP16.16 Both the oxidizing agent and the reducing agent are H_2O_2. The H_2O_2 disproportionates according to the equation

$$2 H_2O_2(aq) \rightarrow 2 H_2O(l) + O_2(g)$$

PP16.17 (a) 2 equiv/mol (reacts with 2 mol of OH^-)
(b) 1 equiv/mol (reacts with 1 mol of OH^-)
(c) 8 equiv/mol (reacts with 8 mol of e^-)

PP16.21 $H_3PO_4(aq) + NaOH(aq) \rightarrow NaH_2PO_4(aq) + H_2O(l)$

$$\left(\frac{1.72 \text{ mol } H_3PO_4}{1 \text{ L}}\right)\left(\frac{1 \text{ equiv } H_3PO_4}{1 \text{ mol } H_3PO_4}\right) = 1.72 \text{ N } H_3PO_4$$

PP16.22 $NaH_2PO_4(aq) + 2 NaOH(aq) \rightarrow Na_3PO_4(aq) + 2 H_2O(l)$

$$90.0 \text{ g } NaH_2PO_4\left(\frac{1 \text{ mol } NaH_2PO_4}{120 \text{ g } NaH_2PO_4}\right)\left(\frac{2 \text{ equiv } NaH_2PO_4}{1 \text{ mol } NaH_2PO_4}\right)$$
$$= 1.50 \text{ equiv } NaH_2PO_4$$

$$\frac{1.50 \text{ equiv}}{0.555 \text{ L}} = 2.70 \text{ N } NaH_2PO_4$$

PP16.23 $N_1V_1 = N_2V_2$

$$N_1(42.8 \text{ mL}) = (4.50 \text{ N})(21.3 \text{ mL})$$
$$N_1 = 2.24 \text{ N}$$

17 Reaction Rates and Chemical Equilibrium

PP17.1 The equilibrium will shift to the left to use up some of the added nitrogen.

PP17.3 The equilibrium will shift to the right.

PP17.9 $K = \dfrac{[N_2][O_2]}{[NO]^2}$

PP17.10 $K = \dfrac{[N_2][O_2]^{1/2}}{[N_2O]}$

PP17.13

	2 SO₃(g)	⇌	2 SO₂(g)	+	O₂(g)
Initial	0.400		0		0.010
Change	−0.075		+0.075		+0.038
Equilibrium	0.325		0.075		0.048

$$K = \frac{[SO_2]^2 [O_2]}{[SO_3]^2} = \frac{(0.075)^2(0.048)}{(0.325)^2} = 0.0026$$
$$= 2.6 \times 10^{-3}$$

The equilibrium concentrations are essentially the same as in Example 17.13, and the value of K is the reciprocal of that in the example because the equation is written in the opposite direction.

PP17.16

	2 A	+	B	⇌	C	+	2 D
Initial	0.25		0.20		0		0
Change	−2x		−x		+x		+2x
Equilibrium	0.25 − 2x		0.20 − x		x		2x

$$K = \frac{[C] [D]^2}{[A]^2 [B]} = \frac{x(2x)^2}{(0.25 - 2x)^2(0.20 - x)} = 1.0 \times 10^{-8}$$
$$4x^3 = 1.2 \times 10^{-10}$$
$$x = 3.1 \times 10^{-4}$$

This value of [C] is very different from that in Example 17.16 because of the $[D]^2$ factor in this equilibrium expression.

18 Acid-Base Theory

PP18.1 $H_3PO_4(aq) + NaOH(aq) \rightarrow NaH_2PO_4(aq) + H_2O(l)$
$NaH_2PO_4(aq) + NaOH(aq) \rightarrow Na_2HPO_4(aq) + H_2O(l)$
$Na_2HPO_4(aq) + NaOH(aq) \rightarrow Na_3PO_4(aq) + H_2O(l)$

The acid salts are acids, but hydrogen is written first only in the formula of the anion, not in that of the compound as a whole.

PP18.2 The nitrate ion, the conjugate of a strong acid, is a feeble base; it does not react with water at all.

PP18.7

	NH₃(aq) + H₂O(l) ⇌	NH₄⁺(aq) +	OH⁻(aq)
Initial	0.120	0	0
Change	−x	+x	+x
Equilibrium	0.120 − x	x	x

$$K_b = \frac{[NH_4^+] [OH^-]}{[NH_3]} = \frac{x^2}{0.120} = 1.8 \times 10^{-5} \quad \text{(From Table 18.2)}$$
$$x = 1.5 \times 10^{-3}$$

The hydroxide ion concentration is 1.5×10^{-3} M.

PP18.9 (a) HCl is a strong acid, so $[H_3O^+]$ is 0.0010 M.
(b) NaOH is a strong, soluble base, so 0.0010 M NaOH is actually 0.0010 M Na^+ and 0.0010 M OH^-. The hydroxide ion concentration is 0.0010 M. To find the hydronium ion concentration, use the K_w expression:

$$K_w = [H_3O^+] [OH^-] = [H_3O^+] (0.0010) = 1.0 \times 10^{-14}$$

Thus, $[H_3O^+] = 1.0 \times 10^{-11}$ M.

PP18.10 1.17×10^{-8} M

PP18.11 $pH = -\log [H_3O^+] = -\log (2.9 \times 10^{-3}) = 2.54$

PP18.12 $pH = 9.19 = -\log [H_3O^+]$

Changing the sign of 9.19 and taking the antilogarithm yields the hydronium ion concentration, 6.5×10^{-10} M.

PP18.13 The pH is 7.00 because the solution is neutral.

PP18.14 The initial concentration of acetic acid (before reaction) is 0.300 M. After it reacts with water, its concentration and those of its products are given in the following table:

	HC₂H₃O₂(aq) + H₂O(l) ⇌	C₂H₃O₂⁻(aq) +	H₃O⁺(aq)
Initial	0.300	0.00	0.00
Change	−x	+x	+x
Equilibrium	0.300 − x	x	x

The equilibrium constant expression and the value of K_b can be used to solve for x:

$$K_b = \frac{[C_2H_3O_2^-] [H_3O^+]}{[HC_2H_3O_2]} = \frac{x^2}{0.300 - x} = 1.8 \times 10^{-5}$$

Neglecting x when it is subtracted from 0.300 yields

$$x^2 = 5.4 \times 10^{-6}$$

Taking the square root of both sides yields

$$x = 2.3 \times 10^{-3}$$

Since $[H_3O^+] = 2.3 \times 10^{-3}$ M,

$$pH = 2.64$$

PP18.16 A strong base is completely ionized, and its cation has no tendency to react. Any additional hydroxide ion added to this solution cannot react with the cation and thus have its concentration reduced.

PP18.17 The base ionizes according to the equation

$$NH_3(aq) + H_2O(l) \rightleftharpoons NH_4^+(aq) + OH^-(aq)$$

The ammonium chloride is a salt, so the NH_4^+ is initially 0.145 M. The ammonium ion affects the position of the equilibrium:

	$NH_3(aq) + H_2O(l) \rightleftharpoons$	$NH_4^+(aq) +$	$OH^-(aq)$
Initial	0.130	0.145	0
Change	$-x$	$+x$	$+x$
Equilibrium	$0.130 - x$	$0.145 + x$	x

Neglecting x when added to or subtracted from larger quantities and using the value of K_b from Table 18.2 yields:

$$K_b = \frac{[OH^-][NH_4^+]}{[NH_3]} = \frac{(x)(0.145)}{0.130} = 1.8 \times 10^{-5}$$

$$x = 1.6 \times 10^{-5}$$
$$[OH^-] = 1.6 \times 10^{-5} \text{ M}$$

The approximations made by neglecting x when added to 0.145 or subtracted from 0.130 are valid.

$$[H_3O^+] = \frac{K_w}{[OH^-]} = \frac{1.00 \times 10^{-14}}{1.6 \times 10^{-5}} = 6.2 \times 10^{-10}$$

$$pH = 9.21$$

19 Organic Chemistry

PP19.1 $C_2H_6(g) + 6\ Cl_2(g) \rightarrow C_2Cl_6(l) + 6\ HCl(g)$
$2\ C_2H_6(g) + 5\ O_2(g) \rightarrow 4\ CO(g) + 6\ H_2O(g)$

PP19.2 (a) $CH_2{=}CHCH_2CH_2CH_2CH_3$
(b) $CH_3CH_2CH{=}CHCH_2CH_3$

PP19.4 (a) $CH_3C{\equiv}CCH_2CH_3$ **(b)** $CH_3CH_2C{\equiv}CCH_2CH_3$

PP19.5 Four. There are two bromine atoms (as you can tell from the prefix *di*-), leaving four positions on the six carbon atoms of benzene for hydrogen atoms.

PP19.6

PP19.7 (a)

If the methyl group is attached to the first carbon atom, the longest continuous chain of carbon atoms includes the methyl group. The compound's name is butane.

(b)

The methyl group is attached to the carbon atom that is third from the left but *second* from the right. The compound's name is 2-methylbutane.

PP19.8 $\underset{1\ \ \ 2\ \ \ \ \ \ \ 3\ \ \ 4\ \ \ \ \ \ 5\ \ \ 6}{CH_3CH(CH_3)CH_2CH(CH_3)CH_2CH_3}$

PP19.9

1-Chloropropane   2-Chloropropane

PP19.12 The five- and six-carbon radicals are called pentyl and hexyl, respectively.

PP19.13 $CH_3CH_2CH_2CH_2OH$ $CH_3CH_2CH(OH)CH_3$
$(CH_3)_2CHCH_2OH$ $(CH_3)_3COH$

PP19.14 In butanone, the carbonyl group must involve one of the middle carbon atoms, and in this compound the methyl group must be on the other.

PP19.15 $\underset{\text{Propanoic acid}}{CH_3CH_2COOH(l)} + \underset{\text{Ethanol}}{HOCH_2CH_3(l)} \rightarrow$

$\underset{\text{Ethyl propanoate}}{CH_3CH_2COOCH_2CH_3(l)} + \underset{\text{Water}}{H_2O(l)}$

20 Nuclear Reactions

PP20.1 $^{214}_{82}Pb \rightarrow X + ^{\ 0}_{-1}\beta$

Thus, $X = ^{214}_{83}Bi$

PP20.3 ^{230}Th has 90 electrons, ^{226}Ra has 88 electrons, and ^4He has 2 electrons. The number of electrons on each side of the equation is the same.

PP20.4 $232 = 232 + 0 = 4(58) + 0 = 4n$
Thus ^{232}Th is a member of the $4n$ series.

PP20.5 ^{235}U loses $235 - 207 = 28$ amu in the change to ^{207}Pb. Since emissions of beta and gamma particles do not change the mass number, there must be $28/4 = 7$ alpha particles emitted.

PP20.6 The atomic number charges by 10 units, from $^{235}_{92}U$ to $^{207}_{82}Pb$. The emission of seven alpha particles (Practice Problem 20.5) lowers the atomic number by 14, so four beta particles must be emitted to reduce the atomic number by only 10 units:

Change in atomic number $= 7(-2) + 4(+1) = -10$

PP20.8 If three-quarters of the sample disintegrates, one-quarter remains. This is essentially the same problem as Example 20.8, in which one-fourth of the 1.00-kg sample remains. The answer is 80 hours.

PP20.10 $\log\left(\dfrac{m_0}{\frac{1}{4}m_0}\right) = \log 4.00 = 0.602$

$$t = \left(\frac{t_{1/2}}{0.301}\right)\log\left(\frac{m_o}{m}\right) = \frac{(12.3 \text{ seconds})(0.602)}{0.301}$$

$$= 2.00(12.3 \text{ seconds}) = 24.6 \text{ seconds}$$

Since $0.602/0.301 = 2$, it will take two half-lives to accomplish this decay. Note that the equation gives the same answer as the simple procedure used earlier for integral numbers of half-lives.

PP20.11 The ratio of masses is equal to the ratio of numbers of atoms, so this is the same problem as Example 20.11:

$$\log\left(\frac{m_o}{m}\right) = \log\left(\frac{100}{30.0}\right) = 0.523$$

PP20.13 The total number of atoms at time zero (when the rock solidified) was

$$9.07 \times 10^{18} + 5.11 \times 10^{17} = 9.58 \times 10^{18}$$

$$\log\left(\frac{9.58 \times 10^{18}}{9.07 \times 10^{18}}\right) = 0.024$$

$$t = \log\left(\frac{N_o}{N}\right)\left(\frac{t_{1/2}}{0.301}\right) = \frac{(0.024)(1.3 \times 10^9 \text{ years})}{0.301}$$

$$= 1.0 \times 10^8 \text{ years}$$

PP20.14 Since the mass numbers are the same, the masses of the isotopes are approximately the same, and the mass ratio is nearly equal to the ratio of numbers of atoms. Half of the ^{40}K atoms have therefore disintegrated, and the age of the rock is equal to the half-life of that isotope, which is 1.3×10^9 years.

PP20.15 $$\log\left(\frac{15.3 \text{ dis/min} \cdot \text{gram}}{x}\right) = \left(\frac{0.301}{5730 \text{ years}}\right)(4000 \text{ years})$$

$$= 0.210$$

$$\frac{15.3 \text{ dis/min} \cdot \text{gram}}{x} = 1.62$$

$$x = 9.44 \text{ dis/min} \cdot \text{gram}$$

PP20.16 $3^{10} = 59\,049$ neutrons

Appendix 1

PPA.6 Dollars/gallon or dollars/liter

PPA.9 $\dfrac{6.00 \text{ m/s}}{0.200 \text{ s}} = 30.0 \text{ m/s}^2$

PPA.18 $\boxed{6}\ \boxed{.}\ \boxed{1}\ \boxed{1}\ \boxed{+/-}\ \boxed{\text{EXP}}\ \boxed{7}\ \boxed{+/-}$

1 Basic Concepts

1.1 Aluminum foil does not shatter the way glass does. Glass is more brittle than aluminum foil.

1.2 C, Cu, and Cr are already used—for carbon, copper, and chromium, respectively.

1.3 (a) IB   (b) IA   (c) IIA   (d) 0   (e) VIIA

1.4 Refer to Figure 1.5 to check your answers.

1.5 Refer to Figure 1.5 to check your answers.

1.6 (a) 1 and 2. No is nobelium; NO has nitrogen and oxygen.
(b) 2 and 1. HF has hydrogen and fluorine; Hf is hafnium.
(c) 2 and 3. $PoCl_2$ has polonium and chlorine; $POCl_3$ has phosphorus, oxygen, and chlorine.
(d) 1 and 2. Si is silicon; SI_2 has sulfur and iodine.

1.7 No. All chemists use all of the branches of chemistry in their work.

1.8 No; the speed at which the person drove does not tell how far or for how long the person drove. Speed is intensive.

1.9 Buy the least expensive brand, if it is pure aspirin.

1.10 Main group elements

1.11 Elements in the same group have similar chemical properties.

1.12 Group VIII has nine elements; the others have three or four each.

1.13 (a) Matter   (b) Matter   (c) Energy   (d) Energy

1.14 The main groups have five to seven elements each, and the typical transition group has only three or four.

1.15 (a) Group IIIA (13)   (b) Third period
(c) A main group element

1.16 Inorganic chemistry

1.17 (a) Physical   (b) Chemical

1.19 The combination is a solution (a homogeneous mixture) because the saltiness persists, the amount of salt in the water can be varied, and the more salt that is added, the saltier the combination becomes. That is, the properties are affected by the percentage of each component. In contrast, a compound would have its own set of properties, and its composition would not be variable.

1.20 Striking a match is one very familiar example.

1.21 (a) Heterogeneous   (b) Heterogeneous
(c) Homogeneous   (d) Homogeneous
(e) Heterogeneous   (f) Homogeneous
(g) Heterogeneous

1.22 (a) Physical; no change in composition occurs.
(b) Chemical; a change in composition occurs.
(c) Physical; no change in composition occurs.
(d) Chemical; a change in composition occurs.
(e) Physical; no change in composition occurs.
(f) Chemical; a change in composition occurs.
(g) Physical; no change in composition occurs.

1.25 (a) Odorless   (b) Ammonia is a compound, with its own set of properties, one of which happens to be a strong smell.

1.26 (a) Mixture (solution)
(b) Compound (a combination with its own set of properties)
(c) Element

1.28 It is a compound, which has its own set of properties, as evidenced by the very high melting point.

1.30 (a) Extensive   (b) Intensive   (c) Intensive
(d) Extensive   (e) Extensive   (f) Intensive
(g) Extensive   (h) Intensive

1.32 It is a compound because it can be decomposed into simpler substances. You know that the products are simpler because each is only *part* of the original sample; each has less mass.

1.33 Iron is heavier (for a given volume), magnetic, rusts more easily, is less easily bent, and is less shiny than aluminum, among many other differences.

1.36 The price is intensive ($600 per ounce, no matter how many ounces you buy), but the cost is extensive (the more you buy, the more you pay).

1.38 One example of each is given.
(a) Alternator   (b) Battery   (c) Battery   (d) Horn
(e) Engine   (f) Starter   (g) Headlight

1.39 Chemical energy to electrical energy to light

1.41 We use batteries for their portability and convenience and as backups in case of power outages.

1.44 $\dfrac{7 \text{ elements}}{103 \text{ elements}} \times 100\% = 6.80\%$

1.50 Fluorine (F)—it is in the same group as chlorine.

1.52 Sodium is a typical metal; it is far from the metal-nonmetal dividing line on the periodic table.

1.54 Hydrogen—it is not a metal at all.

1.59 (a) VIII, 4   (b) IA, 1   (c) IIA, 4

1.61 (a) Ti and Ge   (b) Only Al

1.63 Since the law of conservation of mass states that mass cannot be created or destroyed, it is extremely doubtful that this proposed research would be successful. It would be better to spend the $1 million on a project more likely to succeed.

1.64 (a) Green   (b) No   (c) A solution retains the properties of its components, and yellow plus blue yields green. A compound has a set of properties all its own, and its color cannot be predicted from those of its elements.

1.66 Analytical chemistry

1.69 The minerals required for human health are generally eaten in the form of their compounds, not in elemental form. The patient should eat foods containing compounds of iron but refrain from eating foods containing sodium compounds.

1.71 (a) The statement implies that oil and water do not form a homogeneous mixture (solution) when mixed. *Any* substances can be mixed. A drop of oil placed in a cup of water will float on the water surface. The oil from a tanker spilled on an ocean will also float; getting the oil collected or dissolved is a major project that is too often necessary in the modern world.

(b) The statement means that one should never drive a car after having drunk an alcoholic beverage. People who "mix" gasoline and alcohol imperil their own lives and those of others who use the highways. The statement has nothing to do with the physical mixing of the liquids.

1.72 (b) $NaMnO_4$. Both Cl and Mn are in periodic groups numbered VII, and they have some chemical similarities, especially the type of formulas of their analogous compounds.

2 Measurement

2.1 (a) The dollar is bigger, so it would take more cents to buy any given purchase.

(b) The meter is bigger, so it would take more centimeters to measure any given length.

2.2 (a) The minimum wage is a rate—a ratio of dollars to hours.

(b) The amount of pay is a quantity.

(c) The number of hours worked is a quantity.

2.3 Values less than zero: (b) and (c); magnitudes less than one: (a) and (b); values less than one: (a), (b), and (c)

2.4 The kilogram is bigger, so there are more milligrams in any given mass.

2.5 All the units containing the unit *meter* are units of length: mm (millimeter) and m (meter).

2.6 The answer is 1.

2.7 The answer is 3. Be sure to *divide* by the three in the denominator.

2.8 (a) Chicken

2.9 $\dfrac{1 \text{ mm}}{0.001 \text{ m}}$ or $\dfrac{1000 \text{ mm}}{1 \text{ m}}$

2.10 (a) The final answer is the same as the original value.

(b) The format of the exponential number is changed, but its value is unchanged.

2.11 (a) 10^{-3}　**(b)** 10^3　**(c)** 10^{-2}

The appropriate exponential part may replace a metric prefix. For example,

$3 \text{ mm} = 3 \times 10^{-3} \text{ m}$

The m- for milli- has been replaced by "$\times 10^{-3}$."

2.12 (a) If the company reduces the mass of the coffee in the can without changing the cost of a can, the price effectively goes up.

(b) If you add more black coffee (or water or cream), the concentration of the sugar and therefore the sweetness is reduced.

(c) If you travel for a given time and cover less distance, your speed has gone down.

Each of these processes from everyday life shows that you can change the value of a ratio by changing either its numerator or its denominator (or both).

2.13 1 mg = 1 milligram = 0.001 g
1 Mg = 1 megagram = 1 000 000 g

Be very careful to use the standard abbreviations for units!

2.14 There are 16 cm² in the square, as shown in the accompanying figure (not drawn to scale).

$$(4.0 \text{ cm})^2 = 16 \text{ cm}^2$$

2.15 There is no essential difference. (There is not even any difference in the number of significant digits because all the quantities have two significant digits.)

2.16 (a) 3×10^3　**(b)** 1×10^6
(c) 2×10^8　**(d)** 7×10^9

2.17 (a) $6.00 \text{ dozen}\left(\dfrac{5.00 \text{ dollars}}{1 \text{ dozen}}\right) = 30.00 \text{ dollars}$

(b) $12.50 \text{ dollars}\left(\dfrac{1 \text{ dozen}}{5.00 \text{ dollars}}\right) = 2.50 \text{ dozen}$

(c) $2.50 \text{ dozen}\left(\dfrac{12 \text{ donuts}}{1 \text{ dozen}}\right) = 30 \text{ donuts}$

2.19 $7.75 \text{ hours}\left(\dfrac{60 \text{ minutes}}{1 \text{ hour}}\right)\left(\dfrac{60 \text{ seconds}}{1 \text{ minute}}\right) = 27\,900 \text{ seconds}$

$$= 2.79 \times 10^4 \text{ seconds}$$

2.21 You can change each length to yards first, then multiply:

$12.0 \text{ ft}\left(\dfrac{1 \text{ yd}}{3 \text{ ft}}\right) = 4.00 \text{ yd}$　　$15.0 \text{ ft}\left(\dfrac{1 \text{ yd}}{3 \text{ ft}}\right) = 5.00 \text{ yd}$

$4.00 \text{ yd} \times 5.00 \text{ yd} = 20.0 \text{ yd}^2$

Alternatively, you can multiply the lengths together first and then change the square feet to square yards:

$12.0 \text{ ft} \times 15.0 \text{ ft} = 180 \text{ ft}^2$

$180 \text{ ft}^2\left(\dfrac{1 \text{ yd}}{3 \text{ ft}}\right)^2 = 20.0 \text{ yd}^2$

Note that in each method you divide by 3 *twice*.

2.22 $35 \text{ weeks}\left(\dfrac{15 \text{ hours}}{1 \text{ week}}\right)\left(\dfrac{6.50 \text{ dollars}}{1 \text{ hour}}\right) = 3412.50 \text{ dollars}$

2.24 (a) $1.23 \text{ m}\left(\dfrac{1 \text{ cm}}{0.01 \text{ m}}\right) = 123 \text{ cm} = 1.23 \times 10^2 \text{ cm}$

(b) $1.24 \text{ m}\left(\dfrac{1 \text{ mm}}{0.001 \text{ m}}\right) = 1240 \text{ mm} = 1.24 \times 10^3 \text{ mm}$

(c) $1.25 \text{ m}\left(\dfrac{1 \text{ km}}{1000 \text{ m}}\right) = 0.00125 \text{ km} = 1.25 \times 10^{-3} \text{ km}$

2.27 All the units containing the base unit *liter* are units of volume, as well as those units that are the cube of a length: mL, mm³, m³, and kL.

2.28 (d) 1 km

2.31 (a) $0.0203 \text{ m}^3\left(\dfrac{1000 \text{ L}}{1 \text{ m}^3}\right) = 20.3 \text{ L}$

(b) $303 \text{ cm}^3\left(\dfrac{0.001 \text{ L}}{1 \text{ cm}^3}\right) = 0.303 \text{ L}$

(c) $403 \text{ mL}\left(\dfrac{0.001 \text{ L}}{1 \text{ mL}}\right) = 0.403 \text{ L}$

(d) $503 \text{ mm}^3\left(\dfrac{1 \text{ cm}}{10 \text{ mm}}\right)^3\left(\dfrac{1 \text{ L}}{1000 \text{ cm}^3}\right) = 0.000503 \text{ L}$
$$= 5.03 \times 10^{-4} \text{ L}$$

2.32 The volume is the product of the three dimensions. All the units need to be the same before the multiplication is performed:

$2.5 \text{ cm}\left(\dfrac{0.01 \text{ m}}{1 \text{ cm}}\right) = 0.025 \text{ m}$

$0.000020 \text{ km}\left(\dfrac{1000 \text{ m}}{1 \text{ km}}\right) = 0.020 \text{ m}$

The volume is

$(4.5 \text{ m})(0.025 \text{ m})(0.020 \text{ m}) = 0.0022 \text{ m}^3 = 2200 \text{ cm}^3$

2.35 (a) 145 mL   **(b)** 1450 mL   **(c)** The answers are not the same. The answer in part (b) has a nonsignificant zero to properly express its magnitude.

2.36 (a) Three (107)   **(b)** Four   **(c)** Four   **(d)** Four

2.37 (a) 1.00 mm   **(b)** 0.010 cm   **(c)** 100 m   **(d)** 10.0 km

2.40 (a) 7.65 cm   **(b)** 0.00730 cm   **(c)** 7.20 cm   **(d)** 7.07 cm

2.41 The measurements are 3.4 cm and 3.37 cm.

2.43 (a) 7.00×10^1 m   **(b)** 7.0001×10^2 mg   **(c)** 8.301×10^2 g
(d) 4.05×10^{-3} L

2.44 (a) 7000 cm   **(b)** 7000 cm   **(c)** 7000 cm   **(d)** 7000 cm

2.45 (a) 6.02×10^0 cm   **(b)** 6.13×10^3 cm   **(c)** 6.08×10^6 cm
(d) 4.44×10^2 cm

2.48 (a) 2.723×10^3 **(b)** 4.44×10^{-1}
(c) 6.02×10^{23} **(d)** 2.0×10^{-10}

2.49 (a) 5.0×10^{-1} **(b)** 3.6×10^3 **(c)** 3.5×10^5
(d) 3.0×10^6 **(e)** 4.2×10^{-4}

2.51 $4.50 \times 10^2 \text{ cm}\left(\dfrac{0.01 \text{ m}}{1 \text{ cm}}\right) = 4.50 \text{ m}$

$V = (4.50 \text{ m})^3 = 91.1 \text{ m}^3$

2.54 All digits in the coefficient of a properly reported number in scientific notation are significant:
(a) 1.0×10^2 cm   **(b)** 7.02×10^2 cm   **(c)** 7.0×10^2 cm
(d) 7.00×10^{-2} m   **(e)** 700 mm

2.55 (a) 5.00 mL   **(b)** 5.0 mL   **(c)** 5 mL

2.57 (a) 3.08 kg or 3.08×10^3 g   **(b)** 0.0344 kg or 34.4 g
(c) 0.221 kg or 221 g   **(d)** 2.41×10^3 cm

2.59 (a) 7.22×10^{-3} g/cm³   **(b)** 3.36×10^{-5} cm
(c) 2.582×10^{-4} cm²   **(d)** 3.40×10^{-5} kg/cm³

2.60 (a) 6.0×10^{-1} cm   **(b)** 5.02×10^2 g

Each answer is the same as the larger of the two numbers added, since the smaller value is too small to affect the last significant digit.

2.61 Change the values to liters: **(a)** 4.00 L, **(b)** 4.0 L, **(c)** 4 L

2.64 Aluminum is less dense. Therefore, a given volume (needed to make the airplane) weighs much less.

2.65 $d = \dfrac{m}{V} = \dfrac{32.8 \text{ g}}{4.75 \text{ mL}} = 6.91 \text{ g/mL}$

2.66 $\dfrac{529.8 \text{ g}}{24.70 \text{ cm}^3} = 21.45 \text{ g/cm}^3$ Platinum (Pt)

2.68 (a) 4.899 g/cm³   **(b)** 4.899 g/cm³

2.69 Gasoline is less dense than water. Water is not effective in putting out gasoline fires because the gasoline just floats to the top and burns anyway.

2.70 $709 \text{ mL}\left(\dfrac{13.6 \text{ g}}{1 \text{ mL}}\right) = 9640 \text{ g} = 9.64 \text{ kg}$

2.73 The volume of the water is

$30.0 \times 10^3 \text{ g}\left(\dfrac{1 \text{ mL}}{1.00 \text{ g}}\right) = 30.0 \times 10^3 \text{ mL} = 30.0 \times 10^3 \text{ cm}^3$

The depth of the water is equal to its volume divided by the area of its base:

$\dfrac{30.0 \times 10^3 \text{ cm}^3}{(40.0 \text{ cm})^2} = 18.8 \text{ cm}$

2.76 (a) Density   **(b)** Volume   **(c)** Mass

2.77 $\dfrac{6.25 \text{ kg}}{1 \text{ L}}\left(\dfrac{1 \text{ L}}{1000 \text{ mL}}\right)\left(\dfrac{1000 \text{ g}}{1 \text{ kg}}\right) = \dfrac{6.25 \text{ g}}{1 \text{ mL}}\left(\dfrac{1 \text{ mL}}{1 \text{ cm}^3}\right) = 6.25 \text{ g/cm}^3$

2.83 (a) $(100°\text{F} - 32°\text{F})\frac{5}{9} = 37.8°\text{C}$ **(b)** $-6.67°\text{C}$ **(c)** $25°\text{C}$
(d) $37.0°\text{C}$ **(e)** $-18°\text{C}$ **(f)** $0.0°\text{C}$ **(g)** $100°\text{C}$

2.84 (a) $100°\text{C}(\frac{9}{5}) = 180°$
$180° + 32° = 212°\text{F}$
(b) $32°\text{F}$ **(c)** $91.4°\text{F}$ **(d)** $77°\text{F}$
(e) $50°\text{F}$ **(f)** $-0.4°\text{F}$ **(g)** $-459°\text{F}$

2.85 (a) $90.0°\text{C} + 273° = 363 \text{ K}$
(b) 306 K   **(c)** 263 K   **(d)** 310 K   **(e)** 398 K

2.87 Months have different numbers of days.

2.88 A refrigerator is used to move heat from a cold area (inside) to a warmer area (outside). To move heat from a colder place to a warmer place, energy must be input. Electrical energy is most often used for this purpose, although the required energy may be generated by the combustion of gas.

2.89 (a) 31 cm³   **(b)** 1.2×10^{-4} kg/cm³   **(c)** 0.6780 g/cm³

The subtraction in part (c) yields a difference with four significant digits, which limits the overall answer to four significant digits.

2.95 You must use the same units when comparing densities (and other measured quantities). The density of water in grams per liter is

$\dfrac{1.00 \text{ g}}{1 \text{ mL}}\left(\dfrac{1 \text{ mL}}{0.001 \text{ L}}\right) = 1000 \text{ g/L}$

The oxygen will float on the water.

2.97 Since 1.0 kilometer is 0.62 mile, the kilometer is smaller, and there are more kilometers than miles between any two given points. The multiplication is converting the *number* of kilometers between two points to the number of miles between the points.

2.98 $\ell = \sqrt[3]{V} = \sqrt[3]{8.00 \text{ cm}^3} = 2.00 \text{ cm}$

2.99 (a) The discount is 10.0% of the 16.0% rate, or 1.60%. The actual rate is therefore $16.0\% - 1.60\% = 14.4\%$.
(b) The problem is to find the number of grams of iron in 100 g of the rock:

$\dfrac{21 \text{ g ore}}{100 \text{ g rock}}\left(\dfrac{72 \text{ g iron}}{100 \text{ g ore}}\right) = \dfrac{15 \text{ g iron}}{100 \text{ g rock}}$
$$= 15\% \text{ iron in the rock}$$

This problem emphasizes the importance of asking "percentage of what in what?"

2.104 (a) $2.00 \text{ mg}\left(\dfrac{0.001 \text{ g}}{1 \text{ mg}}\right) = 0.00200 \text{ g active ingredient}$

$\dfrac{0.00200 \text{ g active}}{1.00 \text{ g total}} \times 100\% = 0.200\% \text{ active}$

(b) 0.0200% active   **(c)** 20.0% active

2.109 Consider the rules of addition and subtraction, as well as multiplication and division, when you solve these problems.
(a) $(33°\text{F} - 32°\text{F}) = 1°\text{F}$ $1°\text{F}(5°/9°) = 0.6°\text{C}$
(one significant figure)

(b) $(33.0°F - 32.0°F) = 1.0°F$ $1.0°F(5°/9°) = 0.56°C$
(two significant figures)

(c) $(0°F - 32°F) = -32°F$ $-32°F(5°/9°) = -18°C$
(two significant figures)

2.110 $\ell = \sqrt[3]{V} = \sqrt[3]{73.6 \text{ cm}^3} = 4.19$ cm

To determine the cube root on a calculator, enter the number, press ⎡LOG⎤, divide by 3, press ⎡INV⎤, then ⎡LOG⎤ (or ⎡2nd F⎤ then ⎡LOG⎤). Alternatively, enter the number, press ⎡y^x⎤, enter 3, press ⎡$1/x$⎤, press ⎡=⎤.

2.111 $E = mc^2 = (1.3 \times 10^{-35} \text{ kg})(3.00 \times 10^8 \text{ m/s})^2$
$\qquad = 1.2 \times 10^{-18}$ J

2.112 (a) (Not drawn to scale)

0   1   2   3   4   5   6   7   8   9   10
Centimeters

(b) The line 10.0 cm long can be drawn with the ruler from part (a), but that ruler cannot be used to measure 0.01 cm, much less 0.010 cm. If a better measuring instrument is to be used, it should be used for both lengths, so the first does not have inaccuracies that would make the second useless.

2.113 $15.40 \text{ mL metal}\left(\dfrac{14.7 \text{ g metal}}{1 \text{ mL metal}}\right)\left(\dfrac{62.5 \text{ g gold}}{100 \text{ g metal}}\right) = 141$ g gold

2.116 $V = \dfrac{4\pi}{3}(r^3)$

$r^3 = \dfrac{3V}{4\pi} = \dfrac{3(10.00 \text{ cm}^3)}{4(3.1416)} = 2.387 \text{ cm}^3$

$r = 1.337$ cm   (See the answer to Problem 2.110, if necessary, to calculate a cube root.)

2.124 $121 \text{ mL solution}\left(\dfrac{1.19 \text{ g solution}}{1 \text{ mL solution}}\right) = 144$ g solution

$144 \text{ g solution}\left(\dfrac{25.0 \text{ g salt}}{100 \text{ g solution}}\right) = 36.0$ g salt

2.131 (a) Box is 4.76 cm.

Man is 8.26 cm.

Let x = actual height of man.

Then, $\dfrac{x}{1.00 \text{ m}} = \dfrac{8.26 \text{ cm}}{4.76 \text{ cm}}$

$x = 1.74$ m

(b) Box is $1\frac{7}{8}$ inches.

Man is $3\frac{1}{4}$ inches.

$\dfrac{x}{1.00 \text{ m}} = \dfrac{3\frac{1}{4} \text{ inches}}{1\frac{7}{8} \text{ inches}} = \dfrac{3.25 \text{ inches}}{1.875 \text{ inches}} = 1.73$

$x = 1.73$ m

(c) Decimal fractions (in the metric system) are easier to use than common fractions (in the English system).

3 Atoms and Atomic Masses

3.1 The symbol for the isotope has a superscript on its left side, denoting the mass number.

3.2 Protons and electrons

3.3 (a) $3(93 \text{ lb}) + 2(75 \text{ lb}) = 429$ lb
(b) 429 lb/5 children = 85.8 lb average

(c) $\dfrac{93 \text{ lb} + 75 \text{ lb}}{2} = 84$ lb

3.4 (a) 7   **(b)** 7   **(c)** 7+
The three parts actually ask the same question in different words.

3.5 The mass of ^{12}C

3.6 (a) amu   **(b)** e, the charge on the electron

3.7 Atomic number and atomic mass

3.8 Chemists use atomic mass and atomic weight as synonyms. The other terms have different meanings.

3.9 The mass number is an integer—the number of protons plus neutrons. The mass is an actual measured quantity and is not integral.

3.10 Elements with atomic numbers 61, 84–87, 89, and 94 and higher have mass numbers given because they do not occur naturally.

3.11 (a) K   **(b)** F
(c) K is in group IA, and F is in group VIIA.

3.12 $\dfrac{3}{2} = \dfrac{1.50}{1}$ The gambler wins $1.50 for each dollar bet. Although 1.50/1 is not an integral ratio, it is equal to the integral ratio 3/2.

3.13 According to the law of conservation of mass, 25.3 g of product must result.

3.15 (a) $5.000 \text{ g}\left(\dfrac{79.89 \text{ g C}}{100.0 \text{ g total}}\right) = 3.994$ g C

(b) There is 3.994 g of carbon in any 5.000-g portion because of the law of definite proportions.

3.16 There must be 14.37% hydrogen in this compound:
$100.00\% - 85.63\% = 14.37\%$

$\dfrac{85.63 \text{ g C}}{14.37 \text{ g H}} = \dfrac{5.959 \text{ g C}}{1 \text{ g H}}$ $\dfrac{79.89 \text{ g C}}{20.11 \text{ g H}} = \dfrac{3.973 \text{ g C}}{1 \text{ g H}}$

Per gram of hydrogen:

$\dfrac{\text{C in second compound}}{\text{C in first compound}} = \dfrac{5.959 \text{ g C}}{3.973 \text{ g C}} = \dfrac{1.500 \text{ g C}}{1 \text{ g C}}$

$\qquad\qquad\qquad\qquad\qquad = \dfrac{3.000 \text{ g C}}{2.000 \text{ g C}}$

3.22 No; the law of multiple proportions involves two compounds containing the same elements (such as CO and CO_2), in which the ratio of masses of one element is a small, whole-number ratio (for a given mass of the other element).

3.24 Per gram of oxygen, the following mass of carbon is present in each of the two compounds:

In Carbon Monoxide   *In Carbon Dioxide*

$\dfrac{42.88 \text{ g C}}{57.12 \text{ g O}} = \dfrac{0.7507 \text{ g C}}{1 \text{ g O}}$ $\dfrac{27.29 \text{ g C}}{72.71 \text{ g O}} = \dfrac{0.3753 \text{ g C}}{1 \text{ g O}}$

The ratio of grams of carbon in carbon dioxide (per gram of oxygen) to grams of carbon in carbon monoxide (per gram of oxygen) is

$\dfrac{0.7507 \text{ g C}}{0.3753 \text{ g C}} = \dfrac{2.000}{1.000}$

This ratio is, within limits of experimental error, equal to a small, whole-number ratio.

3.25 (a) $\dfrac{16 \text{ amu}}{12 \text{ amu}} = 1.3$

(b) 1600 amu O; 1200 amu C

(c) $\dfrac{1600 \text{ amu}}{1200 \text{ amu}} = 1.3$

(d) For 1 billion atoms of each:
$\dfrac{16 \times 10^9 \text{ amu}}{12 \times 10^9 \text{ amu}} = 1.3$

(e) For equal numbers of atoms, the ratio is always the same.

3.30 ^{1}H

3.31 (a) ^{1}H   (b) ^{2}H   (c) ^{3}H   (d) ^{3}H   (e) ^{3}He

3.36

	Isotopic Symbol	Atomic Number	Mass Number	No. of Protons	No. of Neutrons	No. of Electrons
(a)	^{19}F	9	19	9	10	9
(b)	^{81}Br	35	81	35	46	35
(c)	^{79}Br	35	79	35	44	35
(d)	^{230}Th	90	230	90	140	90
(e)	$^{56}_{26}$Fe	26	56	26	30	26
(f)	^{27}Al	13	27	13	14	13
(g)	^{37}Cl	17	37	17	20	17

3.37 (a) The atomic number and mass number are given. The atomic number is implied by the symbol for the element.

(b) Since the atomic number is the number of protons, the two numbers are really only one value, which would not yield the mass number or the number of neutrons.

(c) In a neutral atom, the numbers of protons and electrons are the same, and again the number of neutrons and the mass number could not be determined.

3.39 The *average* of the masses of lighter and heavier isotopes is 69.72 amu. (The actual composition is 60.4% ^{69}Ga and 39.6% ^{71}Ga.)

3.41 (a) 4.00 g

(b) $\dfrac{8(5.00\ \text{g}) + 2(3.00\ \text{g})}{10} = 4.60\ \text{g}$

(c) $\dfrac{3(35.0\ \text{amu}) + 37.0\ \text{amu}}{4} = 35.5\ \text{amu}$

3.42 Atomic mass $= (0.99759)(15.9949\ \text{amu})$
$+ (0.00037)(16.9991\ \text{amu})$
$+ (0.00204)(17.9992\ \text{amu})$
$= 15.999\ \text{amu}$

3.43 (a) 4.00 g is much too large to be the mass of an atom.

(b) 16.0 amu is the mass of an oxygen atom.

(c) $2.41 \times 10^{24}\ \text{amu}\left(\dfrac{1\ \text{g}}{6.02 \times 10^{23}\ \text{amu}}\right) = 4.00\ \text{g}$

This is the same as part (a).

(d) This again is much too large to be the mass of an atom.

3.48 That fact was crucial, since atomic numbers were as yet unknown. Because atomic mass rises as atomic number rises, and the periodic table is based on atomic number, Mendeleyev could use atomic mass to build his table.

3.50 NaF   MgF$_2$ AlF$_3$ SiF$_4$ PF$_3$ SF$_2$

3.54 (a) The oxygen in the carbon dioxide has a mass of 5.000 g − 1.364 g = 3.636 g.

(b) The oxygen in the sample of carbon monoxide has a mass half that of the oxygen in the 5.000 g of carbon dioxide: 3.636 g/2 = 1.818 g.

(c) The total mass of carbon monoxide containing 1.364 g of carbon is therefore 1.364 g + 1.818 g = 3.182 g, and the percentage of carbon is

$\left(\dfrac{1.364\ \text{g C}}{3.182\ \text{g total}}\right) \times 100\% = 42.87\%\ \text{C}$

The percentage of oxygen is 100.00% − 42.87% = 57.13%.

3.56 (b) 16.4 g of B (For the fixed mass of element A [5.00 g], there must be a small, whole-number ratio of grams of B [8.20 g to 16.4 g, or 1 g to 2 g].)

3.58 The total mass is 20.0 g. Whatever mass reacts, say x grams, must produce the same mass of products, x grams. The part that did not react, $(20.0 - x)$ grams, plus the x grams of the products gives a total of 20.0 g. The law of conservation of mass gives the same answer.

3.59 The factors are the numbers of significant digits in the percentage of occurrence and in the mass of each isotope. The mass of an isotope can be measured rather precisely with modern mass spectrometers. The percentage of each isotope present can be measured precisely for a given sample but can vary very slightly from sample to sample. Thus, the percentage is the factor that usually limits the precision of an atomic mass.

3.60 You can calculate the mass of each of the other elements per gram of any one of them. For example, per gram of sulfur, the masses of carbon and hydrogen are as follows:

(a) $\dfrac{38.66\ \text{g C}}{51.60\ \text{g S}} = \dfrac{0.7492\ \text{g C}}{1\ \text{g S}}$ $\dfrac{9.734\ \text{g H}}{51.60\ \text{g S}} = \dfrac{0.1886\ \text{g H}}{1\ \text{g S}}$

(b) $\dfrac{47.31\ \text{g C}}{42.10\ \text{g S}} = \dfrac{1.124\ \text{g C}}{1\ \text{g S}}$ $\dfrac{10.59\ \text{g H}}{42.10\ \text{g S}} = \dfrac{0.2515\ \text{g H}}{1\ \text{g S}}$

(c) $\dfrac{53.27\ \text{g C}}{35.55\ \text{g S}} = \dfrac{1.498\ \text{g C}}{1\ \text{g S}}$ $\dfrac{11.18\ \text{g H}}{35.55\ \text{g S}} = \dfrac{0.3145\ \text{g H}}{1\ \text{g S}}$

The gram ratios can be simplified by dividing each set by the smallest value in the set:

Ratios of Masses of Carbon
0.7492 g : 1.124 g : 1.498 g
1.000   : 1.500   : 1.999

Ratios of Masses of Hydrogen
0.1886 g : 0.2515 g : 0.3145 g
1.000   : 1.334   : 1.668

The numbers in the set on the left can be made nearly integral by multiplying each one by 2; those in the set on the right can be made integral by multiplying each by 3:

2.000 : 3.000 : 3.998   3.000 : 4.002 : 5.004

The resulting values are integers within the accuracy of the four significant digits reported. If you chose to calculate the masses per gram of carbon or per gram of hydrogen, the same ratios should have been obtained.

3.63 Since the entire group of noble gases was unknown, it was not apparent to Mendeleyev that any elements were missing.

3.64 $\dfrac{51.82(106.90509\ \text{amu}) + 48.18(108.9047\ \text{amu})}{100} = 107.9\ \text{amu}$

(The number of significant digits is limited by the percentages.)

3.69 It is one-trillionth:

$\dfrac{V_n}{V_a} = \dfrac{r_n^{\ 3}}{r_a^{\ 3}} = (10^{-4})^3 = 10^{-12}$

3.70 HO

4 Electronic Configuration of the Atom

4.1 Lr is in period 7, so it has n values up to 7—1, 2, 3, 4, 5, 6, and 7.

4.2 (a) One   (b) All three   (c) No   (d) Yes

4.3 (a) Since the s subshell contains only one orbital, there is no difference.

(b) A p subshell contains three p orbitals.

4.4 (a) Yes; the sign can be either plus or minus, and since both have equal energy, either is correct.

(b) No; once the first electron has a given sign, the second electron must have the opposite sign so as not to violate the Pauli exclusion principle.

(c) Yes; the sign can be either plus or minus, and since both have equal energy, either is correct.

(d) No; once the other five electrons in the $3p$ subshell have given signs, the last electron must have the opposite sign to the other electron in its orbital so as not to violate the Pauli exclusion principle.

4.5 (a) No   (b) At least one must be different.
(c) No   (d) At least one must be different.

4.6 Hydrogen

4.7 $(-1.634 \times 10^{-18} \text{ J}) + (-3.025 \times 10^{-19} \text{ J}) = -1.936 \times 10^{-18}$ J

The same energy change is accomplished in going from the third shell to the second to the first as in going directly from the third to the first (PP 4.1, part b).

4.8 Helium's outermost shell is complete.

4.9 Electrical energy gives some electrons enough energy to move to higher energy levels; when they fall back to their ground states, the added energy is given off in the form of light.

4.10 Eight paths are possible for an electron descending from the fifth shell (energy level) to the first in a hydrogen atom (see Figure 4.3):

One path:	shell 5 to 4 to 3 to 2 to 1
Another path:	shell 5 to 4 to 3 to 1
A third path:	shell 5 to 4 to 2 to 1
A fourth path:	shell 5 to 4 to 1
A fifth path:	shell 5 to 3 to 2 to 1
A sixth path:	shell 5 to 3 to 1
A seventh path:	shell 5 to 2 to 1
An eighth path:	shell 5 to 1

4.11 Ten distinctly different energies of light would be emitted, corresponding to the transitions from shell 5 to 4, 5 to 3, 5 to 2, 5 to 1, 4 to 3, 4 to 2, 4 to 1, 3 to 2, 3 to 1, and 2 to 1.

4.12 $\ell = 0, 1, 2, 3,$ or 4

4.13 $s = -\frac{1}{2}$ or $+\frac{1}{2}$, no matter what the values of the other quantum numbers are.

4.14 $m = -3, -2, -1, 0, 1, 2,$ or 3

4.18 Only n and ℓ have an effect on energy. Since the n values are the same for the three electrons and the ℓ values are also the same, the electrons are degenerate (their energies are the same).

4.20 The lowest $n + \ell$ values are those of (b) and (d).
   (a) $n = 5, \ell = 1, n + \ell = 6$
   (b) $n = 5, \ell = 0, n + \ell = 5$
   (c) $n = 4, \ell = 2, n + \ell = 6$
   (d) $n = 4, \ell = 1, n + \ell = 5$

Since (d) has a lower n value than (b), it is lowest in energy. Of the other two electrons, with higher $n + \ell$ values, (c) is lower in energy since it has a lower n value. The order of energies is (d) < (b) < (c) < (a).

4.23 (a) 8, oxygen
   (b) 28, nickel
   (c) 33, arsenic

4.24 They are the same; both equal $2\ell + 1$.

4.28
Li	$1s^2\, 2s^1$
Na	$1s^2\, 2s^2\, 2p^6\, 3s^1$
K	$1s^2\, 2s^2\, 2p^6\, 3s^2\, 3p^6\, 4s^1$
Rb	$1s^2\, 2s^2\, 2p^6\, 3s^2\, 3p^6\, 4s^2\, 3d^{10}\, 4p^6\, 5s^1$

Cs and Fr also have one electron in their outermost s subshells:
Cs   [Xe] $6s^1$ Fr   [Rn] $7s^1$

4.32 (a) A p subshell has an ℓ value of 1, so three different m values and two different s values are possible, for a total of six combinations. There may be up to six electrons in a p subshell.
   (b) In any orbital, there can be a maximum of two electrons.

4.34 The Pauli exclusion principle, the $n + \ell$ rule, and the permitted values of the quantum numbers

4.36
N	$1s^2\, 2s^2\, 2p^3$
P	$1s^2\, 2s^2\, 2p^6\, 3s^2\, 3p^3$
As	$1s^2\, 2s^2\, 2p^6\, 3s^2\, 3p^6\, 4s^2\, 3d^{10}\, 4p^3$
Sb	$1s^2\, 2s^2\, 2p^6\, 3s^2\, 3p^6\, 4s^2\, 3d^{10}\, 4p^6\, 5s^2\, 4d^{10}\, 5p^3$

4.37 (a) All three   (b) Two, the d_{z^2} and $d_{x^2-y^2}$

4.38 d_{xz} and d_{yz}

4.41 (a) None   (b) Four   (c) Six

4.42 See the accompanying figures. The numbers of unpaired electrons are
   (a) Three   (b) None   (c) One   (d) Two

Phosphorus   Calcium

Fluorine   Oxygen

4.45

4.46 (a) Four

   (b) Five

   (c) Four (Two are paired.)

4.47 (c) The seven electrons lie in the lowest energy orbitals possible, with the three electrons in the $2p$ subshell in separate orbitals with parallel spins.

4.48 (a) Rb   [Kr] $5s^1$
   (b) Ra   [Rn] $7s^2$
   (c) Sn   [Kr] $5s^2\, 4d^{10}\, 5p^2$

4.50 The differences corresponding to the possible transitions of Problem 4.10 are as follows:

From 5 to 4:	4.898×10^{-20} J
From 5 to 3:	1.549×10^{-19} J
From 5 to 2:	4.574×10^{-19} J
From 5 to 1:	2.091×10^{-18} J
From 4 to 3:	1.059×10^{-19} J
From 4 to 2:	4.084×10^{-19} J
From 4 to 1:	2.042×10^{-18} J
From 3 to 2:	3.025×10^{-19} J
From 3 to 1:	1.936×10^{-18} J
From 2 to 1:	1.634×10^{-18} J

4.51 Three. If there are $2p$ electrons in the ground state, the $1s$ and $2s$ subshells must be full, and there can be only three unpaired electrons maximum.

4.56 Mendeleyev used the chemical properties of the elements to make the periodic table. The chemical properties result from the electronic configurations.

4.59 **(a)** Mn   **(b)** Sn   **(c)** Bi   **(d)** Rb   **(e)** Ga   **(f)** Gd

4.60 No. Every element with an atomic number greater than nine has this *inner* configuration.

4.69 The atom with the most unpaired electrons will be drawn into the magnetic field the most.
(a) V (has three unpaired electrons)
(b) Mn (has five unpaired electrons)
(c) Rh (has three unpaired electrons)
(d) Mn (has five unpaired electrons)

5 Chemical Bonding

5.1 The first is a bromide of cobalt; the second contains carbon and oxygen as well as bromine.

5.2 Ionic: **(b)** $MgCl_2$
Covalent: **(a)** Cl_2 and **(c)** SCl_2

5.3 **(a)** *Diatomic* means two atoms per molecule; *binary* means two elements per compound.
(b) The valence shell is the outermost shell of an uncombined atom. If all the valence electrons have been removed, the outermost shell of the resulting positive ion will be the shell just below the valence shell.

5.4 Eight, seven from the outermost shell of chlorine and one extra transferred from some cation

5.5 **(a)** 0   **(b)** 3+   **(c)** 13+ (equal to the atomic number)
Note the great difference in the meaning of the questions with only a slight difference in the wording.

5.6 **(a)** Hydrogen is a group IA element, but it is not a metal.
(b) Since hydrogen can form H^- ions, only the first statement is correct.

5.7 **(a)** No; it has no electrons. (This ion is not a stable species, but chemists often use the symbol H^+ to represent H_3O^+.)
(b) Yes, it has the electronic structure of He.

5.8 **(a)** The alkali metals, the alkaline earth metals, silver, zinc, cadmium, and aluminum
(b) The group IA and group IB metals

5.9 **(a)** 1+   **(b)** 3+   **(c)** 1−   **(d)** 3−
The metals of (a) and (b) have charges equal to their classical group numbers. For (c) and (d), each monatomic anion has a charge equal to its classical group number minus 8 (or the modern group number minus 18).

5.10 Atoms of the main group elements except for the noble gases have valence electrons equal in number to their (classical) periodic group numbers.

5.11 **(a)** $NiCl_2$ **(b)** Ni^{2+} and Br^-
(c) ZnO   **(d)** Zn^{2+} and S^{2-}

5.12 **(a)** 2+   **(b)** 2+   **(c)** 0   **(d)** 2+   **(e)** 2+
The charge on the calcium ion in each of its compounds is 2+ (equal to its group number). In the free element, as in all free elements, the charge on the atom is zero.

5.13 NaBr in all cases. Note the different ways in which the same problem may be presented.

5.14 The first is a compound, and the second is an ion—part of a compound.

5.15 **(a)** Ionic
(b) Covalent
(c) Covalent

5.16 NH_3 and NH_4^+

5.17 **(a)** Mg: **(b)** Mg^{2+} (no valence electrons)
(c) :Ṡ: **(d)** $\left[:\ddot{S}: \right]^{2-}$

5.18 **(a)** Na^+ :H^-
(b) $H:^-$ Ca^{2+} :H^- or Ca^{2+} 2 :H^-

5.19 The charge on a monatomic anion is equal to the classical group number minus 8. Generally, the charge on a polyatomic anion is even if the central element is in an even periodic group and odd if it is in an odd periodic group.

5.20 **(a)** $H:^-$ **(b)** Li^+ **(c)** He:
All of these species have the same electronic configuration, but since only valence electrons are shown, the electron dot diagram for the lithium ion *looks* different.

5.21 **(a)** None; the two valence electrons of the magnesium atom have been donated to form the ion.
(b) The magnesium atom does not share electrons; it forms ionic compounds.

5.22 **(a)** Ag_2S **(b)** Ag^+ and O^{2-}

5.23 There are two aluminum atoms, three sulfur atoms, and nine oxygen atoms per formula unit of the compound. Moreover, the three oxygen atoms in each SO_3 group are bonded to the sulfur atom in some way, and the SO_3 groups are bonded to the aluminum atoms in some way.

5.24 Since the two mercury atoms are written together, they are bonded in some way. And since they are each part of a cation, the bonding between them is covalent.

5.25 **(a)** 3 Na, 1 As
(b) 1 Al, 3 Cl, 12 O
(c) 4 N, 16 H, 2 P, 7 O

5.27

Symbol	Atomic Number	No. of Protons	No. of Electrons	Net Charge
(a) Na^+	11	11	10	1+
(b) S^{2-}	16	16	18	2−
(c) N^{3-}	7	7	10	3−
(d) Ca^{2+}	20	20	18	2+
(e) Be^{2+}	4	4	2	2+

(a) Sodium has an atomic number of 11 (see the periodic table) and therefore has 11 protons in its nucleus. Since the ion is shown with a 1+ charge, the number of electrons is one less than the number of protons, so 10 electrons are present.
(b) The element with atomic number 16 is sulfur. Since the number of protons is equal to the atomic number and there are two more electrons than protons, the net charge is 2−. The symbol for this ion is therefore S^{2-}. *Do not forget the charge!*
(c) Since the number of protons is always equal to the atomic number, the same reasoning is used here as in part (b).
(d) The atomic number is 20, so the element is calcium, and the number of protons is also 20. Since the ion has a 2+ charge, there must be two fewer electrons than protons. The symbol is Ca^{2+}.
(e) The number of protons is 4, as is the atomic number. The element is beryllium, and the symbol is Be^{2+}.

5.29 There is no real difference, but beginning students find the former easier to interpret.

5.31 **(a)** Br^- $1s^2\,2s^2\,2p^6\,3s^2\,3p^6\,4s^2\,3d^{10}\,4p^6$
(b) N^{3-} $1s^2\,2s^2\,2p^6$
(c) S^{2-} $1s^2\,2s^2\,2p^6\,3s^2\,3p^6$

5.33 In each case, the $4s$ or $5s$ electrons are donated first.
(a) $1s^2\,2s^2\,2p^6\,3s^2\,3p^6\,3d^7$
(b) $1s^2\,2s^2\,2p^6\,3s^2\,3p^6\,4s^2\,3d^{10}\,4p^6\,4d^{10}$
(c) $1s^2\,2s^2\,2p^6\,3s^2\,3p^6\,3d^4$

5.34 **(a)** LiH   **(b)** CaH_2

5.36 **(a)** K_2S (Don't forget that the positive ion is written first.)
(b) Cu_2S **(c)** CuS

5.37 (a) K^+ (b) Cd^{2+} (c) Li^+ (d) Ba^{2+} (e) Al^{3+}

5.38 The net charge on each formula unit must be zero.

	I^-	S^{2-}	P^{3-}
Na^+	NaI	Na_2S	Na_3P
Ba^{2+}	BaI_2	BaS	Ba_3P_2
Al^{3+}	AlI_3	Al_2S_3	AlP

5.40 (a) K^+ Br^- (b) Ca^{2+} O^{2-} (c) Li^+ S^{2-}
(d) Mg^{2+} N^{3-} (e) Na^+ P^{3-} (f) Li^+ N^{3-}
(g) Cr^{3+} F^-

5.42 (a) $1 \ Na^+ \ 1 \ Br^-$ (b) $1 \ Ba^{2+} \ 1 \ S^{2-}$
(c) $1 \ Ba^{2+} \ 2 \ Br^-$ (d) $2 \ Na^+ \ 1 \ S^{2-}$

5.43 Two; there are four valence electrons in the neutral tin atom (group IVA[14]), of which two are donated to some other atom(s) in forming a 2+ ion. (The electronic configuration is $[Kr] \ 5s^2 \ 4d^{10} \ 5p^0$.)

5.46

	Chlorine	*Sulfur*	*Phosphorus*
Potassium	KCl	K_2S	K_3P
Magnesium	$MgCl_2$	MgS	Mg_3P_2
Aluminum	$AlCl_3$	Al_2S_3	AlP

5.49 (a) $\left[:\overset{..}{\underset{..}{P}}:\right]^{3-}$ (b) $\left[:\overset{..}{\underset{..}{S}}:\right]^{2-}$ (c) $\left[:\overset{..}{\underset{..}{Br}}:\right]^{-}$

5.51

	Atoms		*Ions*	
(a)	Li·	:N·	Li^+	$\left[:\overset{..}{\underset{..}{N}}:\right]^{3-}$
(b)	Ba:	:Br:	Ba^{2+}	$\left[:\overset{..}{\underset{..}{Br}}:\right]^{-}$
(c)	·Al:	:S:	Al^{3+}	$\left[:\overset{..}{\underset{..}{S}}:\right]^{2-}$

5.53 Ionic bonding only: (b) $MgCl_2$
Covalent bonding only: (d) I_2 (e) PCl_3 (f) CF_4
Both types: (a) $(NH_4)_2SO_4$ (c) $Ca(ClO_4)_2$

5.55 (a) :N:::N: (b) Lewis structure of PBr_3 (c) Lewis structure of CCl_4

(d) H:N:H with H below   (e) Lewis structure of CBr_2O (f) Lewis structure of SO_2

5.57 The hydrogen atom can have only two electrons in its outermost shell, and with two shared electrons, it can form only one single bond.

5.58 (a) $:\overset{..}{\underset{..}{I}}-\overset{\overset{\textstyle :\overset{..}{\underset{..}{I}}:}{|}}{P}-\overset{..}{\underset{..}{I}}:$ (b) $H-C\equiv N:$ (c) $:\overset{..}{\underset{..}{Cl}}-\overset{\overset{\textstyle :\overset{..}{\underset{..}{O}}:}{|}}{\underset{\underset{\textstyle :\overset{..}{\underset{..}{O}}:}{|}}{S}}-\overset{..}{\underset{..}{Cl}}:$

5.64

Atoms	*Valence Electrons Available*	*Valence Electrons Required*
H	1	2
O	6	8
C	4	8
N	5	8
Total	16	26

The number of shared electrons is $26 - 16 = 10$. Since the hydrogen atom can share a maximum of only two electrons, the structure is

$$H:\overset{..}{\underset{..}{O}}:C:::N: \quad \text{or} \quad H-\overset{..}{\underset{..}{O}}-C\equiv N:$$

5.65

	NO_3^-	SO_4^{2-}	CO_3^{2-}	PO_4^{3-}
K^+	KNO_3	K_2SO_4	K_2CO_3	K_3PO_4
Mg^{2+}	$Mg(NO_3)_2$	$MgSO_4$	$MgCO_3$	$Mg_3(PO_4)_2$
Fe^{2+}	$Fe(NO_3)_2$	$FeSO_4$	$FeCO_3$	$Fe_3(PO_4)_2$
Fe^{3+}	$Fe(NO_3)_3$	$Fe_2(SO_4)_3$	$Fe_2(CO_3)_3$	$FePO_4$

5.67 (a) Na^+ $Cr_2O_7^{2-}$ (b) La^{3+} OH^- (c) Li^+ SO_4^{2-}
(d) VO^{2+} SO_4^{2-} (e) Li^+ NO_3^- (f) Na^+ PO_4^{3-}
(g) K^+ HCO_3^- (h) NH_4^+ BrO_3^- (i) NH_4^+ SO_4^{2-}

5.70 (a) $\overset{..}{\underset{..}{O}}::\overset{..}{O}:\overset{..}{\underset{..}{O}}$ (b) ring structure of S_6

5.71 (a) Li_2S (b) $(NH_4)_3PO_4$ (c) $Cr_2(CO_3)_3$

Note that the cation is written first in each compound, no matter which is stated first in the problem.

5.75 (a) $Na^+\left[:\overset{..}{O}::C::\overset{..}{N}:\right]^-$ or $Na^+\left[:\overset{..}{\underset{..}{O}}:C:::N:\right]^-$ or $Na^+\left[:O:::C:\overset{..}{\underset{..}{N}}:\right]^-$

(b) $H:\overset{H}{\underset{H}{C}}:\overset{H}{N}:H$ (c) $H:\overset{H}{\underset{H}{C}}:\overset{H}{\underset{H}{C}}:\overset{..}{\underset{..}{O}}:H$

5.76 You deduce the number of shared electrons in the usual way:

Atoms	*Valence Electrons Available*	*Valence Electrons Required*
S	6	8
C	4	8
N	5	8
Negative charge	1	
Total	16	24

To be shared: $24 - 16 = 8$

$$\left[:\overset{..}{\underset{..}{S}}::C::\overset{..}{N}:\right]^- \quad \text{or} \quad \left[:\overset{..}{\underset{..}{S}}=C=\overset{..}{\underset{..}{N}}:\right]^-$$

5.77 An atom with a total of eight electrons in its outermost shell (*s* and *p* subshells) is stable.

5.80 (a) An ion is a charged atom or group of atoms.
(b) A cation is a positively charged ion.
(c) Monatomic means "one-atom"; a monatomic ion is composed of one charged atom.
(d) Ozone is an allotropic form of oxygen with formula O_3.
(e) A noble gas configuration has eight (or two for the lightest elements) valence electrons.
(f) A triple bond is a covalent bond formed by six shared electrons.
(g) A lone pair is a pair of electrons not shared between atoms.
(h) An octet is a group of eight electrons in the outermost shell of an atom.

5.81 (a) $:\overset{..}{\underset{..}{O}}=\overset{\underset{\underset{\textstyle :\overset{..}{\underset{..}{O}}:}{|}}{}}{S}-\overset{..}{\underset{..}{O}}:$ (b) $\left[:\overset{..}{\underset{..}{O}}-\overset{\underset{\underset{\textstyle :\overset{..}{\underset{..}{O}}:}{|}}{}}{S}-\overset{..}{\underset{..}{O}}:\right]^{2-}$

(c) $2 \ Na^+ \left[:\overset{..}{\underset{..}{O}}-\overset{\underset{\underset{\textstyle :\overset{..}{\underset{..}{O}}:}{|}}{}}{S}-\overset{..}{\underset{..}{O}}:\right]^{2-}$

(d) $H-\overset{..}{\underset{..}{O}}-\overset{\underset{\underset{\textstyle :\overset{..}{\underset{..}{O}}:}{|}}{}}{S}-\overset{..}{\underset{..}{O}}-H$

5.82 The charge is 3+. No monatomic ion can have a charge of 5+, which might have been expected because bismuth is in periodic group VA (15).

5.87 (a) Eight; one valence electron has been donated to form the ion.
(b) The ion does not share electrons with other ions; it forms ionic compounds by electron transfer.

5.88 Remember that elemental nitrogen and bromine occur in diatomic molecules.
(a) $6 \ Li· + :N:::N: \rightarrow 6 \ Li^+ + 2\left[:\overset{..}{\underset{..}{N}}:\right]^{3-}$
(b) $Ba: + :\overset{..}{\underset{..}{Br}}:\overset{..}{\underset{..}{Br}}: \rightarrow Ba^{2+} + 2\left[:\overset{..}{\underset{..}{Br}}:\right]^-$

5.103 An "infinite" number

5.104 (a) Four   (b) Two

6 Nomenclature

6.1 (a) Carbon monoxide   (b) Cobalt

6.2 The alkali metals, the alkaline earth metals, zinc, cadmium, silver, and aluminum form a single type of ion with charge equal to the metal's classical periodic group number.

6.3 The prefix *bi-* represents hydrogen in an acid salt, as in sodium bicarbonate; the prefix *di-* means two atoms of some element.

6.4 (a) Ionic, lithium chloride
(b) Covalent, iodine monochloride
(c) Covalent, phosphorus trifluoride
(d) Ionic, aluminum fluoride

6.5 (a) Except for the hydride ion, the charge on a monatomic anion is equal to the classical group number minus 8.
(b) The charge on an oxoanion is generally even if the central element is in an even-numbered periodic group and odd if it is in an odd-numbered group.

6.6 The first is a compound, chlorine trioxide, and the second is an ion, chlorate ion. The ion has one more electron.

6.7 (a) Constant type, calcium chloride
(b) Variable type, copper(II) chloride
(c) Constant type, silver chloride
(d) Variable type, iron(II) chloride

6.8 Hydrochloric acid refers to the compound in aqueous solution; hydrogen chloride is used for the pure compound.

6.9 H_2SO_4 is not a *binary* nonmetal-nonmetal compound.

6.10 (a) You can tell the numbers of each type of ion in a formula unit of the compound.
(b) The compound (or ion) is an acid (or acid salt).
(c) There is oxygen in the compound or ion.

6.11 (a) Hydrogen ion is H^+; hydride ion is H^-.
(b) Although hydrogen forms ions with two different charges, the ion with the negative charge is distinguished by the ending *-ide*, so the Roman numeral is unnecessary in the name of H^+.

6.12 (a) Cl^- (b) ClO_3^- (c) ClO_2^-

6.13 (a) *Hydro-* indicates an acid that contains no oxygen.
(b) *Hydrogen* is used in acid salts, but not in covalent acids.
(c) *Hypo-* indicates an oxygen-containing acid with two fewer oxygen atoms than the corresponding acid whose name ends in *-ate.*

6.14 (a) Sodium phosphate   (b) Phosphoric acid

6.15 NH_3 is a compound; NH_4^+ is an ion. The compound is ammonia; the ion is the ammonium ion.

6.16 (a) $1-$ (HCO_3^-)   (b) $1-$ ($H_2PO_4^-$)   (c) $2-$ (HPO_4^{2-})

6.17 (a) Perchloric acid   (b) Chloric acid   (c) Chlorous acid
(d) Hypochlorous acid   (e) Hydrochloric acid

6.18 (a) Iodine pentafluoride   (b) Water
(c) Arsenic pentachloride   (d) Sulfur trioxide
(e) Phosphorus tribromide

6.19 (a) SO_2 (b) CCl_4 (c) PCl_5 (d) AsF_3 (e) NH_3

6.20 (a) Silicon tetrafluoride   (b) Silicon dioxide
(c) Hydrogen chloride
(d) Hydrogen sulfide (rarely, hydrosulfuric acid)
(e) Bromine monochloride   (f) Iodine trifluoride
(g) Nitrogen tribromide

6.21 (a) Sulfur difluoride   (b) Sulfur tetrafluoride
(c) Sulfur hexafluoride

6.26 (a) Potassium ion   (b) Barium ion   (c) Cadmium ion

6.27 (a) Phosphide ion   (b) Oxide ion   (c) Nitride ion
(d) Iodide ion

6.30 (a) Chromate ion   (b) Acetate ion
(c) Dichromate ion   (d) Permanganate ion
(e) Peroxide ion   (f) Cyanide ion

6.31 The word *chloride* refers only to the ion; the word *sodium* can refer to the atom, the element, or the ion.

6.33 (a) Ca^{2+} (b) Mn^{2+} (c) Ag^+ (d) NH_4^+
(e) Hg_2^{2+}

6.35 (a) Potassium sulfate   (b) Aluminum cyanide
(c) Ammonium phosphate

6.37 (a) $Ni(ClO_3)_2$ (b) $Co(OH)_3$ (c) $MgSO_4$
(d) CuO (e) $LiCN$ (f) $(NH_4)_2CO_3$

6.39

	Bromate	*Sulfite*	*Phosphate*	*Acetate*
Sodium	$NaBrO_3$	Na_2SO_3	Na_3PO_4	$NaC_2H_3O_2$
Chromium(II)	$Cr(BrO_3)_2$	$CrSO_3$	$Cr_3(PO_4)_2$	$Cr(C_2H_3O_2)_2$
Iron(III)	$Fe(BrO_3)_3$	$Fe_2(SO_3)_3$	$FePO_4$	$Fe(C_2H_3O_2)_3$
Ammonium	NH_4BrO_3	$(NH_4)_2SO_3$	$(NH_4)_3PO_4$	$NH_4C_2H_3O_2$

6.42

	Hydrogen Sulfate	*Sulfate*	*Acetate*
Mercury(II)	$Hg(HSO_4)_2$	$HgSO_4$	$Hg(C_2H_3O_2)_2$
Cobalt(II)	$Co(HSO_4)_2$	$CoSO_4$	$Co(C_2H_3O_2)_2$
Iron(III)	$Fe(HSO_4)_3$	$Fe_2(SO_4)_3$	$Fe(C_2H_3O_2)_3$

6.43 The instructor would say, "The sodium(II) ion cannot be prepared in a solid."

6.44 There is no oxygen in hydrobromic acid, HBr. Bromic acid is $HBrO_3$.

6.46 (a) HBr   (b) H_3PO_4 (c) $HClO_4$ (d) H_2SO_3

6.47 Phosphorus is the name of the element, and phosphorous is the name of the oxoacid with one fewer oxygen atoms than phosphoric acid.

6.51 (a) Acid salt, sodium hydrogen sulfate
(b) Regular salt, sodium sulfate
(c) Acid, sulfuric acid

6.52 (a) Cadmium sulfate heptahydrate
(b) Calcium bromide hexahydrate

6.53 (a) $CaCO_3\cdot6H_2O$
(b) $Cr(C_2H_3O_2)_3\cdot H_2O$
(c) $Co_3(PO_4)_2\cdot2H_2O$

6.54 The binary nonmetal-nonmetal compounds—(b) phosphorus trichloride and (c) carbon tetrachloride—are named using the prefixes of Table 6.1. Metals having differently charged ions use Roman numerals: (f) cobalt(II) carbonate. Metals forming only one type of ion use neither the prefixes nor Roman numerals: (a) sodium hypochlorite, (d) barium chloride, and (e) cadmium oxide.

6.57 (a) Sulfur trioxide   (b) Sulfite ion   (c) Sodium sulfite
(d) Sulfurous acid   (e) Cobalt(II) sulfite

6.61 (a) IF_5 (b) $Ti_2(SO_3)_3$ (c) HIO
(d) $Cr_2(SO_3)_3$ (e) $BiCl_3$

6.69

	NO_3^-	CO_3^{2-}	PO_4^{3-}
NH_4^+	NH_4NO_3, ammonium nitrate	$(NH_4)_2CO_3$, ammonium carbonate	$(NH_4)_3PO_4$, ammonium phosphate
Mn^{2+}	$Mn(NO_3)_2$, manganese(II) nitrate	$MnCO_3$, manganese(II) carbonate	$Mn_3(PO_4)_2$, manganese(II) phosphate
Fe^{3+}	$Fe(NO_3)_3$, iron(III) nitrate	$Fe_2(CO_3)_3$, iron(III) carbonate	$FePO_4$, iron(III) phosphate

6.73 (a) Sodium hydrogen carbonate
(b) Chromium(II) chloride
(c) Iron(II) nitrate
(d) Cobalt(II) sulfide

6.74 LiHS

6.75 H₂S

6.78 H₂S, HS⁻, and H₂S

6.80

	Formula	Corresponding Acid	Corresponding Anion
(a)	HS⁻	H₂S, hydrosulfuric acid	S²⁻, sulfide ion
(b)	H₂PO₄⁻	H₃PO₄, phosphoric acid	PO₄³⁻, phosphate ion
(c)	HSO₃⁻	H₂SO₃, sulfurous acid	SO₃²⁻, sulfite ion
(d)	HCO₃⁻	H₂CO₃, carbonic acid	CO₃²⁻, carbonate ion

6.83 (a) Sodium dihydrogen phosphate
   (b) Sodium monohydrogen phosphate
   (c) Phosphoric acid
   (d) Sodium phosphate

6.85 (a) Nickel(II) ion
   (b) Gold(III) ion
   (c) Mercury(II) ion
   (d) Chromium(III) ion
   (e) Lead(II) ion

6.88

Formula	Name	Type
BaCO₃	Barium carbonate	IC
PCl₃	Phosphorus trichloride	C
H₂SO₄	Sulfuric acid	A
NH₄NO₃	Ammonium nitrate	IC
MnSO₄	Manganese(II) sulfate	IV
SF₄	Sulfur tetrafluoride	C
Co₃(PO₄)₂	Cobalt(II) phosphate	IV
HNO₃	Nitric acid	A
CCl₄	Carbon tetrachloride	C
FeCl₃	Iron(III) chloride	IV
P₂S₃	Diphosphorus trisulfide	C
Cu₂S	Copper(I) sulfide	IV
Mg(OH)₂	Magnesium hydroxide	IC
BrF₃	Bromine trifluoride	C
HCl	Hydrogen chloride or	C
	Hydrochloric acid	A
BF₃	Boron trifluoride	C
AgBr	Silver bromide	IC
H₃PO₄	Phosphoric acid	A
MnO₂	Manganese(IV) oxide	IV
CoF₃	Cobalt(III) fluoride	IV
KOH	Potassium hydroxide	IC

6.89

Name	Formula	Type
Iron(III) oxide	Fe₂O₃	IV
Nickel(II) sulfide	NiS	IV
Iodine trifluoride	IF₃	C
Lithium hydride	LiH	IC
Gold(I) oxide	Au₂O	IV
Calcium hydroxide	Ca(OH)₂	IC
Nitrous acid	HNO₂	A
Ammonium sulfide	(NH₄)₂S	IC
Magnesium sulfide	MgS	IC
Phosphoric acid	H₃PO₄	A
Hydrochloric acid	HCl	A
Potassium sulfate	K₂SO₄	IC
Ammonium phosphate	(NH₄)₃PO₄	IC

6.93

Formula	Name
Ca(OH)₂	Calcium hydroxide
HNO₂	Nitrous acid
TiO	Titanium(II) oxide
MgCrO₄	Magnesium chromate
Ni(NO₂)₂	Nickel(II) nitrite
Na₂Cr₂O₇	Sodium dichromate
CrCl₂	Chromium(II) chloride
NCl₃	Nitrogen trichloride
SF₆	Sulfur hexafluoride
H₂SO₄	Sulfuric acid

Formula	Name
NaHSO₃	Sodium hydrogen sulfite
BaCl₂	Barium chloride
IF₇	Iodine heptafluoride
OF₂	Oxygen difluoride
LiCl	Lithium chloride
Co(OH)₂	Cobalt(II) hydroxide
AuCl₃	Gold(III) chloride
(NH₄)₂SO₃	Ammonium sulfite
MnO	Manganese(II) oxide
SnSO₄	Tin(II) sulfate
FeBr₂	Iron(II) bromide
NH₃	Ammonia
MgF₂	Magnesium fluoride
Li₃N	Lithium nitride
Ba(OH)₂	Barium hydroxide
FeCl₂	Iron(II) chloride
(NH₄)₂S	Ammonium sulfide
XeF₂	Xenon difluoride
ICl	Iodine monochloride
CaH₂	Calcium hydride
CI₄	Carbon tetraiodide
AgCl	Silver chloride
H₃PO₄	Phosphoric acid
AgC₂H₃O₂	Silver acetate
PbS	Lead(II) sulfide
Zn(ClO₃)₂	Zinc chlorate
CS₂	Carbon disulfide
CoCO₃	Cobalt(II) carbonate
HClO₃	Chloric acid
NaCN	Sodium cyanide

6.94

Name	Formula
Manganese(III) oxide	Mn₂O₃
Cobalt(II) sulfide	CoS
Boron trifluoride	BF₃
Sodium hydride	NaH
Gold(I) bromide	AuBr
Calcium hydride	CaH₂
Carbon tetrachloride	CCl₄
Lithium nitride	Li₃N
Magnesium perchlorate	Mg(ClO₄)₂
Acetic acid	HC₂H₃O₂
Copper(II) sulfate	CuSO₄
Chlorous acid	HClO₂
Copper(I) oxide	Cu₂O
Bromine	Br₂
Bromine trichloride	BrCl₃
Bromic acid	HBrO₃
Sodium hydrogen sulfate	NaHSO₄
Calcium hydrogen sulfate	Ca(HSO₄)₂
Nitric acid	HNO₃
Hydrochloric acid	HCl
Lithium sulfite	Li₂SO₃
Ammonium dihydrogen phosphate	NH₄H₂PO₄
Sodium hydroxide	NaOH
Calcium carbonate	CaCO₃
Nitrogen triiodide	NI₃
Potassium cyanide	KCN
Chloric acid	HClO₃
Sodium arsenate	Na₃AsO₄
Ammonium selenate	(NH₄)₂SeO₄

7 Formula Calculations

7.1 (a) amu   (b) amu   (c) g/mol   (d) grams
   (e) amu

7.2 (a) Multiply by 2 to get 3/2.
   (b) Multiply by 3 to get 4/3.

(c) Multiply by 4 to get 5/4.
(d) Multiply by 3 to get 5/3.
(e) Multiply by 4 to get 7/4.
(f) Multiply by 4 to get 9/4.

7.3 Only C_6H_{13} and CH are empirical formulas. For each of the other formulas, the subscripts can be evenly divided by a small integer.

7.4 The compound contains 87.0% carbon, no matter what the size of the sample is. Percent composition is an intensive property; that is, compounds have definite compositions.

7.5 (a) The formula and the atomic masses of the elements
(b) The percent composition or other mass data and the atomic masses
(c) The mass data (as for empirical formula calculations) and a molecular mass, or the empirical formula itself and the molecular mass

7.6 (a) 36 pairs   (b) 36 socks   (c) 18 pairs
(d) 72 socks   (e) 1.5 dozen pairs
(f) 6 dozen socks

7.7 The parts of this problem mirror those of Problem 7.6. Instead of pairs of socks, there are pairs of hydrogen atoms bonded into molecules; instead of dozens, there are moles. Note that each answer is different, although some of the numbers in the answers are the same.
(a) 1.81×10^{24} H_2 molecules   (b) 1.81×10^{24} H atoms
(c) 9.03×10^{23} H_2 molecules   (d) 3.61×10^{24} H atoms
(e) 1.50 mol H_2 molecules   (f) 6.00 mol H atoms

7.8 Na_2O_2 and $Na_2S_2O_8$ are ionic—they contain Na^+ ions—but the formulas are not empirical formulas. (Each anion has an even charge and an even number of atoms of each element.)

7.9 (a) C_3H_8 (b) C_5H_4O (c) $C_{12}H_{22}O_{11}$ (d) CH_2O

7.10 (a) A dozen watermelons weighs more because each watermelon weighs more than each grape. There are the same number of fruits; a dozen is 12 in each case.
(b) A mole of uranium weighs more because each uranium atom weighs more than each lithium atom. There are the same number of atoms in each case, since a mole of atoms is 6.02×10^{23} atoms.

7.11 (a) The socks weigh the same, whether or not they are rolled together.
(b) The atoms have the same mass, whether or not they are bonded.
(c) The 1.00 mol of nitrogen molecules has double the mass because there are twice as many atoms as 1.00 mol of nitrogen atoms.

7.12 (a) The atomic mass refers to an F atom; the molecular mass refers to an F_2 molecule.
(b) The phrase might refer to the mass of a mole of fluorine atoms or the mass of a mole of fluorine molecules.
(c) It refers to the molar mass of F_2 molecules because fluorine gas exists in the form of diatomic molecules.

7.13 (a) Percent composition
(b) Empirical formula
(c) Molecular formula

7.14 The molar mass (formula mass in grams per mole)

7.15 Avogadro's number is used to convert the number of formula units to moles, and then the molar mass is used to convert the number of moles to grams (mass). (See Figure 7.1.)

7.16 Both NH_3 and S_8 are substances that exist in the form of molecules and thus have molecular masses. Ne can be considered a monatomic molecule, in which case its molecular mass is equal to its atomic mass. All of the substances have molar masses, whether they exist in the form of molecules or not.

7.17 You can do calculations that lead you to the empirical formula.

7.18 The formula yields the following information:

Nitrogen, hydrogen, sulfur, and oxygen are present, and all the atoms are bonded together in some way.

The four hydrogen atoms and the nitrogen atom of each NH_4^+ ion are covalently bonded together.

There are two nitrogen atoms, eight hydrogen atoms, one sulfur atom, and three oxygen atoms per formula unit.

There are two ammonium ions per sulfite ion in each formula unit. The formula mass is 116.14 amu, and the molar mass is 116.14 g/mol.

The percentage of sulfur is $(32.06/116.14) \times 100.0\% = 27.60\%$, the percentage of nitrogen is $[(2 \times 14.007)/116.14] \times 100.0\% = 24.12\%$, and the percentages of the other elements can be calculated similarly.

The ratio of hydrogen atoms to nitrogen atoms is 8 to 2 (and similarly for all the other ratios).

7.19 (a) 1.0 ft

(b) 4.0 ft

$$1 \text{ dozen boxes}\left(\frac{4.0 \text{ inches}}{1 \text{ box}}\right)\left(\frac{1 \text{ foot}}{12 \text{ inches}}\right)\left(\frac{12 \text{ boxes}}{1 \text{ dozen}}\right) = 4.0 \text{ ft}$$

(c) 8.0 ft
(d) The factor (12) that converts inches to feet also converts boxes to dozens of boxes, and the factors cancel each other.

7.20 *Rounding First*
(a) 4(12.01 amu) + 10(1.01 amu) = 58.14 amu
(b) 22.99 amu + 35.45 amu = 58.44 amu
(c) 2(10.81 amu) + 6(1.01 amu) = 27.68 amu

Rounding Last
(a) 4(12.011 amu) + 10(1.0080 amu) = 58.12 amu
(b) 22.9898 amu + 35.453 amu = 58.44 amu
(c) 2(10.81 amu) + 6(1.0080 amu) = 27.67 amu

The rounding can make a small difference.

7.21 (a) 2 Na 2×23.0 amu = 46.0 amu
S   = 32.1 amu
4 O 4×16.0 amu = 64.0 amu
Total = 142.1 amu

(b)   Ca   = 40.1 amu
2 C 2×12.0 amu = 24.0 amu
2 N 2×14.0 amu = 28.0 amu
2 O 2×16.0 amu = 32.0 amu
Total = 124.1 amu

(c) 98.0 amu   (d) 96.0 amu   (e) 114 amu   (f) 7.9 amu

7.22 (a) 116 g/mol   (b) 261 g/mol   (c) 158 g/mol

7.23 (a) 1.008 amu, for a hydrogen atom
(b) 2.016 amu, for an H_2 molecule
(If you ever get a smaller value for an atomic mass or molecular mass, you know that there is a mistake somewhere.)

7.24 3 C 3×12.01 amu = 36.03 amu
5 H 5×1.008 amu = 5.040 amu
3 N 3×14.0067 amu = 42.020 amu
9 O 9×16.00 amu = 144.0 amu
Total = 227.1 amu

$$\% \text{ C} = \left(\frac{36.03 \text{ amu}}{227.1 \text{ amu}}\right) \times 100.0\% = 15.87\%$$

$$\% \text{ H} = \left(\frac{5.040 \text{ amu}}{227.1 \text{ amu}}\right) \times 100.0\% = 2.219\%$$

$$\% \text{ N} = \left(\frac{42.02 \text{ amu}}{227.1 \text{ amu}}\right) \times 100.0\% = 18.50\%$$

$$\% \text{ O} = \left(\frac{144.0 \text{ amu}}{227.1 \text{ amu}}\right) \times 100.0\% = \underline{63.41\%}$$

$$\text{Total} = 100.00\%$$

7.28 Each compound consists of 85.63% carbon and 14.37% hydrogen. They have the same mass ratio of carbon to hydrogen because they have the same mole ratio of carbon to hydrogen; that is, they have the same empirical formula (CH_2).

7.35 (a) 2 mol Al, 6 mol Cr, 21 mol O
(b) 1 mol Ba, 2 mol O, 2 mol H
(c) 2 mol N, 8 mol H, 1 mol S

7.36 (a) $3.000 \text{ g C}_2\text{H}_6\left(\dfrac{1 \text{ mol C}_2\text{H}_6}{30.07 \text{ g C}_2\text{H}_6}\right) = 0.09977 \text{ mol C}_2\text{H}_6$

(b) $0.09977 \text{ mol C}_2\text{H}_6\left(\dfrac{2 \text{ mol C atoms}}{1 \text{ mol C}_2\text{H}_6}\right) = 0.1995 \text{ mol C atoms}$

(c) $0.1995 \text{ mol C atoms}\left(\dfrac{6.022 \times 10^{23} \text{ C atoms}}{1 \text{ mol C atoms}}\right)$
$$= 1.201 \times 10^{23} \text{ C atoms}$$

7.37 $7.30 \text{ g}\left(\dfrac{1 \text{ mol}}{130.1 \text{ g}}\right) = 0.0561 \text{ mol}$

7.39 $17.0 \text{ mol CH}_2\text{O}\left(\dfrac{30.0 \text{ g CH}_2\text{O}}{1 \text{ mol CH}_2\text{O}}\right) = 510 \text{ g CH}_2\text{O}$

7.41 (a) $2.50 \text{ mol Cl}\left(\dfrac{35.5 \text{ g Cl}}{1 \text{ mol Cl}}\right) = 88.8 \text{ g Cl}$

(b) Same as part (a)

(c) $2.50 \text{ mol Cl}_2\left(\dfrac{70.9 \text{ g Cl}_2}{1 \text{ mol Cl}_2}\right) = 177 \text{ g Cl}_2$

7.44 $3.000 \text{ mol C}_4\text{H}_{10}\left(\dfrac{4 \text{ mol C}}{1 \text{ mol C}_4\text{H}_{10}}\right)\left(\dfrac{6.022 \times 10^{23} \text{ C atoms}}{1 \text{ mol C}}\right)$
$$= 7.226 \times 10^{24} \text{ C atoms}$$

7.45 $22.20 \text{ mol NH}_3\left(\dfrac{3 \text{ mol H}}{1 \text{ mol NH}_3}\right)\left(\dfrac{6.02 \times 10^{23} \text{ H atoms}}{1 \text{ mol H}}\right)$
$$= 4.01 \times 10^{25} \text{ H atoms}$$

7.46 $6.78 \times 10^{22} \text{ O atoms}\left(\dfrac{1 \text{ mol O atoms}}{6.02 \times 10^{23} \text{ O atoms}}\right) \times \left(\dfrac{1 \text{ mol H}_2\text{SO}_4}{4 \text{ mol O atoms}}\right)$
$$= 0.0282 \text{ mol H}_2\text{SO}_4$$

7.47 $7.77 \times 10^{26} \text{ H atoms}\left(\dfrac{1 \text{ mol H atoms}}{6.02 \times 10^{23} \text{ H atoms}}\right)\left(\dfrac{1 \text{ mol H}_2\text{O}}{2 \text{ mol H atoms}}\right)$
$$= 645 \text{ mol H}_2\text{O}$$

7.48 $2.30 \text{ mol AgBr}\left(\dfrac{1 \text{ mol Ag}}{1 \text{ mol AgBr}}\right)\left(\dfrac{107.9 \text{ g Ag}}{1 \text{ mol Ag}}\right) = 248 \text{ g Ag}$

7.49 $2.278 \text{ g C}_2\text{H}_6\text{O}\left(\dfrac{1 \text{ mol C}_2\text{H}_6\text{O}}{46.07 \text{ g C}_2\text{H}_6\text{O}}\right) = 0.04945 \text{ mol C}_2\text{H}_6\text{O}$

$0.04945 \text{ mol C}_2\text{H}_6\text{O}\left(\dfrac{2 \text{ mol C atoms}}{1 \text{ mol C}_2\text{H}_6\text{O}}\right) = 0.09890 \text{ mol C atoms}$

$0.09890 \text{ mol C atoms}\left(\dfrac{6.022 \times 10^{23} \text{ C atoms}}{1 \text{ mol C atoms}}\right)$
$$= 5.956 \times 10^{22} \text{ C atoms}$$

7.58 $6.098 \times 10^{24} \text{ C atoms}\left(\dfrac{1 \text{ molecule C}_2\text{H}_4}{2 \text{ C atoms}}\right)$
$$= 3.049 \times 10^{24} \text{ molecules C}_2\text{H}_4$$

7.61 $1 \text{ atom}\left(\dfrac{1.000 \text{ mol}}{6.022 \times 10^{23} \text{ atoms}}\right)\left(\dfrac{12.01 \text{ g}}{1 \text{ mol}}\right) = 1.994 \times 10^{-23} \text{ g}$

7.62 (a) No; both subscripts can be divided by 2.
(b) Yes
(c) Yes
(d) No; both subscripts can be divided by 2.
(e) No; the subscript can be divided by 8.

7.64 $C_2H_5NO_2$

7.65 The empirical formula is SO_3. If you used only one significant digit in the calculations, you might get an incorrect answer, SO_4, because of rounding errors.

7.69 (a) 92.26 g of carbon and 7.74 g of hydrogen
(b) 7.682 mol C and 7.68 mol H
(c) 1:1   (d) CH

7.75 The empirical formula mass of NO_2 is 46.0 g/mol of empirical formula units.

(a) $\dfrac{46.0 \text{ g/mol}}{46.0 \text{ g/mol empirical formula units}}$

$$= \dfrac{1 \text{ mol empirical formula units}}{1 \text{ mol}}$$

The molecular formula is NO_2.

(b) $\dfrac{92.0 \text{ g/mol}}{46.0 \text{ g/mol empirical formula units}}$

$$= \dfrac{2 \text{ mol empirical formula units}}{1 \text{ mol}}$$

The molecular formula is N_2O_4.

Another way to look at this problem is to calculate the molar mass for each of the possible molecular formulas corresponding to the empirical formula, and then simply choose the formula that matches the correct molecular mass:

Possible Molecular Formula	Corresponding Molar Mass
NO_2	46.0 g/mol
N_2O_4	92.0 g/mol
N_3O_6	138 g/mol
⋮	⋮

7.77 Empirical formula: CH_2O; molecular formula: $C_6H_{12}O_6$

7.78 (a) $H_2C_2O_4$ (b) $C_3H_6O_2$ (c) $C_3H_8O_3$ (d) $C_6H_4N_2O_6$

7.82 The mass per mole of empirical formula units is 30.97 g/mol. Therefore,

$$\dfrac{123.9 \text{ g/mol}}{30.97 \text{ g/mol empirical formula units}}$$

$$= \dfrac{4 \text{ mol empirical formula units}}{1 \text{ mol}}$$

The formula is P_4.

7.83 (a) $\% \text{ error} = \dfrac{\text{true value} - \text{value}}{\text{true value}} \times 100\%$

$$= \dfrac{15.9994 \text{ amu} - 16.0 \text{ amu}}{15.9994 \text{ amu}} \times 100\% = 0.004\%$$

(b) 0.083%   (c) 0.13%   (d) 0.20%

7.84 (a) $\% \text{ Co} = \dfrac{58.93 \text{ g/mol Co}}{\text{MM}} \times 100\% = 4.348\% \text{ Co}$

$$\text{MM} = \dfrac{(58.93 \text{ g/mol})(100\%)}{4.348\%} = 1355 \text{ g/mol}$$

(b) $\% \text{ O} = \dfrac{2(16.0 \text{ g/mol}) \text{ O}}{\text{MM}} \times 100\% = 4.03\%$

$$MM = \frac{3200 \text{ g/mol}}{4.03} = 794 \text{ g/mol}$$

7.90 $25.7 \text{ g}\left(\dfrac{60.0 \text{ g water}}{100 \text{ g total}}\right) = 15.4 \text{ g water}$

$15.4 \text{ g}\left(\dfrac{1 \text{ mol}}{18.0 \text{ g}}\right)\left(\dfrac{6.02 \times 10^{23} \text{ molecules}}{1 \text{ mol}}\right)$
$$= 5.15 \times 10^{23} \text{ molecules}$$

7.91 Since the total of the percentages must be 100.00%, the percentage of hydrogen must be

100.00% total − 83.62% C = 16.38% H

First, calculate the mass of carbon and the mass of hydrogen in 1.00 mol of compound:

	Formula Mass	*Percentage*

$1.00 \text{ mol compound}\left(\dfrac{86.2 \text{ g compound}}{1 \text{ mol compound}}\right)\left(\dfrac{83.62 \text{ g C}}{100 \text{ g compound}}\right)$
$$= 72.1 \text{ g C}$$

$1.00 \text{ mol compound}\left(\dfrac{86.2 \text{ g compound}}{1 \text{ mol compound}}\right)\left(\dfrac{16.38 \text{ g H}}{100 \text{ g compound}}\right)$
$$= 14.1 \text{ g H}$$

Now convert the number of grams of each element to moles of the element:

$72.1 \text{ g C}\left(\dfrac{1 \text{ mol C}}{12.01 \text{ g C}}\right) = 6.00 \text{ mol C}$

$14.1 \text{ g H}\left(\dfrac{1 \text{ mol H}}{1.008 \text{ g H}}\right) = 14.0 \text{ mol H}$

The molecular formula is C_6H_{14}. Note that there is no need to try to get integral numbers of moles if an exact molar mass is given; the values come out very nearly integrals. (If an approximate molar mass is given, the numbers of moles of each element must be rounded to the nearest integer. See Problem 7.92.)

7.96 $23.10 \text{ g H}\left(\dfrac{1 \text{ mol H}}{1.008 \text{ g H}}\right)\left(\dfrac{2 \text{ mol C}}{6 \text{ mol H}}\right) = 7.639 \text{ mol C}$

7.97 (a) With percent composition data, you can find the empirical formula.
(b) To get a molecular formula using an empirical formula, we need the molar mass, so the rest of the data must be usable to calculate the molar mass of the compound.

8 Chemical Reactions

8.1 (a) Combination **(b)** Decomposition
(c) Substitution **(d)** Decomposition
(e) Double substitution **(f)** Combination
(g) Combustion

8.2 There is no difference.

8.3 Table 8.2 should be used with substitution reactions, and Table 8.3 should be used with double substitution reactions.

8.4 The more likely product is $Fe(NO_3)_3$; the charge on the cation will not change in a double substitution reaction.

8.5 This is a combustion reaction, for which the complete balanced equation is

$$C_4H_{10}(g) + 6.5 \text{ } O_2(g) \rightarrow 4 \text{ } CO_2(g) + 5 \text{ } H_2O(g)$$

8.6 N_2O_3, SO_2, and Cl_2O_7 are nonmetal oxides and thus are acidic anhydrides; CaO is a metal oxide and therefore a basic anhydride.

8.7 Inactive metal oxides are more likely to decompose. Metals that are classified as active have a great tendency to combine because they are more stable in the combined state.

8.8 $HC_3H_5O_2$ and H_2Se are acids, since their formulas have H written first. Water is not an acid in the Arrhenius sense, even though its formula has H written first.

8.9 An acid salt, such as $NaHSO_4$ (The H is written first in the formula for the *anion,* but the cation is written before the anion in the formula for the compound.)

8.10 Oxygen cannot replace oxygen and yield any net change. However nickel can exist as two different monatomic cations (see Figure 5.10). These reactants undergo a combination reaction:

$$O_2(g) + 2 \text{ NiO}(s) \rightarrow 2 \text{ NiO}_2(s)$$

8.11 (a) The strong acids (HCl, HBr, HI, $HClO_3$, $HClO_4$, HNO_3, and H_2SO_4) react with water to form ions.
(b) None of the common acids is ionic when in the pure form.

8.12 The equations are balanced as written.

8.13 (a) This is a substitution (displacement) reaction, with bromine displacing iodine in the compound.
(b) Br_2, KI, KBr, and I_2 are the formulas.
(c) $Br_2(aq) + 2 \text{ KI}(aq) \rightarrow 2 \text{ KBr}(aq) + I_2(aq)$

8.14 The reaction has two driving forces: the formation of a covalent compound and the formation of an insoluble compound.

8.15 Both cause a reaction to happen without undergoing permanent change themselves.

8.16 (a) $2 \text{ CrF}_2(s) + F_2(g) \rightarrow 2 \text{ CrF}_3(s)$
(b) $4 \text{ NH}_3(g) + 5 \text{ O}_2(g) \rightarrow 4 \text{ NO}(g) + 6 \text{ H}_2O(g)$
(c) $2 \text{ CuCl}(s) + Cl_2(g) \rightarrow 2 \text{ CuCl}_2(s)$
(d) $2 \text{ H}_3PO_4(aq) + 3 \text{ Ba(OH)}_2(aq) \rightarrow Ba_3(PO_4)_2(s) + 6 \text{ H}_2O(l)$

8.17 (a) KCl and O_2 (Since the catalyst does not change its composition permanently, this is a reaction of one compound; it is a decomposition. This decomposition is a common method for preparing oxygen in the laboratory.)
(b) KCl and O_2 (The same products will be produced, since the reactants are the same as in part [a]. It does not matter whether the question identifies MnO_2 as a catalyst or not. This question is harder, since you must remember that the MnO_2 is a catalyst in this reaction.)
(c) KCl and O_2
(d) KCl and O_2 (In the absence of the catalyst, the reaction will proceed more slowly, but the same products are expected.)

8.18 (a) One molecule **(b)** 1 mol of H_2

8.20 (a) 4 Na, 4 Cl, 8 O **(b)** 8 N, 24 H
(c) 2 Co, 2 C, 6 O **(d)** 3 Ba, 6 N, 15 O, 6 H
(e) 6 N, 36 H, 6 P, 18 O

8.21 The process may be started as follows:

$$? \text{ } O_2(g) + ? \text{ } N_2(g) \rightarrow ? \text{ } N_2O(g)$$

With one molecule of each substance, the numbers of oxygen atoms and nitrogen atoms on the two sides of the equation do not match, so the equation is not balanced. You can change the numbers of formula units by substituting numbers for the question marks in front of the formulas. When coefficients are used as follows, the equation is balanced:

$$O_2(g) + 2 \text{ N}_2(g) \rightarrow 2 \text{ N}_2O(g) \quad \text{(Balanced)}$$

The two before N_2O indicates that there are two N_2O molecules containing four N atoms and two O atoms. Since there are four N atoms and two O atoms on each side of the equation, the equation is balanced.

8.22 (a) $2 \text{ C}_3H_7OH(l) + 9 \text{ O}_2(g) \rightarrow 6 \text{ CO}_2(g) + 8 \text{ H}_2O(g)$
(b) $C_3H_8(g) + 5 \text{ O}_2(g) \rightarrow 3 \text{ CO}_2(g) + 4 \text{ H}_2O(g)$
(c) $BiCl_3(aq) + H_2O(l) \rightarrow BiOCl(s) + 2 \text{ HCl}(aq)$
(d) $2 \text{ Sb}_2S_3(s) + 9 \text{ O}_2(g) \rightarrow 2 \text{ Sb}_2O_3(s) + 6 \text{ SO}_2(g)$
(e) $2 \text{ CH}_2O(l) + O_2(g) \rightarrow 2 \text{ CO}(g) + 2 \text{ H}_2O(g)$
(f) $CO_2(g) + H_2(g) \rightarrow CO(g) + H_2O(g)$
(g) $2 \text{ ZnS}(s) + 3 \text{ O}_2(g) \rightarrow 2 \text{ ZnO}(s) + 2 \text{ SO}_2(g)$
(h) $Cu_2S(s) + O_2(g) \rightarrow 2 \text{ Cu}(s) + SO_2(g)$

(i) $4 H_2O(l) + PCl_5(s) \rightarrow 5 HCl(aq) + H_3PO_4(aq)$

(j) $O_2(g) + 6 FeO(s) \rightarrow 2 Fe_3O_4$

(k) $Na_2SO_3(aq) + S(s) \rightarrow Na_2S_2O_3(aq)$

(l) $AlCl_3(aq) + 4 NaOH(aq) \rightarrow NaAl(OH)_4(aq) + 3 NaCl(aq)$

(m) $CuSO_4 \cdot 5H_2O(aq) + 4 NH_3(aq) \rightarrow$
$$CuSO_4 \cdot 4NH_3(aq) + 5 H_2O(l)$$

8.24 $2 CuCl_2(aq) + 4 KI(aq) \rightarrow 2 CuI(s) + I_2(aq) + 4 KCl(aq)$

8.26 **(a)** $4 Li(s) + O_2(g) \rightarrow 2 Li_2O(s)$
(b) $2 Na(s) + O_2(g) \rightarrow Na_2O_2(s)$
(c) $K(s) + O_2(g) \rightarrow KO_2(s)$

8.29 With limited NaOH:

$$H_2SO_4(aq) + NaOH(aq) \rightarrow NaHSO_4(aq) + H_2O(l)$$

With excess NaOH:

$$H_2SO_4(aq) + 2 NaOH(aq) \rightarrow Na_2SO_4(aq) + 2 H_2O(l)$$

8.30 According to the table, alkali metal sulfides and ammonium sulfide are soluble.

8.32 **(a)** $SO_3(g) + CaO(s) \rightarrow CaSO_4(s)$
(b) $SO_2(g) + 2 NaOH(aq) \rightarrow Na_2SO_3(aq) + H_2O(l)$
(c) $SO_2(g) + CaO(s) \rightarrow CaSO_3(s)$

8.33 The products are water and either CO or CO_2.

With limited oxygen:

$$2 C_6H_6(l) + 9 O_2(g) \rightarrow 12 CO(g) + 6 H_2O(g)$$

With excess oxygen:

$$2 C_6H_6(l) + 15 O_2(g) \rightarrow 12 CO_2(g) + 6 H_2O(g)$$

8.35 **(a)** $NaCl(aq) + AgNO_3(aq) \rightarrow AgCl(s) + NaNO_3(aq)$
(b) $2 HCl(aq) + Ba(OH)_2(aq) \rightarrow BaCl_2(aq) + 2 H_2O(l)$
(c) $CH_4(g) + 2 O_2(g) \rightarrow CO_2(g) + 2 H_2O(g)$
(d) $2 Al(s) + 3 Cl_2(g) \rightarrow 2 AlCl_3(s)$
(e) $4 Al(s) + 3 O_2(g) \rightarrow 2 Al_2O_3(s)$

8.37 **(a)** $Fe(s) + 2 HCl(aq) \rightarrow FeCl_2(aq) + H_2(g)$
(b) $2 Fe(s) + 3 Cl_2(g) \rightarrow 2 FeCl_3(s)$

8.38 **(a)** Reducing the relative amount of chlorine
(b) $CrCl_2$
(c) $Cr(s) + 2 CrCl_3(s) \xrightarrow{\text{Heat}} 3 CrCl_2(s)$
(d)

$$\begin{array}{ccc} Cl & & Cl \\ | & & | \\ Cr \; Cl \; Cr \; Cl & Cr & \\ | & & | \\ Cl & & Cl \end{array}$$

8.39 No reaction is expected in any part.
(a) No substitution reaction is possible, and Cl_2 cannot add to $ZnCl_2$.
(b) Ne does not react with anything.
(c) Mg is more active than Al.
(d) H is more active than Cu.

8.43 **(a)** $2 HCl(aq) \xrightarrow{\text{Electricity}} H_2(g) + Cl_2(g)$

(b) $2 H_2O(l) \xrightarrow[\text{Na}_2\text{SO}_4]{\text{Electricity}} 2 H_2(g) + O_2(g)$

(c) $2 NaCl(l) \xrightarrow{\text{Electricity}} 2 Na(l) + Cl_2(g)$

8.44 **(a)** $BaCl_2(aq) + (NH_4)_2SO_4(aq) \rightarrow BaSO_4(s) + 2 NH_4Cl(aq)$
(b) $Ba(ClO_3)_2(aq) + MgSO_4(aq) \rightarrow BaSO_4(s) + Mg(ClO_3)_2(aq)$
(c) $NH_3(g) + HCl(aq) \rightarrow NH_4Cl(aq)$
(d) $3 Ba(s) + 2 FeCl_3(s) \xrightarrow{\text{Heat}} 3 BaCl_2(s) + 2 Fe(s)$
(e) $3 Cl_2(g) + 2 AlI_3(aq) \rightarrow 2 AlCl_3(aq) + 3 I_2(aq)$
(f) $Ba(C_2H_3O_2)_2(aq) + Na_2CO_3(aq) \rightarrow$
$$BaCO_3(s) + 2 NaC_2H_3O_2(aq)$$

8.45 Double substitution reactions

8.46 **(a)** $2 FeCl_2(s) + Cl_2(g) \rightarrow 2 FeCl_3(s)$
(b) $PCl_5(l) + 4 H_2O(l) \rightarrow H_3PO_4(aq) + 5 HCl(aq)$
(c) $PCl_3(l) + Cl_2(g) \rightarrow PCl_5(l)$
(d) $3 Mg(s) + N_2(g) \rightarrow Mg_3N_2(s)$

8.48 No. As soon as it is prepared, it will decompose.

8.50 **(a)** $Zn(s) + 2 HClO_4(aq) \rightarrow H_2(g) + Zn(ClO_4)_2(aq)$
(b) $Cu(s) + HCl(aq) \rightarrow nr$
(c) $2 Al(s) + 6 HI(aq) \rightarrow 3 H_2(g) + 2 AlI_3(aq)$

8.51 **(a)** $NH_4Cl(aq) + NaOH(aq) \rightarrow NH_3(aq) + NaCl(aq) + H_2O(l)$
(b) $NaHCO_3(aq) + NaOH(aq) \rightarrow Na_2CO_3(aq) + H_2O(l)$

8.54 **(a)** $CO_2(g) + H_2O(l) + MgCO_3(s) \rightarrow Mg(HCO_3)_2(aq)$

8.56 **(a)** $2 HClO_4(aq) + CaO(s) \rightarrow Ca(ClO_4)_2(aq) + H_2O(l)$
(b) $N_2O_3(g) + H_2O(l) \rightarrow 2 HNO_2(aq)$
(c) $SO_3(g) + 2 NaOH(aq) \rightarrow Na_2SO_4(aq) + H_2O(l)$

8.59 They are all balanced, but none goes as written: (a) and (b) go in the opposite direction; (c) does not go in either direction.

8.60 $6 KI(aq) + 2 Fe(NO_3)_3(aq) \rightarrow 2 FeI_2(aq) + I_2(aq) + 6 KNO_3(aq)$

8.61 The reaction would produce $CaCO_3(s)$ and NaCl from $CaCl_2(aq)$ and $Na_2CO_3(aq)$, not the products which are wanted.

8.64 Oxygen is one of the reactants in a combustion reaction. Carbon and hydrogen are often components of the other reactant. Note that oxygen can be involved in a displacement reaction:

$$O_2(g) + 2 CCl_4(l) \rightarrow 2 COCl_2(g) + 2 Cl_2(g)$$

Combustion of substances that do not contain carbon *and* hydrogen can often be regarded as either combustion or combination:

$$2 SO_2(g) + O_2(g) \rightarrow 2 SO_3(g)$$
$$P_4(s) + 5 O_2(g) \rightarrow P_4O_{10}(s)$$
$$4 Fe(s) + 3 O_2(g) \rightarrow 2 Fe_2O_3(s)$$
$$2 CO(g) + O_2(g) \rightarrow 2 CO_2(g)$$

8.68 $CaCO_3$ The reaction is

$$Ca(HCO_3)_2(aq) + 2 NH_3(aq) \rightarrow (NH_4)_2CO_3(aq) + CaCO_3(s)$$

8.69 $2 NH_4Cl(aq) + Ba(OH)_2(aq) \rightarrow$
$$BaCl_2(aq) + 2 NH_3(aq) + 2 H_2O(l)$$
The NH_4OH that might have been expected decomposes to NH_3 and H_2O.

8.74 There is no difference in the chemistry, but the questions get progressively harder to answer as more is left for you to interpret.

8.78 **(a)** $HBr(aq) + NaHCO_3(aq) \rightarrow NaBr(aq) + H_2O(l) + CO_2(g)$
(b) $NaOH(aq) + NaHSO_3(aq) \rightarrow Na_2SO_3(aq) + H_2O(l)$
(c) $2 HC_2H_3O_2(aq) + Ba(OH)_2(aq) \rightarrow$
$$Ba(C_2H_3O_2)_2(aq) + 2 H_2O(l)$$
(d) $NaC_2H_3O_2(aq) + HCl(aq) \rightarrow HC_2H_3O_2(aq) + NaCl(aq)$
(The formation of $HC_2H_3O_2$, a covalent compound, drives this reaction.)

9 Net Ionic Equations

9.1 **(a)** Fe^{2+} Cl^- **(b)** NH_4^+ NO_3^- **(c)** Na^+ ClO_4^-
(d) Co^{2+} ClO^- **(e)** Cu^{2+} Cl^- **(f)** Al^{3+} Cl^-

9.2 **(a)** $H^+(aq) + ClO_3^-(aq)$ (Strong acid)
(b) $Ca^{2+}(aq) + 2 Cl^-(aq)$
(c) $H^+(aq) + Cl^-(aq)$ (Strong acid)
(d) $BaSO_4(s)$ (Insoluble)
(e) $Na^+(aq) + OH^-(aq)$
(f) $K^+(aq) + ClO_3^-(aq)$
(g) $Zn(s)$ (Not an ion)
(h) $2 NH_4^+(aq) + SO_4^{2-}(aq)$
(i) $Ni^{2+}(aq) + 2 ClO_3^-(aq)$
(j) $SO_2(aq)$ (Covalent)
(k) $NH_3(aq)$ (Covalent)
(l) $K^+(aq) + OH^-(aq)$
(m) $2 Al^{3+}(aq) + 3 SO_4^{2-}(aq)$
(n) $Mg(OH)_2(s)$ (Insoluble)
(o) $H_2O(l)$ (Covalent)
(p) $CH_3OH(aq)$ (Covalent)

(q) $H_3PO_4(aq)$ — (Weak acid, covalent)
(r) $AgCl(s)$ — (Insoluble)
(s) $PbCl_2(s)$ — (Insoluble)
(t) $Zn^{2+}(aq) + 2\ C_2H_3O_2^-(aq)$
(u) $Ag^+(aq) + ClO_3^-(aq)$
(v) $NH_4^+(aq) + I^-(aq)$
(w) $2\ K^+(aq) + Cr_2O_7^{2-}(aq)$
(x) $CH_2O(aq)$ — (Covalent)
(y) $K^+(aq) + MnO_4^-(aq)$
(z) $H^+(aq) + ClO_4^-(aq)$ — (Strong acid)

9.3 (a) Ionic equation: $Ba^{2+}(aq) + 2\ OH^-(aq) + 2\ H^+(aq) +$
$\qquad 2\ Cl^-(aq) \rightarrow Ba^{2+}(aq) + 2\ Cl^-(aq) + 2\ H_2O(l)$
Net ionic equation: $OH^-(aq) + H^+(aq) \rightarrow H_2O(l)$
(b) $Ba(OH)_2(s) + 2\ H^+(aq) \rightarrow Ba^{2+}(aq) + 2\ H_2O(l)$

9.4 (a)–(c) $Ca(s) + 2\ H^+(aq) \rightarrow Ca^{2+}(aq) + H_2(g)$
See Practice Problem 8.7.

9.5 In Cl_2, the chlorine atoms are covalently bonded to each other. In $SrCl_2$, the chloride ions are ionically bonded to the strontium ion. In SCl_2, the chlorine atoms are covalently bonded to the sulfur atom.

9.6

	Water Soluble?	Ionic?	Written As Separate Ions?
(a) H_2O	Yes	No	No, written $H_2O(l)$
(b) $LiCl$	Yes	Yes	Yes, written $Li^+(aq) + Cl^-(aq)$
(c) $BaSO_4$	No	Yes	No, written $BaSO_4(s)$
(d) $HC_2H_3O_2$	Yes	No	No, written $HC_2H_3O_2(aq)$
(e) $HClO_3$	Yes	Yes	Yes, written $H^+(aq) + ClO_3^-(aq)$
(f) CH_3OH	Yes	No	No, written $CH_3OH(aq)$

9.7 (a) In the pure state, no common acids exist as ions.
(b) In water solution, the strong acids (Table 9.1) exist in the form of their ions.

9.8 It contains Sr^{2+} ions, water, and some type of anion, such as NO_3^-, that forms a soluble compound with strontium ion. No bottle can contain only cations or only anions.

9.9 (a) $H^+(aq) + Cl^-(aq)$ — (HCl is a strong acid.)
(b) $HClO(aq)$ — (HClO is a weak acid.)
(c) $HClO_2(aq)$ — ($HClO_2$ is a weak acid.)
(d) $H^+(aq) + ClO_3^-(aq)$ — ($HClO_3$ is a strong acid.)
(e) $H^+(aq) + ClO_4^-(aq)$ — ($HClO_4$ is a strong acid.)

9.10 All salts are ionic, and all sodium salts are soluble.
(a) $Na^+(aq) + Cl^-(aq)$ (b) $Na^+(aq) + ClO^-(aq)$
(c) $Na^+(aq) + ClO_2^-(aq)$ (d) $Na^+(aq) + ClO_3^-(aq)$
(e) $Na^+(aq) + ClO_4^-(aq)$

9.11 (a) $CO_2(g)$ — (Not ionic)
(b) $BaSO_4(s)$ — (Insoluble)
(c) $Cu^{2+}(aq) + 2\ ClO_3^-(aq)$ — (Soluble and ionic)
(d) $Al^{3+}(aq) + 3\ Cl^-(aq)$ — (Soluble and ionic)
(e) $Ba^{2+}(aq) + 2\ ClO_3^-(aq)$ — (Soluble and ionic)
(f) SF_2 — (Not ionic)

9.15 In each case, a compound that is ionic in solution is required.
(a) NaF yields Na^+ and F^- ions and is thus better. HF ionizes to only a very limited extent.
(b) Since HCl is a *strong* acid, both compounds are fully ionic in solution, and either would be suitable.
(c) NH_4Cl is ionic and yields NH_4^+ ions and Cl^- ions, and therefore it is better. NH_3 reacts with water to form ions only to a slight extent. (It was introduced as a weak base in Chapter 8.)

9.16 Not only the *atoms* but also the *net charges* must be balanced.
(a) $Cu_2O(s) + 2\ H^+(aq) \rightarrow Cu(s) + Cu^{2+}(aq) + H_2O(l)$
(b) $2\ Co^{3+}(aq) + Co(s) \rightarrow 3\ Co^{2+}(aq)$
(c) $2\ Ag^+(aq) + Cd(s) \rightarrow Cd^{2+}(aq) + 2\ Ag(s)$
(d) $2\ I^-(aq) + 2\ Ce^{4+}(aq) \rightarrow 2\ Ce^{3+}(aq) + I_2(aq)$

9.18 (a) $HCO_3^-(aq) + OH^-(aq) \rightarrow CO_3^{2-}(aq) + H_2O(l)$
(b) $HCO_3^-(aq) + H^+(aq) \rightarrow H_2O(l) + CO_2(g)$
(c) $I^-(aq) + Ag^+(aq) \rightarrow AgI(s)$
(d) $OH^-(aq) + H^+(aq) \rightarrow H_2O(l)$
(e) $Ba^{2+}(aq) + SO_4^{2-}(aq) \rightarrow BaSO_4(s)$

(f) $2\ H^+(aq) + Ba(OH)_2(s) \rightarrow Ba^{2+}(aq) + 2\ H_2O(l)$
$Ba(OH)_2$ is only slightly soluble; here, it exists as a solid and is written as a compound.

9.19 $M_2O(s) + 2\ H^+(aq) \rightarrow 2\ M^+(aq) + H_2O(l)$

9.20 $Pb(NO_3)_2(aq) + 2\ LiCl(aq) \rightarrow PbCl_2(s) + 2\ LiNO_3(aq)$
$Pb(NO_3)_2(aq) + 2\ NaCl(aq) \rightarrow PbCl_2(s) + 2\ NaNO_3(aq)$
$Pb(NO_3)_2(aq) + 2\ KCl(aq) \rightarrow PbCl_2(s) + 2\ KNO_3(aq)$
$Pb(NO_3)_2(aq) + 2\ RbCl(aq) \rightarrow PbCl_2(s) + 2\ RbNO_3(aq)$
$Pb(NO_3)_2(aq) + 2\ CsCl(aq) \rightarrow PbCl_2(s) + 2\ CsNO_3(aq)$
$Pb(NO_3)_2(aq) + 2\ FrCl(aq) \rightarrow PbCl_2(s) + 2\ FrNO_3(aq)$

9.24 $Ba(ClO_3)_2(aq) + Na_2SO_4(aq) \rightarrow BaSO_4(s) + 2\ NaClO_3(aq)$
$Ba(NO_3)_2(aq) + Na_2SO_4(aq) \rightarrow BaSO_4(s) + 2\ NaNO_3(aq)$
$Ba(C_2H_3O_2)_2(aq) + Na_2SO_4(aq) \rightarrow BaSO_4(s) + 2\ NaC_2H_3O_2(aq)$
$Ba(ClO_3)_2(aq) + (NH_4)_2SO_4(aq) \rightarrow BaSO_4(s) + 2\ NH_4ClO_3(aq)$
$Ba(NO_3)_2(aq) + (NH_4)_2SO_4(aq) \rightarrow BaSO_4(s) + 2\ NH_4NO_3(aq)$
$Ba(C_2H_3O_2)_2(aq) + (NH_4)_2SO_4(aq) \rightarrow$
$\qquad\qquad\qquad BaSO_4(s) + 2\ NH_4C_2H_3O_2(aq)$
$Ba(ClO_3)_2(aq) + FeSO_4(aq) \rightarrow BaSO_4(s) + Fe(ClO_3)_2(aq)$
$Ba(NO_3)_2(aq) + FeSO_4(aq) \rightarrow BaSO_4(s) + Fe(NO_3)_2(aq)$
$Ba(C_2H_3O_2)_2(aq) + FeSO_4(aq) \rightarrow BaSO_4(s) + Fe(C_2H_3O_2)_2(aq)$

9.25 (a) $H^+(aq) + OH^-(aq) \rightarrow H_2O(l)$
(b) $H^+(aq) + NH_3(aq) \rightarrow NH_4^+(aq)$
(c) $HA(aq) + OH^-(aq) \rightarrow A^-(aq) + H_2O(l)$

9.28 (a) $Fe^{2+}(aq) + S^{2-}(aq) \rightarrow FeS(s)$
(b) $Mn^{2+}(aq) + S^{2-}(aq) \rightarrow MnS(s)$
(c) $Zn^{2+}(aq) + S^{2-}(aq) \rightarrow ZnS(s)$
(d) $Cu^{2+}(aq) + S^{2-}(aq) \rightarrow CuS(s)$
(e) $Ni^{2+}(aq) + S^{2-}(aq) \rightarrow NiS(s)$
(f) $Co^{2+}(aq) + S^{2-}(aq) \rightarrow CoS(s)$

9.30 (a) $H^+(aq) + OH^-(aq) \rightarrow H_2O(l)$
(b) $HClO_2(aq) + OH^-(aq) \rightarrow ClO_2^-(aq) + H_2O(l)$
Note that $HClO_3$ is a strong acid and is completely ionized in water; $HClO_2$ is a weak acid and exists mostly in the form of covalent molecules, so it is written as a compound in the net ionic equation.

9.34 See Figure 8.10.
(a) $CO_3^{2-}(aq) + 2\ H^+(aq) \rightarrow CO_2(g) + H_2O(l)$
(b) $HCO_3^-(aq) + H^+(aq) \rightarrow CO_2(g) + H_2O(l)$
(c) $HCO_3^-(aq) + OH^-(aq) \rightarrow CO_3^{2-}(aq) + H_2O(l)$

9.35 They yield the same quantity of heat because they are essentially the same reaction:

$H^+(aq) + OH^-(aq) \rightarrow H_2O(l)$

The quantity of heat liberated in this reaction does not depend on the anion and the cation that *do not* react.

9.36 CH_3Cl is not ionic. (It also happens to be insoluble in water.) The Ag^+ ion precipitates AgCl in the presence of Cl^- ions, but no Cl^- ions are present in this case.

10 Stoichiometry

10.1

$\dfrac{1\ mol\ PCl_5}{4\ mol\ H_2O}$	$\dfrac{4\ mol\ H_2O}{5\ mol\ HCl}$	$\dfrac{5\ mol\ HCl}{1\ mol\ H_3PO_4}$	$\dfrac{1\ mol\ H_3PO_4}{1\ mol\ PCl_5}$
$\dfrac{1\ mol\ PCl_5}{5\ mol\ HCl}$	$\dfrac{4\ mol\ H_2O}{1\ mol\ H_3PO_4}$	$\dfrac{5\ mol\ HCl}{1\ mol\ PCl_5}$	$\dfrac{1\ mol\ H_3PO_4}{4\ mol\ H_2O}$
$\dfrac{1\ mol\ PCl_5}{1\ mol\ H_3PO_4}$	$\dfrac{4\ mol\ H_2O}{1\ mol\ PCl_5}$	$\dfrac{5\ mol\ HCl}{4\ mol\ H_2O}$	$\dfrac{1\ mol\ H_3PO_4}{5\ mol\ HCl}$

10.2 (a) 2 (b) 6 (c) 100 (d) 1 dozen (e) 1 mol

10.3 All of them; the equation gives the mole ratios only, not the quantities of reactants. (The three different quantities of KOH would require 0.333 mol of H_3PO_4, 0.667 mol of H_3PO_4, and 1.33 mol of H_3PO_4, respectively.)

10.4 The ratio is 1:5. No matter how complicated the equation, once it has been balanced, the mole ratios may be read directly from it.

10.5 (a) 25 sandwiches

(b) 25 sandwiches (This problem involves a limiting quantity; you cannot make 30 sandwiches with only 50 slices of bread.)

(c) 20 sandwiches (You cannot make 25 sandwiches with only 20 patties.)

(d) The quantities of two (or more) reactants are given in the problem statement.

10.6 This problem is analogous to Problem 10.5.

(a) 25 mol

(b) 25 mol [The number of moles of HCl is the limiting quantity, even though it is greater than the number of moles of $Ba(OH)_2$.]

(c) 20 mol [The $Ba(OH)_2$ is present in limiting quantity.]

10.7 (a) $KCl(aq) + AgNO_3(aq) \rightarrow AgCl(s) + KNO_3(aq)$

$$2.50 \text{ mol AgCl}\left(\frac{1 \text{ mol KCl}}{1 \text{ mol AgCl}}\right) = 2.50 \text{ mol KCl}$$

(b) 2.50 mol NH_4Cl (c) 2.50 mol CsCl

10.8 Write formulas from the names given (Chapter 6).

Calculate the numbers of moles of the reactants from the masses given and the masses of products from their numbers of moles (Chapter 7).

Predict the products of a familiar type of reaction (Chapter 8). Know whether or not the excess reactant and the products are volatile from the nature of their ionic or covalent bonding (Chapter 5).

10.9 (a) B   (b) A   (c) A   (d) B   (e) A

10.10 (a) No NaCl can be produced without any chlorine.

(b) 2 mol of NaCl can be produced. The numbers of moles present are exactly the same as the coefficients in the balanced equation.

(c) 2 mol of NaCl can be produced. The sum of the numbers of moles of reactants in parts (a) and (b) produces the sum of the numbers of moles of products. Put another way, after 1 mol of Cl_2 reacts with 2 mol of Na, as in part (b), there is no more Cl_2 left, and we are back to the situation in part (a).

10.11 (a) B   (b) B   (c) B   (d) A   (e) B

10.12 (a) 10 mol

(b) 10 mol [The number of moles of HCl is the limiting quantity, even though it is greater than the number of moles of $La(OH)_3$.]

(c) 5 mol [The $La(OH)_3$ is present in limiting quantity.]

10.13 (a) $2.50 \text{ mol KCl}\left(\dfrac{74.6 \text{ g KCl}}{1 \text{ mol KCl}}\right) = 186 \text{ g KCl}$

(b) $2.50 \text{ mol NH}_4\text{Cl}\left(\dfrac{53.5 \text{ g NH}_4\text{Cl}}{1 \text{ mol NH}_4\text{Cl}}\right) = 134 \text{ g NH}_4\text{Cl}$

(c) $2.50 \text{ mol CsCl}\left(\dfrac{168 \text{ g CsCl}}{1 \text{ mol CsCl}}\right) = 420 \text{ g CsCl}$

10.14 $Ca(s) + 2 HCl(aq) \rightarrow CaCl_2(aq) + H_2(g)$
$2 Al(s) + 6 HCl(aq) \rightarrow 2 AlCl_3(aq) + 3 H_2(g)$
$Zn(s) + 2 HCl(aq) \rightarrow ZnCl_2(aq) + H_2(g)$

The sample of aluminum can produce the most hydrogen because there is a 3:2 mole ratio of hydrogen to aluminum metal in the balanced equation for that reaction and only a 1:1 mole ratio in the other two equations. That is, each mole of aluminum produces more hydrogen.

10.15 $3 Ba(OH)_2(aq) + 2 H_3PO_4(aq) \rightarrow Ba_3(PO_4)_2(s) + 6 H_2O(l)$

$$5.55 \text{ mol Ba(OH)}_2\left(\frac{2 \text{ mol H}_3\text{PO}_4}{3 \text{ mol Ba(OH)}_2}\right) = 3.70 \text{ mol H}_3\text{PO}_4$$

10.19 *Complete combustion* means formation of CO_2 rather than CO. The balanced equation is

$$C_4H_8(g) + 6 O_2(g) \rightarrow 4 CO_2(g) + 4 H_2O(g)$$

$$2.96 \text{ mol C}_4\text{H}_8\left(\frac{6 \text{ mol O}_2}{1 \text{ mol C}_4\text{H}_8}\right) = 17.8 \text{ mol O}_2$$

10.22 $BaCO_3(s) + 2 HCl(aq) \rightarrow BaCl_2(aq) + H_2O(l) + CO_2(g)$

$$9.12 \text{ mol H}_2\text{O}\left(\frac{1 \text{ mol BaCl}_2}{1 \text{ mol H}_2\text{O}}\right) = 9.12 \text{ mol BaCl}_2$$

10.23 (a) $2 KClO_3(s) \rightarrow 2 KCl(s) + 3 O_2(g)$

(b) The number of millimoles *produced* (of O_2) is governed by the equation; the number of millimoles of $KClO_3$ *present initially* is not.

(c) $40.5 \text{ mmol O}_2\left(\dfrac{2 \text{ mmol KClO}_3 \text{ decomposed}}{3 \text{ mmol O}_2}\right)$
$$= 27.0 \text{ mmol KClO}_3 \text{ decomposed}$$

$$\left(\frac{27.0 \text{ mmol KClO}_3 \text{ decomposed}}{40.5 \text{ mmol KClO}_3 \text{ present}}\right) \times 100\%$$
$$= 66.7\% \text{ KClO}_3 \text{ decomposed}$$

10.26 (a) $10.00 \text{ mmol Fe}^{2+}\left(\dfrac{1 \text{ mmol MnO}_4^-}{5 \text{ mmol Fe}^{2+}}\right) = 2.000 \text{ mmol MnO}_4^-$

(b) $10.00 \text{ mmol Fe}^{2+}\left(\dfrac{4 \text{ mmol H}_2\text{O}}{5 \text{ mmol Fe}^{2+}}\right) = 8.000 \text{ mmol H}_2\text{O}$

10.27 (a) $Cu(s) + 2 AgNO_3(aq) \rightarrow 2 Ag(s) + Cu(NO_3)_2(aq)$

(b) $9.97 \text{ g Cu}\left(\dfrac{1 \text{ mol Cu}}{63.5 \text{ g Cu}}\right) = 0.157 \text{ mol Cu}$

(c) $0.157 \text{ mol Cu}\left(\dfrac{2 \text{ mol Ag}}{1 \text{ mol Cu}}\right) = 0.314 \text{ mol Ag}$

(d) $0.314 \text{ mol Ag}\left(\dfrac{108 \text{ g Ag}}{1 \text{ mol Ag}}\right) = 33.9 \text{ g Ag}$

(e) $9.97 \text{ g Cu}\left(\dfrac{1 \text{ mol Cu}}{63.5 \text{ g Cu}}\right)\left(\dfrac{2 \text{ mol Ag}}{1 \text{ mol Cu}}\right)\left(\dfrac{108 \text{ g Ag}}{1 \text{ mol Ag}}\right)$
$$= 33.9 \text{ g Ag}$$

10.29 (a) $Ba(s) + 2 HClO_3(aq) \rightarrow Ba(ClO_3)_2(aq) + H_2(g)$

$$25.0 \text{ g Ba}\left(\frac{1 \text{ mol Ba}}{137 \text{ g Ba}}\right)\left(\frac{1 \text{ mol H}_2}{1 \text{ mol Ba}}\right) = 0.182 \text{ mol H}_2$$

(b) $Mg(s) + 2 HClO_3(aq) \rightarrow Mg(ClO_3)_2(aq) + H_2(g)$

$$25.0 \text{ g Mg}\left(\frac{1 \text{ mol Mg}}{24.3 \text{ g Mg}}\right)\left(\frac{1 \text{ mol H}_2}{1 \text{ mol Mg}}\right) = 1.03 \text{ mol H}_2$$

(c) Although the number of grams of each metal is the same, the numbers of moles are very different because of the great difference in atomic masses. Therefore, the number of moles of H_2 produced differs greatly.

10.31 $FeCl_3(aq) + 3 AgNO_3(aq) \rightarrow Fe(NO_3)_3(aq) + 3 AgCl(s)$

$$17.2 \text{ g AgCl}\left(\frac{1 \text{ mol AgCl}}{143 \text{ g AgCl}}\right)\left(\frac{1 \text{ mol FeCl}_3}{3 \text{ mol AgCl}}\right)\left(\frac{162 \text{ g FeCl}_3}{1 \text{ mol FeCl}_3}\right)$$
$$= 6.50 \text{ g FeCl}_3$$

10.32 (a) $2 Na(s) + S(s) \rightarrow Na_2S(s)$

$$20.0 \text{ g Na}\left(\frac{1 \text{ mol Na}}{23.0 \text{ g Na}}\right)\left(\frac{1 \text{ mol Na}_2\text{S}}{2 \text{ mol Na}}\right)\left(\frac{78.1 \text{ g Na}_2\text{S}}{1 \text{ mol Na}_2\text{S}}\right)$$
$$= 34.0 \text{ g Na}_2\text{S}$$

(b) $Mg(s) + F_2(g) \rightarrow MgF_2(s)$

$$30.0 \text{ g Mg}\left(\frac{1 \text{ mol Mg}}{24.3 \text{ g Mg}}\right)\left(\frac{1 \text{ mol MgF}_2}{1 \text{ mol Mg}}\right)\left(\frac{62.3 \text{ g MgF}_2}{1 \text{ mol MgF}_2}\right)$$
$$= 76.9 \text{ g MgF}_2$$

(c) $4\ Al(s) + 3\ O_2(g) \rightarrow 2\ Al_2O_3(s)$

$$40.0\ g\ Al\left(\frac{1\ mol\ Al}{27.0\ g\ Al}\right)\left(\frac{2\ mol\ Al_2O_3}{4\ mol\ Al}\right)\left(\frac{102\ g\ Al_2O_3}{1\ mol\ Al_2O_3}\right)$$
$$= 75.6\ g\ Al_2O_3$$

10.33 $1.50 \times 10^6\ g\ Al\left(\frac{1\ mol\ Al}{27.0\ g\ Al}\right)\left(\frac{1\ mol\ Al_2O_3}{2\ mol\ Al}\right)\left(\frac{102\ g\ Al_2O_3}{1\ mol\ Al_2O_3}\right)$
$$= 2.83 \times 10^6\ g\ Al_2O_3$$

10.36 $5.95 \times 10^5\ g\ H_3PO_4\left(\frac{1\ mol\ H_3PO_4}{98.0\ g\ H_3PO_4}\right)\left(\frac{1\ mol\ (NH_4)_3PO_4}{1\ mol\ H_3PO_4}\right) \times$

$$\left(\frac{149\ g\ (NH_4)_3PO_4}{1\ mol\ (NH_4)_3PO_4}\right) = 9.05 \times 10^5\ g\ (NH_4)_3PO_4$$

10.37 **(a)** and **(b)** $25.0\ g\ NaHCO_3\left(\frac{1\ mol\ NaHCO_3}{84.0\ g\ NaHCO_3}\right) \times$

$$\left(\frac{1\ mol\ Na_2CO_3}{2\ mol\ NaHCO_3}\right)\left(\frac{106\ g\ Na_2CO_3}{1\ mol\ Na_2CO_3}\right) = 15.8\ g\ Na_2CO_3$$

(c) Parts (a) and (b) are the same problem, since Na_2CO_3 is the only solid product.

10.42 $225\ g\ Zn\left(\frac{1\ mol\ Zn}{65.4\ g\ Zn}\right)\left(\frac{1\ mol\ NH_4NO_3}{4\ mol\ Zn}\right)\left(\frac{80.0\ g\ NH_4NO_3}{1\ mol\ NH_4NO_3}\right)$
$$= 68.8\ g\ NH_4NO_3$$

10.43 $Zn(s) + 2\ AgNO_3(aq) \rightarrow 2\ Ag(s) + Zn(NO_3)_2(aq)$

$$424\ g\ Zn\left(\frac{1\ mol\ Zn}{65.4\ g\ Zn}\right)\left(\frac{2\ mol\ Ag}{1\ mol\ Zn}\right)\left(\frac{108\ g\ Ag}{1\ mol\ Ag}\right) = 1400\ g\ Ag$$

10.44 $35.9\ g\ HNO_3\left(\frac{1\ mol\ HNO_3}{63.0\ g\ HNO_3}\right)\left(\frac{2\ mol\ NO_2}{4\ mol\ HNO_3}\right)\left(\frac{46.0\ g\ NO_2}{1\ mol\ NO_2}\right)$
$$= 13.1\ g\ NO_2$$

10.47 $NiCl_2(aq) + 2\ AgNO_3(aq) \rightarrow Ni(NO_3)_2(aq) + 2\ AgCl(s)$

$$13.7\ g\ AgCl\left(\frac{1\ mol\ AgCl}{143\ g\ AgCl}\right)\left(\frac{1\ mol\ NiCl_2}{2\ mol\ AgCl}\right)\left(\frac{130\ g\ NiCl_2}{1\ mol\ NiCl_2}\right)$$
$$= 6.23\ g\ NiCl_2$$

10.48 $6.93\ g\ H_2\left(\frac{1\ mol\ H_2}{2.016\ g\ H_2}\right)\left(\frac{1\ mol\ C_8H_{18}}{2\ mol\ H_2}\right)\left(\frac{114\ g\ C_8H_{18}}{1\ mol\ C_8H_{18}}\right)$
$$= 196\ g\ C_8H_{18}$$

10.52 $2\ CrCl_2(s) + Cl_2(g) \rightarrow 2\ CrCl_3(s)$

$$72.4\ g\ CrCl_2\left(\frac{1\ mol\ CrCl_2}{123\ g\ CrCl_2}\right)\left(\frac{1\ mol\ Cl_2}{2\ mol\ CrCl_2}\right)\left(\frac{70.9\ g\ Cl_2}{1\ mol\ Cl_2}\right)$$
$$= 20.9\ g\ Cl_2$$

10.53 $Fe(s) + Sn(NO_3)_2(aq) \rightarrow Sn(s) + Fe(NO_3)_2(aq)$

$$27.6\ g\ Fe\left(\frac{1\ mol\ Fe}{55.8\ g\ Fe}\right)\left(\frac{1\ mol\ Sn}{1\ mol\ Fe}\right)\left(\frac{119\ g\ Sn}{1\ mol\ Sn}\right) = 58.9\ g\ Sn$$

10.54 $2.50\ mol\ (NH_4)_2CO_3\left(\frac{2\ mol\ NH_3}{1\ mol\ (NH_4)_2CO_3}\right)\left(\frac{17.0\ g\ NH_3}{1\ mol\ NH_3}\right)$
$$= 85.0\ g\ NH_3$$

10.55 $P_4(s) + 5\ O_2(g) \rightarrow P_4O_{10}(s)$

$$6.173\ g\ P_4\left(\frac{1\ mol\ P_4}{123.9\ g\ P_4}\right)\left(\frac{1\ mol\ P_4O_{10}}{1\ mol\ P_4}\right)\left(\frac{283.9\ g\ P_4O_{10}}{1\ mol\ P_4O_{10}}\right)$$
$$= 14.14\ g\ P_4O_{10}$$

10.59 $1055\ g\ C\left(\frac{1\ mol\ C}{12.01\ g\ C}\right)\left(\frac{1\ mol\ Fe_2O_3}{3\ mol\ C}\right)\left(\frac{159.7\ g\ Fe_2O_3}{1\ mol\ Fe_2O_3}\right)$
$$= 4676\ g\ Fe_2O_3$$

10.60 $22.7\ g\ (NH_4)_2SO_3\left(\frac{1\ mol\ (NH_4)_2SO_3}{116\ g\ (NH_4)_2SO_3}\right)\left(\frac{2\ mol\ NH_3}{1\ mol\ (NH_4)_2SO_3}\right)$
$$= 0.391\ mol\ NH_3$$

10.61 $H_2SO_4(aq) + 2\ KOH(aq) \rightarrow K_2SO_4(aq) + 2\ H_2O(l)$

$$29.7\ g\ H_2SO_4\left(\frac{1\ mol\ H_2SO_4}{98.1\ g\ H_2SO_4}\right)\left(\frac{2\ mol\ KOH}{1\ mol\ H_2SO_4}\right)\left(\frac{56.1\ g\ KOH}{1\ mol\ KOH}\right)$$
$$= 34.0\ g\ KOH$$

10.63 **(a)** $7.22 \times 10^{24}\ Cl\ atoms\left(\frac{1\ mol\ Cl}{6.02 \times 10^{23}\ Cl\ atoms}\right)$
$$= 12.0\ mol\ Cl$$

(b) $12.0\ mol\ Cl\left(\frac{1\ mol\ PCl_5}{5\ mol\ Cl}\right) = 2.40\ mol\ PCl_5$

(c) $2.40\ mol\ PCl_5\left(\frac{2\ mol\ POCl_3}{2\ mol\ PCl_5}\right) = 2.40\ mol\ POCl_3$

(d) $2.40\ mol\ POCl_3\left(\frac{153\ g\ POCl_3}{1\ mol\ POCl_3}\right) = 367\ g\ POCl_3$

10.64 $2\ CrCl_2(s) + Cl_2(g) \rightarrow 2\ CrCl_3(s)$

$$2.68\ g\ Cl_2\left(\frac{1\ mol\ Cl_2}{70.9\ g\ Cl_2}\right)\left(\frac{2\ mol\ CrCl_3}{1\ mol\ Cl_2}\right)\left(\frac{3\ mol\ Cl}{1\ mol\ CrCl_3}\right)$$
$$= 0.227\ mol\ Cl$$

$$0.227\ mol\ Cl\left(\frac{6.02 \times 10^{23}\ atoms\ Cl}{1\ mol\ Cl}\right) = 1.37 \times 10^{23}\ atoms\ Cl$$

10.65 $227\ mL\left(\frac{0.997\ g}{1\ mL}\right)\left(\frac{1\ mol\ H_2O}{18.0\ g}\right)\left(\frac{2\ mol\ H}{1\ mol\ H_2O}\right) = 25.1\ mol\ H$

10.69 $2\ SO_2(g) + O_2(g) \rightarrow 2\ SO_3(g)$

$$72.9\ g\ O_2\left(\frac{1\ mol\ O_2}{32.0\ g\ O_2}\right)\left(\frac{2\ mol\ SO_3}{1\ mol\ O_2}\right)\left(\frac{3\ mol\ O}{1\ mol\ SO_3}\right)$$
$$= 13.7\ mol\ O$$

$$13.7\ mol\ O\left(\frac{6.02 \times 10^{23}\ atoms\ O}{1\ mol\ O}\right) = 8.25 \times 10^{24}\ atoms\ O$$

10.70 $75.0\ g\ NH_3\left(\frac{1\ mol\ NH_3}{17.0\ g\ NH_3}\right)\left(\frac{1\ mol\ NH_4NO_3}{1\ mol\ NH_3}\right) \times$

$$\left(\frac{2\ mol\ N}{1\ mol\ NH_4NO_3}\right)\left(\frac{6.02 \times 10^{23}\ atoms\ N}{1\ mol\ N}\right) = 5.31 \times 10^{24}\ atoms\ N$$

10.74 $C(s) + 2\ S(s) \rightarrow CS_2(l)$

$$5.00 \times 10^{22}\ S\ atoms\left(\frac{1\ mol\ S}{6.02 \times 10^{23}\ S\ atoms}\right)\left(\frac{1\ mol\ CS_2}{2\ mol\ S}\right) \times$$

$$\left(\frac{76.1\ g\ CS_2}{1\ mol\ CS_2}\right) = 3.16\ g\ CS_2$$

10.75 **(a)** The ratios are

Present		Required
$\dfrac{27\ bases}{10\ plates} = \dfrac{2.70\ bases}{1\ plate}$		$\dfrac{3\ bases}{1\ plate}$

Since the ratio present is smaller, the bases are in limiting quantity. Therefore, the number of fields is calculated using the number of bases:

$$27\ bases\left(\frac{1\ field}{3\ bases}\right) = 9\ fields$$

(b) There will be extra home plates:

$$27\ bases\left(\frac{1\ plate}{3\ bases}\right) = 9\ plates\ required$$

$$10\ plates\ present - 9\ plates\ required = 1\ plate\ extra$$

10.76 (a) First, calculate the number of moles of each reactant:

$$273.8 \text{ g C}_3\text{H}_4 \left(\frac{1 \text{ mol C}_3\text{H}_4}{40.06 \text{ g C}_3\text{H}_4} \right) = 6.835 \text{ mol C}_3\text{H}_4$$

$$40.18 \text{ g H}_2 \left(\frac{1 \text{ mol H}_2}{2.016 \text{ g H}_2} \right) = 19.93 \text{ mol H}_2$$

Since the mole ratio of H_2 to C_3H_4 is 2:1 in the balanced equation, and the ratio of H_2 to C_3H_4 present is greater than that, C_3H_4 is in limiting quantity. The 6.835 mol of C_3H_4 present will react with 2(6.835 mol) of H_2 to yield 6.835 mol of C_3H_8.

$$6.835 \text{ mol C}_3\text{H}_8 \left(\frac{44.09 \text{ g C}_3\text{H}_8}{1 \text{ mol C}_3\text{H}_8} \right) = 301.4 \text{ g C}_3\text{H}_8$$

(b) Since H_2 is in excess, not all of it will react. The mass of C_3H_8 produced is *not* the sum of the masses of H_2 and C_3H_4 originally present.

10.78 The numbers of moles present are calculated first:

$$6.22 \times 10^{23} \text{ molecules H}_2\text{SO}_4 \left(\frac{1 \text{ mol H}_2\text{SO}_4}{6.02 \times 10^{23} \text{ molecules}} \right)$$
$$= 1.03 \text{ mol H}_2\text{SO}_4$$

$$179 \text{ g Ba(OH)}_2 \left(\frac{1 \text{ mol Ba(OH)}_2}{171 \text{ g Ba(OH)}_2} \right) = 1.05 \text{ mol Ba(OH)}_2$$

$$\text{Ba(OH)}_2(aq) + \text{H}_2\text{SO}_4(aq) \rightarrow \text{BaSO}_4(s) + 2 \text{ H}_2\text{O}(l)$$

The ratio of moles present is

$$\frac{1.05 \text{ mol Ba(OH)}_2}{1.03 \text{ mol H}_2\text{SO}_4} = \frac{1.02 \text{ mol Ba(OH)}_2}{1 \text{ mol H}_2\text{SO}_4}$$

This ratio is greater than that required, so $Ba(OH)_2$ is in excess. The number of moles of H_2SO_4 is the limiting quantity for the complete neutralization.

$$1.03 \text{ mol H}_2\text{SO}_4 \left(\frac{2 \text{ mol H}_2\text{O}}{1 \text{ mol H}_2\text{SO}_4} \right) = 2.06 \text{ mol H}_2\text{O}$$

10.81 $\text{MgO}(s) + 2 \text{ HNO}_3(aq) \rightarrow \text{Mg(NO}_3)_2(aq) + \text{H}_2\text{O}(l)$

$$115 \text{ g HNO}_3 \left(\frac{1 \text{ mol HNO}_3}{63.0 \text{ g HNO}_3} \right) = 1.83 \text{ mol HNO}_3$$

$$52.0 \text{ g MgO} \left(\frac{1 \text{ mol MgO}}{40.3 \text{ g MgO}} \right) = 1.29 \text{ mol MgO}$$

Present		*Required*
$\dfrac{1.83 \text{ mol HNO}_3}{1.29 \text{ mol MgO}} = \dfrac{1.42 \text{ mol HNO}_3}{1 \text{ mol MgO}}$		$\dfrac{2 \text{ mol HNO}_3}{1 \text{ mol MgO}}$

The HNO_3 is in limiting quantity.

$$1.83 \text{ mol HNO}_3 \left(\frac{1 \text{ mol MgO}}{2 \text{ mol HNO}_3} \right) = 0.915 \text{ mol MgO reacts}$$

$$1.29 \text{ mol MgO} - 0.915 \text{ mol MgO} = 0.38 \text{ mol MgO excess}$$

10.82 $25.0 \text{ g SCl}_4 \left(\dfrac{1 \text{ mol SCl}_4}{174 \text{ g SCl}_4} \right) = 0.144 \text{ mol SCl}_4$

$$5.00 \text{ g H}_2\text{O} \left(\frac{1 \text{ mol H}_2\text{O}}{18.0 \text{ g H}_2\text{O}} \right) = 0.278 \text{ mol H}_2\text{O}$$

Present		*Required*
$\dfrac{0.278 \text{ mol H}_2\text{O}}{0.144 \text{ mol SCl}_4} = \dfrac{1.93 \text{ mol H}_2\text{O}}{1 \text{ mol SCl}_4}$		$\dfrac{2 \text{ mol H}_2\text{O}}{1 \text{ mol SCl}_4}$

The H_2O is in limiting quantity, so the mass of SCl_4 that reacts can be calculated from that quantity:

$$0.278 \text{ mol H}_2\text{O} \left(\frac{1 \text{ mol SCl}_4}{2 \text{ mol H}_2\text{O}} \right) \left(\frac{174 \text{ g SCl}_4}{1 \text{ mol SCl}_4} \right)$$
$$= 24.2 \text{ g SCl}_4 \text{ reacts}$$

The mass in excess (which remains unreacted) is then:

$$25.0 \text{ g SCl}_4 - 24.2 \text{ g SCl}_4 = 0.8 \text{ g SCl}_4$$

10.83 First, calculate the number of moles of each reactant:

$$2.50 \text{ g (NH}_4)_2\text{S} \left(\frac{1 \text{ mol (NH}_4)_2\text{S}}{68.1 \text{ g (NH}_4)_2\text{S}} \right) = 0.0367 \text{ mol (NH}_4)_2\text{S}$$

$$12.3 \text{ g AgNO}_3 \left(\frac{1 \text{ mol AgNO}_3}{170 \text{ g AgNO}_3} \right) = 0.0724 \text{ mol AgNO}_3$$

The balanced equation is

$$\text{(NH}_4)_2\text{S}(aq) + 2 \text{ AgNO}_3(aq) \rightarrow \text{Ag}_2\text{S}(s) + 2 \text{ NH}_4\text{NO}_3(aq)$$

The ratios are

Present		*Required*
$\dfrac{0.0724 \text{ mol AgNO}_3}{0.0367 \text{ mol (NH}_4)_2\text{S}} = \dfrac{1.97 \text{ mol AgNO}_3}{1 \text{ mol (NH}_4)_2\text{S}}$		$\dfrac{2 \text{ mol AgNO}_3}{1 \text{ mol (NH}_4)_2\text{S}}$

Since the ratio of moles present is less than that required, $AgNO_3$ is in limiting quantity.

$$0.0724 \text{ mol AgNO}_3 \left(\frac{1 \text{ mol Ag}_2\text{S}}{2 \text{ mol AgNO}_3} \right) \left(\frac{248 \text{ g Ag}_2\text{S}}{1 \text{ mol Ag}_2\text{S}} \right)$$
$$= 8.98 \text{ g Ag}_2\text{S}$$

10.86 (a) $2.27 \times 10^{23} \text{ molecules HCl} \left(\dfrac{1 \text{ mol HCl}}{6.02 \times 10^{23} \text{ molecules HCl}} \right)$
$$= 0.377 \text{ mol HCl}$$

$$\text{Ba(OH)}_2(aq) + 2 \text{ HCl}(aq) \rightarrow \text{BaCl}_2(aq) + 2 \text{ H}_2\text{O}(l)$$

Present		*Required*
$\dfrac{0.377 \text{ mol HCl}}{0.197 \text{ mol Ba(OH)}_2} = \dfrac{1.91 \text{ mol HCl}}{1 \text{ mol Ba(OH)}_2}$		$\dfrac{2 \text{ mol HCl}}{1 \text{ mol Ba(OH)}_2}$

Since the ratio present is lower than that required, HCl is in limiting quantity.

$$0.377 \text{ mol HCl} \left(\frac{2 \text{ mol H}_2\text{O}}{2 \text{ mol HCl}} \right) = 0.377 \text{ mol H}_2\text{O}$$

(b) $0.377 \text{ mol HCl} \left(\dfrac{1 \text{ mol Ba(OH)}_2}{2 \text{ mol HCl}} \right)$

$$= 0.188 \text{ mol Ba(OH)}_2 \text{ reacts}$$
$$0.197 \text{ mol Ba(OH)}_2 \text{ present} - 0.188 \text{ mol Ba(OH)}_2 \text{ reacts}$$
$$= 0.009 \text{ mol Ba(OH)}_2 \text{ excess}$$

10.87 Since two quantities of reactants are stated, this problem involves a limiting quantity. The masses are first converted to moles:

$$72.7 \text{ g Zn} \left(\frac{1 \text{ mol Zn}}{65.4 \text{ g Zn}} \right) = 1.11 \text{ mol Zn}$$

$$95.1 \text{ g HBr} \left(\frac{1 \text{ mol HBr}}{80.9 \text{ g HBr}} \right) = 1.18 \text{ mol HBr}$$

The balanced equation is

$$\text{Zn}(s) + 2 \text{ HBr}(aq) \rightarrow \text{ZnBr}_2(aq) + \text{H}_2(g)$$

The ratio of HBr present to Zn present, compared to the ratio required by the balanced chemical equation, is

Present		*Required*
$\dfrac{1.18 \text{ mol HBr}}{1.11 \text{ mol Zn}} = \dfrac{1.06 \text{ mol HBr}}{1.00 \text{ mol Zn}}$		$\dfrac{2 \text{ mol HBr}}{1 \text{ mol Zn}}$

Since the ratio of reactants present is lower than the ratio required, Zn (in the denominator) is in excess, and HBr is in limiting quantity. Use the limiting quantity to calculate the mass of $ZnBr_2$ that can be produced:

$$1.11 \text{ mol Zn} \left(\frac{1 \text{ mol ZnBr}_2}{1 \text{ mol Zn}} \right) \left(\frac{225 \text{ g ZnBr}_2}{1 \text{ mol ZnBr}_2} \right) = 250 \text{ g ZnBr}_2$$

10.88 (a) $AlCl_3(aq) + 3 AgNO_3(aq) \rightarrow 3 AgCl(s) + Al(NO_3)_3(aq)$

$$1.25 \text{ mol AgNO}_3\left(\frac{1 \text{ mol AlCl}_3}{3 \text{ mol AgNO}_3}\right) = 0.417 \text{ mol AlCl}_3$$

(b) Since 0.417 mol of $AlCl_3$ will react with 1.25 mol of $AgNO_3$, as shown in part (a), 0.417 mol of $AlCl_3$ will react, even though more than that is present. Alternatively, the systematic method may be used. The ratios are

Present		Required
$\dfrac{1.25 \text{ mol AgNO}_3}{1.11 \text{ mol AlCl}_3} =$	$\dfrac{1.13 \text{ mol AgNO}_3}{1 \text{ mol AlCl}_3}$	$\dfrac{3 \text{ mol AgNO}_3}{1 \text{ mol AlCl}_3}$

Since the ratio present is less than that required, $AgNO_3$ is in limiting quantity.

$$1.25 \text{ mol AgNO}_3\left(\frac{1 \text{ mol AlCl}_3}{3 \text{ mol AgNO}_3}\right) = 0.417 \text{ mol AlCl}_3$$

(c) Since 0.417 mol of $AlCl_3$ would react but 0.417 mol is *not* present, $AlCl_3$ is in limiting quantity. All the $AlCl_3$ that is present will react—0.411 mol. Alternatively, the systematic method may be used. The ratios are

Present		Required
$\dfrac{1.25 \text{ mol AgNO}_3}{0.411 \text{ mol AlCl}_3} =$	$\dfrac{3.04 \text{ mol AgNO}_3}{1 \text{ mol AlCl}_3}$	$\dfrac{3 \text{ mol AgNO}_3}{1 \text{ mol AlCl}_3}$

Since the ratio present is greater than that required, $AlCl_3$ is in limiting quantity. All the $AlCl_3$ present—0.411 mol—will react.

10.90 $SO_2(g) + Cl_2(g) \rightarrow SO_2Cl_2(l)$

(a) $2.48 \text{ mol SO}_2\left(\dfrac{1 \text{ mol SO}_2Cl_2}{1 \text{ mol SO}_2}\right) = 2.48 \text{ mol SO}_2Cl_2$

(b) $\left(\dfrac{2.46 \text{ mol}}{2.48 \text{ mol}}\right) \times 100\% = 99.2\%$

10.91 $5.89 \text{ g SO}_2\left(\dfrac{1 \text{ mol SO}_2}{64.1 \text{ g SO}_2}\right)\left(\dfrac{1 \text{ mol SOCl}_2}{1 \text{ mol SO}_2}\right)\left(\dfrac{119 \text{ g SOCl}_2}{1 \text{ mol SOCl}_2}\right)$
$$= 10.9 \text{ g SOCl}_2$$

$$\left(\frac{5.99 \text{ g}}{10.9 \text{ g}}\right) \times 100\% = 55.0\%$$

10.94 The balanced net ionic equation is

$HCO_3^-(aq) + H^+(aq) \rightarrow CO_2(g) + H_2O(l)$

The number of moles of CO_2 is given by

$$0.575 \text{ mol HCO}_3^-\left(\frac{1 \text{ mol CO}_2}{1 \text{ mol HCO}_3^-}\right) = 0.575 \text{ mol CO}_2$$

10.95 $CO_3^{2-}(aq) + 2 H^+(aq) \rightarrow H_2O(l) + CO_2(g)$

$$1.23 \text{ mol H}^+\left(\frac{1 \text{ mol CO}_3^{2-}}{2 \text{ mol H}^+}\right) = 0.615 \text{ mol CO}_3^{2-}$$

10.98 (a) $0.654 \text{ mol NaOH}\left(\dfrac{1 \text{ mol Na}^+}{1 \text{ mol NaOH}}\right) = 0.654 \text{ mol Na}^+$

(b) The full equation is

$NaOH(aq) + HCl(aq) \rightarrow NaCl(aq) + H_2O(l)$

Since the acid and base react in a 1:1 mole ratio, HCl is present in limiting quantity, and 0.407 mol of NaCl will be produced.

(c) The excess will be 0.247 mol of NaOH, equal to the 0.654 mol initially present minus the 0.407 mol that reacts.

(d) There are 0.407 mol of Na^+ from the NaCl and 0.247 mol of Na^+ from the excess NaOH for a total of 0.654 mol of Na^+.

(e) The 0.654 mol of Na^+ initially present (see part [a]) is still present after the reaction, consistent with Na^+ being a spectator ion in the reaction:
$H^+(aq) + OH^-(aq) \rightarrow H_2O(l)$

10.99 The balanced net ionic equation is

$Pb^{2+}(aq) + 2 Br^-(aq) \rightarrow PbBr_2(s)$

The number of grams of $PbBr_2$ is given by

$$0.506 \text{ mol Pb}^{2+}\left(\frac{1 \text{ mol PbBr}_2}{1 \text{ mol Pb}^{2+}}\right)\left(\frac{367 \text{ g PbBr}_2}{1 \text{ mol PbBr}_2}\right) = 186 \text{ g PbBr}_2$$

10.102 The balanced net ionic equation is

$Ba^{2+}(aq) + SO_4^{2-}(aq) \rightarrow BaSO_4(s)$

Since the reacting ratio is 1:1, the limiting quantity is SO_4^{2-}. The number of grams of $BaSO_4$ is given by

$$1.74 \text{ mol SO}_4^{2-}\left(\frac{1 \text{ mol BaSO}_4}{1 \text{ mol SO}_4^{2-}}\right)\left(\frac{233 \text{ g BaSO}_4}{1 \text{ mol BaSO}_4}\right)$$
$$= 405 \text{ g BaSO}_4$$

10.103 What mass of octane, C_8H_{18}, is required to produce 983 g of carbon dioxide on complete combustion?

10.104 (a) $2 LiOH(aq) + H_2SO_4(aq) \rightarrow Li_2SO_4(aq) + 2 H_2O(l)$

$$50.0 \text{ g LiOH}\left(\frac{1 \text{ mol LiOH}}{23.9 \text{ g LiOH}}\right)\left(\frac{1 \text{ mol Li}_2SO_4}{2 \text{ mol LiOH}}\right) \times$$
$$\left(\frac{110 \text{ g Li}_2SO_4}{1 \text{ mol Li}_2SO_4}\right) = 115 \text{ g Li}_2SO_4$$

(b) $KOH(aq) + HC_2H_3O_2(aq) \rightarrow KC_2H_3O_2(aq) + H_2O(l)$

$$50.0 \text{ g HC}_2H_3O_2\left(\frac{1 \text{ mol HC}_2H_3O_2}{60.0 \text{ g HC}_2H_3O_2}\right)\left(\frac{1 \text{ mol KC}_2H_3O_2}{1 \text{ mol HC}_2H_3O_2}\right) \times$$
$$\left(\frac{98.0 \text{ g KC}_2H_3O_2}{1 \text{ mol KC}_2H_3O_2}\right) = 81.7 \text{ g KC}_2H_3O_2$$

(c) $MgO(s) + 2 HNO_3(aq) \rightarrow Mg(NO_3)_2(aq) + H_2O(l)$

$$50.0 \text{ g MgO}\left(\frac{1 \text{ mol MgO}}{40.3 \text{ g MgO}}\right)\left(\frac{1 \text{ mol Mg(NO}_3)_2}{1 \text{ mol MgO}}\right) \times$$
$$\left(\frac{148 \text{ g Mg(NO}_3)_2}{1 \text{ mol Mg(NO}_3)_2}\right) = 184 \text{ g Mg(NO}_3)_2$$

(d) $2 LiOH(aq) + N_2O_5(g) \rightarrow 2 LiNO_3(aq) + H_2O(l)$

$$50.0 \text{ g N}_2O_5\left(\frac{1 \text{ mol N}_2O_5}{108 \text{ g N}_2O_5}\right)\left(\frac{2 \text{ mol LiNO}_3}{1 \text{ mol N}_2O_5}\right)\left(\frac{68.9 \text{ g LiNO}_3}{1 \text{ mol LiNO}_3}\right)$$
$$= 63.8 \text{ g LiNO}_3$$

10.107 The balanced equation is

$$NaOH(aq) + HC_2H_3O_2(aq) \rightarrow NaC_2H_3O_2(aq) + H_2O(l)$$

Since the acid and the base react in a 1:1 mole ratio, 1.50 mol of NaOH will react with 1.50 mol of $HC_2H_3O_2$, leaving an excess of 1.50 mole of $HC_2H_3O_2$ in solution, Therefore, at the end of the reaction, there will be 1.50 mole of $NaC_2H_3O_2$ and 1.50 mol of $HC_2H_3O_2$ in the solution. In summary (with all quantities in moles):

$$NaOH(aq) + HC_2H_3O_2(aq) \rightarrow NaC_2H_3O_2(aq) + H_2O(l)$$

	NaOH	$HC_2H_3O_2$	$NaC_2H_3O_2$
Present initially	1.50	3.00	0.00
Change due to reaction	−1.50	−1.50	+1.50
Present finally	0.00	1.50	1.50

10.108 $2 NaHCO_3(s) \xrightarrow{\text{Heat}} Na_2CO_3(s) + CO_2(g) + H_2O(g)$

The word *solid* in the statement of this problem is equivalent to Na_2CO_3 because that is the only solid product.

$$275 \text{ g NaHCO}_3\left(\frac{1 \text{ mol NaHCO}_3}{84.0 \text{ g NaHCO}_3}\right)\left(\frac{1 \text{ mol Na}_2CO_3}{2 \text{ mol NaHCO}_3}\right) \times$$
$$\left(\frac{106 \text{ g Na}_2CO_3}{1 \text{ mol Na}_2CO_3}\right) = 174 \text{ g Na}_2CO_3$$

10.109 The balanced equation is

$$Li_2SO_4(aq) + Ba(NO_3)_2(aq) \rightarrow BaSO_4(s) + 2\ LiNO_3(aq)$$

$$71.2\ g\ Li_2SO_4\left(\frac{1\ mol\ Li_2SO_4}{110\ g\ Li_2SO_4}\right) = 0.647\ mol\ Li_2SO_4$$

$$244\ g\ Ba(NO_3)_2\left(\frac{1\ mol\ Ba(NO_3)_2}{261\ g\ Ba(NO_3)_2}\right) = 0.935\ mol\ Ba(NO_3)_2$$

The equation indicates that a 1:1 mole ratio of reactants is required, and the ratio of barium nitrate to lithium sulfate present is greater than 1:1; thus, the Li_2SO_4 is in limiting quantity.

$$0.647\ mol\ Li_2SO_4\left(\frac{1\ mol\ BaSO_4}{1\ mol\ Li_2SO_4}\right)\left(\frac{233\ g\ BaSO_4}{1\ mol\ BaSO_4}\right)$$
$$= 151\ g\ BaSO_4$$

10.111 $2\ KClO_3(s) \rightarrow 2\ KCl(s) + 3\ O_2(g)$
The number of moles of KCl produced is

$$1.01\ g\ KCl\left(\frac{1\ mol\ KCl}{74.5\ g\ KCl}\right) = 0.0136\ mol\ KCl$$

(a) $0.0136\ mol\ KCl\left(\frac{3\ mol\ O_2}{2\ mol\ KCl}\right)\left(\frac{32.0\ g\ O_2}{1\ mol\ O_2}\right)$
$$= 0.653\ g\ O_2\ produced$$

(b) $0.0136\ mol\ KCl\left(\frac{2\ mol\ KClO_3}{2\ mol\ KCl}\right)\left(\frac{122.5\ g\ KClO_3}{1\ mol\ KClO_3}\right)$
$$= 1.67\ g\ KClO_3\ decomposed$$

The $KClO_3$ did not all decompose.

10.113 This reaction proceeds because of the formation of the (largely covalent) acetic acid. Since $NaC_2H_3O_2$ and HCl react in a 1:1 mole ratio, 3.00 mol of $NaC_2H_3O_2$ will react with 3.00 mol of HCl, leaving an excess of 2.00 mol of HCl in solution. Therefore, at the end of the reaction, there will be 3.00 mol of $HC_2H_3O_2$, 3.00 mol of NaCl, and 2.00 mol of HCl in the aqueous solution. In summary (all quantities in moles):

$$NaC_2H_3O_2(aq) + HCl(aq) \rightarrow NaCl(aq) + HC_2H_3O_2(aq)$$

Present initially	3.00	5.00	0.00	0.00
Change due to reaction	−3.00	−3.00	+3.00	+3.00
Present finally	0.00	2.00	3.00	3.00

10.118 $Na_2CO_3(aq) + 2\ HCl(aq) \rightarrow 2\ NaCl(aq) + CO_2(g) + H_2O(l)$

$$14.5\ g\ Na_2CO_3\left(\frac{1\ mol\ Na_2CO_3}{106\ g\ Na_2CO_3}\right) = 0.137\ mol\ Na_2CO_3$$

$$10.0\ g\ HCl\left(\frac{1\ mol\ HCl}{36.5\ g\ HCl}\right) = 0.274\ mol\ HCl$$

Present *Required*
$$\frac{0.274\ mol\ HCl}{0.137\ mol\ Na_2CO_3} = \frac{2.00\ mol\ HCl}{1\ mol\ Na_2CO_3} \qquad \frac{2\ mol\ HCl}{1\ mol\ Na_2CO_3}$$

Both reactants are in limiting quantity, so either quantity may be used to calculate the mass of NaCl, the only solid produced:

$$0.137\ mol\ Na_2CO_3\left(\frac{2\ mol\ NaCl}{1\ mol\ Na_2CO_3}\right)\left(\frac{58.5\ g\ NaCl}{1\ mol\ NaCl}\right)$$
$$= 16.0\ g\ NaCl$$

10.119 (a)

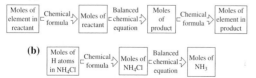

11 Molarity

11.1 **(a)** $\dfrac{2.00\ mol}{1.00\ L} = 2.00\ M$

(b) $\dfrac{2.00\ mol}{2.00\ L} = 1.00\ M$

(c) $\dfrac{2.00\ mol}{4.00\ L} = 0.500\ M$

11.2 **(a)** Less than 3 L of water is used.
(b) Exactly 3.00 L of water is used, and more than 3 L of solution is produced.

11.3 The concentration of the portion is 2.52 M, the same as the concentration of the original sample. (Tea would taste as sweet if sipped from a 100-mL portion or from 25.0 mL poured directly from that 100-mL portion.)

11.4 The two solutions have the same initial concentration. When they are combined, the final concentration is also the same. (The two initial solutions and the final solution would all taste equally sweet. Try it.) As an analogy, if you travel 50 miles per hour during the first half of a trip and 50 miles per hour during the second half of the trip, your overall speed is 50 miles per hour. Alternatively, you may calculate the final concentration:

$$0.050\ L\left(\frac{3.00\ mol}{1\ L}\right) = 0.150\ mol$$

$$0.100\ L\left(\frac{3.00\ mol}{1\ L}\right) = 0.300\ mol$$

$$\frac{0.450\ mol}{0.150\ L} = 3.00\ M$$

11.5 **(a)** 1.000 mL
(b) 4.000 mmol (one-thousandth of the 4.000 mol in the sample)
(c) 4.000 M (The small sample is the same concentration as the original solution, since it has not been changed.)
(d) 4.000 M = 4.000 mmol/mL

11.6 **(a)** A chemical reaction occurs.
(b) The number of moles of sodium ion in the final solution must be calculated from the numbers of moles in the two solutions.
(c) Each solution merely dilutes the concentrations of the ions in the other.

11.7 **(a)** The final volume is about 4.0 L.
(b) The final volume is 2.5 L. Note the small difference in the wording of the parts of this problem, which makes a large difference in the answer.

11.8 **(a)** 1.0 M K^+ and 1.0 M Cl^-
(b) 1.0 M Mg^{2+} and 2.0 M Cl^-
(c) 1.0 M Al^{3+} and 3.0 M Cl^-
(d) 1.0 M K^+ and 1.0 M ClO_3^-
(e) 1.0 M Mg^{2+} and 2.0 M ClO_3^-
(f) 1.0 M Al^{3+} and 3.0 M ClO_3^-
Note that the subscript after the O in the formulas in parts (d)–(f) represents the number of oxygen atoms per chlorate ion, not the number of chlorate ions per formula unit.

11.9 **(a)** 2 dozen brides, 2 dozen grooms
(b) 2.0 M Na^+, 2.0 M Cl^-

11.10 $\dfrac{2.00\ mol}{0.250\ L} = 8.00\ M$

11.12 $40.0\ g\ CH_2O\left(\dfrac{1\ mol\ CH_2O}{30.0\ g\ CH_2O}\right) = 1.33\ mol\ CH_2O$

$$\frac{1.33\ mol}{0.100\ L} = 13.3\ M$$

11.14 $\dfrac{125.0 \text{ mmol}}{400.0 \text{ mL}} = 0.312 \text{ M}$

Note that molarity is the number of millimoles per milliliter, as well as the number of moles per liter.

11.16 $1.270 \text{ L}\left(\dfrac{1.755 \text{ mol}}{1 \text{ L}}\right) = 2.229 \text{ mol}$

11.17 $3.000 \text{ L}\left(\dfrac{4.000 \text{ mol NaBr}}{1 \text{ L}}\right)\left(\dfrac{102.9 \text{ g NaBr}}{1 \text{ mol NaBr}}\right) = 1235 \text{ g NaBr}$

11.20 $1.19 \text{ mol}\left(\dfrac{1 \text{ L}}{1.27 \text{ mol}}\right) = 0.937 \text{ L}$

11.21 $7.14 \text{ g}\left(\dfrac{1 \text{ mol}}{58.5 \text{ g}}\right)\left(\dfrac{1000 \text{ mmol}}{1 \text{ mol}}\right)\left(\dfrac{1 \text{ mL}}{2.25 \text{ mmol}}\right) = 54.2 \text{ mL}$

11.22 The initial solution contains

$2.0 \text{ L}\left(\dfrac{1.5 \text{ mol}}{1 \text{ L}}\right) = 3.0 \text{ mol}$

(a) The final volume is about 5.0 L, so the final concentration is

$\dfrac{3.0 \text{ mol}}{5.0 \text{ L}} = 0.60 \text{ M}$

(b) The final volume is 3.0 L, so the final concentration is

$\dfrac{3.0 \text{ mol}}{3.0 \text{ L}} = 1.0 \text{ M}$

11.25 The initial solution contains

$25.0 \text{ mL}\left(\dfrac{1.15 \text{ mmol}}{1 \text{ mL}}\right) = 28.8 \text{ mmol}$

The final solution contains the same number of moles of solute, so its concentration is

$\dfrac{28.8 \text{ mmol}}{125.0 \text{ mL}} = 0.230 \text{ M}$

11.26 The initial solution contains

$25.0 \text{ mL}\left(\dfrac{1.15 \text{ mmol}}{1 \text{ mL}}\right) = 28.8 \text{ mmol}$

The final solution contains the same number of moles of solute, so its volume is

$28.8 \text{ mmol}\left(\dfrac{1 \text{ mL}}{0.800 \text{ mmol}}\right) = 36.0 \text{ mL}$

11.27 The numbers of moles of solute are the same in the initial and final solutions. Use the known volume and molarity of the final solution to calculate that number of moles, and then use the number of moles to find the volume of the initial solution:

$4.00 \text{ L}\left(\dfrac{1.75 \text{ mol}}{1 \text{ L}}\right) = 7.00 \text{ mol}$

$7.00 \text{ mol}\left(\dfrac{1 \text{ L}}{4.50 \text{ mol}}\right) = 1.56 \text{ L}$

11.31 (a) 5.21 M Na^+ and 5.21 M Cl^-

(b) 3.45 M K^+ and 1.15 M PO_4^{3-} (Note that the phosphorus and oxygen atoms are covalently bonded, and are not dissociated into monatomic ions.)

(c) 3.58 M NH_4^+ and 1.79 M SO_4^{2-}

(d) 1.50 M Al^{3+} and 2.25 M SO_4^{2-}

(e) 0.200 M Na^+ and 0.200 M ClO_2^-

11.33 $25.0 \text{ mL CoCl}_2\left(\dfrac{3.00 \text{ mmol CoCl}_2}{1 \text{ mL CoCl}_2}\right) = 75.0 \text{ mmol CoCl}_2$

The 75.0 mmol of $CoCl_2$ consists of 75.0 mmol of Co^{2+} and 150 mmol of Cl^-. The concentrations are

$\dfrac{75.0 \text{ mmol Co}^{2+}}{80.0 \text{ mL}} = 0.938 \text{ M Co}^{2+}$

$\dfrac{150 \text{ mmol Cl}^-}{80.0 \text{ mL}} = 1.88 \text{ M Cl}^-$

11.35 (a) There are ions in common. The concentration of the chloride ion is the sum of that provided by KCl and that provided by $CuCl_2$.

(b) There is no interaction between the ions in these compounds.

(c) The Ag^+ ions react with the Cl^- ions, forming a solid and leaving fewer ions in solution.

(d) The Ba^{2+} ions react with the SO_4^{2-} ions, forming a solid and leaving fewer ions in solution. In addition, the H^+ ions react with the OH^- ions, forming water and leaving still fewer ions in solution.

(e) The H^+ ions react with the CO_3^{2-} ions, forming water and carbon dioxide, two covalent compounds, and leaving fewer ions in solution.

(f) There is no interaction between the ions in these compounds.

11.36 (a) 0.200 M (0.100 M Na^+ plus 0.100 M Cl^-)

(b) 0.300 M (0.100 M Mg^{2+} plus 0.200 M Cl^-)

(c) $21.6 \text{ g (NH}_4)_2SO_4\left(\dfrac{1 \text{ mol (NH}_4)_2SO_4}{132 \text{ g (NH}_4)_2SO_4}\right)$
$= 0.164 \text{ mol (NH}_4)_2SO_4$

$\dfrac{0.164 \text{ mol (NH}_4)_2SO_4}{0.252 \text{ L}} = 0.651 \text{ M (NH}_4)_2SO_4$

1.30 M NH_4^+ plus 0.651 M SO_4^{2-} = 1.95 M

11.37 The 0.250 mol of K_2SO_4 consists of

$0.250 \text{ mol K}_2SO_4\left(\dfrac{2 \text{ mol K}^+}{1 \text{ mol K}_2SO_4}\right) = 0.500 \text{ mol K}^+$

and

$0.250 \text{ mol K}_2SO_4\left(\dfrac{1 \text{ mol SO}_4^{2-}}{1 \text{ mol K}_2SO_4}\right) = 0.250 \text{ mol SO}_4^{2-}$

The 0.250 mol of Na_2SO_4 consists of

$0.250 \text{ mol Na}_2SO_4\left(\dfrac{2 \text{ mol Na}^+}{1 \text{ mol Na}_2SO_4}\right) = 0.500 \text{ mol Na}^+$

and

$0.250 \text{ mol Na}_2SO_4\left(\dfrac{1 \text{ mol SO}_4^{2-}}{1 \text{ mol Na}_2SO_4}\right) = 0.250 \text{ mol SO}_4^{2-}$

The total number of moles of SO_4^{2-} is

0.250 mol + 0.250 mol = 0.500 mol

The concentrations are

$\dfrac{0.500 \text{ mol K}^+}{0.250 \text{ L}} = 2.00 \text{ M K}^+$

$\dfrac{0.500 \text{ mol Na}^+}{0.250 \text{ L}} = 2.00 \text{ M Na}^+$

$\dfrac{0.500 \text{ mol SO}_4^{2-}}{0.250 \text{ L}} = 2.00 \text{ M SO}_4^{2-}$

11.38 The compound consists of Hg^{2+} ions and NO_3^- ions. There are 0.100 mol of Hg^{2+} ions per liter and twice that concentration of NO_3^- ions—0.200 M NO_3^-.

11.39 $48.27 \text{ mL}\left(\dfrac{1.738 \text{ mmol}}{1 \text{ mL}}\right) = 83.89 \text{ mmol HClO}_3$

$$50.00 \text{ mL}\left(\frac{0.5000 \text{ mmol}}{1 \text{ mL}}\right) = 25.00 \text{ mmol NaOH}$$

The net ionic equation is

$$H^+(aq) + OH^-(aq) \rightarrow H_2O(l)$$

The OH^- ion is in limiting quantity, and there is 58.89 mmol of H^+ ion in excess. The final concentrations are

$$\frac{58.89 \text{ mmol H}^+}{98.27 \text{ mL}} = 0.5993 \text{ M H}^+$$

$$\frac{83.89 \text{ mmol ClO}_3^-}{98.27 \text{ mL}} = 0.8537 \text{ M ClO}_3^-$$

$$\frac{25.00 \text{ mmol Na}^+}{98.27 \text{ mL}} = 0.2544 \text{ M Na}^+$$

11.40 $272 \text{ mL}\left(\dfrac{1.42 \text{ mmol NaCl}}{1 \text{ mL}}\right) = 386 \text{ mmol NaCl}$

$432 \text{ mL}\left(\dfrac{1.19 \text{ mmol AlCl}_3}{1 \text{ mL}}\right) = 514 \text{ mmol AlCl}_3$

386 mmol NaCl consists of 386 mmol Na^+ and 386 mmol Cl^-
514 mmol $AlCl_3$ consists of 514 mmol Al^{3+} and 1540 mmol Cl^-
There is a total of 1930 mmol of Cl^- in the final solution. The final concentrations are

$$\frac{386 \text{ mmol Na}^+}{1000 \text{ mL}} = 0.386 \text{ M Na}^+$$

$$\frac{514 \text{ mmol Al}^{3+}}{1000 \text{ mL}} = 0.514 \text{ M Al}^{3+}$$

$$\frac{1930 \text{ mmol Cl}^-}{1000 \text{ mL}} = 1.93 \text{ M Cl}^-$$

11.41 The NaCl solution contains 100 mmol of NaCl, or 100 mmol of Na^+ and 100 mmol of Cl^-. The NaBr solution contains 50.0 mmol of NaBr, or 50.0 mmol of Na^+ and 50.0 mmol of Br^-. The total number of millimoles of Na^+ is 50.0 mmol + 100 mmol = 150 mmol. These values may be tabulated for clarity:

Solute	Millimoles of Compound	Millimoles of Na^+	Millimoles of Cl^-	Millimoles of Br^-
NaCl	100	100	100	
NaBr	50.0	50.0		50.0
Totals for ions:		150	100	50.0

The concentrations are

$$\frac{150 \text{ mmol Na}^+}{100 \text{ mL}} = 1.50 \text{ M Na}^+$$

$$\frac{100 \text{ mmol Cl}^-}{100 \text{ mL}} = 1.00 \text{ M Cl}^-$$

$$\frac{50.0 \text{ mmol Br}^-}{100 \text{ mL}} = 0.500 \text{ M Br}^-$$

11.47 $H_2SO_4(aq) + 2 NaOH(aq) \rightarrow Na_2SO_4(aq) + 2 H_2O(l)$

The number of millimoles of NaOH is

$$46.19 \text{ mL}\left(\frac{1.500 \text{ mmol}}{1 \text{ mL}}\right) = 69.28 \text{ mmol NaOH}$$

$$69.28 \text{ mmol NaOH}\left(\frac{1 \text{ mmol H}_2SO_4}{2 \text{ mmol NaOH}}\right) = 34.64 \text{ mmol H}_2SO_4$$

The concentration of the H_2SO_4 is

$$\frac{34.64 \text{ mmol}}{25.00 \text{ mL}} = 1.386 \text{ M}$$

11.48 1.148 M. This problem is essentially the same as Problem 11.47, except that names are used instead of formulas.

11.51 $6.593 \text{ g KHPh}\left(\dfrac{1 \text{ mol KHPh}}{204.2 \text{ g KHPh}}\right)\left(\dfrac{1 \text{ mol KOH}}{1 \text{ mol KHPh}}\right)$
$\qquad\qquad\qquad\qquad\qquad = 3.229 \times 10^{-2} \text{ mol KOH}$

$$\frac{3.229 \times 10^{-2} \text{ mol KOH}}{41.99 \times 10^{-3} \text{ L}} = 0.7690 \text{ M KOH}$$

11.52 $HCl(aq) + NaHCO_3(s) \rightarrow NaCl(aq) + CO_2(g) + H_2O(l)$

$22.0 \text{ g NaHCO}_3\left(\dfrac{1 \text{ mol NaHCO}_3}{84.0 \text{ g NaHCO}_3}\right)\left(\dfrac{1 \text{ mol HCl}}{1 \text{ mol NaHCO}_3}\right) \times$

$\qquad\qquad\qquad\left(\dfrac{1 \text{ L}}{4.47 \text{ mol HCl}}\right) = 0.0586 \text{ L HCl}$

The tablet can neutralize 58.6 mL of stomach acid.

11.54 $NaOH(aq) + HCl(aq) \rightarrow NaCl(aq) + H_2O(l)$

$13.87 \text{ mL}\left(\dfrac{6.000 \text{ mmol HCl}}{1 \text{ mL}}\right)\left(\dfrac{1 \text{ mmol NaOH}}{1 \text{ mmol HCl}}\right)$
$\qquad\qquad\qquad\qquad\qquad = 83.22 \text{ mmol NaOH}$

11.56 $Ba(OH)_2(aq) + 2 HClO_3(aq) \rightarrow Ba(ClO_3)_2(aq) + 2 H_2O(l)$

$24.17 \text{ mL}\left(\dfrac{6.000 \text{ mmol HClO}_3}{1 \text{ mL}}\right)\left(\dfrac{1 \text{ mmol Ba(OH)}_2}{2 \text{ mmol HClO}_3}\right)$
$\qquad\qquad\qquad\qquad\qquad = 72.51 \text{ mmol Ba(OH)}_2$

11.58 $1.91 \text{ L}\left(\dfrac{2.71 \text{ mol NaCl}}{1 \text{ L}}\right) = 5.18 \text{ mol NaCl}$

$2.27 \text{ L}\left(\dfrac{0.985 \text{ mol AlCl}_3}{1 \text{ L}}\right) = 2.24 \text{ mol AlCl}_3$

5.18 mol NaCl consists of 5.18 mol Na^+ and 5.18 mol Cl^-
2.24 mol $AlCl_3$ consists of 2.24 mol Al^{3+} and 6.72 mol Cl^-
There is a total of 11.90 mol of Cl^- ions in the final solution. The final concentrations are

$$\frac{5.18 \text{ mol Na}^+}{5.00 \text{ L}} = 1.04 \text{ M Na}^+$$

$$\frac{2.24 \text{ mol Al}^{3+}}{5.00 \text{ L}} = 0.448 \text{ M Al}^{3+}$$

$$\frac{11.90 \text{ mol Cl}^-}{5.00 \text{ L}} = 2.38 \text{ M Cl}^-$$

11.59 $1.79 \text{ L}\left(\dfrac{2.22 \text{ mol NaOH}}{1 \text{ L}}\right) = 3.97 \text{ mol NaOH}$

$2.19 \text{ L}\left(\dfrac{0.505 \text{ mol H}_2SO_4}{1 \text{ L}}\right) = 1.11 \text{ mol H}_2SO_4$

3.97 mol NaOH consists of 3.97 mol Na^+ and 3.97 mol OH^-
1.11 mol H_2SO_4 consists of 1.11 mol SO_4^{2-} and 2.22 mol H^+
The OH^- ions react with the H^+ ions according to the net ionic equation

$$OH^-(aq) + H^+(aq) \rightarrow H_2O(l)$$

There is a limiting quantity of H^+, so 3.97 mol $-$ 2.22 mol = 1.75 mol of OH^- will remain in the solution. The final concentrations are

$$\frac{3.97 \text{ mol Na}^+}{5.00 \text{ L}} = 0.794 \text{ M Na}^+$$

$$\frac{1.11 \text{ mol SO}_4{}^{2-}}{5.00 \text{ L}} = 0.222 \text{ M SO}_4{}^{2-}$$

$$\frac{1.75 \text{ mol OH}^-}{5.00 \text{ L}} = 0.350 \text{ M OH}^-$$

11.62 (a) The NaCl consists of 40.50 mmol Na^+ and 40.50 mmol Cl^-.
The NaOH consists of 14.55 mmol Na^+ and 14.55 mmol OH^-.
The total number of millimoles of Na^+ is 55.05 mmol.
The concentrations are

$$\frac{55.05 \text{ mmol Na}^+}{82.05 \text{ mL}} = 0.6709 \text{ M Na}^+$$

$$\frac{14.55 \text{ mmol OH}^-}{82.05 \text{ mL}} = 0.1773 \text{ M OH}^-$$

$$\frac{40.50 \text{ mmol Cl}^-}{82.05 \text{ mL}} = 0.4936 \text{ M Cl}^-$$

(b) $NaOH(aq) + HCl(aq) \rightarrow NaCl(aq) + H_2O(l)$
When the

$$27.00 \text{ mL}\left(\frac{1.500 \text{ mmol}}{1 \text{ mL}}\right) = 40.50 \text{ mmol of HCl}$$

reacts with the excess NaOH, 40.50 mmol of NaCl is formed, and 14.55 mmol of NaOH remains unreacted because it is in excess. The same concentrations are obtained as in part (a).

(c) Since both part (a) and part (b) have the same volumes and the same quantities of the same compounds, the results are the same.

11.64 The HCl solution contains

$$47.57 \text{ mL}\left(\frac{1.000 \text{ mmol}}{1 \text{ mL}}\right)$$
$$= 47.57 \text{ mmol Cl}^- \text{ and } 47.57 \text{ mmol H}^+$$

The NaOH solution contains

$$23.46 \text{ mL}\left(\frac{1.527 \text{ mmol}}{1 \text{ mL}}\right)$$
$$= 35.82 \text{ mmol Na}^+ \text{ and } 35.82 \text{ mmol OH}^-$$

The H^+ ions react with OH^- ions:

$$H^+(aq) + OH^-(aq) \rightarrow H_2O(l)$$

The limiting quantity is the quantity of OH^- ions, therefore 35.82 mmol of OH^- reacts with 35.82 mmol of H^+ to produce water. This leaves 47.57 mmol − 35.82 mmol = 11.75 mmol of H^+ ions remaining. The chloride and sodium ions are spectator ions; they do not react. Therefore, there are 47.57 mmol of Cl^- and 35.82 mmol of Na^+ in the final solution. The final volume is assumed to be 47.57 mL + 23.46 mL = 71.03 mL, and so the final concentrations are

$$\frac{11.75 \text{ mmol H}^+}{71.03 \text{ mL}} = 0.1654 \text{ M H}^+$$

$$\frac{35.82 \text{ mmol Na}^+}{71.03 \text{ mL}} = 0.5043 \text{ M Na}^+$$

$$\frac{47.57 \text{ mmol Cl}^-}{71.03 \text{ mL}} = 0.6697 \text{ M Cl}^-$$

11.65 $Li_2O(s) + H_2O(l) \rightarrow 2 \text{ OH}^-(aq) + 2 \text{ Li}^+(aq)$

$$0.400 \text{ mol Li}_2O\left(\frac{2 \text{ mol OH}^-}{1 \text{ mol Li}_2O}\right) = 0.800 \text{ mol OH}^-$$

The 0.800 mol of OH^- in 0.500 L of solution represents a 1.60 M OH^- solution.

11.66

$$\frac{0.400 \text{ mol Cu}^{2+}}{1 \text{ L}}\left(\frac{2 \text{ mol H}^+}{1 \text{ mol Cu}^{2+}}\right) = \frac{0.800 \text{ mol H}^+}{1 \text{ L}}$$
$$= 0.800 \text{ M H}^+$$

11.68 The solution contains CH_2O from two sources:

$$2.50 \text{ L}\left(\frac{1.28 \text{ mol}}{1 \text{ L}}\right) = 3.20 \text{ mol}$$

$$1.70 \text{ L}\left(\frac{1.33 \text{ mol}}{1 \text{ L}}\right) = 2.26 \text{ mol}$$

$$3.20 \text{ mol} + 2.26 \text{ mol} = 5.46 \text{ mol}$$

$$\text{Molarity} = \frac{5.46 \text{ mol}}{5.00 \text{ L}} = 1.09 \text{ M}$$

11.70 $100 \text{ mL HC}_2\text{H}_3\text{O}_2\left(\frac{2.00 \text{ mmol HC}_2\text{H}_3\text{O}_2}{1 \text{ mL HC}_2\text{H}_3\text{O}_2}\right)$
$$= 200 \text{ mmol HC}_2\text{H}_3\text{O}_2$$

$$100 \text{ mL OH}^-\left(\frac{1.00 \text{ mmol OH}^-}{1 \text{ mL OH}^-}\right) = 100 \text{ mmol OH}^-$$

The $HC_2H_3O_2$ is in excess, so 100 mmol of $HC_2H_3O_2$ will react, producing 100 mmol of $C_2H_3O_2^-$ and leaving 100 mmol of $HC_2H_3O_2$ in excess. The final concentrations of $C_2H_3O_2^-$ and $HC_2H_3O_2$ are

$$\frac{100 \text{ mmol C}_2\text{H}_3\text{O}_2{}^-}{200 \text{ mL}} = 0.500 \text{ M C}_2\text{H}_3\text{O}_2{}^-$$

$$\frac{100 \text{ mmol HC}_2\text{H}_3\text{O}_2}{200 \text{ mL}} = 0.500 \text{ M HC}_2\text{H}_3\text{O}_2$$

(Note that the concentration of $HC_2H_3O_2$ was halved by reaction with OH^- and halved again by the doubling of the original volume of the solution.)

11.73 The numbers of millimoles of acid and base are calculated from the volumes and molarities:

$$25.00 \text{ mL H}_3\text{PO}_4\left(\frac{2.000 \text{ mmol H}_3\text{PO}_4}{1 \text{ mL H}_3\text{PO}_4}\right) = 50.00 \text{ mmol H}_3\text{PO}_4$$

$$30.95 \text{ mL NaOH}\left(\frac{3.231 \text{ mmol NaOH}}{1 \text{ mL NaOH}}\right) = 100.0 \text{ mmol NaOH}$$

The ratio of base to acid is

$$\frac{100.0 \text{ mmol NaOH}}{50.00 \text{ mmol H}_3\text{PO}_4} = \frac{2 \text{ mol NaOH}}{1 \text{ mol H}_3\text{PO}_4}$$

The balanced equation is therefore

$$2 \text{ NaOH}(aq) + H_3PO_4(aq) \rightarrow Na_2HPO_4(aq) + 2 \text{ H}_2O(l)$$

11.75 Complete neutralization would require three times the volume of NaOH solution, or 83.88 mL, because 3 mol of NaOH is being used per mole of H_3PO_4 instead of only 1 mol:

$$3 \text{ NaOH}(aq) + H_3PO_4(aq) \rightarrow Na_3PO_4(aq) + 3 \text{ H}_2O(l)$$

11.76 Represent the acids as HX. Then

$$HX(aq) + NaOH(aq) \rightarrow NaX(aq) + H_2O(l)$$

$$19.73 \text{ mL NaOH}\left(\frac{0.1000 \text{ mmol NaOH}}{1 \text{ mL NaOH}}\right)\left(\frac{1 \text{ mmol HX}}{1 \text{ mmol NaOH}}\right)$$
$$= 1.973 \text{ mmol HX}$$

$$\frac{1.973 \text{ mmol HX}}{10.0 \text{ mL HX}} = 0.197 \text{ M HX}$$

The total concentration of all acids is 0.197 M.

11.77 $Na_2CO_3(s) + 2 \text{ HCl}(aq) \rightarrow 2 \text{ NaCl}(aq) + CO_2(g) + H_2O(l)$

$$27.16 \text{ mL HCl}\left(\frac{6.000 \text{ mmol HCl}}{1 \text{ mL HCl}}\right) = 163.0 \text{ mmol HCl added}$$

$$2.471 \text{ mL NaOH}\left(\frac{1.000 \text{ mmol NaOH}}{1 \text{ mL NaOH}}\right) = 2.471 \text{ mmol NaOH}$$

Of the 163.0 mmol of HCl added, 2.471 mmol was in excess and was neutralized by the NaOH. The rest, 163.0 mmol − 2.471 mmol = 160.5 mmol, reacted with the Na_2CO_3:

$$160.5 \text{ mmol HCl}\left(\frac{1 \text{ mmol Na}_2\text{CO}_3}{2 \text{ mmol HCl}}\right)\left(\frac{106.0 \text{ mg Na}_2\text{CO}_3}{1 \text{ mmol Na}_2\text{CO}_3}\right)$$
$$= 8506 \text{ mg Na}_2\text{CO}_3 = 8.506 \text{ g Na}_2\text{CO}_3$$

11.79 $0.04318 \text{ L}\left(\dfrac{4.000 \text{ mol}}{1 \text{ L}}\right) = 0.1727 \text{ mol NaOH}$

Since the HA and the NaOH react in a 1:1 ratio, the sample contained 0.1727 mol of acid:

$$\text{MM} = \frac{8.153 \text{ g}}{0.1727 \text{ mol}} = 47.21 \text{ g/mol}$$

11.80 $2.169 \text{ g}\left(\dfrac{1 \text{ mol}}{204.2 \text{ g}}\right)\left(\dfrac{1 \text{ mol NaOH}}{1 \text{ mol}}\right) = 0.01062 \text{ mol NaOH}$

$$\frac{0.01062 \text{ mol}}{0.04191 \text{ L}} = 0.2534 \text{ M NaOH}$$

12 Gases

12.1 (a) $x = 0.214$ (b) $V = 0.214$ (c) $x = 26.96$
     (d) $x = 2.333$ (e) $x = 2.61$ (f) $x = 45.29$
     (g) $x = 8.337$ (h) $x = 8.95$

Note that parts (a) and (b) are the same, despite the difference in variable identity. Part (d) could be done in your head, and part (e) should have a similar result because each factor is approximately the same. You should always check your calculations to see that they are approximately correct.

12.2 Three significant digits: 5°C + 273° = 278 K

12.3 The Kelvin (absolute) scale

12.4 Since the temperature and volume do not change, the pressure also does not change.

12.5 The gas laws apply only to gases; under the given conditions, only CO and N_2 are gases.

12.6 The gas laws apply only to gases; under the given conditions, all four substances are gases. (Water does evaporate even at 0°C.)

12.7 The combined gas law equation is

$$\frac{P_1 V_1}{T_1} = \frac{P_2 V_2}{T_2}$$

(a) If the temperature is constant, $T_1 = T_2$. Multiplying both sides of the combined gas law equation by this equation yields $P_1 V_1 = P_2 V_2$, which is the Boyle's law equation.
(b) If the pressure is constant, $P_1 = P_2$. Dividing both sides of the combined gas law equation by this equation yields the Charles' law equation:

$$\frac{V_1}{T_1} = \frac{V_2}{T_2}$$

12.8 (a) The final volume is 5.00 L.
     (b) The final volume is 3.00 L.
     (c) The final volume is 5.00 L.

12.9 The volume of a gas may be increased by allowing it to expand into an evacuated vessel, as shown in part (a) of the problem, by withdrawing a piston in a cylinder, as shown in part (b), or by other means.
     (a) If a 2.00-L sample is allowed to expand into a 3.00-L evacuated vessel, its final volume is 5.00 L. The gas starts out in the left vessel; there is nothing in the right vessel. The gas is expanded 3.00 L, or expanded *by* 3.00 L, finally occupying both vessels with a volume of 5.00 L.
     (b) If a 2.00-L sample is expanded in a cylinder to a total volume of 3.00 L, its final volume is 3.00 L. It is expanded *to* 3.00 L.

12.10 The volume will go down because the temperature is going down. Celsius temperature is *not* directly proportional to volume.

12.11 The average distance between molecules in a liquid is very much smaller than the average distance between molecules in a gas, as illustrated in Example 12.28.

12.12 The molecules get farther apart on the average.

12.13 The tennis ball must move much faster to have the same kinetic energy as the much more massive bowling ball.

12.14 (a) 76.0 cm × 1.00 cm² = 76.0 cm³

$$76.0 \text{ cm}^3\left(\frac{13.6 \text{ g}}{1 \text{ cm}^3}\right) = 1.03 \times 10^3 \text{ g}\left(\frac{1 \text{ kg}}{1000 \text{ g}}\right) = 1.03 \text{ kg}$$

(b) The mass is twice as great, 2.06 kg.
(c) The weight of the wider column is twice as great because its mass is twice as great. The force pushing it up is also twice as great, however, because its area is twice as great, and the force is equal to pressure times area:

$$f = PA$$

Thus, the height of mercury in a simple barometer is independent of the cross-sectional area of the tube.

12.15 The gas will push the piston back until the pressures are equal.

12.16 (a) $100.0 \text{ kPa}\left(\dfrac{1.000 \text{ atm}}{101.3 \text{ kPa}}\right)\left(\dfrac{760.0 \text{ torr}}{1 \text{ atm}}\right) = 750.2 \text{ torr}$

(b) $1.00 \text{ Pa}\left(\dfrac{1 \text{ kPa}}{1000 \text{ Pa}}\right)\left(\dfrac{1.000 \text{ atm}}{101.3 \text{ kPa}}\right)\left(\dfrac{760.0 \text{ torr}}{1 \text{ atm}}\right)$
$$= 0.00750 \text{ torr}$$

(c) 205 torr

12.17 (a) $4.00 \text{ atm}\left(\dfrac{760 \text{ torr}}{1 \text{ atm}}\right) = 3040 \text{ torr}$

(b) $1.23 \text{ atm}\left(\dfrac{760 \text{ mm Hg}}{1 \text{ atm}}\right) = 935 \text{ mm Hg}$

(c) $720 \text{ torr}\left(\dfrac{1 \text{ atm}}{760 \text{ torr}}\right) = 0.947 \text{ atm}$

(d) $920 \text{ torr}\left(\dfrac{1 \text{ mm Hg}}{1 \text{ torr}}\right) = 920 \text{ mm Hg}$

12.18 (a) $0.395 \text{ L}\left(\dfrac{1 \text{ mL}}{0.001 \text{ L}}\right) = 395 \text{ mL}$

(b) $P_2 = \dfrac{P_1 V_1}{V_2} = \dfrac{(1.14 \text{ atm})(395 \text{ mL})}{197 \text{ mL}} = 2.29 \text{ atm}$

12.20 $P_2 = \dfrac{P_1 V_1}{V_2} = \dfrac{(781 \text{ torr})(2.71 \text{ L})}{1.17 \text{ L}} = 1810 \text{ torr}$

12.22 $P_1 V_1 = P_2 V_2$

(a) $P_2 = \dfrac{P_1 V_1}{V_2} = \dfrac{(770 \text{ torr})(1.75 \text{ L})}{2.26 \text{ L}} = 596 \text{ torr}$

(b) $P_1 = \dfrac{P_2 V_2}{V_1} = \dfrac{(770 \text{ torr})(1.75 \text{ L})}{2.26 \text{ L}} = 596 \text{ torr}$

Note that parts (a) and (b) have the same answer. It does not matter whether you are solving for the initial or the final pressure, as long as the 770 torr and 1.75 L conditions represent the same state.

12.24 Since the temperature does not change (it is constant), this is a Boyle's law problem. Change the second volume to milliliters to match the units of the first volume:

$$0.899 \text{ L}\left(\frac{1 \text{ mL}}{0.001 \text{ L}}\right) = 899 \text{ mL}$$

Tabulate the data:

State	P	V
1	777 torr	716 mL
2	P_2	899 mL

Rearrange the equation and solve:

$$P_1V_1 = P_2V_2$$

$$P_2 = \frac{P_1V_1}{V_2} = \frac{(777 \text{ torr})(716 \text{ mL})}{899 \text{ mL}} = 619 \text{ torr}$$

12.26 (a) 1.33 L

(b) Estimating the volume at a pressure of 12.00 atm is difficult, since it is not easy to estimate how much the line will curve at a point past the experimental data.

12.28 (a) $V_2 = 2.00 \text{ L} + 4.00 \text{ L} = 6.00 \text{ L}$

$$P_2 = \frac{P_1V_1}{V_2} = \frac{(1.00 \text{ atm})(2.00 \text{ L})}{6.00 \text{ L}} = 0.333 \text{ atm}$$

(b) $V_2 = 4.00 \text{ L}$

$$P_2 = \frac{(1.00 \text{ atm})(2.00 \text{ L})}{4.00 \text{ L}} = 0.500 \text{ atm}$$

12.29 As shown in the accompanying figure, absolute zero is about $-273°C$.

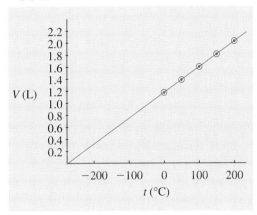

12.31 (a) No; the volume is *not* proportional to the Celsius temperature.

(b) The ratio is equal to the ratio of absolute (Kelvin) temperatures:

$$\frac{V_2}{V_1} = \frac{T_2}{T_1} = \frac{276 \text{ K}}{274 \text{ K}} = 1.01$$

12.33

State	V	T
1	6.11 L	303 K
2	V_2	441 K

$$\frac{V_1}{T_1} = \frac{V_2}{T_2}$$

$$V_2 = \frac{V_1T_2}{T_1} = \frac{(6.11 \text{ L})(441 \text{ K})}{303 \text{ K}} = 8.89 \text{ L}$$

12.35 $T_2 = 344 + 273 = 617 \text{ K}$

$T_1 = 108 + 273 = 381 \text{ K}$

This problem is the same as Problem 12.34, once the Celsius temperatures have been changed to the Kelvin scale.

$$V_2 = \frac{V_1T_2}{T_1} = \frac{(7.17 \text{ L})(617 \text{ K})}{(381 \text{ K})} = 11.6 \text{ L}$$

12.36

State	V	T
1	V_1	317 K
2	1.18 L	302 K

$$\frac{V_1}{T_1} = \frac{V_2}{T_2}$$

$$V_1 = \frac{V_2T_1}{T_2} = \frac{(1.18 \text{ L})(317 \text{ K})}{302 \text{ K}} = 1.24 \text{ L}$$

12.39 $\dfrac{P_1V_1}{T_1} = \dfrac{P_2V_2}{T_2}$

(a) $T_2 = \dfrac{P_2V_2T_1}{P_1V_1} = \dfrac{(1.71 \text{ atm})(3.00 \text{ L})(303 \text{ K})}{(1.33 \text{ atm})(1.11 \text{ L})} = 1050 \text{ K}$

(b) $P_2 = \dfrac{P_1V_1T_2}{V_2T_1} = \dfrac{(762 \text{ torr})(6.17 \text{ L})(310 \text{ K})}{(10.5 \text{ L})(310 \text{ K})} = 448 \text{ torr}$

(c) $V_2 = \dfrac{P_1V_1T_2}{P_2T_1} = \dfrac{(6.11 \text{ atm})(1.79 \text{ L})(331 \text{ K})}{(6.11 \text{ atm})(300 \text{ K})} = 1.97 \text{ L}$

(d) $V_2 = \dfrac{P_1V_1T_2}{P_2T_1} = \dfrac{(777 \text{ torr})(559 \text{ mL})(303 \text{ K})}{(1520 \text{ torr})(303 \text{ K})} = 286 \text{ mL}$

(e) $P_2 = \dfrac{P_1V_1T_2}{V_2T_1} = \dfrac{(1.14 \text{ atm})(1.92 \text{ L})(314 \text{ K})}{(1.18 \text{ L})(288 \text{ K})} = 2.02 \text{ atm}$

(f) $T_1 = \dfrac{P_1V_1T_2}{P_2V_2} = \dfrac{(2.11 \text{ atm})(1.97 \text{ L})(292 \text{ K})}{(3.09 \text{ atm})(0.918 \text{ L})} = 428 \text{ K}$

(g) $T_2 = \dfrac{P_2V_2T_1}{P_1V_1} = \dfrac{(836 \text{ torr})(4.10 \text{ L})(298 \text{ K})}{(767 \text{ torr})(10.1 \text{ L})} = 132 \text{ K}$

(h) $P_1 = \dfrac{P_2V_2T_1}{V_1T_2} = \dfrac{(821 \text{ torr})(1.92 \text{ L})(312 \text{ K})}{(0.973 \text{ L})(270 \text{ K})}$

$$= 1.87 \times 10^3 \text{ torr}$$

12.42 The 785 torr is converted to atmospheres for each part:

$$785 \text{ torr}\left(\frac{1 \text{ atm}}{760 \text{ torr}}\right) = 1.03 \text{ atm}$$

(a) $V_2 = \dfrac{P_1V_1T_2}{T_1P_2} = \dfrac{(1.00 \text{ atm})(2.00 \text{ L})(373 \text{ K})}{(273 \text{ K})(1.03 \text{ atm})} = 2.65 \text{ L}$

(b) $V = \dfrac{nRT}{P} = \dfrac{(2.00 \text{ mol})(0.0821 \text{ L·atm/mol·K})(373 \text{ K})}{1.03 \text{ atm}}$

$$= 59.5 \text{ L}$$

12.43 The number of moles of gas before the change may be calculated as follows:

$$n = \frac{PV}{RT} = \frac{(1.00 \text{ atm})(2.00 \text{ L})}{(0.0821 \text{ L·atm/mol·K})(273 \text{ K})} = 0.0892 \text{ mol}$$

The same number of moles is present after the change, since it is the same sample of gas. The new pressure in atmospheres is

$$785 \text{ torr}\left(\frac{1 \text{ atm}}{760 \text{ torr}}\right) = 1.03 \text{ atm}$$

The new volume is

$$V = \frac{nRT}{P} = \frac{(0.0892 \text{ mol})(0.0821 \text{ L·atm/mol·K})(373 \text{ K})}{1.03 \text{ atm}}$$

$$= 2.65 \text{ L}$$

12.45 $785 \text{ torr}\left(\dfrac{1 \text{ atm}}{760 \text{ torr}}\right) = 1.03 \text{ atm}$

$$n = \frac{PV}{RT} = \frac{(1.03 \text{ atm})(2.33 \text{ L})}{(0.0821 \text{ L·atm/mol·K})(339 \text{ K})} = 0.0862 \text{ mol}$$

12.46 $781 \text{ torr}\left(\dfrac{1 \text{ atm}}{760 \text{ torr}}\right) = 1.03 \text{ atm}$

$$V = \frac{nRT}{P} = \frac{(0.114 \text{ mol})(0.0821 \text{ L·atm/mol·K})(303 \text{ K})}{1.03 \text{ atm}}$$

$$= 2.75 \text{ L}$$

12.49 $10.2 \text{ g CO}_2\left(\dfrac{1 \text{ mol CO}_2}{44.0 \text{ g CO}_2}\right) = 0.232 \text{ mol}$

$$P = \frac{nRT}{V} = \frac{(0.232 \text{ mol})(0.0821 \text{ L·atm/mol·K})(299 \text{ K})}{17.7 \text{ L}}$$

$$= 0.322 \text{ atm}$$

12.50 $n = \dfrac{PV}{RT} = \dfrac{(1.04 \text{ atm})(2.92 \text{ L})}{(0.0821 \text{ L·atm/mol·K})(319 \text{ K})} = 0.116 \text{ mol}$

12.51 (a) $T = \dfrac{PV}{nR} = \dfrac{(1.01 \text{ atm})(9.93 \text{ L})}{(1.22 \text{ mol})(0.0821 \text{ L·atm/mol·K})} = 100 \text{ K}$

(b) The gas is more likely to be He, since H_2O would turn to a solid (ice) at that temperature.

12.55 The gas laws apply to moles of *molecules*, whether the molecules are monatomic, diatomic, or polyatomic. In any case, the correct number of atoms per molecule is necessary to calculate masses of gaseous elements from the numbers of their moles.

12.56 $n = \dfrac{PV}{RT} = \dfrac{(1.00 \text{ atm})(0.750 \text{ L})}{(0.0821 \text{ L·atm/mol·K})(296 \text{ K})} = 0.0309 \text{ mol O}_2$

Be careful; this is the number of moles of O_2 gas, not the number of moles of HgO! You must determine the number of moles of HgO from the balanced chemical equation.

$2 \text{ HgO(s)} \rightarrow 2 \text{ Hg(l)} + \text{O}_2\text{(g)}$

$0.0309 \text{ mol O}_2\left(\dfrac{2 \text{ mol HgO}}{1 \text{ mol O}_2}\right) = 0.0618 \text{ mol HgO}$

12.57 $\text{Zn(s)} + 2 \text{ HCl(aq)} \rightarrow \text{H}_2\text{(g)} + \text{ZnCl}_2\text{(aq)}$

$4.50 \text{ g Zn}\left(\dfrac{1 \text{ mol Zn}}{65.4 \text{ g Zn}}\right)\left(\dfrac{1 \text{ mol H}_2}{1 \text{ mol Zn}}\right) = 0.0688 \text{ mol H}_2$

The volume occupied by 0.0688 mol of H_2 under the conditions given is

$$V = \frac{nRT}{P} = \frac{(0.0688 \text{ mol})(0.0821 \text{ L·atm/mol·K})(301 \text{ K})}{1.00 \text{ atm}}$$

$$= 1.70 \text{ L}$$

12.64 $2 \text{ NaHCO}_3\text{(s)} \xrightarrow{\text{Heat}} \text{Na}_2\text{CO}_3\text{(s)} + \text{CO}_2\text{(g)} + \text{H}_2\text{O(g)}$

$2.75 \text{ g NaHCO}_3\left(\dfrac{1 \text{ mol NaHCO}_3}{84.0 \text{ g NaHCO}_3}\right)\left(\dfrac{1 \text{ mol CO}_2}{2 \text{ mol NaHCO}_3}\right)$

$$= 0.0164 \text{ mol CO}_2$$

$$V = \frac{nRT}{P} = \frac{(0.0164 \text{ mol})(0.0821 \text{ L·atm/mol·K})(298 \text{ K})}{1.00 \text{ atm}}$$

$$= 0.401 \text{ L}$$

12.67 $2 \text{ H}_2\text{(g)} + \text{O}_2\text{(g)} \rightarrow 2 \text{ H}_2\text{O(g)}$

The number of moles of water produced may be calculated from the ideal gas law:

$$n = \frac{PV}{RT} = \frac{(0.998 \text{ atm})(16.0 \text{ L})}{(0.0821 \text{ L·atm/mol·K})(395 \text{ K})}$$

$$= 0.492 \text{ mol H}_2\text{O(g)}$$

$0.492 \text{ mol H}_2\text{O}\left(\dfrac{1 \text{ mol O}_2}{2 \text{ mol H}_2\text{O}}\right)\left(\dfrac{32.0 \text{ g O}_2}{1 \text{ mol O}_2}\right) = 7.87 \text{ g O}_2$

12.69 (a) $3 \text{ H}_2\text{(g)} + \text{N}_2\text{(g)} \rightarrow 2 \text{ NH}_3\text{(g)}$

(b) $V_{\text{H}_2} = \dfrac{nRT}{P}$

$$= \frac{(3.00 \text{ mol H}_2)(0.0821 \text{ L·atm/mol·K})(273 \text{ K})}{1.00 \text{ atm}}$$

$$= 67.2 \text{ L H}_2$$

$$V_{\text{N}_2} = \frac{(1.00 \text{ mol N}_2)(0.0821 \text{ L·atm/mol·K})(273 \text{ K})}{1.00 \text{ atm}}$$

$$= 22.4 \text{ L N}_2$$

$$V_{\text{NH}_3} = \frac{(2.00 \text{ mol NH}_3)(0.0821 \text{ L·atm/mol·K})(273 \text{ K})}{1.00 \text{ atm}}$$

$$= 44.8 \text{ L NH}_3$$

(c) You can see in the setup for part (b) that the volumes of the gases are in the same ratio as their numbers of moles because R, T, and P are the same for all three gases.

12.71 No matter what temperature and what pressure you choose, as long as all three substances are gases, their volume ratios are the same as their mole ratios. The volumes of gases involved in a chemical reaction are governed by the balanced chemical equation as long as they are all at the same temperature and pressure.

12.72 The numbers of moles of all the components are also the same. The volumes and temperatures of all gases in any gaseous mixture are the same, and in this mixture, the pressures are also the same. Since $n = PV/RT$ and all the factors on the right side of the equation are the same for all the gases, the value of n must also be the same for all the gases.

12.73 The total number of moles is 3.00 mol. The ideal gas law works for both the gas mixture and the individual components, so

$$P_{\text{total}} = n_{\text{total}}RT/V \quad \text{and} \quad P_{\text{He}} = n_{\text{He}}RT/V$$

Dividing the second of these equations by the first yields

$$\frac{P_{\text{He}}}{P_{\text{total}}} = \frac{n_{\text{He}}RT/V}{n_{\text{total}}RT/V}$$

R, T, and V all cancel out, leaving

$$\frac{P_{\text{He}}}{P_{\text{total}}} = \frac{n_{\text{He}}}{n_{\text{total}}}$$

The partial pressures are directly proportional to the numbers of moles present.

$$P_{\text{He}} = \frac{n_{\text{He}}P_{\text{total}}}{n_{\text{total}}} = \frac{(1.00 \text{ mol})(1.50 \text{ atm})}{3.00 \text{ mol}} = 0.500 \text{ atm}$$

The pressures of the other gases can be calculated in the same way:

$$P_{\text{Ne}} = P_{\text{Ar}} = 0.500 \text{ atm}$$

As a check, calculate the total pressure:

$$0.500 \text{ atm} + 0.500 \text{ atm} + 0.500 \text{ atm} = 1.50 \text{ atm}$$

12.75 As shown in Problem 12.73, the number of moles of a gas in a mixture is directly proportional to the partial pressure of that gas. Therefore,

$$\frac{P_{\text{H}_2}}{n_{\text{H}_2}} = \frac{P_{\text{N}_2}}{n_{\text{N}_2}}$$

$$P_{\text{N}_2} = \frac{P_{\text{H}_2}n_{\text{N}_2}}{n_{\text{H}_2}} = \frac{(0.118 \text{ atm})(0.173 \text{ mol})}{0.616 \text{ mol}} = 0.0331 \text{ atm}$$

$$\frac{P_{\text{H}_2}}{n_{\text{H}_2}} = \frac{P_{\text{Ne}}}{n_{\text{Ne}}}$$

$$P_{\text{Ne}} = \frac{P_{\text{H}_2}n_{\text{Ne}}}{n_{\text{H}_2}} = \frac{(0.118 \text{ atm})(0.291 \text{ mol})}{0.616 \text{ mol}} = 0.0557 \text{ atm}$$

12.77 The partial pressure of water vapor in this system is 770 torr − 743 torr = 27 torr. The temperature is determined from Table 12.3 to be 27°C.

12.79 The number of moles of hydrogen is

$$2.00 \text{ g } H_2 \left(\frac{1 \text{ mol } H_2}{2.016 \text{ g } H_2} \right) = 0.992 \text{ mol } H_2$$

The pressure of the hydrogen is

$$P_{H_2} = P_{total} - P_{H_2O} = 760 \text{ torr} - 24 \text{ torr} = 736 \text{ torr}$$

$$736 \text{ torr} \left(\frac{1 \text{ atm}}{760 \text{ torr}} \right) = 0.968 \text{ atm}$$

The volume is

$$V = \frac{nRT}{P} = \frac{(0.992 \text{ mol})(0.0821 \text{ L·atm/mol·K})(298 \text{ K})}{0.968 \text{ atm}}$$

$$= 25.1 \text{ L}$$

12.81 The partial pressure of the oxygen is

$$P_{O_2} = P_{total} - P_{H_2O} = 771 \text{ torr} - 20 \text{ torr} = 751 \text{ torr}$$

$$751 \text{ torr} \left(\frac{1 \text{ atm}}{760 \text{ torr}} \right) = 0.988 \text{ atm}$$

$$n_{O_2} = \frac{P_{O_2}V}{RT} = \frac{(0.988 \text{ atm})(1.00 \text{ L})}{(0.0821 \text{ L·atm/mol·K})(295 \text{ K})}$$

$$= 0.0408 \text{ mol } O_2$$

$$2 \text{ KClO}_3(s) \xrightarrow{\text{Heat}} 2 \text{ KCl}(s) + 3 \text{ O}_2(g)$$

$$0.0408 \text{ mol } O_2 \left(\frac{2 \text{ mol KClO}_3}{3 \text{ mol } O_2} \right) \left(\frac{122 \text{ g KClO}_3}{1 \text{ mol KClO}_3} \right) = 3.32 \text{ g KClO}_3$$

12.82 The equation for the reaction is

$$2 \text{ KClO}_3(s) \xrightarrow{\text{Heat}} 2 \text{ KCl}(s) + 3 \text{ O}_2(g)$$

$$0.255 \text{ g KClO}_3 \left(\frac{1 \text{ mol KClO}_3}{122 \text{ g KClO}_3} \right) \left(\frac{3 \text{ mol } O_2}{2 \text{ mol KClO}_3} \right)$$

$$= 3.14 \times 10^{-3} \text{ mol } O_2$$

The oxygen pressure is the total pressure minus the water vapor pressure (from Table 12.3):

$$775 \text{ torr} - 24 \text{ torr} = 751 \text{ torr} \left(\frac{1 \text{ atm}}{760 \text{ torr}} \right) = 0.988 \text{ atm}$$

$$V = \frac{nRT}{P}$$

$$= \frac{(3.14 \times 10^{-3} \text{ mol } O_2)(0.0821 \text{ L·atm/mol·K})(298 \text{ K})}{0.988 \text{ atm}}$$

$$= 0.0778 \text{ L}$$

12.85 $18.8 \text{ mL} \left(\frac{1 \text{ L}}{1000 \text{ mL}} \right) = 0.0188 \text{ L}$

$$\left(\frac{0.0188 \text{ L}}{30.6 \text{ L}} \right) \times 100\% = 0.0614\%$$

The molecules themselves occupy less than 0.10% of the volume of the gas, in accord with the kinetic molecular theory.

12.86 The more moles of the gas there are, the more molecules bombard the walls and, therefore, the greater the force and pressure, all other factors being equal.

12.89 Parts (b) and (d) may be solved with Boyle's law, since the temperature is constant in each. Part (c) may be solved with Charles' law, since the pressure is constant.

12.90 The final pressure will be half the initial pressure $(0.5P_1)$, and the final absolute temperature will be 1.20 times the initial temperature $(1.20T_1)$.

$$V_2 = \frac{V_1 P_1 T_2}{P_2 T_1} = \frac{(1.00 \text{ L})(P_1)(1.20T_1)}{(0.5P_1)(T_1)} = \frac{(1.00 \text{ L})(1.20)}{0.5}$$

$$= 2.40 \text{ L}$$

12.93 This problem is identical to Problem 12.92.

12.94 $n = \dfrac{PV}{RT} = \dfrac{(2.20 \text{ atm})(1.38 \text{ L})}{(0.0821 \text{ L·atm/mol·K})(285 \text{ K})} = 0.130 \text{ mol}$

$$\frac{4.16 \text{ g}}{0.130 \text{ mol}} = 32.0 \text{ g/mol}$$

12.96 $752 \text{ torr} \left(\dfrac{1 \text{ atm}}{760 \text{ torr}} \right) = 0.989 \text{ atm}$

$$n = \frac{PV}{RT} = \frac{(0.989 \text{ atm})(0.869 \text{ L})}{(0.0821 \text{ L·atm/mol·K})(296 \text{ K})} = 0.0354 \text{ mol}$$

$$\frac{1.06 \text{ g}}{0.0354 \text{ mol}} = 29.9 \text{ g/mol}$$

The empirical formula is calculated as in Section 7.4:

$$79.89 \text{ g C} \left(\frac{1 \text{ mol C}}{12.01 \text{ g C}} \right) = 6.652 \text{ mol C}$$

$$20.11 \text{ g H} \left(\frac{1 \text{ mol H}}{1.008 \text{ g H}} \right) = 19.95 \text{ mol H}$$

$$\frac{19.95 \text{ mol H}}{6.652 \text{ mol C}} = \frac{3 \text{ mol H}}{1 \text{ mol C}}$$

The empirical formula is CH_3. The empirical formula mass is therefore 15.0 amu per empirical formula unit. The number of empirical formula units per molecule is given by

$$\frac{29.9 \text{ amu/molecule}}{15.0 \text{ amu/empirical formula unit}} = \frac{2 \text{ empirical formula units}}{1 \text{ molecule}}$$

The molecular formula is $(CH_3)_2$, or C_2H_6.

12.98 (a) The volume of each gas is 1.00 L, and the pressures of the two gases are given by the ideal gas law equation:

$$P_{He} = \frac{nRT}{V} = 4.89 \text{ atm} \qquad P_{Ne} = \frac{nRT}{V} = 2.45 \text{ atm}$$

There is no mixture, so there is no total pressure. The total volume is 2.00 L.

(b) The volume is 1.00 L, and the pressures of the two gases are the same as in part (a). The total pressure is the sum of those two partial pressures:

$$P_{total} = 2.45 \text{ atm} + 4.89 \text{ atm} = 7.34 \text{ atm}$$

(Note: If gases are mixed, their volumes are the same and their pressures are added. If they are not mixed, their volumes are added.)

12.99 (a) At constant pressure, the volume of a gas is proportional to its absolute temperature (Charles' law). After the volume increases with temperature, it can be brought back to the original volume by increasing the pressure (Boyle's law) by the same factor. Thus, at constant volume, the pressure is directly proportional to the absolute temperature. For example, if we heat a gas to double its original absolute temperature at constant pressure, the volume will double. Doubling the pressure will halve the volume to its original value. The gas is now at double the absolute temperature and double the original pressure at the same original volume. Thus, the pressure is directly proportional to the absolute temperature at constant volume.

(b) Divide the combined gas law equation, $P_1V_1/T_1 = P_2V_2/T_2$, by an equation that states that the volume is constant, $V_1 = V_2$. The result is

$$\frac{P_1}{T_1} = \frac{P_2}{T_2}$$

This equation states that the ratio P/T is a constant under the given conditions ($V_1 = V_2$)—that is, that the pressure is directly proportional to the absolute temperature at constant volume.

13 Atomic and Molecular Properties

13.1

		Number of Protons	Number of Electrons	Size
(a)	Li	3	3	Larger
	Li$^+$	3	2	Smaller
(b)	F	9	9	Smaller
	F$^-$	9	10	Larger
(c)	Ne	10	10	Larger
	Na$^+$	11	10	Smaller
(d)	Ne	10	10	Smaller
	F$^-$	9	10	Larger

13.2 (a) IA, alkali metals
(b) IA, alkali metals

13.3 (a) F (b) F (c) No difference

13.4 Polyatomic ions are larger.

13.5 (a) No (b) Ionization energies are all positive; that is, it takes energy to remove the electron.

13.6 The bond in H_2 is a covalent chemical bond. The hydrogen bond is an intermolecular force between molecules containing hydrogen atoms and nitrogen, oxygen, or fluorine atoms.

13.7 A polar bond is a covalent bond that exists between two atoms having an electronegativity difference greater than 0.2. A polar molecule results if one or more polar bonds in a molecule is not balanced by other polar bonds in the molecule.

13.8 Cea (Size diminishes toward the right in the periodic table.)

13.11 F$^-$ (It has the greatest nuclear charge.)

13.15 H$^-$

13.16 (a) IA, alkali metals (b) Group IIIA

13.18 Noble gases

13.19 (a) He (b) Fr

13.21 Groups IIA and 0. (Their electron affinities are negative.)

13.22 (a) Ionic (b) Polar covalent (c) Nonpolar covalent

13.24 (a) Nonpolar covalent (b) Nonpolar covalent

13.27 (a), (b), and (c) Tetrahedral

13.30 Nonpolar bonds have atoms of equal electronegativity that share electrons equally. Nonpolar molecules either have all bonds nonpolar or have any polar bonds oriented so that they cancel out each other's effects.

13.31 Yes, if the orientations of the polar bonds in the molecule cancel each other's effects.

13.34 (a) No, the molecules are nonpolar.
(b) Yes, the molecules are polar.

13.38 (a) Hydrogen bonding
(b) Dipole moments but no hydrogen bonding
(c) Van der Waals forces only
(d) Dipole moments but no hydrogen bonding
(e) Van der Waals forces only

13.41 (a) Na 5058 kJ/mol Al 5140 kJ/mol
(b) It takes less energy to produce Na^{2+}(g) from Na(g) than it takes to form Al^{3+}(g) from Al(g).
(c) Al^{3+} is the familiar ion in the solid state and in solution; Na^{2+} is not a stable species.

13.43 (a) Trigonal planar (b) Tetrahedral

13.46 N^{3-} is larger because it has eight electrons in the second shell, while Li$^+$ has none.

13.47 (a) It is angular. (b) The angle in NO_2 is less because the single electron does not repel the other electron groups as well as the electron pair in NO_2^-.

13.52 CFClBrI, CF_2Br_2, or many other possibilities

13.54 (a) Trigonal planar (b) Trigonal pyramidal

14 Solids and Liquids, Energies of Physical and Chemical Changes

14.1 Gases have very small attractive forces between the particles (Section 12.8). An ionic substance has strong ionic bonds between its ions. If an ionic substance were in the gaseous state at room temperature, it would condense into a solid very quickly.

14.2 The liquid phase

14.3

Forces	Example
Chemical bonds	
Ionic bonds	NaCl
Covalent bonds	Graphite
Metallic bonds	Iron
Intermolecular forces	
Van der Waals forces	Br_2
Dipolar attractions	ICl
Hydrogen bonding	Ice

14.4 When metal is placed in water, both the metal and the water will finally come to the same temperature. The final temperature of the metal is thus 22.4°C.

14.5

	Initial (°C)	Final (°C)	Change (°C)
(a)	10.0	24.1	14.1
(b)	10.0	34.1	24.1
(c)	11.6	35.7	24.1
(d)	11.6	35.7	24.1
(e)	24.1	59.8	35.7
(f)	24.1	10.0	−14.1

14.6 It takes −5440 J. The same quantity of energy is involved, but in the cooling process, the energy is *removed*, and its sign is minus.

14.7 (a) 30.0°C (b) 10.0°C

14.8 (a) One is the reverse of the other. (See Figure 14.4.)
(b) Vaporization is endothermic.
(c) They have the same magnitude but opposite signs:
$$\text{Heat}_{vap} = -\text{Heat}_{cond}$$

14.9 You would do a specific heat calculation for the warming of the ice, a heat of fusion calculation for the melting of the ice, and another specific heat calculation for the warming of the resulting liquid water. Finally, you would add the three values together.

14.10 (a) Heat = $mc\Delta t$ = (50.0 g)(2.09 J/g·°C)(10.0°C) = 1040 J
(b) Heat = mc_{fusion} = (50.0 g)(335 J/g) = 16 800 J
(c) Heat = $mc\Delta t$ = (50.0 g)(4.184 J/g·°C)(25.0°C) = 5230 J
(d) Heat = 1040 J + 16 800 J + 5230 J = 23 100 J = 23.1 kJ

14.11 (a) One is the reverse of the other. (See Figure 14.4.)
(b) Fusion is endothermic.
(c) They have the same magnitude but opposite signs:
$$\text{Heat}_{fusion} = -\text{Heat}_{solidification}$$

14.12 CO_2 is a molecular solid; SiO_2 is a network solid.

14.13 They have about the same van der Waals forces. Neither substance can exhibit hydrogen bonding, since neither contains hydrogen. ICl has a dipole and Br_2 does not, so ICl should have the greater intermolecular forces. Since it has greater intermolecular forces, ICl boils at a higher temperature (ICl, 97.4°C; Br_2, 58.78°C).

14.14 PCl_3 should have the lowest melting point because it has the smallest forces holding its components together. NaCl has ionic bonds holding its ions in a solid lattice, and SiO_2 has covalent bonds holding the silicon and oxygen atoms in a three-dimensional network.

14.15 P_2O_3 is a molecular solid; Al_2O_3 is an ionic solid.

14.16 SO_3 has the greater molar mass and the greater number of electrons. Therefore it has greater intermolecular forces, it boils at a higher temperature (SO_2, $-10°C$; SO_3, $44.8°C$).

14.17 He should have the lowest melting point, because it has the smallest forces holding its atoms together.

14.18 The larger molecules have more electrons and, therefore, greater van der Waals forces to hold them to one another. The lighter F_2 and Cl_2 molecules have insufficient intermolecular forces to keep their molecules from being in the gaseous state under ordinary conditions of temperature and pressure.

14.19 NaCl has strong bonds holding its ions together; CCl_4 has only van der Waals forces holding its molecules to one another.

14.20 The substance is a crystalline solid. The regularity of the positions of the particles is a characteristic of crystalline solids.

14.21 The compound with the doubly charged ions will melt at a higher temperature. The stronger attractions of the dipositive and dinegative charges require higher temperatures to disrupt. For example, NaF melts at $993°C$ and MgO melts at $2800°C$.

14.22 SCl_2 has only dipolar attractions between its molecules and is lowest melting (at $-78°C$). The molecules of H_2O are attracted to each other by hydrogen bonding, and it is next highest melting (at $0°C$). $MgCl_2$ has strong ionic bonds linking its ions and is the highest melting (at $714°C$).

14.23 No; the pressure will be equal to the vapor pressure if the liquid and vapor phases are both present in equilibrium, but if all the liquid is vaporized, the pressure may be below the vapor pressure of water at that temperature.

14.24 (a) The water cools.
(b) The water freezes (at $0°C$) until it is all solid.

14.25 Only (d), increasing the temperature, affects the vapor pressure of a given pure substance.

14.26 (a) The water boils.
(b) The water warms up.

14.27 There will be no change. Vapor pressure depends on temperature, not surface area.

14.28 (a) Heat $= mc\Delta t$
$= (20.0 \text{ g})(4.184 \text{ J/g·°C})(53.3°C - 22.2°C)$
$= 2600 \text{ J} = 2.60 \text{ kJ}$
(b) Heat $= mc\Delta t$
$= (20.0 \text{ g})(4.184 \text{ J/g·°C})(22.2°C - 53.3°C)$
$= -2600 \text{ J} = -2.60 \text{ kJ}$
In part (b), 2.60 kJ of heat must be *removed* from the water.

14.29 (a) Heat $= mc\Delta t$
$= (10.0 \text{ g})(2.089 \text{ J/g·°C})[(-5.0°C) - (-19.0°C)]$
$= 292 \text{ J}$
(b) Heat $= mc\Delta t$
$= (10.0 \text{ g})(2.089 \text{ J/g·°C})[(-19.0) - (-5.0)]$
$= -292 \text{ J}$
In part (b), 292 J of heat must be *removed* from the ice.

14.30 The specific heat of zinc, listed in Table 14.4, is 0.388 J/g·°C. The change in temperature is given by

$$\Delta t = \frac{\text{heat}}{mc} = \frac{273 \text{ J}}{(25.0 \text{ g})(0.388 \text{ J/g·°C})} = 28.1°C$$

The initial temperature was $22.7°C$, and the change in temperature is $28.1°C$, so the final temperature is $22.7°C + 28.1°C = 50.8°C$.

14.31 The specific heat of zinc, listed in Table 14.4, is 0.388 J/g·°C. The heat required is given by

Heat $= mc\Delta t = (49.8 \text{ g})(0.388 \text{ J/g·°C})(14.7°C) = 284 \text{ J}$

14.32 The specific heat of aluminum, from Table 14.4, is 0.90 J/g·°C. The change in temperature is given by

$$\Delta t = \frac{\text{heat}}{mc} = \frac{155 \text{ J}}{(40.0 \text{ g})(0.90 \text{ J/g·°C})} = 4.3°C$$

The initial temperature was $22.7°C$ and the change in temperature is $4.3°C$, so the final temperature is $22.7°C + 4.3°C = 27.0°C$.

14.33 The specific heat of cobalt, listed in Table 14.4, is 0.46 J/g·°C. The heat required is given by

Heat $= mc\Delta t = (150 \text{ g})(0.46 \text{ J/g·°C})(10.6°C) = 730 \text{ J}$

14.34 The specific heat of cobalt, listed in Table 14.4, is 0.46 J/g·°C. The heat required is given by

Heat $= mc\Delta t = (150 \text{ g})(0.46 \text{ J/g·°C})(29.3°C - 18.7°C) = 730 \text{ J}$
(This problem is the same as Problem 14.33.)

14.35 The total heat added to the metal and water is zero. Thus,

$$0 = (80.5 \text{ g})(c_{\text{metal}})(23.0°C - 75.0°C)$$
$$+ (150 \text{ g})(4.184 \text{ J/g·°C})(23.0°C - 20.0°C)$$

$c_{\text{metal}} = 0.45 \text{ J/g·°C}$

The metal is most likely chromium.

14.36 The heat required to heat the liquid water to $100°C$ is

Heat $= mc\Delta t = (10.0 \text{ g})(4.184 \text{ J/g·°C})(8.0°C) = 330 \text{ J}$

The heat required to vaporize the water is

Heat $= (10.0 \text{ g})(2260 \text{ J/g}) = 22\,600 \text{ J}$

The heat required to heat the water vapor to $115°C$ is

Heat $= (10.0 \text{ g})(2.042 \text{ J/g·°C})(15°C) = 306 \text{ J}$

The total heat is 23 200 J.

14.40 Heat $= 0 = (m_{\text{water}})(c_{\text{water}})(\Delta t_{\text{water}}) + (m_{\text{metal}})(c_{\text{metal}})(\Delta t_{\text{metal}})$
$0 = (102 \text{ g})(4.184 \text{ J/g·°C})(t_2 - 19.7°C)$
$+ (39.1 \text{ g})(0.0650 \text{ J/g·°C})(t_2 - 65.7°C)$
$0 = 427t_2 - 8410°C + 2.54t_2 - 167°C$
$430t_2 = 8580°C$
$t_2 = 20.0°C$

14.41 Heat $= mc\Delta t$
$1000 \text{ J} = (100.0 \text{ g})(4.184 \text{ J/g·°C})(\Delta t)$
$\Delta t = 2.39°C$

14.42 (a) -20.0 kJ (b) $+10.0 \text{ kJ}$

14.43 (a) $H_2O(s) \rightarrow H_2O(l)$
(b) $\frac{1}{2} H_2(g) + \frac{1}{2} Cl_2(g) \rightarrow HCl(g)$
(c) $CH_4(g) + 2 O_2(g) \rightarrow CO_2(g) + 2 H_2O(l)$
(d) $H_2O(l) \rightarrow H_2O(g)$
(e) $C(s) + \frac{1}{2} O_2(g) \rightarrow CO(g)$
(f) $CO(g) + \frac{1}{2} O_2(g) \rightarrow CO_2(g)$

14.44 $C_2H_4(g) + 3 O_2(g) \rightarrow 2 CO_2(g) + 2 H_2O(l)$
$\Delta H = 2\Delta H_f(CO_2) + 2\Delta H_f(H_2O) - \Delta H_f(C_2H_4) - 3\Delta H_f(O_2)$
$-1410 \text{ kJ} = 2(-393.5 \text{ kJ}) + 2(-285.9 \text{ kJ}) - \Delta H_f(C_2H_4)$
$\Delta H_f(C_2H_4) = 51 \text{ kJ}$

14.48 (a) $+10.0 \text{ kJ}$ (the reverse of the first equation)
(b) $+20.0 \text{ kJ}$ (double the answer in part [a])
(c) $+5.0 \text{ kJ}$ (the sum of the two given equations)

14.49 $CO(g) + \frac{1}{2} O_2(g) \rightarrow CO_2(g)$

For 1 mol of CO:

$\Delta H = (-393.5 \text{ kJ}) - (-110.5 \text{ kJ}) = -283.0 \text{ kJ}$

For 50.0 g of CO:

$$50.0 \text{ g CO}\left(\frac{1 \text{ mol CO}}{28.0 \text{ g CO}}\right)\left(\frac{-283.0 \text{ kJ}}{1 \text{ mol CO}}\right) = -505 \text{ kJ}$$

14.51 $2 \text{ NO}_2(g) \rightarrow \text{N}_2\text{O}_4(g)$

$\Delta H = -(67.70 \text{ kJ}) + 9.67 \text{ kJ} = -58.03 \text{ kJ}$

Since the N_2O_4 is much more stable (the ΔH is very negative), the reaction is apt to be spontaneous.

14.52 $\Delta H = (-1305 \text{ kJ}) - (-1541 \text{ kJ}) + 2(-286 \text{ kJ}) = -336 \text{ kJ}$

14.54 $\text{C}_8\text{H}_{18}(l) + \frac{25}{2} \text{O}_2(g) \rightarrow 8 \text{ CO}_2(g) + 9 \text{ H}_2\text{O}(l)$

For 1 mol of C_8H_{18}:

$$\begin{aligned}\Delta H &= \Delta H_f(\text{products}) - \Delta H_f(\text{reactants}) \\ &= 8\Delta H_f(\text{CO}_2) + 9\Delta H_f(\text{H}_2\text{O}) - \Delta H_f(\text{C}_8\text{H}_{18}) \\ &= 8(-393.5 \text{ kJ}) + 9(-285.9 \text{ kJ}) - (-208 \text{ kJ}) = -5513 \text{ kJ}\end{aligned}$$

For 0.100 mol of C_8H_{18}:

$0.100 \text{ mol } (-5513 \text{ kJ/mol}) = -551 \text{ kJ}$

14.58 Note that $\Delta H_{\text{combustion}}$, not ΔH_f values are given.

$\text{C}_2\text{H}_2(g) + \frac{5}{2} \text{O}_2(g) \rightarrow 2 \text{ CO}_2(g) + \text{H}_2\text{O}(l)$　　$\Delta H = -1305 \text{ kJ}$

$\text{C}_6\text{H}_6(l) + \frac{15}{2} \text{O}_2(g) \rightarrow 6 \text{ CO}_2(g) + 3 \text{ H}_2\text{O}(l)$　　$\Delta H = -3273 \text{ kJ}$

$\Delta H = -(-3273 \text{ kJ}) + 3(-1305 \text{ kJ}) = -642 \text{ kJ}$

14.60 (a) $(4.184 \text{ J/g·°C})(18.016 \text{ g/mol}) = 75.38 \text{ J/mol·°C}$
　　　(b) $(0.442 \text{ J/g·°C})(55.85 \text{ g/mol}) = 24.7 \text{ J/mol·°C}$

14.61 CaO has doubly charged ions, as opposed to singly charged ions in LiF. Since CaO has greater forces holding the particles together, it should have the higher melting point. The actual values are 2580°C for CaO and 845°C for LiF.

14.63 The sudden expansion of liquid CO_2 does work pushing back the atmosphere and overcoming intermolecular forces in the liquid. The energy to do that work comes from the molecules themselves, so the average energy of the molecules is lowered. The CO_2 condenses to a solid because of this loss of energy.

14.64

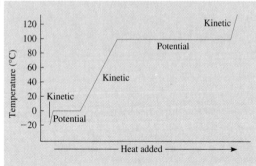

14.65 (a) Since 104.0 g − 100.0 g = 4.0 g, that amount of water is present at the end of the process and was not present initially. It has condensed from the steam.
　　　(b) The water has warmed from 19.0°C to 42.9°C; the heat required for this process is

$$\begin{aligned}\text{Heat} &= mc\Delta t = (100.0 \text{ g})(4.184 \text{ J/g·°C})(23.9°C) \\ &= 10\,000 \text{ J} = 10.0 \text{ kJ}\end{aligned}$$

　　　(c) The water condensed from steam has cooled from 100°C to 42.9°C. The heat liberated is

$$\begin{aligned}\text{Heat} &= mc\Delta t = (4.0 \text{ g})(4.184 \text{ J/g·°C})(-57.1°C) \\ &= -960 \text{ J}\end{aligned}$$

(d) The total heat supplied is −10 000 J. If you subtract the heat supplied by the cooling water, −960 J, you get the heat supplied by the condensation process:

$-10\,000 \text{ J} - (-960 \text{ J}) = -9000 \text{ J} = -9.0 \text{ kJ}$

The heat is negative, since the condensation process supplies heat to the cold water.

(e) The heat of vaporization per gram is the heat of part (d) divided by the number of grams, but changed in sign because the heat of vaporization is heat *added to the system:*

$$\frac{9.0 \text{ kJ}}{4.0 \text{ g}} = 2.2 \text{ kJ/g}$$

14.68 HF and HI each has greater intermolecular forces than does HCl. (HF has hydrogen bonding, and HI has greater van der Waals forces.)

14.70 (a) No
　　　(b) Since energy is produced, ΔH is negative for the reaction.
　　　(c) Since energy is absorbed, ΔH is positive for the warming process.
　　　(d) The overall enthalpy change is zero. Since the energy from the reaction is added to the solution, no energy is gained or lost from the system.

14.72 For hydrogen:

$(14.4 \text{ J/g·°C})(2.016 \text{ g/mol}) = 29.0 \text{ J/mol·°C}$

The molar heat capacities of the other elements are calculated in the same manner, giving the following results (in J/mol·°C):

H_2	29.0	Al	24	Fe	24.7
O_2	29.5	Ag	26	Mg	24
N_2	29.1	Au	25.4	Pb	27
		Co	27	Sn	26
		Cr	23	Zn	25.4
		Cu	24.5		

Despite a wide range of *specific* heat capacities, the diatomic gases have *molar* heat capacities of about 29 J/mol·°C, and the *molar* heat capacities of all the metallic elements are close to 26 J/mol·°C. The latter generalization is known as the law of Dulong and Petit.

14.73 Since the solid does not have regularly repeating units, it is an amorphous solid. Amorphous solids do not have distinct melting points but become soft over a range of temperatures; examples include glass and a chocolate bar.

14.74 It went into potential energy—the energy required to disorder the molecules or ions from their regular arrangement by lessening the forces holding them together.

14.76 A plot of these data shows that water should boil at about −75°C. Since it actually boils 175°C higher than that, the effects of hydrogen bonding are obviously very important.

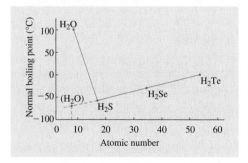

14.77 A plot of these data shows that ammonia should boil at about $-105°C$. Since it actually boils 70°C higher than that, the effects of hydrogen bonding are obviously very important.

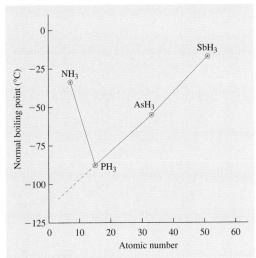

14.78 Since the concentration of water that can be held by the air increases with increasing temperature but the actual concentration of water in the air does not change, the relative humidity decreases.

14.79 The water will be heated to 100.0°C, at which point it will vaporize; then the vapor will be heated to 121.0°C. The heat required for each of these three steps is calculated separately, and the total heat required is the sum of the quantities required for the individual steps. For heating the water:

Heat = $mc\Delta t$ = (153 g)(4.184 J/g·°C)(15.0°C) = 9600 J

For vaporizing the water:

Heat = (153 g)(2260 J/g) = 346 000 J

For heating the vapor:

Heat = $mc\Delta t$ = (153 g)(2.042 J/g·°C)(21.0°C) = 6560 J

The total heat required is

Total heat = 9600 J + 346 000 J + 6560 J = 362 000 J = 362 kJ

14.80 This problem may be thought of as involving limiting quantities, in that the heat available can be compared with the heat required for a certain purpose. The heat required to raise the temperature to 0°C is given by

Heat = $mc\Delta t$ = (22.2 g)(2.089 J/g·°C)(15°C) = 700 J

More heat than that is available. After the ice has been raised to the melting point, there will still be available

3021 J − 700 J = 2320 J

The ice will start to melt. The heat required to melt all the ice is

(22.2 g)(335 J/g) = 7440 J

Since not that much heat is still available, not all the ice will melt. At the end, there will be a mixture of water and ice, so the temperature will be 0°C.

14.83

CO(g) + $\frac{1}{2}$ O$_2$(g) 25°C	→ 3 →	CO$_2$(g) 25°C	
↑ 1	↑ 2	↓ 4	
CO(g) + $\frac{1}{2}$ O$_2$(g) 125°C	→ ΔH →	CO$_2$(g) 125°C	

$\Delta H_1 = mc\Delta t = (28.0\ g)(1.04\ J/g\cdot°C)(-100°C) = -2910\ J$
$\Delta H_2 = mc\Delta t = \frac{1}{2}(32.0\ g)(0.922\ J/g\cdot°C)(-100°C) = -1480\ J$
$\Delta H_3 = \Delta H_{combustion} = -283.0\ kJ = -283\ 000\ J$
$\Delta H_4 = mc\Delta t = (44.0\ g)(0.852\ J/g\cdot°C)(+100°C) = 3750\ J$
$\Delta H = \Delta H_1 + \Delta H_2 + \Delta H_3 + \Delta H_4$
$= (-2.91\ kJ) + (-1.48\ kJ) + (-283.0\ kJ) + (3.75\ kJ)$
$= -283.6\ kJ$

14.84 **(a)** $200\ mL\left(\dfrac{1.01\ g}{1\ mL}\right) = 202\ g$

Heat = $mc\Delta t$ = 5520 J = (202 g)(4.10 J/g·°C)Δt

$\Delta t = 6.67°C$

$t_f = 25.0°C + 6.67°C = 31.7°C$

(b) NaOH(aq) + HCl(aq) → NaCl(aq) + H$_2$O(l)

0.100 mol H$_2$O(−55.2 kJ/mol H$_2$O) = −5.52 kJ

(c) The final volume is 200 mL; the final concentration of NaCl is (0.100 mol)/(0.200 L) = 0.500 M. The heat of reaction in part (b) is −5.52 kJ, which is provided to the solution. Thus, heat added to the solution is +5.52 kJ. Since all the factors are the same as in part (a), the final temperature is the same, 31.7°C.

14.86 $1.000\ gal\left(\dfrac{3.785\ L}{1\ gal}\right)\left(\dfrac{10^3\ mL}{1\ L}\right)\left(\dfrac{0.7025\ g}{1\ mL}\right)\left(\dfrac{1\ mol\ C_8H_{18}}{114.2\ g\ C_8H_{18}}\right)$
$= 23.28\ mol\ C_8H_{18}$

For 1 mol C$_8$H$_{18}$:

C$_8$H$_{18}$(l) + $\frac{17}{2}$ O$_2$(g) → 8 CO(g) + 9 H$_2$O(l)	ΔH = ?
C$_8$H$_{18}$(l) + $\frac{25}{2}$ O$_2$(g) → 8 CO$_2$(g) + 9 H$_2$O(l)	ΔH = −5450 kJ
8 CO(g) + 4 O$_2$(g) → 8 CO$_2$(g)	ΔH = 8(−283 kJ)

Subtracting the last equation from the one before yields the desired equation, with $\Delta H = -3190$ kJ per mol of C$_8$H$_{18}$. For 23.28 mol:

23.28 mol(−3190 kJ/mol) = −74 300 kJ = −7.43 × 10^4 kJ

14.87 C(s) + $\frac{1}{2}$ O$_2$(g) → CO(g)   −110.5 kJ
C(s) + O$_2$(g) → CO$_2$(g)   −393.5 kJ

Let x = number of moles of CO formed.

Then $1.00 - x$ = number of moles of CO$_2$ formed.

Heat = x(−110.5 kJ) + (1.00 − x)(−393.5 kJ) = −313.8 kJ
−110.5x − 393.5 + 393.5x = −313.8
283.0x = 79.7
x = 0.282 mol CO formed
1.00 − x = 0.718 mol CO$_2$ formed

From the balanced equations, 0.141 mol O$_2$ was used to form the CO and 0.718 mol O$_2$ was used to form the CO$_2$, for a total of 0.859 mol O$_2$. The mass of O$_2$ is

0.859 mol (32.0 g/1 mol) = 27.5 g O$_2$.

15 Solutions

15.1 **(a)** The first solution is unsaturated, since it could hold at least 3.00 g more of the solute at the temperature specified.

(b) There is not enough information to tell if the final solution is unsaturated or saturated. It might be holding as much as is stable at 20°C. It cannot be supersaturated because more solute would not simply dissolve at 20°C than is stable at that temperature.

15.2 The names differ by only one letter. The symbol for molality is lowercase *m;* for molarity, it is capital *M*. Kilograms rather than liters are involved in molality. A quantity of solvent rather than a quantity of solution appears in the denominator of the expression for molality.

15.3 They are similar in that they both are ratios of each component to the total and in that the total is specified in each case (100% for percent composition and 1 mole for mole fraction). They differ in the units used—moles for mole fraction and mass for percent composition.

15.4 By definition, the sum of the mole fractions must be 1.000; therefore,

$$X_{CH_2O} = 1.000 - 0.850 = 0.150$$

15.5 The identity of the molecular solute does not affect the freezing-point depression (as long as the solute does not ionize); it is the number of molecules per unit mass of solvent that affects the extent of the freezing-point depression. Therefore, for all three parts of the problem,

$$\Delta t = k_f m = (1.86°C/m)(0.100\ m) = 0.186°C$$

15.6 By definition, the sum of the mole fractions must be 1.000; therefore,

$$X_{CH_3OH} = 1.000 - 0.650 - 0.100 = 0.250$$

15.7 (a) The greater the molality, the greater is the freezing-point depression, so the 0.30 m solution has the greater depression.
  (b) The lower the freezing-point depression, the higher is the freezing point, so the 0.15 m solution has the higher freezing point.

15.8 (a) The greater the molality, the greater is the boiling-point elevation, so the 0.20 m solution has the greater elevation.
  (b) The greater the boiling-point elevation, the higher is the boiling point, so the 0.20 m solution has the higher boiling point.

15.9 (a) Mole fraction (b) Molality (c) Molality (d) Molarity

15.10 (a) KBr is more soluble at 20°C.
  (b) KNO_3 is more soluble at 80°C.

15.11 Both CH_3OH and H_2O consist of polar molecules, and they both form hydrogen bonds. They are so similar that they readily dissolve in one another. C_6H_6 is not polar and does not form hydrogen bonds; it is so different from water that it is not expected to dissolve appreciably in water, and it does not.

15.13 Both $C_{10}H_8$ and C_6H_6 are nonpolar hydrocarbons. They are so similar that they readily dissolve in one another. C_6H_6 is nonpolar, and NaCl is ionic. They are so different that NaCl is not expected to dissolve in C_6H_6, and it does not.

15.14 (a) The negative end (b) The positive end
  (c) The positive end, since the oxygen atom has a small negative charge. Hydrogen bonding is expected between the hydrogen atoms of water and the oxygen atoms of formaldehyde.

15.15 The —OH groups are very similar to water in their polarity and in their ability to form hydrogen bonds: "Like dissolves like."

15.16 About 115 g of KNO_3 will dissolve in 100 g of water at 60°C.

15.18 (a) Unsaturated (b) Saturated (c) Unsaturated

15.20 Of these compounds, KNO_3 changes solubility most with increasing temperature.

15.22 (a) The excess solute would crystallize out.
  (b) The added crystal would dissolve.
  (c) The added crystal would not dissolve, nor would any more solid form.

15.23 Add excess salt to about 150 mL of water, and allow the mixture to stand until no more dissolves. Carefully pour off 100 mL of the solution (or filter off the remaining solid).

15.24 The number of moles of $C_{12}H_{22}O_{11}$ and the number of kilograms of H_2O are calculated first:

$$7.12\text{ g }C_{12}H_{22}O_{11}\left(\frac{1\text{ mol }C_{12}H_{22}O_{11}}{344\text{ g }C_{12}H_{22}O_{11}}\right) = 0.0207\text{ mol }C_{12}H_{22}O_{11}$$

$$10.0\text{ g}\left(\frac{1\text{ kg}}{1000\text{ g}}\right) = 0.0100\text{ kg }H_2O$$

The molality is

$$m = \frac{0.0207\text{ mol}}{0.0100\text{ kg}} = 2.07\text{ m }C_{12}H_{22}O_{11}$$

15.25 The molality is the number of moles of solute per kilogram of solvent:

$$m = \frac{0.2083\text{ mol}}{0.1000\text{ kg}} = 2.083\text{ m CsCl}$$

15.26 $50.0\text{ g }H_2O\left(\dfrac{1\text{ kg }H_2O}{1000\text{ g }H_2O}\right)\left(\dfrac{2.00\text{ mol solute}}{1\text{ kg }H_2O}\right)$
$$= 0.100\text{ mol solute}$$

15.27 $7.00\text{ mol solute}\left(\dfrac{1\text{ kg solvent}}{4.00\text{ mol solute}}\right) = 1.75\text{ kg solvent}$

15.33 (a) $X_{H_2O} = \dfrac{10.0\text{ mol }H_2O}{30.0\text{ mol total}} = 0.333$

  (b) The numbers of moles are calculated first:

$$10.0\text{ g }C_2H_5OH\left(\frac{1\text{ mol }C_2H_5OH}{46.0\text{ g }C_2H_5OH}\right) = 0.217\text{ mol }C_2H_5OH$$

$$10.0\text{ g }CH_2O\left(\frac{1\text{ mol }CH_2O}{30.0\text{ g }CH_2O}\right) = 0.333\text{ mol }CH_2O$$

$$10.0\text{ g }H_2O\left(\frac{1\text{ mol }H_2O}{18.0\text{ g }H_2O}\right) = 0.556\text{ mol }H_2O$$

The total number of moles in the solution is 1.106 mol. The mole fraction of water is

$$X_{H_2O} = \frac{0.556\text{ mol }H_2O}{1.106\text{ mol total}} = 0.503$$

15.34 Any size sample of a given solution will have the same concentration, whether it is expressed as a molality or a mole fraction. It is convenient to work with a sample that has 1.00 kg of water, in which there are 3.00 mol of ammonia and

$$1000\text{ g }H_2O\left(\frac{1\text{ mol }H_2O}{18.0\text{ g }H_2O}\right) = 55.6\text{ mol }H_2O$$

The mole fraction of ammonia is therefore

$$X_{NH_3} = \frac{3.00\text{ mol}}{3.00\text{ mol} + 55.6\text{ mol}} = 0.0512$$

15.36 In a solution containing 1.00 mol total, there are 0.150 mol alcohol and 0.850 mol water. The mass of water is

$$0.850\text{ mol}\left(\frac{18.0\text{ g}}{1\text{ mol}}\right)\left(\frac{1\text{ kg}}{1000\text{ g}}\right) = 0.0153\text{ kg}$$

The molality is (0.150 mol)/(0.0153 kg) = 9.80 m

15.38 $P_{C_6H_6} = X_{C_6H_6}P°_{C_6H_6} = \left(\dfrac{2.50\text{ mol }C_6H_6}{4.25\text{ mol total}}\right)96.0\text{ torr} = 56.5\text{ torr}$

15.40 (a) The temperature is below the freezing point of water, so the sample is solid.
  (b) The temperature is above the freezing point, so the sample is liquid.
  (c) Lowering the freezing point (without changing the temperature of the sample) can cause melting, which is the rationale for salting icy roads and sidewalks.

15.41 The solution would freeze below −11.5°C, since the presence of a solute—the water—depresses the freezing point, even of antifreeze. It should be noted that some mixtures of antifreeze and water freeze far below −11.5°C (equal to 11.3°F).

15.42 No matter what the molecular solute is, the freezing-point depression is determined using the constant for the solvent (Table 15.2). Thus, for (a)–(c),

$$\Delta t = k_f m = (1.86°C/m)(0.150\ m) = 0.279°C$$

15.43 (a) $\Delta t = k_f m = (1.86°C/m)(0.750\ m) = 1.40°C$

$$t_f = 0.000°C - 1.40°C = -1.40°C$$

(b) $\Delta t = k_b m = (0.512°C/m)(0.750\ m) = 0.384°C$

$$t_b = 100.000°C + 0.384°C = 100.384°C$$

15.45 Use the k_f value and the freezing point for the *solvent* (Table 15.2) in each case.

(a) $\Delta t = k_f m = (6.85°C/m)(0.100\ m) = 0.685°C$

$$t_f = 80.22°C - 0.685°C = 79.54°C$$

(b) $\Delta t = k_f m = (5.12°C/m)(0.100\ m) = 0.512°C$

$$t_f = 5.5°C - 0.512°C = 5.0°C$$

15.47 $m = \dfrac{\Delta t}{k_f} = \dfrac{0.457°C}{1.86°C/m} = 0.246\ m$

15.49 The freezing-point depression is $5.5°C - 3.00°C = 2.5°C$.

$$m = \frac{\Delta t}{k_f} = \frac{2.5°C}{5.12°C/m} = 0.49\ m$$

15.51 $m = \dfrac{\Delta t}{k_f} = \dfrac{1.22°C}{1.86°C/m} = 0.656\ m = \dfrac{0.656\ mol}{1\ kg\ water}$

$$\frac{13.00\ g\ solute}{0.2500\ kg\ water} = \frac{52.00\ g\ solute}{1\ kg\ water}$$

Since 0.656 mol of solute is 52.00 g, the molar mass is

$$\frac{52.00\ g}{0.656\ mol} = 79.3\ g/mol$$

15.53 $\pi V = nRT$

$$\pi = \frac{nRT}{V} = \left(\frac{n}{V}\right)RT$$

$$= \left(\frac{0.250\ mol}{1\ L}\right)(0.0821\ L\cdot atm/mol\cdot K)(298\ K)$$

$$= 6.12\ atm$$

15.54 The freezing-point depression is 1.20°C, and therefore, the molality of the solution is

$$m = \frac{\Delta t}{k_f} = \frac{1.20°C}{1.86°C/m} = 0.645\ m$$

The number of grams of solute in 1 kg of solvent is

$$1\ kg\ solvent\left(\frac{5.00\ g\ solute}{0.0200\ kg\ solvent}\right) = 250\ g\ solute$$

The molar mass is the number of grams in a kilogram of solvent divided by the number of moles in a kilogram of solvent:

$$Molar\ mass = \frac{250\ g}{0.645\ mol} = 388\ g/mol$$

15.57 $25.0\ g\ C_6H_6\left(\dfrac{1\ mol\ C_6H_6}{78.0\ g\ C_6H_6}\right) = 0.321\ mol\ C_6H_6$

$$25.0\ g\ C_2H_5OH\left(\frac{1\ mol\ C_2H_5OH}{46.0\ g\ C_2H_5OH}\right) = 0.543\ mol\ C_2H_5OH$$

The mole fractions are

$$X_{C_6H_6} = \frac{0.321\ mol\ C_6H_6}{0.864\ mol\ total} = 0.372\ \text{and therefore,}\ X_{C_2H_5OH} = 0.628$$

$$P_{C_6H_6} = P°_{C_6H_6}X_{C_6H_6} = (100\ torr)(0.372) = 37.2\ torr$$

$$P_{C_2H_5OH} = P°_{C_2H_5OH}X_{C_2H_5OH} = (44.0\ torr)(0.628) = 27.6\ torr$$

The total pressure is 37.2 torr + 27.6 torr = 64.8 torr

15.58 $P = P°X$

$$X = P/P° = (26.5\ torr)/(96.0\ torr) = 0.276$$

15.59 $P_{benzene} = \left(\dfrac{1.75\ mol}{3.00\ mol}\right)105\ torr = 61.2\ torr$

$$P_{toluene} = \left(\frac{1.25\ mol}{3.00\ mol}\right)34.0\ torr = 14.2\ torr$$

The total pressure is 61.2 torr + 14.2 torr = 75.4 torr

15.61 Assume any mass for each substance, such as 10.0 g. Calculate the number of moles of each substance:

$$10.0\ g\ CH_3OH\left(\frac{1\ mol\ CH_3OH}{32.0\ g\ CH_3OH}\right) = 0.312\ mol\ CH_3OH$$

$$10.0\ g\ CH_2O\left(\frac{1\ mol\ CH_2O}{30.0\ g\ CH_2O}\right) = 0.333\ mol\ CH_2O$$

The total number of moles is 0.645 mol, and the mole fraction of CH_3OH is

$$\frac{0.312\ mol}{0.645\ mol} = 0.484$$

15.63 From the boiling-point elevation, the molality of the solution can be calculated:

$$m = \frac{\Delta t}{k_b} = \frac{2.1°C}{0.512°C/m} = 4.1\ m$$

From that molality, the freezing-point depression can be calculated:

$$\Delta t = k_f m = (1.86°C/m)(4.1\ m) = 7.6°C$$

The freezing point is $-7.6°C$.

15.64 The number of moles of gas in a liter is given by the ideal gas law:

$$n = \frac{PV}{RT} = \frac{(1.00\ atm)(1.00\ L)}{(0.0821\ L\cdot atm/mol\cdot K)(298\ K)} = 0.0409\ mol$$

The number of moles of ammonia in a liter of 12 M solution is 12 mol. The ratio of numbers of moles is

$$\frac{12\ mol}{0.0409\ mol} = \frac{290}{1}$$

15.65 The boiling-point elevation is 1.22°C; therefore, the molality is

$$m = \frac{1.22°C}{0.512°C/m} = 2.38\ m$$

The solution contains 14.19 g of solute in 100.0 g of solvent, or 141.9 g per kilogram of solvent. Thus, 2.38 mol of solute has a mass of 141.9 g, and

$$Molar\ mass = \frac{141.9\ g}{2.38\ mol} = 59.6\ g/mol$$

15.67 The number of moles of Na_2SO_4 is

$$0.650\ kg\ H_2O\left(\frac{0.200\ mol\ Na_2SO_4}{1\ kg\ H_2O}\right) = 0.130\ mol\ Na_2SO_4$$

The number of moles of Na^+ ions is

$$0.130\ mol\ Na_2SO_4\left(\frac{2\ mol\ Na^+}{1\ mol\ Na_2SO_4}\right) = 0.260\ mol\ Na^+$$

15.68 The empirical formula is determined as in Section 7.4:

$$40.0\ g\ C\left(\frac{1\ mol\ C}{12.0\ g\ C}\right) = 3.33\ mol\ C$$

$$6.7\ g\ H\left(\frac{1\ mol\ H}{1.0\ g\ H}\right) = 6.7\ mol\ H$$

$53.3 \text{ g O}\left(\dfrac{1 \text{ mol O}}{16.0 \text{ g O}}\right) = 3.33 \text{ mol O}$

The ratio is 1 mol of carbon to 2 mol of hydrogen to 1 mol of oxygen, and the empirical formula is CH_2O. The freezing-point depression is 0.41°C, and therefore, the molality of the solution is

$m = \dfrac{\Delta t}{k_f} = \dfrac{0.41°C}{1.86°C/m} = 0.22 \text{ m}$

The number of grams of solute in 1.00 kg of solvent is

$1.00 \text{ kg solvent}\left(\dfrac{2.90 \text{ g solute}}{0.106 \text{ kg solvent}}\right) = 27.4 \text{ g solute}$

The molar mass is the number of grams in a kilogram of solvent, divided by the number of moles in a kilogram of solvent:

$\text{Molar mass} = \dfrac{27.4 \text{ g solute}}{0.22 \text{ mol solute}} = 120 \text{ g/mol}$

The empirical formula mass for CH_2O is 12.0 amu + 2.0 amu + 16.0 amu = 30.0 amu. The number of moles of empirical formula units per mole of molecules is

$\dfrac{120 \text{ g/mol}}{30.0 \text{ g/mol empirical formula units}}$

$= \dfrac{4 \text{ empirical formula units}}{1 \text{ molecule}}$

The molecular formula is $C_4H_8O_4$.

15.70 The number of moles per *liter of solution* is known, and the number of moles per *kilogram of solvent* is to be found. Assume that there is 1.00 L of solution. That volume of solution contains 0.324 mol of $CaCl_2$ and has the following mass:

$1.000 \text{ L solution}\left(\dfrac{1000 \text{ mL}}{1 \text{ L}}\right)\left(\dfrac{1.029 \text{ g}}{1 \text{ mL}}\right) = 1029 \text{ g solution}$

The mass of the $CaCl_2$ is

$0.324 \text{ mol CaCl}_2\left(\dfrac{111.0 \text{ g}}{1 \text{ mol}}\right) = 36.0 \text{ g CaCl}_2$

The mass of the solvent is

$1029 \text{ g solution} - 36.0 \text{ g CaCl}_2 = 993 \text{ g solvent}$

Therefore, the molality is $m = \dfrac{0.324 \text{ mol CaCl}_2}{0.993 \text{ kg solvent}} = 0.326 \text{ m}$

15.71 $P = P°X = 177.5 \text{ torr} = (268 \text{ torr})X$

$X = \dfrac{177.5 \text{ torr}}{268 \text{ torr}} = 0.662$

Let y = number of moles of solute per mole of benzene. Then

$X = 0.662 = \dfrac{1.00 \text{ mol benzene}}{1.00 \text{ mol} + y}$

$0.662(1.00 + y) = 1.00 \text{ mol}$
$0.662y = 0.338 \text{ mol}$
$y = 0.511 \text{ mol}$

Check: $\dfrac{1.00 \text{ mol}}{1.511 \text{ mol}} = 0.662$

15.74 The mole fraction of water must be 0.877, since the total of both mole fractions must equal 1.000. In a solution containing 1.000 mol total, there are 0.123 mol of methyl alcohol and 0.877 mol of water. The mass of the water is

$0.877 \text{ mol H}_2O\left(\dfrac{18.0 \text{ g H}_2O}{1 \text{ mol H}_2O}\right) = 15.8 \text{ g H}_2O$

Thus,

$\text{Molality} = \dfrac{0.123 \text{ mol alcohol}}{0.0158 \text{ kg water}} = 7.78 \text{ m}$

15.76 For 1.00 kg of water:

$1.35 \times 10^2 \text{ mol CH}_3OH\left(\dfrac{32.0 \text{ g CH}_3OH}{1 \text{ mol CH}_3OH}\right)\left(\dfrac{1 \text{ kg}}{1000 \text{ g}}\right)$
$= 4.32 \text{ kg CH}_3OH$

$1000 \text{ g H}_2O\left(\dfrac{1 \text{ mol H}_2O}{18.0 \text{ g H}_2O}\right) = 55.6 \text{ mol H}_2O$

$\text{Molality} = \dfrac{55.6 \text{ mol H}_2O}{4.32 \text{ kg CH}_3OH} = 12.9 \text{ m H}_2O$

15.78 The 75.0 g of H_2O is 0.0750 kg of H_2O and also

$75.0 \text{ g H}_2O\left(\dfrac{1 \text{ mol H}_2O}{18.0 \text{ g H}_2O}\right) = 4.17 \text{ mol H}_2O$

The 22.7 g of C_2H_5OH is 0.0227 kg of C_2H_5OH and also

$22.7 \text{ g C}_2H_5OH\left(\dfrac{1 \text{ mol C}_2H_5OH}{46.0 \text{ g C}_2H_5OH}\right) = 0.493 \text{ mol C}_2H_5OH$

(a) $\dfrac{0.493 \text{ mol C}_2H_5OH}{0.0750 \text{ kg H}_2O} = 6.57 \text{ m C}_2H_5OH$

(b) $\dfrac{4.17 \text{ mol H}_2O}{0.0227 \text{ kg C}_2H_5OH} = 184 \text{ m H}_2O$

(c) $X_{H_2O} = \dfrac{4.17 \text{ mol H}_2O}{4.66 \text{ mol total}} = 0.895$

$X_{C_2H_5OH} = \dfrac{0.493 \text{ mol C}_2H_5OH}{4.66 \text{ mol total}} = 0.106$

(d) Which component is regarded as the solvent makes no difference in determining the mole fractions.

15.80 KNO_3 changes solubility the most with increasing temperature, so it would be purified most effectively by this technique.

15.81 $\Delta P_{\text{solvent}} = P°_{\text{solvent}} - P_{\text{solvent}}$
$= P°_{\text{solvent}} - X_{\text{solvent}}P°_{\text{solvent}}$
$= P°_{\text{solvent}}(1 - X_{\text{solvent}})$
$= P°_{\text{solvent}}X_{\text{solute}}$

16 Oxidation Numbers

16.1 (a) M_2A (b) Y_2Z

16.2 (a) The oxidation number of each sodium atom is +1; that of the sulfur atom is −2.
(b) The oxidation number of phosphorus is +5; that of each chlorine atom is −1.

16.3 (a) The oxidation number of copper is +1; that of the oxygen atom is −2.
(b) The oxidation number of carbon is +4; that of each oxygen atom is −2.

16.4 The sum of the oxidation numbers of all the atoms is equal to the charge on the species (rule 1).
(a) −2   (b) 0   (c) −1   (d) 0

16.5 The sum of the oxidation numbers of all the atoms is equal to the charge on the species (rule 1).
(a) −1   (b) +1   (c) 0   (d) −1

16.6 (a) +2   (b) −1   (c) −1

16.7 The elemental sodium (on the left) has an oxidation number of 0 (rule 2), and the sodium in Na_2S has an oxidation number of +1 (rule 5). Note that sodium always has +1 as its oxidation number *in its compounds*.

16.8 The element that changed oxidation number has already been balanced, so hydrogen needs to have an oxidation number of +1.

16.9 Six electrons must be added to the right side.

16.10 **(a)** Double substitution is one reaction type described in Chapter 8 in which oxidation numbers never change. Combinations and decompositions of *compounds* to form new compounds (for example, calcium oxide plus carbon dioxide yielding calcium carbonate, or vice versa) also have no changes of oxidation numbers.
(b) No. We balanced simple oxidation-reduction equations in Chapter 8.

16.11 :$\ddot{\text{C}}\text{l}$:)C($\ddot{\ddot{\text{C}}\text{l}}$:
:S:

The oxidation number of carbon is $4 - 0 = +4$; that of the sulfur atom is $6 - 8 = -2$; and that of each of the chlorine atoms is $7 - 8 = -1$. Thus, the sulfur in $CSCl_2$ is like the oxygen in $COCl_2$, and the carbon has the same oxidation number in both compounds.

16.13 **(a)** +5 **(b)** +3 **(c)** +1 **(d)** −1

16.15 **(a)** −4 **(b)** −2 **(c)** 0 **(d)** +2 **(e)** +4

16.17 **(a)** +6 **(b)** +6

16.19 **(a)** +4 **(b)** +4 **(c)** +4

16.21 **(a)** −2 **(b)** −3 **(c)** +1 **(d)** 0
(e) $-\frac{1}{3}$ **(f)** −1 **(g)** +3 **(h)** +2
(i) +4 **(j)** +5 **(k)** +4

16.22 **(a)** +1 **(b)** +3 **(c)** +5 **(d)** +7

16.25 **(a)** +2 **(b)** +3 **(c)** +2.67

16.27 **(a)** +1 **(b)** −1 **(c)** −1 **(d)** −1

16.29 **(a)** Cu_2O **(b)** CuO **(c)** Hg_2Cl_2

16.31 **(a)** Vanadium pentachloride **(b)** Carbon tetrafluoride
(c) Phosphorus pentoxide, or diphosphorus pentoxide

16.33 **(a)** Lead(II) sulfide
(b) Chromium(III) fluoride
(c) Iron(III) nitrate

16.35 **(a)** Nitrogen(III) oxide
(b) Chlorine(V) fluoride
(c) Nitrogen(III) chloride

16.36 +1 (in most of its compounds); −1 (in its compounds with active metals)

16.38 The highest oxidation numbers are associated with covalent bonding. The four oxygen atoms in OsO_4 are covalently bonded to the osmium atom, and no ions are formed. For any monatomic ion, 4+ is the maximum charge.

16.39 **(a)** 0 **(b)** −3 **(c)** 0

16.40 Since phosphorus is on the right side of the main group section of the periodic table, its oxidation numbers vary in steps of 2. Its maximum oxidation number is +5, so its other positive oxidation numbers are most likely +3 and +1. Phosphorus exhibits these oxidation numbers in phosphorous acid and hypophosphorous acid, respectively (as well as in the salts of these acids and in other compounds). The formulas are H_3PO_4, H_3PO_3, and H_3PO_2.

16.41 The group IB elements have oxidation numbers that exceed the group number. Most group VIII elements have maximum oxidation numbers that are not as large as the group number.

16.45 Since fluorine in its compounds exhibits only a negative oxidation number, the sulfur must have positive oxidation numbers in all three compounds. It can exhibit a +6 oxidation number (its maximum), as well as +4 and +2 (steps of 2). The formulas are therefore SF_6, SF_4, and SF_2.

16.46 In the anions with the prefix *per,* the halogens have the oxidation number +7, their highest oxidation number.

16.49 The suffix *-ic* signifies the higher oxidation number.

16.52

	Oxidizing Agent	Reducing Agent	Element Oxidized	Element Reduced
(a)	CuO	CuO	O^{2-}	Cu^{2+}
(b)	$PbCl_2$	Mg	Mg	Pb^{2+}
(c)	$KMnO_4$	KCl	Cl^-	Mn^{VII}
(d)	$FeCl_3$	Fe	Fe	Fe^{3+}

16.54

	Oxidizing Agent	Reducing Agent	Element Oxidized	Element Reduced
(a)	HNO_3	Cu	Cu	N^V
(b)	NO_3^-	Cu	Cu	N^V

Usually, the net ionic equation (b) is easier to work with. For example, in the total equation for the reaction, two nitrogen atoms of the four HNO_3 molecules are actually reduced, and two remain as nitrate ions with the Cu^{2+} ion. Only the nitrogen atoms that change oxidation number are included in the net ionic equation.

16.55 $C_2O_4^{2-}(aq) \rightarrow 2\ CO_2(g)$
$C_2O_4^{2-}(aq) \rightarrow 2\ CO_2(g) + 2\ e^-$

$Pb^{4+}(aq) \rightarrow Pb^{2+}(aq)$
$2e^- + Pb^{4+}(aq) \rightarrow Pb^{2+}(aq)$

Combining these equations yields:

$C_2O_4^{2-}(aq) + Pb^{4+}(aq) \rightarrow Pb^{2+}(aq) + 2\ CO_2(g)$

16.56 **(a)** $3\ Cu(s) + 8\ H^+(aq) + 2\ NO_3^-(aq) \rightarrow$
$3\ Cu^{2+}(aq) + 2\ NO(g) + 4\ H_2O(l)$
(b) $H_2O_2(aq) + 2\ Co^{3+}(aq) \rightarrow 2\ Co^{2+}(aq) + O_2(g) + 2\ H^+(aq)$
(c) $5\ BiO_3^-(aq) + 2\ Mn^{2+}(aq) + 14\ H^+(aq) \rightarrow$
$2\ MnO_4^-(aq) + 5\ Bi^{3+}(aq) + 7\ H_2O(l)$
(d) $3\ H_2SO_4 \text{ (concentrated)} + 8\ Al(s) + 24\ H^+(aq) \rightarrow$
$3\ H_2S(g) + 8\ Al^{3+}(aq) + 12\ H_2O(l)$
(e) $2\ MnO_4^-(aq) + 5\ Pb^{2+}(aq) + 16\ H^+(aq) \rightarrow$
$2\ Mn^{2+}(aq) + 5\ Pb^{4+}(aq) + 8\ H_2O(l)$

16.57 The atoms and charges in this equation can be balanced by inspection:

$2\ Br^-(aq) + 2\ Ce^{4+}(aq) \rightarrow Br_2(aq) + 2\ Ce^{3+}(aq)$

Alternatively, the equation can be balanced by the systematic method:

$e^- + Ce^{4+}(aq) \rightarrow Ce^{3+}(aq)$
$2\ Br^-(aq) \rightarrow Br_2(aq) + 2\ e^-$

Doubling the first equation and then combining these equations yields the overall equation:

$2\ Br^-(aq) + 2\ Ce^{4+}(aq) \rightarrow Br_2(aq) + 2\ Ce^{3+}(aq)$

16.58 *Step 1:* $2\ I^-(aq) \rightarrow I_2(aq)$
Step 5: $2\ I^-(aq) \rightarrow I_2(aq) + 2\ e^-$
Step 1: $Cu^{2+}(aq) \rightarrow CuI(s)$
Step 2: $I^-(aq) + Cu^{2+}(aq) \rightarrow CuI(s)$
Step 5: $e^- + I^-(aq) + Cu^{2+}(aq) \rightarrow CuI(s)$

Doubling the reduction half-reaction:

$2\ e^- + 2\ I^-(aq) + 2\ Cu^{2+}(aq) \rightarrow 2\ CuI(s)$

Combining:

$4\ I^-(aq) + 2\ Cu^{2+}(aq) \rightarrow 2\ CuI(s) + I_2(aq)$

16.59 **(a)** $4\ Zn(s) + 10\ H^+(aq) + NO_3^-(aq) \rightarrow$
$4\ Zn^{2+}(aq) + NH_4^+(aq) + 3\ H_2O(l)$
(b) $H_2SO_4(aq) + Zn(s) \rightarrow H_2(g) + ZnSO_4(aq)$
(c) $2\ MnO_4^-(aq) + 10\ Cl^-(aq) + 16\ H^+(aq) \rightarrow$
$2\ Mn^{2+}(aq) + 5\ Cl_2(g) + 8\ H_2O(l)$
(d) $10\ H^+(aq) + 3\ H_2O_2(aq) + 2\ CrO_4^{2-}(aq) \rightarrow$
$2\ Cr^{3+}(aq) + 3\ O_2(g) + 8\ H_2O(l)$

16.61 **(a)** 1 equivalent per mole
(b) 1 equivalent per mole (2 mol of HCl reacts with the 2 mol of OH^- in each mole of $Ba(OH)_2$; thus, the 2 mol of HCl corresponds to 2 equivalents of HCl: 2 equiv/2 mol = 1 equiv/mol.

(c) 1 equivalent per mole (This is an oxidation-reduction reaction: $10 \, Cl^-(aq) \rightarrow 5 \, Cl_2(g) + 10 \, e^-$. Since 10 mol of Cl^- produces 10 mol of e^-, there are 10 equivalents of HCl in the 10 mol of HCl.)

16.63 The reaction is

$$H_2SO_4(aq) + 2 \, OH^-(aq) \rightarrow SO_4^{2-}(aq) + 2 \, H_2O(l)$$

There are 2 equivalents per mole, so

$$(5.00 \text{ mol})(2 \text{ equiv/mol}) = 10.0 \text{ equiv}$$

16.65 Since the concentration is given in *normality*, it does not make any difference which base is used. Thus, for both parts, the number of equivalents of base is

$$(0.03835 \text{ L})(0.2193 \text{ equiv/L}) = 0.008410 \text{ equiv}$$

The number of equivalents of acid is equal to the number of equivalents of base, and the normality of the acid is therefore

$$\frac{0.008410 \text{ equiv acid}}{0.02500 \text{ L}} = 0.3364 \text{ N acid}$$

16.67 The number of equivalents of acid (and therefore of base) is

$$(0.03197 \text{ L})(1.500 \text{ equiv/L}) = 0.04796 \text{ equiv}$$

The equivalent mass is the number of grams per equivalent:

$$\frac{2.135 \text{ g}}{0.04796 \text{ equiv}} = 44.52 \text{ g/equiv}$$

16.69 $(31.49 \text{ mL NaOH})(3.127 \text{ mequiv/mL}) = 98.47 \text{ mequiv NaOH}$

The number of milliequivalents of acid is equal to this number of milliequivalents of base. There is 98.47 mequiv of acid.

16.70 $N_1V_1 = N_2V_2$

$$N_2 = \frac{(0.1000 \text{ N})(25.00 \text{ mL})}{44.16 \text{ mL}} = 0.05661 \text{ N}$$

16.73 $\left(\dfrac{145 \text{ g}}{1 \text{ mol}}\right)\left(\dfrac{1 \text{ mol}}{2 \text{ equiv}}\right) = \dfrac{72.5 \text{ g}}{1 \text{ equiv}}$

16.74 **(a)** Na_2S and many others
(b) Cl_2O, H_2S, or many others, including Na_2S

16.75 $:C\!\!\left(\!:::N\!:\right)^-$

The oxidation number of carbon is $+2$; that of nitrogen is -3. The carbon in CO has an oxidation number of $+2$, since it retains control of only two electrons. The oxygen has an oxidation number of -2. There is a difference because there is no charge on the CO molecule.

16.76 The ion whose central element has the higher oxidation number is named with the ending *-ate*.

16.78 Tin(IV) chloride and tin tetrachloride

16.80 Since electrons are included in balanced half-reactions, there is a charge on one side of the equation, which must be balanced by the charge(s) on ions.

16.81 **(a)** $HX(aq) + NaOH(aq) \rightarrow NaX(aq) + H_2O(l)$

$$0.04273 \text{ L NaOH}\left(\frac{1.000 \text{ mol NaOH}}{1 \text{ L NaOH}}\right)\left(\frac{1 \text{ mol HX}}{1 \text{ mol NaOH}}\right)$$
$$= 0.04273 \text{ mol HX}$$

$$\text{Molar mass} = \frac{4.54 \text{ g HX}}{0.04273 \text{ mol HX}} = 106 \text{ g/mol HX}$$

(b) $H_2X(aq) + 2 \, NaOH(aq) \rightarrow Na_2X(aq) + 2 \, H_2O(l)$

$$0.04273 \text{ L NaOH}\left(\frac{1.000 \text{ mol NaOH}}{1 \text{ L NaOH}}\right)\left(\frac{1 \text{ mol H}_2X}{2 \text{ mol NaOH}}\right)$$
$$= 0.02136 \text{ mol H}_2X$$

$$\text{Molar mass} = \frac{4.54 \text{ g H}_2X}{0.02136 \text{ mol H}_2X} = 213 \text{ g/mol H}_2X$$

(c) There is no way to tell from the data about the base how many ionizable hydrogens are on each acid molecule.

16.83 The oxidation number of the alkali metals is $+1$ in each of their compounds, so the oxygen has an unusual oxidation number in parts (a) and (b). To have peroxides and superoxides, the other elements present must be in their highest oxidation states. Thus, lead must be present as lead(IV) in part (c).
(a) -1 (a peroxide) **(b)** $-\frac{1}{2}$ (a superoxide)
(c) -2 (a normal oxide)

16.89 There is no change in oxidation number. Chromium is in the $+6$ oxidation state on both sides.

16.90 **(a)** $6 \, e^- + 14 \, H^+(aq) + Cr_2O_7^{2-}(aq) \rightarrow 2 \, Cr^{3+}(aq) + 7 \, H_2O(l)$
For the oxidation half-reaction:

Step 1: $\qquad\qquad\qquad CN^-(aq) \rightarrow CO_2(g) + NO_2(g)$
Step 3: $\quad 4 \, H_2O(l) + CN^-(aq) \rightarrow CO_2(g) + NO_2(g)$
Step 4: $\quad 4 \, H_2O(l) + CN^-(aq) \rightarrow$
$$CO_2(g) + NO_2(g) + 8 \, H^+(aq)$$
Step 5: $\quad 4 \, H_2O(l) + CN^-(aq) \rightarrow$
$$CO_2(g) + NO_2(g) + 8 \, H^+(aq) + 9 \, e^-$$
Step 6: $\quad 8 \, H_2O(l) + 2 \, CN^-(aq) \rightarrow$
$$2 \, CO_2(g) + 2 \, NO_2(g) + 16 \, H^+(aq) + 18 \, e^-$$
$$18 \, e^- + 42 \, H^+(aq) + 3 \, Cr_2O_7^{2-}(aq) \rightarrow$$
$$6 \, Cr^{3+}(aq) + 21 \, H_2O(l)$$

The whole equation:
$$26 \, H^+(aq) + 3 \, Cr_2O_7^{2-}(aq) + 2 \, CN^-(aq) \rightarrow$$
$$6 \, Cr^{3+}(aq) + 2 \, CO_2(g) + 2 \, NO_2(g) + 13 \, H_2O(l)$$

(b) $62 \, H^+(aq) + 7 \, Cr_2O_7^{2-}(aq) + 6 \, CNO^-(aq) \rightarrow$
$$14 \, Cr^{3+}(aq) + 6 \, CO_2(g) + 6 \, NO_2(g) + 31 \, H_2O(l)$$

(c) $38 \, H^+(aq) + 5 \, Cr_2O_7^{2-}(aq) + 2 \, CNS^-(aq) \rightarrow$
$$10 \, Cr^{3+}(aq) + 2 \, CO_2(g) + 2 \, NO_2(g) + 2 \, SO_4^{2-}(aq) + 19 \, H_2O(l)$$

17 Reaction Rates and Chemical Equilibrium

17.1 The equation regulates the ratio that reacts, not how much can be placed in a vessel. The reacting ratio does not control the extent of the reaction. Therefore, statements (a), (b), and (c) are incorrect. Statement (d) is also incorrect, since the reaction is an equilibrium and does not go to completion to give 2 mol of CO_2. Statement (e) is the only correct one.

17.2 **(a)** At the start of the reaction, the rate of combination is zero, since there is no nitrogen or hydrogen present.
(b) The rate of combination increases as the decomposition reaction proceeds, producing more of the elements. At the point when the rate of decomposition is equal to the rate of combination, equilibrium is established.

17.3 The concentration ratio is the same as the mole ratio because the volume of every gas in a mixture is the same.

$$\frac{[A]}{[B]} = \frac{(\text{mol A})/V_A}{(\text{mol B})/V_B} = \frac{\text{mol A}}{\text{mol B}}$$

17.4 **(a)** $K = \dfrac{[HI][HI]}{[H_2][I_2]}$ **(b)** $K = \dfrac{[HI]^2}{[H_2][I_2]}$

Note that the two chemical equations state the same facts, as do the two equilibrium constant expressions. This pair of equations shows *why* the coefficient in the balanced chemical equation becomes an exponent in the equilibrium constant expression.

17.5 The equilibrium constant expression is the same for both parts; the heat is not included in the equilibrium constant expression.

$$K = \frac{[NH_3]^2}{[N_2][H_2]^3}$$

17.6 **(a)** K for the first equation is the square root of that for the second equation.

(b) K for the first equation is the reciprocal of that for the second equation.

17.7 The equilibrium constant expression is the same for both parts; the heat is not included in the equilibrium constant expression.

$$K = \frac{[CO_2][H_2]}{[CO][H_2O]}$$

17.8 **(a)** A reaction rate increases with increasing temperature, so the combination reaction will go fastest at 500°C.
(b) The decomposition reaction also will go fastest at 500°C.

17.9 You could increase the temperature by about 10°C.

17.13 **(a)** Since the system is at equilibrium, the rate of decomposition is the same as the rate of combination. Thus, 9.72×10^{-4} mol/L·s of CO is combining with O_2.
(b) Since this is the start of the reaction, there is no CO or O_2 present. Thus, their rate of combination must be zero at this point.

17.15 Since CO is combining at the rate of 4.00×10^{-4} mol/L·s, and the balanced equation shows a 2:1 ratio, the O_2 must be combining at 2.00×10^{-4} mol/L·s.

17.16 The equilibrium will shift to the left, to use up some of the heat added to raise the temperature.

17.18 **(a)** Increasing the pressure by decreasing the volume shifts the equilibrium to the right, to produce fewer moles of gas.
(b) Increasing the pressure by decreasing the volume shifts the equilibrium to the left, to produce fewer moles of *gas*. Note that the carbon is not a gas.
(c) Since the same number of moles of gas is on both sides, there is no shift.

17.19 **(a)** Shift to the right **(b)** Shift to the left
(c) Shift to the right **(d)** Shift to the right
(e) No shift **(f)** Shift to the right
(g) Shift to the left

17.21 **(a)** Right shift **(b)** Left shift
(c) Right shift (to lower the number of moles of gas)
(d) No shift **(e)** Left shift **(f)** Right shift

17.23 **(a)** $K = \dfrac{[SO_2Cl_2]}{[SO_2][Cl_2]}$ **(b)** $K = \dfrac{[PF_5]}{[PF_3][F_2]}$

17.25 For the reaction given,

$$\frac{[N_2]^{1/2}[O_2]}{[NO_2]} = K_1 = 0.010$$

(a) $\dfrac{[N_2][O_2]^2}{[NO_2]^2} = K_1{}^2 = 0.00010 = 1.0 \times 10^{-4}$

(b) $\dfrac{[NO_2]}{[N_2]^{1/2}[O_2]} = 1/K_1 = 100$

(c) $\dfrac{[NO_2]^2}{[N_2][O_2]^2} = 1/K_1{}^2 = 1.0 \times 10^4$

17.27 The equilibrium constant expression is

$$K = \frac{[C][D]}{[A]^2[B]^3}$$

The initial numbers of moles are converted to concentrations (in moles/liter) and tabulated as follows:

	2 A	+	3 B	⇌	C	+	D
Initial	1.00		1.50		0		0
Change							
Equilibrium	0.10						

It is apparent that 0.90 mol/L of A has been used up by the reaction, which also uses up 1.35 mol/L of B and produces 0.45 mol/L each of C and D. These values are added to the table:

	2 A	+	3 B	⇌	C	+	D
Initial	1.00		1.50		0		0
Change	−0.90		−1.35		+0.45		+0.45
Equilibrium	0.10						

The values involved in the change are added to or subtracted from the initial values to yield the equilibrium values:

	2 A	+	3 B	⇌	C	+	D
Initial	1.00		1.50		0		0
Change	−0.90		−1.35		+0.45		+0.45
Equilibrium	0.10		0.15		0.45		0.45

The value of the equilibrium constant is given by

$$K = \frac{[C][D]}{[A]^2[B]^3} = \frac{(0.45)(0.45)}{(0.10)^2(0.15)^3} = 6.0 \times 10^3$$

17.29 **(a)**

	2 T	⇌	R	+	Z
Initial	4.00		0		0
Change	−2x		x		x
Equilibrium	4.00 − 2x		x		x
	≈ 4.00				

$$K = \frac{[R][Z]}{[T]^2} = \frac{x^2}{(4.00)^2} = 1.0 \times 10^{-8}$$

$$\frac{x}{4.00} = 1.0 \times 10^{-4}$$

$$x = 4.0 \times 10^{-4}$$

(b)

	T	⇌	½ R	+	½ Z
Initial	4.00		0		0
Change	−2y		y		y
Equilibrium	4.00 − 2y		y		y
	≈ 4.00				

$$K = \frac{[R]^{1/2}[Z]^{1/2}}{[T]} = \frac{(y)^{1/2}(y)^{1/2}}{4.00} = 1.0 \times 10^{-4}$$

$$y = 4.0 \times 10^{-4}$$

The concentration of R is the same no matter which equation is used.

17.31 The equilibrium constant expression is

$$K = \frac{[C][D]}{[A]^2[B]^2} = 2.50 \times 10^{-9}$$

The initial numbers of moles are converted to concentrations (in moles/liter) and then all values are tabulated:

	2 A	+	2 B	⇌	C	+	D
Initial	1.00		1.00		0		0
Change	−2x		−2x		+x		+x
Equilibrium	1.00 − 2x		1.00 − 2x		x		x

Neglecting $2x$ when subtracted from the large concentrations yields:

$$K = \frac{[C][D]}{[A]^2[B]^2} = \frac{x^2}{(1.00)^2(1.00)^2} = 2.50 \times 10^{-9}$$

$$x^2 = 2.50 \times 10^{-9}$$
$$x = 5.00 \times 10^{-5}$$

The value of $2x$ is small enough to be neglected when subtracted from the much larger quantity 1.00.

17.33 The initial numbers of moles are divided by 2.00 L to get the initial concentrations:

	$Br_2(g)$	+	$Cl_2(g)$	⇌	2 BrCl(g)
Initial	0.100		0.200		0.000
Change	−0.055		−0.055		+0.110
Equilibrium	0.045		0.145		0.110

$$K = \frac{[BrCl]^2}{[Br_2][Cl_2]} = \frac{(0.110)^2}{(0.045)(0.145)} = 1.9$$

17.34 **(a)** Since O_2 is reacting at 4.07×10^{-3} mol/L·s, that reaction is producing SO_3 at 8.14×10^{-3} mol/L·s. Since the rate of decomposition of SO_3 is less than the rate of its production, some net quantity of SO_3 will be produced after this point. There will be more present at equilibrium than at this point.

(b) Since SO_3 is being produced at the point in question, it must have been the product since the start of the experiment. Reactions never go past the equilibrium point (unless a stress is applied), so SO_2 and O_2 are the starting materials.

17.36 The added nitrogen will cause the equilibrium to shift to the left, producing some heat. The added heat causes the temperature to rise.

17.38

	$N_2(g)$	$+$	$2\,O_2(g)$	$\rightleftharpoons$	$2\,NO_2(g)$
Initial	0		0		1.000
Change	$(1.000 - x)/2$		$1.000 - x$		$1.000 - x$
Equilibrium	$(1.000 - x)/2$		$1.000 - x$		x

$$K = \frac{x^2}{[(1.000 - x)/2](1.000 - x)^2} = 5.00 \times 10^{-6}$$

Neglecting x when subtracted from 1.000 yields

$$K = \frac{x^2}{(0.500)(1.000)^2} = 5.00 \times 10^{-6}$$

$$x^2 = 2.50 \times 10^{-6}$$
$$x = 1.58 \times 10^{-3} = [NO_2]$$

$[O_2] = 1.000 - x = 0.998$
$[N_2] = (1.000 - x)/2 = 0.499$

Check: $\dfrac{(1.58 \times 10^{-3})^2}{(0.499)(0.998)^2} = 5.02 \times 10^{-6}$

17.40 (a) Each of these stresses will shift the equilibrium to the left, so the combination will shift it to the left.

(b) and (d) Each stress will shift the equilibrium in a different direction, but we cannot tell by how much, so no prediction can be made about the combined effect.

(c) A catalyst has no effect on the position of the equilibrium and the removal of CO_2 will shift it to the right, so the combination will shift it to the right.

17.42 LeChâtelier's principle applies only to systems already at equilibrium. It cannot predict anything about this initial reaction mixture.

17.44 Since the NH_3 is decomposing at 2.00×10^{-3} mol/L·s, it must be producing 3.00×10^{-3} mol/L·s of H_2, since they react in a 2:3 ratio. Since the rate of production of H_2 is less than its rate of reaction, less is being produced than is being used up, and the elements are the original reactants. The reaction is proceeding to the left.

17.46 Since an equilibrium system is *dynamic* (that is, reaction continues in both directions), you expect both the combination reaction and the decomposition reaction to proceed. Therefore, some ND_3 should react with some NH_3, and some HD will be formed. The reaction of the D_2 and the HD with N_2 will result in the formation of some ND_3, NHD_2, and NH_2D molecules. The existence of molecules with both ^1H and D in them proves that both reactions are still occurring, even though at equilibrium no *net reaction* occurs.

17.48

	$2\,NO_2(g)$	$+$	$Cl_2(g)$	$\rightleftharpoons$	$2\,NO_2Cl(g)$
Initial	0.200		0.100		0.000
Change	-0.054		-0.027		$+0.054$
Equilibrium	0.146		0.073		0.054

$$K = \frac{[NO_2Cl]^2}{[NO_2]^2[Cl_2]} = \frac{(0.054)^2}{(0.146)^2(0.073)} = 1.9$$

18 Acid-Base Theory

18.1 (a) Weak (b) Weak (c) Strong (d) Weak (e) Strong

18.2 (a) Strong (b) Weak (c) Strong (d) Strong (e) Weak

18.3 (a) The salt is completely ionized. There are no NaCl molecules in solution. The ions Na^+ and Cl^- are present.

(b) Ammonia is weak, and only a small percentage of the molecules react with water to form ions. NH_3 is the major solute, with only a little NH_4^+ and OH^- present.

(c) K^+ and SO_4^{2-} are present. The salt is ionic even when pure, and it stays ionic in solution.

(d) NH_4^+ and SO_4^{2-} are present both in the solid state and in solution. (NH_4^+ does react with water to a slight extent to form NH_3 and H_3O^+ [Example 18.4].)

18.4 (a) The acid is weak and ionizes very little. There are mostly HNO_2 molecules in solution. The ions H_3O^+ and NO_2^- are present to a very small extent.

(b) Na^+ and NO_3^- are present. The salt is soluble. It is ionic even when pure, and it stays ionic in solution.

18.5 Compounds (a), (d), and (e) exist essentially as ions.
(a) KBr is a salt.
(b) $HClO_2$ is a weak acid.
(c) H_2SO_3 is a weak acid.
(d) $Ca(HCO_3)_2$ is a salt.
(e) NH_4ClO_3 is a salt.
(f) CH_3OH is a soluble covalent compound.

18.6 $HF(aq) + H_2O(l) \rightleftharpoons F^-(aq) + H_3O^+(aq)$
  Acid   Base   Base   Acid

18.7 $NH_3(aq) + H_2O(l) \rightleftharpoons NH_4^+(aq) + OH^-(aq)$
  Base   Acid   Acid   Base

18.8 (a) $pH = -\log[H_3O^+] = -\log(0.300) = 0.523$
(b) $pH = 0.523$ (The HCl ionizes completely, yielding 0.300 M H_3O^+.)

18.9 (a) 3.016 (b) 8.642 (c) 3.294 (d) 6.157 (e) 0.000

18.10 (a) 6.110 (b) 10.044 (c) 1.650 (d) 11.301 (e) -1.000

18.11 (a) 6.21×10^{-13} (b) 7.78×10^{-7} (c) 7.76×10^{-13} (d) 1.00×10^{-7} (e) 1.26×10^{-3}

18.12 The solution has a pH of 7.00. Since the salt is composed of two ions, neither of which reacts with water, the solution is neutral. The pH of a neutral solution is 7.00.

18.13 (a) Neutral (b) Basic (c) Acidic (d) Acidic (e) Basic (f) Basic

18.14 (a) The solution is 0.200 M $HC_2H_3O_2$ and 0.130 M $C_2H_3O_2^-$.

(b) The solution is 0.200 M $HC_2H_3O_2$ and 0.130 M $C_2H_3O_2^-$. The $C_2H_3O_2^-$ ions resulted from the partial neutralization of the $HC_2H_3O_2$.

18.15 The cations are Li^+, Na^+, K^+, Rb^+, Cs^+, Fr^+, (and Ag^+). These are the cations of the strong bases containing one OH^- ion per formula unit.

18.16 Any strong acid could be used to provide the H_3O^+ ions, and any soluble metal chlorite could be used to provide the anion. For example HCl and $NaClO_2$ could be used.

18.17 The salt consists of the conjugate acid of a base (the cation) and the conjugate base of an acid (the anion).

18.18 (a) ClO_2^- (b) $CH_3NH_3^+$ (c) $H_2PO_4^-$

18.20 (a) H_2O (b) $HClO_2$ (c) NH_3 (d) H_2O

18.22 $H_2SO_4(aq) + OH^-(aq) \rightleftharpoons HSO_4^-(aq) + H_2O(l)$
  Acid   Base   Base   Acid

$HSO_4^-(aq) + OH^-(aq) \rightleftharpoons SO_4^{2-}(aq) + H_2O(l)$
  Acid   Base   Base   Acid

18.24 (a) Acidic (The solution contains a weak acid, NH_4^+, and an almost feeble base.)
   (b) Basic (The solution contains a feeble acid and a weak base, F^-.)
   (c) Basic (The solution contains a feeble acid and a weak base, ClO_2^-.)
   (d) Neutral
   (e) Basic (The solution contains a feeble acid and a weak base, $C_2H_3O_2^-$.)

18.26 (a) H_2SO_4 and SO_4^{2-} **(b)** H_3O^+ and OH^-

18.28 $HClO_2(aq) + H_2O(l) \rightleftharpoons ClO_2^-(aq) + H_3O^+(aq)$
   Acid   Base   Base   Acid

$C_5H_5N(aq) + H_2O(l) \rightleftharpoons C_5H_5NH^+(aq) + OH^-(aq)$
   Base   Acid   Acid   Base

18.30 The hydronium ion concentration at equilibrium is equal to the concentration of A^-, and the concentration of HA is not affected significantly. Therefore,

$$K_a = \frac{[H_3O^+][A^-]}{[HA]} = \frac{(4.79 \times 10^{-5})^2}{0.100} = 2.29 \times 10^{-8}$$

18.32 Since $NaClO_2$ is a salt, it is present in solution in the form of Na^+ and ClO_2^- ions. The concentration 0.200 M means that the solution was made up with 0.200 mol of $NaClO_2$ per liter of solution; it does not mean that un-ionized molecules of $NaClO_2$ are present in the solution.

18.34 $HCHO_2(aq) + H_2O(l) \rightleftharpoons CHO_2^-(aq) + H_3O^+(aq)$
Initial   0.250   0   0
Change $-x$ $+x$ $+x$
Equilibrium $0.250 - x$ x x

$$K_a = \frac{[CHO_2^-][H_3O^+]}{[HCHO_2]} = \frac{x^2}{0.250} = 1.7 \times 10^{-4}$$
$$x^2 = 4.2 \times 10^{-5}$$
$$x = 6.5 \times 10^{-3}$$
$$[H_3O^+] = 6.5 \times 10^{-3} \text{ M}$$

18.35 The equation for the reaction of a general base, B, with water is

$$B(aq) + H_2O(l) \rightleftharpoons BH^+(aq) + OH^-(aq)$$

The equilibrium constant expression is

$$K_b = \frac{[BH^+][OH^-]}{[B]}$$

From this expression, you can see that the smaller the value of K_b, the lower is the concentration of hydroxide ion. Of the bases in Table 18.2, therefore, pyridine has the lowest hydroxide ion concentration, since it has the lowest value of K_b.

18.37 (a) $CH_3NH_2(aq) + H_2O(l) \rightleftharpoons CH_3NH_3^+(aq) + OH^-(aq)$
Initial   0.333   0   0
Change $-x$ $+x$ $+x$
Equilibrium $0.333 - x$ x x

$$K_b = \frac{[CH_3NH_3^+][OH^-]}{[CH_3NH_2]} = \frac{x^2}{0.333}$$
$$= 4.4 \times 10^{-4} \text{ (From Table 18.2)}$$
$$x^2 = 1.5 \times 10^{-4}$$
$$x = 1.2 \times 10^{-2}$$

$$\text{Percent ionization} = \left(\frac{1.2 \times 10^{-2}}{0.333}\right) \times 100\% = 3.6\%$$

(b) $K_b = \dfrac{x^2}{0.100} = 4.4 \times 10^{-4}$

$$x = 6.6 \times 10^{-3}$$

$$\text{Percent ionization} = \left(\frac{6.6 \times 10^{-3}}{0.100}\right) \times 100\% = 6.6\%$$

(c) $K_b = \dfrac{x^2}{0.0333} = 4.4 \times 10^{-4}$

$$x = 3.8 \times 10^{-3}$$

$$\text{Percent ionization} = \left(\frac{3.8 \times 10^{-3}}{0.0333}\right) \times 100\% = 11\%$$

In summary,

Initial $[CH_3NH_2]$	$[OH^-]$	*Percent Ionization*
0.333 M	1.2×10^{-2}	3.6
0.100 M	6.6×10^{-3}	6.6
0.0333 M	3.8×10^{-3}	11

With increasing dilution of the base, the hydroxide ion concentration decreases. However, because the concentration of the un-ionized base decreases more, the percent ionization increases.

18.38 (a) 1.000   **(b)** 13.000   **(c)** 7.000

18.40 (a) 0.100 M   **(b)** 1.00×10^{-13} M   **(c)** 1.00×10^{-7} M
   (d) 1.00   **(e)** 1.00×10^{-14}

18.41 (a) 1.762   **(b)** 11.627
   (c) 10.425 [Two moles of OH^- ions are present per mole of $Ba(OH)_2$.]

18.43 (a) 7.46×10^{-8} M
   (b) 8.07×10^{-14} M
   (c) 7.9×10^{-11} M

18.45 From each $[OH^-]$, the corresponding $[H_3O^+]$ is calculated. From that, the pH is obtained.

   (a) $K_w = [H_3O^+][OH^-] = 1.00 \times 10^{-14}$

$$[H_3O^+] = \frac{K_w}{[OH^-]} = \frac{1.00 \times 10^{-14}}{6.39 \times 10^{-3}} = 1.56 \times 10^{-12} \text{ M}$$

   pH = 11.806

   (b) pH = 10.862   **(c)** pH = 7.338

18.46 From the pH, you can calculate the equilibrium hydronium ion concentration:

$$[H_3O^+] = 10^{-4.93} = 1.2 \times 10^{-5} \text{ M}$$

That concentration is equal to the concentration of A^-, and the concentration of HA does not change significantly. Therefore,

$$K_a = \frac{[H_3O^+][A^-]}{[HA]} = \frac{(1.2 \times 10^{-5})^2}{0.100} = 1.4 \times 10^{-9}$$

18.49 From the pH, you can calculate the equilibrium hydronium ion concentration:

$$[H_3O^+] = 10^{-12.79} = 1.6 \times 10^{-13} \text{ M}$$

The OH^- concentration is thus

$$\frac{1.0 \times 10^{-14}}{1.6 \times 10^{-13}} = 6.2 \times 10^{-2}$$

which is also equal to the BH^+ concentration. The concentration of B is equal to 0.200 M $-$ 0.062 M = 0.138 M. Therefore,

$$K_b = \frac{[BH^+][OH^-]}{[B]} = \frac{(6.2 \times 10^{-2})^2}{0.138} = 2.8 \times 10^{-2}$$

Note that, in this case, the quantity of B that ionized is not negligible, relative to the initial quantity.

18.50 (a) 1.34×10^{-8} M   **(b)** 1.62×10^{-3} M
   (c) 8.17×10^{-12} M

18.52 (a) pH = 11.534   **(b)** pH = 11.656   **(c)** pH = 6.857

18.55 From the pH, you can calculate the equilibrium hydronium ion concentration, and from that the hydroxide ion concentration:

$$[H_3O^+] = 10^{-12.15} = 7.1 \times 10^{-13} \text{ M}$$
$$[OH^-] = K_w/[H_3O^+] = 1.4 \times 10^{-2} \text{ M}$$

The concentration of BH^+ is equal to 0.140 M, and the concentration of B is 0.130 M. Thus,

$$K_b = \frac{[BH^+][OH^-]}{[B]} = \frac{(0.140)(1.4 \times 10^{-2})}{0.130} = 1.5 \times 10^{-2} \text{ M}$$

18.57 From the pH, you can calculate the equilibrium hydronium ion concentration:

$$[H_3O^+] = 10^{-4.27} = 5.4 \times 10^{-5} \text{ M}$$

The concentration of A^- is equal to 0.311 M, and the concentration of un-ionized HA is 0.213 M. Thus,

$$K_a = \frac{[H_3O^+][A^-]}{[HA]} = \frac{(5.4 \times 10^{-5})(0.311)}{0.213} = 7.9 \times 10^{-5}$$

18.59 **(a)** Formic acid ionizes according to the equation

$$HCHO_2(aq) + H_2O(l) \rightleftharpoons CHO_2^-(aq) + H_3O^+(aq)$$

The acid dissociation constant is 1.7×10^{-4}. (The formate ion from the sodium formate shifts this equilibrium to the left; it does not react directly with the formic acid.) Start by assuming that no hydronium ion is present:

$$HCHO_2(aq) + H_2O(l) \rightleftharpoons CHO_2^-(aq) + H_3O^+(aq)$$

Initial	0.130	0.120	0
Change	$-x$	$+x$	$+x$
Equilibrium	$0.130 - x$	$0.120 + x$	x

Neglecting x when subtracted from 0.130 or added to 0.120 yields

$$K_a = \frac{[CHO_2^-][H_3O^+]}{[HCHO_2]} = \frac{(0.120)(x)}{0.130} = 1.7 \times 10^{-4}$$
$$x = 1.8 \times 10^{-4}$$

$[H_3O^+] = 1.8 \times 10^{-4}$ M

(b) The NaOH reacts with some of the $HCHO_2$, leaving 0.130 mol of $HCHO_2$ and producing 0.120 mol of CHO_2^-. Therefore,

$$HCHO_2(aq) + H_2O(l) \rightleftharpoons CHO_2^-(aq) + H_3O^+(aq)$$

Initial	0.130	0.120	0
Change	$-x$	$+x$	$+x$
Equilibrium	$0.130 - x$	$0.120 + x$	x

Neglecting x when subtracted from or added to 0.100 yields

$$K_a = \frac{[CHO_2^-][H_3O^+]}{[HCHO_2]} = \frac{(0.120)(x)}{0.130} = 1.7 \times 10^{-4}$$
$$x = 1.8 \times 10^{-4}$$

$[H_3O^+] = 1.8 \times 10^{-4}$ M

The solution formed in part (b) is the same as that in part (a).

18.62 The HCl ionizes completely. The acetic acid is the only acid for which an equilibrium calculation need be done. The initial concentration of H_3O^+ is 0.140 M from the ionization of the HCl.

$$HC_2H_3O_2(aq) + H_2O(l) \rightleftharpoons C_2H_3O_2^-(aq) + H_3O^+(aq)$$

Initial	0.120	0	0.140
Change	$-x$	$+x$	$+x$
Equilibrium	$0.120 - x$	x	$0.140 + x$

Neglecting x when subtracted from or added to larger values and using the value of K_a from Table 18.2 yields

$$K_a = \frac{[C_2H_3O_2^-][H_3O^+]}{[HC_2H_3O_2]} = \frac{(x)(0.140)}{0.120} = 1.8 \times 10^{-5}$$
$$x = 1.5 \times 10^{-5}$$

$[C_2H_3O_2^-] = 1.5 \times 10^{-5}$ M

You can readily see that the excess H_3O^+ from the HCl has repressed the ionization of the $HC_2H_3O_2$ (compare with Example 18.7). Any weaker acid is affected in this way in the presence of a stronger acid.

18.64 The solution has a very weak acid (phenol) and a moderately weak base (phenolate ion). The weak base is stronger than the weak acid, so the solution is basic.

18.66 **(a)** Yes **(b)** Yes **(c)** No
(d) Yes **(e)** No **(f)** Yes

18.68 The neutralization reaction is

$$NaOH(aq) + HCHO_2(aq) \rightarrow NaCHO_2(aq) + H_2O(l)$$

Exactly halfway through the titration, half of the acid will have been neutralized, and half will remain. The concentrations of $HCHO_2$ and CHO_2^- will be equal, and the hydronium ion concentration will equal K_a (whose value is given in Table 18.2):

$$K_a = \frac{[H_3O^+][CHO_2^-]}{[HCHO_2]} = [H_3O^+] = 1.7 \times 10^{-4} \text{ M}$$

The hydronium ion concentration is 1.7×10^{-4} M, so the pH is 3.77.

18.70 **(a)** No **(b)** No **(c)** No **(d)** No
(e) No **(f)** Yes **(g)** No

19 Organic Chemistry

19.1 There are five carbons in each. The stem *penta* means "five."

19.2 **(a)** $HOCH_2CH_3$ **(b)** $HCOCH_3$ **(c)** $HOCOCH_3$
(d) NH_2CH_3

19.3 **(a)** An ammonia molecule with one hydrogen replaced by a radical, CH_3—
(b) A water molecule with both hydrogen atoms replaced by radicals, CH_3—
(c) An ammonia molecule with two hydrogen atoms replaced by radicals, CH_3—
(d) Hydrogen chloride, HCl, with the hydrogen replaced by a radical, CH_3CH_2—

19.4 82 $[2n + 2 = 2(40) + 2 = 82]$

19.5 **(a)** A ketone cannot have its carbonyl group at the end of the carbon chain, and there is only one carbon atom not on the end in a three-carbon chain.
(b) An aldehyde must have its carbonyl group at the end of the carbon chain, and in this compound, it does not make any difference which end.

19.6 **(a)** Water **(b)** Water **(c)** Water

19.7 The radical cannot be hydrogen in organic halides, alcohols, ethers, or amines (which would be hydrogen halides, water, or ammonia if R were hydrogen). In esters, the radical denoted R′ in Table 19.4 cannot be hydrogen, but the one denoted R can. In ketones, neither radical can be hydrogen (if it were, the compound would be an aldehyde).

19.8 **(a)**   **(b)**

19.9 **(a)** Eight (There are seven in the heptane chain and one additional in the methyl side chain.)
(b) Nine (There are two in the side chain.)
(c) Nine (There is one in each of the two side chains.)

19.10 **(a)** Eighteen **(b)** Twenty **(c)** Twenty

19.11 **(a)** Alkene **(b)** Alkyne **(c)** Alkane

19.15 **(a)** $CH_3CH_2CH=CHCH_2CH_3$
(b) $CH_3CH(CH_3)_2$ **(c)** $(CH_3)_2CHCH(CH_3)_2$
(d) $(CH_3)_2CHCH_2CH=C(CH_3)CH_2CH_3$

19.17

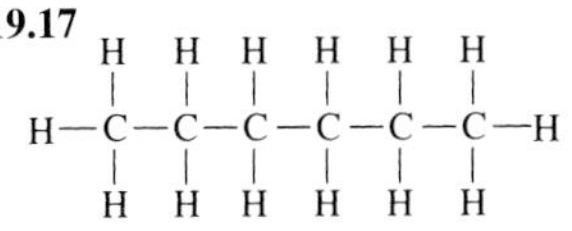

Hexane

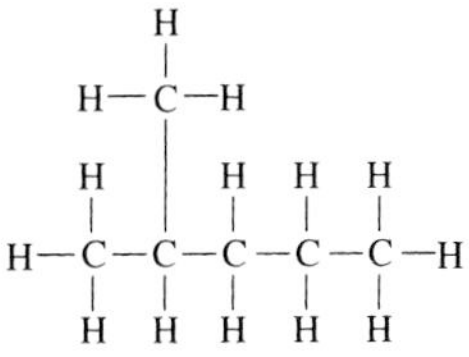

2-Methylpentane

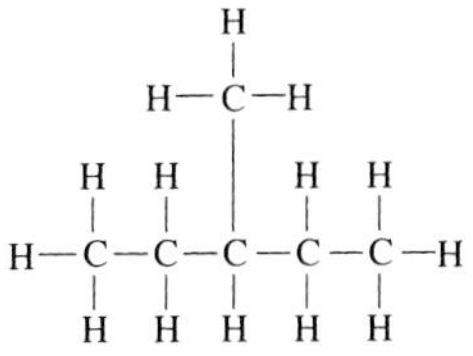

3-Methylpentane

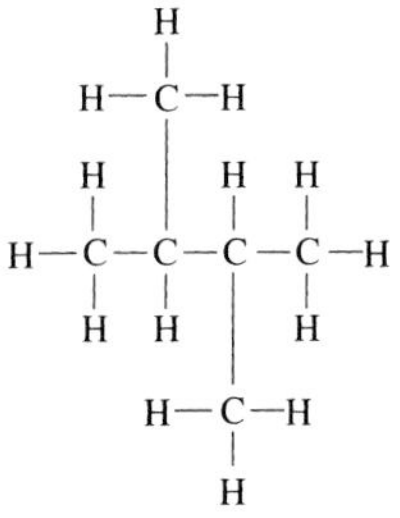

2,3-Dimethylbutane

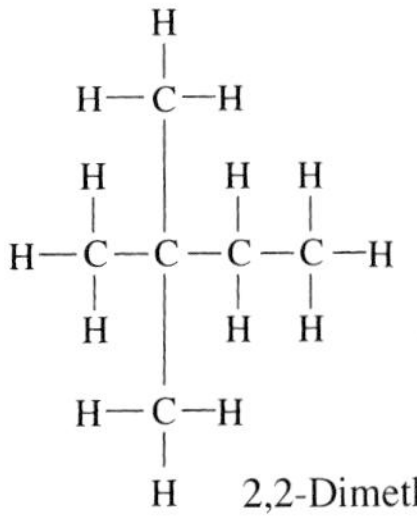

2,2-Dimethylbutane

19.19 (a) $CH_3CHCH_2CH_2CH_2CH_3$
              |
              CH_3

(b) $CH_3CH_2CHCH_2CH_2CH_3$
              |
              CH_3

(c) If you placed a methyl group on the fourth carbon in the chain, it would be the third carbon from the right end, and the formula would represent 3-methylhexane.

19.21 (a)

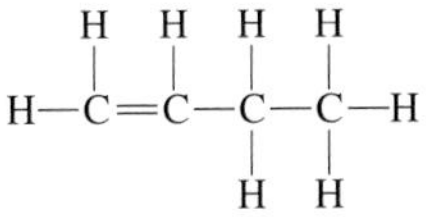

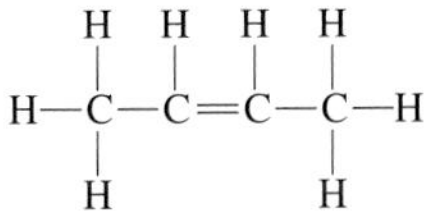

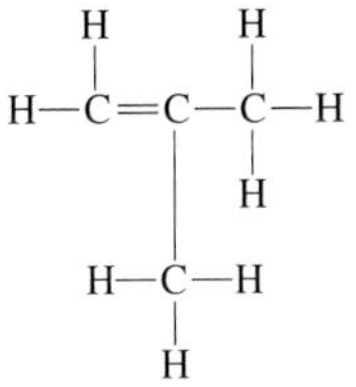

(b)

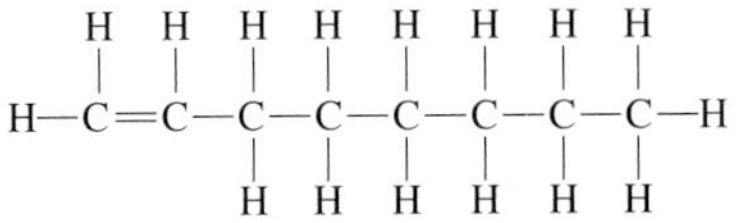

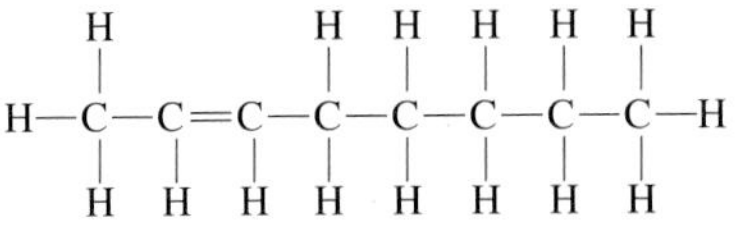

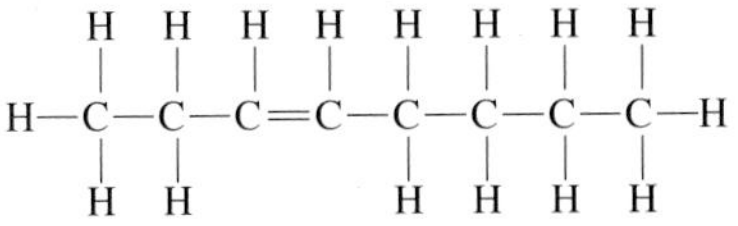

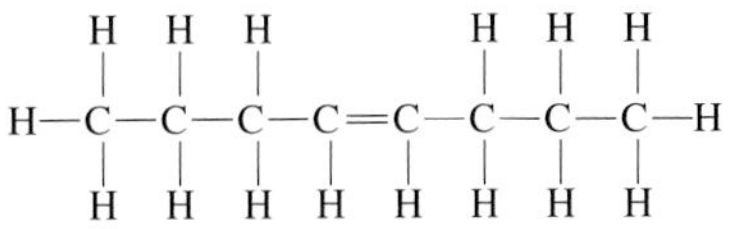

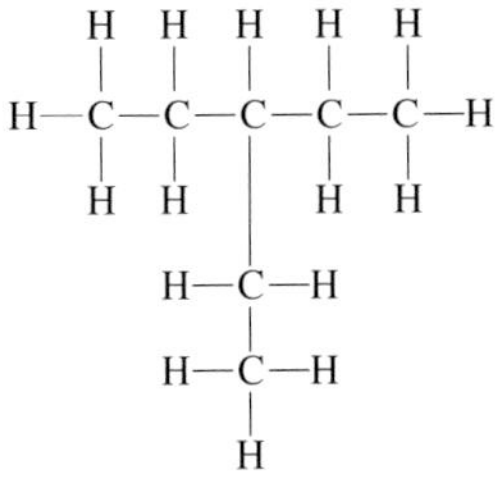

(The branch must be on the middle carbon atom, or the longest chain will have more than five carbon atoms and the branch will not be $-C_2H_5$.)

19.23

19.24

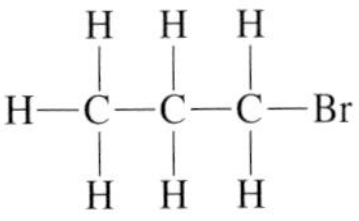

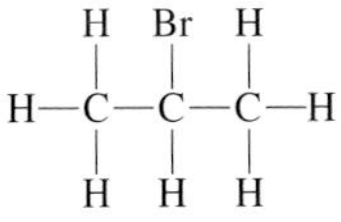

19.26 (a) $CH\equiv CH$ (b) $CH_2=CH_2$ (c) CH_4 (d) $HCOOH$
(e) CH_3Cl (f) CH_3OH (g) CH_3COCH_3

19.27 (a) Ethyne (acetylene)
(b) Ethene (ethylene)
(c) Methane
(d) Methanoic acid (formic acid)
(e) Chloromethane (methyl chloride)
(f) Methanol (methyl alcohol)
(g) Propanone (acetone)

19.29 (a) Amine   (b) Alcohol   (c) Aldehyde
(d) Organic acid   (e) Ether   (f) Ketone

19.31 (a) Methyl alcohol   (b) Formaldehyde   (c) Formic acid

19.32 (a)   (b)

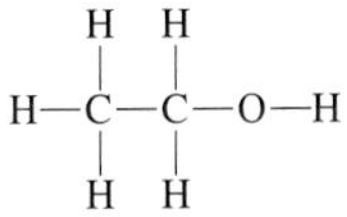

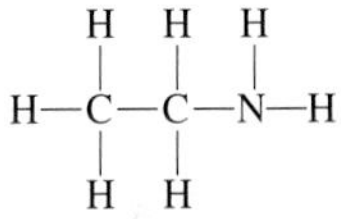

(c) 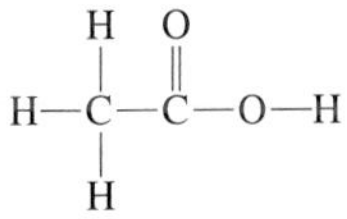   (d)

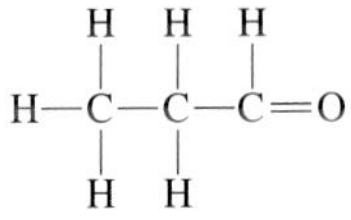

19.33 The compound given in the problem statement is identified first:
(a) Ether and alcohol
(b) Amine (secondary) and amine (primary)
(c) Ester and organic acid
(d) Ketone and aldehyde

19.34 The compound given in the problem statement is named first:
(a) Dimethyl ether and ethanol
(b) Dimethyl amine and ethyl amine
(c) Methyl formate and ethanoic acid (acetic acid)
(d) Propanone (acetone) and propanal (propionaldehyde)

19.41 An amide and an amine cannot be isomers. An amide molecule contains oxygen, but an amine does not. However, alcohols and ethers can be isomers of each other, as can aldehydes and ketones, and also organic acids and esters.

19.44

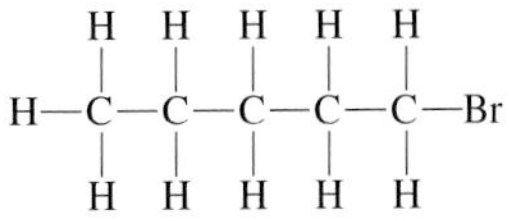

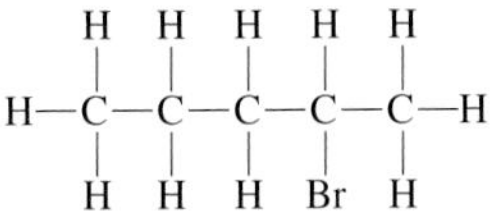

(structural formulas)

$$H-\overset{\overset{\displaystyle H}{|}}{\underset{\underset{\displaystyle H}{|}}{C}}-\overset{\overset{\displaystyle H}{|}}{\underset{\underset{\displaystyle H}{|}}{C}}-\overset{\overset{\displaystyle H}{|}}{\underset{\underset{\displaystyle Br}{|}}{C}}-\overset{\overset{\displaystyle H}{|}}{\underset{\underset{\displaystyle H}{|}}{C}}-\overset{\overset{\displaystyle H}{|}}{\underset{\underset{\displaystyle H}{|}}{C}}-H$$

19.46   (Each R stands for C_6H_5.)

19.48   Gases have little intermolecular forces between their molecules. The molecules of polymers are very large, and intermolecular forces, even in nonpolar polymers, are considerable. You should not expect polymers to be gases under the conditions given.

19.50   An amino acid has both an amine group and an acid group in the same molecule; an amide is the product of the reaction of separate molecules having those two groups:

NH_2CH_2COOH $CH_3CONHCH_3$
An amino acid   An amide

19.52   (a) Carbon, hydrogen, oxygen, and nitrogen
(b) Carbon, hydrogen, and oxygen
(c) Carbon, hydrogen, and oxygen

19.53   (a) For CH_3CH_2OH and CH_3OCH_3:
$$2x + 6(+1) + (-2) = 0$$
$$x = -2$$
For CH_3CH_2CHO and CH_3COCH_3:
$$3x + 6(+1) + (-2) = 0$$
$$x = -\tfrac{4}{3}$$
For $HCOOCH_3$ and CH_3COOH:
$$2x + 4(+1) + 2(-2) = 0$$
$$x = 0$$
(b) The isomeric molecules have the same oxidation number for carbon. (The oxidation number increases in the order given as fewer hydrogen atoms or more oxygen atoms appear in the molecule.)

19.55

Number of Carbon Atoms	Alkane	Alkene	Alkyne	Alkyl Radical
1	CH_4	None	None	$—CH_3$
2	C_2H_6	C_2H_4	C_2H_2	$—C_2H_5$
3	C_3H_8	C_3H_6	C_3H_4	$—C_3H_7$
4	C_4H_{10}	C_4H_8	C_4H_6	$—C_4H_9$
5	C_5H_{12}	C_5H_{10}	C_5H_8	$—C_5H_{11}$

19.57 $^+NH_3CH_2COO^-$

19.60 $C_6H_{12}O_6 + C_6H_{12}O_6 \rightarrow C_{12}H_{22}O_{11} + H_2O$

19.62 (a)

1,1-Di(4-chlorophenyl)-2,2,2-trichloroethane

(b)

Di(2-chlorophenyl)methane

19.64 The amide group, $—\overset{\overset{\displaystyle O}{\|}}{C}—NH—$ (known as the peptide linkage in proteins)

19.67 Aldehydes and acids do not need positional numbers for their functional groups. In these compounds, the functional groups are always at the end of the carbon chain. (Esters and amides do not need positional numbers for the carbon chain of the acid part but might need them for the other part.)

19.69 A fully hydrogenated fat is a saturated fat, whether it started out unsaturated or not. You should not pay extra for it.

20 Nuclear Reactions

20.1 There is no difference. All four are representations for an alpha particle (a helium nucleus).

20.2 The symbol Co stands for the naturally occurring mixture of isotopes of cobalt; ^{60}Co stands for only one isotope—the one used in nuclear medicine—and represents the nucleus, rather than the entire atom.

20.3 You can look at a periodic table to find the atomic numbers, which are the subscripts. You cannot look at the periodic table to find the superscripts because mass numbers are not generally given there. (Mass numbers for the few elements that do not occur naturally are given in parentheses in most periodic tables.) The atomic mass given in a periodic table can give a clue to the mass number of the most abundant isotope for many but not all elements. For example, the atomic mass of bromine is almost exactly 80, but the two isotopes that form the naturally occurring mixture have mass numbers 79 and 81. (They occur in almost a 1 : 1 ratio.)

20.4 When half of the sample has disintegrated, the rate of disintegration is half of what it was before, and only half of what is left will disintegrate in the next 10.0 minutes.

20.5 (a) $_0^0\gamma$ (gamma particle)

(b) $_1^2H$ (deuteron)

(c) $_1^3H$ (tritium nucleus)

(d) $_{-1}^0e$ (electron) or $_{-1}^0\beta$ (beta particle)

(e) $_{+1}^0\beta$ (positron)   (f) $_2^4He$ (alpha particle)

(g) $_0^1n$ (neutron)   (h) $_1^1H$ (proton)

20.6 N_0 is the number of atoms of a particular isotope at the start of the time period, N is the number of atoms of that isotope remaining at time t, and $t_{1/2}$ is the half-life. Note that N is *not* the number of atoms that have disintegrated.

20.7 (a) Fusion   (b) Spontaneous   (c) Fission   (d) Fusion

20.8 (a) $_{85}^{217}At \rightarrow _2^4He + _{83}^{213}Bi$

(b) $_{89}^{227}Ac \rightarrow _{-1}^0\beta + _{90}^{227}Th$

(c) $_{50}^{119}Sn \rightarrow _0^0\gamma + _{50}^{119}Sn$

20.9 (a) $_{88}^{225}Ra \rightarrow _{89}^{225}Ac + _{-1}^0\beta$

(b) $_{87}^{221}Fr \rightarrow _{85}^{217}At + _2^4\alpha$

(c) $_{83}^{211}Bi \rightarrow _{81}^{207}Tl + _2^4\alpha$

(d) $_{83}^{212}Bi \rightarrow _{81}^{208}Tl + _2^4\alpha$

20.12 The nuclear reactions of radon produce subatomic particles that cause damage to humans.

20.14 After the first 6.80 years, half of the 60-mg sample will still be the original isotope. After 6.80 more years, half of that remaining 30 mg will still be the original isotope. That is, 15 mg will remain after 13.6 years.

20.16 $\log\left(\dfrac{N_0}{N}\right) = \left(\dfrac{0.301}{t_{1/2}}\right)t$

(a) $\log\left(\dfrac{6.21\ g}{0.881\ g}\right) = 0.848$

$t_{1/2} = \dfrac{(0.301)(6.42\ days)}{0.848} = 0.228\ day$

(b) Note the wording of the problem. Here 0.881 g disintegrates, and 6.21 g − 0.881 g = 5.33 g remains.

$\log\left(\dfrac{6.21\ g}{5.33\ g}\right) = 0.0664$

$t_{1/2} = \dfrac{(0.301)(6.42\ days)}{0.0664} = 29.1\ days$

20.18

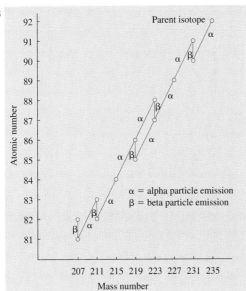

α = alpha particle emission
β = beta particle emission

20.19 The number of atoms of an isotope (N) is equal to Avogadro's number times the mass of the isotope (m) divided by its molar mass (MM), and the same relationship applies for the initial number of atoms (N_0). Thus,

$$\dfrac{N_0}{N} = \dfrac{(m_0/MM)(\text{Avogadro's number})}{(m/MM)(\text{Avogadro's number})} = \dfrac{m_0}{m}$$

20.21 The total number of atoms at time zero (when the rock solidified) was

$(8.96 \times 10^{21}) + (6.44 \times 10^{22}) = 7.34 \times 10^{22}$

$\log\left(\dfrac{7.34 \times 10^{22}}{8.96 \times 10^{21}}\right) = 0.913$

$t = \log\left(\dfrac{N_0}{N}\right)\left(\dfrac{t_{1/2}}{0.301}\right) = (0.913)\left(\dfrac{1.3 \times 10^9\ years}{0.301}\right)$

$= 3.9 \times 10^9\ years$

20.23 (a) $_{92}^{235}U + _0^1n \rightarrow _{54}^{143}Xe + _{38}^{90}Sr + 3\ _0^1n$

(b) $_{13}^{27}Al + _2^4\alpha \rightarrow _{15}^{30}P + _0^1n$

(c) $_7^{14}N + _2^4\alpha \rightarrow _8^{17}O + _1^1H$

(d) $_{92}^{235}U + _0^1n \rightarrow _{56}^{140}Ba + _{36}^{94}Kr + 2\ _0^1n$

20.25 (a) $_4^9Be + _2^4He \rightarrow _6^{12}C + _0^1n$

(b) $_{92}^{238}U + _0^1n \rightarrow _{92}^{239}U + _0^0\gamma$

(c) $_{94}^{239}Pu + _0^1n \rightarrow _{94}^{240}Pu$

20.27 (a) $_6^{12}C + _1^1H \rightarrow _7^{13}N$

(b) $_7^{13}N \rightarrow _6^{13}C + _{+1}^0\beta$ (The positron produced can react with an electron, which converts all the matter of both to energy. See Example 20.18.)

(c) $_6^{13}C + _1^1H \rightarrow _7^{14}N$ (d) $_7^{14}N + _1^1H \rightarrow _8^{15}O$

(e) $_8^{15}O \rightarrow _7^{15}N + _{+1}^0\beta$ (f) $_7^{15}N + _1^1H \rightarrow _6^{12}C + _2^4He$

20.28 The net reaction, when the annihilation of positrons and electrons is included, is

$4\ ^1H \rightarrow {}^4He + energy$

20.29 (a) $_1^2H + _1^2H \rightarrow _1^3H + _1^1H$ (b) $_1^2H + _1^2H \rightarrow _2^3He + _0^1n$

20.32 Since 90.0% disintegrates, 10.0% is left.

$\log\left(\dfrac{100.0\%}{10.0\%}\right) = \left(\dfrac{0.301}{7.50\ min}\right)t$

$t = 24.9\ min$

20.34 (a) Natural   (b) Artificially induced

(c) Artificially induced

20.35 The number of moles of each isotope is calculated first:

$4.79 \times 10^{-3}\ g\ ^{238}U\left(\dfrac{1\ mol}{238\ g}\right) = 2.01 \times 10^{-5}\ mol\ ^{238}U$

$2.93 \times 10^{-4}\ g\ ^{206}Pb\left(\dfrac{1\ mol}{206\ g}\right) = 1.42 \times 10^{-6}\ mol\ ^{206}Pb$

The total number of moles of ^{238}U when the rock solidified is the sum of these.

$\log\left(\dfrac{n_0}{n}\right) = \left(\dfrac{0.301}{t_{1/2}}\right)t$

$\log\left(\dfrac{2.15 \times 10^{-5}\ mol}{2.01 \times 10^{-5}\ mol}\right) = \left(\dfrac{0.301}{4.51 \times 10^9\ years}\right)t$

$t = 4.38 \times 10^8\ years$

20.37 Boiling water is a technique used to kill harmful bacteria—living organisms. Since the chemical environment of a radioactive isotope does not affect its radioactivity, boiling alone will not make water containing a radioactive isotope safe to drink. (Distillation, a process involving boiling and condensation in another place, may separate the isotope from the water, making the water safe to drink.)

20.39 $E = mc^2$

$m = E/c^2 = (1.00 \times 10^{-11} \text{ J})/(3.00 \times 10^8 \text{ m/s})^2$
$= 1.11 \times 10^{-28}$ kg

20.41 No, there is no obvious relationship.

20.43 **(a)** At least half of the neutrons must escape or be absorbed.
(b) At least two-thirds of the neutrons must escape or be absorbed.

Appendix 1

A.1 Be sure to distinguish between equations and expressions when simplifying.
(a) $x = 13$ **(b)** $5x/y$
(c) $x = 0.200$ **(d)** $0.286x$
(e) $x = 72$ **(f)** $x - 2y + 4$

A.2 **(a)** T_2
(b) The expression cannot be simplified.
(c) $P_1 = 3P_2$

A.3 $x = +4$

A.4 $x = -1$

A.5 The exponent 3 *outside* the parentheses indicates that the 3.0 is to be cubed, as well as the unit centimeter.

A.6 **(a)** Both expressions are equal.
(b) All three expressions are equal.
(c) None of the expressions is equal to any of the others.

A.7 **(a)** Not equal: $\frac{3}{4} \neq \frac{1}{6}$
(b) Not equal: $-\frac{1}{4} \neq \frac{1}{2}$
(c) Equal: $\frac{1}{8} = \frac{1}{8}$

A.8 **(a)** They are not equal: $-d$ times $-f$ is equal to $+df$, not $-df$.
(b) They are equal.

A.9 **(a)** g/mL   **(b)** g/cm³   **(c)** g/m³

A.10 **(a)** kg/L   **(b)** g/L   **(c)** mg/L

A.11 **(a)** 3.60 pounds   **(b)** 5.51 gallons

A.12 For $(1.23)^2$, the $\boxed{x^2}$ key should be pressed after entering 1.23.
For 1.23×10^2, the $\boxed{\text{EXP}}$ or $\boxed{\text{EE}}$ key followed by 2 should be pressed after entering 1.23.

A.13 The reciprocal key, $\boxed{1/x}$

A.14 $2.5 \times 10^8 \text{ m}^{-1} = 2.5 \times 10^8/\text{m}$

Note that the unit for the reciprocal is different from the unit for the given value.

A.15 **(a)** 25   **(b)** 19   **(c)** -16

A.16 **(a)** -21 **(b)** -1 **(c)** $+8$

A.17 **(a)** 1.23×10^6 **(b)** 1.23×10^{-6} **(c)** -1.23×10^6
(d) -1.23×10^{-6} **(e)** 1×10^{11}

A.18 The answer is 81. The 3 is squared the first time $\boxed{x^2}$ is pressed, yielding 9 in the display. That 9 is squared when $\boxed{x^2}$ is pressed again.

A.19 **(a)** 7.71   03   **(b)** 7.7 − 03   **(c)** −7.71   03
(d) −7.71 − 03   **(e)** 2.2   22

A.20 **(a)** $R = \dfrac{PV}{nT}$ **(b)** $n = MV$ **(c)** $V = m/d$

(d) $P_2 = \dfrac{P_1T_2}{T_1}$ **(e)** $T = \dfrac{\pi V}{nR}$

A.22 **(a)** cm   **(b)** cm²   **(c)** cm³   **(d)** cm

A.24 **(a)** Dollars   **(b)** cm²   **(c)** cm³   **(d)** cm

A.26 Specific gravity has *no units*. The units of the density of the substance cancel the units of the density of water. For example, the specific gravity of mercury is

$$\frac{13.6 \text{ g/mL}}{1.00 \text{ g/mL}} = 13.6$$

The units cancel each other.

A.27 **(a)** $P = \dfrac{nRT}{V} = \dfrac{(3.00 \text{ mol})(0.0821 \text{ L·atm/mol·K})(295 \text{ K})}{6.00 \text{ L}}$

$= 12.1$ atm

(b) $V = n/M = 0.317 \text{ mol}\left(\dfrac{1 \text{ L}}{0.953 \text{ mol}}\right) = 0.333$ L

(c) $V = m/d = 122 \text{ g}\left(\dfrac{1 \text{ mL}}{4.73 \text{ g}}\right) = 25.8$ mL

(d) $T_1 = \dfrac{P_1T_2}{P_2} = \dfrac{(750 \text{ torr})(273 \text{ K})}{760 \text{ torr}} = 269$ K

(e) $\pi = \dfrac{nRT}{V}$

$= \dfrac{(0.0100 \text{ mol})(0.0821 \text{ L·atm/mol·K})(293 \text{ K})}{3.00 \text{ L}}$

$= 0.0802$ atm

(f) $P_2 = \dfrac{P_1V_1}{V_2} = \dfrac{(0.989 \text{ atm})(425 \text{ mL})}{722 \text{ mL}} = 0.582$ atm

(g) $\nu = E/h = (4.45 \times 10^{-19} \text{ J})/(6.63 \times 10^{-34} \text{ J·s})$
$= 6.71 \times 10^{14}/\text{s}$

(h) $k_b = \Delta t/m = (0.300°\text{C})/(0.400 \text{ m}) = 0.750°\text{C/m}$

A.29 **(a)** $(a - b) - c$ **(b)** $(a/b) + c$
(c) $a/(bc)$ **(d)** $(ab)/c$ or $a(b/c)$

A.30 **(a)** $\boxed{2}\ \boxed{\times}\ \boxed{3}\ \boxed{x^2}\ \boxed{=}$

(b) $\boxed{2}\ \boxed{\times}\ \boxed{3}\ \boxed{=}\ \boxed{x^2}$

(c) $\boxed{2}\ \boxed{x^2}\ \boxed{+/-}$

(d) $\boxed{2}\ \boxed{x^2}\ \boxed{\times}\ \boxed{2}\ \boxed{=}$

A.32 Part (c) is different. On the calculator, the quotient a/b is multiplied by c, so the expression $(a/b) \times c$ is the same as the value without the parentheses.

A.34 **(a)** 1.2×10^{12} **(b)** 3×10^{-1} **(c)** 1.5×10^{-4}
(d) 2.8×10^5 **(e)** -5.0×10^8 **(f)** -5.0×10^{-6}
(g) 4.0×10^1

A.35 **(a)** -21.5 **(b)** 1.95×10^{-3} **(c)** -33.1
(d) 3.92×10^5 **(e)** -1.84×10^4 **(f)** -578
(g) 4.64   **(h)** 72.0   **(i)** -0.986 **(j)** -43.5

A.38 **(a)** 10.2   **(b)** 0.00707   **(c)** 0.279
(d) 44.8   **(e)** -121 **(f)** -34.3
(g) 2.7×10^2 **(h)** 81.0   **(i)** -0.952 **(j)** 178.4

A.40 The calculator does not keep track of significant digits, so it does not differentiate between 4.00 and 4.

A.42 **(a)** -8.3×10^3 **(b)** -8.8×10^{12} **(c)** 9.62×10^7
(d) 6.02×10^{-2} **(e)** 4×10^2 **(f)** 0.1667
(g) 1.00006

A.43 Exponentiation is done right to left. Thus, first square x, yielding 9, and then raise the 4 to the ninth power, yielding 262 144.

A.44 $t = 420$ years

A.46

	a	b	c	x
(a)	1	2	1	-1 and -1
(b)	1	-2	1	1 and 1
(c)	1	0	-4	-2 and 2

(d) 1   −4   4   2 and 2
(e) 4   4   1 $-\frac{1}{2}$ and $-\frac{1}{2}$

The equations of parts (f)–(j) may be rearranged to give the same equations as in parts (a)–(e), respectively, and so the answers are the same.

A.48 **(a)** −2.356666667 − 10   **(b)** −2.27 − 10
(c) 8.51851394   **(d)** 0.2
(e) 2.   **(f)** 1.   11
(g) 1.301029996   **(h)** 64.

Glossary

The section numbers in which the terms are explained are given in parentheses.

absolute temperature (12.3) A temperature, measured on the Kelvin scale, that must be used for gas law and certain other types of calculations.

absolute zero (12.3) 0 K, the lowest possible temperature.

accuracy (2.3) The closeness of a measurement or set of measurements to the correct value.

acetone (19.3) CH_3COCH_3, propanone, the simplest ketone.

acetylene (19.1) $HC \equiv CH$, ethyne, the simplest alkyne.

acid (6.3, 8.4) A compound that provides H^+ ions to water solutions. (See also *Brønsted acid* and *organic acid*.)

acid carbonate (8.4) A compound containing the HCO_3^- ion.

acid dissociation constant (18.2) The equilibrium constant that controls the extent of ionization of a weak acid in water; symbolized by K_a.

acid salt (6.3, 8.4) A partially neutralized acid that originally contained two or more ionizable hydrogen atoms per molecule; for example, HSO_4^- is partially neutralized H_2SO_4.

acid strength (18.1) The tendency for a certain percentage of acid molecules to ionize in water; the higher the acid strength, the higher is the percent ionization.

acidic anhydride (8.4) A nonmetal oxide that can react with water to form an acid.

active (8.3) Having a high tendency to react; reactive.

alcohol (19.3) An organic compound with the —OH functional group.

aldehyde (19.3) An organic compound with the —CHO functional group.

alkali metal (1.5) A metal in group IA of the periodic table—Li, Na, K, Rb, Cs, or Fr.

alkaline earth metal (1.5) A metal in group IIA of the periodic table—Be, Mg, Ca, Sr, Ba, or Ra.

alkane (19.1) A hydrocarbon containing only single bonds.

alkene (19.1) A hydrocarbon containing one double bond per molecule.

alkyl radical (19.3) A portion of an alkane formed by loss of one hydrogen atom.

alkyne (19.1) A hydrocarbon containing one triple bond per molecule.

allotrope (5.5) One of two or more forms of an uncombined element; for example, diamond and graphite are allotropes of carbon.

alpha particle (20.1) A helium nucleus generated in a nuclear reaction; a stream of such particles is referred to as an alpha ray.

amide (19.3) An organic compound with the —CONH— functional group.

amine (19.3) An organic compound with the —NR_2 functional group, where either or both Rs can be hydrogen atoms.

ammonia (6.1) The covalent compound NH_3.

ammonium ion (6.2) The NH_4^+ ion.

amorphous solid (14.1) A solid with an internal structure that does not repeat over many molecules and that therefore has a large melting range.

amphiprotic (18.1) The ability of a molecule or ion to react with both acids and bases.

analytical chemistry (Chapter 1, Introduction) The branch of chemistry dealing with the composition of samples.

angular molecule (13.4) A three-atom molecule in which the atoms do not lie on a line.

angular momentum quantum number (4.2) The quantum number (ℓ) that governs the shape of the space occupied by an electron in an atom.

anhydride (8.4) A compound resulting from loss of water by another compound; for example, the CaO resulting from strong heating of $Ca(OH)_2$.

anhydrous (5.1) Without water; for example, $CuSO_4$ resulting from loss of water from $CuSO_4 \cdot 5H_2O$ is said to be anhydrous.

anion (5.2) A negatively charged ion.

anode (5.2) The electrode where oxidation occurs in an electrochemical reaction.

aqueous solution (8.2) Any solution in which water is the solvent.

aromatic hydrocarbon (19.1) A hydrocarbon containing one or more benzene rings.

Arrhenius acid (8.4) A compound that provides H^+ ions to water.

Arrhenius base (8.4) A compound that provides OH^- ions to water.

Arrhenius theory (8.4) The fundamental theory of acids and bases. (Contrast with *Brønsted theory*.)

atmosphere (12.1) A unit equal to 760 torr and abbreviated atm that is the pressure of the atmosphere on a "normal" day at sea level; the envelope of air surrounding the earth.

atmospheric pressure (12.1) The pressure of the atmosphere.

atom (3.2) The smallest particle that retains the characteristic composition of an element.

atom smasher (20.3) A machine designed to initiate nuclear reactions.

atomic bomb (20.3) A nuclear bomb; a bomb in which matter is converted to energy for destructive purposes.

atomic mass (3.4, 7.1) The weighted average of the masses of the naturally occurring isotopes of an element, compared to one-twelfth of the mass of a ^{12}C atom.

atomic mass scale (3.4) A relative scale of masses based on the mass of ^{12}C being the standard and having a mass defined as exactly 12 amu.

atomic mass unit (3.4) A mass equal to one-twelfth of the mass of a ^{12}C atom; abbreviated amu.

atomic number (3.3) The number of protons in the nucleus of each atom of an element.

atomic size (13.1) The size of an atom.

atomic theory See *Dalton's atomic theory.*

atomic weight See *atomic mass.*

autoionization (18.3) A reaction of molecules of a single substance to produce both positive and negative ions; for example, water molecules react to produce both H_3O^+ and OH^- ions.

auto-oxidation reduction See *disproportionation.*

average kinetic energy (12.8) The total kinetic energy of all the molecules of a sample, divided by the number of molecules; the average kinetic energy is dependent on temperature only.

Avogadro's number (7.3) 6.02×10^{23}, which is the number of ^{12}C atoms in exactly 12 g of ^{12}C (as well as the number of atomic mass units in exactly 1 g).

balanced equation (8.1) A written representation of a chemical reaction in which formulas of the reactants and products appear on two sides of an arrow with coefficients to make the numbers of atoms of each element the same on both sides.

barometer (12.1) An instrument for measuring the pressure of a gas, especially the atmosphere.

barometric pressure (12.1) The pressure of the atmosphere.

base (6.3, 8.4) A compound that provides OH^- ions to water solutions. (See also *Brønsted base.*)

base (2.4) The number that is raised to a power and multiplied by a coefficient in an exponential number; for example, the 10 in 1.5×10^2.

base dissociation constant (18.2) The equilibrium constant that controls the extent of ionization of a weak base in water; symbolized K_b.

base strength (18.1) The tendency for a certain percentage of molecules of a base to ionize in water; the higher the base strength, the higher is the percent ionization.

basic anhydride (8.4) A metal oxide that can react with water to form a base.

bent molecule (13.4) See *angular molecule.*

benzene (19.1) C_6H_6, the simplest aromatic hydrocarbon.

beta particle (20.1) A high-energy electron emitted from a nucleus in a nuclear reaction; a stream of such particles is called a beta ray.

binary compound (5.1) A compound composed of two elements.

biochemistry (Chapter 1, Introduction) The branch of chemistry dealing with living things.

Bohr theory (4.1) The first theory of the atom to propose that electrons in atoms were in definite energy levels.

boiling point (14.2) The temperature at which a liquid changes to a gas at the prevailing pressure.

boiling-point elevation (15.5) An increase in the boiling point of a solvent due to the presence of a solute.

bond See *covalent bond* and *ionic bond.*

bond order See *total bond order.*

Boyle's law (12.2) At constant temperature, the volume of a given sample of gas is inversely proportional to its pressure.

Brønsted acid (18.1) A proton donor.

Brønsted base (18.1) A proton acceptor.

Brønsted theory (18.1) A theory of acids and bases that is broader than the Arrhenius theory in that it does not limit acids and bases to water solutions and it defines ions as acids or bases depending on their tendency to react with water.

buffer solution (18.4) A solution of a weak acid and its conjugate base or a weak base and its conjugate acid, which resists changes in its acidity even on addition of a moderate quantity of strong acid or strong base.

buildup principle (4.3) The addition of an electron to the configuration of an atom of the preceding element on the periodic table allows determination of the electronic configuration of an atom of a given element.

buret (11.3) A piece of laboratory glassware calibrated for measuring the volume of liquid delivered.

calorie (2.6) A unit of energy equal to the quantity of heat necessary to raise the temperature of 1 g of water by 1°C; equivalent to 4.184 J.

carbohydrate (19.5) A compound containing carbon plus hydrogen and oxygen in a mole ratio of 2:1; the general formula is $C_xH_{2y}O_y$.

carbonate (8.4) An ionic compound containing the CO_3^{2-} ion.

catalyst (8.3, 17.1) A material that affects the speed of a chemical reaction without any permanent change in its own composition.

cathode (5.2) The electrode where reduction occurs in an electrochemical reaction.

cation (5.2) A positively charged ion.

Celsius scale (2.6) The temperature scale defined with 0° as the freezing point of water and 100° as the normal boiling point of water; the centigrade scale.

centered dot (5.1) An indication that water of hydration is part of a compound, as in $CuSO_4 \cdot 5H_2O$.

centigrade scale (2.6) The Celsius scale.

central atom (13.4) The atom in a molecule or ion to which (most of) the other atoms are covalently bonded.

chain reaction (20.3) A nuclear reaction whose products cause the same reaction to occur again.

Charles' law (12.3) At constant pressure, the volume of a given sample of gas is directly proportional to its absolute temperature.

chemical change (1.1) A chemical reaction.

chemical reaction (1.1) A change in which the composition (or structure) of one or more substances is altered.

chemistry (1.3) The study of the interaction of matter and energy and the changes that matter undergoes.

coefficient (2.4) The number that is multiplied by a base raised to a certain power in a number written in exponential notation, such as the 1.5 in 1.5×10^2.

coefficient (8.1) The number placed before a formula in a chemical equation to balance the equation.

coinage metal (1.5) An element of periodic group IB—Cu, Ag, or Au.

colligative properties (15.5) Characteristics of solutions that are dependent on the concentration of solute particles and the nature of the solvent but not on the nature of the solute.

combination reaction (8.3) A reaction in which two substances combine to form one.

combined gas law (12.4) For a given sample of gas, the volume is directly proportional to the absolute temperature and inversely proportional to the pressure.

combustion reaction (8.3) A reaction of an element or compound with oxygen at high temperature.

completion (10.4) The condition a reaction has reached when the limiting quantity of reactant has been exhausted.

compound (1.1) A chemical combination of elements that has a definite composition and its own set of properties.

concentration (11.1) The quantity of solute per unit volume of solution or per unit mass of solvent.

condensation (14.2) A change of phase from gaseous to liquid or solid.

conjugate acid (18.1) The cation (or molecule) that results from the reaction of a base with a proton.

conjugate acid-base pair (18.1) A Brønsted acid and base that differ from each other by H^+.

conjugate base (18.1) The anion (or molecule) that results from the loss of a proton by an acid.

constant composition See *law of constant composition.*

"control" of shared electrons (16.1) Assignment of the electrons in a covalent bond to the more electronegative atom of the bond.

control rod (20.3) A neutron-absorbing rod used to slow down nuclear reactions in commercial reactors.

conversion factor (2.1) A ratio equal to 1, which can be multiplied by a quantity to change its form without changing its value.

covalent bond (5.5) A bond resulting from electron sharing.

critical mass (20.3) The smallest mass capable of sustaining a nuclear chain reaction.

crystalline solid (14.1) A solid with a regular internal structure of repeating units and a definite melting point.

cubic meter (2.2) The volume of a cube that measures 1 m on each edge; the basic unit of volume in SI.

Dalton's atomic theory (3.2) The theory that matter is made up of small particles (atoms) that have properties characteristic of an element.

Dalton's law of partial pressures (12.7) The total pressure of a gas mixture is equal to the sum of the partial pressures of its components.

daughter isotope (20.1) The large product of a natural radioactive decay event.

decay See *radioactive decay.*

decomposition reaction (8.3) A reaction in which one compound is broken down into two or more substances.

definite composition (1.1) The given ratio by mass of each element in a compound to any other element in the compound.

definite proportions See *law of definite proportions.*

degenerate (4.2) Having the same energy.

density (2.5) The mass per unit volume of a sample of matter.

deuterium (20.4) The isotope of hydrogen with mass number 2; symbolized ^2H or D.

deuteron (20.4) A deuterium (^2H) nucleus; symbolized d.

diatomic molecule (5.1) A molecule containing two atoms.

dimensional analysis See *factor label method.*

dipole (13.5) See *dipole moment.*

dipole moment (13.5) An unequal distribution of charge in a molecule resulting from unsymmetrical orientation of polar bonds; also referred to as a dipole.

direct proportionality (12.3) The relationship in which one variable changes by the same factor as another.

disaccharide (19.5) A sugar containing two simple sugar molecules.

discrete energy levels (4.1) Atomic energy levels that have specific energies.

disintegration (20.1) A reaction of an unstable nucleus.

displacement reaction (8.3) The reaction of an element with a compound to produce another element and another compound.

disproportionation (16.4) The reaction of a species with itself to produce products having higher and lower oxidation numbers; also called auto-oxidation-reduction.

dissociation (Chapter 9, Introduction) Separation of ions from their close proximity in a solid lattice to a distance when dissolved in a solvent.

dissolve (1.2) To go into solution, making a homogeneous mixture.

distillation (14.2) The conversion of a liquid to a gas and then back to a liquid to separate it from impurities.

double bond (5.5) The sharing of two pairs of electrons between atoms.

double displacement reaction (8.3) The reaction of two compounds in which two other compounds are produced, as a result of a trade of the anions by the cations.

double substitution reaction (8.3) A double displacement reaction.

ductile (1.5) Capable of being drawn into a wire.

duet (5.2) A pair of electrons associated with a hydrogen, helium, lithium, or beryllium atom, resulting in the stable configuration of a noble gas (helium).

Einstein's equation (20.3) The equation $E = mc^2$, which relates the energy produced to the mass of matter converted to energy in a nuclear reaction.

electrode (5.2) The solid portion of an electrochemical reaction apparatus at which a change from electron conduction to ion conduction, or vice versa, takes place.

electromagnetic radiation (20.1) Any form of light energy—visible light, ultraviolet light, X rays, gamma rays, infrared radiation, or microwaves.

electron (3.3) A negatively charged subatomic particle; a fundamental particle of nature.

electron affinity (13.2) The energy liberated when a gaseous atom acquires an electron to form a gaseous anion.

electron dot diagram (5.4) A pictorial model of an atom, molecule, or ion consisting of the symbol(s) for element(s) and dots representing the valence electrons.

electron group (13.4) A single electron pair, the four or six electrons in a double or triple bond, or rarely, a single electron, which are attached to a central atom of a molecule or polyatomic ion.

electron sharing (5.5) The sharing of electrons between atoms to form covalent bonds.

electronegativity (5.1, 13.3) The tendency of atoms involved in covalent bonds to attract electrons.

electronic charge (3.3) The charge on one electron, 1.60×10^{-19} coulomb.

electronic configuration (4.4) The arrangement of the electrons in an atom, ion, or molecule.

electronic structure See *electronic configuration.*

electropositive (5.1) Having only a small tendency to attract the electrons in a covalent bond.

element (1.1) A substance that cannot be broken down into simpler substances by chemical means; one of the basic building blocks of which all matter is composed.

elemental (5.1) Existing as an element; not combined into any compound.

empirical formula (7.4) The simplest formula for a compound that shows the atomic ratio of elements in the compound.

end point (11.3) The point in a titration when the indicator signals that the reaction is complete.

endothermic process (14.3) A process in which energy is absorbed from outside the system.

energy (1.3) The capacity to do work.

energy level diagram (4.6) A graph in which horizontal lines represent the orbitals of an atom, those with higher energies toward the top, and in which arrows may be used to represent electrons.

enthalpy change (14.4) The heat (under certain familiar conditions) involved in a process; symbolized ΔH.

enthalpy of formation (14.4) The enthalpy change in the formation of a substance from its elements in their standard states; symbolized ΔH_f.

enthalpy of fusion See *heat of fusion.*

enthalpy of vaporization See *heat of vaporization.*

enzyme (17.1) A biochemical catalyst.

equation (8.1) A written representation of a chemical reaction, using formulas for the reactants and products and coefficients to indicate the mole ratios involved.

equilibrium (17.2) The situation in which the reactants and the products are producing each other *at the same rate.*

equilibrium constant (17.4) A constant that tells how far a reaction will proceed until it reaches equilibrium.

equilibrium constant expression (17.4) The ratio of the product of the concentrations of the products divided by the product of the concentrations of the reactants, each raised to the power corresponding to its coefficient in the balanced equation: for $a\,A + b\,B \rightleftharpoons c\,C + d\,D$,

$$K = \frac{[C]^c[D]^d}{[A]^a[B]^b}$$

equivalent (16.5) The quantity of a substance that will react with or produce 1 mole of electrons in an oxidation-reduction reaction *or* 1 mole of hydrogen ions or hydroxide ions in an acid-base reaction.

equivalent mass (16.5) The mass in grams of 1 equivalent of a substance.

Erlenmeyer flask (11.3) A flask designed to allow swirling of the liquid contents without spillage.

ester (19.3) An organic compound with the —COO—R functional group.

ether (19.3) An organic compound with the —O— functional group.

ethylene (19.1) Ethene, $H_2C{=}CH_2$.

evaporation (14.2) A change of phase from liquid to gaseous.

event (20.1) A nuclear reaction involving one nucleus.

excess (10.4) The quantity of a reactant that exceeds that which can react with the limiting quantity of another reactant.

exothermic process (14.3) A process in which energy is transferred to the surroundings.

exponent (2.4) The number of times the base is multiplied by the coefficient; for example, in 1.5×10^3, the 1.5 is multiplied by 10 three times, where 3 is the exponent.

exponential notation (2.4) The format for writing large or small numbers that consists of the coefficient times a base raised to a power; for example, 1.5×10^3.

exponential part (2.4) The base raised to a power in an exponential number.

extensive property (1.2) A characteristic that depends on the quantity of the sample.

factor label method (2.1) A system that involves the use of units to indicate the proper arithmetic operation to perform; also called dimensional analysis.

Fahrenheit scale (2.6) A temperature scale in common use in the United States on which the freezing point of water is defined as 32°F and the normal boiling point of water is defined as 212°F.

family (1.5) In the periodic table, a column that includes elements with similar chemical properties; a periodic group.

fat (19.5) A compound formed by the reaction of glycerine with long-chain fatty acids.

feeble acid (18.1) An acid that has no tendency to react with water; the conjugate of a strong base.

feeble base (18.1) A base that has no tendency to react with water; the conjugate of a strong acid.

fluidity (14.1) The characteristic of a gas or liquid that allows it to flow.

force (12.1) A push or a pull.

formaldehyde (19.3) Methanal, HCHO.

formula (1.4, 5.1) A combination of symbols and subscripts that identifies the composition of an element, compound, or ion.

formula mass (7.1) The relative mass of one formula unit compared to the mass of a ^{12}C atom, which is defined exactly as 12 amu.

formula unit (5.1) The collection of atoms described by a chemical formula—an atom of an uncombined element, a molecule of a molecular compound, or the set of ions in the formula of an ionic compound.

formula weight (7.1) Formula mass.

freezing (14.2) Changing from a liquid to a solid.

freezing-point depression (15.5) A lowering of the freezing point of a solvent due to the presence of a solute.

functional group (19.3) The characteristic group of atoms attached to a radical that gives a class of organic compounds its characteristic properties.

fusion (14.2) Melting, or changing from a solid to a liquid; see also *nuclear fusion.*

gamma particle (20.1) A particle (photon) of high-energy electromagnetic radiation (light) emitted from a nucleus in a nuclear reaction; a stream of such particles is often referred to as a gamma ray.

gas (12.1) A state of matter; a sample of matter that has its volume and shape determined by the volume and shape of its container.

Gay-Lussac's law of combining volumes (12.6) The pressure of a given sample of gas at constant volume is directly proportional to its absolute temperature.

Geiger counter (20.1) An instrument for measuring radioactivity.

glycerine (19.5) The trialcohol with the formula $CH_2OHCHOHCH_2OH$.

gram (2.2) The basic unit of mass in the metric system; one-thousandth of the SI standard mass—the kilogram.

ground state (4.2) The lowest energy state of the set of electrons in an atom.

group (1.5) In the periodic table, a column that includes elements with similar chemical properties; a family.

halate (6.2) A chlorate, bromate, or iodate.

half-life (20.2) The period of time it takes for half of a radioactive sample to disintegrate naturally.

half-reaction (16.4) The oxidation or reduction half of a redox reaction.

half-reaction method (16.4) The method of balancing oxidation-reduction equations that involves the completion and balancing of the oxidation and reduction half-reactions separately, followed by the combining of the two.

halogen (1.5) An element of periodic group VIIA—F, Cl, Br, I, or At.

heat (2.6, 14.4) A form of energy.

heat capacity See *specific heat.*

heat of formation See *enthalpy of formation.*

heat of fusion (14.3) The heat required to change a given quantity of solid to the liquid state.

heat of vaporization (14.3) The heat required to change a given quantity of liquid to the gas state.

heating curve (14.3) A graph showing the temperature changes of a substance as a function of the quantity of heat added.

Heisenberg uncertainty principle (4.5) It is impossible to know exactly both the energy and the momentum of a subatomic particle at the same time.

Hess's law (14.4) When two or more processes combine to give a resulting process, their enthalpy changes add up to give the enthalpy change for the resulting process.

heterogeneous mixture (1.1) A physical combination of substances having distinguishable parts.

homogeneous mixture (1.1) A physical combination of substances whose parts are not distinguishable, even with the best optical microscope; a solution.

Hund's rule (4.6) The electrons in a partially filled subshell in an atom will occupy the orbitals singly as far as possible.

hydrate (5.1) A compound that has water molecules bonded in it.

hydrocarbon (19.1) A compound containing only carbon and hydrogen.

hydrogen (1.5) The first element on the periodic table.

hydrogen (6.3) Word used in names of acid salts to indicate the hydrogen remaining, as in $NaHCO_3$, sodium hydrogen carbonate.

hydrogen bomb (20.4) A thermonuclear device in which the nuclear fusion of a hydrogen isotope is the source of energy.

hydrogen bonding (13.6) The intermolecular force resulting from the attraction of a hydrogen atom on one molecule to a small, highly electronegative atom (F, O, or N) on another molecule (or the same molecule).

hydronium ion (18.1) The H_3O^+ ion.

hypothesis (1.6) A proposed explanation for a body of observed facts.

ideal gas law (12.5) The pressure, volume, number of moles, and temperature of a sample of gas can be related by the equation $PV = nRT$, where $R = 0.0821$ L · atm/mol · K.

ideal solution (15.5) A solution that obeys Raoult's law exactly.

indicator (8.4) A compound that has different colors in solutions of different acidities and that is used to signal the end of a titration.

inner transition element (1.5) An element with atomic number 58–71 or 90–103; a member of the lanthanide or actinide series.

inorganic chemistry (Chapter 1, Introduction) The branch of chemistry dealing with compounds other than those containing C—C and/or C—H bonds.

intensive property (1.2) A characteristic such as color that does not depend on the quantity of material present.

intermolecular force (13.6) An attraction between molecules.

inverse proportionality (12.2) A relationship in which one variable gets smaller by the same factor as another gets larger, or vice versa.

ion (5.2) A charged atom or group of atoms.

ionic bond (5.2) The attractive force between oppositely charged ions.

ionic equation (9.2) A chemical equation in which soluble, ionic substances are written with their ions separated.

ionic size (13.1) The size of an ion.

ionic solid (14.1) A solid consisting of ions.

ionizable hydrogen atom (6.3) Any of the hydrogen atoms in an acid that are capable of reacting with water to form ions.

ionization (18.2) Reaction of a covalent substance with solvent to produce ions, as, for example, the reaction of a strong acid with water.

ionization constant (18.2) The equilibrium constant for the reaction of a weak acid or base with water to form ions in solution; symbolized K_i.

ionization energy (13.2) The energy required to remove an electron from a gaseous atom to form a gaseous cation.

isomers (19.2) Different compounds having the same molecular formula.

isotope (3.3, 20.1) A form of an element whose atoms all have the same number of protons and the same number of neutrons.

joule (2.6) The SI unit of energy; it takes 4.184 J to raise the temperature of 1.000 g of water 1.000°C.

kelvin (12.3) The unit of the Kelvin temperature scale; abbreviated K.

Kelvin scale (2.6, 12.3) The temperature scale with 273 as the freezing point of water and 373 as the normal boiling point of water; the scale required for gas law and certain other scientific calculations.

ketone (19.3) A class of organic compounds with the

$$\text{C}{=}\text{O}$$ functional group not at the end of a carbon chain.

kilogram (2.2) 1000 grams; the standard mass in SI.

kinetic energy (12.8) Energy of motion; KE $= \frac{1}{2}mv^2$.

kinetic molecular theory (12.8) The theory that explains the gas laws (and other phenomena) in terms of the motions and characteristics of the molecules of a gas.

law (1.6) A generalized statement that summarizes a collection of observations.

law of conservation of energy (1.3) Energy can be neither created nor destroyed in any chemical or physical process.

law of conservation of mass (1.6) Mass can be neither created nor destroyed in any chemical or physical process.

law of constant composition (3.1) The composition of a compound is fixed; also called law of definite proportions.

law of definite proportions (3.1) The composition of a compound is fixed; also called law of constant composition.

law of multiple proportions (3.1) When two or more elements combine to form more than one compound, for a fixed mass of one element, the masses of the other element(s) in the compounds are in a small, whole-number ratio.

LeChâtelier's principle (17.3) When a stress is applied to a system at equilibrium, the equilibrium shifts so as to tend to reduce the stress.

length (2.2) The distance from one end of an object to the other.

light absorption (4.1) Process in which the energy of certain wavelengths of light is converted to energy of electrons in atoms, increasing their energy.

light emission (4.1) Process in which light of specific wavelengths is produced as electrons in atoms fall to lower energy levels.

limiting quantity (10.4) The quantity of the reactant that will be exhausted first in a chemical reaction, limiting the quantity of products that can be produced.

line formula (19.2) A formula for an organic compound that is written on a single line, such as CH_3CH_2COOH.

linear molecule (13.4) A molecule whose atoms all lie on a line.

liquid (12.1) A state of matter; a sample of matter that has a definite volume but assumes the shape of its container.

liter (2.2) The basic unit of volume of the metric system, equal to 1 dm^3.

lobe (4.5) One portion of an atomic orbital.

lone pair (5.5) An unshared pair of electrons on a bonded atom.

macromolecule (5.5) A molecule containing hundreds of thousands to millions of covalently bonded atoms.

magnetic properties (4.6) The characteristics of substances that make them attracted to or repelled by magnetic fields.

magnetic quantum number (4.2) The quantum number (m) that designates the orientation in space of the volume occupied by an electron in an atom.

main group element (1.5) An element of one of the periodic groups labeled A, which extend up to the first or second period of the periodic table.

malleable (1.2) Capable of being pounded into various shapes.

mass (1.3) A fundamental characteristic of a sample of matter that quantifies its attraction to the earth (weight) and its resistance to a change in its motion (inertia).

mass number (3.3) The sum of the number of protons and the number of neutrons in an atom; the distinguishing difference among isotopes of a given element.

matter (1.3) Anything that has mass and occupies space.

melting (14.2) A change of phase from solid to liquid.

metal (1.5) An element on the left in the periodic table, or a mixture of such elements.

metallic solid (14.1) A solid consisting of atoms of one or more elemental metals.

metalloid (1.5) An element near the dividing line between metals and nonmetals on the periodic table and that has properties of both metals and nonmetals.

metathesis reaction (8.3) A double displacement reaction.

meter (2.2) The basic unit of length in the metric system and SI.

metric system (2.2) A system of measurement whose units have subunits and multiples that are based on powers of 10.

metric ton (10.2) 1 000 000 grams.

mixture (1.1) A physical combination of substances that has an arbitrary composition and properties characteristic of its components.

molal (15.3) The unit of molality, abbreviated m.

molality (15.3) A measure of concentration defined as the number of moles of solute per kilogram of solvent; symbolized m.

molar (11.1) The unit of molarity; abbreviated M.

molar mass (7.3) The mass in grams of 1 mol of a substance.

molarity (11.1) A measure of concentration defined as the number of moles of solute per liter of solution; symbolized M.

mole (7.3) The chemical unit of quantity for any substance; equal to 6.02×10^{23} individual atoms, molecules, or formula units of the substance; abbreviated mol.

mole fraction (15.4) The ratio of the number of moles of a component (A) of a solution to the total number of moles in the solution; symbolized X_A.

molecular formula (7.5) The formula of a molecular substance that gives the ratio of atoms of each element to the substance's molecules.

molecular mass (7.1) The relative mass of a molecule of a substance compared to the mass of a ^{12}C atom.

molecular shape (13.4) The spatial arrangement of the atoms in a molecule.

molecular solid (14.1) A solid consisting of (relatively small) molecules.

molecular weight (7.1) Molecular mass.

molecule (3.2, 5.5) An uncharged, covalently bonded group of atoms.

molten (8.3) Melted; in a liquid state without being dissolved.

monatomic ion (5.2) An ion consisting of one atom only.

monomer (19.4) A molecule that is capable of reacting with other similar molecules to form a polymer.

monosaccharide (19.5) A simple sugar, such as glucose, $C_6H_{12}O_6$.

multiple bond (5.5) A double or triple bond.

$n + \ell$ rule (4.3) The electron with the lower value for the sum of quantum numbers n and ℓ is lower in energy.

net ionic equation (9.2) An ionic equation in which the ions that do not change (the spectator ions) are omitted.

network solid (14.1) A solid consisting of macromolecules.

neutral (3.3) Neither positively nor negatively charged.

neutral (8.4) Neither acidic nor basic.

neutralization reaction (8.4) A reaction of an acid and a base.

neutron (3.3) A subatomic particle that has no charge and a mass slightly greater than 1 amu.

noble gas (1.5) An element of periodic group 0—He, Ne, Ar, Kr, Xe, or Rn.

noble gas configuration (5.2) An electronic configuration like that of a noble gas, with eight electrons (or two for very light elements) in the outermost shell.

nomenclature (Chapter 6, Introduction) The systematic naming of chemical substances.

non-octet structure (5.5) Any electronic structure in which an atom contains more or fewer than eight electrons in its valence shell.

nonlinear molecule (13.4) See *angular molecule*.

nonmetal (1.5) Hydrogen or any element on the right in the periodic table.

nonpolar bond (13.3) A covalent bond in which the electrons are shared equally.

nonpolar molecule (13.5) A molecule in which the centers of positive and negative charge lie at the same place.

nonvolatile (15.5) Not easily vaporized.

normal (16.5) The unit of normality; symbolized N.

normal boiling point (14.2) The boiling point of a substance at a pressure of 1 atm.

normality (16.5) A measure of concentration equal to the number of equivalents of solute per liter of solution; symbolized N.

nuclear fission (20.3) The splitting of a nucleus into two more or less equally sized smaller nuclei, plus some subatomic particles.

nuclear fusion (20.4) Combination of nuclei in a nuclear reaction.

nuclear radiation (20.1) Alpha, beta, and/or gamma particles; particles emitted in nuclear reactions.

nucleus (3.3) The center of an atom, consisting of the protons and neutrons.

octet (5.2) A set of eight electrons in the outermost shell of an atom or ion.

octet rule (5.2) Atoms or ions with an octet are stable.

orbital (4.4) A part of a subshell of an atom having a given set of n, ℓ, and m quantum numbers.

orbital shape (4.5) The shape of the volume of space that can be occupied by an electron in an atom.

organic acid (19.3) An organic compound containing the —COOH functional group.

organic chemistry (Chapter 1, Introduction) The branch of chemistry dealing with organic compounds.

organic compound (Chapter 19, Introduction) A compound containing at least one carbon-carbon or carbon-hydrogen bond.

organic halide (19.3) An organic compound that has at least one halogen atom per molecule.

osmotic pressure (15.5) The difference in pressure between a solution and the pure solvent separated by a semipermeable membrane.

outermost shell (5.2) The shell of highest principal quantum number that contains electrons.

overall equation See *total equation*.

oxidation (16.4) An increase in oxidation number.

oxidation number (16.1) The difference in the number of electrons in a neutral atom and the number "controlled" by the atom in a compound, where control is assigned to the more electronegative atom of a covalent bond.

oxidation-reduction reaction (16.4) A reaction in which the oxidation number of one element is raised and the oxidation number of another element (or the same element) is lowered.

oxidation state (16.1) The oxidation number.

oxidizing agent (16.4) A species that can increase the oxidation number of another reactant.

oxoacid (6.3) An acid containing oxygen.

oxoanion (6.2) An anion containing oxygen covalently bonded to another element.

ozone (5.5) O_3, an allotrope of oxygen.

parent isotope (20.1) The isotope that is the reactant in natural radioactive decay.

partial pressure (12.7) The pressure of one gas in a mixture of gases.

Pauli exclusion principle (4.2) The rule that prohibits more than one electron in an atom from having the same set of four quantum numbers.

percent (2.1) Parts per hundred parts.

percent composition (7.2) The percentages by mass of all elements in a compound.

percent yield (10.5) The ratio of actual yield to theoretical yield, expressed as a percentage.

perfectly elastic collision (12.8) A collision that occurs with no loss of kinetic energy.

period (1.5) One of the seven horizontal rows of the periodic table.

periodic table (1.5) An assemblage of elements in order of atomic number, with elements having similar chemical properties aligned in vertical columns.

periodicity of electronic configuration (4.7) The regularity of the change in the outermost electronic configuration of the elements in the groups of the periodic table.

peroxide ion (6.2) The O_2^{2-} ion.

pH (18.3) A measure of the acidity of a solution, defined as $-\log [H_3O^+]$.

phase (12.1) A state of matter—solid, liquid, or gas.

phase change (14.3) A transition from one to another of the three states of matter—for example, from gas to liquid.

physical change (1.1) A process in which no change in composition occurs.

physical chemistry (Chapter 1, Introduction) The branch of chemistry dealing with the properties of substances.

physical equilibrium (14.2) A situation in which two exactly opposite physical processes are occurring at the same rate in the same system.

pipet (11.3) A piece of laboratory glassware designed to deliver an exact volume of liquid.

polar bond (13.3) A covalent bond in which there is unequal sharing of electrons.

polar molecule (13.5) A molecule that has a permanent dipole.

polyatomic ion (5.5) An ion composed of two or more atoms.

polymer (19.4) A molecule built from many (thousands or even more) smaller molecules (monomers) or parts of molecules.

polysaccharide (19.5) A complex carbohydrate composed of more than one monosaccharide.

positron (20.3) A subatomic particle created in a nuclear reaction that has the same properties as an electron, except for being *positively* charged.

postulate (3.2) A proposed explanation for an observation or set of observations.

precipitate (8.3) A solid formed from substances in solution.

precision (2.3) The reproducibility of measurements.

prefix (6.1) A word fragment placed before another part of a word to impart a special meaning.

pressure (12.1) Force divided by area.

principal quantum number (4.2) The main quantum number (*n*) of an electron in an atom, which is the major factor determining the energy of the electron and its mean distance from the nucleus.

proceed to the left (17.2) A phrase used to indicate that the products of a chemical equation (as written) will yield the reactants.

proceed to the right (17.2) A phrase used to indicate that the reactants of a chemical equation (as written) will yield the products.

product (8.1) Any substance produced in a reaction and appearing on the right-hand side of a chemical equation.

property (1.2) A characteristic of a substance.

propylene (19.1) Propene, $CH_2{=}CHCH_3$.

protein (19.4) A polymer formed from amino acids.

proton (3.3) A subatomic particle with a mass slightly greater than 1 amu and a charge of $1+$.

proton (18.1) In the Brønsted sense, the nucleus of the hydrogen atom.

proton acceptor (18.1) A molecule or ion that accepts a proton; a Brønsted base.

proton donor (18.1) A molecule or ion that donates a proton; a Brønsted acid.

pure substance (1.1) An element or compound.

quantitative property (1.2) A property that is measurable.

quantum numbers (4.2) Four numbers assigned to each electron in an atom that describe the energy and other properties of the electron.

radiation See *electromagnetic radiation* or *nuclear radiation*.

radical (19.3) The hydrocarbon portion of an organic molecule that has a functional group.

radioactive dating (20.2) The determination of the age of an object from the measurement of the radioactivity of an isotope it contains.

radioactive decay (20.1) Disintegration of nuclei of an isotope.

radioactive series (20.1) A parent isotope and the successive daughters of its natural decay.

radioactivity (20.1) Spontaneous reaction of nuclei.

random motion (12.8) Motion of molecules in arbitrary directions.

Raoult's law (15.5) The vapor pressure of a solute in a solution is equal to its mole fraction in the solution times the vapor pressure of the pure solute: $P_A = X_A P_A^\circ$.

rate (17.1) The number of moles per liter produced or used up in a chemical reaction per unit time.

reactant (8.1) Any substance that undergoes a reaction and thus appears on the left-hand side of a chemical equation.

reacting ratio (10.1) A ratio of coefficients from a balanced equation, which represents the ratio of moles of reactants and/or products involved in the reaction.

reaction See *chemical reaction*.

reactive (8.3) Having a high tendency to undergo chemical reaction.

reagent (8.1) A reactant.

redox reaction (16.4) An abbreviation for *oxidation-reduction reaction*.

reducing agent (16.4) A species that reduces the oxidation number of another reactant.

reduction (16.4) The lowering of oxidation number.

relative scale (3.4) A scale based on an arbitrarily chosen standard; the atomic mass scale is a relative scale of masses based on the mass of ^{12}C.

rounding (2.3) Reducing the number of digits in a calculated result to the proper number of significant digits.

salt (6.3, 8.4) An ionic compound that does not contain H^+, OH^-, or O^{2-} ions.

saturated fat (19.5) A fat that contains no carbon-carbon multiple bonds.

saturated hydrocarbon (19.1) A compound containing only carbon and hydrogen linked only by single bonds.

saturated solution (15.2) A solution that holds as much solute as it is capable of holding stably at a given temperature.

scientific notation (2.4) A format for writing large and small numbers, using a coefficient with one (nonzero) integer digit times 10 to an integral exponent.

second (2.6) The basic unit of time in SI.

second ionization energy (13.2) The energy required to remove an electron from a gaseous, monopositive ion to form a gaseous, dipositive ion.

shell (4.1) An energy level for electrons in an atom that is characterized by a given value of the principal quantum number, n.

shift (17.3) A net reaction of a system at equilibrium in response to a stress—a change in the conditions on the system.

SI (2.2) Système International d'Unités; the modern form of the metric system.

significant digit (2.3) Any digit that reflects the accuracy with which a measurement was made.

significant figure (2.3) A significant digit.

single bond (5.5) A covalent bond formed by a single pair of shared electrons.

soap (19.5) A salt of a fatty acid, such as sodium stearate, $Na^+C_{17}H_{35}COO^-$.

sodium chloride structure (5.2) The three-dimensional lattice containing alternating positive and negative ions that is characteristic of sodium chloride and many other ionic compounds.

solid (12.1) A state of matter; a sample of matter that has a definite shape and volume.

solubility (8.3, 15.2) The concentration of a saturated solution at a given temperature.

solute (11.1) The component of a solution that is dissolved in another component—the solvent.

solution (1.1) A homogeneous mixture.

solvent (11.1) The component of a solution that does the dissolving.

specific heat (14.3) The quantity of heat required to raise the temperature of 1 g of a substance by 1°C; the specific heat capacity, symbolized c.

spectator ion (9.2) An ion that is present but does not change during a chemical reaction.

spin quantum number (4.2) The quantum number (s) that determines electron pairings.

stability (8.3) Resistance to reaction; relative lack of reactivity.

standard (2.2) A basis for comparison, such as the mass of ^{12}C.

standard atmosphere (12.1) A pressure of 760 torr.

standard exponential form (2.4) A format in which a number is expressed as a coefficient with one (nonzero) integer digit times 10 to an integral exponent.

standard state (14.4) The normal state for a substance of 1 atm and the temperature involved; for example, the standard state for oxygen at 25°C is gaseous O_2 molecules.

standard temperature and pressure (12.4) 0°C and 1 atm, abbreviated STP; used in gas law problems.

state (8.2, 12.1) A phase in which matter exists—solid, liquid, or gas.

state function (14.4) A variable that depends only on the initial and final states and not on the path between them.

state of a system (12.2) The condition of a system.

state of subdivision (17.1) The particle size of a sample of a solid.

Stock system (6.2, 16.2) The nomenclature system for inorganic compounds in which the oxidation state (or charge for a monatomic cation) is represented as a Roman numeral in the name of the compound.

stoichiometry (10.1) The determination of how much a reactant can produce or how much of a product can be produced from a given quantity of another substance in a reaction.

stress (17.3) Any change in conditions affecting a system at equilibrium.

strong acid (8.4) An acid that reacts completely with water to form ions in solution.

strong base (8.4) A base that is fully ionized in water solution.

structural formula (5.5) A formula in which lines represent electron pairs shared by covalently bonded atoms.

subatomic particle (3.3) A proton, neutron, electron, deuteron, positron, alpha particle, etc.

sublimation (13.6) A phase change in which a solid goes directly into the gas phase.

subscript (5.1) A number following the symbol of an element (or a closing parenthesis) that denotes the number of atoms of the element (or the number of groups) in the formula unit.

subshell (4.4) The portion of a shell characterized by the same principal quantum number and the same angular momentum quantum number.

substance See *pure substance*.

substitution reaction (8.3) A reaction in which an uncombined element reacts with a compound to produce a new compound and a different uncombined element.

sugar (19.5) A carbohydrate consisting of a monosaccharide or a combination of them—for example, sucrose.

superoxide ion (16.1) The O_2^- ion.

supersaturated solution (15.2) A solution holding more solute than it can hold stably at a given temperature.

surroundings (12.2) Anything outside of the system under investigation.

symbol (1.4) A one- or two-letter representation of an element.

synthetic fiber (19.4) A polymer, such as nylon, that has the form of a fiber.

system (12.2) The portion of the universe under investigation.

Système International d'Unités See *SI*.

temperature (2.6) The intensity of heat in a body.

ternary compound (8.3) A compound consisting of three elements.

tetrahedral (13.4) A molecule with electrons oriented toward the corners of a tetrahedron (a solid object with four sides, all of which are identical equilateral triangles).

theoretical yield (10.5) The calculated quantity of product that would result from a chemical reaction based on the laws of stoichiometry.

theory (1.6) A generally accepted explanation for a law (or series of observations).

third ionization energy (13.2) The energy required to remove an electron from a gaseous, dipositive ion to form a gaseous, tripositive ion.

titration (11.3) An experimental technique used to determine the concentration of a solution of unknown concentration or the number of moles in an unknown sample of a substance.

torr (12.1) A unit of pressure equal to 1 mm Hg or $\frac{1}{760}$ atm.

total bond order (19.3) The number of electron pairs shared by an atom in a molecule or ion.

total equation (9.2) The equation representing the complete compounds undergoing reaction, as opposed to a net ionic equation.

tracer (20.1) A radioactive isotope of an element used to determine what happens to the element, often in some biochemical system.

transition element (1.5) Any element in the groups that start in the fourth period of the periodic table, having atomic numbers 21–30, 39–48, 57, 72–80, or 104 or higher.

transmutation (20.3) The conversion of one element into another by a nuclear reaction.

trigonal planar molecule (13.4) A molecule with atoms oriented toward the corners of an equilateral triangle, with the central atom in the same plane.

trigonal pyramidal molecule (13.4) A molecule with atoms oriented toward the corners of an equilateral triangle, with the central atom out of that plane.

triple bond (5.5) A covalent bond consisting of three pairs of electrons shared between two atoms.

tritium (20.4) The isotope ^{3}H.

uncertainty principle See *Heisenberg uncertainty principle.*

unit (2.1) A standard division of measure having a certain value; for example, the meter is the basic metric unit of length.

unsaturated fat (19.5) A fat that contains one or more carbon-carbon multiple bonds per molecule.

unsaturated hydrocarbon (19.1) A compound containing only carbon and hydrogen and having one or more multiple bonds per molecule.

unsaturated solution (15.2) A solution that contains less solute than it could hold stably at a given temperature.

unshared pair (5.5) A pair of electrons in a molecule or ion that is not shared between atoms.

urea (Chapter 19, Introduction) NH_2CONH_2 (an organic compound despite its lack of carbon-carbon or carbon-hydrogen bonds).

valence electron (5.2) An electron in or from the outermost electron-containing shell of an uncombined atom.

valence shell (5.2) The outermost shell containing electrons in an uncombined atom, or that same shell even when the atom is combined in a compound.

van der Waals force (13.6) An intermolecular force resulting from instantaneous dissymmetry of charge in otherwise nonpolar molecules.

vapor (14.2) A gas in contact with its liquid (or solid) phase.

vapor pressure (12.7, 14.2) The pressure of the vapor over a liquid in equilibrium with its vapor.

vapor-pressure lowering (15.5) A decrease in the vapor pressure of a solvent due to the presence of a solute.

vaporization (14.2) A phase change from liquid to gas (vapor).

volatile (15.5) Easily vaporized.

volume (2.2) The extent of space occupied by a sample of matter.

volume ratio (12.6) The ratio of volumes of gases involved in a chemical reaction.

water ionization constant (18.3) The equilibrium constant for the autoionization of water; symbolized K_w; at 25°C, $K_w = [H_3O^+][OH^-] = 1.0 \times 10^{-14}$.

weak acid (8.4) An acid that reacts only partially with water to form ions.

weak base (8.4) A base that reacts only partially with water to form ions.

weighted average (3.4) The average value of several types of items, taking into account the number of individual items of each type.

work (14.4) All forms of energy except heat.

Photo Credits

Fundamental Photographs; 15.4: © The McGraw-Hill Companies, Inc./Bob Coyle, photographer

Chapter 16

Opener: © Richard Megna/ Fundamental Photographs

Chapter 17

Opener: © Richard Magna/ Fundamental Photographs

Chapter 18

Opener, 18.1: © The McGraw-Hill Companies, Inc./Terry Wild Studio; 18.2: © The McGraw-Hill Companies, Inc./Bob Coyle, photographer

Chapter 19

Opener: Courtesy of Molecular Model Co., P.O. Box 250, Edgerton, WI 53534; 19.1,

19.3-19.5: © The McGraw-Hill Companies, Inc./Terry Wild Studio

Chapter 20

Opener: © Paolo Koch, Science Source/Photo Researchers, Inc.; 20.4: Portraits by Florence

Index

Pages on which key terms are introduced are printed in **bold**. The letter f with a page number indicates a figure, and the letter t a table.

Abbreviations, for units, 478t
Absolute temperature, **258,** 260, 276
Absolute zero, **259**
Absorption of light, **79**
Acceleration, 462
Accuracy, of scientific measurement, **36**–37
Acetic acid, 181, 198, 390, 393, 398, 422
Acetone, 421
Acetylene, 158f, 411
Acid(s), **136, 180.** *See also* Acid-base theory; Acid salts; Strong acids; Weak acids
  naming of, 137–38, 141
  organic acids, **422,** 423, 429
  properties of, 181–83, 187
Acid-base reaction, 359, 360
Acid-base theory
  answers to practice problems, 492–93
  autoionization of water, 394–98, 402
  Brønsted theory, 388–91, 402
  buffer solutions, 398–401, 402
  ionization constants, 392–94, 402
Acid-base titration, 242–46
Acid carbonates, **185,** 186–87
Acid dissociation constant, **392**
Acidic anhydrides, **184**
Acid rain, 184
Acid salts, **136**
  naming of, 140, 141
  properties of, 184–85, 187
Acid strength, **390**
Actinide series, 13, 353
Activity, **174**
Addition, of exponential numbers, 49
Addition polymerization, 425
Air pollution, 450

Alcohols, **419,** 420, 429
Aldehydes, **421,** 429
Algebra, scientific, **49,** 457–64
Alkali metals, 12, 92, 93, 123, 348
Alkaline earth metals, 12, 92, 93, 123, 348
Alkanes, **408,** 409, 429
Alkenes, **409**–11, 429
Alkynes, **411,** 412, 429
Allotropes, 114
Alpha particles, **435,** 437, 440, 453
Alum, 131
Aluminum, 51t, 71, 94, 110, 113, 206–7, 214, 218–19
Amides, **423,** 424, 429
Amines, **423,** 424, 429
Amino acids, 425
Ammonia, 101, **129,** 183, 290, 350, 370–71, 372–73, 389
Ammonium, 118, 131
Ammonium carbonate, 267
Ammonium chloride, 391
Ammonium fluoride, 391
Ammonium hydroxide, 177
Ammonium phosphate, 213–14
Amorphous solids, **304**
Amphiproticity, **389**
Analytical chemistry, 2
Angular molecule, **291**
Angular momentum quantum number, **81,** 95
Anhydrides, **184,** 187
Anhydrous compound, **104**
Anions, **107,** 108
  naming of, 134–35
  net ionic equations, 223
Anode, **107**
Anthracene, 337
Antifreeze, 336, 338, 339, 420
Antilogarithms, 471–72

Aqueous solution, 194–95, 196, 209, 238, 246, 390, 391, 398
Archaeology, 446, 447–48, 453
Argon, 85–86, 447
Aromatic hydrocarbons, **412,** 413, 429
Arrhenius theory, **180**
Arsenic, 71
Arsine, 129
Aspirin, 158, 399
Atmosphere (atm), **254**
Atmospheric pressure, **254**
Atom(s). *See also* Nuclear reactions
  answers to practice problems, 484, 490
  Bohr theory, 79–81, 95
  Dalton's atomic theory, 67, 73
  electronegativity and bond polarity, 288–89, 299
  energy level diagrams, 89–92, 95
  history of theory of, 64
  ionization energy and electron affinity, 286–88, 299
  laws of chemical combination, 64–67
  periodic variation of electronic configuration, 92–95
  quantum numbers, 81–83, 95
  relative energies of electrons, 83–86, 95
  shells, subshells, and orbitals, 86–88, 89, 95
  sizes of, 283–86, 299
  subatomic particles, 68–69
Atomic bomb, **449**
Atomic mass, 70–71, 73, 160, 484
Atomic mass scale, **70**
Atomic mass units, **70,** 151
Atomic number, **68,** 73, 286f, 436, 437
Atomic weight, **70.** *See also* Atomic mass

Atomic weight scale, **70**
Atom smasher, **448**
Autoionization, of water, 394–98
Auto-oxidation-reduction, 358
Average kinetic energy, **274**
Avogadro's number, **151,** 152, 160

Balanced equations, **166**–71, 187
   limiting quantity, 241
   mole calculations, 205, 246
   volume ratios, 267–68
Barium bromide, 169
Barium carbonate, 168–69
Barium chloride, 168–69
Barium hydroxide, 169
Barium iodide, 166, 167f
Barium sulfate, 135, 178
Barium sulfite, 135
Barometer, **252,** 253f, 276
Barometric pressure, **254,** 271
Base, in exponential number, **43**
Base(s), **136, 180,** 181–83, 187. *See also* Acid-base theory; Acid(s); Strong base; Weak base
Base dissociation constant, **392**
Base strength, **391**
Basic anhydrides, **184**
Bent molecule, **291**
Benzene, 158f, 336, 337, **412,** 429
Beryllium, 290
Beta particles, **435,** 437, 440, 453
Binary compounds, **101,** 121, 122, 129–31, 141
Binary nonmetal-nonmetal compounds, 129–31, 141
Biochemistry, 2
Biological systems, and osmotic pressure, 340
Bohr, Niels, 79, 95
Bohr theory, 79–81, 95
Boiling point, **310,** 322, 338–39, 341
Boiling-point elevation, **338,** 339, 341
Bond(s) and bonding
   covalent, **113,** 121, 122, 299, 346–47
   double, **115,** 123
   ionic, **107,** 122
   polar, **289,** 299
   single, **113,** 123
   triple, **115,** 123
Bond polarity, 288–89
Boric acid, 182, 390, 391
Boron, 121
Boyle, Robert, 254
Boyle's law, 254–58, 271, 275–76
Bromide, 108
Bromine, 106f, 130, 358

Bromobenzene, 412
Brønsted, J., 388
Brønsted acid, **388**
Brønsted base, **388**
Brønsted theory, **388**–91, 402
Buffer solutions, 398–401, 402
Buildup principle, **85**
Buret, **242,** 243
Butane, 408, 411, 414f

Calcium, 132, 172, 287
Calcium carbonate, 186f
Calculators, electronic. *See also* Scientific calculations
   exponential notation, 45
   mathematics and, 464–72
   radioactive disintegration time, 443–44
   significant digits, 39–41
Calorie, **55**
Cancer, 445
Capitalization, in chemical formulas, 11
Carbohydrates, **427,** 428, 429
Carbon
   allotropes of, 114
   electron dot diagrams, 117, 118
   energy level diagrams, 91
   macromolecules, 116f
   oxidation number, 351, 354
   radioactive dating and carbon-14, 447–48
   60-carbon molecule, 306
   tetrahedron structure, 413
Carbonates, **185**–87
Carbon dioxide
   carbonates and acid carbonates, 185, 186
   combination reaction, 172
   combustion reaction, 179
   enthalpy of formation, 320–22
   equilibrium and, 377, 378
   molecular shape, 291
   pure sample of, 65
   unbalanced equation, 168
Carbon disulfide, 6–8
Carbon-14 dating, 447–48
Carbonic acid, 177
Carbon monoxide
   chemical reactions, 168, 172, 173
   combustion reactions, 179
   equilibrium state, 374, 377
   Hess's law, 320–22
   nomenclature, 130, 351
   pure sample of, 65
   safety and, 409
Carbon tetrachloride, 419

Catalyst, **174, 369,** 381
Cathode, **107**
Cations, **107**–9, 131–33. *See also* Polyatomic ions
Caves, formation of, 186f, 369
Cellulose, 429
Celsius temperature scale, **54, 258,** 260, 276
Centered dot, **104**
Centi-, as metric prefix, **28**
Centigrade scale, **54**
Central atom, **290**
Chain reactions, 449–50, 453
Change sign key, on scientific calculator, 468–69
Changes in concentration, 378
Charge(s). *See* Electrons; Ions
Charles, J. A. C., 258
Charles' law, 258–61, 276
Chemical bonding
   answers to practice problems, 484
   chemical formulas, 101–5, 121
   covalent bonding, 113–21
   electron dot diagrams, **112,** 113, 122
   formulas for ionic compounds, 109–12, 122
   ionic bonding, 105–9, 122
Chemical change, **2**
Chemical combination, law of, 64–67
Chemical equation, 166–67. *See also* Balanced equation
Chemical formulas, **11, 101**–5, 109–12, 121, 122
   answers to practice problems, 485–86
   empirical formulas, 155–58, 160
   formula masses, 149–50, 160
   mole, 151–55, 160
   molecular formulas, 158–59, 160
   naming of ionic compounds and writing of, 135–36, 141
   percent composition, 150–51, 160
Chemical reactions, **2.** *See also* Molarity; Reaction rates; Stoichiometry
   acids and bases, 180–87
   answers to practice problems, 486
   balanced equations, 167–71, 187
   chemical equation, 166–67, 187
   enthalpy changes, 316–22
   gases in, 265–70
   mass calculations for, 207–10
   mole calculations for, 205–7
   nuclear reactions and, 434
   predicting products of, 171–80, 187
Chemistry, 2, **9,** 16. *See also* Organic chemistry
Chloric acid, 137
Chloride ion, 123, 390

Chlorine
    atomic mass of, 70
    balanced equation, 170
    electron affinity, 288
    electron dot diagrams, 117
    ionic compounds, 109
    molecule, 102f
    oxidation number, 347, 351, 353,
        354, 358–59
    sodium metal and, 205–6, 208,
        216–17
Chloroform, 419
Chromate ion, 354, 357
Chromium, 150, 349
Citric acid, 181
Classical system, of nomenclature, 133
Coal, 370, 450
Cobalt, 132, 445–46
Coefficients, **43, 166–67,** 168, 187
Coinage metals, 12, 351
Cold fusion, 452
Colligative properties, **335,** 335–39
Combination reactions, **171**–73,
        179, 187
Combined gas law, 261–62, 264, 276
Combustion, enthalpy change for,
        319–20
Combustion reactions, **179**–80, 187
Completion, **214**
Composition
    definite, **3**
    percent, **150,** 151, 160
Compounds, **3,** 16
    oxidation number and naming of,
        350–51
    properties of, 6, 8
Concentration, **232,** 234–35, 240, 246,
        369–70. *See also* Molality; Molar-
        ity; Mole fraction; Normality
    binary, **101,** 121, 122, 129–31, 141
    binary metal-nonmetal, 129–31, 141
    equilibrium constants, 376–78, 381
Condensation, **309**
Condensation polymerization, 425
Conjugate acid, **389**
Conjugate acid-base pair, **389**
Conjugate base, **389**
Conservation of energy, law of, 9
Conservation of mass, law of, 16, 17,
        64f, 67
Constant(s)
    equilibrium, **375,** 375–81, 402
    ionization, **392,** 392–94
    symbols for, 477–78t
Constant composition, law of, 64–65
Control rods, **450**
Control of shared electrons, 347

Conversion factor, **23**
Copper
    density of, 51t
    industrial processing of, 210
    ionic compounds, 111–12, 132
    net ionic equation, 198–99
    oxidation number, 351, 352–53
    stability, 176
Covalent bond, **113,** 121, 122, 299,
        346–47
Covalent bonding, 113–21, 122
Covalent compounds
    double substitution reactions, 183
    molarities of, 238
    net ionic equations, 197
    oxidation numbers, 354
Cracking process, for petroleum, 417
Critical mass, **450**
Crystalline solids, **304,** 305, 322, 336
Cube, volume of, 33–34, 460
Cubic meter, **28,** 34
Curie, Marie, 438
Cyanide, 158

Dalton, John, 16, 64, 67, 69, 73
Dalton's atomic theory, 64, 67, 73
Dalton's law of partial pressures,
        270–73
Daughter isotope, **436,** 453
DDT, 419
Decay, radioactive, **435**
Decimal fraction numbers, conversion
        to integral ratios, 463–64
Decimal place digits, pH values
        as, 396
Decomposition reactions, **173**–74, 187,
        370–71
Definite composition, 3
Definite proportions, law of, 64–65,
        67, 73
Degeneration, of electrons, **83**
Density, **51**
    of common substance, 51t
    measurement of, 51–53, 56
    scientific algebra, 458
Desalinization, of seawater, 340
Deuterons, **452**
Diamond, 115, 116f, 305, 307
Diatomic molecules, **103,** 121,
        122, 293
Dichromate ion, 354, 357
Diethyl ether, 420
Diethylpropane, 416
Dimensional analysis. *See* Factor label
        method
Dipolar attraction, 295–96, 299

Dipole(s), **292,** 293, 295–96, 299
Dipole moment, **292,** 293, 299
Direct proportionality, **258**
Disaccharides, **428**
Discrete energy levels, **79**
Disintegration, radioactive, **435,** 438f,
        439f, 441–48, 453
Disintegration time, calculation of,
        443–44
Disodium monohydrogen
        phosphate, 170
Dispersion forces, 296
Displacement reaction, **174**
Disproportionation, **358**
Dissociation, 194
Distance, in metric system, 32
Distillation, **309**
Division, in scientific algebra, 46,
        467–68
DNA, hydrogen bonding in double
        helix, 298f
Double bond, **115,** 123
Double displacement reaction, **176**
Double substitution reaction, **176**–79,
        183, 187
Ductile, **14**
Duet, **106**

Einstein, Albert, 9, 451, 453. *See also*
        Relativity, theory of
Electricity
    conduction of by ions, 107f
    decomposition reactions, 173
    nuclear energy and generation
        of, 450
Electrodes, **107**
Electrolysis, 173f, 209
Electromagnetic radiation, **435**
Electron(s), **68,** 73
    affinity, **287,** 288, 299
    Bohr theory, 79–81, 95
    energy level diagrams, 89–92, 95
    groups, **290**
    oxidation half-reaction, 356–57
    oxidation number, 346
    relative energies of, 83–86, 95
    sharing of, **113**
    shells, subshells, and orbitals, 86–88,
        89, 95
    sizes of atoms, 285
Electron dot diagrams, **112,** 113, 114,
        115–18, 122, 123, 347
Electronegativity, **101, 288,** 289, 299
Electronic configuration, **88,** 92–95,
        108–9, 484
Electron sharing, **113**

Element(s), **2**, 16. *See also* Periodic tables; specific elements
diatomic molecules, 103f
electron affinities, 288t
electronegativities of main group, 102f
monatomic ions, 110f
names and symbols beginning with different letters, 11t
names and symbols to be learned, 10f
periodic table and, 11–15
relative reactivities of uncombined, 175t
valence electrons of main group, 106f
Elemental sulfur, 103
Emission of light, **79**, 80f
Empirical formulas, **155**–58, 159, 160
Endothermic process, **313**
End point, **242**, 243
Energy, **9**. *See also* Electricity; Heat
enthalpy changes in chemical reactions, 316–22
forms of, 9t
matter and, 8–9, 16
measurement of, 55, 56, 310–16
nuclear energy and generation of electricity, 450
of sun and stars, 452
temperature and, 53f
Energy level diagrams, 89–92, 95
English system of measurement, metric conversion factors, 29–30, 31t
Enthalpy
changes in chemical reactions, 316–22, 324
of formation, 318–20
of fusion, **314**, 322
of sublimation, 323
of vaporization, **314**, 317–18, 322–23
Enzymes, **369**
Equations, basic mathematical, 481–82t. *See also* Balanced equations; Chemical equations; Net ionic equations
Equilibrium, condition of, 370–72, 381, 382, 402, 492
Equilibrium constant expression, **375**, 382, 392, 402
Equilibrium constants, **375**–81, 402
Equivalent mass, **362**, 363
Equivalents, **359**–62, 363
Equivalent weight. *See* Equivalent mass
Erlenmeyer flask, **242**, 243
Esters, organic, **422**, 423, 429
Ethane, 409

Ethanol, 419, 420, 423–24
Ethers, **420**, 429
Ethyl acetate, 422
Ethyl alcohol, 334–35
Ethyl benzene, 157
Ethylene, 409, 425–26
Ethylene glycol, 339, 420
Ethyne, 411
Evaporation, **308**
Event, radioactive decay, **435**
Excess, **214**
Exothermic process, **313**
Expanded octets, 121
Exponent, **44**
Exponential notation, **43**, 56
Exponential numbers, 43–51, 464, 469–70
Exponential part, **43**, 49
Extensive properties, 6, 16

Factor label method, 22–27, 56
Fahrenheit temperature scale, **54**
Family, in periodic table, 12
Fats, dietary, **426**, 427, 429
Fatty acids, 427
Feeble acid, **390**
Feeble base, **390**
Fertilizer, 213–14
First transition series, 13
Fission reaction, 451–52
Fluidity, **308**
Fluorinated polymers, 426
Fluorine, 101, 130, 154, 172, 351, 353
Foods, and organic chemistry, 426–28. *See also* Minerals; Vitamins
Force, **252**
Form, in factor label method, 25
Formaldehyde, 117, 421, 424
Format, of exponential numbers, 46
Formation, enthalpy of, 318–20
Formic acid, 424
Formula(s). *See* Chemical formulas
Formula mass, **149**, 150, 160
Formula unit, **103**, 106f, 121
Formula weight, **149**, 160
Fossil fuels, 450. *See also* Petroleum
Fragrance, and esters, 422
Francium, 101
Free elements, 101–2
Freezing, **309**
Freezing point, 336, 341
Freezing-point depression, 336–38, 341
Freons, 419
Fructose, 428

Functional group, **418**
Fusion, **308**, 314, 322
Fusion reaction, 451–52

Gallium, 72
Gamma particles, **435**, 437, 453
Gamma ray, 435
Gas(es), **252**. *See also* Noble gases
answers to practice problems, 489–90
Boyle's law, 254–58, 276
Charles' law, 258–61, 276
chemical reactions, 265–70
combined gas law, 261–62, 276
Dalton's law of partial pressures, **270**–73
ideal gas law, 263–65
intermolecular forces, 304, 308
kinetic molecular theory, 273–76
partial pressure and equilibrium, 374, 381
physical properties of, 276, 304t
pressure and, 252–54
reaction rates and concentration of, 370
Gay-Lussac's law of combining volumes, **267**
Geiger counter, **440**, 441
Geology, 446, 453
Germanium, 72
Glucose
boiling-point elevation, 338–39
empirical formula of, 155
osmotic pressure, 340
reaction rates, 369
structure of, 428
Glycerine, 426–27
Gold, 51t, 52, 176, 351
Grain alcohol, 423
Gram, **28**
Graphite, 115, 116f, 305, 307
Gravity, law of, 17, 22, 273
Greek letters, as symbols, 479t
Ground state, **81**
Groups, in periodic table, 12

Haber, Fritz, 371
Haber process, 371
Half-life, of radioactivity, **441**–48, 453
Half-reaction, **355**, 356
Half-reaction method, **356**, 363
Halides, organic, **418**, 419, 429
Halogenated hydrocarbons, 409, 429
Halogens, 12, 348, 353

Heat, **53, 316.** *See also* Energy;
   Specific heat; Temperature
   equilibrium and, 373–74
   of fusion, **314**
   measurement of energy changes,
      310–16
   of phase change, 323
   of sublimation, 323
   of vaporization, **314**
Heating curve, **315**
Heisenberg uncertainty principle, **89**
Helium
   discovery of, 72
   electronic configuration, 93
   emission of light by atoms of, 80f
   ionic bonding, 105
   quantum number, 83–84
Hess's law, 320–22, 324
Heterogeneous mixture, 3, 16
Homogeneous mixture, 3, **4**, 16
Homogenization, **4**
Hund's rule, **90–91**, 95
Hydrate, **104**, 121, 140, 141
Hydrazoic acid, 350
Hydrobromic acid, 169
Hydrocarbons, 159, 179, 319t,
   407–13, 429
Hydrochloric acid
   balanced equations, 171
   chemical formula, 101
   ionization constants, 393
   limiting quantity and, 214, 215
   mass calculations, 210
   mole calculations, 206–7
   net ionic equations, 197
   nomenclature, 137
   titration, 242–44
Hydrogen
   acids, 136–37, 180–81
   acid salts, 140, 185
   balanced equation for reaction of
      oxygen and, 166f, 167
   binary compounds, 101, 129
   covalent bonding, 113–14
   electron dot diagrams, 117, 120
   electronic configuration, 93
   emission of light by atoms of, 80f
   energy levels and electron
      transitions, 80f
   equilibrium and, 370–71, 377
   ionic compounds, 110
   nomenclature and, 141
   oxidation number, 348, 349–50
   periodic table, 15
   quantum numbers, 83
   reactivity, 175
Hydrogenation, of fats, 427

Hydrogen bomb, **451,** 452
Hydrogen bonding, **297**–99
Hydrogen chloride, 137, 390
Hydrogen fluoride, 297, 388–89
Hydroiodic acid, 138
Hydronium ion, 131, **388,** 390,
   394–95, 397
Hydroxide ion, 394–95
Hypochlorous acid, 138
Hypothesis, **16,** 17

Ice, 297. *See also* Freezing;
   Freezing point
Ideal gas law, **263**–65, 268–70, 271
Ideal solution, **335**
Indicator, **182,** 245, 246
Inner transition elements, 13
Inorganic chemistry, 2
Insecticides, 419
Integer, and atomic number, 72
Integral mole ratio, 157, 160
Integral ratios, conversion of decimal
   fraction numbers to, 463–64
Intensive properties, 6, 16, 52
Intermolecular forces, 295–99, 304,
   308, 322
Inverse proportionality, **254,** 255
Iodine, 72, 129, 306, 441
Ion(s), **106.** *See also* Anions; Cations;
   Ionic bonding; Ionic compounds
   charges on and chemical
      formulas, 122
   conduction of electricity by, 107f
   mobility of, 195f
   molarities of, 238–42, 246
   monatomic, **107,** 110, 122, 123, 131,
      346, 348, 350, 363
   polyatomic, **118**–21, 122, 123, 131,
      141, 170, 346–47
   sizes of, 283–86
Ionic bonding, 105–9, 122
Ionic bonds, **107,** 122
Ionic compounds
   chemical reactions, 183, 187
   formulas for, 109–12, 122, 158
   melting points, 307
   molarity and, 238–42, 246
   nomenclature, 131–36, 141
   properties of in aqueous solution,
      194–95
Ionic equation, **196,** 200. *See also* Net
   ionic equations
Ionic lattice, 330
Ionic solids, **304,** 307, 329, 330, 340
Ionizable hydrogen atoms, **136**
Ionization, 194

Ionization constants, **392**–94
Ionization energy, **286**–88, 299
Iron, 6–8, 51t, 109, 199
Iron pyrite, 52
Iron-sulfur compound, 7t, 8
Isobutane, 424
Isomerism, 413–17
Isomers, **415,** 429
Isotope(s), **68,** 73
   daughter, **436,** 453
   half-life of, 441–48
   natural radioactivity and, **434,**
      435–41
   parent, **436,** 453

Joliot-Curie, Irène, 448
Joule, **55**

K (equilibrium constant), 375, 382
Kelvin, Lord, 259
Kelvin temperature scale, **54, 259,** 276
Kelvin unit, **259,** 260, 276
Ketones, **421,** 429
Kilo-, as metric prefix, **29**
Kilocalorie, 55
Kilogram, 28, 33
Kinetic energy, 55, 274, 275
Kinetic molecular theory of gases,
   273–76

Lanthanide series, 13, 94
Lavoisier, Antoine, 64, 73
Law(s), **15,** 17. *See also* Hess's law
   of chemical combination, 64–67
   of combined gas, 261–62, 264, 276
   of conservation of energy, 9
   of conservation of mass, 16, 17,
      64f, 67
   of constant composition, 64–65
   of definite proportions, 64–65, 67, 73
   of multiple proportions, 65–66,
      67, 73
   of partial pressure, **270,** 374
Lead, 51t, 65, 109, 446
LeChâtelier's principle, 372–74, 382,
   399, 402
Length, in metric system, **32**
Life insurance, and statistics, 441
Light, absorption and emission of,
   79, 80f
Limestone, 186, 369
Limiting quantity, **214,** 214–21, 241
Linear molecule, **291**
Line formula, **414**

Liquids, **252**
  answers to practice problems, 490–91
  intermolecular forces, 304
  properties of, 304t, 307–8, 322
  ranges of solvents and solutions, 338f
Liter, **28, 34**
Lithium, 84, 111, 351
Lithium hydroxide, 184
Litmus, 182
Lobes, orbital electrons, **89**
Logarithms, 395, 471–72
London forces, 296
Lone pair, **113**
Lowry, T. M., 388
LSD (lysergic acid diethylamide), 182

Macromolecular solids, 305
Macromolecule, **115,** 116f, 122, 307
Magnesium, 51t, 109
Magnesium chloride, 171, 346
Magnetic properties, of atoms, **91**
Magnetic quantum number, 81t, **82,** 95
Main group elements, 13, 353
Malleability, **7**
Manganese, 65–66, 350, 357–58
Mass, **8,** 16
  calculations for chemical reactions, 207–10
  density and equation for, 458
  law of conservation of, 16, 17, 64f, 67
  metric system and measurement of, 33
  of solvent, 341
Mass number, **68,** 437
Mathematics. *See* Algebra; Calculators
Matter, **8,** 16
  classification of, 2–5
  energy and, 8–9
  states of, **252**
Maximum molarity, 238
Maximum oxidation number, 351
Measurement, 22, 56
  answers to practice problems, 483–84
  of density, 51–53
  exponential numbers, 43–51
  factor label method, 22–27
  metric system, 27–35
  significant digits, 35–43
  of time, temperature, and energy, 53–55
Medicine
  blood and osmotic pressure, 340
  radiation therapy, 445
  radioactive tracers, 440–41, 453

Melting, **308**
Melting point
  intermolecular forces, 322
  ionic compounds, 307
  solids, 308t
Mendeleyev, Dmitry, 71, 73
Mercury, 51t, 131, 211, 253f, 350
Metal(s). *See also* Alkali metals; Alkaline earth metals; Coinage metals; Inner transition elements; Transition metals
  electronegativity, 101
  ionic bonding, 106, 121
  ionic compounds, 110
  naming ions of, 131
  oxidation numbers, 352, 354
  properties of, 14
  relative reactivities, 175
  specific heat, 313–14
Metallic solids, **304,** 305, 307
Metalloids, 14
Metathesis reactions, **176**
Meter, **28**
Methanal, 421, 424
Methane, 179, 180, 212, 290, 319, 408
Methanoic acid, 424
Methanol, 420, 423–24
Methyl alcohol, 334–35, 336
Methylbutane, 416
Methylbutanone, 421
Methylpropane, 410–11, 415–16, 424
Metric system, 27–35, 56
Meyer, Lothar, 71
Milli-, as metric prefix, **29**
Milliliters, 234, 236–37, 246
Millimoles, 234, 236–37, 246
Minerals, in diet, 3
Minimum oxidation number, 352
Mixtures, **3,** 16
  of carbon monoxide and carbon dioxide, 65
  heterogeneous, 3, 16
  homogeneous, 3, **4,** 16
  properties of, 6, 8
Molal, **332**
Molality, 332–34, 337, 341
Molar (M), **233**
Molarity, **233**
  answers to practice problems, 488–89
  definition and uses of, 232–38, 246
  of ions, 238–42, 246
  molality, 332
  osmotic pressure, 340
  titration, 242–46
Molar mass, **151,** 153, 154, 159, 160, 268–70

Mole(s), **151**–55, 160. *See also* Mole fraction; Mole ratio
  chemical reactions and, 205–7
  conversions of, 211–14, 362f
  ideal gas law, 264, 265, 271
Molecular compounds, 159
Molecular equation, 196
Molecular formula, **158,** 159, 160, 268–70
Molecular mass, **149,** 160
Molecular shape, 289–92
Molecular solids, **304,** 305, 307, 308
Molecular weight, **149**
Molecule(s), **67, 113**
  answers to practice problems, 490
  intermolecular forces, 295–99
  molecular shape, 289–92, 299
  polar and nonpolar, 292–94, 299
  structural formulas, 413–17
Mole fraction, **334,** 335
Mole ratio, 153, 155, 156, 167, 223, 224
Monatomic ions, **107,** 110, 122, 123, 131, 346, 348, 350, 363
Monomers, **425**
Monosaccharides, **428**
Multiple proportions, law of, 65–66, 67, 73
Multiplication, of exponential numbers, 46

Natural radioactivity, 434–41, 453
Neon, 80f, 85t, 172
Net ionic equations, 200, 223
  answers to practice problems, 486
  ionic compounds in aqueous solutions, 194–95
  mole and mass ratios, 224
  writing of, 196–99
Network solids, **304,** 307
Neutrality, **181**
Neutral atoms, **68,** 73
Neutralization reaction, **181,** 187
Neutrons, **68,** 73, 285, 449–50, 453
Newton, Sir Isaac, 22
Nitric acid, 182
Nitrogen
  binary compounds, 129
  covalent bonding, 114–15
  electron dot diagrams, 118
  energy level diagrams, 91
  equilibrium and combination with hydrogen, 370–71
  fertilizer and quantity of, 213–14
  ionic compounds, 111
  oxidation number, 349, 353

Nitrogen monoxide, 374
Noble gas(es), 12
electronegativity, 288
electronic configurations, 92–94
ionic bonding, 105, 121
maximum oxidation numbers, 351
minimum oxidation numbers, 352
periodic table, 72
van der Waals forces, 297
Noble gas configuration, **108**
Nomenclature, **129**. *See also*
Abbreviations; Prefixes; Stock
system; Subscripts; Suffixes;
Superscripts; Symbols
of acids and acid salts, 136–40, 141
of alcohols, 419
of aldehydes and ketones, 421
of alkanes, 408
of alkenes, 410
of alkynes, 411
of amines and amides, 423–24
answers to practice problems, 485
of binary nonmetal-nonmetal
compounds, 129–31, 141
of hydrates, 140, 141
of ionic compounds, 131–36, 141
organic acids and esters, 422
outline for, 138t
oxidation number and compounds,
350–51
of radicals, 418
summary of for compounds and
ions, 139f
Nonlinear molecule, **291**
Nonmetals
binary nonmetal-nonmetal
compounds, 129–31
electronegativity, 101
ionic bonding, 106, 121–22
properties of, 14
relative reactivities, 175
Non-octet structures, **121**
Nonpolar bond, **289**, 299
Nonpolar molecules, 292–94, 329
Nonpolar solutes, 329–30, 340
Nonvolatile solute, **336**
Nonzero concentrations, 400
Normal boiling point, **310**
Normality, of solution, **360**–62, 363
Nuclear disintegration, 435–36
Nuclear energy, 440, 450, 451f, 453
Nuclear fission, 448–51, 453
Nuclear fusion, 451–52
Nuclear radiation, **435**
Nuclear reactions
answers to practice problems, 493–94
energetics of, 451, 453

fission, 448–51, 453
fusion, 451–52, 453
half-life, 441–48, 453
natural radioactivity, 434–41, 453
Nuclear weapons, 449, 451–52
Nucleus, of atom, **68**, 73, 435
Nylon, 424, 429

Octane, 51t, 179, 407, 417
Octet, **105**, 114
Octet rule, **105**, 121, 122, 123
Odometer, 35
Odors, and esters, 422
Orbitals, **86**, 89, 95
Organic acids, **422**, 423, 429
Organic chemistry, 2, **407**
Organic compounds
acids and esters, 422–23, 429
alcohols and ethers, 419–20, 429
aldehydes and ketones, 421, 429
amines and amides, 423–24, 429
answers to practice problems, 493
classes of organic compounds, 424t
foods, 426–28, 429
halides, 418–19, 429
hydrocarbons, 407–13, 429
isomerism, 413–17, 429
polymers, 424–26, 429
Organic halides, **418**, 419, 429
Osmosis, 339–41
Osmotic pressure, **339**, 341
Outermost shell, **105**
Oxidation, **354**, 363
Oxidation number(s), **346**
answers to practice problems,
491–92
assigning of, 346–50, 363
equivalents and normality,
359–62, 363
naming of compounds, 350–51, 363
oxidation-reduction equations,
354–59, 363
periodic variation of, 351–53, 363
Oxidation-reduction equations, 354–59
Oxidation-reduction reaction, 359, 363
Oxidation state, **346**
Oxide, 108
Oxidizing agent, **354**
Oxoacids, **137**, 141
Oxoanions, **134**
Oxygen. *See also* Carbon dioxide;
Carbon monoxide; Ozone
allotropes of, 114
balanced equation for reaction of
hydrogen and, 166f, 167
binary compounds, 129
compound of lead and, 65

compound of manganese and, 65–66
electron dot diagrams, 117, 120
electronic configuration of, 88, 108
energy level diagrams, 91
macromolecules, 116f
mercury liquid and, 211
oxidation number, 348, 349–50
oxoanions, 134
Ozone, 114, 222
Ozone layer, depletion of, 419

Paraffins, 408
Parentheses, and precedence rules,
466, 468
Parent isotope, **436**, 453
Partial pressures, **270**, 374
Pauli exclusion principle, **82**, 87, 95
Penicillin, 159
Pentanal, 421
Pentane, 416
Pentanone, 421
Percent, **25**
Percent composition, **150**, 151, 160
Percent yield, **221**, 222, 224
Perchloric acid, 138, 220–21
Perfectly elastic collision, **274**
Period, **11**
Periodicity of electronic configuration,
92–95
Periodic table, 11–15, 16
atomic mass, 71–73
electronic configurations, 93f, 95
ionization energy, 287f
nomenclature of binary
compounds, 129
oxidation numbers, 351–53, 363
sizes of atoms and ions, 284f, 285f
transition metals, 86
Permanganate ion, 354, 358
Peroxides, 348
Petroleum, 417. *See also* Fossil fuels
pH scale, **395**–98, 399–401, 402
Phase, of matter, **252**
Phase change, **308**–10, 322
Phenolphthalein, 182
Phosphate, electron dot diagram,
119–20
Phosphine, 129
Phosphoric acid, 170, 219
Phosphorus
atomic mass, 71
binary compounds, 129, 130
electron dot diagram, 116–17, 121
molecule, 102f
nuclear fission, 448
oxidation number, 350, 351, 352

Phosphorus trichloride, 348
Physical changes, **2**
Physical chemistry, 2
Physical equilibrium, **310**
Pipet, **242,** 243
Platinum, 51t
Polar bond, **289,** 299
Polar molecules, 292–94, 329
Polar solutes, 329, 330, 340
Polyatomic ions, **118**–21, 122, 123, 131, 141, 170, 346–47
Polyester, 424, 425
Polyethylene, 409, 424, 426, 429
Polymerization process, 425–26
Polymers, 424–26
Polypropylene, 409, 426t
Polysaccharides, **428**
Polystyrene, 426t
Polyvinylchloride (PVC), 426t
Positron, **449**
Postulates, Dalton's, 67
Potassium, 111, 172
Potassium chlorate, 174, 207
Potassium chloride, 194, 196, 330
Potassium cyanide, 158
Power, in factor label method, 26
Precedence rules, and scientific calculators, 465–67
Precipitate, **177**
Precision, of scientific measurement, **35,** 36–37
Prediction, of oxidation numbers, 351–53
Prefixes, and nomenclature, 130t, 132, 134, 137, 138, 140, 141, 479–80t
Pressure, **252**–54, 276
    atmospheric, **254**
    barometric, **254,** 271
    calculation of, 257–58
    chemical reactions and, 381
    graphs of volume and, 255–56
    osmotic, **339,** 341
    partial, **270**
Priestley, Joseph, 173
Primary amine, 423
Principal quantum number, **81**
Proceed to the right or left, in chemical reaction, **372**
Products, of chemical reactions, **166,** 171–80
Propane, 319, 408
Propanol, 420
Propanone, 421
Properties, **5**
    as basic concept, 5–8, 16
    colligative, **335**–39
    electronic configuration, 95

extensive, 6, 16
intensive, 6, 16, 52
magnetic, **91**
quantitative, 6
of subatomic particles, 68t
Propylene, 409
Protein, **425,** 429
Proton(s), **68,** 73, 285, 388
Proton acceptor, **388**
Proton donor, **388**
Pure substance, 3

Quadratic equations, 462–63
Quadratic formula, 462
Quantitative property, 6
Quantity, unit of, 461
Quantum numbers, 81–83, 95

Radiation therapy, 445
Radicals, **418,** 429
Radioactive dating, **446**
Radioactive decay, **435**
Radioactive disintegration, **437**–40, 441–48
Radioactivity, 434–41, 449t, 453
Radium, 438
Radon, 297
Random motion, **274**
Raoult's law, **335,** 341
Rate, in factor label method, 24. *See also* Reaction rates
Ratio, in factor label method, 24. *See also* Mole ratio
Reactant, **166**
Reaction(s), 2. *See also* Chemical reactions; Nuclear reactions
    acid-base, 359, 360
    chemical vs. Nuclear, 434
    combination, **171**–73, 179, 187
    combustion, **179**–80, 187
    decomposition, **173**–74, 187, 370–71
    displacement, **174**
    double displacement, **176**
    double substitution, **176**–79, 183, 187
    gases in, 265–70
    mass calculations, 207–10
    metathesis, **176**
    neutralization, **181,** 187
    oxidation-reduction, 359, 363
    rates of, **369**–72, 381, 492
    redox, 355
    substitution, **174**–76
Reaction rates, **369**–72, 381, 382, 492
Reactivity, **174,** 175t, 187

Reciprocal key, on scientific calculator, 470–71
Recrystallization, 331
Redox reaction, 355
Reducing agents, **354,** 355
Reduction, **354,** 363
Relative density, 52
Relative energies, of electrons, 83–86
Relative reactivities, 175
Relative scale, of atomic mass, 70
Relativity, theory of, 9, 16, 273, 451, 453
Reverse osmosis, 340
Rhombic sulfur, 102f
Rounding off, 41–43, 160
Rutherford, Ernest, 448

Salts, **181**
    acid, **136,** 140, 141, 184–85, 187
    net ionic equation, 197
Saturated fats, 427, 429
Saturated hydrocarbons, 408
Saturated solutions, **330,** 331, 332
Scientific calculations
    basic mathematical equations, 481–82t
    calculator mathematics, 467–72
    scientific algebra, **49,** 457–64
Scientific notation, **44,** 56
Second, as unit of time, **53**
Secondary amine, 423
Second ionization energy, **287**
Second transition series, 13
Selenium, 129
Semipermeability, 339
Shell, atomic, **86,** 95. See also Outermost shell
Shift, of equilibrium, **372**
SI (*Système International d'Unités*), **27**
Significant digits, 35–43, 56
Significant figures, **37,** 472
Silica, 115, 116f, 305
Silicon, 71, 116f
Silver, 176, 198–99, 223
Silver chloride, 194, 196, 200, 223
Silver fluoride, 166, 167f
Silver nitrate, 171, 223
Single bond, **113,** 123
60-carbon molecule, 306
Soap, **427**
Sodium, 123
    chlorine gas and, 205–6, 208, 216–17
    electron dot diagrams, 112, 113
    ionic compound, 110–11
    ionization energy, 288   *(cont.)*

Sodium (*cont.*)
  sulfur and, 217
  valence shell, 106–7
Sodium acetate, 331–32, 391, 400
Sodium bicarbonate, 140, 185
Sodium carbonate, 168–69, 171, 210, 220–21
Sodium chlorate, 118–19
Sodium chloride
  balanced equations, 168–69, 170–71
  ionic compounds, 194
  limiting quanity, 216–17
  mass calculations, 208, 209
  mole calculations, 205–6
  solution process, 330
  structure of, 107, 305, 306f
Sodium hydroxide, 170, 197, 242–44
Sodium perchlorate, 220–21
Sodium stearate, 427
Sodium sulfide, 170–71, 206, 217
Solar energy, 450
Solids, 252
  amorphous, **304**
  answers to practice problems, 490–91
  crystalline, **304**, 305, 322, 336
  ionic, **304**, 307, 329, 330, 340
  macromolecules, 305
  melting points, 308t
  metallic, **304**, 305, 307
  molecular, **304**, 305, 307, 308
  network, **304**, 307
  properties of, 304–7, 322
  reaction rates and subdivision of, 370, 381
Solubility, **177**, 178, 187, 200, **330**, 331f
Soluble metal hydroxides, 183
Solute, **232**
Solutions, **3**, 4–5, 16. *See also* Molarity
  answers to practice problems, 491
  buffer, 398–401, 402
  colligative properties, 335–39, 341
  molality, 332–34, 341
  mole fraction, 334–35, 341
  normality, **360**–62, 363
  reaction rates and concentration of, 369
  saturated, unsaturated, and supersaturated, 330–32, 340–41
  solution process, 329–30, 340
  types and examples of, 329t
Solvay process, 178, 179
Solvent, **232**
Specific heat, **311**, 312, 313–14, 322
Spectator ions, **196**, 197, 200
Spectrum, of sunlight, 72

Spin quantum number, 81t, **82**, 87
Square root, 50, 394
Stability, **176**
Stalactite, 186
Stalagmite, 186
Standard atmosphere, **254**
Standard exponential form, **44**
Standard mass, 28, 33
Standard state, **318**
Standard temperature and pressure (STP), **262**, 276
Starches, dietary, 429
Stars, and nuclear fusion, 452
State. *See also* Gas(es); Liquids; Solids
  of matter, **252**
  of reactant or product, **171**
  of system, **254**
State function, **317**
Stearic acid, 427
Stock system, **131**, 133, 350, 363
Stoichiometry, **205**
  answers to practice problems, 486–88
  calculations involving conversions, 211–14, 224
  gases in chemical reactions, 265
  limited quantities and problems of, 214–21, 224
  mass calculations for chemical reactions, 207–10, 224
  mole calculations for chemical reactions, 205–7, 224
  net ionic equations and, 223, 224
  theoretical yield and percent yield, 221–22, 224
Stress, **372**
Strong acids, **183**, 187, 195, 197, 200, 242, 390, 399, 400, 402
Strong bases, **183**, 195t, 197, 242, 399, 400
Structural formula, **114**, 413–17
Subatomic particles, 68–69, 73, 434t, 478t
Subdivision, of solid, **370**, 381
Sublimation, **296**, 323
Subscripts. *See also* Nomenclature
  alpha, beta, and gamma particles, 435
  in chemical formulas, **103–4**, 121
Subshell, **86**, 89–95
Substance, **3**
Substitution reactions, **174**–76
Subtraction, of exponential numbers, 49
Sucrose, 428f. *See also* Sugar
Suffixes, and nomenclature, 133, 134, 137, 138, 141, 479–80t
Sugars
  density of, 51t
  organic chemistry of food, **427**–29

percent composition of, 155
reaction of with excess oxygen, 180
Sulfur
  ionic compounds, 132
  molecule, 102f
  oxidation number, 347, 349, 352, 353
  properties of, 6–8
  sodium metal and, 217
Sulfur dichloride, 351
Sulfur dioxide, 184, 351
Sulfuric acid, 137, 138, 182, 184, 209, 218–19
Sun
  hydrogen and helium as component of, 80f
  nuclear fusion, 452, 453
  spectrum of sunlight, 72
Superoxides, 348
Supersaturated solutions, **331**, 332, 340, 341
Superscripts, and alpha, beta, and gamma particles, 435. *See also* Nomenclature
Surroundings, **254**
Symbol(s), chemical. *See also* Nomenclature; Periodic table
  atom of element and, 68–69
  basic concept of, 9, 10f, 11, 16
  Greek letters as, 479t
  in scientific algebra, 458, 460
  for subatomic particles, 478t
  for units, 478t
  for variables and constants, 477–78t
System, **254**
Système International d'Unités (SI), **27**

Teflon, 424, 426
Tellurium, 72
Temperature, **53**. *See also* Boiling point; Freezing point; Heat
  equilibrium, 373–74, 381
  measurement of, 54, 56
  reaction rates, 369, 381
  scales of, 258–61
  solubility and, 331f
  vapor pressure of liquid, 310
Ternary compound, **173**
Tertiary amine, 423
Tetrahedral molecule, **290**, 291, 413
Tetraphosphorus decoxide, 130
Thallium, 94, 108
Theoretical yield, **221**, 222, 224
Theory, **16**, 17. *See also* Relativity, theory of
Third ionization energy, **287**

Third transition series, 13
Thomson, William, 259
Thyroid gland, 441
Time, measurement of, 53, 56
Titanium, 88, 354
Titration, **242**, 242–46
TNT, 413
Torr, **252**, 276
Torricelli, Evangelista, 252
Total bond order, **418**
Total equation, **196**
Total pressure, 374
Trace elements, in diet, 3
Tracers, radioactive, **440**, 441, 453
Transition group elements, 13, 353.
    See also Transition metals
Transition metals, 86, 109, 354
Transmutation, **449**
Tree ring dating, 448
Trigonal planar molecule, **290**
Trigonal pyramidal molecule, **291**
Trinitrotoluene, 413
Triple bond, **115**, 123
Tritium, **452**

Uncertainty principle, **89**
Unit, **22**
    in scientific algebra, 459–62
    symbols for, 478t
Unit cell, 50
Unsaturated fats, 427, 429
Unsaturated fatty acids, 427
Unsaturated hydrocarbons, **409**
Unsaturated solution, **331**, 332, 340
Unshared pair, **113**
Uranium, 438f, 440, 445, 446, 450

Valence electron, **106**, 111, 112, 119,
    346, 347
Valence shell, **106**
Value, in factor label method, 25
Value, of exponential numbers, 46
Vanadium, 133
van der Waals, Johannes D., 296
van der Waals forces, 296–97, 299
Vapor, **308**
Vaporization, **308**, 317–18, 322–23
Vapor pressure, **272**, **310**, 322
Vapor-pressure lowering,
    335–36, 341
Variables
    in scientific algebra, 457–59
    symbols for, 477–78t
Velocity, 462
Vinegar, 422
Vitamins, 3, 159
Volatile solvent, **336**
Volume
    calculation of gas pressure and,
        257–58
    density and, 52, 458
    graphs of gas pressure and, 255–56
    measurement of in metric system,
        33–34
Volume ratios, in chemical reactions,
    267, 268

Water
    acid-base theory and, 389
    atomic mass of, 70
    autoionization of, 394–98
    binary compounds of
        hydrogen, 129

density of, 51t, 52
desalinization of, 340
heating curve, 315f
hydrates, 140, 141
hydrogen bonding, 297
ionic solids, 330f
net ionic equation, 197
osmotic pressure, 339f
solubility, 177t
vapor pressure, 272t, 272f, 310
vapor and solubility of gases, 271–72
volume of liquid, 275
Water ionization constant, **394**
Wavelengths
    emission of light by gaseous
        atoms, 80f
    visible spectrum of sunlight, 72
Weak acids, **183**, 187, 198, 390, 402
Weak bases, **183**, 402
Weight, **8–9**, 16
Weighted average, **70**, 73
Welding torch, 103, 411
White phosphorus, 102f
Wind energy, 450
Wöhler, Friedrich, 407
Wood, dating of samples, 448
Wood alcohol, 423
Work, **316**

Xenon, 130
X rays, and barium sulfate, 135

Zinc, 71, 174, 175f, 213, 215, 219
Zinc chloride, 213

Table of the Elements

Element	Symbol	Atomic Number	Atomic Mass	Element	Symbol	Atomic Number	Atomic Mass
Actinium	Ac	89	(227)	Europium	Eu	63	151.96
Aluminum	Al	13	26.9815	Fermium	Fm	100	(253)
Americium	Am	95	(243)	Fluorine	F	9	18.9984
Antimony	Sb	51	121.75	Francium	Fr	87	(223)
Argon	Ar	18	39.948	Iron	Fe	26	55.847
Arsenic	As	33	74.9216	Gadolinium	Gd	64	157.25
Astatine	At	85	(210)	Gallium	Ga	31	69.72
Silver	Ag	47	107.868	Germanium	Ge	32	72.59
Gold	Au	79	196.9665	Gold	Au	79	196.9665
Barium	Ba	56	137.34	Hafnium	Hf	72	178.49
Berkelium	Bk	97	(249)	Helium	He	2	4.00260
Beryllium	Be	4	9.01218	Holmium	Ho	67	164.9303
Bismuth	Bi	83	208.9806	Hydrogen	H	1	1.0080
Boron	B	5	10.81	Mercury	Hg	80	200.59
Bromine	Br	35	79.904	Indium	In	49	114.82
Cadmium	Cd	48	112.40	Iodine	I	53	126.9045
Calcium	Ca	20	40.08	Iridium	Ir	77	192.22
Californium	Cf	98	(251)	Iron	Fe	26	55.847
Carbon	C	6	12.011	Krypton	Kr	36	83.80
Cerium	Ce	58	140.12	Potassium	K	19	39.102
Cesium	Cs	55	132.9055	Lanthanum	La	57	138.9055
Chlorine	Cl	17	35.453	Lawrencium	Lr	103	(257)
Chromium	Cr	24	51.996	Lead	Pb	82	207.2
Cobalt	Co	27	58.9332	Lithium	Li	3	6.941
Copper	Cu	29	63.546	Lutetium	Lu	71	174.97
Curium	Cm	96	(247)	Magnesium	Mg	12	24.305
Dysprosium	Dy	66	162.50	Manganese	Mn	25	54.9380
Einsteinium	Es	99	(254)	Mendelevium	Md	101	(256)
Erbium	Er	68	167.26				